Advances in Magnetic Resonance Technology and Applications
Volume 7

Magnetic Resonance Image Reconstruction

Theory, Methods, and Applications

Advances in Magnetic Resonance Technology and Applications Series

Series Editors

In-Young Choi, PhD
Department of Neurology, Department of Radiology, Department of Molecular & Integrative Physiology, Hoglund Biomedical Imaging Center, University of Kansas Medical Center, Kansas City, KS, United States

Peter Jezzard, PhD
Wellcome Centre for Integrative Neuroimaging, Nuffield Department of Clinical Neurosciences University of Oxford, Oxford, United Kingdom

Brian Hargreaves, PhD
Department of Radiology, Department of Electrical Engineering, Department of Bioengineering Stanford University, Stanford, CA, United States

Greg Zaharchuk, MD, PhD
Department of Radiology, Stanford University, Stanford, CA, United States

Titles published:

Volume 1 – Quantitative Magnetic Resonance Imaging – Edited by Nicole Seiberlich and Vikas Gulani

Volume 2 – Handbook of Pediatric Brain Imaging – Edited by Hao Huang and Timothy Roberts

Volume 3 – Hyperpolarized Carbon-13 Magnetic Resonance Imaging and Spectroscopy – Edited by Peder Larson

Volume 4 – Advanced Neuro MR Techniques and Applications – Edited by In-Young Choi and Peter Jezzard

Volume 5 – Breast MRI – Edited by Katja Pinker, Ritse Mann and Savannah Partridge

Volume 6 – Motion Correction in MR – Edited by André Van Der Kouwe and Jalal B. Andre

Volume 7 – Magnetic Resonance Image Reconstruction – Edited by Mehmet Akçakaya, Mariya Doneva and Claudia Prieto

Visit the Series webpage at https://www.elsevier.com/books/book-series/advances-in-magnetic-resonance-technology-and-applications

Advances in Magnetic Resonance
Technology and Applications
Volume 7

Magnetic Resonance Image Reconstruction

Theory, Methods, and Applications

Edited by

Mehmet Akçakaya
University of Minnesota
Minneapolis, MN, United States

Mariya Doneva
Philips Research
Hamburg, Germany

Claudia Prieto
King's College London
London, United Kingdom

Academic Press is an imprint of Elsevier
125 London Wall, London EC2Y 5AS, United Kingdom
525 B Street, Suite 1650, San Diego, CA 92101, United States
50 Hampshire Street, 5th Floor, Cambridge, MA 02139, United States
The Boulevard, Langford Lane, Kidlington, Oxford OX5 1GB, United Kingdom

ISBN: 978-0-12-822726-8
ISSN: 2666-9099

For information on all Academic Press publications
visit our website at https://www.elsevier.com/books-and-journals

Publisher: Mara E. Conner
Acquisitions Editor: Tim Pitts
Editorial Project Manager: Fernanda A. Oliveira
Production Project Manager: Nirmala Arumugam
Cover Designer: Vicky Pearson Esser

Typeset by VTeX

Contents

Contributors

Mehmet Akçakaya
Department of Electrical and Computer Engineering, University of Minnesota, Minneapolis, MN, United States
Center for Magnetic Resonance Research, University of Minnesota, Minneapolis, MN, United States

M. Salman Asif
Department of Electrical and Computer Engineering, University of California, Riverside, CA, United States

Suchandrima Banerjee
GE Healthcare, Menlo Park, CA, United States

Berkin Bilgic
Athinoula A. Martinos Center for Biomedical Imaging, Charlestown, MA, United States
Harvard Medical School, Boston, MA, United States
Harvard/MIT Health Sciences and Technology, Cambridge, MA, United States

Christof Boehm
Klinikum rechts der Isar, Technical University of Munich, Department of Diagnostic and Interventional Radiology, School of Medicine, Munich, Germany

Itthi Chatnuntawech
National Nanotechnology Center, Pathum Thani, Thailand

Anthony G. Christodoulou
Biomedical Imaging Research Institute, Cedars-Sinai Medical Center, Los Angeles, CA, United States

Chiara Coletti
Department of Imaging Physics, Delft University of Technology, Delft, The Netherlands

Gastao Cruz
School of Biomedical Engineering and Imaging Sciences, King's College London, London, United Kingdom

Evan Cummings
Department of Biomedical Engineering, University of Michigan, Ann Arbor, MI, United States

Maximilian N. Diefenbach
Klinikum rechts der Isar, Technical University of Munich, Department of Diagnostic and Interventional Radiology, School of Medicine, Munich, Germany

Mariya Doneva
Philips Research, Hamburg, Germany

Holger Eggers
Philips Research, Hamburg, Germany

Li Feng
Department of Radiology, Icahn School of Medicine at Mount Sinai, New York, NY, United States

Justin P. Haldar
University of Southern California, Signal and Image Processing Institute, Ming Hsieh Department of Electrical and Computer Engineering, Los Angeles, CA, United States

Kerstin Hammernik
Technical University Munich, AI in Healthcare and Medicine, Klinikum Rechts der Isar, Munich, Germany

Rakib Hyder
Department of Electrical and Computer Engineering, University of California, Riverside, CA, United States

Mathews Jacob
University of Iowa, Iowa City, IA, United States

Ulugbek S. Kamilov
Department of Electrical and Systems Engineering, Washington University in St. Louis, St. Louis, MO, United States
Department of Computer Science and Engineering, Washington University in St. Louis, St. Louis, MO, United States

Dimitrios C. Karampinos
Klinikum rechts der Isar, Technical University of Munich, Department of Diagnostic and Interventional Radiology, School of Medicine, Munich, Germany

Kirsten Kerkering
Physikalisch-Technische Bundesanstalt (PTB), Braunschweig and Berlin, Germany

Melanie Kircheis
Technische Universität Chemnitz, Faculty of Mathematics, Chemnitz, Germany

Christoph Kolbitsch
Physikalisch-Technische Bundesanstalt (PTB), Braunschweig and Berlin, Germany

Thomas Küstner
University Hospital Tübingen, Medical Image and Data Analysis, Department of Radiology, Tübingen, Germany

Fan Lam

Department of Bioengineering, University of Illinois Urbana Champaign, Urbana, IL, United States

Beckman Institute for Advanced Science and Technology, University of Illinois Urbana Champaign, Urbana, IL, United States

Zhi-Pei Liang

University of Illinois at Urbana-Champaign, Department of Electrical and Computer Engineering, Urbana, IL, United States

Sajan Goud Lingala

University of Iowa, Iowa City, IA, United States

Jiaming Liu

Department of Electrical and Systems Engineering, Washington University in St. Louis, St. Louis, MO, United States

Jacob A. Macdonald

Department of Radiology, Duke University, Durham, NC, United States

Department of Radiology, University of Michigan, Ann Arbor, MI, United States

Merry Mani

University of Iowa, Iowa City, IA, United States

Steen Moeller

Center for Magnetic Resonance Research, University of Minnesota, Minneapolis, MN, United States

Freddy Odille

IADI U1254, Inserm and University of Lorraine, Nancy, France

CIC-IT 1433, Inserm, Université de Lorraine and CHRU Nancy, Nancy, France

Daniel Polak

Siemens Healthcare GmbH, Erlangen, Germany

Daniel Potts

Technische Universität Chemnitz, Faculty of Mathematics, Chemnitz, Germany

Claudia Prieto

School of Biomedical Engineering and Imaging Sciences, King's College London, London, United Kingdom

Daniel Rueckert

Technical University Munich, AI in Healthcare and Medicine, Klinikum Rechts der Isar, Munich, Germany

Imperial College London, Department of Computing, London, United Kingdom

Stefan Ruschke
Klinikum rechts der Isar, Technical University of Munich, Department of Diagnostic and Interventional Radiology, School of Medicine, Munich, Germany

Tobias Schaeffter
Physikalisch-Technische Bundesanstalt (PTB), Braunschweig and Berlin, Germany

Nicole Seiberlich
Department of Biomedical Engineering, University of Michigan, Ann Arbor, MI, United States
Department of Radiology, University of Michigan, Ann Arbor, MI, United States

Bradley P. Sutton
Department of Bioengineering, University of Illinois Urbana Champaign, Urbana, IL, United States
Beckman Institute for Advanced Science and Technology, University of Illinois Urbana Champaign, Urbana, IL, United States

Joao Tourais
Department of Imaging Physics, Delft University of Technology, Delft, The Netherlands

Sebastian Weingärtner
Department of Imaging Physics, Delft University of Technology, Delft, The Netherlands

Burhaneddin Yaman
Department of Electrical and Computer Engineering, University of Minnesota, Minneapolis, MN, United States
Center for Magnetic Resonance Research, University of Minnesota, Minneapolis, MN, United States

Christoph Zoellner
Klinikum rechts der Isar, Technical University of Munich, Department of Diagnostic and Interventional Radiology, School of Medicine, Munich, Germany

Editor Biographies

Mehmet Akçakaya

Dr. Akçakaya received the B. Eng. degree from McGill University, Montreal, QC, Canada, in 2005, and the S.M. and Ph.D. degrees from Harvard University, Cambridge, MA, USA, in 2010. From 2010 to 2015, he was with the Harvard Medical School. He is currently an Associate Professor at the University of Minnesota, MN, USA. His works on accelerated MRI has received a number of international recognitions. His research interests include image reconstruction, machine learning, MRI physics, inverse problems, and image processing.

Mariya Doneva

Dr. Doneva received her B.Sc. and M.Sc. degrees in physics from the University of Oldenburg in 2006 and 2007, respectively, and her Ph.D. degree in physics from the University of Lübeck in 2010. She joined Philips Research in 2010 and was a research associate in the Electrical Engineering and Computer Sciences Department at the University of California (UC), Berkeley, during 2015 and 2016. She is currently a senior scientist and a team leader at Philips Research, Hamburg, Germany. Her research interests include methods for efficient data acquisition, image reconstruction, and quantitative (multi)-parameter mapping in the context of MR imaging.

Claudia Prieto

Dr. Prieto received her B.Sc. and M.Sc. degrees in electrical engineering from the Pontificia Universidad Católica de Chile in 2005 and the PhD degree from the same institution in 2007. She was a postdoctoral fellow with the Imaging Sciences Division, King's College London, from 2008 to 2011. She is currently Full Professor at the School of Biomedical Engineering and Imaging Sciences, King's College London, London, UK. Her research interests include methods for efficient volumetric and free-breathing data acquisition, undersampled and motion-corrected image reconstruction, and quantitative multiparametric mapping in the context of cardiovascular magnetic resonance imaging.

Introduction

1 MRI reconstruction and its role in clinical practice

The goal of Magnetic Resonance Imaging (MRI) reconstruction is to convert the measured raw MR data into images that can be clinically interpreted, and thus it plays a fundamental role in the clinical use of MRI.

In the early days of MRI, most of the research focused on hardware improvements and the development of new pulse sequences aiming at faster acquisition and/or improved contrast, while the image reconstruction was performed via a simple Fourier transform. Interestingly, the Fourier transform was not the first approach to MRI reconstruction. In 1973, Paul Lauterbur published his seminal paper showing the first MR image of two water-filled glass tubes. This image was obtained by applying magnetic field gradients at several different directions and performing a reconstruction similar to the filtered back projection used in computed tomography. Although this was very exciting, the reconstruction was very time consuming and the images were prone to artifacts, mainly due to the main magnetic field inhomogeneities. The introduction of spin warp imaging, in which data are acquired on a uniform grid in k-space (Cartesian sampling), enabled the application of the fast Fourier transform (FFT) for MR image reconstruction. This made MR reconstruction faster and less sensitive to field inhomogeneities, which were crucial for the clinical adoption of MRI.

The fast Fourier transform has well served the MR community as the conventional image reconstruction method for fully sampled Cartesian k-space. However, it is important to mention that it is an approximation that does not always describes the MR physics very well, which may lead to artifacts in the reconstructed images. Moreover, it requires the acquisition of a fully sampled Cartesian k-space, which may lead to very long scan times. In other instances, such as when only undersampled k-space data are available, when data are acquired with multiple coils or on a non-Cartesian grid, or when physical effects like field inhomogeneities are important, the simple FFT image reconstruction becomes suboptimal. In these scenarios, advanced reconstruction methods that use more accurate physics models for the data acquisition or apply prior knowledge about the image properties are more appropriate.

An important conceptual shift that facilitated the development of such advanced MR reconstruction methods has been the move towards model-based reconstruction that uses an inverse problem formulation to provide a unified view of different reconstruction problems. This also enables the application of algorithms for solving these inverse problems with proven convergence guarantees. It also allows the use of more complicated MR physics models to either reduce artifacts or extract physical properties of the tissue from the measurement data (quantitative MRI). In this optimization framework, it is easy to include additional constraints, e.g., priors describing image properties or structure. This framework is very powerful for the development of new reconstruction approaches and understanding the limitations arising from various approximations.

Although advanced MR reconstruction techniques have a long history, the last two decades marked a fast development of this research field, as well as the adoption of advanced reconstruction methods by vendors and subsequent integration into clinical practice. Parallel imaging has revolutionized MRI by providing typically a two–three-fold acceleration, reducing the risk of motion artifacts, improving patient comfort, and enabling new applications. With the introduction of compressed sensing, an additional acceleration was achieved with virtually no loss in image quality which has already helped many hospitals to improve their productivity by scanning more patients a day. More recently, deep learning-based reconstruction approaches, using data adapted priors that can push the acceleration even a little further, have gained a lot of attention These advances were due to a combination of factors, including the advances in computation speed, development of the necessary mathematical frameworks, and advances in deep learning, as well as the clinical need for improved imaging speed.

2 Organization of the book

This book covers the fundamental concepts of MR image reconstruction, including its formulation as an inverse problem and the most common models and optimization methods used to reconstruct MR images nowadays. The most recent developments, such as compressed sensing, tensor-based reconstruction, and machine learning-based reconstruction are also included. Approaches for specific applications such as non-Cartesian imaging, undersampled reconstruction, motion correction, dynamic imaging, and quantitative MRI are discussed. Code examples and tutorials are provided for some of the MRI reconstruction methods discussed in this book. The tutorials can be accessed at https://www.elsevier.com/books-and-journals/book-companion/9780128227268.

Chapters 1–4 describe the basic principles of MR image reconstruction and are complemented by Appendix A, in which the necessary mathematical tools are explained in more detail. Chapter 1 gives a brief introduction to MR physics and the image formation process. Chapter 2 formulates MRI reconstruction as an inverse problem and discusses some basic considerations for solving this inverse problem. Chapter 3 gives an overview of some commonly used algorithms used for MRI reconstruction. Chapter 4 considers sampling on a non-Cartesian grid and the corresponding reconstruction methods.

Chapters 5–11 describe more advanced reconstruction methods to speed up MR acquisition. Chapter 5 provides an overview of early constrained image reconstruction methods used in MRI. Chapter 6 explains the basics of parallel imaging reconstruction and discusses several approaches for volumetric, dynamic, and non-Cartesian acquisitions. In Chapter 7 simultaneous multislice (SMS) MRI reconstruction that enables acquisition of multiple slices at the same time is discussed. Chapter 8 presents an overview of compressed sensing and its application in rapid MRI by exploiting image sparsity. Low-rank matrix and tensor approaches to MR image reconstruction are discussed in Chapter 9. In Chapter 10 undersampled MRI reconstruction methods that approximate groups of signals by subspaces, dictionary learning methods, and smooth manifold based approaches are described. Chapter 11 provides the basics of machine learning for MRI reconstruction and discusses the use of various neural networks for single and multicoil MRI reconstruction.

Model-based MRI reconstruction approaches are discussed in Chapters 12–16. Chapter 12 explains the imaging impacts of magnetic field inhomogeneities and describes approaches that incorporate models of field inhomogeneity in the image reconstruction to correct for it. Similarly, since motion can introduce artifacts in the reconstructed images, Chapter 13 describes how to incorporate motion into the reconstruction process to correct for it. In Chapter 14, chemical-shift encoding methods, together with reconstruction approaches to separate water and fat signals, are described. Chapter 15 discusses various reconstruction approaches for model-based quantitative parameter mapping, while Chapter 16 describes specific approaches for quantitative susceptibility mapping reconstruction.

PART

Basics of MRI Reconstruction

1

CHAPTER

1 Brief Introduction to MRI Physics

Joao Tourais[a], Chiara Coletti[a], and Sebastian Weingärtner
Department of Imaging Physics, Delft University of Technology, Delft, The Netherlands

1.1 A brief history of MRI

In the 1970s, the State University of New York decided against supporting a patent application, with the rationale that the resulting patent would be unlikely to recoup the patent application costs. The patent application in question was written by Paul C. Lauterbur, describing his technique for creating an image from nuclear magnetic resonance (NMR) using field gradients. Some 30 years later, this invention would be recognized by the Nobel Committee as fundamental to the development of what is now known as magnetic resonance imaging (MRI)—and by the State University of New York, Lauterbur's employer at the time, as a huge missed opportunity. Ever since Lauterbur created his early predecessor to MR images by rotating a probe with deuterium and water in an NMR spectrometer [11], MRI has come a long way. With nearly 100 million scans performed each year worldwide, MRI is one of the backbones of modern image-based diagnosis. A plethora of key innovations by brilliant minds has facilitated the techniques that now enable unique insights into the human body.

MRI is based on the phenomenon of NMR, which was first described by Isidor I. Rabi in the 1930s [17] and later confirmed by experimental validation in 1938 [18]. The next step was taken by Felix Bloch and Edward M. Purcell, who independently demonstrated NMR in liquids and solids in 1946 [2,16]. These discoveries marked the beginning of NMR spectroscopy, which in the following years proved to be a transformative technique for the study of molecules. However, further leaps were required to enable the formation of images.

The first step in this direction was taken by Hermann Y. Carr, who reported in his 1952 Ph.D. thesis the use of spatially varying magnetic fields, so-called gradient fields, to encode the spatial location of magnetization in NMR experiments [4]. In his landmark paper from 1973, Lauterbur built on this idea to create a two-dimensional image, introducing the fundamental concept of image formation used in modern MRI machines [11]. Following Lauterbur's seminal work, Sir Peter Mansfield in Nottingham, UK, introduced the mathematical characterization of the newly formed imaging methodology [13]. The contributions of Paul C. Lauterbur and Sir Peter Mansfield that enabled the development of MRI were honored with the Nobel Prize for Physiology and Medicine in 2003. These developments sparked the rapidly evolving field of MRI research, which remains an active area of research to date.

In this chapter, we describe the basic physical principles that underpin MRI and that are necessary to understand the properties of MR images, data acquisition techniques, and reconstruction methods.

[a] Those authors contributed equally to the work.

Advances in Magnetic Resonance Technology and Applications, Volume 7, ISSN 2666-9099. https://doi.org/10.1016/B978-0-12-822726-8.00010-5

We take an explorative approach to derive six principles that aim to enable the reader to follow in-depth discussions of MRI: 1) the formation of a magnetization based on nuclear magnetism; 2) the dynamics of this magnetization in an external magnetic field; 3) the use of rotating magnetic fields to create a detectable magnetization component; 4) the use of induction to measure this component; 5) the evolution of this signal in the presence of relaxation mechanisms, and finally 6) the use of gradient fields to enable image formation. By the end of this chapter, the reader will hopefully have attained a basic understanding of an MRI experiment and the various components involved. This forms the foundation necessary to understand and discuss advanced MRI reconstructions.

1.2 Nuclear magnetism

The signal in MRI originates from magnetization that forms within the sample or the subject that is being imaged. The source of this magnetization can be found in the nuclei of certain atoms within the sample, due to the physical phenomenon known as nuclear magnetism. To understand nuclear magnetism and the resulting MRI signal, the basic properties giving rise to this phenomenon and the physical principles describing its behavior are laid out in this section.

1.2.1 Spin

Particles and nuclei are described by several characteristics, for example, their mass or their charge. Similar to mass or charge, *spin* also describes a fundamental property intrinsic to particles and nuclei. As hinted at by its terminology, a spin is related to an angular momentum, similar to the rotational momentum of a spinning top. However, unlike the classical angular momentum of the spinning top, the spin angular momentum is quantized, i.e., it can only assume certain discrete states.

Think of this as the difference between a bicycle wheel and a fortune wheel: classical angular momentum can align in any direction, similar to a bicycle wheel that can stop at any angle. A spin angular momentum, however, has only a discrete set of directions, similar to the stopping position of the fortune wheel. The number of possible directions, also known as spin states, is characterized by the *spin quantum number*, s. By conventional definition, s is a half or full integer ($s = 0, 1/2, 1, 3/2...$). Using the spin quantum number, the number of different spin states is given by $2s + 1$. The spin quantum number is a constant property that does not change over time and cannot be altered without modifying the particle or the nucleus. The spin state, on the other hand, can change over time and may be different for otherwise identical particles or nuclei.

Going back to our analogy of the spinning top, if additionally an electric charge is placed on the edge of the spinning top, it rotates due to the rotational momentum. Classical electrodynamics tells us that the rotation of the charge gives rise to a magnetic moment oriented along the rotational axis. Similarly, nuclei with nonzero spin develop a magnetic moment. This nuclear magnetic moment is primarily a magnetic dipole moment, which means its magnetic field is comparable to the one created by an infinitely small bar magnet. Thus, a useful analogy of nuclei for the understanding of MRI is a spinning bar magnet as illustrated in Fig. 1.1.[1]

[1] The analogy has naturally certain limitations. For example, nuclei with asymmetric charge distribution also exhibit the weak quadrupole moments associated with different resulting magnetic fields. These effects can have relevant consequences on the signal when imaged with MRI. For more details, the reader is referred to [28,21].

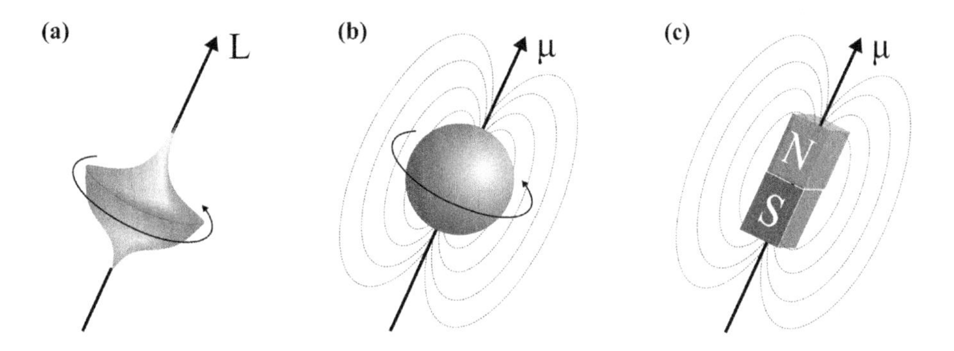

FIGURE 1.1

(a) Spinning top with angular momentum L, (b) hydrogen nucleus (proton) with magnetic moment μ, and (c) spinning bar magnet with magnetic dipole moment μ.

1.2.2 Net magnetization

When a nucleus that exhibits a nuclear magnetic moment is placed into an external magnetic field $\mathbf{B}$, different orientations of the magnetic moment are associated with different magnetic potential energies. For the discrete spin states, this entails an energy gap between the various states. This splitting of the energy levels is called the *Zeeman effect*. The gap between the energy states depends on the strength of the external magnetic field B, the magnetic moment (μ), and the spin (s). A useful quantity to describe the behavior of nuclei in external magnetic fields is the gyromagnetic ratio that is obtained as the ratio between the magnetic and the angular momentum $\gamma = \mu/(s\hbar)$, where $\hbar$ is the reduced Planck constant. The energy gap between the spin states can then be expressed as

$$\Delta E = \frac{\mu}{s} B = \gamma \hbar B. \tag{1.1}$$

When placing a multitude of nuclei into the external magnetic field, they will naturally distribute across the spin states following the principles of thermodynamics. This distribution can be described using Boltzmann statistics as a function of the energy gap ΔE and the temperature T as

$$\frac{N^+}{N^-} = e^{-\frac{\Delta E}{kT}}, \tag{1.2}$$

with N^+ and N^- being the number of nuclei in the higher and lower energy spin state, respectively. Here, k denotes the Boltzmann constant. The uneven distribution of the orientations of the magnetic moments entails that, when summing all magnetic moments in a volume, a nonzero magnetization is formed.

MRI most commonly focuses on the hydrogen nucleus ^1H which comprises only a single proton. A proton is a fermion, which means it has a half-integer spin, in this case, $s = 1/2$. Accordingly, the hydrogen nucleus has only $2s + 1 = 2$ spin states, which are often referred to as *spin-up* and *spin-down*, respectively (Fig. 1.2). In this case, the net magnetization is easily derived by subtracting the

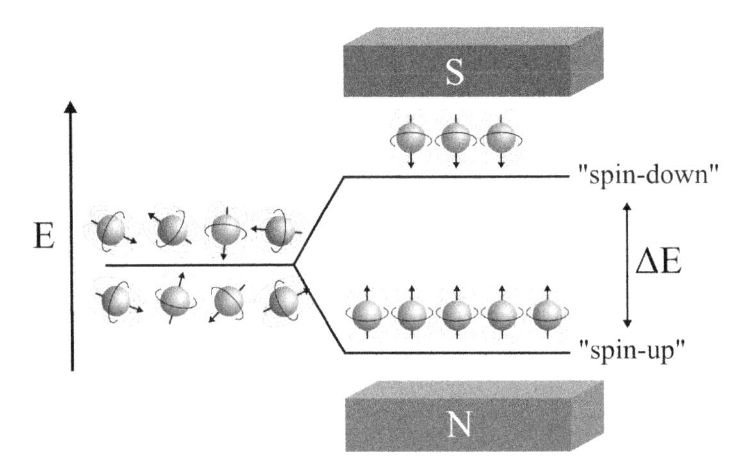

FIGURE 1.2

Zeeman splitting effect: When no external magnetic field is present, spins with randomly distributed magnetic moments are characterized by the same energy level. In the presence of an external magnetic field, the spin-up and spin-down states are separated by a potential energy gap.

anti-parallel (spin-down) magnetic moments from the parallel magnetic moment (spin-up).

$$M_0 = N^+\mu - N^-\mu = N^-\mu\left(e^{-\frac{\Delta E}{kT}} - 1\right) \tag{1.3}$$

$$= N\mu\frac{e^{-\frac{\Delta E}{kT}} - 1}{e^{-\frac{\Delta E}{kT}} + 1} = N\mu\tanh\left(\frac{\Delta E}{2kT}\right) \tag{1.4}$$

$$\approx \frac{N\mu\Delta E}{2kT} = \frac{N\mu\gamma\hbar B}{2kT}. \tag{1.5}$$

Here, $N^- = N/(e^{-\frac{\Delta E}{kT}} + 1)$, with $N = N^+ + N^-$, was used in (1.4), which can be easily derived from Eq. (1.2). Further, $\tanh(x) \approx x$ was approximated under the assumption of $x \ll 1$. Eq. (1.5) can be similarly derived for other spin values and is known as Curie's law. As illustrated in Fig. 1.3, this describes the first fundamental step in forming a magnetization that can later be used for imaging.

Principle 1: Net magnetization

If samples containing nuclei with nonzero spin s are brought into a magnetic field $\mathbf{B}$, a net-magnetization $\mathbf{M}_0$ forms. This magnetization is formed parallel to the external field ($\mathbf{M}_0 \parallel \mathbf{B}$), and its magnitude is characterized as

$$\mathbf{M}_0 = \frac{N\gamma s(s+1)\hbar^2}{3kT}\mathbf{B}, \tag{1.6}$$

with the gyromagnetic ratio γ, the spin quantum number s, the number of nuclei per volume N, the temperature T, and the Boltzmann constant k.

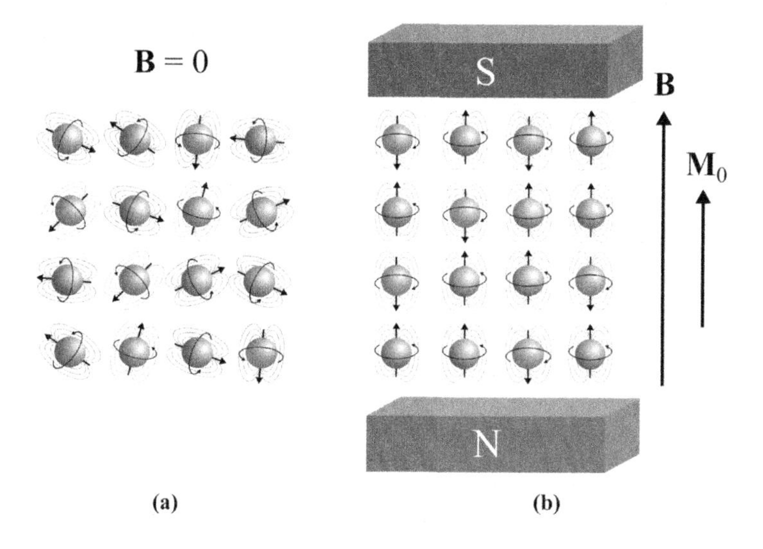

FIGURE 1.3

(a) Spins and corresponding magnetic moments are randomly oriented in the absence of an external magnetic field ($\mathbf{B} = 0$). (b) In the presence of an external magnetic field $\mathbf{B}$, however, the majority of the spins occupy the lower energy configuration, the spin-up state, generating a net magnetization $\mathbf{M}_0$ aligned with the magnetic field.

1.2.3 Magnetization dynamics

Forming a magnetization along the external magnetic field proves to be not very fruitful by itself. Instead, it helps to consider the magnetization dynamics over time. Any sample with a magnetic dipole moment $\mathbf{M}(t)$ experiences a torque $\mathbf{T}(t)$ in the presence of a static external field $\mathbf{B}$:

$$\mathbf{T}(t) = \mathbf{M}(t) \times \mathbf{B}. \tag{1.7}$$

Accordingly, the same happens to a sample with nonzero spin nuclei. However, as initially explained, the spin is also associated with an angular momentum. Thus, this torque affects the cumulative angular momentum $\mathbf{L}_s$ of the sample as[2]

$$\dot{\mathbf{L}}_s(t) = \mathbf{M}(t) \times \mathbf{B}. \tag{1.8}$$

The gyromagnetic ratio γ links the magnetic and the angular moment of our spin ensemble. Hence, we can link the dynamics of the angular momentum to that of the net magnetization as presented in the following principle.

[2] For convenience, we use Newton's notation for differentiation throughout this chapter. If f is a function of t, then the derivative of f with respect to t is denoted as $\dot{f}$.

Principle 2: Larmor precession

The net magnetization $\mathbf{M}(t)$ described in Principle 1 undergoes the following dynamics in a static external magnetic field $\mathbf{B}$.[3]

$$\dot{\mathbf{M}}(t) = \gamma \mathbf{M}(t) \times \mathbf{B} + \text{relaxation terms}, \tag{1.9}$$

with the gyromagnetic ratio γ. These dynamics are known as Larmor precession [10].

This enables us to gain some important insights into the magnetization. First, if the magnetization is parallel to the external magnetic field, no dynamic behavior is observed

$$\mathbf{M}(t) \parallel \mathbf{B} \implies \mathbf{M}(t) \times \mathbf{B} = 0 \implies \dot{\mathbf{M}} = 0 \implies \mathbf{M}(t) = \mathbf{M}(0). \tag{1.10}$$

However, if the magnetization is skewed away from the external field by an angle θ, Principle 2 dictates that the gradient direction is tangential to a circle around the main magnetic field, and its magnitude is found as

$$\|\dot{\mathbf{M}}(t)\| = \gamma \sin(\theta) M B, \tag{1.11}$$

with the (static) magnitude of the magnetization $M = \|\mathbf{M}(t)\|$ and the field strength $B = \|\mathbf{B}\|$.

Thus, the magnetization vector describes a cone around the external magnetic field, with the tip lying on a circle. This type of motion is called precession and can again be likened to our spinning top. If the spinning top is nudged so that its rotational axis is angled with respect to the gravitational force, it precesses comparably around this axis.

The circle prescribed in this motion has radius $\sin(\theta)M$. Thus, the frequency of the rotation can be obtained as

$$\frac{\gamma \sin(\theta) M B}{\sin(\theta) M} = \gamma B = \omega. \tag{1.12}$$

This field-dependent angular frequency is also known as the *Larmor frequency* and will prove important for our understanding of the MRI signal.

1.3 NMR/MRI signal

1.3.1 Signal creation and reception

There is little use in knowing the magnetization dynamics if we cannot manipulate or detect them. We will now explore how the resonance phenomenon can be employed to change the magnetization and how this can be read out into a detectable signal.

[3] The relevant relaxation terms will be introduced in Subsection 1.3.2.

1.3.1.1 Radiofrequency pulses

In MRI a signal can only be created if the magnetization is nutated away from the external magnetic field. For this purpose, the special case of rotating external magnetic fields is particularly relevant. To this end, we start by defining the overall magnetic field $\mathbf{B}_{tot}(t)$ as

$$\mathbf{B}_{tot}(t) = \begin{pmatrix} B_1\cos(\omega_1 t) \\ B_1\sin(\omega_1 t) \\ B_0 \end{pmatrix}, \tag{1.13}$$

with a stationary magnetic field with strength B_0 along the z-axis and a secondary magnetic field with amplitude B_1 rotating with angular frequency ω_1 in the transverse plane, i.e., the plane orthogonal to the $\mathbf{B}_0$ field.

To better understand the effects of rotating magnetic fields, it is useful to consider a rotating frame of reference (RFR). RFRs are a familiar concept. The reader is likely observing this chapter in an RFR: In a galactic reference frame, this book would rotate around the axis of the globe. However, the reference frame we observe is spinning with the globe and makes the book appear stationary. Similarly, we can define an RFR to rotate with the rotating magnetic field.[4] For the stationary or laboratory frame of reference (x, y, z) and the RFR (x', y', z'), we can define the following reference frame transformation

$$\begin{pmatrix} x \\ y \\ z \end{pmatrix} \mapsto \begin{pmatrix} x' \\ y' \\ z' \end{pmatrix} : \mathbf{R}(t) = \begin{pmatrix} \cos(\omega_1 t) & \sin(\omega_1 t) & 0 \\ -\sin(\omega_1 t) & \cos(\omega_1 t) & 0 \\ 0 & 0 & 1 \end{pmatrix}, \tag{1.14}$$

with $\mathbf{R}(t)$ being a proper rotation matrix, i.e., $\mathbf{R}(t)\mathbf{R}^T(t) = \mathbf{I}$ and $\det(\mathbf{R}(t)) = 1$.

Observing the magnetization dynamics of a magnetization $\mathbf{M}'(t) = \mathbf{R}(t)\mathbf{M}(t)$ in an RFR in the presence of a rotating dynamic magnetic field $\mathbf{B}_{tot}(t)$ such as defined in Eq. (1.13) yields

$$\dot{\mathbf{M}}'(t) = \frac{d}{dt}(\mathbf{R}(t)\mathbf{M}(t)) = \dot{\mathbf{R}}(t)\mathbf{M}(t) + \mathbf{R}(t)\dot{\mathbf{M}}(t) \tag{1.15}$$

$$= \dot{\mathbf{R}}(t)\mathbf{R}^T(t)\mathbf{M}'(t) + \mathbf{R}(t)\gamma\mathbf{M}(t) \times \mathbf{B}_{tot}(t). \tag{1.16}$$

By multiplying out the matrices, it can quickly be concluded that

$$\dot{\mathbf{R}}(t)\mathbf{R}^T(t)\mathbf{M}'(t) = -\omega_1\mathbf{M}'(t) \times \begin{pmatrix} 0 \\ 0 \\ 1 \end{pmatrix} \tag{1.17}$$

and

$$\mathbf{M}(t) \times \mathbf{B}(t) = \mathbf{R}^T(t)\mathbf{M}'(t) \times \mathbf{B}_{tot}(t), \tag{1.18}$$

[4] An extensive explanation of the RFR systems in MRI can be found in [33].

due to the rotation properties of the cross product. It follows that

$$\dot{\mathbf{M}}'(t) = -\omega_1 \mathbf{M}'(t) \times \begin{pmatrix} 0 \\ 0 \\ 1 \end{pmatrix} + \mathbf{R}(t)\gamma \mathbf{R}^T(t)\mathbf{M}'(t) \times \mathbf{B}_{tot}(t) \tag{1.19}$$

$$= \gamma \left(\mathbf{B}'_{tot}(t) - \begin{pmatrix} 0 \\ 0 \\ B_0 \frac{\omega_1}{\omega_0} \end{pmatrix} \right), \tag{1.20}$$

with the Larmor frequency $\omega_0 = \gamma B_0$ and

$$\mathbf{B}'_{tot}(t) = \mathbf{R}(t)\mathbf{B}(t) = \begin{pmatrix} B_1 \\ 0 \\ B_0 \end{pmatrix}. \tag{1.21}$$

This yields the following principle:

Principle 3: Rotating frame excitation

For a magnetic field with a rotational component in the xy-plane $\mathbf{B}_{tot}(t) = \begin{pmatrix} B_1 \cos(\omega_1 t) & B_1 \sin(\omega_1 t) & B_0 \end{pmatrix}^T$, the net magnetization $\mathbf{M}'(t)$ in a RFR rotating about the z-axis with frequency ω_1 can be described as

$$\dot{\mathbf{M}}'(t) = \gamma \mathbf{M}'(t) \times \mathbf{B}_{\mathbf{eff}}(t) + \text{relaxation terms}, \tag{1.22}$$

with the gyromagnetic ratio γ, the Larmor frequency $\omega_0 = \gamma B_0$ and

$$\mathbf{B}_{\mathbf{eff}}(t) = \begin{pmatrix} B_1 \\ 0 \\ B_0 \left(1 - \frac{\omega_1}{\omega_0}\right) \end{pmatrix}. \tag{1.23}$$

Again, this principle allows us to derive some important properties. Namely, if the secondary magnetic field $\mathbf{B}_1$ is applied at the resonance frequency, i.e., $\omega_1 = \omega_0$, the magnetization vector in the RFR rotates about the secondary magnetic field only, as shown in Fig. 1.4. The speed of this rotation is determined by the strength of the secondary magnetic field. Hence, the angle accumulated by the secondary magnetic field of time-varying amplitude $B_1(t)$ can be computed as

$$\alpha = \int_0^\infty \gamma B_1(t) dt, \tag{1.24}$$

and is called the *flip angle*. Accordingly, for the simple case of a constant magnetic field B_1 applied during a time τ, the flip angle $\alpha = \gamma B_1 \tau$. A flip angle of 90° tips the magnetization into the transverse plane and yields the strongest magnetization component orthogonal to the main magnetic field. The application of $\alpha = 180°$ returns no transverse magnetization, essentially inverting the magnetization.

On the other hand, for $\omega_1 \neq \omega_0$ the magnetization precesses around an angled effective field, where the angle between $\mathbf{B}_{eff}$ and the z-axis amounts to $\theta = \text{atan}\left(B_1 / \left(B_0 \left(1 - \omega_1/\omega_0\right)\right)\right)$. In MRI, the field

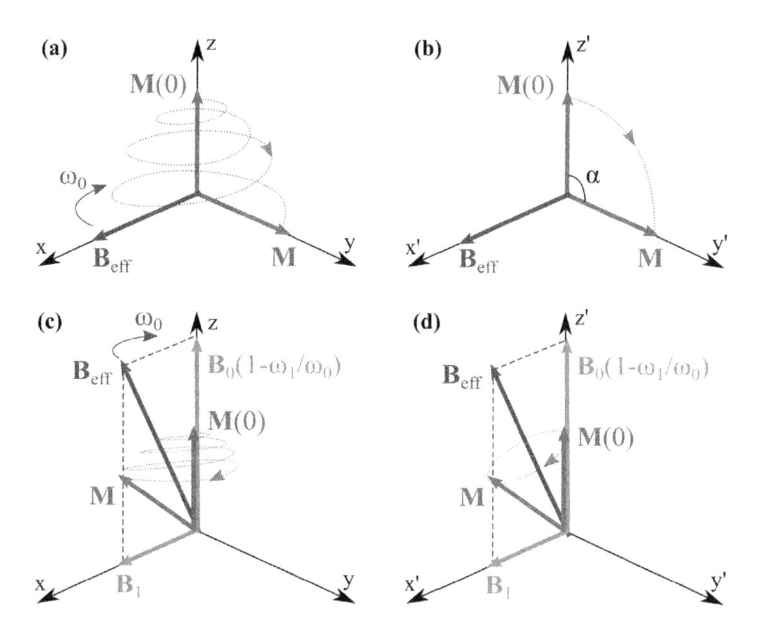

FIGURE 1.4

On-resonance 90° excitation pulse ($\mathbf{B}_{eff} = \mathbf{B}_1$, $\omega_1 = \omega_0$) tipping the magnetization to the transverse plane, as viewed in the (a) laboratory and (b) rotating frame of reference (RFR). (c, d) Off-resonance excitation pulse ($\omega_1 \neq \omega_0$), causing $\mathbf{B}_{eff}$ to be angled towards the z-axis, preventing full excitation.

strength of the secondary magnetic field is much smaller than the main magnetic field ($B_1 \ll B_0$), thus, to induce a precession that rotates the magnetization away from the z-axis, it is important to closely match the resonance frequency.

In clinical MRI systems, the field strength is often around $B_0 = 1.5T$. At this field strength, resonance is achieved at the Larmor frequency of $\omega_0/(2\pi) \approx 63.9$ MHz (for the hydrogen nucleus). As these frequencies are in the range of radio waves, the rotating magnetic field $\mathbf{B}_1$ may be referred to as the radiofrequency (RF) field. Because the B_1 field is employed for a short duration, it is also often commonly referred to as an RF pulse. An RF pulse creating a flip angle of 90° is (conveniently) called a 90° pulse.

1.3.1.2 Signal detection

To better understand how to create a detectable signal from the magnetization $\mathbf{M}$ in MRI, it is useful to consider the individual magnetization components

$$\mathbf{M}(t) = \begin{pmatrix} M_x(t) \\ M_y(t) \\ M_z(t) \end{pmatrix}. \tag{1.25}$$

$M_z(t)$ describes the magnetization component along the z-axis that is aligned with the external magnetic field $\mathbf{B}_0 = \begin{pmatrix} 0 & 0 & B_0 \end{pmatrix}^T$, and is called the longitudinal magnetization. The vector $\mathbf{M}_{xy} = \begin{pmatrix} M_x & M_y \end{pmatrix}^T$ is referred to as the transverse magnetization that lies in the xy-plane, orthogonal to the main magnetic field $\mathbf{B}_0$.

Using this notation, the net magnetization $\mathbf{M}$ that is formed, following Principle 1, is aligned with the external magnetic field $\mathbf{B}_0$

$$\mathbf{M}(0) = \begin{pmatrix} 0 \\ 0 \\ M_0 \end{pmatrix}. \tag{1.26}$$

Using Principle 3, we can calculate the effect of an RF pulse played along the y'-axis with a duration τ and strength B_1 chosen to achieve a flip angle $\alpha = \gamma \tau B_1 = 90°$. When neglecting the relaxation terms, it follows for the magnetization that

$$\mathbf{M}'(\tau) = \begin{pmatrix} M_0 \\ 0 \\ 0 \end{pmatrix} \Longrightarrow \mathbf{M}(\tau + t) = \begin{pmatrix} \cos(\omega_0 t) M_0 \\ \sin(\omega_0 t) M_0 \\ 0 \end{pmatrix}, \tag{1.27}$$

in the RFR and the stationary frame, respectively.

This means that a time-varying magnetization component has been created in the transverse plane. The time-varying magnetic field originating in the precession of the magnetization vectors can be picked up using an inductive coil near the sample. In its most simplistic form, this coil is an individual loop of wire. Faraday's law of induction declares that a time-varying magnetic flux through a wire loop produces a current in the loop and induces a voltage ϕ over the edges of the loop.

Principle 4: Faraday's law of induction

If a coil is placed such that a magnetic field has a temporally varying component normal to the coil area, a voltage is induced in the coil. The produced voltage E is proportional to the negative of the rate of temporal variation of the flux ϕ that in turn is proportional to the magnetic field strength normal to the coil ($\phi \propto \mathbf{M}_{xy}$):

$$E \propto -\frac{d\phi}{dt}. \tag{1.28}$$

The demand for a time-varying magnetic flux is the reason solely precessing magnetization in the xy-plane yields an increase to an NMR signal: the z-component does not precess and, hence, generates no voltage. At greater strengths of the $\mathbf{B}_0$ field, the magnetization precesses at a higher frequency, and so the value of $d\phi/dt$ rises. Thus, stronger magnetic fields yield better signal strength not only because of the larger nuclear polarization (Principle 1) but also because of the additional increase in magnetic flux.

1.3.2 Signal relaxation and decay

When researchers first created NMR signals from the principles just cited, such as Erwin Hahn with his pulsed NMR experiments in 1950 [8], they made the sobering observation that the carefully created signal would quickly decay. The signal that was induced in a nearby coil in this experiment was termed free induction decay (FID). It showed an oscillating behavior at the Larmor frequency, but also decay of the amplitude of those oscillations was observed. It was Felix Bloch who phenomenologically concluded that the created NMR signal is subject to multiple relaxation and decay processes, which prevent a transverse magnetization from indefinite signal induction [2].

To create a detectable signal, following Principle 3, an RF pulse needs to be applied at the Larmor frequency. Note that this frequency corresponds to the energy gap between the spin states

$$\Delta E = \omega_0 h = \gamma h \mathbf{B}_0, \tag{1.29}$$

where h is the Planck constant. Thus, when a secondary magnetic field is applied, energy is transmitted into the system. This excess energy excites the system above the thermal equilibrium, to which it naturally relaxes back. In the vector model, the M_z component has been lowered from its equilibrium value of M_0, and the M_x and/or M_y components may have a nonzero value. Each of the magnetization components M_z, M_x, and M_y returns to the thermal equilibrium value over time. To better understand this relaxation, it is useful to consider two different processes: 1) The recovery of the M_z component to the thermal equilibrium value M_0, also known as longitudinal relaxation, and 2) The decay of the transverse magnetization towards zero, also known as transverse relaxation.

1.3.2.1 Longitudinal relaxation

During the recovery to thermal equilibrium, the spin system dissipates energy to the surrounding lattice. Thus, the recovery of the longitudinal magnetization is also known as spin-lattice relaxation. The closer the system gets to thermal equilibrium, the closer the longitudinal magnetization is to the equilibrium magnetization M_0. The rate of recovery depends on the excess of energy in the spin system, as well as a sample-specific constant T_1. This allows us to describe the dynamics of longitudinal magnetization, including magnetization recovery, as

$$\dot{\mathbf{M}}_z(t) = \gamma(\mathbf{M}(t) \times \mathbf{B}(t))_z - \frac{M_z(t) - M_0}{T_1}, \tag{1.30}$$

where $(\mathbf{x})_z = x_z$, for $\mathbf{x} = \begin{pmatrix} x_x & x_y & x_z \end{pmatrix}^T$.

In MRI, observing the longitudinal magnetization following an RF pulse of flip angle α is of particular interest. Following Principle 3, we see that, promptly after the application of the RF pulse, the longitudinal magnetization M_z can be described by $M_0 \cos(\alpha)$. For a time t after the RF pulse, Eq. (1.30) gives the value of M_z as

$$M_z(t) = M_0 \cos(\alpha) + (M_0 - M_0 \cos(\alpha)) \left(1 - e^{-\frac{t}{T_1}}\right). \tag{1.31}$$

For instance, following a 90° pulse, the value of M_z is given by

$$M_z(t) = M_0 \left(1 - e^{-\frac{t}{T_1}}\right). \tag{1.32}$$

This recovery and an example of longitudinal magnetization regrowth following a 180° pulse can be seen in Fig. 1.5.

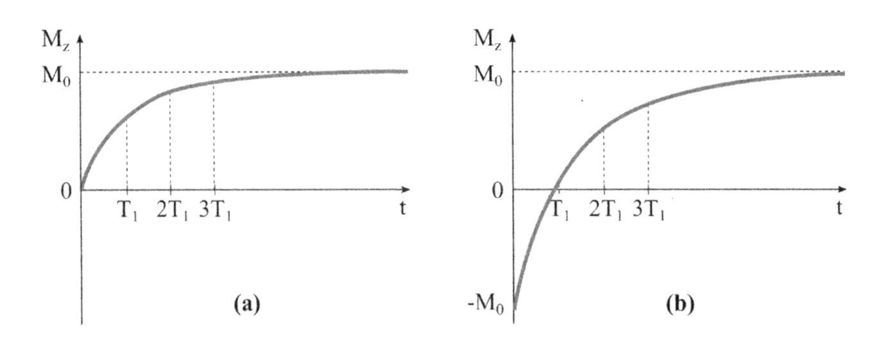

FIGURE 1.5

Exponential regrowth, over time, of the longitudinal component of the magnetization $M_z(t)$ to the equilibrium value M_0 after a 90° (a) or 180° (b) excitation pulse.

1.3.2.2 Transverse relaxation

The physical basis for the second relaxation mechanism, the transverse magnetization decay, differs from the T_1 relaxation process. A coherent magnetization in the transverse plane can be maintained only as long as all magnetic moments contributing to the transverse magnetization precess at the same frequency. However, even in a perfectly homogeneous magnetic field, magnetic moments precess at marginally different frequencies. This is due to the magnetic moments of neighboring nuclei that influence the observed net magnetic field strength at any location and in a time-varying manner, due to the molecular motion. Transverse magnetization decay is caused by losing the *phase coherence* among the magnetic moments. As a result, the net magnetization decays over time (Fig. 1.6): The magnetic moments fan out, and consequently, the magnitude of the net sum decreases. Thus, the transverse magnetization unwinds back to the thermal equilibrium value of zero with a sample-specific time constant called the spin–spin relaxation time, or T_2:

$$\dot{M}_x(t) = \gamma(\mathbf{M}(t) \times \mathbf{B}(t))_x - \frac{M_x(t)}{T_2}, \tag{1.33}$$

$$\dot{M}_y(t) = \gamma(\mathbf{M}(t) \times \mathbf{B}(t))_y - \frac{M_y(t)}{T_2}. \tag{1.34}$$

Together these relaxation processes constitute the Bloch equations, one of the most important tools for MRI researchers.

Principle 5: Bloch equations

The time evolution of M_z, M_x, and M_y can be described by differential equations, recognized as the Bloch equations:

$$\dot{\mathbf{M}}(t) = \frac{d}{dt}\begin{pmatrix} M_x \\ M_y \\ M_z \end{pmatrix} = \gamma(\mathbf{M}(t) \times \mathbf{B}(t)) - \begin{pmatrix} \frac{M_x(t)}{T_2} \\ \frac{M_y(t)}{T_2} \\ \frac{M_z(t)-M_0}{T_1} \end{pmatrix} \tag{1.35}$$

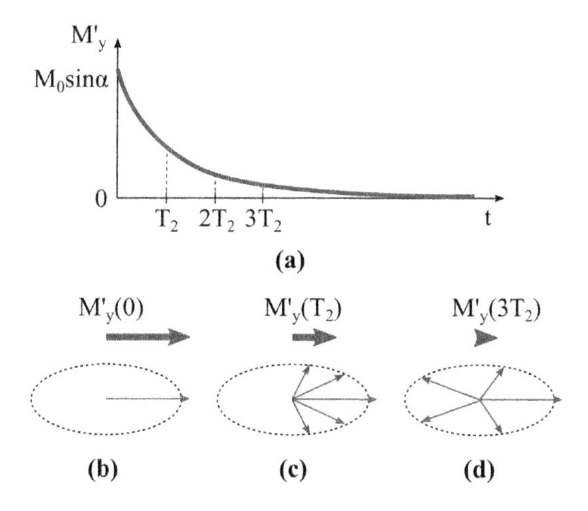

FIGURE 1.6

(a) Exponential decay of the transverse component of the magnetization ($M_y'(t)$) in the RFR over time. The transverse component recovers to the equilibrium value zero after excitation with an arbitrary flip angle α. The coherence loss of magnetic moments precessing in the transverse plane and the corresponding net transverse magnetization M_y is shown right after the excitation pulse (b) and after one (c) or several (d) spin–spin relaxation periods T_2.

Two mechanisms that contribute to the loss of phase coherence of the transverse magnetization can be differentiated. The first mechanism is the "real" T_2 decay originating in spin–spin interaction, as described before. The second mechanism derives from the spatial disparities of the magnetic field strength inside the body or across the sample. There are, in turn, two notable causes for these differences. The first is the inherent magnet design: It is impracticable to design a magnet generating a perfectly homogeneous magnetic field throughout the whole subject. The second cause is given by regional changes in the magnetic field due to the distinct magnetic susceptibilities of different tissues. This effect is particularly pronounced at the air/tissue and bone/tissue boundaries. Collectively, these effects create a decline in phase coherence, which is described by a relaxation time T_2'. The effective relaxation time that characterizes the decay of transverse magnetization is a combination of signal loss due to T_2 and T_2' effects. This is defined by T_2^*, and it can be expressed by

$$\frac{1}{T_2^*} = \frac{1}{T_2'} + \frac{1}{T_2}. \tag{1.36}$$

Considering these decay mechanisms, the detectable magnetization in the transverse plane from Eq. (1.27), which is formed after the application of a 90° RF pulse, can be described as

$$\mathbf{M}(t) = \begin{pmatrix} \cos(\omega_0 t) \\ \sin(\omega_0 t) \\ 0 \end{pmatrix} M_0 e^{-\frac{t}{T_2^*}}. \tag{1.37}$$

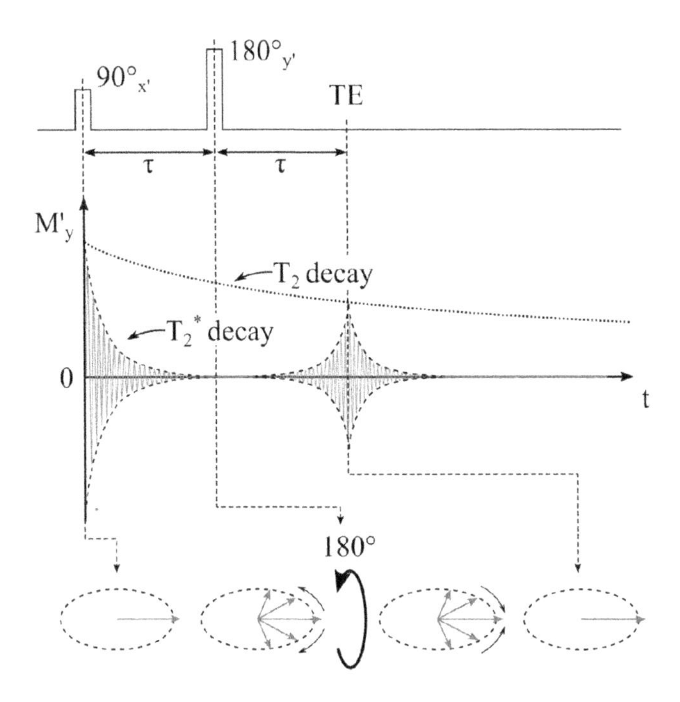

FIGURE 1.7

Signal evolution during a spin-echo experiment in the RFR (green, solid) and the laboratory frame of reference (gray, dashed). After the first 90° pulse, spins start dephasing, generating the FID signal, characterized by the T_2^* decay. The application of a 180° pulse inverts the process and generates an echo at $TE = 2\tau$, recovering T_2' losses. Overall, the spin-echo signal follows a T_2 decay.

In high-resolution NMR spectroscopy for chemical examination, the sample is small and spatially uniform. Thus, there are only small field inhomogeneities leading to long T_2' times. In this case, the value of T_2^* is adequately approximated by T_2, and Principle 5 is reasonably accurate for representing the decay of transverse magnetization in the entire sample. In MRI, however, the value of T_2' can be up to 10–100 times lower than that of T_2. Thus, T_2' is often the main driver of T_2^* decay.

In practice, it can be important to discriminate between T_2 and T_2^*. The field variations contributing to T_2 are temporally variant, but the effects contributing to T_2^* are static over time. Fortunately, a mechanism can also be used to recover the signal loss due to temporally invariant causes. If, at a time τ, following the 90° pulse, an additional 180° pulse is applied, the magnetization is flipped. As shown in Fig. 1.7, phase dispersion that has been caused by temporally invariant inhomogeneities, is then rewound. Thus, this pulse is also known as *refocusing pulse*. At the time point *echo time* ($TE = 2\tau$), when the phase coherence is maximally restored, a *spin-echo* is formed, which compensated for T_2' decay, and T_2 can be measured [8][22].

1.4 Image formation

The NMR signal defined so far is the gross summation of all net magnetic moments in the sample. It was not before 1973 when the contributions of Paul C. Lauterbur enabled to make spatial sense of the signal [11]. The key innovation was the introduction of *gradient fields*, magnetic fields whose strength is a function of the spatial location. It was discovered that, in the presence of a gradient field, the spatial distribution of the protons gives rise to a variety of proton resonant frequencies, each dependent upon the location of the proton in the sample.

To understand image formation using magnetic field gradients, consider the most common form, linear field gradients. Let $\mathbf{G} = \nabla \mathbf{B}$ denote the *gradient* of the magnetic field strength of this gradient field. The total magnetic field strength at a position $\mathbf{r}$ can then be obtained as

$$B(\mathbf{r}) = B_0 + \mathbf{G}^T \mathbf{r}. \tag{1.38}$$

It should be noted that the orientation of the gradient fields (but not the direction of change) is always parallel to the static field, in order to alter the longitudinal component of the overall magnetic field.

In the presence of a field gradient, the Larmor frequency of the magnetization becomes a function of its spatial location,

$$\omega(\mathbf{r}) = \gamma \mathbf{G}^T \mathbf{r} \tag{1.39}$$

In other words, the gradient fields allow linking the Larmor frequency to the spatial location inside the magnet.

To see how the difference in precession frequency can be used for encoding, consider the application of gradient fields after a transverse magnetization has been created. For this purpose, the transverse magnetization is best described as a complex function of the time and the location $M_{xy}(t, \mathbf{r}) = M_x(t, \mathbf{r}) + i M_y(t, \mathbf{r})$. With this definition, the complex plane of $M_{xy}(t, \mathbf{r})$ coincides with the xy-plane. Following Principle 2, and assuming no relaxation effects due to $t \ll T_2^*$, it is easily verified that the precession can be described as

$$M_{xy}(t, \mathbf{r}) = e^{-i\gamma B(\mathbf{r})t} M_{xy}^0(\mathbf{r}). \tag{1.40}$$

Combining Eqs. (1.40) and (1.38), yields

$$M_{xy}(t, \mathbf{r}) = e^{-i\gamma B_0 t} e^{-i\gamma \mathbf{G}^T \mathbf{r} t} M_{xy}^0(\mathbf{r}). \tag{1.41}$$

The MRI signal can then be obtained as the spatial sum across the imaging volume

$$S'(t) \propto \int e^{-i\gamma B_0 t} e^{-i\gamma \mathbf{G}^T \mathbf{r} t} M_{xy}^0(\mathbf{r}) d\mathbf{r} \tag{1.42}$$

$$= e^{-i\omega_0 t} \int e^{-i\gamma \mathbf{G}^T \mathbf{r} t} M_{xy}^0(\mathbf{r}) d\mathbf{r}. \tag{1.43}$$

After demodulation of the signal with the Larmor frequency ω_0,

$$S(t) \propto \int e^{-i\gamma \mathbf{G}^T \mathbf{r} t} M_{xy}^0(\mathbf{r}) d\mathbf{r}. \tag{1.44}$$

The well-versed reader might recognize the resemblance to a Fourier transform. Indeed, to connect to the Fourier transform, consider a gradient $\mathbf{G} = \begin{pmatrix} G_x & G_y & G_z \end{pmatrix}^T$ after time t. The gradient moments can be defined as $m_x = G_x t$, $m_y = G_y t$ and $m_z = G_z t$ and wavenumbers as $k_x = \gamma m_x$, $k_y = \gamma m_y$ and $k_z = \gamma m_z$ to yield

$$S(k_x, k_y, k_z) \propto \int\int\int e^{-i(k_x x + k_y y + k_z z)} M^0_{xy}(x, y, z)\, dx\, dy\, dz. \tag{1.45}$$

It follows that the Fourier transform of the signal S recovers the spatial information of the magnetization.

Principle 6: Gradient encoding

If gradient fields with gradient moments m_x, m_y and m_z are applied after the creation of a transverse magnetization $M^0_{xy}(x, y, z)$, the resulting signal can be described as

$$S(k_x, k_y, k_z) \propto \int\int\int e^{-i(k_x x + k_y y + k_z z)} M_{xy}(x, y, z)\, dx\, dy\, dz, \tag{1.46}$$

using the wavenumbers $k_x = \gamma m_x$, $k_y = \gamma m_y$, and $k_z = \gamma m_z$, and the gyromagnetic ratio γ.

Of note, the Fourier transform of this signal recovers the spatial information about the magnetization

$$\mathbf{F}(S)(x, y, z) \propto M^0_{xy}(x, y, z). \tag{1.47}$$

In the remainder of this section it will be explored, how this principle can be used to encode an image in MRI. Commonly, three steps can be differentiated in this process: frequency encoding, phase encoding, and slice selection. Principle 6 is commonly considered in a stationary reference frame aligned with the image orientation to ease notation. Thus, we can assume that frequency encoding is performed along the x-axis, phase encoding along the y-axis, and slice selection along the z-axis.

1.4.1 Frequency encoding

Frequency encoding is the most straightforward way to apply Principle 6 for spatial encoding. In frequency encoding, a gradient G_x is turned on and the signal is continuously sampled, while the gradient is applied. The acquired signal can thus be described as

$$S(\gamma G_x t, 0, 0) \propto \int e^{-ik_x x} \int\int M_{xy}(x, y, z)\, dy\, dz\, dx. \tag{1.48}$$

Accordingly, one spatial dimension can be fully characterized using frequency encoding.

Since the frequency encoding gradient G_x remains turned on, the spins precess at different frequencies, while the signal is acquired. Thus, the received composite signal is a summation of the contributions at many frequencies. This situation can be compared to a glass harp (Fig. 1.8a). Each spin is a different wine glass, and the filling line that determines the frequency is a function of the position along the frequency encoding direction. In this analogy, frequency encoding is obtained by playing all glasses at the same time, and a frequency analysis can be used to recover the position of the individual glasses.

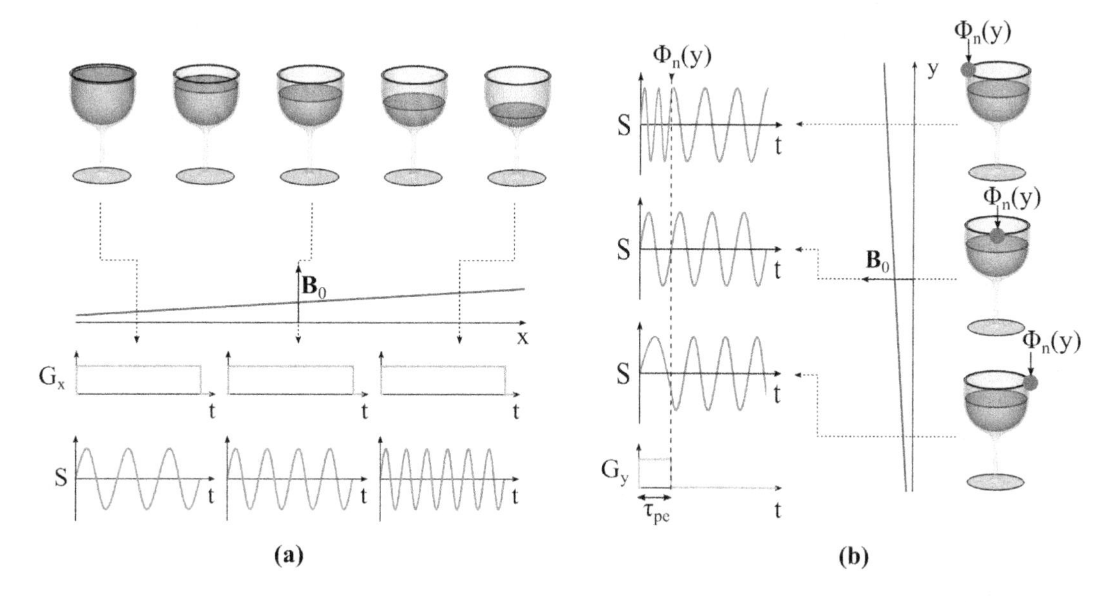

FIGURE 1.8

(a) Frequency encoding: when G_x is applied, it induces a spatially dependent shift in the B_0 field and causes spins, at different locations along the x-axis, to precess at different frequencies. This enables signal localization in the x direction, similar to playing glass harp, with wine glasses filled to different water levels, where the higher the water level, the lower the pitch. (b) Phase encoding: When G_y is applied for a duration τ_{pe}, the spins accumulate a spatially dependent phase value ϕ, precessing with the same frequency but different phases once G_y is turned off. This can be compared to partially water-filled glasses, with identical water levels, being played at different initial phases.

1.4.2 Phase encoding

In order to encode additional spatial dimensions, phase encoding can be applied. In phase encoding, a gradient is turned on and off *before* the frequency encoding and data acquisition begins. A number N_y of distinct values of this phase encoding gradient with stepwise amplitude increment ΔG_y are used for a fixed duration τ_{pe}, to obtain different gradient moments $m_y(n) = n\Delta G_y \tau_{pe}$, $-N_y < n \leq N_y$.

In combination with frequency encoding, we obtain the following signal:

$$S(\gamma G_x t, \gamma n \Delta G_y \tau_{pe}, 0) \propto \int\int e^{-ik_x x + k_y y} \int M_{xy}^0(x, y, z)\, dz\, dy\, dx. \tag{1.49}$$

During the application of G_y with moment $m_y(n)$, spins accumulate a spatially dependent phase shift $\phi_n(y) = \gamma m_y(n) y$. So, while all spins along the phase encode dimension precess at the same frequency, the phase between the signals is shifted. Since only one phase shift can be obtained in the phase encoding direction, the application of many gradient amplitudes is necessary to spatially resolve this information. In the glass-harp analogy, phase encoding is represented by glasses with the same

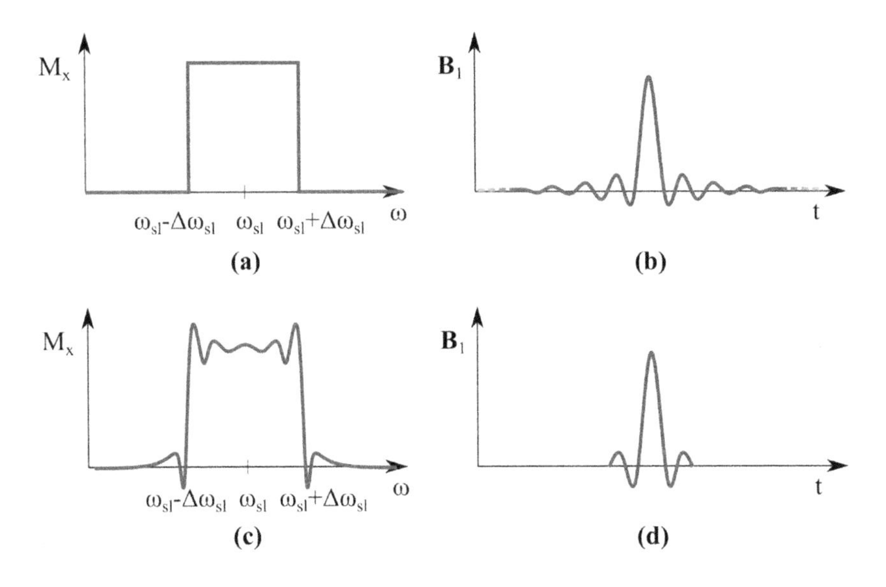

FIGURE 1.9

(a) Ideal slice profile of an excitation RF pulse centered at frequency ω_{sl}, with an excitation bandwidth $\pm\Delta\omega_{sl}$, and (b) its corresponding temporal profile, an infinitely long sinc pulse. (d) Example of a time-limited RF pulse, obtained by truncating a sinc pulse with a rectangular window. This causes ripples at the edge of the slice profile, the so-called Gibbs ringing (c).

filling line, thus producing the same frequency, but they are played with a different phase, as illustrated in Fig. 1.8b.

Phase encoding, as just described, can be applied along multiple dimensions. For example, additional phase encoding can be used in the slice dimension to enable the acquisition of a 3D image. To this end, a second phase encoding gradient is applied such that, for each gradient step in the new axis, a whole set of phase and frequency encoding steps are obtained. However, the number of phase encoding steps required grows exponentially with the number of dimensions encoded. Because phase encoding is the major time-limiting factor in MRI, 3D imaging is often time-consuming, and the acquisition of (multiple) 2D slices remains standard in clinical use.

1.4.3 Slice selection

To acquire a 2D slice in MRI, slice selection can be accomplished by a frequency-selective RF pulse employed together with a magnetic field gradient G_z. During the application of G_z, the Larmor frequency varies as a function of the location along the slice-encoding direction. Thus, if the selective RF pulse is administered at a frequency ω_{sl} with an excitation bandwidth of $\pm\Delta\omega_{sl}$ (Fig. 1.9), then following Principle 3 only protons precessing at frequencies between $\omega_{sl} + \Delta\omega_{sl}$ and $\omega_{sl} - \Delta\omega_{sl}$ are flipped into the transverse plane; those with resonant frequencies outside this range are not effectively nutated and remain largely aligned with the z-direction. In combination with frequency and phase encoding,

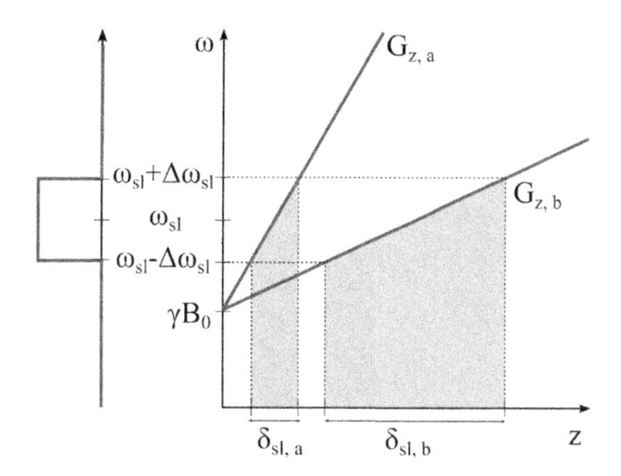

FIGURE 1.10

Illustration of two different gradient strengths, $G_{z,a} > G_{z,b}$, and their effect on slice selection. Considering an RF pulse with an excitation bandwidth of $2\Delta\omega_{sl}$, the slice thickness δ_{sl} can be varied by changing the gradient strength G_z.

slice selection enables two-dimensional imaging of the slice of interest, as follows:

$$S(\gamma G_x t, \gamma n \Delta G_y \tau_{pe}) \propto \int_{slice} \int \int e^{-ik_x x + k_y y} M_{xy}^0(x, y, z)\, dx\, dy\, dz. \quad (1.50)$$

The thickness δ_{sl} of the acquired slice can be characterized by the bandwidth of the RF pulse, $2\Delta\omega_{sl}$, and the value of the slice-selection gradient:

$$\delta_{sl} = \frac{2\Delta\omega_{sl}}{\gamma G_z}. \quad (1.51)$$

The slice thickness can thus be increased either by reducing the strength of G_z or by broadening the frequency bandwidth of the RF pulse, as illustrated in Fig. 1.10. The ideal slice-excitation profile of an RF pulse would cause the magnetization to be excited with the target flip angle inside the slice thickness and with a zero flip angle outside. This corresponds to a rectangular slice profile. For small flip angles, the achieved slice profile can be approximated by the frequency spectrum of the RF pulse. Thus, the commonly chosen RF pulse shapes are an approximation to sinc pulses, whose frequency spectrum is rectangular (Fig. 1.9a–b). However, to obtain a perfectly rectangular slice profile, an infinitely long RF pulse would be required. In practice, truncated sinc pulses are employed (Fig. 1.9c–d). For a fixed pulse shape, increasing the pulse duration results in a narrowing of the frequency bandwidth and, consequently, smaller slice thickness. Shifting the center frequency ω_{sl} of the RF pulse, on the other hand, alters the slice position. Thus, the excitation pulse properties can be used to select the volume of interest.

The application of a slice-selection gradient also induces a shift of Larmor frequency within the excited slice. Hence, after slice selection, the phase is spatially inhomogeneous across the slice thickness, which results in signal loss. To reduce this undesired loss of phase coherence, a rephasing gradient

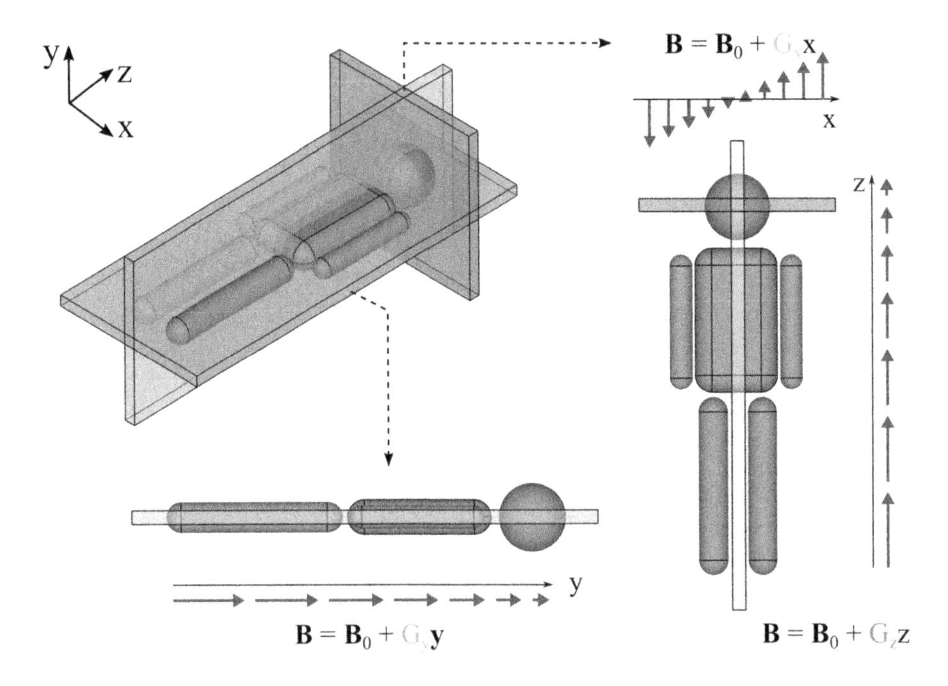

FIGURE 1.11

Sagittal, coronal, and axial slices acquired by choosing the slice-selection direction along the x, y, or z-axis (in the laboratory frame), respectively. Gradients induce spatially dependent shifts in the magnetic field strength B_0 along the direction of their application, allowing the excitation of narrow bands of spins in this direction.

with opposite polarity is employed for half the duration to match the gradient moment. The area of the rephasing lobe is half of the slice selection gradient lobe since dephasing only affects the transverse magnetization M_{xy}. A viable approximation is to assume that, during the RF excitation, the magnetization is instantaneously tipped into the transverse plane at the center of the pulse, thus dephasing only occurs during the second half of the RF pulse duration.

1.4.4 Sequence diagram

So far, all encoding steps were considered in the image reference system. However, if frequency, phase, or slice encoding is angled with respect to the laboratory frame, multiple physical gradient coils are employed for each of the encoding steps. The three standard orientations, sagittal, coronal, and axial (Fig. 1.11) can be obtained by performing slice selection along with the laboratory frame x, y, or z gradient axis, respectively. To acquire an oblique slice, two or three of the gradients are employed concurrently with suitably weighted strengths to achieve slice selection.

Fig. 1.12 depicts a simple MRI pulse sequence diagram. At this point, the reader should be equipped with all the tools required to interpret the sequence diagrams. The imaging process starts with the application of an RF pulse with a flip angle α. The shape of the RF pulse is chosen as a truncated sinc, as illustrated in Fig. 1.9. At the same time, a gradient for slice selection G_z is applied, followed by the

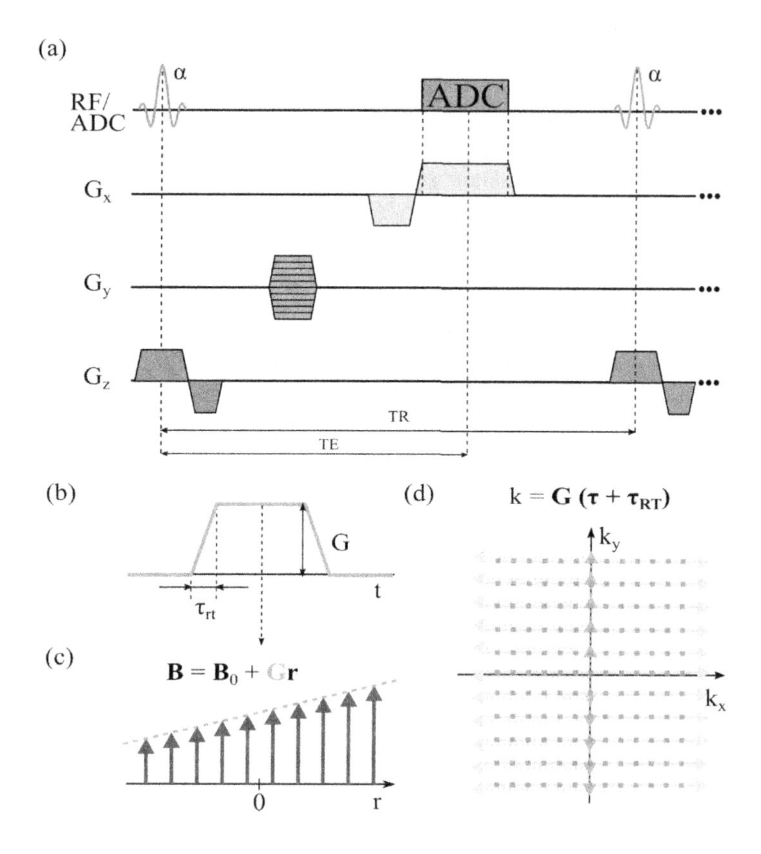

FIGURE 1.12

(a) Pulse sequence diagram example, showing an α-RF excitation pulse and a simultaneous slice-selection gradient with subsequent refocusing lobe, followed by a phase-encoding step and, finally, the frequency-encoding gradient applied during signal acquisition. (b) Trapezoidal gradient shape, highlighting rise time τ_{rt}. (c) B_0 dependence on spatial location, with gradient strength determining the field-intensity slope. (d) k-space trajectory of the pulse sequence. Every time the pulse sequence snippet in *a)* is executed, a single line in k-space is acquired (yellow (light gray in print version)). Multiple repetitions with different phase encoding steps (orange (mid gray in print version)) enable the sampling of the 2D k-space in parallel lines. This sampling trajectory in k-space is known as Cartesian sampling.

slice rewinder. Next, the phase-encoding gradient G_y is played with one of many stepwise incremented amplitudes. To readout the frequencies along one line, a prewinder is played before the frequency encoding, such that, during frequency encoding, k_x is sampled from $-k_{x,max}$ to $k_{x,max}$. Finally, the frequency encoding gradient G_x is applied concurrently with the image readout, where the analog-to-digital converter (ADC) is switched on and the signal induced by the transverse magnetization M_{xy} is captured. As denoted in the sequence diagram, the most basic form of gradient field has a trapezoidal shape over time. The duration of the ramp is called rise time τ_{rt} that describes how long it takes for the gradient field to be ramped up to its maximum value G, as shown in Fig. 1.12b.

When characterizing pulse sequences, two timing parameters are particularly useful: 1) The duration between the creation of the transverse magnetization and the time point when the sampling along the frequency encoding direction accumulates zero moment from the pre- and rewinding. This time is commonly known as echo time or TE; 2) the time between the application of two excitation RF pulses in the same slice selection. This time is known as repetition time or TR. TE and TR are illustrated in Fig. 1.12a.

1.4.5 *k*-space formalism

Due to the naming of the wavenumbers k_x, k_y, and k_z, MRI is often considered to collect image data in "k-space". Following Principle 6 it is easily seen that k-space is indeed the Fourier domain of the MR image. The concept of k-space formalism is thoroughly discussed in [32,35].

For 2D imaging, in practical terms, k-space is a data matrix of size $N_r \times N_{pe}$. It can be derived from the properties of the Fourier transform that the distance between the k-space points determines the field-of-view (FOV) of the image, which describes the extent of the object to be imaged:

$$FOV_x = \frac{1}{\Delta k_x} \tag{1.52}$$

$$FOV_y = \frac{1}{\Delta k_y}. \tag{1.53}$$

In a dual principle, the extent to which k-space is sampled determines the image resolution

$$\Delta x = \frac{1}{2k_{x,max}} \tag{1.54}$$

$$\Delta y = \frac{1}{2k_{y,max}}. \tag{1.55}$$

1.4.6 *k*-space trajectories

The k-space can be traversed in multiple ways to acquire sufficient data for image reconstruction. k-space trajectory refers to the path along which this data is acquired. Most imaging sequences in clinical MRI sample k-space using a *Cartesian* sampling scheme (Fig. 1.12d and Fig. 1.13b). Here, a rectangular grid with equidistant k-space sampling in each of the two directions (k_x and k_y) is filled. In Cartesian sampling, one row of k-space is acquired at a time using frequency encoding, before navigating to the next row employing phase encoding. By collecting data in this fashion, high and low spatial frequencies are sampled uniformly. After the acquisition of all rows required to achieve the desired FOV and resolution, the k-space can be converted to an image using the Fourier transform. This will be discussed in further detail in Chapter 2 of this book.

1.4.6.1 *Echo-planar imaging*

A particular example of Cartesian k-space sampling is the so-called echo-planar imaging (EPI) [12]. In EPI, the collection of all the data necessary to reconstruct an image is performed after a single excitation of the magnetization in one set of echoes, commonly referred to as an echo train. Following the creation of the transverse magnetization, multiple k-space lines are sampled using frequency encoding. These

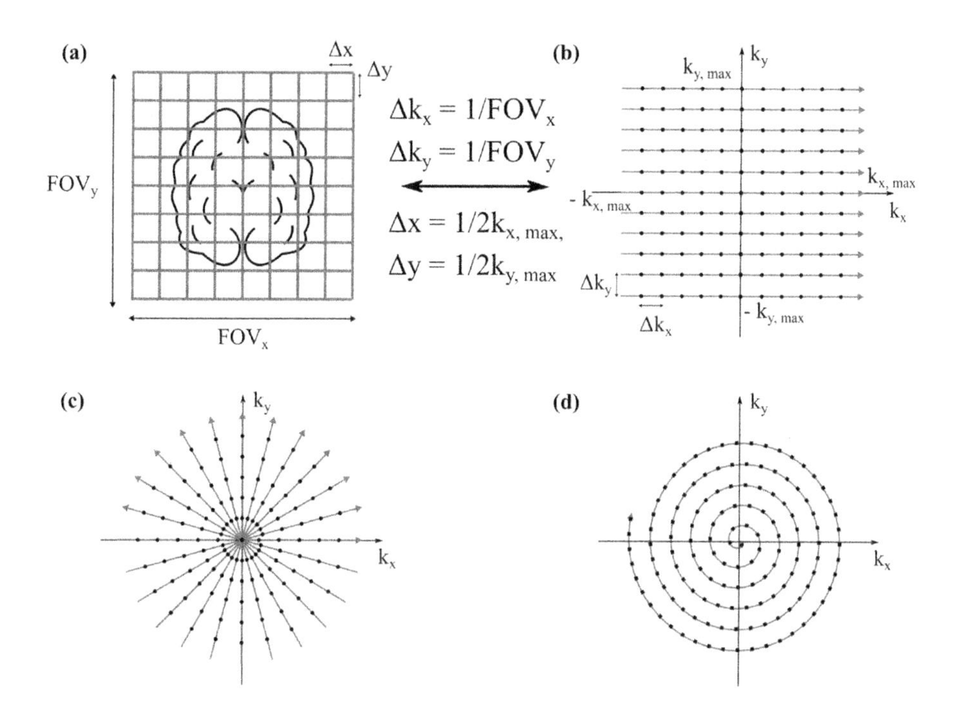

FIGURE 1.13

Image-space representation, illustrating the concepts of resolution (Δx, Δy) and field of view (FOV) (a) and their relationship with k-space for the example of Cartesian sampling (b). Examples of radial (c) and spiral (d) trajectories for k-space sampling.

are separated by a short phase-encoding gradient pulse to traverse along the phase-encoding direction. To allow for symmetric k-space readouts, this echo train is preceded by frequency- and phase-encoding prewinders. EPI is one of the fastest imaging pulse sequences, allowing image acquisition in the order of a few tens of ms. Thus, it plays an essential role in many advanced MR applications, such as diffusion, perfusion, neuro-functional, and dynamic imaging. However, EPI sequences are prone to artifacts due to T_2^* decay, eddy currents, flow, and off-resonance spins (more details in Chapter 12).

1.4.6.2 *Non-Cartesian trajectories*

There are advantages in collecting data using non-Cartesian (non-rectangular) k-space trajectories. Radial imaging, or projection reconstruction (PR), was the original MR k-space trajectory used by Paul C. Lauterbur to make the first MR image [11]. To perform the radial projection pattern, frequency encoding is employed together on multiple axes (e.g., x and y for transverse slices) with various amplitudes to attain a rotational pattern of radial spokes (Fig. 1.13c) [6].

Another imaging trajectory is commonly known as spiral imaging (Fig. 1.13d) [1,15]. In spiral imaging, long readout trains allow frequency encoding of a spiral arm. This bears potential advantages. First, as in radial imaging, low spatial frequencies are sampled more densely because the spiral density is highest near the origin of the k-space axis, leading to a higher image signal-to-noise ratio (SNR) and intrinsic robustness against motion. Second, the gradient *slew rate*, the rate at which the gradient strength has to be altered, is low for spiral scanning. However, errors in the gradient accuracy can lead to deviations from the prescribed non-Cartesian trajectories, with the risk of deteriorating the image quality.

Since the radial and spiral k-space samples do not lie on a regular grid, it is not possible to directly employ the fast Fourier transform, and advanced reconstruction methods are necessary. Further details regarding non-Cartesian trajectories and reconstruction will be discussed in Chapter 4 of this book.

1.4.7 Pulse sequence types

In most cases, numerous excitations are necessary to harvest sufficient information for image reconstruction, as just described. Thus, to allow for efficient imaging, while creating strong magnetization signals with the desired contrast, several pulse-sequence schemes, such as the one shown in Fig. 1.12a, have been devised. In this subsection, we will introduce the most commonly used pulse sequences, divided into two main families: *spin echo* and *gradient echo*. For an in-depth discussion about MRI pulse sequences, the reader is referred to [22].

The difference between *spin echo* and *gradient echo* sequences is most easily explained when considering how a strong signal is being attained: In *spin echo* (SE) sequences the transverse magnetization M_{xy} is being retained, while in *gradient echo* (GRE) sequences, the strong longitudinal magnetization M_z is preserved, as explained now in more detail.

1.4.7.1 Spin echo

In spin echo (SE) sequences, the idea is to keep the transverse magnetization for as long as possible, by refocussing the magnetization dephasing incurred due to the reversible signal decay (T_2', see Section 1.3.2.2). Thus, for an SE experiment, at least two RF pulses are used for echo generation. First, the transverse magnetization is created with a 90° pulse, which then decays due to dephasing. After time τ, a 180° pulse is played to recover the signal losses incurred due to reversible dephasing, such as those related to inhomogeneities in the main magnetic field. This leads to maximum rephasing at time point 2τ, creating the so-called *spin-echo*. SE sequences are characterized by excellent image quality precisely because the effects of static field inhomogeneities are eliminated. The trade-off is long scan times, which makes the sequence highly sensitive to motion artifacts. For a review of spin-echo basic concepts, the reader is referred to [31].

Fast spin-echo (FSE) sequences have been introduced to considerably shorten the scan times. This is accomplished by delivering several 180° refocusing RF pulses after a single magnetization excitation, as depicted in its sequence diagram in Fig. 1.14a [9]. Multiple phase-encoding gradients with different amplitudes enable the readout of multiple k-space lines for each excitation. This series of spin echoes is called an echo train, and the number of echoes sampled is the echo train length. FSE sequences have a longer TR to deliver as many refocusing pulses as possible before the transverse magnetization is considerably decayed due to T_2 relaxation. FSE sequences find use in a broad spectrum of MRI applications and are consequently a workhorse in today's clinical MRI.

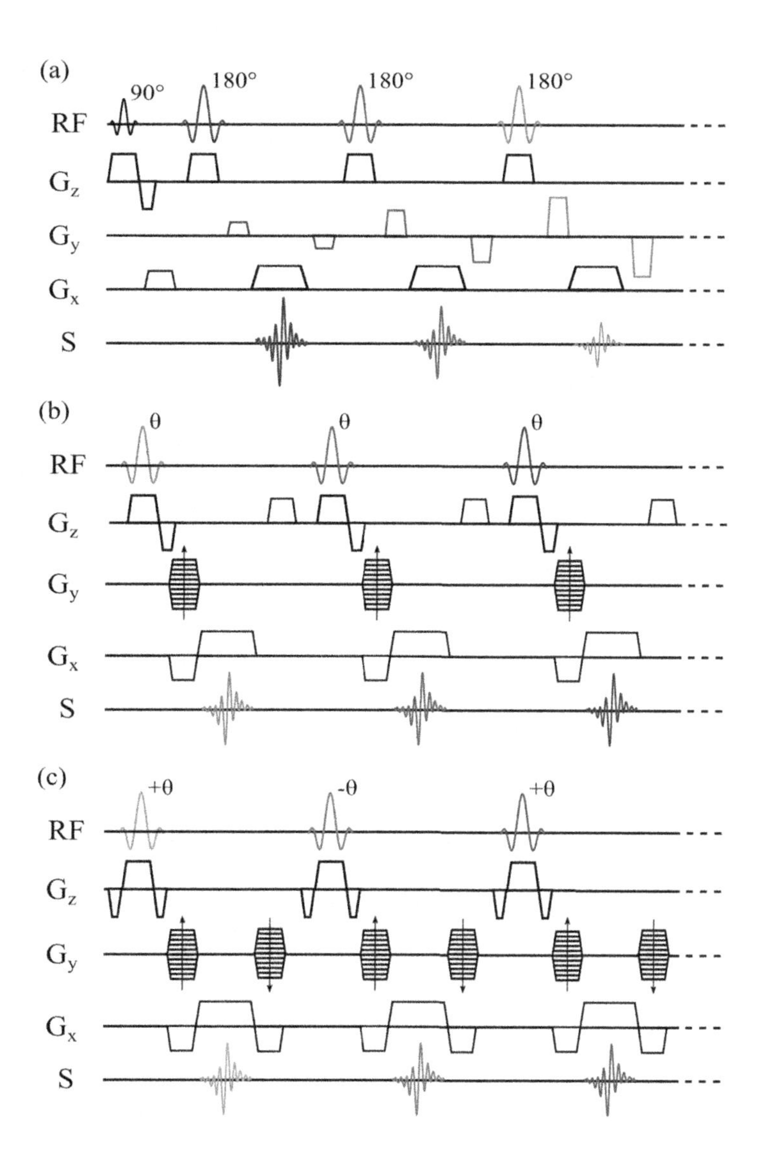

FIGURE 1.14

Pulse sequence diagrams of (a) a fast spin echo (FSE), (b) a fast gradient echo, and (c) balanced steady-state free precession (bSSFP) sequence.

1.4.7.2 Gradient echo

In gradient echo (GRE) imaging, the idea is to only use a part of the longitudinal magnetization to generate a transverse magnetization. This leaves substantial longitudinal magnetization, which can then be used to quickly generate a new transverse magnetization.

In GRE, a single RF pulse is used and gradient encoding is applied during the free induction decay. The time point when the gradient prewinder and rewinder amount to a zero moment is often referred to as the *gradient echo* [14,5]. Every RF excitation chips away from the longitudinal magnetization that can be used for the following TRs. Simultaneously, T_1 relaxation during this time leads to the rebuilding of the longitudinal magnetization. This way, a steady-state magnetization (M_{ss}) is being approached after a number of TRs.

Since no rephasing of the transverse magnetization is applied in GRE sequences, the signal intensity decays much faster than in SE sequences, and the TE has to be much shorter to yield sufficient signal intensity.

The main advantage of GRE sequences is the ability to achieve short TR and, therefore, faster image acquisition. With very short TRs, however, part of the signal will be "left over" from cycle to cycle. This residual magnetization needs to be accounted for to avoid image artifacts. Two basic concepts are often used for this purpose: 1) destroying (described in the next subsection) or 2) recycling (described in Section 1.4.7.3) the left-over magnetization.

The purposeful destruction of the residual MR signal is called spoiling. Spoiling can be accomplished by turning on additional gradients, often along the slice encoding direction, to dephase the magnetization before the next RF pulse is applied. Alternatively, cycling of the RF phase, i.e., the direction of the B_1 field in the RFR, can be used to achieve spoiling [20]. Both yield a T_1-weighted image since the transverse magnetization decay does not contribute to the steady-state magnetization. Spoiled GRE sequences are widely used in the clinical setting, but small flip angles, typically much less than 45°, need to be applied for sufficient SNR.

1.4.7.3 Balanced steady-state free precession

The idea of recycling the transverse magnetization in a GRE sequence is called a balanced steady-state free precession (bSSFP) sequence [3]. In bSSFP sequences, it is ensured that, at the end of each cycle, the phase incurred by all gradients is fully rewound. This is achieved by a symmetric pulse-sequence diagram as shown in Fig. 1.14c, with identical gradients before and after the gradient echo. This way, a second gradient echo is created at the time point of the RF pulse. Additionally, the RF pulses are played with alternating polarity. In the steady-state, this leads to alternating periods of dephasing and rephasing. As such, in an in-depth analysis, bSSFP sequences can be considered a hybrid of GRE and SE sequences, since both the longitudinal and the transverse magnetization are retained. For more intuition on this, the reader is referred to [23,25].

In bSSFP, the generated contrast is determined by the T_2/T_1 ratio (e.g., blood has a high T_2/T_1 ratio and therefore appears bright in bSSFP images). Due to the recycling of the transverse magnetization, bSSFP images have a higher SNR than spoiled GRE images with comparable TRs. bSSFP sequences are characterized by very short scan times and are thus well-suited for vascular imaging and real-time imaging of moving organs, such as the heart. However, the recycling of the transverse magnetization is prone to differences in the precession frequency of the spins and, thus, off-resonance artifacts are common at high field strengths.

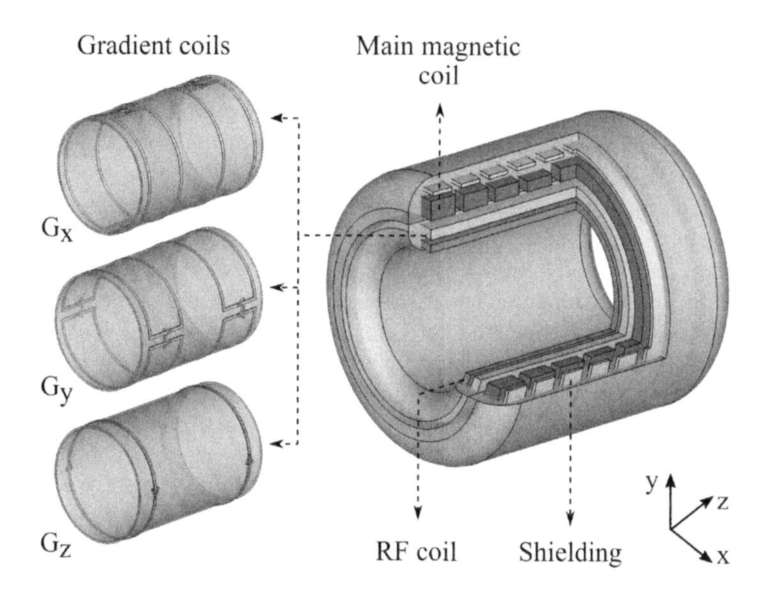

FIGURE 1.15

3D cross-sectional view of a conventional clinical MR scanner, with the various major components. The scanner bore comprises a transmitting body coil (RF coil) and gradient coils, surrounded by a superconducting magnet, responsible for the main magnetic field. The shielding procedure is achieved by placing permanent magnets and shield coils. On the left, the three different types of gradient coils are depicted.

1.5 Components of an MRI scanner

Modern MRI systems can be described by three major components, corresponding to three integral parts of the image formation:

1. The superconducting magnet, responsible for the main magnetic field $\mathbf{B}_0$ that facilitates the formation of the net magnetization [26,24];
2. Radiofrequency (RF) coils, placed around and/or on the subject, inducing secondary magnetic fields to create and detect transverse magnetization [27];
3. Three sets of magnetic field gradient coils, enabling spatial encoding of the signal [34,30].

Fig. 1.15 contains a schematic representation of these components.

1.5.1 Magnet

The design of a magnet has to consider several factors to create a magnetic field $\mathbf{B}_0$ with maximal spatial homogeneity and temporal stability in the subject [30]. These factors are important to avoid inhomogeneity-related artifacts and to achieve the best possible SNR. Over the years, magnet design has evolved to enable stronger and more homogeneous magnetic fields. A wide variety of designs is available for a vast range of field strength (0.02–10.5T, for whole-body magnets). The magnets can be divided into three types: resistive, permanent, and superconducting.

Resistive magnets are no longer used for clinical MR scanners because the resulting magnetic field is greatly temperature-dependent and the maximum attainable field strength is very low (about 0.2T). In resistive magnets, the magnetic field is created by the passage of an electric current through an inductive loop (typically, a copper wire or aluminum band) and is proportional to the magnitude of the current. However, due to the electrical resistance in the loop, a large amount of electrical power is dissipated in the form of heat, requiring active water-cooling systems with high power consumption to lower the temperature of the magnet.

Permanent magnets are typically employed in the so-called *open bore* systems, which reduce the subject's stress caused by claustrophobia and have lower costs for the installation and maintenance of the system. These magnets use rare earth alloys (like samarium-cobalt and neodymium-iron-boron) in which large magnetic fields are induced during manufacturing. However, the maximum achievable field strength is lower than 1T. The magnetic field is generally vertically oriented with reduced stray fields. However, the field homogeneity and stability are also temperature-dependent, for example, a 1 °C increase will cause a 1,000 ppm decrease in the magnetic field strength of neodymium-iron.

In clinical scanners, which generally operate at a field strength of 1.5T or 3T, superconducting magnets are employed. These magnets use the properties of superconductive materials (like those made from niobium-titanium). Superconductors have zero electrical resistance below a certain material-dependent critical temperature (e.g., −263.96°C (9.2 K) for niobium-titanium) and an associated critical magnetic field (e.g., 10T for niobium-titanium). Thus, once a current is fed into a loop of superconducting wire, it will circulate indefinitely. In comparison with other common types of magnets, superconducting magnets can achieve a much higher field strength (up to 10.5T whole-body, currently) and feature excellent field uniformity.

The superconducting material, which is designed as a conducting copper matrix with multi-stranded filaments, is surrounded by liquid helium at an extremely low temperature (4.2 K). Heat transfer slowly boils off helium, making regular refills necessary (commonly, every 3–18 months). Modern MRI scanners use increasingly effective cryogenic cooling of the helium chamber with the effect of minimizing the boil-off.

To further improve the magnet homogeneity and to minimize any field distortions, fixed shimming can be performed. Permanent magnets and shim coils, an arrangement of independently wired coils, can be employed either passively, actively, or as a combination of both. A more thorough discussion of MRI magnets can be found in [26].

1.5.2 Gradient coils

As discussed in Section 1.4, to spatially encode the MRI signal in three spatial dimensions, three linear magnetic field gradients are used. In reality, these are generated by using *magnetic field gradient coils* or in short *gradient coils*. The gradient coils are placed inside the bore of the magnet, on a cylindrical former. By design, the center of the gradient coils, and also the center of the magnet, is placed at the *isocenter* ($z = 0$, $y = 0$, $x = 0$).

The design of the gradient coils needs to balance four major objectives: good linearity over the imaging field-of-view (FOV), maximum achievable amplitude, minimal rise time, and minimization of eddy-current effects induced in the magnet.

Contrary to the magnet design, where the goal is to obtain a uniform magnetic field, the magnetic field produced by the gradient coils should be as linear as possible. Although the method of producing the magnetic field is the same—current passing through conducting wires—the chosen material, the cooling system, and the geometrical configuration differ from the main magnet. For the selected material of the three gradient coils, copper at room temperature produces a sufficiently strong gradient (40–80 mT/m for clinical scanners). To minimize the heat losses in this resistive coil design and to maintain the desired temperature, water cooling is used. Although the gradient coils are placed in a cylindrical form inside the bore of the magnet, the geometrical configuration is different for the longitudinal (z) and the transverse (x and y) fields as seen in Fig. 1.15. The z gradient is typically arranged in a *Maxwell pair*, where a pair of wire loops with counter-rotating currents are separated by $r\sqrt{3}$, with r being the coil radius. The x and y gradients use a different layout known as the *saddle coil* arrangement or the *Golay* configuration. However, the coil configuration in x is rotated by 90° with respect to y. With those design considerations, all three gradient fields are approximately parallel to $\mathbf{B}_0$.

For gradient design, as important as the gradient linearity is the minimization of the gradient slew rate, which relates to the times at which a gradient can be turned on and off. The slew rate is commonly defined as the maximum gradient strength of the gradient divided by the rise time. When the gradients are rapidly turned on and off, eddy currents will be induced in the nearby conducting components of the scanner and may be manifested in the final images as specific artifacts and/or signal loss. To reduce the eddy currents, clinical MRI scanners make use of active shielding, where additional coils are placed in the surroundings of the gradient coils, to minimize the stray gradient fields. This, however, results in reduced bore space and additional power consumption. For an in-depth discussion on gradient coil design, the reader is referred to [24,30,34].

1.5.3 Radiofrequency coils

Radiofrequency (RF) coils are the components responsible for generating the rotating magnetic field $\mathbf{B}_1$ and for picking up the signal induced by the rotating magnetic fields in the sample. The RF coils can be divided into two types: *transmit* and *receive*.

Transmit coils

In most clinical scanners, the principal transmitting coil is known as the body coil, which encompasses the full subject volume and is built into the scanner bore. This coil typically uses a saddle or birdcage design, with the latter generally being able to create a more homogeneous distribution of the transmitted energy throughout the object. The transmit RF coils commonly extend over a large area to optimize the uniformity of the $\mathbf{B}_1$ field and to cover a substantial volume. As a disadvantage, if used as a receiver, such coils will have low sensitivity to the MR signal. When scanning certain anatomies, e.g., the head or the knee, specifically designed coils are used for transmission where less power is required, often at the expense of less uniform fields.

At high field strength, the wavelength of the transmit field approximates the size of the object, which can lead to inadvertent energy distribution, e.g., due to the formation of standing waves. Therefore, in modern MRI systems at high (3T) and, in particular, ultra-high field strengths (7T and above), multiple transmit channels are used simultaneously to circumvent this problem. In a process called B_1^+ shimming, those transmit channels are combined to create optimal homogeneity in the transmit-field amplitude throughout the object and are essential to achieve reliable image quality.

Receive coils

In contrast to transmit coils, dedicated receive coils are designed to better match the anatomy of interest and are placed close to the sample of interest. This will result in increased sensitivity and noise minimization. Generally, two types of receive coils can be differentiated: *volume* and *surface* coils.

Volume coils, which are typically combined transmit and receive (Tx/Rx), cover the full area of interest. In contrast, surface coils are placed on top of the patient's anatomy of interest (e.g., a flexible surface coil is positioned on the chest when imaging the cardiac or abdominal region) to detect signals from a patient's specific anatomical region. Generally, surface coils are only used for signal reception because the RF field response with this coil design is highly inhomogeneous.

To circumvent the limited FOV, an array of multiple surface coils can be combined and used simultaneously to receive the MR signal, while maintaining the high sensitivity of each element. To eliminate the interaction between the different individual elements, these elements need to be decoupled. Most commonly, the coil elements are decoupled by design (i.e., there is a geometrical overlap between neighboring elements). Since each element acts as an independent receiver, the coil noise[5] is uncorrelated with the other elements in the coil. Images from multiple coil elements are then combined in the reconstruction to yield coverage of the entire imaging volume. Additional details about multiple coil reconstruction will be described in Chapter 6. In-depth analysis of MRI RF coils can be found in [27].

1.5.4 Noise properties

As with any measurement, the quality of the MRI signal and the resulting images are curtailed by noise in the measurement process [29]. Multiple sources of noise taint the signal quality. The overall measurement is usually characterized by the dimensionless signal-to-noise ratio (SNR). To maximize SNR, either the signal strength needs to be improved or the noise contributions need to be mitigated.

Even though the MR signal is complex, most commonly the magnitude images are used for clinical diagnosis (Fig. 1.16). The mapping from real and imaginary components to their magnitude is nonlinear. Therefore, even though noise in the real and imaginary components can be assumed to follow a Gaussian distribution, noise in the magnitude images is non-Gaussian. For the simplest case of receiving the signal with one channel, and without using advanced reconstruction methods, the noise distribution for the measured pixel intensity M is given by the Rice distribution

$$p_M(M) = \frac{M}{\sigma^2} e^{\frac{-(M^2+A^2)}{2\sigma^2}} I_0\left(\frac{A \cdot M}{\sigma^2}\right), \tag{1.56}$$

where A is the image pixel intensity without noise, I_0 is the modified zeroth-order Bessel function of the first kind, and σ is the standard deviation of the Gaussian noise in the real and imaginary images, which can be assumed to be identical [7]. This corresponds to the Rice distribution and can be obtained by taking the square root of the sum of squares of two independent and identically distributed Gaussian variables [19].

With Rician noise, the SNR can be defined as the ratio of A over σ. For high SNR (≥ 3), the noise distribution can be well approximated with a simple Gaussian distribution [7]. However, for SNR

[5] Unfortunately, this is only one of two types of noise, as elaborated in the next section.

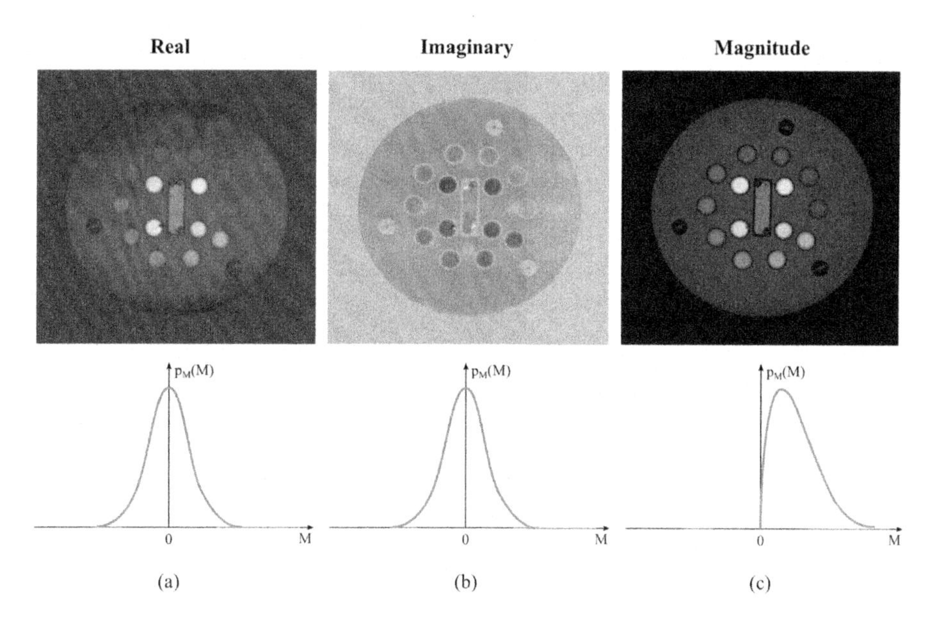

FIGURE 1.16

Real (a), imaginary (b), and magnitude (c) images of the NIST/ISMRM system phantom, with their corresponding noise distributions.

values ≤ 1, the noise distribution is substantially skewed. In regions where only noise is presented (i.e., $A = 0$ in Eq. (1.56)), $p_M(M)$ simplifies to a Rayleigh distribution. Note that multicoil reconstructions or nonlinear reconstructions further influence the noise distribution and can lead to different noise properties.

As described in Eq. (1.5), the net magnetization is proportional to the value of the static magnetic field $\mathbf{B}_0$ since stronger $\mathbf{B}_0$ fields lead to an increase in the nuclear polarization and, thus, an increase in signal. Also, at higher field strength, higher voltage is produced in the coil by the precessing magnetization, which further improves the signal strength. However, when $\mathbf{B}_0$ increases, the value of T_1 increases and T_2^* decreases, resulting in a decreased signal strength. Thus, the choice of an optimal field strength remains highly application-dependent.

Besides the signal strength, two important sources of noise need to be considered: first, the human body, which is conducting and, thus, produces thermal noise. This is picked up by the receiving RF coil during imaging. Second, random voltage variations in the copper conductors in the RF coil itself will also increase the received noise. The SNR can be considered proportional to the sensitivity of the RF coil.

Several other imaging parameters affect the SNR. If multiple images are obtained using equal parameters, averaging can be performed to enhance the SNR at the cost of extra scan time. Since the MRI signal is coherent but the noise is incoherent, the SNR increases with $\sqrt{N_a}$, where N_a is the number of signal averages. Also, as the imaging signal is proportional to the number of hydrogen nuclei per voxel, an increase in voxel size proportionally increases the SNR. Accordingly, for a given FOV, if the amount of phase-encoding steps N_{pe} is doubled, then the image SNR is decreased by a factor of two.

Similarly, if the number of points collected in the frequency-encoding direction N_r is doubled for a given FOV, the SNR decreases by a factor of two. If frequency-encoding is performed with twice the bandwidth, the SNR is decreased by a factor of two, leading to shorter readouts but more noise in the signal. Finally, increasing the slice thickness yields a proportional improvement in the image SNR as the imaging volume and the number of protons in the volume are increased.

1.6 Summary

This chapter introduces the basic aspects of MRI physics and the basic concepts involved in image acquisition. This will hopefully enable the reader to understand advanced MRI methods and image-reconstruction techniques as described in the remainder of this book. It describes the physics of MRI around six central principles:

1. *Origin of the net magnetization*: Certain nuclei are characterized by a physical property, a nonzero spin, which causes them to form a magnetic moment when placed into an external magnetic field. Many of these nuclei together form a net magnetization, which is the basis of all MRI signals.
2. *The dynamics of this magnetization in an external magnetic field*: The net magnetization is related to both a magnetic and angular momentum. Thus, if the net magnetization is angled with respect to the external effective field, it prescribes a precession motion around the axis of the effective field, with a characteristic field strength-dependent resonance frequency, the so-called Larmor frequency.
3. *Using rotating magnetic fields to create a detectable magnetization component*: While a strong static magnetic field is necessary to create net magnetization, a much weaker rotating magnetic field can be used to change the orientation of the effective field. Specifically, when this second magnetic field is applied near the resonance frequency, the net magnetization can be tipped away from the direction of the main magnetic field.
4. *Using induction to measure this component*: A magnetization component orthogonal to the main magnetic field precesses and can be used to create a signal. Specifically, this temporally varying magnetic field induces a current in the nearby receiver coils via induction, which can be recorded as the imaging signal.
5. *The evolution of this signal in the presence of relaxation mechanisms*: If the net magnetization is altered from its equilibrium state, e.g., by nutating away from the main magnetic field, multiple recovery processes are at play to return the magnetization back to the thermal equilibrium. The longitudinal magnetization, i.e., the component along the axis of the main magnetic field, undergoes an exponential recovery. The transverse magnetization, which is the component orthogonal to the main magnetic field, experiences an exponential decay.
6. *Gradient fields to enable image formation*: To spatially locate the acquired signal, gradient fields are used to create a spatially dependent phase in the signal. The signal is then acquired in the so-called k-space. To fully resolve the signal in all spatial dimensions, frequency, phase, and slice-encoding gradients can be used.

Based on these basic principles, a wide variety of pulse sequences that allow for acquisitions with different properties can be designed. Most fundamentally, gradient-echo and spin-echo sequences have been introduced along with their variations to make possible faster imaging. Image acquisition can also be performed by sampling various trajectories in k-space to further speed up the scan.

For the practical realization of the imaging process, several hardware components relating to these physical principles are necessary. Superconducting electromagnets are most widely used to create the primary magnetic field. Different coil arrangements enable spatially varying gradient fields and the creation of a secondary magnetic field to excite the magnetization. Anatomy-specific coil arrays are used to receive the signal at maximum strength.

The basic properties of the magnetization signal, its dynamics, and relaxation help to understand the limitations in MRI pulse-sequence design. The fundamental spatial encoding principles and the related hardware further explain the bottlenecks in the image acquisition speed. In recent years, advanced reconstruction methods have been taking center stage in the acceleration of MRI scans. Craftily reducing the number of samples, or supplementing the acquired information with a-priori knowledge has enabled unprecedented acquisition times. However, in MRI, the development of reconstruction techniques, as described throughout the remainder of the book, goes hand in hand with a thorough understanding of the physical principles of the magnetization signal. This chapter is intended to serve as a reference for understanding the limitations and opportunities in the signal properties of MRI that are relevant for image reconstruction.

References

[1] Ahn CB, Kim JH, Cho ZH. High-speed spiral-scan echo planar nmr imaging-I. IEEE Trans Med Imaging 1986;5(1):2–7.
[2] Bloch F, Hansen WW, Packard M. Nuclear induction. Phys Rev 1946;69:127.
[3] Carr H. Steady-state free precession in nuclear magnetic resonance. Phys Rev 1958;112(5):1693.
[4] Carr HY. Free precession techniques in nuclear magnetic resonance. PhDT. 1953.
[5] Elster AD. Gradient-echo mr imaging: techniques and acronyms. Radiology 1993;186(1):1–8.
[6] Glover GH, Pauly JM. Projection reconstruction techniques for reduction of motion effects in mri. Magn Reson Med 1992;28(2):275–89.
[7] Gudbjartsson H, Patz S. The Rician distribution of noisy MRI data. Magn Reson Med 1995;34(6):910–4.
[8] Hahn EL. Spin echoes. Phys Rev 1950;80:580–94.
[9] Hennig J, Nauerth A, Friedburg H. Rare imaging: a fast imaging method for clinical mr. Magn Reson Med 1986;3(6):823–33.
[10] Larmor J. LXIII. On the theory of the magnetic influence on spectra; and on the radiation from moving ions. Philos Mag 1897;44(271):503–12.
[11] Lauterbur PC. Image formation by induced local interactions: examples employing nuclear magnetic resonance. Nature 1973;242(5394):190–1.
[12] Mansfield P. Multi-planar image formation using NMR spin echoes. J Phys C, Solid State Phys 1977;10(3):L55–8.
[13] Mansfield P, Grannell PK. NMR diffraction in solids. J Phys C, Solid State Phys 1973;6(22):L422.
[14] Markl M, Leupold J. Gradient echo imaging. J Magn Reson Imaging 2012;35(6):1274–89.
[15] Meyer CH, Hu BS, Nishimura DG, Macovski A. Fast spiral coronary artery imaging. Magn Reson Med 1992;28(2):202–13.
[16] Purcell EM, Torrey HC, Pound RV. Resonance absorption by nuclear magnetic moments in a solid. Phys Rev 1946;69:37–8.
[17] Rabi II. Space quantization in a gyrating magnetic field. Phys Rev 1937;51:652–4.
[18] Rabi II, Zacharias JR, Millman S, Kusch P. A new method of measuring nuclear magnetic moment. Phys Rev 1938;53:318.

[19] Rice SO. Mathematical analysis of random noise. Bell Syst Tech J 1944;23(3):282–332.
[20] Zur Y, Wood M, Neuringer L. Spoiling of transverse magnetization in steady-state sequences. Magn Reson Med 1991;21(2):251–63.

Suggested readings

[21] Abragam A. The principles of nuclear magnetism. Oxford University Press; 1961. p. 32.
[22] Bernstein MA, King KF, Zhou XJ. Handbook of MRI pulse sequences. Elsevier; 2004.
[23] Bieri O, Scheffler K. Fundamentals of balanced steady state free precession mri. J Magn Reson Imaging 2013;38(1):2–11.
[24] Chapman B. Shielded gradients. And the general solution to the near field problem of electromagnet design. Magn Reson Mater Phys Biol Med 1999;9(3):146–51.
[25] Chavhan GB, Babyn PS, Jankharia BG, Cheng H-LM, Shroff MM. Steady-state mr imaging sequences: physics, classification, and clinical applications. Radiographics 2008;28(4):1147–60.
[26] Cosmus TC, Parizh M. Advances in whole-body MRI magnets. IEEE Trans Appl Supercond 2010;21(3):2104–9.
[27] Gruber B, Froeling M, Leiner T, Klomp DW. RF coils: a practical guide for nonphysicists. J Magn Reson Imaging 2018;48(3):590–604.
[28] Hanson LG. Is quantum mechanics necessary for understanding magnetic resonance?. Concepts Magn Reson, Part A. Educ J 2008;32(5):329–40.
[29] Henkelman RM. Measurement of signal intensities in the presence of noise in MR images. Med Phys 1985;12(2):232–3.
[30] Hidalgo-Tobon S. Theory of gradient coil design methods for magnetic resonance imaging. Concepts Magn Reson, Part A 2010;36(4):223–42.
[31] Jung BA, Weigel M. Spin echo magnetic resonance imaging. J Magn Reson Imaging 2013;37(4):805–17.
[32] Ljunggren S. A simple graphical representation of Fourier-based imaging methods. J Magn Reson (1969) 1983;54(2):338–43.
[33] Rabi II, Ramsey NF, Schwinger J. Use of rotating coordinates in magnetic resonance problems. Rev Mod Phys 1954;26(2):167.
[34] Turner R. Gradient coil design: a review of methods. Magn Reson Imaging 1993;11(7):903–20.
[35] Twieg DB. The k-trajectory formulation of the NMR imaging process with applications in analysis and synthesis of imaging methods. Med Phys 1983;10(5):610–21.

CHAPTER

MRI Reconstruction as an Inverse Problem

2

Gastao Cruz[a], Burhaneddin Yaman[b,c], Mehmet Akçakaya[b,c], Mariya Doneva[d], and Claudia Prieto[a]

[a]*School of Biomedical Engineering and Imaging Sciences, King's College London, London, United Kingdom*
[b]*Department of Electrical and Computer Engineering, University of Minnesota, Minneapolis, MN, United States*
[c]*Center for Magnetic Resonance Research, University of Minnesota, Minneapolis, MN, United States*
[d]*Philips Research, Hamburg, Germany*

2.1 Inverse problems

An inverse problem arises when we have access to the effects or observed measurements of a certain physical phenomenon and would like to determine its causes. This is the opposite of a forward problem that predicts the effects of certain phenomenon or system given knowledge of their causes. For example, anticipating that you will suffer fever, headache, and fatigue if you have gotten the flu is a forward problem. On the other hand, guessing the specific disease from its symptoms is an inverse problem. From this example, where different diseases may cause the same/similar symptoms, we can notice that inverse problems are more difficult to solve than forward or direct problems.

The difficulty of solving an inverse problem can be characterized by its well-posedness. Specifically, a problem is said to be *well-posed* if it satisfies the three Hadamard conditions: 1) solution existence, 2) solution uniqueness, and 3) solution stability (Fig. 2.1). The first condition is evident: The problem must have a solution. The second condition states that, if a solution exists, it must be unique. The third condition states that the solution should be stable, i.e., small perturbations to the acquired data should produce small perturbations to the solution. If the problem violates one or more of these requirements, it is said to be *ill-posed*. The difficulty is that inverse problems are typically ill-posed. Therefore, a reformulation is needed to find a suitable solution for ill-posed inverse problems. This reformulation often involves including additional requirements or constraints to the solution.[1]

MR reconstruction, and indeed most medical image reconstruction problems, can be treated as an inverse problem since we would like to reconstruct the unknown image from the (noisy) sampled measurements. Specifically, as discussed in Chapter 1, MR reconstruction corresponds to the problem of estimating the transverse macroscopic magnetization (i.e., the image) from the electric potential induced in the receiver coil(s) (i.e., the k-space data). The MR reconstruction problem is typically ill-posed. If the forward model of the reconstruction does not correctly capture the phenomena of the acquisition process, then a solution may not exist (no solution). Since we would (ideally) like to reconstruct the continuous (infinite) distribution of the object's magnetization from a discrete (finite) number of measurements, it is also clear that there are infinite solutions to this problem. For example, the

[1] For more details, the reader is referred to [[22]: "Discrete Inverse Problems: Insight and Algorithms" by Christian Hansen].

Advances in Magnetic Resonance Technology and Applications, Volume 7, ISSN 2666-9099. https://doi.org/10.1016/B978-0-12-822726-8.00011-7

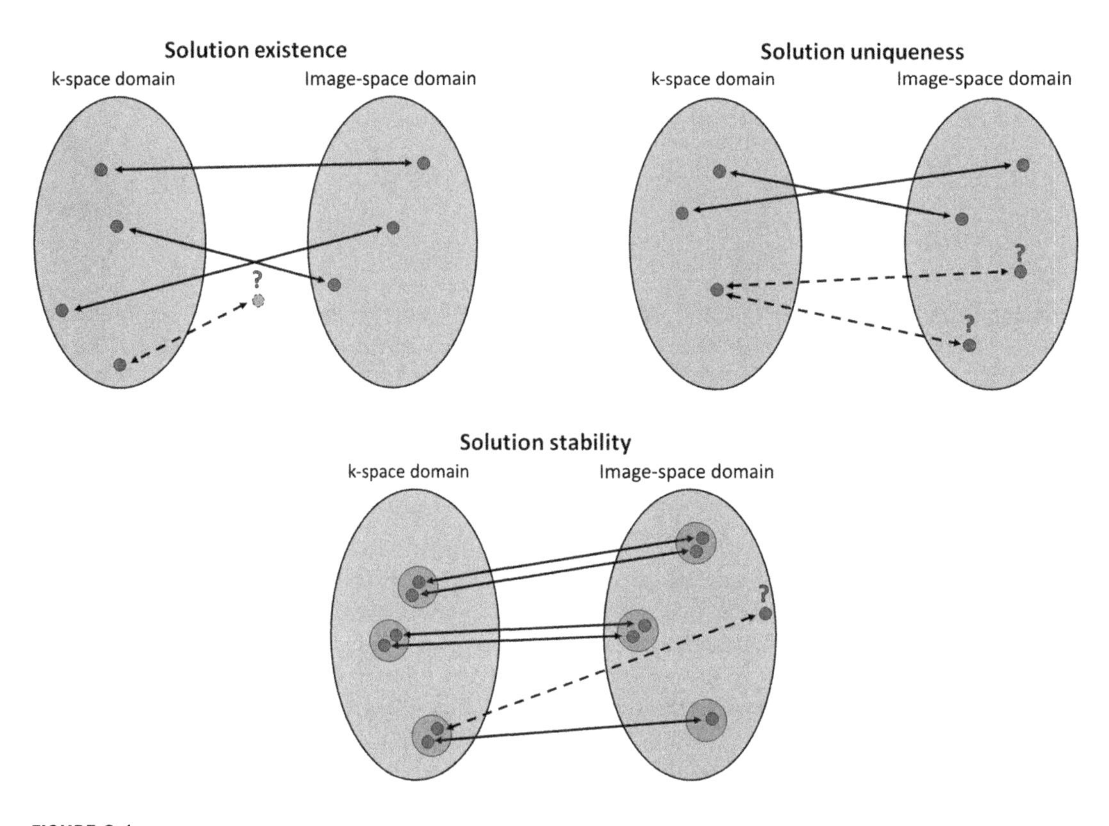

FIGURE 2.1

Representative diagram of the Hadamard conditions that should be satisfied for a well-posed problem. First, a solution must exist according to the underlying mathematical model. Second, the solution must be unique, i.e., a set of k-space measurements should a single solution in the image space domain. Third, the solution must be stable, meaning that small perturbations to the measured data should cause only small perturbations to the image solution.

acquired band-limited signal will impose resolution limitations: there are infinite arrangements of the magnetization at higher resolutions, all of them consistent with the acquired data. If not enough samples are collected (i.e., undersampling), the Fourier transform-based MRI reconstruction problem will be ill-posed and lack a unique solution. This results in a violation of the second Hadamard condition. The third condition, stability, relates to how much perturbations to the acquired data affect the solution. The Fourier transform is in fact sensitive to noise amplification since large perturbations to high-frequency information (e.g., $\sin(kx\pi)$) will amount to small changes to the acquired data $\Delta S(k)$, but can produce arbitrarily large changes in the image $\Delta M(x)$. In other words, $\lim_{k \longrightarrow \infty} \mathcal{F}\{\sin(kx\pi)\} = 0$, which is an instance of the Riemann–Lebesque lemma [1]. This sensitivity to high-frequency perturbations contributes to the instability of the Fourier transform and often manifests itself as noise amplification.

Luckily, the MR reconstruction problem can be reformulated, incorporating additional constraints and regularization terms, to find an appropriate solution to this otherwise ill-posed problem. Hereafter, we will formulate the general MR reconstruction problem, starting with the discretization of the continuous MR signal, and discuss approaches to solve this problem. The formulation and solution to specific MRI applications will be discussed in the remaining chapters of this book.

2.2 Discretization of the MR signal

Chapter 1 introduced the relationship between the measured signal $S(\boldsymbol{k})$ and the transverse magnetization of the underlying object $M_{xy}(\boldsymbol{r})$. If we neglect relaxation effects and receiver coil sensitivity profile, the signal equation is expressed by Principle 6 (Eq. (1.46)):

$$S(\boldsymbol{k}) = \int_V M_{xy}(\boldsymbol{r}) e^{(-2\pi i \boldsymbol{k}\cdot\boldsymbol{r})} d\boldsymbol{r} = \mathcal{F}\{M_{xy}(\boldsymbol{r})\} \tag{2.1}$$

where $\mathcal{F}$ is the Fourier transform and $\boldsymbol{r} = (x, y, z)$ denotes the spatial coordinate; $\boldsymbol{k} = (k_x, k_y, k_z)$ denotes the k-space coordinate and is determined by the magnetic field gradients $\boldsymbol{G}$, $\boldsymbol{k} = \int_0^t \gamma (2\pi)^{-1} \times \boldsymbol{G}(t')dt'$. This equation is the forward model that describes the MR acquisition process. As also discussed in Chapter 1, Eq. (2.1) informs that the measured MR signal is given by the Fourier transform of the magnetization; therefore, if $S(\boldsymbol{k})$ is sampled in a continuous (and infinite) fashion, the original magnetization can be obtained via $\mathcal{F}^{-1}\{S(\boldsymbol{k})\}$. In practice, however, MR acquisition requires the discretization of the acquired (and finite) signal. For example, considerable discretization is required along the phase-encoding direction since data along this dimension is populated along several repetition times (TR), which are limited by scan time. From the full spectrum of $S(\boldsymbol{k})$, only a discrete and band-limited subset of samples $S(\boldsymbol{k_n})$ are acquired in practice.

To investigate the impact of signal discretization on MR reconstruction, we can evaluate $\mathcal{F}^{-1}\{S(\boldsymbol{k_n})\}$. Starting with the definition of $S(\boldsymbol{k_n})$, the discretized sampling of $S(\boldsymbol{k})$ within a spectral band that is limited by $\boldsymbol{k_{max}}$ (since the signal is finite) is

$$S(\boldsymbol{k_n}) = S(\boldsymbol{k})\, III(\boldsymbol{k}) = S(\boldsymbol{k}) \sum_{-\boldsymbol{k}_{max}}^{+\boldsymbol{k}_{max}} \delta(\boldsymbol{k} - n\boldsymbol{\Delta_k}), \tag{2.2}$$

where III is the comb function, defined as a set of Dirac deltas δ, separated by $\boldsymbol{\Delta_k}$, the space between samples. Rewriting the previous equation as an infinite summation bounded by the *rect* function Π, we arrive at:

$$\mathcal{F}^{-1}\{S(\boldsymbol{k_n})\} = \mathcal{F}^{-1}\left\{\Pi(\frac{\boldsymbol{k}}{\boldsymbol{k}_{max}}) S(\boldsymbol{k}) \sum_{-\infty}^{+\infty} \delta(\boldsymbol{k} - n\boldsymbol{\Delta_k})\right\}. \tag{2.3}$$

Following the convolution theorem,

$$\begin{aligned} &\mathcal{F}^{-1}\left\{\Pi\left(\frac{\boldsymbol{k}}{\boldsymbol{k}_{max}}\right) S(\boldsymbol{k}) \sum_{-\infty}^{+\infty} \delta(\boldsymbol{k} - n\boldsymbol{\Delta_k})\right\} = \\ &= \mathcal{F}^{-1}\left\{\Pi\left(\frac{\boldsymbol{k}}{2\boldsymbol{k}_{max}}\right)\right\} * \mathcal{F}^{-1}\{S(\boldsymbol{k})\} * \mathcal{F}^{-1}\left\{\sum_{-\infty}^{+\infty} \delta(\boldsymbol{k} - n\boldsymbol{\Delta_k})\right\} = \end{aligned} \tag{2.4}$$

$$= 2k_{max} sinc\left(\boldsymbol{r} 2\pi \boldsymbol{k}_{max}\right) * M_{xy}(\boldsymbol{r}) * (\boldsymbol{\Delta}_k)^{-1} \sum_{-\infty}^{+\infty} \delta\left(\boldsymbol{r} - \frac{n}{\boldsymbol{\Delta}_k}\right).$$

Therefore, the reconstruction from the acquired k-space data can be written as:

$$\mathcal{F}^{-1}\{S(\boldsymbol{k}_n)\} = sinc\left(\boldsymbol{r} 2\pi \boldsymbol{k}_{max}\right) * \frac{2\boldsymbol{k}_{max}}{\boldsymbol{\Delta}_k} \sum_{-\infty}^{+\infty} M_{xy}\left(\boldsymbol{r} - \frac{n}{\boldsymbol{\Delta}_k}\right). \tag{2.5}$$

The result in Eq. (2.5) is proportional to $\sum_{-\infty}^{+\infty} M_{xy}\left(\boldsymbol{r} - \frac{n}{\boldsymbol{\Delta}_k}\right)$, indicating that a reconstruction from discrete samples will not only reconstruct the original object $M_{xy}(\boldsymbol{r})$, but also infinite replicas $M_{xy}(\boldsymbol{r} - \frac{n}{\boldsymbol{\Delta}_k})$ separated by $\boldsymbol{\Delta}_k^{-1}$ (Fig. 2.2). Therefore, the largest $\boldsymbol{FOV} = (\boldsymbol{x}_{max}, \boldsymbol{y}_{max}, \boldsymbol{z}_{max})$ admissible is determined by $\boldsymbol{\Delta}_k^{-1}$ (introduced without derivation in Chapter 1, Eq. (1.52) and Eq. (1.53)). If the object extends beyond this range, then there will be regions where $M_{xy}(\boldsymbol{r})$ and $M_{xy}(\boldsymbol{r} \pm \frac{n}{\boldsymbol{\Delta}_k})$ overlap, creating *aliasing* artifacts. In this case, $\boldsymbol{\Delta}_k^{\circ -1} < \boldsymbol{FOV}$ (where $^{\circ -1}$ denotes the Hadamard inverse), and the reconstruction is characterized as *undersampled* since the sampling of the measurements in k-space does not satisfy the Nyquist-Shannon criteria. If $\boldsymbol{\Delta}_k^{\circ -1} = \boldsymbol{FOV}$, then the reconstruction is characterized as *fully-sampled*, whereas, if $\boldsymbol{\Delta}_k^{\circ -1} > \boldsymbol{FOV}$, then it is referred to as *oversampled*. The acceleration (or undersampling, sometimes also called reduction) factor is commonly used to describe the degree of undersampling: $Acc = N_{acquired}/N_{fully-sampled}$, where $N_{acquired}$ and $N_{fully-sampled}$ are the number of acquired samples (in a given accelerated acquisition) and the number of required samples for a fully-sampled acquisition, respectively. The description so far has focused on a quasiuniform degree of undersampling everywhere in k-space, however some k-space sampling strategies can *oversample* and *undersample* different regions of k-space.

Returning to Eq. (2.5), we note further that the result is convolved with $sinc\left(\boldsymbol{r} 2\pi \boldsymbol{k}_{max}\right)$ as a result of having a band-limited signal. This convolution will effectively blur the underlying object M_{xy}, and the resulting spatial resolution $\boldsymbol{\Delta}_r$ can be estimated via

$$\boldsymbol{\Delta}_r = \int_{-\infty}^{+\infty} sinc\left(\boldsymbol{r} 2\pi \boldsymbol{k}_{max}\right) d\boldsymbol{r} = (2\boldsymbol{k}_{max})^{-1}. \tag{2.6}$$

This relationship was also introduced (without derivation) in Chapter 1, Eqs. (1.54) and (1.55).

Moreover, the convolution of a *sinc* with any sharp edges present in the image will produce intensity ripples around said edges, known as *Gibbs ringing* (Fig. 2.2). The relationship between k-space sampling and field-of-view ($\boldsymbol{\Delta}_k^{-1} = \boldsymbol{FOV}$), along with the relationship between the resolution and the highest frequency sampled in k-space ($\boldsymbol{\Delta}_r = (2\boldsymbol{k}_{max})^{\circ -1}$), imposes some fundamental trade-offs in MR acquisition. Noting that the total number of acquired samples $\boldsymbol{K}_{samples} = 2\boldsymbol{k}_{max} \oslash \boldsymbol{\Delta}_k$ (which is proportional to total scan time) where $\oslash$ denotes Hadamard division, the following relationship can be established:

$$\boldsymbol{K}_{samples} = \boldsymbol{FOV} \oslash \boldsymbol{\Delta}_r. \tag{2.7}$$

Eq. (2.7) notes that there is an inherent trade-off between the field-of-view and resolution. To increase either the FOV or resolution, additional samples must be acquired, extending scan time. The triangular

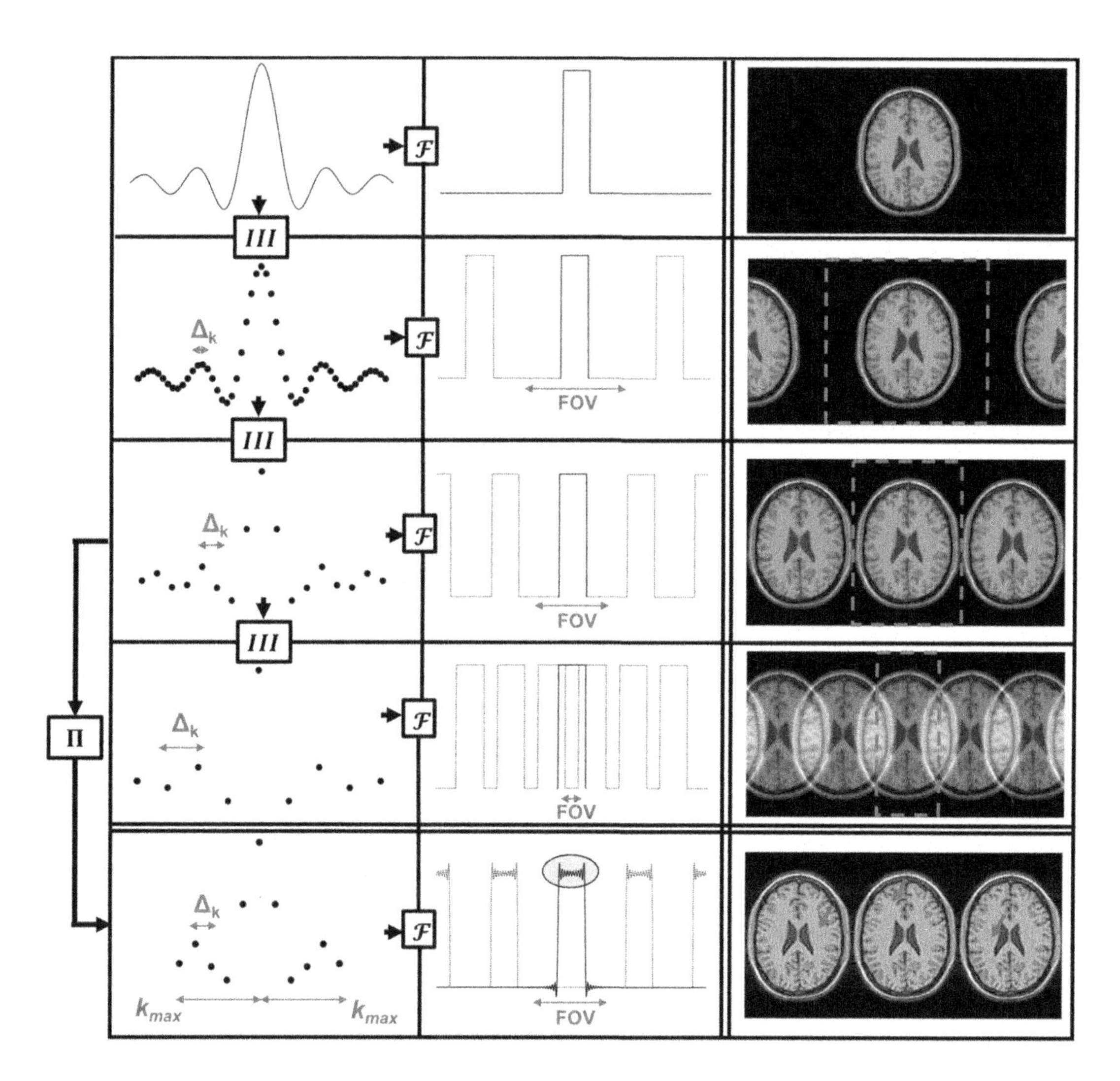

FIGURE 2.2

Practical reconstruction consequences of discrete and band-limited sampling. The first column denotes a set of measurements in k-space domain; the second column contains the corresponding the representation in image-space domain; the third column is an example of MR reconstruction. Discretization of measurements constrains the effective FOV of the reconstruction (red-dashed lines), whereas band-limiting imposes resolution limits and potentially Gibbs ringing (blue arrows).

relationship between resolution, field-of-view, and scan time is a key property of MR and often an important consideration in various MR applications.

Considering that the acquired data is finite, it seems unrealistic that the reconstruction of a continuous (i.e., infinite) object is guaranteed to be correct. As discussed in the previous section, this observation is true, and such a problem is fundamentally ill-posed since we have an infinite number of unknowns to be solved with a finite number of equations. Specifically, there will be an infinite number of solutions to this reconstruction problem, i.e., the solution is not unique. This issue is resolved by assuming that the object $M_{xy}(\boldsymbol{r})$ also lies in a finite-dimensional subspace. Specifically, we will be assuming that the object $M_{xy}(\boldsymbol{r})$ to be reconstructed has finite support (i.e., it is bounded by some $\boldsymbol{FOV}$) and finite resolution (i.e., it is discretized in space, commonly by pixels/voxels). Under these conditions, the following relationship between the continuous ($M_{xy}(\boldsymbol{r})$) and discrete ($M_{xy}(\boldsymbol{r}_n)$) object can be established:

$$M_{xy}(\boldsymbol{r}) = \sum_{n=1}^{N_{pixels}} M_{xy}(\boldsymbol{r}_n)\, \boldsymbol{b}(\boldsymbol{r} - \boldsymbol{r}_n), \tag{2.8}$$

where $\boldsymbol{b}(\boldsymbol{r} - \boldsymbol{r}_n)$ are the chosen basis functions for discretization. The *rect* function is commonly used for this purpose, discretizing the image into pixels, i.e., $\boldsymbol{b}(\boldsymbol{r}) = \Pi\left(\frac{\boldsymbol{r}}{\boldsymbol{\Delta_r}}\right)$.

If the acquisition is *fully-sampled* and the reconstructed image is bounded by $\boldsymbol{FOV}$ (assuming that the object lies on a finite-dimension subspace), then only the original object in Eq. (2.5) remains, i.e., $\sum_{-\infty}^{+\infty} M_{xy}\left(\boldsymbol{r} - \frac{n}{\boldsymbol{\Delta_k}}\right) = M_{xy}(\boldsymbol{r})$. Combining Eq. (2.5) and Eq. (2.8), we obtain

$$\mathcal{F}^{-1}\{S(\boldsymbol{k}_n)\} = sinc\left(\frac{\boldsymbol{r}\pi}{\boldsymbol{\Delta_r}}\right) * N_{pixels} \sum_{n=1}^{N_{pixels}} M_{xy}(\boldsymbol{r}_n)\, \Pi\left(\frac{\boldsymbol{r} - \boldsymbol{r}_n}{\boldsymbol{\Delta_r}}\right). \tag{2.9}$$

Eq. (2.9) relates the reconstruction of a discrete, band-limited signal $S(\boldsymbol{k}_n)$ to a discrete image $M_{xy}(\boldsymbol{r}_n)$ blurred by a $sinc$ function if the image is bounded by $\boldsymbol{FOV}$. We have not, up to this point, considered that the acquisition of $S(\boldsymbol{k}_n)$ may be affected by noise, and said noise may affect the reconstructed image.

2.3 MR reconstruction as a linear inverse problem

The discretized (single-channel) signal $S(\boldsymbol{k}_n)$ can be expressed by replacing the continuous object's magnetization ($M_{xy}(\boldsymbol{r})$) in Eq. (2.1) with the relationship derived in Eq. (2.8) and noting the quadrature rule, which results in

$$S(\boldsymbol{k}_n) = \sum_{n=1}^{N_{pixels}} M_{xy}(\boldsymbol{r}_n)\, \boldsymbol{b}(\boldsymbol{r} - \boldsymbol{r}_n) e^{(-2\pi i \boldsymbol{k}_n \cdot \boldsymbol{r}_n)} \boldsymbol{\Delta_r}. \tag{2.10}$$

This is the discretized forward model of the MR acquisition. Assuming that the pixels are small enough that magnetization changes within a voxel are negligible, a center voxel approximation can be reasonably adopted. Eq. (2.10) can be stated as the linear inverse problem

$$\boldsymbol{s} = \boldsymbol{E}\boldsymbol{x}, \tag{2.11}$$

where $\boldsymbol{s}$ corresponds to the (vectorized) k-space data, $\boldsymbol{E}$ is the encoding operator that describes the acquisition process, and $\boldsymbol{x}$ is the (vectorized) image. The encoding operator $\boldsymbol{E}$ includes the Fourier transform and other operations that can be included in the forward model, such as the image-space basis $\boldsymbol{b}$, MR physics effects (e.g., relaxation), or the spatial sensitivity profile of the receiving coil array (commonly used for parallel imaging reconstructions, as we will see in Chapter 6). Formulating the MR reconstruction as the linear inverse problem given by Eq. (2.11) is advantageous since we can leverage the many tools of linear algebra to analyze its properties and solve the problem. Further discussion of this topic can be found in the appendix.

As in any measurement system, MRI samples are affected by noise during the measurement process. As discussed in Chapter 1, the noise distribution can (generally) be well-approximated by a simple Gaussian distribution [2]. Therefore, the noise corrupting the MR signal can be modeled by additive white Gaussian noise $\boldsymbol{\varepsilon}$ (following the central limit theorem), and Eq. (2.11) becomes

$$s = \boldsymbol{E}\boldsymbol{x} + \boldsymbol{\varepsilon}. \tag{2.12}$$

Aiming to estimate the image $\boldsymbol{x}$ in the presence of Gaussian noise $\boldsymbol{\varepsilon}$, we turn to Maximum Likelihood Estimation (MLE). Under the MLE framework, we seek to find the best (unbiased) estimate of the image $\hat{\boldsymbol{x}}$ that maximizes the conditional probability that the k-space data $\boldsymbol{s}$ would be measured, in other words,

$$\hat{\boldsymbol{x}} = argmax_{\boldsymbol{x}}\, p\left(\boldsymbol{s}|\boldsymbol{x}\right). \tag{2.13}$$

Assuming that the noise is modeled by an independent, identically distributed (i.i.d.) Gaussian distribution, we can write the noise term as

$$\boldsymbol{\varepsilon} \sim \mathrm{N}\left(z;\mu,\sigma^2\right) = \frac{1}{\sqrt{\pi 2\sigma^2}} e^{-\frac{(z-\mu)^2}{2\sigma^2}}, \tag{2.14}$$

where $\mu = 0$ is the mean and σ is the standard deviation. Since the acquired (single-channel) data $\boldsymbol{s}$ is affected by i.i.d. Gaussian noise (where $\boldsymbol{\varepsilon} = \boldsymbol{s} - \boldsymbol{E}\boldsymbol{x}$), the conditional probability of simultaneously observing all the data points in $\boldsymbol{s}$ is given by

$$p\left(\boldsymbol{s}|\boldsymbol{x}\right) = \prod_{n=1}^{N} \frac{1}{\sqrt{\pi 2\sigma^2}} e^{-\frac{(s_i-(\boldsymbol{E}\boldsymbol{x})_i)^2}{2\sigma^2}}, \tag{2.15}$$

where N is the total number of samples (i.e. $N = K_{samples}$). Noting that log is a monotonically increasing function, the $\hat{\boldsymbol{x}}$ that maximizes $f(x)$ is the same that maximizes $\log f(x)$, i.e.,

$$\hat{\boldsymbol{x}} = argmax_x \log p\left(\boldsymbol{s}|\boldsymbol{x}\right) = N \log \frac{1}{\sqrt{\pi 2\sigma^2}} - \sum_{n=1}^{N} \frac{(s_i - (\boldsymbol{E}\boldsymbol{x})_i)^2}{2\sigma^2}. \tag{2.16}$$

Maximizing a function is equivalent to minimizing the corresponding negative function, i.e., $argmax_x f(x) = argmin_x - f(x)$. Moreover, constants do not affect the computation of the minimum/maximum argument, i.e., $argmin_x f(x) = argmin_x \alpha f(x) + \beta$. Leveraging these two properties, the best image estimate $\hat{\boldsymbol{x}}$ under independent, identically distributed Gaussian noise can be written as:

$$\hat{\boldsymbol{x}} = argmin_x \sum_{n=1}^{N} (s_i - (\boldsymbol{E}\boldsymbol{x})_i)^2. \tag{2.17}$$

In other words, the reconstructed image will be the one that best fits the acquired data under a least squares cost function. This can be written in corresponding vector-matrix notation as

$$\hat{x} = argmin_{x} \|s - Ex\|_2^2. \tag{2.18}$$

Note that the least squares formulation results from the noise model, i.e., a different noise model (e.g., Rician, as discussed in Chapter 1) would result in a different formulation. In the following section, we will provide the theoretical background to compute the minimum of this least squares functional to determine the solution of the reconstruction problem, whereas practical algorithms to solve this problem will be discussed in Chapter 3.

2.4 Solution of the MR reconstruction problem

In the previous section, we established that, under i.i.d. Gaussian noise, the linear inverse problem of MR reconstruction amounts to minimizing a least squares cost function:

$$\hat{x} = argmin_{x} f(x) = argmin_{x} \|s - Ex\|_2^2. \tag{2.19}$$

The solution image $\hat{x}$ is a minimum argument of $f(x)$ if it satisfies the first-order condition

$$\nabla f = \left[\frac{\partial f}{\partial x_1}, \dots, \frac{\partial f}{\partial x_N}\right]^T = \mathbf{0}. \tag{2.20}$$

Let's consider the computation of $\frac{\partial f}{\partial x_n}$, starting by rewriting the cost function

$$f(x) = (s - Ex)^H (s - Ex), \tag{2.21}$$

where H is the conjugate transpose (Hermitian transpose). For simplicity, we will use Einstein notation, i.e., the entry in matrix $C = AB$ corresponding to the *i-th* row and *k-th* column is written as $C^i_k = A^i_{\ j} B^j_k$, where the repeated index implies a summation. In this notation, the least squares cost function can be written as

$$f(x) = \left(s_j - x_i E^i_j\right)\left(s^j - E^j_i x^i\right). \tag{2.22}$$

The partial derivative of the cost function with respect to the *n-th* entry of x is

$$\frac{\partial f}{\partial x_n} = \left(\frac{\partial s_j}{\partial x_n} - \frac{\partial x_i E^i_j}{\partial x_n}\right)\left(s^j - E^j_i x^i\right) + \left(s_j - x_i E^i_j\right)\left(\frac{\partial s^j}{\partial x_n} - \frac{\partial E^j_i x^i}{\partial x_n}\right), \tag{2.23}$$

where the product rule was used. The derivative between two entries of $\boldsymbol{x}$ is zero unless the entry is the same, leading to

$$\begin{aligned}\frac{\partial f}{\partial x_n} &= \left(0-\delta_{in}E_j^i\right)\left(s^j-E_i^j x^i\right)+\left(s_j-x_i E_j^i\right)\left(0-E_i^j\delta_{in}\right)= \\ &= -E_j^n s^j+E_j^n E_i^j x^i - s_j E_n^j + x_i E_j^i E_n^j = \\ &= -2E_j^n\left(s^j-E_i^j x^i\right),\end{aligned} \tag{2.24}$$

where δ_{in} is the Kronecker delta. Generalizing this last result to every entry in the gradient of the cost function and returning to vector-matrix notation, we obtain:

$$\nabla f = -2\boldsymbol{E}^H\left(\boldsymbol{s}-\boldsymbol{E}\boldsymbol{x}\right). \tag{2.25}$$

Enforcing the first-order condition,

$$\begin{aligned}\nabla f &= \boldsymbol{0} \\ -2\boldsymbol{E}^H\left(\boldsymbol{s}-\boldsymbol{E}\hat{\boldsymbol{x}}\right) &= \boldsymbol{0} \\ \hat{\boldsymbol{x}} &= \left(\boldsymbol{E}^H\boldsymbol{E}\right)^{-1}\boldsymbol{E}^H\boldsymbol{s} = \boldsymbol{E}^\dagger\boldsymbol{s}.\end{aligned} \tag{2.26}$$

The last line defines the solution of the MR reconstruction, where $\boldsymbol{E}^\dagger$ is known as the Moore–Penrose generalized inverse. If the acquisition is at least *fully-sampled* (meaning $\boldsymbol{\Delta}_k^{-1} \geq \boldsymbol{FOV}$, i.e., there are at least as many data points in $\boldsymbol{s}$ as unknowns in $\boldsymbol{x}$) and the noise level is adequate, then we can generally expect a well-posed problem. If a single coil fully sampled Cartesian acquisition is considered, then the forward model simplifies to a Fourier matrix, which has a condition number of 1. In such a case, $\boldsymbol{E} = \boldsymbol{F}$ and a very fast reconstruction is possible since

$$\hat{\boldsymbol{x}} = \left(\boldsymbol{F}^H\boldsymbol{F}\right)^{-1}\boldsymbol{F}^H\boldsymbol{s} = \boldsymbol{F}^{-1}\boldsymbol{s}, \tag{2.27}$$

where $\boldsymbol{F}$ can be efficiently computed using the fast Fourier transform (FFT) [3]. Returning to our sampling considerations, if the data is *under-sampled* ($\boldsymbol{\Delta}_k^{-1} < \boldsymbol{FOV}$, there are fewer points in $\boldsymbol{s}$ than unknowns in $\boldsymbol{x}$), then the condition number is $\sim\infty$ and infinitely many solutions exist. Naturally, the second Hadamard condition does not apply in this case. Consequently, Eq. (2.27) is not guaranteed to produce a correct (or even useful) reconstruction.

Since MR data is commonly acquired with an array of coils (multiple channels) with unique spatially varying sensitivities, these can also easily be incorporated into the *fully sampled* reconstruction (improving SNR due to the spatial proximity to the measured object) as $\hat{\boldsymbol{x}} = \left(\boldsymbol{C}^H\boldsymbol{I}\boldsymbol{C}\right)^{-1}\boldsymbol{C}^H\boldsymbol{F}^{-1}\boldsymbol{s}$, where $\boldsymbol{C}$ are the coil sensitivity maps. This is a type of parallel imaging [4–7] reconstruction and is discussed in detail in Chapter 6.

A point of note is that the computation of $\boldsymbol{E}^\dagger$ in the general case requires the inversion of the matrix $\boldsymbol{E}^H\boldsymbol{E}$. In practice, for even small 2D image reconstructions, the size of the matrix $\boldsymbol{E}$ can easily exceed 10,000 × 10,000, meaning that direct matrix inversion is in general not computationally feasible. Since the least squares cost function is convex and differentiable, we can attempt to reach the solution $\hat{\boldsymbol{x}}$ by iteratively searching for a local minimum (which will also be the global minimum). One of the simplest algorithms to find $\hat{x}$ is the gradient descent method. The field of optimization is rich in methods to solve problems of the type $Ax = b$, of which the gradient descent is one the simplest approaches. Another popular (and much more efficient) algorithm to solve this problem is the conjugate gradient. These and other algorithms will be described in detail in Chapter 3.

As discussed earlier, one of the properties that characterize the well-posedness of the MR reconstruction problem is its sensitivity to noise. This can be captured by the *condition number* of the forward model matrix, $\kappa(\boldsymbol{E})$, that provides a theoretical bound for error propagation in $\boldsymbol{s}$ to the solution $\hat{\boldsymbol{x}}$. In other words, the condition number predicts how small perturbations in the measured signal $\boldsymbol{s}$ affect the solution of the inverse problem $\hat{\boldsymbol{x}}$, i.e., a measure of *stability* of the solution. The condition number dictates the following relationship of (relative) error propagation from $\boldsymbol{s}$ to $\boldsymbol{x}$

$$\begin{aligned} \frac{\|\boldsymbol{x}-\hat{\boldsymbol{x}}\|_2}{\|\boldsymbol{x}\|_2} &\leq \kappa(\boldsymbol{E}) \frac{\|\boldsymbol{\varepsilon}\|_2}{\|\boldsymbol{s}\|_2} \\ \kappa(\boldsymbol{E}) &= \frac{\tau_{max}}{\tau_{min}}, \end{aligned} \tag{2.28}$$

where τ_{max} and τ_{min} correspond to the maximum and minimum singular values of $\boldsymbol{E}$. From this definition, it can be seen that $\kappa(\boldsymbol{E}) \geq 1$. The matrix is considered well-conditioned if the condition number $\kappa(\boldsymbol{E}) \approx 1$; in this case, the solution is stable to small data perturbations. As the condition number increases, the matrix $\boldsymbol{E}$ becomes increasingly ill-conditioned, meaning that small perturbations of the right-hand side ($\boldsymbol{s}$) can lead to large perturbations of the solution ($\hat{\boldsymbol{x}}$). The condition number κ effectively becomes infinite when $\boldsymbol{E}$ is singular, i.e., the problem is underdetermined. For example, if the acquired data $\boldsymbol{s}$ is *undersampled* (i.e., $\boldsymbol{\Delta}_k^{-1} < \boldsymbol{FOV}$), then the reconstruction problem is ill-posed. However, it is still possible to solve ill-posed problems by incorporating additional information in the form of regularization or constraints.

If the reconstruction problem is ill-posed, applying the least squares solution as previously described can produce different artifacts as can be seen in Fig. 2.3. Low levels of noise have a minor effect on the image, however larger noise levels can obscure smaller image structures and lead to an apparent loss of resolution, even though contrast is mostly unaffected. Acquiring only (approximately) half of the k-space, known as the partial Fourier [8,9], can create blurring/smearing artifacts if Hermitian symmetry is not considered. A uniform undersampling pattern (often referred to as Cartesian undersampling) will generate Acc overlapping replicas of the object (often referred to as ghosting), separated by a distance of FOV/Acc. These can be seen in Fig. 2.2 for 1D Cartesian undersampling and in Fig. 2.3 for 2D Cartesian undersampling. Other types of undersampling patterns also generate specific artifacts. Undersampling with a radial trajectory generates streaking artifacts, and undersampling with a spiral trajectory generates spiral-shaped aliasing, as we will see later in Chapter 4. Nonetheless, it is still possible to reconstruct the correct image in all these cases by incorporating additional information into the problem.

2.5 Regularizing the MR reconstruction problem

In the case of an ill-posed MR reconstruction problem (e.g., due to undersampling), the cost function outlined in Eq. (2.18) and its subsequent solution provided in Eq. (2.25) may not produce an adequate

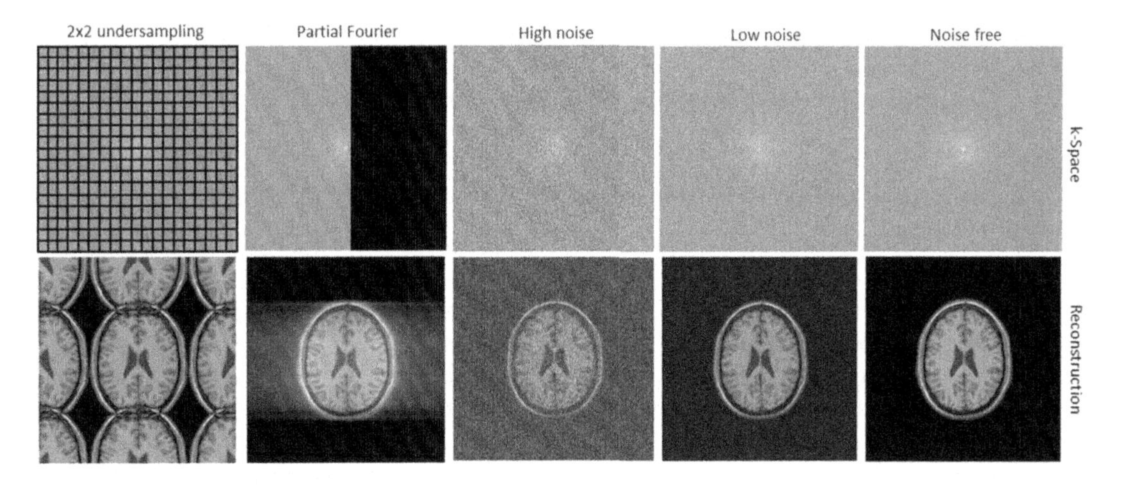

FIGURE 2.3

Reconstructed images from measurements with various sampling properties. Cartesian undersampling leads to aliasing, partial Fourier sampling leads to blurring, and white Gaussian noise perturbations in k-space lead to noise amplification.

image reconstruction of the underlying object. In this case, we can include additional *assumptions* or *prior information* into the cost function to isolate (in the case of multiple solutions) or to stabilize (in the case of unstable solutions) the solution.

These assumptions can be incorporated in the reconstruction via a cost function using a constrained formulation, as shown here:

$$\begin{aligned} \hat{\boldsymbol{x}} &= argmin_{\boldsymbol{x}} R(\boldsymbol{x}) \\ &subject\ to\ \ \|\boldsymbol{s} - \boldsymbol{E}\boldsymbol{x}\|_2^2 < \varepsilon. \end{aligned} \tag{2.29}$$

This means that, among all feasible solutions, we want to choose the one with the smallest cost function $R(\boldsymbol{x})$ for a given assumption or prior information, while, enforcing consistency between the acquired data and the image we want to reconstruct (i.e., data consistency, $\|\boldsymbol{s} - \boldsymbol{E}\boldsymbol{x}\|_2^2 < \varepsilon$).

This approach can also be cast as the regularization problem

$$\hat{\boldsymbol{x}} = argmin_{\boldsymbol{x}} \|\boldsymbol{s} - \boldsymbol{E}\boldsymbol{x}\|_2^2 + \lambda R(\boldsymbol{x}), \tag{2.30}$$

where the first term enforces consistency between the forward model and the acquired data and the second term $R(\boldsymbol{x})$ adds an assumption or prior information about the image that we want to reconstruct, whereas the regularization parameter λ controls the trade-off between data consistency and faithfulness to the optimization criterion.

For example, say we have an estimate of the image $\boldsymbol{x}_0$ (e.g., obtained from a previous examination of the same patient). We could leverage this prior information by stating that the reconstructed image should be similar (in the least squares sense) to $\boldsymbol{x}_0$, corresponding to the following constrained problem:

$$\begin{aligned} \hat{\boldsymbol{x}} &= argmin_{\boldsymbol{x}} \|\boldsymbol{x} - \boldsymbol{x}_0\|_2^2 \\ subject\ to\ &\|\boldsymbol{s} - \boldsymbol{E}\boldsymbol{x}\|_2^2 < \varepsilon. \end{aligned} \tag{2.31}$$

For purposes of optimization, it can be more practical to formulate Eq. (2.31) as an unconstrained regularized problem, i.e.,

$$\hat{\boldsymbol{x}} = argmin_{\boldsymbol{x}} f(\boldsymbol{x}) = argmin_{\boldsymbol{x}} \|\boldsymbol{s} - \boldsymbol{E}\boldsymbol{x}\|_2^2 + \lambda \|\boldsymbol{x} - \boldsymbol{x}_0\|_2^2, \tag{2.32}$$

where the regularization term is $\|\boldsymbol{x} - \boldsymbol{x}_0\|_2^2$ in this case. For an appropriate choice of λ, Eqs. (2.31) and (2.32) have the same solution. This specific type of regularization is known as Tikhonov regularization [10]. The choice of an L_2-norm ($\|.\|_2$) for the regularizer is practical because it will still lead to a linear problem in $\boldsymbol{x}$ with a closed-form solution. Following the same analysis as outlined in Eqs. (2.22)–(2.26), it can be shown that:

$$\begin{aligned} \frac{\partial f}{\partial x_n} &= -2E_j^n \left(s^j - E_i^j x^i\right) + 2\lambda I_j^n I_i^j (x^i - x_0^i) \\ \nabla f &= -2\boldsymbol{E}^H (\boldsymbol{s} - \boldsymbol{E}\boldsymbol{x}) + 2\lambda \boldsymbol{I}^H \boldsymbol{I} (\boldsymbol{x} - \boldsymbol{x}_0), \end{aligned} \tag{2.33}$$

where $\boldsymbol{I}$ corresponds to the identity matrix. Setting $\nabla f = \boldsymbol{0}$ leads to the following solution of the Tikhonov regularized problem

$$\hat{\boldsymbol{x}} = \left(\boldsymbol{E}^H \boldsymbol{E} + \lambda \boldsymbol{I}\right)^{-1} \left(\boldsymbol{E}^H \boldsymbol{s} + \lambda \boldsymbol{x}_0\right). \tag{2.34}$$

Increasing λ will make the solution further insensitive to the noise affecting the measurement; however, increasing λ will also bias the solution towards $\boldsymbol{x}_0$, potentially producing a solution $\hat{\boldsymbol{x}}$ that is not consistent with the acquired data. The latter scenario is important to note since regularization can potentially suppress noise and image features within the noise level alike. Let's consider an example where we have a low resolution prior $\boldsymbol{x}_0$ of the image and we wish to incorporate it as regularization in a noisy, undersampled reconstruction (Fig. 2.4). The reconstruction will be under-regularized if λ is too low, leading to similar quality as in the absence of regularization, presenting residual aliasing and noise amplification. In this case, the term $\|\boldsymbol{s} - \boldsymbol{E}\boldsymbol{x}\|_2^2$ in Eq. (2.32) dominates the cost function. On the other extreme, we have over-regularization (λ is too high); some aliasing and noise are suppressed, however resolution loss is instead observed as the reconstruction converges to the prior. In this case, the term $\lambda \|\boldsymbol{x} - \boldsymbol{x}_0\|_2^2$ dominates the cost function in Eq. (2.32). The ideal choice of λ achieves a good balance between both terms, allowing aliasing/noise reduction without compromising fidelity to the ground-truth image (Fig. 2.4).

The choice of the regularizer when some prior estimate $\boldsymbol{x}_0$ of the image is known is not limited to this example; any useful linear operator $\boldsymbol{L}$ can also be included in the Tikhonov regularization term. In that case its corresponding solution is given by

$$\begin{aligned} \hat{\boldsymbol{x}} &= argmin_{\boldsymbol{x}} \|\boldsymbol{s} - \boldsymbol{E}\boldsymbol{x}\|_2^2 + \lambda \|\boldsymbol{L}\boldsymbol{x} - \boldsymbol{x}_0\|_2^2 \\ \hat{\boldsymbol{x}} &= \left(\boldsymbol{E}^H \boldsymbol{E} + \lambda \boldsymbol{L}^H \boldsymbol{L}\right)^{-1} \left(\boldsymbol{E}^H \boldsymbol{s} + \lambda \boldsymbol{L}^H \boldsymbol{x}_0\right). \end{aligned} \tag{2.35}$$

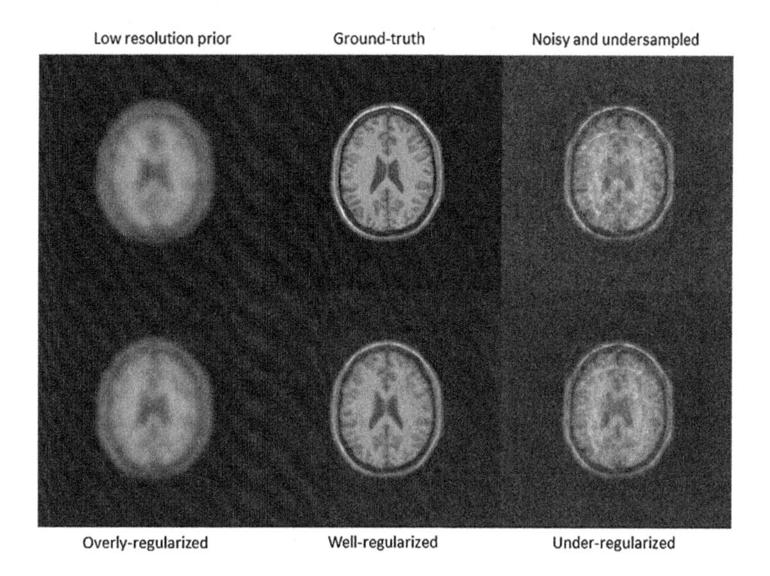

FIGURE 2.4

Example of a regularized reconstruction using a low resolution image as prior information. Image artifacts arise due to undersampling and noisy measurements; however, these can be adequately suppressed, granted the regularization strength is correctly set.

Furthermore, the regularization term does not have to necessarily feature a L_2-norm, for example, compressed sensing [11–14] employs an L_1-norm (as discussed in Chapter 8 in greater detail). Minimizing the L_1-norm of some vector $\boldsymbol{x}$ corresponds to enforcing the prior information that $\boldsymbol{x}$ should be sparse, i.e., have only a few nonzero elements. While there are only a few MR applications where the image itself is sparse, the sparsity constraint can be enforced in some domain (other than the image space) where the transformed $\boldsymbol{x}$ becomes sparse. The general formulation for L_1 regularization is written as

$$\hat{\boldsymbol{x}} = argmin_{\boldsymbol{x}} f(\boldsymbol{x}) = argmin_{\boldsymbol{x}} \|\boldsymbol{s} - \boldsymbol{E}\boldsymbol{x}\|_2^2 + \lambda \|\boldsymbol{\Phi}\boldsymbol{x}\|_1, \tag{2.36}$$

where $\boldsymbol{\Phi}$ is referred to as a sparsifying transform (e.g., finite differences or a wavelet transform), i.e., some linear operator that projects the nonsparse image $\boldsymbol{x}$ into another domain where $\boldsymbol{\Phi x}$ is sparse. Computing the gradient of this cost function requires evaluating $\nabla \|\boldsymbol{\Phi x}\|_1$ and consequently $\frac{\partial |x_i|}{\partial x_n}$, which is not well defined. Specifically, the derivative of $|x|$ approaching from 0^- is different if approaching from 0^+ (-1 and $+1$, respectively). One approach is to "round out" the "kink" at $x = 0$ by redefining $|x| \approx \sqrt{x^2 + \epsilon}$. Noting that the partial derivative $\frac{\partial |x_i|}{\partial x_n} = \delta_{in} \frac{x_i}{|x_n|}$ and also that L_1-norm $\|\boldsymbol{\Phi x}\|_1 = \left|\Phi_i^j x^i\right|$, we can compute the partial derivative of the L_1-norm

$$\frac{\partial \left|\Phi_i^j x^i\right|}{\partial x_n} = {\Phi^i}_j \delta_{in} \frac{\Phi_i^j x^i}{\left|\Phi_n^j x^n\right|} = \frac{\Phi_j^n \Phi_n^j x^n}{\sqrt{x_n \Phi_j^n \Phi_n^j x^n + \epsilon}}. \tag{2.37}$$

We can then evaluate the gradient of the sparsity-regularized reconstruction

$$\frac{\partial f}{\partial x_n} = -2E_j^n\left(s^j - E_i^j x^i\right) + 2\lambda \frac{\Phi_j^n \Phi_n^j x^n}{\sqrt{x_n \Phi_j^n \Phi_n^j x^n + \epsilon}}$$
$$\nabla f = -2\boldsymbol{E}^H(\boldsymbol{s} - \boldsymbol{E}\boldsymbol{x}) + 2\lambda \boldsymbol{\Phi}^H \boldsymbol{W}\boldsymbol{\Phi}\boldsymbol{x}, \tag{2.38}$$

where $W_n^n = \left(x_n \Phi_j^n \Phi_n^j x^n + \epsilon\right)^{-1/2}$ is a diagonal matrix that includes a term with $x^{-1/2}$. As we can see, the L_1-norm regularization creates a problem that is no longer linear in $\boldsymbol{x}$. Consequently, algorithms designed to solve linear problems will not be suitable for this optimization process and more advanced nonlinear methods must be employed. Suitable algorithms to solve this problem will be discussed in Chapter 3.

Finally, we note that many other forms of regularization are also possible, and a plethora of MR reconstruction methods exist that incorporate additional information into the process to improve performance. The main strategies are explored in Chapters 6–11.

2.6 Nonlinear inverse problems in MR

All reconstruction problems considered up to this point assume a known linear forward model, resulting in a linear inverse problem. However, there are applications in MR where the forward model $\boldsymbol{E}$ depends nonlinearly on parameters $\boldsymbol{\theta}$ that need to be estimated, in addition to the image pixels in $\boldsymbol{x}$. These cases lead to nonlinear inverse problems that can be severely underdetermined due to the increased number of unknowns. The general problem may be stated as

$$\left\{\hat{\boldsymbol{x}}, \hat{\boldsymbol{\theta}}\right\} = argmin_{\boldsymbol{x},\boldsymbol{\theta}}\, g(\boldsymbol{x}, \boldsymbol{\theta}) =$$
$$= argmin_{\boldsymbol{x},\boldsymbol{\theta}} \|\boldsymbol{s} - \boldsymbol{E}(\boldsymbol{\theta})\boldsymbol{x}\|_2^2 + \lambda_x \boldsymbol{R}_{\boldsymbol{x}}(\boldsymbol{x}) + \lambda_\theta \boldsymbol{R}_{\boldsymbol{\theta}}(\boldsymbol{\theta}). \tag{2.39}$$

Simultaneously, while estimating the image and the forward model can lead to improved results, it results in a more challenging problem than previously discussed in Section 2.3. Evaluating the partial derivatives $\partial g/\partial \boldsymbol{x}$ and $\partial g/\partial \boldsymbol{\theta}$ to identify potential minima reveals a nonlinear solution on $\left\{\hat{\boldsymbol{x}}, \hat{\boldsymbol{\theta}}\right\}$ that is not immediately easy to compute, nor is it necessarily unique. This function may admit multiple (possibly local) minima, and consequently the converged solution is dependent on the initial conditions of the chosen algorithm. Furthermore, even when the data acquisition is fully-sampled, the corresponding nonlinear problem can be underdetermined, as $K_{samples} < N_{pixels} + \theta_{parameters}$. Despite all these challenges, several nonlinear MR forward models have been developed in recent years to avoid assumptions in the forward model, increase robustness to errors in the model, or directly estimate parameters of interest that are embedded in the model. General strategies to overcome these challenges include parametrization to reduce the number of unknowns, splitting the reconstruction into multiple subproblems and linearizing the model around a neighborhood of existing solutions. We will briefly discuss three examples of nonlinear problems in MR: parallel imaging, motion estimation/correction, and MR parameter reconstruction. Detailed analysis of various nonlinear models used in MR can be found in Chapters 12–16.

2.6.1 Nonlinear parallel imaging

A common *image-based* formulation of parallel imaging includes the coil sensitivities $\boldsymbol{C}$ (complex, block diagonal matrix) into the forward model as $\boldsymbol{E} = \boldsymbol{BFC}$, where $\boldsymbol{B}$ is a k-space sampling operator (logical, diagonal sparse matrix). The coil sensitivities can be estimated a priori from calibration data; in that case, the following parallel imaging reconstruction becomes a simple linear inverse problem. Errors in the initial estimation of $\boldsymbol{C}$ can propagate into reconstruction errors in $\boldsymbol{x}$. Recent methods [15,16] have been proposed to simultaneously solve for the image $\boldsymbol{x}$ and the coil maps $\boldsymbol{C}$:

$$\left\{\hat{\boldsymbol{x}}, \hat{\boldsymbol{C}}\right\} = argmin_{\boldsymbol{x},\boldsymbol{C}} \|\boldsymbol{s} - \boldsymbol{E}(\boldsymbol{C})\boldsymbol{x}\|_2^2 + \sum_i \lambda_i \boldsymbol{R}_i(.). \tag{2.40}$$

The operator $\boldsymbol{C} = [\boldsymbol{C}_1, \ldots, \boldsymbol{C}_s]$ is a block-diagonal matrix, where the nonzero entries to be estimated are $c_{r,s}$: one parameter per image location r and per coil channel s. Consequently, even for *fully sampled* data, this problem will be underdetermined since $K_{samples} < N_{pixels} + N_{coils}N_{pixels}$ ($K_{samples} = N_{coils}N_{pixels}$ for *fully sampled* data). Furthermore, the solution is nonlinear for the parameters of interest, requiring more complex optimization algorithms. One approach [15] to this problem is to employ Gauss–Newton algorithms. The strategy here to linearize the problem around the current solution at the *i-th* iteration $\boldsymbol{y}^{[i]}$ and to solve for the small increment to the solution $d\boldsymbol{y}$. Reformulating the problem as $\boldsymbol{E}(\boldsymbol{y}) = \boldsymbol{s}$, where $\boldsymbol{y} = \left[x_1, \ldots, x_r, c_{1,1}, \ldots, c_{r,s}\right] = [\boldsymbol{x}, \boldsymbol{C}_1, \ldots, \boldsymbol{C}_s]$, the forward model can be approximated by the first-order Taylor expansion

$$\boldsymbol{E}\left(\boldsymbol{y}^{[i]} + d\boldsymbol{y}\right) \approx \boldsymbol{E}\left(\boldsymbol{y}^{[i]}\right) + \mathcal{D}\boldsymbol{E}\left(\boldsymbol{y}^{[i]}\right) d\boldsymbol{y} = \boldsymbol{s}, \tag{2.41}$$

where $\mathcal{D}$ is the Jacobian. We can then formulate an approximate linearized inverse problem on $d\boldsymbol{y}$ as

$$d\hat{\boldsymbol{y}}^{[i+1]} = argmin_{d\boldsymbol{y}} \left\|\boldsymbol{s} - \boldsymbol{E}\left(\boldsymbol{y}^{[i]}\right) - \mathcal{D}\boldsymbol{E}\left(\boldsymbol{y}^{[i]}\right) d\boldsymbol{y}\right\|_2^2 + \lambda \boldsymbol{R}(d\boldsymbol{y}), \tag{2.42}$$

for which there are efficient solvers like the conjugate gradient. Iteratively solving for $d\boldsymbol{y}$ and updating $\boldsymbol{y}^{[i+1]} = \boldsymbol{y}^{[i]} + d\boldsymbol{y}^{[i]}$ will generally converge to a solution, however, the final solution of $\hat{\boldsymbol{y}}$ can vary depending on the choice of the initial guess $\boldsymbol{y}^{[0]}$. Note that, before solving the linear inverse problem in Eq. (2.42), the partial derivatives of the encoding operator must be computed, requiring additional computations. Considering $\boldsymbol{E}(\boldsymbol{y}) = [\boldsymbol{E}_1(\boldsymbol{y}), \ldots, \boldsymbol{E}_s(\boldsymbol{y})]^T$, $\partial \boldsymbol{E}_s\left(\boldsymbol{y}^{[i]}\right)/\partial \boldsymbol{x} = \boldsymbol{BFC}_s$ and $\partial \boldsymbol{E}_s\left(\boldsymbol{y}^{[i]}\right)/\partial \boldsymbol{C}_s = \boldsymbol{BF}\boldsymbol{y}^{[i]}$ must be evaluated every time $\boldsymbol{y}$ is updated.

Another approach to tackle the problem of nonlinear parallel imaging is to parametrize the coil sensitivities in order to reduce the number of unknowns, as proposed in [16]. Coil sensitivities are known to be smooth, and therefore can be parametrized by low-order polynomials such as $c_{r,s} = \sum_k^P \alpha_{k,s} r^{\langle k \rangle}$, where P is the order of the polynomial (and $.^{\langle k \rangle}$ denotes the exponential). Instead of solving for $c_{r,s}$ we can solve for $\alpha_{i,s}$, which significantly reduces the number of unknowns since $P \ll N_{coils}N_{pixels}$. A further property of this formulation is that appropriate selection of P will inherently enforce a smooth solution. The corresponding joint minimization problem then becomes

$$\left\{\hat{\boldsymbol{x}}, \hat{\boldsymbol{\theta}}\right\} = argmin_{\boldsymbol{x},\boldsymbol{a}} \|\boldsymbol{s} - \boldsymbol{BFC}(\boldsymbol{\theta})\boldsymbol{x}\|_2^2, \tag{2.43}$$

where $\boldsymbol{\theta}$ contains the vectorized $\alpha_{i,s}$. A common approach for these types of problems is alternating minimization [17]: solving for one variable while fixing the other variable (iteratively). This amounts to solving the following subproblems

$$\hat{x}^{[i+1]} = argmin_x \left\| s - BFC\left(\hat{\theta}^{[i]}\right)x \right\|_2^2$$
$$\hat{\theta}^{[i+1]} = argmin_a \left\| s - BFC(\theta)\hat{x}^{[i]} \right\|_2^2, \tag{2.44}$$

while fixing one variable at a time. The first problem is a simple linear inverse problem. Since C is block diagonal, the second problem can be rewritten as

$$\hat{\theta}^{[i+1]} = argmin_a \left\| s - BF\tilde{C}(\hat{x}^{[i]})\theta \right\|_2^2, \tag{2.45}$$

where each row i of $\tilde{C}(x)$ is given by $\left[r_i^{\langle 1\rangle}x_i, r_i^{\langle 2\rangle}x_i, \ldots, r_i^{\langle P\rangle}x_i\right]$. Under this formulation, estimation of θ becomes a linear inverse problem as well. Further discussion of various (linear and nonlinear) methods for parallel imaging can be found in Chapter 6.

2.6.2 Nonlinear motion estimation/correction

MR acquisition times are typically in the order of minutes, making the scan sensitive to motion. Reconstructing data from a moving object will lead to artifacts similar to aliasing artifacts (in addition to blurring artifacts) because the reconstructed image will correspond to aliased views of the object in the various motion states. One way to correct for these artifacts is to incorporate motion into the forward model E

$$\hat{x} = argmin_x \|s - BFCMx\|_2^2, \tag{2.46}$$

where $B = [B_1, \ldots, B_n]^T$ is the sampling for each motion state n (logical, block diagonal sparse matrix) and $M = [M, \ldots, M_n]^T$ are the corresponding motion fields (real sparse matrix). If the motion is known (along with coil sensitivities and motion state information), the motion corrected image can be easily obtained since this is a linear inverse problem (note that the problem of motion corrected reconstruction is generally ill-posed since rotation or scaling motion can open gaps in k-space, effectively inducing undersampling). A nonlinear problem arises when we attempt to simultaneously estimate the motion corrected image and the motion

$$\left\{\hat{x}, \hat{\theta}\right\} = argmin_{x,\theta} \|s - BFCM(\theta)x\|_2^2, \tag{2.47}$$

where θ are the motion parameters (we assume B is known). This problem is generally underdetermined due to the large number of unknowns, $K_{samples} < N_{pixels} + DN_{pixels}N_{shots}$, where $D = 2$ (3) for 2D (3D) imaging. If the motion can be captured by a simple model, e.g., rigid motion, then the conditioning of the problem can improve significantly [18]. Considering 2D rigid motion, we can reduce the number of motion parameters from $DN_{pixels}N_{shots}$ to $3N_{shots}$ since $\theta_{(n)} = \left[t_{u(n)}, t_{v(n)}, \alpha_{(n)}\right]^T$, where $t_{u(n)}$ and $t_{v(n)}$ are global translations (along two dimensions, $\cdot_{u,v}$) and $\alpha_{(n)}$ is the rotation angle (for the nth motion state). This can be a very useful motion model for brain imaging, for example. While the dependency on the translational component of motion is linear, it is not the case for the rotational component that involves sin(α) and cos(α). As seen in the previous section, one way to tackle joint estimation is to iteratively alternate between two subproblems

$$\begin{aligned}\hat{\boldsymbol{x}}^{[i+1]} &= argmin_{\boldsymbol{x}} \left\| \boldsymbol{s} - \boldsymbol{BFCM}(\hat{\boldsymbol{\theta}}^{[i]})\boldsymbol{x} \right\|_2^2 \\ \hat{\boldsymbol{\theta}}^{[i+1]} &= argmin_{\boldsymbol{\theta}} \left\| \boldsymbol{s} - \boldsymbol{BFCM}(\boldsymbol{\theta})\hat{\boldsymbol{x}}^{[i]} \right\|_2^2 .\end{aligned} \tag{2.48}$$

Once again, the first problem is a straightforward linear inverse problem. For the second problem, one approach is to employ the Newton method by performing a second-order Taylor expansion

$$\boldsymbol{E}\left(\boldsymbol{\theta}^{[i]} + d\boldsymbol{\theta}\right) \approx \boldsymbol{E}\left(\boldsymbol{\theta}^{[i]}\right) + \mathcal{D}\boldsymbol{E}\left(\boldsymbol{\theta}^{[i]}\right) d\boldsymbol{\theta} + \frac{1}{2}\mathcal{H}\boldsymbol{E}\left(\boldsymbol{\theta}^{[i]}\right) d\boldsymbol{\theta}^2, \tag{2.49}$$

where $\mathcal{H}$ is the Hessian. It can be shown that $d\boldsymbol{\theta} = -\left[\mathcal{H}\boldsymbol{E}\left(\boldsymbol{\theta}^{[i]}\right)\right]^{-1} \mathcal{D}\boldsymbol{E}\left(\boldsymbol{\theta}^{[i]}\right)$, making the corresponding Newtown update as follows: $\boldsymbol{\theta}^{[i+1]} = \boldsymbol{\theta}^{[i]} - \left[\mathcal{H}\boldsymbol{E}\left(\boldsymbol{\theta}^{[i]}\right)\right]^{-1} \mathcal{D}\boldsymbol{E}\left(\boldsymbol{\theta}^{[i]}\right)$. Since the Hessian is not guaranteed to be positive definite, not all iterations are guaranteed to reduce the cost function. The Levenberg–Marquardt is a common variant of the algorithm that ensures convergence at every step; in this case, the update is given by $\boldsymbol{\theta}^{[i+1]} = \boldsymbol{\theta}^{[i]} - \left[\mathcal{H}\boldsymbol{E}\left(\boldsymbol{\theta}^{[i]}\right) + \alpha^{[i]}\boldsymbol{I}\right]^{-1} \mathcal{D}\boldsymbol{E}\left(\boldsymbol{\theta}^{[i]}\right)$, with $\alpha \geq 0$. If a rigid (or affine) motion model is considered, then the first and second derivatives, $\mathcal{D}\boldsymbol{M}(\boldsymbol{\theta})$ and $\mathcal{H}\boldsymbol{M}(\boldsymbol{\theta})$, are practical to compute.

A more challenging problem arises when affine motion models are not sufficient to accurately describe the underlying motion (e.g., of a respiratory or cardiac nature). In this case, the motion is modeled as a deformation field ($\boldsymbol{w}$) with two parameters (with directions denoted as $\cdot_{u,v}$), per pixel ($\cdot^{(x_i)}$), per motion state ($\cdot_{(n)}$), i.e., $\boldsymbol{\theta}_{(n)} = \left[w_{u(n)}^{x_1}, w_{v(n)}^{x_1}, \ldots, w_{u(n)}^{x_N}, w_{v(n)}^{x_N}\right]^T$. One approach to this problem is to establish a link between residual errors in the image $\boldsymbol{x}$ and small perturbations in the motion model $\boldsymbol{M}(\boldsymbol{\theta})$ [19]. In the presence of errors in the model (namely, in the motion operator), there will be a residual error in the reconstruction, $\boldsymbol{s} = \boldsymbol{E}(\hat{\boldsymbol{\theta}})\boldsymbol{x} + \mathcal{E}$, where $\mathcal{E} = \boldsymbol{BFC}\left[\boldsymbol{M}(\boldsymbol{\theta}_{true}) - \boldsymbol{M}(\hat{\boldsymbol{\theta}})\right]\boldsymbol{x}_{true}$, $\boldsymbol{x}_{true}$ is the motion free image and $\boldsymbol{\theta}_{true}$ are the correct motion parameters. A key assumption based on optical flow makes possible the following approximation: $\left[\boldsymbol{M}(\boldsymbol{\theta}_{true}) - \boldsymbol{M}(\hat{\boldsymbol{\theta}})\right]\boldsymbol{x}_{true} \cong -\nabla_s\hat{\boldsymbol{x}}\, d\boldsymbol{\theta}$, where ∇_s denotes the spatial gradient. $\nabla_s\hat{\boldsymbol{x}}$ is a block (for each motion state $\cdot_{(n)}$), bidiagonal matrix, where the ith row of the nth motion state is given by $\nabla_s\hat{\boldsymbol{x}}_{i(n)} = \left[0, \ldots, 0, \partial x_{i(n)}/\partial u, \partial x_{i(n)}/\partial v, 0, \ldots, 0\right]$. Under this approximation, we can establish the following relationship

$$\mathcal{E} = \boldsymbol{s} - \boldsymbol{BFCM}\left(\hat{\boldsymbol{\theta}}\right)\boldsymbol{x} = -\boldsymbol{BFC}\nabla_s\hat{\boldsymbol{x}}d\boldsymbol{\theta}. \tag{2.50}$$

Consequently, we can write two coupled problems to be solved by alternate minimization as

$$\begin{aligned}\hat{\boldsymbol{x}}^{[i+1]} &= argmin_{\boldsymbol{x}} \left\| \boldsymbol{s} - \boldsymbol{BFCM}(\hat{\boldsymbol{\theta}}^{[i]})\boldsymbol{x} \right\|_2^2 \\ d\hat{\boldsymbol{\theta}}^{[i+1]} &= argmin_{d\boldsymbol{\theta}} \left\| \boldsymbol{s} - \boldsymbol{BFCM}\left(\hat{\boldsymbol{\theta}}^{[i]}\right)\hat{\boldsymbol{x}}^{[i]} + \boldsymbol{BFC}\nabla_s\hat{\boldsymbol{x}}^{[i]}d\boldsymbol{\theta} \right\|_2^2 ,\end{aligned} \tag{2.51}$$

with the motion model parameters being updated by $\boldsymbol{\theta}^{[i+1]} = \boldsymbol{\theta}^{[i]} + d\boldsymbol{\theta}^{[i]}$. Note that both problems in Eq. (2.51) are linear inverse problems and have practical solutions. The second problem, however, is highly underdetermined ($DN_{pixels}N_{shots}$ unknowns). Parametrization of the motion and further regularization can be employed to improve the condition of the second problem. The general problem of motion estimation and motion correction in MRI is discussed in greater detail in Chapter 13.

2.6.3 Nonlinear parameter reconstruction

One of the strengths of MRI is the capability of measuring a plethora of physical tissue and system properties, such as relaxation times, diffusion, susceptibility, temperature, water-fat separation, magnetization transfer, and magnetic field (B0 or B1) mapping. Conventional methods to estimate these parameters rely on a two-step approach: i) reconstruct a set of images with different contrasts, weighted by the parameter of interest; ii) perform a pixel-wise fit of the reconstructed images to the theoretical MR physics model to estimate the parameter of interest. Note that the reconstruction step in i) requires $K_{samples} = N_{contrasts}N_{pixels}$ to produce alias-free images, enabling the fit of a parametric map of N_{pixels}. On the other hand, reconstructing the parameter map directly from the acquired k-space data can reduce the number of unknowns, improving performance. Similar to previous examples, the MR physics parameters of interest will be embedded into the encoding operator, and their direct estimation will require challenging nonlinear reconstructions.

The initial forward model outlined at the start of the chapter (Eq. (2.1)) neglects the fact that the transverse magnetization M_{xy} is a function tissue parameters $\boldsymbol{\theta}$ such as the proton density ρ, relaxation times T_1, T_2 or T_2^* (and/or other MR physics parameters), and sequence parameters $\boldsymbol{\kappa}$ that control the RF excitation field $\mathbf{B}_1$ and the magnetic field gradients $\mathbf{G}$. In other words, $M_{xy}(\boldsymbol{r}, t) = Z(\boldsymbol{r}, t, \boldsymbol{B}_1, \boldsymbol{G}, \boldsymbol{\theta})$, where Z models the Bloch equations. For simplicity, we will consider steady-state applications, where the signal model is usually in the form $M_{xy}(\boldsymbol{r}) = Z_{\kappa}(\boldsymbol{r}, \boldsymbol{\theta})\, \rho(\mathbf{r})$, where Z is the signal model for the MR sequence determined by sequence parameters $\boldsymbol{\kappa}$. Inserting our new signal model into Eq. (2.1) leads to the following inverse problem:

$$\left\{\hat{\rho}, \hat{\boldsymbol{\theta}}\right\} = argmin_{\rho,\boldsymbol{\theta}} \|s - \boldsymbol{F}\boldsymbol{C}\boldsymbol{Z}_{\kappa}(\boldsymbol{\theta})\boldsymbol{\rho}\|_2^2 + \sum_i \lambda_i \boldsymbol{R}_i(.), \tag{2.52}$$

which is an instance of Eq. (2.39). Despite being linear on $\boldsymbol{\rho}$, this problem is not linear on $\boldsymbol{\theta}$. As shown in the previous sections, a Taylor expansion can be performed to linearize the problem around an existing solution [20]

$$\boldsymbol{E}\left(\boldsymbol{y}^{[i]} + d\boldsymbol{y}\right) \approx \boldsymbol{E}\left(\boldsymbol{y}^{[i]}\right) + \mathcal{D}\boldsymbol{E}\left(\boldsymbol{y}^{[i]}\right) d\boldsymbol{y} = \boldsymbol{s}, \tag{2.53}$$

with $\boldsymbol{E} = \boldsymbol{F}\boldsymbol{C}\boldsymbol{Z}_{\kappa}(\boldsymbol{\theta})$ and $\boldsymbol{y} = [\boldsymbol{\rho}, \boldsymbol{\theta}]$. This leads to the same strategy outlined previously in nonlinear parallel imaging reconstruction and can be solved by some Gauss–Newton variant. As before, the initialization of $\boldsymbol{y}^{[0]}$ can determine the convergence of the algorithm. Moreover, the scales of the partial derivatives of $\boldsymbol{Z}_{\kappa}$ can vary by several orders of magnitude (depending on the MR physics model and parameters under consideration), which can significantly increase the condition number of the problem. This issue is addressed by normalizing these partial derivatives (a form of preconditioning); the normalizing scales are often experimentally determined. If the partial derivatives are not normalized, the iterative process may take a very large number of iterations to converge or, possibly, diverge. Finally, a variety of regularization approaches can be enforced for any of the parameters being reconstructed; since the parameter maps often have similar structure to common contrast-weighted MR images, similar regularization strategies apply.

A similar approach [21] exploits the fact that, for a given $\boldsymbol{\theta}$, the linear term $\boldsymbol{\rho}$ can be directly obtained by solving the corresponding linear inverse problem $\boldsymbol{\rho} = [\boldsymbol{F C Z}_\kappa(\boldsymbol{\theta})]^{\boldsymbol{H}} s$. Reinserting this into Eq. (2.52) leads to:

$$\hat{\boldsymbol{\theta}} = argmin_{\boldsymbol{\theta}} \left\| \left\{ \boldsymbol{I} - \boldsymbol{F C Z}_\kappa(\boldsymbol{\theta}) \left[\boldsymbol{F C Z}_\kappa(\boldsymbol{\theta})\right]^{\boldsymbol{H}} \right\} s \right\|_2^2 + \sum_i \lambda_i \boldsymbol{R}_i(.). \tag{2.54}$$

Recasting the problem in this manner reduces the total number of unknowns and is expected to have better convergence properties (i.e., avoid local minima). The same strategy as outlined in Eq. (2.48) can be used here, solving the nonlinear problem with some Newton-variant algorithm. As seen in previous examples, these will require the Jacobian (and potentially the Hessian) of the encoding operator, which may incur considerable computation costs. Notably, in nonlinear parameter reconstruction, $\partial \boldsymbol{Z}_k(\boldsymbol{\theta})/\partial \boldsymbol{\theta}$ will need to be computed. For most simple cases, $\boldsymbol{Z}_k$ is an exponential of some sort, and an analytical solution exists. On the other hand, if a general (nonsteady state) MR sequence is considered, then $\boldsymbol{Z}_\kappa(\boldsymbol{\theta}, t)$ may not have a closed-form solution, requiring a full evaluation of the Bloch equations instead. In such cases, finite difference methods may be employed to approximate the required partial derivatives of the MR sequence forward model $\boldsymbol{Z}_\kappa(\boldsymbol{\theta}, t)$. Calculating the derivative via finite differences will require repeated evaluations of $\boldsymbol{Z}_\kappa(\boldsymbol{\theta}, t)$ and $\boldsymbol{Z}_\kappa(\boldsymbol{\theta} + \boldsymbol{\Delta\theta}, t)$, which can add to the computational burden, particularly as the complexity of the MR physics model $\boldsymbol{Z}_\kappa$ increases and, correspondingly, the number of parameters in $\boldsymbol{\theta}$. A thorough description of various linear and nonlinear approaches to parametric mapping is provided in Chapter 15.

2.7 Summary

In this chapter we covered the fundamentals of reconstruction in MRI, how it can be formulated as a linear inverse problem, and some of the tools we have available to solve this problem. MR reconstruction is generally an ill-posed problem (in the Hadamard sense), with several considerations that underpin the process. The required discretization and finite sampling during acquisition impose limitations on the reconstructed alias-free FOV, resolution, and potential *Gibbs-ringing* artifacts. The presence of Gaussian noise during the acquisition also dictates that the best image estimate will be given by the minimization of a L_2 data consistency term. In the case of a *fully sampled* acquisition, the process requires only an inverse Fourier transform; however, in the general *undersampled* case, the problem is underdetermined, and additional information is required. We explored various types of regularization for *undersampled* reconstructions, using L_1 and L_2 norms, and how one goes about solving these problems. In the last part of this chapter, we briefly introduced some nonlinear inverse problems in the context of MR reconstruction. These usually arise when the unknown is not (just) the image, but some parameters that nonlinearly characterize the forward model. Nonlinear inverse problems are present in many MR applications and have been tackled with a myriad of different strategies, but we introduced only a few in this chapter. In the remaining chapters of the book we will explore all of the key methods in MR reconstruction, from seminal to state-of-the-art, for linear and nonlinear inverse problems.

References

[1] Rudin W. Real and complex analysis; 1987.

[2] Haacke EM, Brown RW, Thompson MR, Venkatesan R. Magnetic resonance imaging: physical principles and sequence design. John Wiley & Sons, Ltd; 1999.

[3] Cooley JW, Tukey JW. An algorithm for the machine calculation of complex Fourier series. Math Comput 1965;19:297–301.

[4] Sodickson DK, Manning WJ. Simultaneous acquisition of spatial harmonics (SMASH): fast imaging with radiofrequency coil arrays. Magn Reson Med 1997;38:591–603. https://doi.org/10.1002/mrm.1910380414.

[5] Griswold MA, Jakob PM, Heidemann RM, et al. Generalized autocalibrating partially parallel acquisitions (GRAPPA). Magn Reson Med 2002;1210:1202–10. https://doi.org/10.1002/mrm.10171.

[6] Pruessmann KP, Weiger M, Scheidegger MB, Boesiger P. SENSE: sensitivity encoding for fast MRI. Magn Reson Med 1999;42:952–62. https://doi.org/10.1002/(SICI)1522-2594(199911)42:5<952::AID-MRM16>3.0.CO;2-S.

[7] Uecker M, Lai P, Murphy MJ, et al. ESPIRiT - an eigenvalue approach to autocalibrating parallel MRI: where SENSE meets GRAPPA. Magn Reson Med 2014;71:990–1001. https://doi.org/10.1002/mrm.24751.

[8] Noll DC, Nishimura DG, Macovski A. Homodyne detection in magnetic resonance imaging. IEEE Trans Med Imaging 1991;10:154–63. https://doi.org/10.1109/42.79473.

[9] Blaimer M, Gutberlet M, Kellman P, Breuer FA, Köstler H, Griswold MA. Virtual coil concept for improved parallel MRI employing conjugate symmetric signals. Magn Reson Med 2009;61:93–102. https://doi.org/10.1002/mrm.21652.

[10] Tikhonov A. On the stability of inverse problems. Proc USSR Acad Sci 1943;39:195–8.

[11] Lustig M, Donoho D, Pauly JM. Sparse MRI: the application of compressed sensing for rapid MR imaging. Magn Reson Med 2007;58:1182–95. https://doi.org/10.1002/mrm.21391.

[12] Block KT, Uecker M, Frahm J. Undersampled radial MRI with multiple coils. Iterative image reconstruction using a total variation constraint. Magn Reson Med 2007;57:1086–98. https://doi.org/10.1002/mrm.21236.

[13] Candès EJ, Romberg J, Tao T. Robust uncertainty principles: exact signal frequency information. IEEE Trans Inf Theory 2006;52:489–509. https://doi.org/10.1109/TIT.2005.862083.

[14] Donoho DL. Compressed sensing. IEEE Trans Inf Theory 2006;52:1289–306. https://doi.org/10.1109/TIT.2006.871582.

[15] Uecker M, Hohage T, Block KT, Frahm J. Image reconstruction by regularized nonlinear inversion - joint estimation of coil sensitivities and image content. Magn Reson Med 2008;60:674–82. https://doi.org/10.1002/mrm.21691.

[16] Ying L, Sheng J. Joint image reconstruction and sensitivity estimation in SENSE (JSENSE). Magn Reson Med 2007;57:1196–202. https://doi.org/10.1002/mrm.21245.

[17] Sutton BP, Noll DC, Fessler JA. Dynamic field map estimation using a spiral-in/spiral-out acquisition. Magn Reson Med 2004;51:1194–204. https://doi.org/10.1002/mrm.20079.

[18] Cordero-Grande L, Teixeira RPAG, Hughes EJ, Hutter J, Price AN, Hajnal JV. Sensitivity encoding for aligned multishot magnetic resonance reconstruction. IEEE Trans Comput Imaging 2016;2:266–80. https://doi.org/10.1109/tci.2016.2557069.

[19] Odille F, Vuissoz PA, Marie PY, Felblinger J. Generalized reconstruction by inversion of coupled systems (GRICS) applied to free-breathing MRI. Magn Reson Med 2008;60:146–57. https://doi.org/10.1002/mrm.21623.

[20] Roeloffs V, Wang X, Sumpf TJ, Untenberger M, Voit D, Frahm J. Model-based reconstruction for T1 mapping using single-shot inversion-recovery radial FLASH. Int J Imaging Syst Technol 2016;26:254–63. https://doi.org/10.1002/ima.22196.

[21] Sbrizzi A, van der Heide O, Cloos M, van der Toorn A, Van den Berg CA. Fast quantitative MRI as a nonlinear tomography problem. Magn Reson Imaging 2018;46:56–63. https://doi.org/10.1007/s10393-014-0979-y.

Suggested readings

[22] Liang Zhi-Pei, Lauterbur PC. Principles of magnetic resonance imaging: a signal processing perspective. The Institute of Electrical and Electronics Engineers Press; 2000.
[23] Nishimura DG. Principles of magnetic resonance imaging. Stanford University; 1996.
[24] Hansen PC. Discrete inverse problems: insight and algorithms. Society for Industrial and Applied Mathematics; 2010 Jan 1.
[25] Aster RC, Borchers B, Thurber CH. Parameter estimation and inverse problems. Elsevier; 2018 Oct 16.
[26] Fessler JA. Model-based image reconstruction for MRI. IEEE Signal Process Mag 2010 Jun 14;27(4):81–9.
[27] Wang X, Tan Z, Scholand N, Roeloffs V, Uecker M. Physics-based reconstruction methods for magnetic resonance imaging. Philos Trans R Soc A 2021 Jun 28;379(2200):20200196.

CHAPTER

Optimization Algorithms for MR Reconstruction

3

Jiaming Liu[a], Rakib Hyder[c], M. Salman Asif[c], and Ulugbek S. Kamilov[a,b]

[a]*Department of Electrical and Systems Engineering, Washington University in St. Louis, St. Louis, MO, United States*
[b]*Department of Computer Science and Engineering, Washington University in St. Louis, St. Louis, MO, United States*
[c]*Department of Electrical and Computer Engineering, University of California, Riverside, CA, United States*

3.1 Introduction

As discussed in Chapters 1 and 2, the data collection in MRI is performed in the *k-space*, which corresponds to the Fourier transform of the image. When data is fully sampled, the relationship between the vectorized form of the k-space data $\mathbf{s} \in \mathbb{C}^p$ and the vectorized form of the underlying 2D image $\mathbf{x} \in \mathbb{C}^n$ in MRI can be modeled as

$$\mathbf{s} = \mathbf{F}\mathbf{x}, \tag{3.1}$$

where $\mathbf{F}$ represents the 2D Fourier transform with $F_{kr} = \exp(-i2\pi \mathbf{k}_k^{\mathsf{T}} \mathbf{r}_r)$, $\mathbf{k}_k \in \mathbb{R}^2$ denotes the location of the kth k-space sample, and $\mathbf{r}_r \in \mathbb{R}^2$ denotes the spatial coordinates of the rth pixel. When both $\{\mathbf{k}_k\}$ and $\{\mathbf{r}_r\}$ are on the Cartesian grid with $n = p$, the k-space measurements $\mathbf{s}$ can be computed using a 2D *fast Fourier transform* (FFT), and the image $\mathbf{x}$ can be easily reconstructed by applying a 2D inverse FFT. In other words, we never have to explicitly store or apply the matrix $\mathbf{F}$. This is true also for the non-Cartesian sampling described in Chapter 4.

Parallel MRI uses a set of coil elements for data acquisition, each of which comes with an independent receiver [1,2]. Each coil element is more sensitive to the magnetization closest to it and less sensitive to magnetization further away, which results in different coil sensitivity profiles in the spatial domain. The fully sampled k-space data recorded by each coil can be represented as

$$\mathbf{s}_c = \mathbf{F}\mathbf{C}_c\mathbf{x}, \tag{3.2}$$

where $\mathbf{C}_c$ is a diagonal matrix that contains the spatial sensitivity map for the cth coil as its diagonal. Note that, even though we represent $\mathbf{C}_c$ as a diagonal matrix, we do not need to store the coil sensitivity maps as a diagonal matrix. Data acquisition with a coil array placed close to the body provides better SNR compared to data acquisition with the body coil, which is integrated in the scanner bore and further away from the measured subject. Parallel imaging refers to applying the coil array for acceleration. This is explained in detail in the chapter on parallel imaging.

As described in Chapter 2, it is desirable in many applications to *accelerate* the MR data acquisition by collecting $p < n$ measurements in the k-space [3,4]. The measurements are typically collected in parallel using $T \geq 1$ receiver coils [1], leading to the model

$$\mathbf{s}_c = \mathbf{B}\mathbf{F}\mathbf{C}_c\mathbf{x} + \boldsymbol{\epsilon}_c,$$

Advances in Magnetic Resonance Technology and Applications, Volume 7, ISSN 2666-9099. https://doi.org/10.1016/B978-0-12-822726-8.00012-9

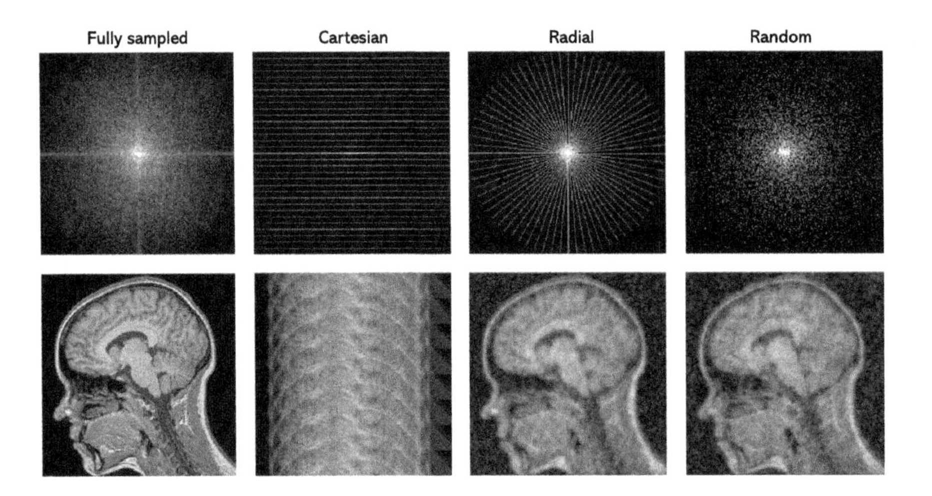

FIGURE 3.1 Sampling strategies in MRI.

Fully sampled k-space can be converted back to the image domain by a simple inverse Fourier transform. *Cartesian*, *radial*, and *random* sampling strategies in this figure correspond to the k-space undersampling by a factor of 8, leading to imaging artifacts in the spatial domain. Note how Cartesian sampling leads to structured artifacts that are more difficult to mitigate computationally.

where $\mathbf{B} \in \mathbb{R}^{p \times n}$ denotes the sampling matrix that contains p rows of the $n \times n$ identity matrix and $\boldsymbol{\epsilon}_c \in \mathbb{C}^p$ is the measurement noise. Note that $\mathbf{BF}$ does not depend on the coil since all coils see the same k-space sampling pattern. By concatenating the measurements from all coils into the vector $\mathbf{s} = (\mathbf{s}_1, \ldots, \mathbf{s}_T)$, we obtain the conventional MRI *forward model*

$$\mathbf{s} = \mathbf{E}\mathbf{x} + \boldsymbol{\epsilon}, \tag{3.3}$$

where $\mathbf{E} \in \mathbb{C}^{m \times n}$, with $m = pT$, is the *measurement operator* representing the physical information and $\boldsymbol{\epsilon} \in \mathbb{C}^m$ is the *measurement noise*, which is often assumed to be additive white Gaussian. The measurement operator $\mathbf{E}$ is assumed to be known during image reconstruction with the sampling matrix $\mathbf{B}$ determined by the location of the k-space samples $\{\mathbf{k}_k\}$ acquired and the sensitivity maps $\{\mathbf{C}_c\}$ determined through a dedicated calibration procedure [5,6]. Note that we do not need $\mathbf{E}$ as an explicit matrix; we can apply $\mathbf{E}$ as a function that computes element-wise multiplication of coil sensitivity maps and image, followed by 2D FFT and subsampling.

The data in MRI is collected using a sequence of measurement trajectories through k-space. The most commonly used trajectories include *Cartesian*, *radial*, and *spiral* (see Fig. 3.1). Additionally, randomized sampling strategies have also been explored in the context of compressive sensing [3,4]. While non-Cartesian sampling generally provides more efficient coverage of k-space and yields less structured artifacts conducive for subsequent image reconstruction, Cartesian sampling is still the most widely-used choice in clinical practice because of its simplicity.

This chapter reviews several widely used optimization strategies for MR image reconstruction. The natural starting point for our discussion is the formulation of MR image reconstruction as an *inverse problem*, where the acquired MRI data $\mathbf{s}$ is combined and complemented by prior knowledge of the unknown image. Different optimization methods arise naturally based on the properties of the functions used for imposing consistency with the measured data and the image priors. Additionally, modern optimization methods can be conceptually divided into purely model-based reconstruction methods and those that can leverage data-driven priors of unknown images. Throughout the chapter, we seek to explicitly state both the assumptions and results characterizing the convergence properties of discussed iterative optimization algorithms. The readers interested in further exploring the topic of MRI optimization algorithms are encouraged to read a recent review by Fessler [7] that provides a complementary view of the material presented here.

3.2 Least squares reconstruction

Consider the problem of reconstructing an unknown image $\mathbf{x} \in \mathbb{C}^n$ from a set of noisy measurements $\mathbf{s} \in \mathbb{C}^m$ in (3.3). When the matrix $\mathbf{E}$ is well-conditioned with $m \geq n$, the *maximum likelihood (ML)* estimator provides a reasonable approach for reconstructing $\mathbf{x}$. For the *Additive White Gaussian Noise (AWGN)* model, the ML estimator reduces to the *least squares (LS)* problem that has the closed-form solution

$$\widehat{\mathbf{x}} = argmin_{\mathbf{x}} \frac{1}{2} \|\mathbf{s} - \mathbf{E}\mathbf{x}\|_2^2 = (\mathbf{E}^{\mathbf{H}}\mathbf{E})^{-1}(\mathbf{E}^{\mathbf{H}}\mathbf{s}), \tag{3.4}$$

where $(\cdot)^{\mathbf{H}}$ denotes the Hermitian transpose.

Even though we can represent the solution for the LS problem in closed form, computing the solution can pose challenges in terms of computational and storage complexity if the problem dimension is large. Note that we can apply $\mathbf{E}$ as a computationally efficient operator using element-wise multiplication of vectors and FFTs, but computing $\mathbf{E}^{\mathbf{H}}\mathbf{E} = \sum_{c=1}^{T} \mathbf{C}_c^{\mathbf{H}}\mathbf{F}^{\mathbf{H}}\mathbf{B}^T\mathbf{B}\mathbf{F}\mathbf{C}_c$ and its inverse is not always efficient.

Suppose we are given fully sampled k-space data from all the receiver coils (i.e., $\mathbf{B}$ is an $n \times n$ identity matrix and $m = Tn$). In such a case, we can simplify $\mathbf{E}^{\mathbf{H}}\mathbf{E} = \sum_{c=1}^{T} \mathbf{C}_c^{\mathbf{H}}\mathbf{C}_c$ and write the solution in (3.4) as

$$\widehat{\mathbf{x}} = \left(\sum_{c=1}^{T} \mathbf{C}_c^{\mathbf{H}}\mathbf{C}_c \right)^{-1} \sum_{c=1}^{T} \mathbf{C}_c^{\mathbf{H}}\mathbf{F}^{\mathbf{H}}\mathbf{s}_c, \tag{3.5}$$

where $\sum_{c=1}^{T} \mathbf{C}_c^{\mathbf{H}}\mathbf{C}_c$ is a diagonal matrix and its inverse can be computed as element-wise inverse of the diagonal entries.

In the case of accelerated MRI, $\mathbf{B}$ is a subsampling matrix with $p < n$, and we cannot simplify the expression for $\mathbf{E}^{\mathbf{H}}\mathbf{E}$ or its inverse. Instead of explicitly computing $\mathbf{E}$, $\mathbf{E}^{\mathbf{H}}\mathbf{E}$ or its inverse, we compute the solution of the LS problem in (3.4) using some iterative method such as the gradient descent or conjugate gradient method [8].

The *gradient descent (GD)* method (which is also called steepest descent) is one of the simplest iterative methods for solving the least squares problem in (3.4). The basic principle in GD method is to iteratively compute the gradient of the LS objective at the current estimate and update the estimate in the negative direction of the gradient with an appropriate step size. Mathematically, we can write the

GD update at iteration t as

$$\mathbf{x}^t = \mathbf{x}^{t-1} - \gamma_t \mathbf{E}^{\mathbf{H}}(\mathbf{E}\mathbf{x}^{t-1} - \mathbf{s}), \tag{3.6}$$

where $\gamma_t > 0$ denotes the step size, which can be adjusted at every iteration or kept fixed. The rate of convergence for GD depends on the choice of step size and the eigenvalues of $\mathbf{E}^{\mathbf{H}}\mathbf{E}$ [9,10].

We can show that the iterative update in (3.6) converges to the LS estimate given in (3.4). Assuming that the gradient descent is initialized with a zero vector $\mathbf{x}^0 = 0$ and $\gamma_t = \gamma$ for all t, we can write

$$\mathbf{x}^t = (I - \gamma \mathbf{E}^{\mathbf{H}}\mathbf{E})\mathbf{x}^{t-1} + \gamma \mathbf{E}^{\mathbf{H}}\mathbf{s} = \gamma \sum_{k=0}^{t-1} (I - \gamma \mathbf{E}^{\mathbf{H}}\mathbf{E})^k \mathbf{E}^{\mathbf{H}}\mathbf{s}. \tag{3.7}$$

The matrix geometric series $\sum_{k=0}^{t-1}(I - \gamma \mathbf{E}^{\mathbf{H}}\mathbf{E})^k$ converges to $(\gamma \mathbf{E}^{\mathbf{H}}\mathbf{E})^{-1}$ if the spectral norm $\|I - \gamma \mathbf{E}^{\mathbf{H}}\mathbf{E}\| < 1$ [11]. This condition implies that, if

$$0 < \gamma \leq \frac{2}{\lambda_{\max}(\mathbf{E}^{\mathbf{H}}\mathbf{E})}, \tag{3.8}$$

then the gradient descent converges to the LS solution in (3.4), where $\lambda_{\max}(\mathbf{E}^{\mathbf{H}}\mathbf{E})$ represents the largest eigenvalue of $\mathbf{E}^{\mathbf{H}}\mathbf{E}$. Depending on the scaling of $\mathbf{E}$ and the choice of γ, gradient descent may not converge or converge very slowly.

The *conjugate gradient (CG)* method is a popular method for solving a system of linear equations when the system matrix is symmetric and positive definite [12,8]. At every iteration, the CG method updates the estimate in a direction that is *conjugate* to the update directions in all the previous iterations. This property ensures that the CG method will converge in at most n iterations while solving a system with n equations. Suppose we are given a linear system as

$$\mathbf{y} = \mathbf{A}(\mathbf{x}), \tag{3.9}$$

where $\mathbf{A}(\cdot)$ is an operator defined by an $n \times n$ symmetric positive definite matrix. We use the operator notation to highlight that we do not need an explicit matrix form for the operator. A pseudocode with all the steps of the CG method is given in Algorithm 1. Each iteration of CG requires a single application of the system operator $\mathbf{A}(\cdot)$. Note that the LS solution given in (3.4) can be written as a system of linear equations as

$$\underbrace{\mathbf{E}^{\mathbf{H}}\mathbf{s}}_{\mathbf{y}} = \underbrace{\mathbf{E}^{\mathbf{H}}\mathbf{E}\mathbf{x}}_{\mathbf{A}(\mathbf{x})}, \tag{3.10}$$

which can be solved using the CG method given in Algorithm 1. Note that we do not need to compute or apply $\mathbf{E}^{\mathbf{H}}\mathbf{E}$ as an explicit matrix; we can compute the effect of matrix multiplication using a function that involves element-wise multiplication and FFTs.

From the discussion so far, one can consider the CG as a minimizer for the quadratic function, $f(\mathbf{x}) = \|\mathbf{A}(\mathbf{x}) - \mathbf{y}\|^2$. However, the CG can be generalized to minimizing any continuous function, $f(\mathbf{x})$, for which we can calculate the gradient. For the nonlinear CG method, we follow the Algorithm 2. Unlike the linear version of CG, several expressions for calculating β have been developed for nonlinear

Algorithm 1 Conjugate gradient method for solving $\mathbf{y} = \mathbf{A}(\mathbf{x})$

1: **input:** initial estimate $\mathbf{x}^0$ and residual $\mathbf{r}_0 = \mathbf{A}(\mathbf{x}^0) - \mathbf{y}$.
2: $t = 0$
3: $\mathbf{p}_0 = \mathbf{r}_0$
4: **repeat**
5: $\quad \alpha_t = \dfrac{\|\mathbf{r}_t\|_2^2}{\mathbf{p}_t^T \mathbf{A}(\mathbf{p}_t)}$
6: $\quad \mathbf{x}^{t+1} = \mathbf{x}^t + \alpha_t \mathbf{p}_t$
7: $\quad \mathbf{r}_{t+1} = \mathbf{r}_t - \alpha_t \mathbf{A}(\mathbf{p}_t)$
8: $\quad$ **if** $\|\mathbf{r}_t\|_2^2 \leq$ `threshold` **then**
9: $\quad\quad$ `break`
10: $\quad$ **end if**
11: $\quad \beta_t = \dfrac{\|\mathbf{r}_{t+1}\|_2^2}{\|\mathbf{r}_t\|_2^2}$
12: $\quad \mathbf{p}_{t+1} = \mathbf{r}_{t+1} + \beta_t \mathbf{p}_t$
13: $\quad t \leftarrow t + 1$
14: **until** convergence/termination criteria satisfied

Algorithm 2 Conjugate gradient method for minimizing continuous $f(x)$

1: **input:** initial value $\mathbf{x}^0$
2: $t = 0$
3: $\mathbf{p}_0 = \mathbf{r}_0 = -\nabla f(x^0)$.
4: **repeat**
5: $\quad \alpha_t = \arg\min_{\alpha_t} f(\mathbf{x}^t + \alpha_t \mathbf{p}_t)$ ▷ via line search
6: $\quad \mathbf{x}^{t+1} = \mathbf{x}^t + \alpha_t \mathbf{p}_t$
7: $\quad \mathbf{r}_{t+1} = -\nabla f(x^{t+1})$
8: $\quad$ **if** $\|\mathbf{r}_t\|_2^2 \leq$ `threshold` **then**
9: $\quad\quad$ `break`
10: $\quad$ **end if**
11: $\quad \beta_{t+1} = \beta_{t+1}^{formula}$ ▷ Using FR, PR, HS, DY etc formula
12: $\quad \mathbf{p}_{t+1} = \mathbf{r}_{t+1} + \beta_{t+1} \mathbf{p}_t$
13: $\quad t \leftarrow t + 1$
14: **until** convergence/termination criteria satisfied

CG over the years. Four well-known formulas are given by Fletcher–Reeves (FR) [13], Polak–Ribière (PR) [14], Hestenes–Stiefel (HS) [15], and Dai–Yuan (DY) [16]:

$$\beta_{t+1}^{FR} = \frac{\mathbf{r}_{t+1}^T \mathbf{r}_{t+1}}{\mathbf{r}_t^T \mathbf{r}_t}, \beta_{t+1}^{PR} = \frac{\mathbf{r}_{t+1}^T (\mathbf{r}_{t+1} - \mathbf{r}_t)}{\mathbf{r}_t^T \mathbf{r}_t}, \beta_{t+1}^{HS} = \frac{\mathbf{r}_{t+1}^T (\mathbf{r}_{t+1} - \mathbf{r}_t)}{-\mathbf{p}_t^T (\mathbf{r}_{t+1} - \mathbf{r}_t)}, \beta_{t+1}^{DY} = \frac{\mathbf{r}_{t+1}^T \mathbf{r}_{t+1}}{-\mathbf{p}_t^T (\mathbf{r}_{t+1} - \mathbf{r}_t)}.$$

Since the general function f may have many local minima, there is no guarantee that the CG will converge to a global minima. Line search can be used to find α_t in CG as this operation requires zero finding. Two well-known iterative line-search methods for zero-finding are the Newton–Raphson method and the secant method. Both methods require the existence of Hessian matrix of f with respect to α. However, the Newton–Raphson method uses second derivative, whereas the secant method approximates the second derivative using the first derivative. Although the Newton–Raphson method converges much faster, it is computationally expensive.

3.3 Model-based reconstruction

The least squares problem is known to be suboptimal, when $m < n$, because of the ill-posed nature of the inverse problem. We can pose a general recovery problem as the following regularized optimization problem:

$$\widehat{\mathbf{x}} = argmin_{\mathbf{x}} f(\mathbf{x}) \quad \text{with} \quad f(\mathbf{x}) = g(\mathbf{x}) + h(\mathbf{x}), \tag{3.11}$$

where g is the *data-fidelity term* that quantifies the consistency with the observed data $\mathbf{s}$ and h is a *regularizer* that encodes the prior knowledge. The optimization problem in Eq. (3.11) can be recognized as the *Maximum A Posteriori probability (MAP)* estimator of $\mathbf{x}$ for the terms

$$g(\mathbf{x}) = -\log(p_{\mathbf{s}|\mathbf{x}}(\mathbf{s}|\mathbf{x})) \quad \text{and} \quad h(\mathbf{x}) = -\log(p_{\mathbf{x}}(\mathbf{x})),$$

where $p_{\mathbf{s}|\mathbf{x}}$ is the likelihood and $p_{\mathbf{x}}$ is the image prior. Naturally, the LS in Eq. (3.4) is a special case of MAP corresponding to AWGN with a uniform image prior.

In this section, we discuss optimization methods for solving problems of form (3.11). We start by discussing the traditional *smooth optimization* methods applicable to differentiable functions f. We then consider *nonsmooth optimization* methods applicable to problems where the regularizers are not differentiable.

3.3.1 Smooth optimization

Tikhonov regularization is one of the oldest and most well-known techniques for stabilizing ill-posed inverse problems [17,18]. In its classical form, it seeks to minimize the sum of the LS data-fidelity term and a quadratic regularizer

$$g(\mathbf{x}) = \frac{1}{2}\|\mathbf{s} - \mathbf{E}\mathbf{x}\|_2^2 \quad \text{and} \quad h(\mathbf{x}) = \frac{\lambda}{2}\|\mathbf{\Phi}\mathbf{x}\|_2^2,$$

where $\lambda > 0$ denotes a regularization parameter and $\mathbf{\Phi}$ denotes a sparsifying transform. The Tikhonov solution reduces to the following closed-form expression:

$$\widehat{\mathbf{x}} = (\mathbf{E}^{\mathbf{H}}\mathbf{E} + \lambda\mathbf{\Phi}^{\mathbf{H}}\mathbf{\Phi})^{-1}(\mathbf{E}^{\mathbf{H}}\mathbf{s}),$$

which can be calculated using the *conjugate gradient (CG)* method in Algorithm 1 [19–21].

In a more general setting, the regularizer h corresponds to a nonquadratic differentiable function. One widely used algorithm for minimizing such functions is the *Accelerated Gradient Method (AGM)* [22,11], whose iterates can be written as

$$\mathbf{x}^t = \mathbf{z}^{t-1} - \gamma_t \nabla f(\mathbf{z}^{t-1}) \tag{3.12a}$$

$$\mathbf{z}^t = \mathbf{x}^t + ((q_{t-1} - 1)/q_t)(\mathbf{x}^t - \mathbf{x}^{t-1}), \tag{3.12b}$$

with $t = 1, 2, 3, \ldots,$ where $\mathbf{x}^0 = \mathbf{z}^0 \in \mathbb{C}^n$ is the initialization and $\gamma_t > 0$ is the step parameter. Note that, when the *overrelaxation parameters* $q_t = 1$ for all $t \geq 1$, AGM reduces to the traditional gradient descent. However, by adapting $\{q_t\}$ at every iteration, AGM seeks to accelerate the convergence relative to GD. A widely used choice for $\{q_t\}$ is to set it as

$$q_t = \frac{1}{2}\left(1 + \sqrt{1 + 4q_{t-1}^2}\right) \quad \text{starting from} \quad q_0 = 1. \tag{3.13}$$

The step-size parameter of AGM is generally either fixed at a sufficiently small value $\gamma > 0$ or adapted via *backtracking line search* [9]. Backtracking line search seeks to ensure a sufficient decrease in the objective at every iteration. Given the gradient vector $\nabla f(\mathbf{z}^{t-1})$, initial step $\gamma_t > 0$, and parameters $\theta \in (0, 1/2]$ and $\beta \in (0, 1)$, backtracking is performed as follows:

$$\begin{aligned} &\text{while } f(\mathbf{z}^{t-1} - \gamma_t \nabla f(\mathbf{z}^{t-1})) > f(\mathbf{z}^{t-1}) - \theta \gamma_t \|\nabla f(\mathbf{z}^{t-1})\|_2^2 \\ &\text{do } \gamma_t = \beta \gamma_t. \end{aligned}$$

For smooth functions f, this while-loop is guaranteed to terminate for a sufficiently small value of γ_t.

The convergence rate of AGM can be precisely characterized for convex and smooth functions. Consider the following conditions on the optimization problem.

Assumption 1. *The function $f = g + h$ is continuously differentiable and convex. Additionally, the gradient ∇f is Lipschitz continuous with a constant $L > 0$.*

This is a widely-used assumption in the optimization literature that simply asserts the convexity and smoothness of the objective. When f is twice continuously differentiable, Assumption 1 is equivalent to the condition $0 \preceq \nabla^2 f(\mathbf{x}) \preceq L\mathbf{I}$, where $\nabla^2 f(\mathbf{x}) \in \mathbb{R}^{n \times n}$ is the Hessian matrix of f.

Assumption 2. *The function f has a finite minimum $f^* = f(\mathbf{x}^*)$ attained at some $\mathbf{x}^* \in \mathbb{C}^n$ such that $\|\mathbf{x}^0 - \mathbf{x}^*\|_2 \leq R$.*

The second assumption simply asserts the existence of a minimizer whose distance to the initialization $\mathbf{x}^0$ is bounded. We can now state the following well-known convergence result [11].

Theorem 1. *Run AGM for $T \geq 1$ iterations under Assumptions 1–2 using a fixed step size $0 < \gamma \leq 1/L$. Then, the iterates of the method satisfy*

$$f(\mathbf{x}^{t-1}) - f^* \leq \frac{2R^2}{\gamma(t+1)^2}.$$

Algorithm 3 PGM/APGM

1: **input:** initial values $\mathbf{x}^0 = \mathbf{z}^0$ and step-size $\gamma > 0$.
2: **for** $t = 1, 2, \ldots$ **do**
3: $\quad \mathbf{u}^t \leftarrow \mathbf{z}^{t-1} - \gamma \nabla g(\mathbf{z}^{t-1})$ ▷ more data-consistency
4: $\quad \mathbf{x}^t \leftarrow \mathrm{prox}_{\gamma h}(\mathbf{u}^t)$ ▷ more regularization
5: $\quad \mathbf{z}^t \leftarrow \mathbf{x}^t + ((q_{t-1} - 1)/q_t)(\mathbf{x}^t - \mathbf{x}^{t-1})$
6: **end for**

As has been shown by Nesterov [11], AGM achieves the optimal $O(1/t^2)$ convergence rate for first-order methods in smooth convex optimization. The *optimized gradient method (OGM)* [23] is a recent alternative to AGM that has a similar per-iteration complexity but leads to a better constant factor LR^2 compared to $2LR^2$ of Theorem 1, achievable using the step parameter $\gamma = 1/L$.

3.3.2 Nonsmooth optimization

Many widely used regularizers $h(\cdot)$ in MRI are nonsmooth. Some well-known examples include the *total variation (TV)* [24,6], *total generalized variation (TGV)* [25,26], and *Hessian Schatten-norm* [27, 28]. For example, the anisotropic TV-regularized MRI reconstruction can be formulated using the terms

$$g(\mathbf{x}) = \frac{1}{2}\|\mathbf{s} - \mathbf{E}\mathbf{x}\|_2^2 \quad \text{and} \quad h(\mathbf{x}) = \lambda \|\mathbf{\Phi}\mathbf{x}\|_1,$$

where $\lambda > 0$ is the regularization parameter and $\mathbf{\Phi}$ represents finite differences computed along every dimension of the image $\mathbf{x}$.

Nondifferentiability of widely-used regularizers has led to the widespread adoption of *proximal algorithms* [29] for image reconstruction without differentiation of h. Two widely used algorithms in this context are the *proximal gradient method (PGM)* [30–33] and *alternating direction method of multipliers (ADMM)* [34–36]. The key concept enabling the elegant handling of nonsmooth regularizers is the *proximal operator* [37], defined as

$$\mathrm{prox}_{\lambda h}(\mathbf{z}) = \underset{\mathbf{x} \in \mathbb{C}^n}{\arg\min} \left\{ \frac{1}{2}\|\mathbf{x} - \mathbf{z}\|_2^2 + \lambda h(\mathbf{x}) \right\}, \quad \lambda > 0, \quad \mathbf{z} \in \mathbb{C}^n. \tag{3.14}$$

The *projection* onto a subset $\mathcal{X} \subseteq \mathbb{C}^n$ is the proximal operator for $h(\cdot) = \mathbb{1}_{\mathcal{X}}(\cdot)$, where $\mathbb{1}_{\mathcal{X}}(\cdot)$ is the indicator function that takes the value 0 for any $\mathbf{x} \in \mathcal{X}$ and $+\infty$ for $\mathbf{x} \notin \mathcal{X}$. Although the proximal operator can be defined for nonconvex functions, its existence and uniqueness for every $\mathbf{z} \in \mathbb{C}^n$ is guaranteed when h is proper, closed, and convex [38]. The proximal operator can additionally be interpreted as a MAP estimator for the AWGN denoising problem

$$\mathbf{z} = \mathbf{x} + \eta, \quad \text{where} \quad \mathbf{x} \sim p_{\mathbf{x}}, \quad \eta \sim \mathcal{CN}(\mathbf{0}, \lambda \boldsymbol{I}),$$

by setting $h(\mathbf{x}) = -\log(p_{\mathbf{x}}(\mathbf{x}))$.

Algorithm 3 summarizes both PGM and *accelerated PGM (APGM)* (also known as *Fast Iterative/Shrinkage Thresholding Algorithm (FISTA)*), where the latter is obtained by simply including

Algorithm 4 ADMM

1: **input:** initial value $\mathbf{x}^0$ and penalty parameter $\gamma > 0$.
2: **for** $t = 1, 2, 3, \ldots$ **do**
3: $\quad \mathbf{u}^t \leftarrow \text{prox}_{\gamma g}(\mathbf{x}^{t-1} + \mathbf{z}^{t-1})$ ▷ more data-consistency
4: $\quad \mathbf{x}^t \leftarrow \text{prox}_{\gamma h}(\mathbf{u}^t - \mathbf{z}^{t-1})$ ▷ more regularization
5: $\quad \mathbf{z}^t \leftarrow \mathbf{z}^{t-1} + \mathbf{x}^t - \mathbf{u}^t$
6: **end for**

the Nesterov acceleration into PGM. In order to implement APGM, one needs to update $\{q_t\}$ as in Eq. (3.13). On the other hand, when $q_t = 1$ for all iterations, the algorithms corresponds to the conventional PGM.

The ADMM-based image reconstruction is obtained by reformulating the inverse problem (3.11) as an equivalent constrained optimization problem. Algorithm 4 summarizes ADMM for the constrained formulation

$$\text{minimize } g(\mathbf{z}) + h(\mathbf{x}) \text{ subject to } \mathbf{z} = \mathbf{x}. \tag{3.15}$$

This constrained problem can then be addressed by forming the *Augmented Lagrangian (AL)* [39] with the dual variable s and optimizing in an alternating fashion over $\mathbf{x}$, z, and s. One can consider other similar *variable splitting* strategies for obtained different variants of ADMM [40].

Careful inspections of PGM/APGM and ADMM reveal a number of similarities and differences between those two classes of algorithms. Both algorithms alternate between improving data consistency by applying the operators $\mathrm{I} - \gamma \nabla g$ and $\text{prox}_{\gamma g}$, respectively, and imposing the prior knowledge on the image by applying the operator $\text{prox}_{\gamma h}$. The per-iteration complexities of PGM/APGM and ADMM can be dramatically different depending on the difficulty of evaluating corresponding operations. For example, for the LS term, we have

$$\nabla g(\mathbf{x}) = \mathbf{E}^{\mathbf{H}}(\mathbf{E}\mathbf{x} - \mathbf{s}) \tag{3.16}$$

and

$$\text{prox}_{\gamma g}(\mathbf{x}) = \underset{\mathbf{z} \in \mathbb{C}^n}{\arg\min} \left\{ \frac{1}{2}\|\mathbf{z} - \mathbf{x}\|_2^2 + \frac{\gamma}{2}\|\mathbf{E}\mathbf{z} - \mathbf{s}\|_2^2 \right\} \tag{3.17a}$$

$$= \left(\boldsymbol{I} + \gamma \mathbf{E}^{\mathbf{H}}\mathbf{E}\right)^{-1} \left(\mathbf{x} + \gamma \mathbf{E}^{\mathbf{H}}\mathbf{s}\right), \tag{3.17b}$$

which is typically solved with a few iterations of CG.

The convergence behavior of APGM on convex functions is identical to that for AGM. To state it explicitly, we assume the following properties:

Assumption 3. *The function g is continuously differentiable and convex. Additionally, the gradient ∇g is Lipschitz continuous with a constant $L > 0$.*

Assumption 4. *The function h is proper, closed, and convex.*

Assumption 5. *The function $f = g + h$ has a finite minimum $f^* = f(\mathbf{x}^*)$ attained at some $\mathbf{x}^* \in \mathbb{C}^n$ such that $\|\mathbf{x}^0 - \mathbf{x}^*\|_2 \leq R$.*

We can now state the following well-known result in optimization (see proof in [33]).

Theorem 2. *Run APGM for $T \geq 1$ iterations under Assumptions 3–5 using a fixed step size $0 < \gamma \leq 1/L$. Then, the iterates of the method satisfy*

$$f(\mathbf{x}^t) - f^* \leq \frac{2R^2}{\gamma(t+1)^2}.$$

APGM is thus a generalization of AGM that maintains the same optimal worst-case convergence rate. Note that, when the proximal operator corresponds to a projection, APGM reduces to the accelerated variant of the projected gradient method.

The convergence of ADMM has also been extensively investigated in the literature [34–36,40]. In order to state one result, we adopt the following assumptions:

Assumption 6. *The functions g and h are both proper, closed, and convex.*

Assumption 7. *Strong duality holds for the constrained problem in Eq.* (3.15).

Assumption 6 simply states that we are considering a convex optimization problem. A sufficient condition for Assumption 7 is that the interiors of the domains of g and h have common vectors (i.e., their intersection is not empty). We can now state the following classical result (see [40] for a proof).

Theorem 3. *Under Assumptions 6 and 7, the ADMM iterates satisfy the following conditions:*

- Residual convergence: *The iterates approach feasibility*

$$\lim_{t\to\infty} \left(\mathbf{x}^t - \mathbf{z}^t\right) = \mathbf{0}.$$

- Objective convergence: *The objective of the iterates approaches the optimum*

$$\lim_{t\to\infty} \left\{g(\mathbf{z}^t) + h(\mathbf{x}^t)\right\} = f^*.$$

In practice, proximal methods offer faster convergence compared to the traditional techniques based on subgradients or smoothing. Both APGM and ADMM are significantly faster than PGM, which is illustrated on the problem of TV-regularized image reconstruction in MRI in Fig. 3.2. Note the typical behavior of ADMM (discussed in [40]), where it converges to moderate accuracy within a few tens of iterations, but is slow to converge to high accuracy. The rapid convergence of ADMM to moderate accuracy makes it suitable for large-scale image reconstruction, where one is only interested in an approximately good solution.

The primal-dual algorithm can also be used for solving various optimization problems that arise in MRI. A typical primal objective function can be written as

$$\min_{\mathbf{x}} g(\mathbf{E}\mathbf{x}) + h(\mathbf{x}) \quad \text{(primal)}, \tag{3.18}$$

where $g(\mathbf{E}\mathbf{x})$ is the data fidelity and $h(\mathbf{x})$ is the regularization. The Chambolle–Pock (CP) algorithm, as proposed in [41], is a first-order primal-dual method that solves the following saddle-point problem:

$$\min_{\mathbf{x}} \max_{\mathbf{v}} \langle \mathbf{E}\mathbf{x}, \mathbf{v}\rangle + h(\mathbf{x}) - g^*(\mathbf{v}) \quad \text{(primal-dual)}, \tag{3.19}$$

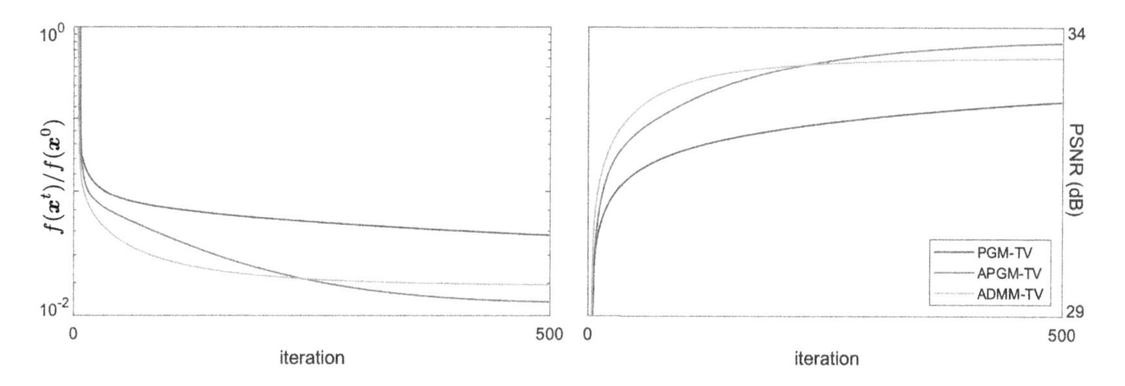

FIGURE 3.2 Convergence behavior of PGM, APGM, and ADMM.

The evolution of the relative objective (left) and PSNR (dB) (right) for MR image reconstruction with the TV regularizer. Both APGM and ADMM are known to lead to faster convergence compared to the traditional PGM.

Algorithm 5 Primal-dual method in [41,42]

1: **input:** initial value $\mathbf{x}^0$, $\bar{\mathbf{x}}^0$, $\mathbf{v}^0$
2: $t = 0$
3: Choose step size $\sigma > 0$ and $\tau > 0$, so that $\sigma\tau L^2 < 1$, where $L = \|\mathbf{E}\|$ is the induced norm, and $\theta \in [0, 1]$.
4: **repeat**
5: $\quad \mathbf{v}^{t+1} = \text{prox}_{\sigma g^*}(\mathbf{v}^t + \sigma \mathbf{E}\bar{\mathbf{x}}^t)$ ▷ dual proximal
6: $\quad \mathbf{x}^{t+1} = \text{prox}_{\tau h}(\mathbf{x}^t - \tau \mathbf{E}^* \mathbf{v}^{t+1})$ ▷ primal proximal
7: $\quad \bar{\mathbf{x}}^{t+1} = \mathbf{x}^{t+1} + \theta(\mathbf{x}^{t+1} - \mathbf{x}^t)$ ▷ extrapolation
8: $\quad t \leftarrow t + 1$
9: **until** convergence/termination criteria satisfied

---

where $g^*(\mathbf{x})$ is the convex conjugate of $g(\mathbf{x})$ [9]. The reasoning behind moving towards the primal-dual approach instead of solving the primal function lies directly in the nontrivial formulation of the proximal operator for $g(\mathbf{Ex})$. We can formulate proximal operator for $g^*(\mathbf{x})$ and $h(\mathbf{x})$ easily. We alternately do proximal gradient descent for $\mathbf{x}$ and $\mathbf{v}$ to solve the optimization in (3.19) according to Algorithm 5. Over the years, many variations of the primal-dual approach have been proposed to solve smooth and nonsmooth optimization [41–48].

$g(\mathbf{Ex})$ can be a nonsmooth function, such as hinge loss [49], generalized hinge loss [50], absolute loss [51], or ϵ-insensitive loss [52]. $h(\mathbf{x})$ can be a nonsmooth regularizer, such as lasso [53], group lasso [54], sparse group lasso [55], exclusive lasso [56], $l_{1,\infty}$ regularizer [57], or trace-norm regularizer [58].

3.3.3 Stochastic gradient-based approaches

Stochastic versions of gradient descent methods have become popular with the rise of neural networks. Stochastic gradient descent (SGD) methods have been extremely successful in solving various nonconvex optimization problems in recent years. Multiple variations of the GD and SGD methods have been proposed over the years to ensure *better* convergence. Such approaches include Nesterov accelerated gradient (NAG) [59], Adagrad [60], RMSprop [61], Adadelta [62], adaptive moment estimation (ADAM) [63], and Nesterov-accelerated adaptive moment estimation (NADAM) [64]. Nesterov accelerated gradient (NAG) incorporates the momentum term to the GD update steps. As the GD iterations approach a local minima, the updates usually become gradually smaller, otherwise we will jump from the vicinity of one local minima to another. To ensure that, the momentum term is introduced. The momentum term increases for parameters whose gradients point in the same directions and reduces updates for parameters whose gradients change directions. Adagrad adapts the learning rate of the parameters based on the frequencies of the features. Adagrad has diminishing learning rate for infrequent features. RMSprop and Adadelta have both been developed to resolve Adagrad's diminishing learning rate. Adam uses the first and second moments of the gradients for GD updates. NADAM incorporates Adam and NAG by modifying the momentum term.

3.4 Summary

This chapter reviewed several widely used optimization strategies in the context of MRI image reconstruction. The key property desired in such algorithms is the ability to leverage useful prior information about the unknown image $\mathbf{x}$. We have discussed algorithms for handling smooth and nonsmooth regularizers.

References

[1] Pruessmann KP, Weiger M, Scheidegger MB, Boesiger P. SENSE: sensitivity encoding for fast MRI. Magn Reson Med 1999;42(5):952–62.

[2] Griswold MA, Jakob PM, Heidemann RM, Nittka M, Jellus V, Wang J, et al. Generalized autocalibrating partially parallel acquisitions (GRAPPA). Magn Reson Med 2002;47:1202–10.

[3] Lustig M, Donoho DL, Pauly JM, Sparse MRI. The application of compressed sensing for rapid MR imaging. Magn Reson Med 2007;58(6):1182–95.

[4] Lustig M, Donoho DL, Santos JM, Pauly JM. Compressed sensing MRI. IEEE Signal Process Mag 2008;25(2):72–82.

[5] Ying J, Sheng J. Joint image reconstruction and sensitivity estimation in SENSE (JSENSE). Magn Reson Med 2007;6(57):1196–202.

[6] Block KT, Uecker M, Frahm J. Undersampled radial MRI with multiple coils. iterative image reconstruction using a total variation constraint. Magn Reson Med 2007;6(57):1086–98.

[7] Fessler JA. Optimization methods for magnetic resonance image reconstruction. IEEE Signal Process Mag 2020;1(37):33–40.

[8] Golub GH, Van Loan CF. Matrix computations, vol. 3. JHU Press; 2013.

[9] Boyd S, Vandenberghe L. Convex optimization. Cambridge Univ. Press; 2004.

[10] Bertsekas DP. Incremental proximal methods for large scale convex optimization. Math Program Ser B 2011;129:163–95.

[11] Nesterov Y. Introductory lectures on convex optimization: a basic course. Kluwer Academic Publishers; 2004.

[12] Shewchuk JR, et al. An introduction to the conjugate gradient method without the agonizing pain; 1994.

[13] Fletcher R, Reeves CM. Function minimization by conjugate gradients. Comput J 1964;7(2):149–54.

[14] Polak E, Ribiere G. Note sur la convergence de méthodes de directions conjuguées. Math Model Numer Anal 1969;3(R1):35–43.
[15] Hestenes MR, Stiefel E, et al. Methods of conjugate gradients for solving linear systems, vol. 49. Washington, DC: NBS; 1952.
[16] Dai YH, Yuan Y. A nonlinear conjugate gradient method with a strong global convergence property. SIAM J Optim 1999;10(1):177–82.
[17] Tikhonov AN, Arsening VY. Solution of ill-posed problems. Winston-Wiley; 1977.
[18] Ribés A, Schmitt F. Linear inverse problems in imaging. IEEE Signal Process Mag 2008;25(4):84–99.
[19] Pruessmann KP, Weiger M, Boernert P, Boesiger P. Advances in sensitivity encoding with arbitrary k-space trajectories. Magn Reson Med 2001;4(46):638–51.
[20] Wajer F, Pruessmann KP. Major speedup of reconstruction for sensitivity encoding with arbitrary trajectories. In: Proc. Int. Soc. magnetic resonance medicine; 2001. p. 767.
[21] Sutton BP, Noll DC, Fessler JA. Fast, iterative image reconstruction for mri in the presence of field inhomogeneities. IEEE Trans Med Imaging 2003;2(22):178–88.
[22] Nesterov YE. A method for solving the convex programming problem with convergence rate $O(1/k^2)$. Dokl Akad Nauk SSSR 1983;269:543–7 (in Russian).
[23] Kim D, Fessler JA. Optimized first-order methods for smooth convex minimization. Math Program 2016;159:81–107.
[24] Rudin LI, Osher S, Fatemi E. Nonlinear total variation based noise removal algorithms. Physica D 1992;60(1–4):259–68.
[25] Bredies K, Kunisch K, Pock T. Total generalized variation. SIAM J Imaging Sci 2010;3(3):492–526.
[26] Knoll F, Brendies K, Pock T, Stollberger R. Second order total generalized variation (TGV) for MRI. Magn Reson Med 2011;65(2):480–91.
[27] Lefkimmiatis S, Bourquard A, Unser M. Hessian-based norm regularization for image restoration with biomedical applications. IEEE Trans Image Process 2012;21(3):983–95.
[28] Lefkimmiatis S, Ward JP, Unser M. Hessian Schatten-norm regularization for linear inverse problems. IEEE Trans Image Process 2013;22(5):1873–88.
[29] Parikh N, Boyd S. Proximal algorithms. Found Trends Optim 2014;1(3):123–231.
[30] Figueiredo MAT, Nowak RD. An EM algorithm for wavelet-based image restoration. IEEE Trans Image Process 2003;12(8):906–16.
[31] Daubechies I, Defrise M, Mol CD. An iterative thresholding algorithm for linear inverse problems with a sparsity constraint. Commun Pure Appl Math 2004;57(11):1413–57.
[32] Bect J, Blanc-Feraud L, Aubert G, Chambolle A. A ℓ_1-unified variational framework for image restoration. In: Proc. ECCV, vol. 3024. New York: Springer; 2004. p. 1–13.
[33] Beck A, Teboulle M. A fast iterative shrinkage-thresholding algorithm for linear inverse problems. SIAM J Imaging Sci 2009;2(1):183–202.
[34] Eckstein J, Bertsekas DP. On the Douglas-Rachford splitting method and the proximal point algorithm for maximal monotone operators. Math Program 1992;55:293–318.
[35] Afonso MV, MBioucas-Dias J, Figueiredo MAT. Fast image recovery using variable splitting and constrained optimization. IEEE Trans Image Process 2010;19(9):2345–56.
[36] Ng MK, Weiss P, Yuan X. Solving constrained total-variation image restoration and reconstruction problems via alternating direction methods. SIAM J Sci Comput 2010;32(5):2710–36.
[37] Moreau JJ. Proximité et dualité dans un espace hilbertien. Bull Soc Math Fr 1965;93:273–99.
[38] Beck A. First-order methods in optimization. MOS-SIAM series on optimization. SIAM; 2017.
[39] Nocedal J, Wright SJ. Numerical optimization. 2nd ed. Springer; 2006.
[40] Boyd S, Parikh N, Chu E, Peleato B, Eckstein J. Distributed optimization and statistical learning via the alternating direction method of multipliers. Found Trends Mach Learn 2011;3(1):1–122.
[41] Chambolle A, Pock T. A first-order primal-dual algorithm for convex problems with applications to imaging. J Math Imaging Vis 2011;40(1):120–45.

[42] Pock T, Cremers D, Bischof H, Chambolle A. Global solutions of variational models with convex regularization. SIAM J Imaging Sci 2010;3(4):1122–45.
[43] Yang T, Mahdavi M, Jin R, Zhu S. An efficient primal dual prox method for non-smooth optimization. Mach Learn 2015;98(3):369–406.
[44] Tang YC, Zhu CX, Wen M, Peng JG. A splitting primal-dual proximity algorithm for solving composite optimization problems. Acta Math Sin Engl Ser 2017;33(6):868–86.
[45] Dupé FX, Fadili MJ, Starch JL. Inverse problems with Poisson noise: primal and primal-dual splitting. In: 2011 18th IEEE international conference on image processing. IEEE; 2011. p. 1901–4.
[46] Yurtsever A, Tran Dinh Q, Cevher V. A universal primal-dual convex optimization framework. Adv Neural Inf Process Syst 2015;28:3150–8.
[47] Valkonen T. First-order primal–dual methods for nonsmooth non-convex optimisation. In: Handbook of mathematical models and algorithms in computer vision and imaging: mathematical imaging and vision; 2021. p. 1–42.
[48] Alacaoglu A, Tran-Dinh Q, Fercoq O, Cevher V. Smooth primal-dual coordinate descent algorithms for nonsmooth convex optimization. In: NeurIPS; 2017.
[49] Vapnik V. Statistical learning theory. Wiley series on adaptive and learning systems for signal processing, communications and control; 1998.
[50] Bartlett PL, Wegkamp MH. Classification with a reject option using a hinge loss. J Mach Learn Res 2008;9(8).
[51] Friedman JH. The elements of statistical learning: data mining, inference, and prediction. Springer Open; 2017.
[52] Rosasco L, De Vito E, Caponnetto A, Piana M, Verri A. Are loss functions all the same? Neural Comput 2004;16(5):1063–76.
[53] Zhu J, Rosset S, Tibshirani R, Hastie TJ. 1-norm support vector machines. In: Advances in neural information processing systems. Citeseer; 2003.
[54] Yuan M, Lin Y. Model selection and estimation in regression with grouped variables. J R Stat Soc, Ser B, Stat Methodol 2006;68(1):49–67.
[55] Yang H, Xu Z, King I, Lyu MR. Online learning for group lasso. In: ICML; 2010.
[56] Zhou Y, Jin R, Hoi SCH. Exclusive lasso for multi-task feature selection. In: Proceedings of the thirteenth international conference on artificial intelligence and statistics, JMLR workshop and conference proceedings; 2010. p. 988–95.
[57] Quattoni A, Carreras X, Collins M, Darrell T. An efficient projection for $\ell 1, \infty$, infinity regularization. In: International conference on machine learning; 2009.
[58] Rennie JD, Srebro N. Fast maximum margin matrix factorization for collaborative prediction. In: Proceedings of the 22nd international conference on machine learning; 2005. p. 713–9.
[59] Nesterov Y. A method for unconstrained convex minimization problem with the rate of convergence o (1/k^ 2). Dokl Akad Nauk USSR 1983;269:543–7.
[60] Duchi J, Hazan E, Singer Y. Adaptive subgradient methods for online learning and stochastic optimization. J Mach Learn Res 2011;12(7).
[61] Hinton G. Neural networks for machine learning, lecture 6a; 2019.
[62] Zeiler MD. Adadelta: an adaptive learning rate method. Available from: arXiv:1212.5701, 2012.
[63] Kingma DP, Ba J. Adam: a method for stochastic optimization. In: ICLR; 2015.
[64] Dozat T. Incorporating Nesterov momentum into Adam; 2016.

CHAPTER

Non-Cartesian MRI Reconstruction

4

Holger Eggers[a], Melanie Kircheis[b], and Daniel Potts[b]
[a]*Philips Research, Hamburg, Germany*
[b]*Technische Universität Chemnitz, Faculty of Mathematics, Chemnitz, Germany*

4.1 Introduction

As elaborated in Chapter 1, signal localization in MRI is predominantly achieved by superposing a comparatively weak magnetic field gradient onto the strong and homogeneous main magnetic field. The magnetic field gradient ideally produces a linear increase in the magnetic field strength in one direction. Since the frequency of the precession of the nuclear spins and of the signal this precession induces in the receive coils is basically proportional to the local magnetic field strength experienced by the nuclear spins, the magnetic field gradient linearly encodes the origin of the signal in the frequency. In this way, the signal is effectively measured in the spatial frequency domain, the so-called k-space.

By allowing rapid changes of the magnetic field gradient in strength and direction over time, MRI provides a remarkable flexibility in sampling k-space. The sampling position $\mathbf{k}$ at time t is simply given by the integral of the magnetic field gradient $\mathbf{G}$ over time

$$\mathbf{k}(t) = \frac{\gamma}{2\pi} \int_0^t \mathbf{G}(\tau)\,\mathrm{d}\tau, \tag{4.1}$$

assuming the signal to be generated by radiofrequency excitation at $t = 0$. $\mathbf{k}$ and $\mathbf{G}$ are vectors in typically two-dimensional (2D) or three-dimensional (3D) spaces and are measured in units of $[\mathrm{m}^{-1}]$ and $[\mathrm{T/m}]$, respectively. The gyromagnetic ratio γ links the local magnetic field strength to the angular frequency of the precession and depends on the nucleus selected for imaging. The temporal variation of $\mathbf{G}$, commonly referred to as the gradient waveform, thus determines the k-space trajectory, along which samples are collected after a single excitation. It is only restricted by hardware constraints and more fundamentally by safety constraints, which mainly limit the velocity and acceleration in k-space. Using multiple excitations, or shots, permits segmenting the k-space trajectory by restarting in the center of k-space.

MRI was first demonstrated with projection reconstruction in 1973 [1]. This choice was presumably inspired by computed tomography (CT), with which the first clinical images had recently been obtained at that time. The data acquisition proceeded along lines through the origin and covered a polar grid in k-space, as illustrated in the bottom left plot in Fig. 4.1. According to the Fourier slice theorem, a one-dimensional (1D) inverse Fourier transform turns the samples along one of these lines into samples of a projection perpendicular to this line. This allowed employing backprojection for image reconstruction, as in CT.

Advances in Magnetic Resonance Technology and Applications, Volume 7, ISSN 2666-9099. https://doi.org/10.1016/B978-0-12-822726-8.00013-0

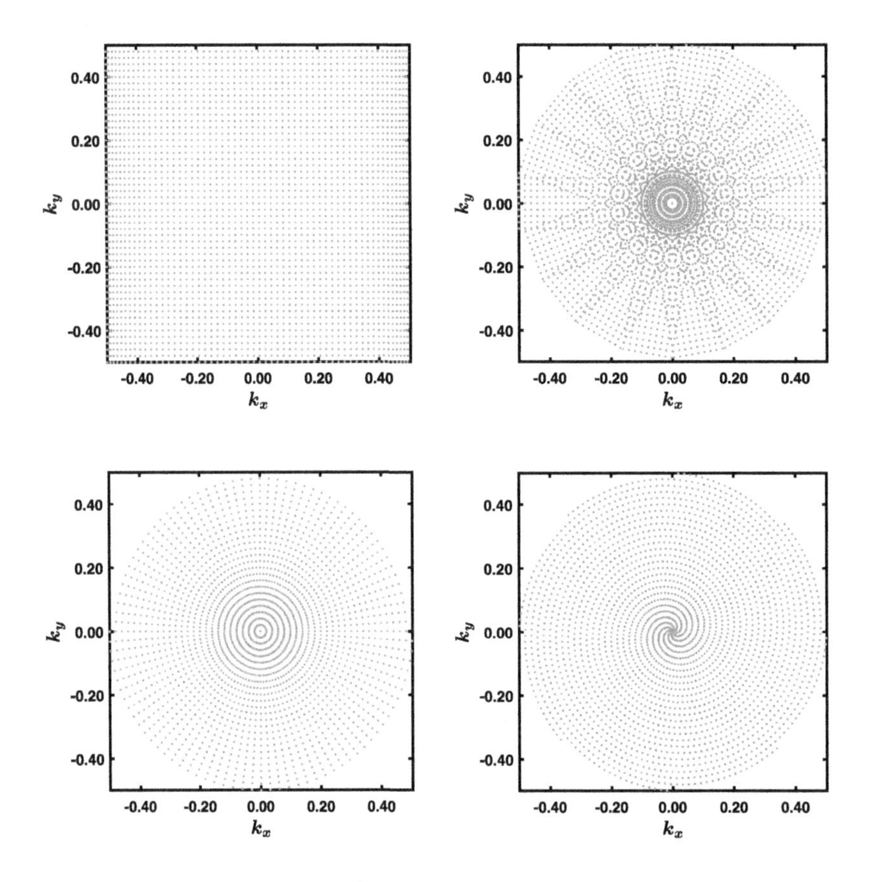

FIGURE 4.1 2D Non-Cartesian Imaging — Sampling Patterns

Nonequidistant sampling of k-space in at least one dimension is involved in 2D echo-planar (top, left), PROPELLER (top, right), radial (bottom, left), and spiral (bottom, right) imaging.

Two years later, sampling k-space on a Cartesian grid along parallel lines was proposed as a simpler alternative for MRI [2]. The application of different magnetic field gradients before and during data acquisition was introduced, establishing the distinction between phase and frequency encoding, and the use of a multidimensional inverse fast Fourier transform (FFT) for image reconstruction was suggested. As later remarked by one of the authors of this seminal work, this was driven by the conviction that projections cannot efficiently cover a 2D plane or 3D volume due to the resulting strong variation in sampling density between the center and the periphery of k-space [3]. The combination of sampling k-space on a Cartesian grid and reconstructing images based on an inverse FFT, known as Cartesian imaging, soon developed into the prevalent method for MRI. In particular in clinical practice, this still holds today.

Nevertheless, a considerable variety of methods that deliberately acquires data not on a Cartesian grid has been devised and explored over the last decades. The sampling patterns obtained in k-space

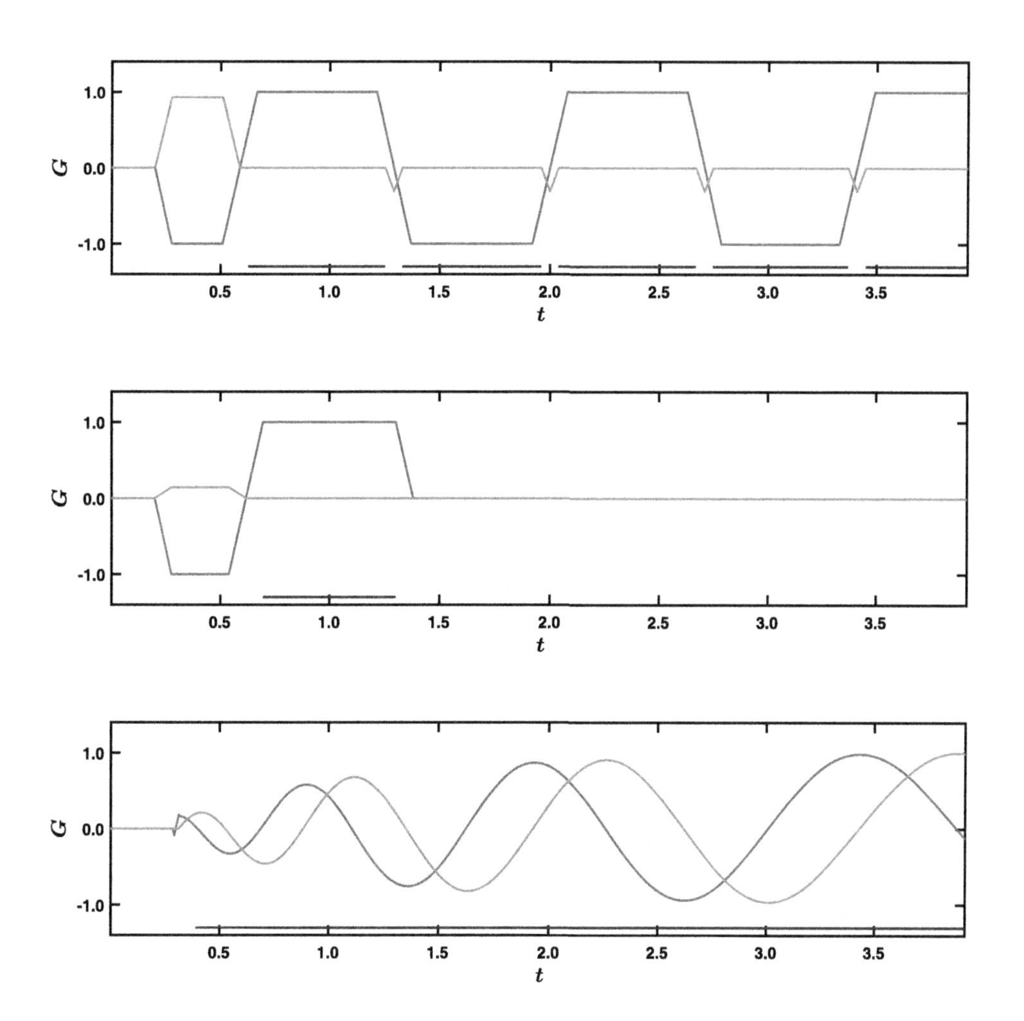

FIGURE 4.2 2D Non-Cartesian Imaging — Gradient Waveforms

2D echo-planar (top), PROPELLER or radial (middle), and spiral (bottom) imaging rely on different gradient waveforms $G(t)$ in the frequency- and phase-encoding directions (red, green (mid gray, gray in print version)) and on different acquisition windows (blue (dark gray in print version)) to generate the sampling patterns in Fig. 4.1. For the sake of simplicity, they are all assumed to be based on gradient echoes.

with four of the most popular of these methods in two dimensions, including projection reconstruction, or radial imaging, are plotted in Fig. 4.1.

Echo-planar imaging (EPI) is often not even considered as a non-Cartesian imaging method [4]. It employs a gradient waveform consisting of a train of trapezoids with alternating polarity in the frequency-encoding direction. The beginning of such a train is shown in the top plot in Fig. 4.2, where G and t are both normalized. In its purest form, the whole k-space is traversed after a single excitation along parallel lines in opposite directions. To increase speed and efficiency further, the first samples along each line are already collected while the magnetic field gradient is still ramped up, and the last samples are still collected while the magnetic field gradient is already ramped down again. This so-called ramp sampling leads to an increasing sampling density towards the edges of k-space in the frequency-encoding direction, while a constant sampling density is preserved everywhere else. The data acquisition is only interrupted for switching between the parallel lines by application of a magnetic field gradient in the phase-encoding direction and is not started before a corner of k-space has been reached by means of the respective first trapezoid in each direction. EPI is in widespread use today, especially for diffusion-weighted imaging (DWI) and functional imaging (fMRI).

Periodically rotated overlapping parallel lines with enhanced reconstruction (PROPELLER) imaging is also a hybrid Cartesian and non-Cartesian imaging method [5]. The data acquisition is split up into blades, with each of which a rectangular k-space area is covered. Each blade supports the reconstruction of an intermediate image with anisotropic spatial resolution based on an inverse FFT. These intermediate images enable detection and correction of motion between the acquisition of the individual blades, which are rotated about the k-space origin with respect to each other. However, the reconstruction of the final image has to cope with the highly nonequidistant sampling of k-space by all blades together. The gradient waveforms are identical to those in Cartesian imaging, except that the frequency- and phase-encoding directions are rotated with the blades. PROPELLER imaging is of particular relevance in applications prone to motion, including head, neck, spine, and shoulder imaging.

Radial imaging is primarily employed in areas in which the strong variation in sampling density between the center and the periphery of k-space actually promises advantages. Above all, the oversampling in the center leads to insensitivity to motion by averaging out inconsistencies in the data to a large extent. Moreover, residual artifacts are spread in two dimensions and appear as streaks, which are usually more benign than the coherent or incoherent ghosts arising along one dimension in Cartesian imaging without variation in sampling density. Additionally, the repeated coverage of the center enables image reconstruction at various spatial and temporal resolutions in dynamic imaging, as well as retrospective self-gating in abdominal and cardiac imaging during free breathing without interspersed navigators or dedicated sensors. Besides, radial imaging allows dispensing with phase encoding by starting the data acquisition in the center, permitting very short echo times to capture rapidly decaying signal, as required for lung and bone imaging. Compared to PROPELLER imaging, the gradient waveform in the phase-encoding direction is eliminated, and the gradient waveform in the frequency-encoding direction is rotated with the projections. The first trapezoid in the frequency-encoding direction is kept to obtain full projections or is removed to obtain half projections, as demonstrated in the later examples.

Spiral imaging first and foremost provides a fast and flexible coverage of k-space [6]. It supports single-shot and multishot imaging, uniform sampling density in all but the very center of k-space and variable sampling density across k-space, as well as traversing k-space from the center outward and from the periphery inward. Unlike in EPI, the magnetic field gradient continuously changes during the data acquisition, and dead times for switching its polarity are eliminated. Furthermore, motion and aliasing artifacts are rather incoherent, and especially flow artifacts are less pronounced. Spiral imaging is mainly of interest in a wide range of applications demanding high speed and efficiency, but is very susceptible to an imperfect main magnetic field or magnetic field gradient.

All of these 2D methods can be extended to three dimensions, either by the so-called stacking, the addition of an equidistant sampling along the third dimension, or by a suitable generalization.

Due to the diversity of sampling patterns in MRI, general algorithms are favored for the reconstruction in non-Cartesian imaging. In view of the limited signal-to-noise ratio (SNR) attainable with MRI, their accuracy only has to be moderate, in particular in comparison with CT. An essential part of such algorithms, a nonequispaced fast Fourier transform (NFFT), is described in the next section. Based on it, ubiquitous gridding is introduced in the following section.

4.2 NFFT

The signal s measured in k-space is, as in Chapter 2, modeled by

$$s(k) = \int_{-N/2}^{N/2} x(r)\, \mathrm{e}^{-\mathrm{i}2\pi kr}\, \mathrm{d}r, \tag{4.2}$$

where x denotes the transverse magnetization of an object of maximum extent $[-\frac{N}{2}, \frac{N}{2})$, from now on simply referred to as the object and the field of view (FOV).

An FFT of length N efficiently converts N equidistant samples of x into N equidistant samples of $\tilde{s}$ defined by

$$\tilde{s}_m = \sum_{n=1}^{N} x_n\, \mathrm{e}^{-\mathrm{i}2\pi k_m r_n}. \tag{4.3}$$

Here, r and k are unitless and normalized to the range $[-\frac{N}{2}, \frac{N}{2})$ and $[-\frac{1}{2}, \frac{1}{2})$, respectively. Assuming N to be an even positive integer, the sampling positions in space are then located at $r_n = -\frac{N}{2}, -\frac{N}{2} + 1, ..., \frac{N}{2} - 1$. $\tilde{s}$ is linked to s by

$$\tilde{s}(k) = \sum_{p=-\infty}^{\infty} s(k+p) \tag{4.4}$$

due to the limited coverage of k-space by the acquisition and the spatial discretization of the object in the reconstruction. Thus, the $\tilde{s}_m$ in Eq. (4.3) are not samples of s, but of s turned into a periodic signal $\tilde{s}$ with a period of 1. To reduce the computational complexity from $\mathcal{O}(N^2)$ of a direct evaluation of Eq. (4.3) to $\mathcal{O}(N\log N)$, an FFT exploits symmetries in the exponential factors, which are lost if the r_n or k_m are not equidistant anymore.

An NFFT generalizes an FFT to nonequidistant samples of x, $\tilde{s}$, or both [7,8]. It relies on an FFT for the actual transformation between the spatial and the spatial frequency domain and, therefore, has to convert nonequidistant to equidistant samples intermediately. This involves an approximation, the accuracy of which can be traded off for computational complexity. In the following, the case of equidistant samples of x and nonequidistant samples of $\tilde{s}$ is considered. It is of relevance for an efficient evaluation of Eq. (4.3), the forward model, which requires in non-Cartesian imaging the calculation of M samples of $\tilde{s}$ from N samples of x. The other two cases are discussed at the end of this chapter.

To convert equidistant into nonequidistant samples in k-space, a convolution with a window function c, turned into a periodic function $\tilde{c}$ with a period of 1 by

$$\tilde{c}(k) = \sum_{p=-\infty}^{\infty} c(k+p), \tag{4.5}$$

is performed. $\tilde{c}$ is assumed to have a uniformly convergent Fourier series, which allows writing it as

$$\tilde{c}(k) = \sum_{r=-\infty}^{\infty} \tilde{c}_r \, \mathrm{e}^{\mathrm{i}2\pi kr} \tag{4.6}$$

with the Fourier coefficients

$$\tilde{c}_r = \int_{-1/2}^{1/2} \tilde{c}(k) \, \mathrm{e}^{-\mathrm{i}2\pi kr} \mathrm{d}k. \tag{4.7}$$

Substituting k by $k - k'$ results in

$$\tilde{c}_r = \int_{-1/2}^{1/2} \tilde{c}(k - k') \, \mathrm{e}^{-\mathrm{i}2\pi\,(k-k')\,r} \mathrm{d}k', \tag{4.8}$$

which is approximated by

$$\tilde{c}_r \approx \frac{1}{fN} \sum_{n'=-fN/2}^{fN/2-1} \tilde{c}\left(k - \tfrac{n'}{fN}\right) \mathrm{e}^{-\mathrm{i}2\pi\,\left(k - \frac{n'}{fN}\right) r}. \tag{4.9}$$

The oversampling factor $f \geq 1$ is introduced to increase accuracy, where fN is also assumed to be an even positive integer [9,10]. If the periodic function $\tilde{c}$ is bandlimited with a maximum frequency of $\frac{fN}{2}$, Eq. (4.9) holds exactly. Otherwise, aliasing occurs, and the estimation of the $\tilde{c}_r$ is confounded by higher frequencies. More specifically, the frequency bands $[fNp - \frac{N}{2}, fNp + \frac{N}{2})$, for any integer $p \neq 0$, fold back onto the baseband $[-\frac{N}{2}, \frac{N}{2})$. Rearranging Eq. (4.9) leads to

$$\mathrm{e}^{\mathrm{i}2\pi kr} \approx \frac{1}{fN\tilde{c}_r} \sum_{n'=-fN/2}^{fN/2-1} \tilde{c}\left(k - \tfrac{n'}{fN}\right) \mathrm{e}^{\mathrm{i}2\pi \frac{n'}{fN} r}, \tag{4.10}$$

provided that $\tilde{c}_r \neq 0$, and substituting r by $-r$ to

$$\mathrm{e}^{-\mathrm{i}2\pi kr} \approx \frac{1}{fN\tilde{c}_{-r}} \sum_{n'=-fN/2}^{fN/2-1} \tilde{c}\left(k - \tfrac{n'}{fN}\right) \mathrm{e}^{-\mathrm{i}2\pi \frac{n'}{fN} r}, \tag{4.11}$$

which is inserted into Eq. (4.3) to obtain

$$\tilde{s}_m \approx \sum_{n'=-fN/2}^{fN/2-1} \tilde{c}\left(k_m - \tfrac{n'}{fN}\right) \sum_{n=1}^{N} \frac{x_n}{fN\tilde{c}_{-r_n}} \mathrm{e}^{-\mathrm{i}2\pi \frac{n'}{fN} r_n}. \tag{4.12}$$

The inner sum amounts to an FFT of length fN, transforming N equidistant samples of x in space, after weighting and zero padding to fN equidistant samples in space, into fN equidistant samples in k-space. The outer sum converts the latter by a convolution into M nonequidistant samples in k-space.

Choosing

$$\tilde{c}_r = \begin{cases} \frac{1}{fN}, & -\frac{fN}{2} < r \leq \frac{fN}{2} \\ 0, & \text{otherwise} \end{cases} \tag{4.13}$$

allows dispensing with the weighting, usually referred to as deapodization, and results with Eq. (4.6) in

$$\tilde{c}(k) = \frac{1}{fN} \sum_{r=-fN/2+1}^{fN/2} \mathrm{e}^{\mathrm{i}2\pi kr}. \tag{4.14}$$

Here, the shift in the range of r arises from substituting r by $-r$ in Eq. (4.11). $\tilde{c}(k)$ is a complex trigonometric polynomial of degree $\frac{fN}{2}$ in this case, for which Eq. (4.9) holds exactly. It is written as

$$\tilde{c}(k) = \begin{cases} \frac{\sin(\pi k f N)}{fN \sin(\pi k)} \mathrm{e}^{\mathrm{i}\pi k}, & k \neq 0 \\ 1, & \text{otherwise} \end{cases} \tag{4.15}$$

and approximated by

$$\tilde{c}(k) \approx \operatorname{sinc}(\pi k f N) \tag{4.16}$$

for small πk using

$$\operatorname{sinc}(x) = \begin{cases} \frac{\sin(x)}{x}, & x \neq 0 \\ 1, & \text{otherwise} \end{cases}. \tag{4.17}$$

According to the sampling theorem, s can be recovered from equidistant samples of s by

$$s(k) = \sum_{n=-\infty}^{\infty} s\left(\frac{n}{fN}\right) \operatorname{sinc}\left(\pi\left(k - \frac{n}{fN}\right) fN\right). \tag{4.18}$$

Consequently, the sinc function of infinite extent is the ideal window function in the hypothetical case of unlimited coverage of k-space by the acquisition. In the practicable case of limited coverage of k-space, however, it is, according to Eq. (4.15), better replaced by the sinc function turned into a periodic function, multiplied by the phasor $\mathrm{e}^{\mathrm{i}\pi k}$ arising from the asymmetry in the r_n for even fN, which is known as the Dirichlet kernel.

To reduce the computational complexity of the convolution, the window function is preferably real and limited in its extent, either by design or by truncation, to $[-\frac{K}{2fN}, \frac{K}{2fN}]$ with a positive integer K, the kernel size. A truncation leads to

$$c'(k) = \begin{cases} c(k), & |k| \leq \frac{K}{2fN} \\ 0, & \text{otherwise} \end{cases} \tag{4.19}$$

and a corresponding periodic function $\tilde{c}'$ with a period of 1, and it changes Eq. (4.12) to

$$\tilde{s}_m \approx \sum_{n'=-fN/2}^{fN/2-1} \tilde{c}'\left(k_m - \frac{n'}{fN}\right) \sum_{n=1}^{N} \frac{x_n}{fN\tilde{c}_{-r_n}} \mathrm{e}^{-\mathrm{i}2\pi \frac{n'}{fN} r_n}. \tag{4.20}$$

Prolate spheroidal wave functions maximize the relative energy of a function of given extent in one domain in an interval of given extent in the other domain [11]. The Kaiser–Bessel window is, in the continuous case, a simple approximation to them [12]. Thus, it is particularly well-suited to minimize aliasing errors in Eq. (4.12) and truncation errors in Eq. (4.20) [10,13]. If aliasing errors are to be minimized, the deapodization is defined by the Kaiser–Bessel window

$$\tilde{c}_r = \begin{cases} \frac{1}{fN} I_0\left(b\sqrt{1-\left(\frac{\pi K}{fNb}r\right)^2}\right), & |r| \leq \frac{fNb}{\pi K} \\ 0, & \text{otherwise} \end{cases}, \tag{4.21}$$

where I_0 is the zeroth order modified Bessel function of the first kind. By setting the shape parameter b to

$$b = \pi K\left(1 - \frac{1}{2f}\right), \tag{4.22}$$

the Fourier coefficients drop to zero for $|r| > fN - \frac{N}{2}$ [14]. The window function, also referred to as the convolution kernel, is then given by

$$c(k) = \begin{cases} \frac{2b}{\pi K}\operatorname{sinhc}\left(b\sqrt{1-\left(\frac{2fN}{K}k\right)^2}\right), & |k| \leq \frac{K}{2fN} \\ \frac{2b}{\pi K}\operatorname{sinc}\left(b\sqrt{\left(\frac{2fN}{K}k\right)^2 - 1}\right), & \text{otherwise} \end{cases} \tag{4.23}$$

using

$$\operatorname{sinhc}(x) = \begin{cases} \frac{\sinh(x)}{x}, & x \neq 0 \\ 1, & \text{otherwise} \end{cases}. \tag{4.24}$$

If truncation errors are to be minimized instead, the convolution kernel is defined by the Kaiser–Bessel window

$$c(k) = \begin{cases} \frac{1}{K} I_0\left(b\sqrt{1-\left(\frac{2fN}{K}k\right)^2}\right), & |k| \leq \frac{K}{2fN} \\ 0, & \text{otherwise} \end{cases}, \tag{4.25}$$

and the deapodization is given by

$$\tilde{c}_r = \begin{cases} \frac{1}{fN}\operatorname{sinhc}\left(b\sqrt{1-\left(\frac{\pi K}{fNb}r\right)^2}\right), & |r| \leq \frac{fNb}{\pi K} \\ \frac{1}{fN}\operatorname{sinc}\left(b\sqrt{\left(\frac{\pi K}{fNb}r\right)^2 - 1}\right), & \text{otherwise} \end{cases}. \tag{4.26}$$

Both choices lead to similar convolution kernels and deapodizations, as illustrated in Fig. 4.3 and Fig. 4.4 with $f = 2.0$, $K = 5$, $N = 800$, and a normalization of the values of the functions to the range

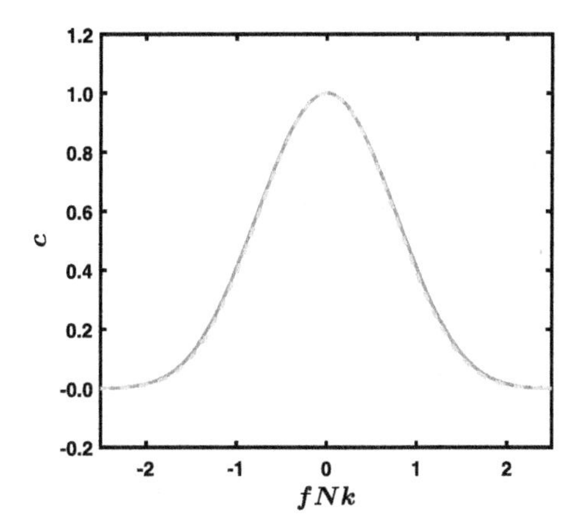

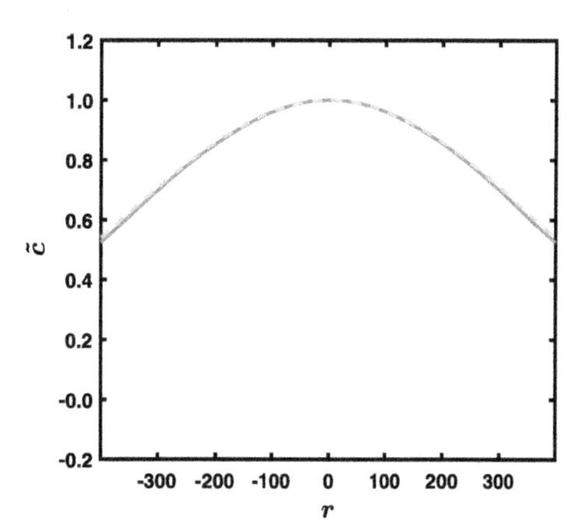

FIGURE 4.3 Convolution Kernels and Deapodizations

The Fourier transform of a Kaiser–Bessel window is employed as a convolution kernel in the conversion of equidistant into nonequidistant samples in k-space (solid red (mid gray in print version), left). To compensate for the inherent low-pass filtering, a deapodization is performed in space by dividing by the Kaiser–Bessel window (solid red (mid gray in print version), right). Alternatively, a Kaiser–Bessel window serves as convolution kernel (dashed green (gray in print version), left) and dividing by the inverse Fourier transform of the Kaiser–Bessel window as deapodization (dashed green (gray in print version), right).

[0, 1]. It is worth noting that the second case in Eq. (4.23) and Eq. (4.26) is only provided for completeness. Due to the truncation of the convolution kernel and the limited extent of the object, only the first case is of relevance in practice. Moreover, by subtracting an offset of $\frac{1}{fN}I_0(b)$ and $\frac{1}{K}I_0(b)$ from the first case of Eq. (4.21) and Eq. (4.25), the Kaiser–Bessel window smoothly drops to zero already at $|r| = \frac{fNb}{\pi K}$ and $|k| = \frac{K}{2fN}$, which enables eliminating aliasing errors and preventing the convolution kernel from extending over $K+1$ equidistant samples in k-space in the worst case, respectively.

In general, the accuracy and the computational complexity of an NFFT increase with the oversampling factor and the kernel size. Error bounds have been established for various window functions to guide the selection of minimal values for f and K given a target accuracy [8]. Since the running time of an FFT does not monotonically increase with the length of the FFT, the choice of f is preferably made in view of the running time of an FFT of length fN [15]. The complexity of the window function is less of a concern because it is typically calculated only once and stored in a look-up table [16].

Eq. (4.20) can be rewritten as

$$\tilde{\mathbf{s}} \approx \mathbf{C}'\mathbf{F}\mathbf{D}\mathbf{x} \tag{4.27}$$

with a diagonal N x N deapodization matrix $\mathbf{D}$

$$[\mathbf{D}]_{n,n} = \frac{1}{fN\tilde{c}_{-r_n}}, \tag{4.28}$$

an fN x N Fourier transform matrix $\mathbf{F}$

$$[\mathbf{F}]_{n',n} = \mathrm{e}^{-\mathrm{i}2\pi\left(\frac{n'-1}{fN}-\frac{1}{2}\right)r_n} \tag{4.29}$$

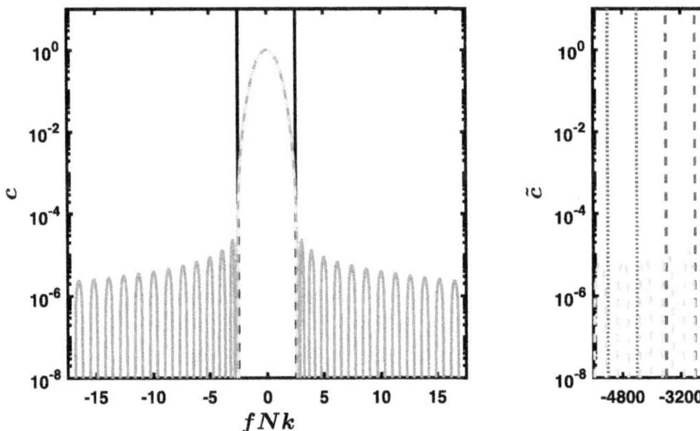

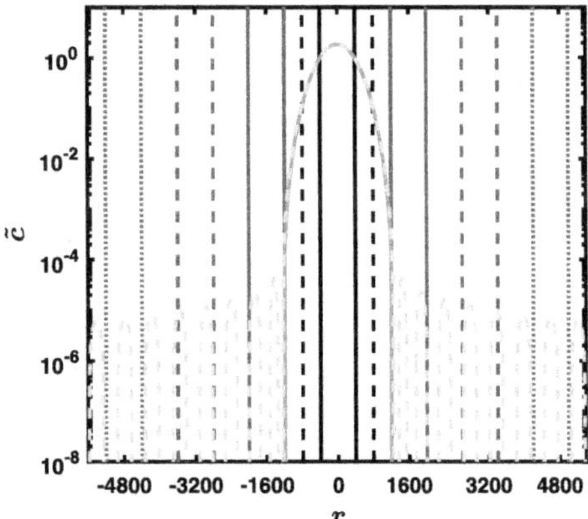

FIGURE 4.4 Aliasing and Truncation Errors

The convolution kernels from Fig. 4.3 and their inverse Fourier transforms are displayed with a logarithmic scale over a larger range of k and r, respectively. While a Kaiser–Bessel window is a function with compact support, its Fourier transform and inverse Fourier transform are not. Consequently, truncation errors occur in k-space using a Kaiser–Bessel window for deapodization in space (solid red (mid gray in print version)), whereas aliasing errors occur in space using it as convolutional kernel in k-space (dashed green (gray in print version)). The prescribed kernel size and FOV (solid black, left and right), the FOV extended by oversampling (dashed black, right), and the first three frequency bands folding back onto the prescribed FOV (solid, dashed, and dotted blue (dark gray in print version), right) are indicated.

and a sparse M x fN convolution matrix $\mathbf{C}'$

$$[\mathbf{C}']_{m,n'} = \tilde{c}'\left(k_m - \tfrac{n'-1}{fN} + \tfrac{1}{2}\right). \tag{4.30}$$

Here, the entries in the columns and rows of the matrices are numbered starting at 1, which complicates the expressions in the parentheses in Eq. (4.29) and Eq. (4.30) compared to Eq. (4.20). Eq. (4.27) is an approximation of

$$\tilde{\mathbf{s}} = \mathbf{E}\mathbf{x} \tag{4.31}$$

with an M x N encoding matrix $\mathbf{E}$

$$[\mathbf{E}]_{m,n} = \mathrm{e}^{-\mathrm{i}2\pi k_m r_n}. \tag{4.32}$$

Throughout this chapter, M and N are assumed to be of the same order of magnitude, ignoring Fourier interpolation for the moment. Undersampling is the subject of subsequent chapters.

Extending an NFFT to multiple dimensions is straightforward by using a tensor product approach. While separable convolution kernels, defined by the product of 1D convolution kernels for each dimension, are convenient and prevailing in practice, nonseparable convolution kernels, such as circularly and spherically symmetric ones, promise to enhance accuracy given a target computational complexity [17]. However, a theoretical comparison of them still seems to be missing.

4.3 Gridding

The object x is ideally recovered from the signal s by

$$x(r) = \int_{-\infty}^{\infty} s(k)\, \mathrm{e}^{\mathrm{i}2\pi kr} \mathrm{d}k. \tag{4.33}$$

Due to the limited coverage of k-space by the acquisition, s is only known for k in $[-\frac{1}{2}, \frac{1}{2})$ and, without prior knowledge, disregarded elsewhere, which leads to the approximation

$$\hat{x}(r) = \int_{-1/2}^{1/2} s(k)\, \mathrm{e}^{\mathrm{i}2\pi kr} \mathrm{d}k. \tag{4.34}$$

In addition, s is sampled with finite resolution, limited by hardware constraints in the frequency-encoding direction and more fundamentally by time constraints in the phase-encoding direction, resulting in the further approximation

$$\hat{x}_n = \frac{1}{M} \sum_{m=1}^{M} s_m\, \mathrm{e}^{\mathrm{i}2\pi k_m r_n} \tag{4.35}$$

or

$$\hat{\mathbf{x}} = \frac{1}{M} \mathbf{E}^{\mathrm{H}} \mathbf{s}. \tag{4.36}$$

An inverse FFT of length M efficiently converts M equidistant samples of s into M equidistant samples of $\hat{x}$. If the object is to be reconstructed with higher spatial resolution, that is to be enlarged by Fourier interpolation, the number of equidistant samples of $\hat{x}$, and with it the length of the inverse FFT, is increased here to $N > M$, whereas the spacing of the equidistant samples of s is decreased to $\frac{1}{N}$. In analogy with Eq. (4.14) and Eq. (4.15), the point spread function (PSF), which links x to $\hat{x}$, is then given by

$$\begin{aligned} \mathrm{PSF}(r) &= \frac{1}{M} \sum_{m=-M/2}^{M/2-1} \mathrm{e}^{\mathrm{i}2\pi \frac{m}{N} r} \\ &= \frac{\sin\left(\pi \frac{r}{N} M\right)}{M \sin\left(\pi \frac{r}{N}\right)} \mathrm{e}^{-\mathrm{i}\pi \frac{r}{N}}, \end{aligned} \tag{4.37}$$

where the scaling is chosen such that $\mathrm{PSF}(0) = 1$. Thus, the equidistant samples of $\hat{x}$ are approximately related by a sinc interpolation. Only in the case of $N = M$, which is exceptional, because Fourier interpolation is almost always applied in clinical routine, they are uncorrelated.

An inverse FFT is both the adjoint and the inverse of the corresponding FFT. For an NFFT, this does not hold anymore. Gridding was developed in radioastronomy from simpler interpolation methods and adopted in medical imaging before NFFTs were proposed and thoroughly analyzed in mathematics [18,9]. It aims at reconstructing N equidistant samples of x from M nonequidistant samples of s. To provide an approximate explicit inverse NFFT, gridding relies on an adjoint NFFT and introduces an additional weighting of the nonequidistant samples of s, the sampling density compensation. It is described by

$$\hat{\mathbf{x}} \approx \mathbf{D}^{\mathrm{H}}\, \mathbf{F}^{\mathrm{H}}\, \mathbf{C}'^{\mathrm{H}}\, \mathbf{W}\, \mathbf{s} \tag{4.38}$$

with an M x M diagonal matrix $\mathbf{W}$ of a vector of weights $\mathbf{w}$. Accordingly, it involves, after the sampling density compensation, a convolution with a window function, an inverse FFT, and a deapodization. Compared to the NFFT derived in the previous section, these three steps are performed in reverse order. Moreover, the convolution is evaluated on an oversampled Cartesian grid, the FFT is replaced by an inverse FFT, and the extended FOV is cropped by the deapodization.

Various approaches to deriving a suitable sampling density compensation have been proposed. One class of approaches considers each s_m as representative for a certain neighborhood in k-space and turns Eq. (4.35) into the Riemann sum

$$\hat{x}_n = \sum_{m=1}^{M} w_m s_m \, \mathrm{e}^{\mathrm{i}2\pi k_m r_n}. \tag{4.39}$$

Here, the sum of the positive weights that correspond to lengths in one dimension is normalized to a value of 1. The calculation of the weights is based on either a continuous or a discrete model.

For simple non-Cartesian sampling patterns, the weights can analytically be obtained from the Jacobian determinant of a differentiable coordinate transformation. In the case of 2D radial imaging, the coordinate transformation from a polar grid to a Cartesian grid is defined by

$$\begin{aligned} k_x &= k_r \cos\left(2\pi k_\phi\right) \\ k_y &= k_r \sin\left(2\pi k_\phi\right) \end{aligned} \tag{4.40}$$

with the radial direction k_r and the azimuthal direction k_ϕ normalized to the range $[0, \frac{1}{2}]$ and $[-\frac{1}{2}, \frac{1}{2})$, respectively. The determinant of the Jacobian matrix

$$J_R = \begin{bmatrix} \frac{\partial k_x}{\partial k_r} & \frac{\partial k_x}{\partial k_\phi} \\ \frac{\partial k_y}{\partial k_r} & \frac{\partial k_y}{\partial k_\phi} \end{bmatrix} \tag{4.41}$$

is then equal to $2\pi k_r$. Extending the integration in Eq. (4.34) to a circular area R of radius $\frac{1}{2}$ in two dimensions results in

$$\hat{x}(r_x, r_y) = \iint_R s(k_x, k_y)\, \mathrm{e}^{\mathrm{i}2\pi (k_x r_x + k_y r_y)}\, \mathrm{d}k_x\, \mathrm{d}k_y \tag{4.42}$$

and a change in variables in

$$\hat{x}(r_x, r_y) = 2\pi \int_0^{1/2} \int_{-1/2}^{1/2} s(k_r, k_\phi)\, k_r\, \mathrm{e}^{\mathrm{i}2\pi k_r (\cos(2\pi k_\phi) r_x + \sin(2\pi k_\phi) r_y)}\, \mathrm{d}k_\phi\, \mathrm{d}k_r. \tag{4.43}$$

Accordingly, $s(k_r, k_\phi)$ is weighted by $2\pi k_r$, which corresponds to the ideal ramp filter used in backprojection.

In the case of 2D spiral imaging, a coordinate transformation to a polar grid exists for some common k-space trajectories, which can readily be extended to a coordinate transformation to a Cartesian grid [19]. For a segmented, or interleaved, variable angular speed k-space trajectory, as illustrated in the bottom right plot in Fig. 4.1, it is defined by

$$\begin{aligned} k_r &= \frac{k_s}{2\sqrt{\alpha + (1-\alpha)\, k_s}} \\ k_\phi &= \frac{N}{M_I} \cdot \frac{k_s}{2\sqrt{\alpha + (1-\alpha)\, k_s}} + k_i\,, \end{aligned} \tag{4.44}$$

where k_s parameterizes the sample, or time, along an interleaf, normalized to the range $[0, 1]$, and k_i parameterizes the interleaf, or its rotation about the k-space origin, normalized to the range $[-\frac{1}{2}, \frac{1}{2})$. Here, both k_s and k_i are considered as continuous. The range of k_ϕ is extended to $[-\frac{1}{2}, \frac{1}{2} + \frac{N}{2M_I})$, and M is subdivided into M_I interleaves with M_S samples each. Since

$$k_r = \frac{M_I}{N}\left(k_\phi - k_i\right), \tag{4.45}$$

this is an Archimedean spiral with a parameter α in the range $[0, 1]$. For $\alpha = 0$ and $\alpha = 1$, a constant linear velocity spiral and a constant angular velocity spiral is obtained, respectively. The determinant of the Jacobian matrix

$$J_S = \begin{bmatrix} \frac{\partial k_r}{\partial k_s} & \frac{\partial k_r}{\partial k_i} \\ \frac{\partial k_\phi}{\partial k_s} & \frac{\partial k_\phi}{\partial k_i} \end{bmatrix} \tag{4.46}$$

is then given by

$$|J_S| = \frac{2\alpha + (1-\alpha)\, k_s}{4\left(\alpha + (1-\alpha)\, k_s\right)^{\frac{3}{2}}} \tag{4.47}$$

and the determinant of the Jacobian matrix of the composite coordinate transformation to a Cartesian grid by

$$|J_S J_R| = \frac{\pi}{4} \cdot \frac{2\alpha k_s + (1-\alpha)\, k_s^2}{\left(\alpha + (1-\alpha)\, k_s\right)^2}. \tag{4.48}$$

For more complex non-Cartesian sampling patterns, the weights can geometrically be obtained from the discrete k-space sampling positions by constructing a Voronoi diagram and calculating the area of each Voronoi cell [20].

Another class of approaches solves a minimization problem. For example, replacing $\mathbf{s}$ in Eq. (4.38) by $\tilde{\mathbf{s}}$ from Eq. (4.27) leads to

$$\hat{\mathbf{x}} \approx \mathbf{D}^{\mathrm{H}}\,\mathbf{F}^{\mathrm{H}}\,\mathbf{C}'^{\mathrm{H}}\,\mathbf{W}\,\mathbf{C}'\,\mathbf{F}\,\mathbf{D}\,\mathbf{x}. \tag{4.49}$$

In analogy with Eq. (4.13) and Eq. (4.14), turning the sparse convolution matrix $\mathbf{C}'$ into a dense convolution matrix $\mathbf{C}$ allows eliminating the deapodization by setting $\mathbf{D} = \mathbf{I}$. Choosing additionally $f = 1.0$ then simplifies Eq. (4.49) to

$$\mathbf{F}\,\hat{\mathbf{x}} \approx N\,\mathbf{C}^{\mathrm{H}}\,\mathbf{W}\,\mathbf{C}\,\mathbf{F}\,\mathbf{x} \tag{4.50}$$

since $\mathbf{F}\,\mathbf{F}^{\mathrm{H}} = N\,\mathbf{I}$. Ideally, applying the forward model to an image and then gridding to the simulated signal recovers the image. Independent of $\mathbf{x}$, this would be ensured by the condition

$$N\,\mathbf{C}^{\mathrm{H}}\,\mathbf{W}\,\mathbf{C} = \mathbf{I}. \tag{4.51}$$

However, this matrix equation is normally overdetermined. It can be rewritten as a linear system of N^2 equations and multiplied on the left with the system matrix, or the Frobenius norm of the matrix $N\,\mathbf{C}^{\mathrm{H}}\,\mathbf{W}\,\mathbf{C} - \mathbf{I}$ can be minimized instead [21,22]. Both approaches lead to the same linear system of M equations

$$\mathbf{A}\,\mathbf{w} = \mathbf{b} \tag{4.52}$$

with

$$\begin{aligned} [\mathbf{A}]_{m,m'} &= \left| \left[\mathbf{C}\,\mathbf{C}^{\mathrm{H}} \right]_{m,m'} \right|^2 \\ [\mathbf{b}]_m &= \frac{1}{N} \left[\mathbf{C}\,\mathbf{C}^{\mathrm{H}} \right]_{m,m}. \end{aligned} \tag{4.53}$$

Due to the often poor condition and the high computational complexity, solving Eq. (4.52) is only practicable for non-Cartesian sampling patterns with small number of samples. It is worth noting that the weights obtained by the least squares approximation of Eq. (4.51) are not necessarily positive anymore.

Similarly,

$$N\,\mathbf{C}\,\mathbf{C}^{\mathrm{H}}\,\mathbf{W} = \mathbf{I} \tag{4.54}$$

is derived by demanding consistency of the signal in k-space rather than of the image in space. If this condition is relaxed by considering the diagonal elements of the resulting matrices on the left and on the right only, it reduces to

$$N\,\mathbf{C}\,\mathbf{C}^{\mathrm{H}}\,\mathbf{w} = \mathbf{1}. \tag{4.55}$$

The fixed-point iteration

$$\left[\mathbf{w}^{[i+1]} \right]_m = \frac{\left[\mathbf{w}^{[i]} \right]_m}{N \left[\mathbf{C}\,\mathbf{C}^{\mathrm{H}} \mathbf{w}^{[i]} \right]_m} \tag{4.56}$$

was suggested for the solution of Eq. (4.55), which remains practicable for non-Cartesian sampling patterns with higher number of samples using a truncated convolution kernel [23].

A generalization of gridding is obtained by replacing the sampling density compensation with a spatially variant convolution kernel. It promises better accuracy, but requires an optimization of the convolution kernel once per sampling pattern [22].

4.4 Iterative reconstruction

Mainly as a reference for gridding in the later examples, an iterative reconstruction for non-Cartesian imaging based on an NFFT and an adjoint NFFT is derived in this section.

Replacing $\tilde{\mathbf{s}}$ by $\mathbf{s}$ in Eq. (4.31) leads to the forward model

$$\mathbf{s} = \mathbf{E}\,\mathbf{x}. \tag{4.57}$$

The inverse problem is solved by the least squares approximation

$$\min_{\hat{\mathbf{x}}} \left\| \mathbf{s} - \mathbf{E}\,\hat{\mathbf{x}} \right\|_2^2 \tag{4.58}$$

using the normal equations of the first kind

$$\mathbf{E}^{\mathrm{H}}\,\mathbf{E}\,\hat{\mathbf{x}} = \mathbf{E}^{\mathrm{H}}\,\mathbf{s} \tag{4.59}$$

or by the weighted least squares approximation

$$\min_{\hat{\mathbf{x}}} \sum_{m=1}^{M} [\mathbf{w}]_m \left| \left[\mathbf{s} - \mathbf{E}\,\hat{\mathbf{x}} \right]_m \right|^2 \tag{4.60}$$

using the modified normal equations of the first kind

$$\mathbf{E}^{\mathrm{H}}\,\mathbf{W}\,\mathbf{E}\,\hat{\mathbf{x}} = \mathbf{E}^{\mathrm{H}}\,\mathbf{W}\,\mathbf{s}. \tag{4.61}$$

Here, Eq. (4.57) is assumed to be overdetermined. Undersampling is considered in subsequent chapters, as well as multiple receive coils. Moreover, the vector $\mathbf{w}$ and the diagonal matrix $\mathbf{W}$ are specifically those introduced in Eq. (4.38) for sampling density compensation.

To determine $\hat{\mathbf{x}}$ in Eq. (4.59) or Eq. (4.61), iterative methods, such as the conjugate gradient (CG) method described in Chapter 3, are preferably employed. They typically involve a repeated calculation of products of the system matrix and a vector $\hat{\mathbf{p}}$. These products can efficiently be computed either with an NFFT and an adjoint NFFT or with a fast convolution [24]. For the latter, the entries of the system matrix are written as

$$\left[\mathbf{E}^{\mathrm{H}}\,\mathbf{E} \right]_{n,n'} = \sum_{m=1}^{M} \mathrm{e}^{\mathrm{i}2\pi k_m (r_n - r_{n'})} \tag{4.62}$$

or

$$\left[\mathbf{E}^{\mathrm{H}}\,\mathbf{W}\,\mathbf{E} \right]_{n,n'} = \sum_{m=1}^{M} w_m\, \mathrm{e}^{\mathrm{i}2\pi k_m (r_n - r_{n'})} \tag{4.63}$$

and referred to as $q(r_n - r_{n'})$. It is worth noting that $q(r)$ closely corresponds to the PSF of a reconstruction with an adjoint NFFT or gridding. The products then reduce to the convolution

$$\sum_{n'=1}^{N} \hat{p}_{n'}\, q(r_n - r_{n'}) \tag{4.64}$$

in space, which can be evaluated by a multiplication in k-space. In this way, an NFFT and an adjoint NFFT are essentially replaced by an FFT and an inverse FFT. The Fourier transform of q has to be

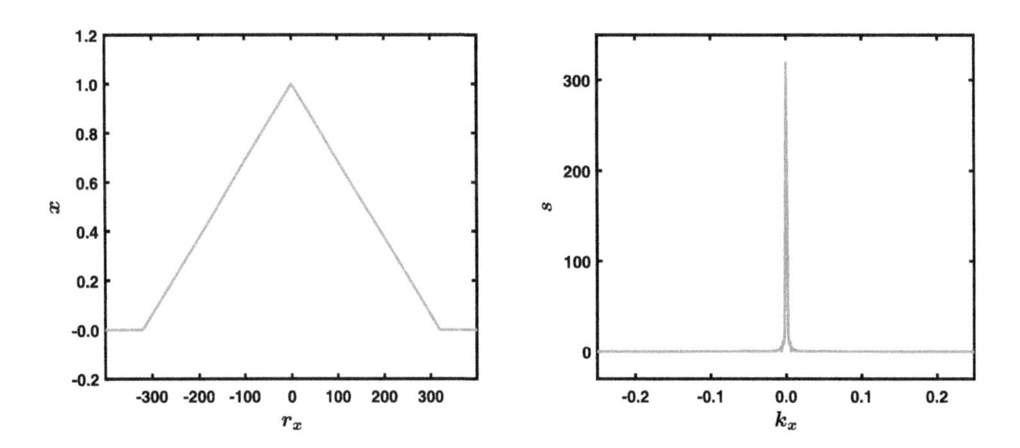

FIGURE 4.5 1D Object

A continuous triangular function, covering 80% of the FOV, is chosen as 1D object (left). Its frequency spectrum is real and nonnegative (right).

computed only once. To prevent backfolding, the fast convolution requires a two-fold oversampling. Nevertheless, it usually provides a substantial acceleration for $M \geq N$ [25].

Iterative methods can synthesize signal outside the k-space area covered by the acquisition [26]. This concerns in particular the corners of a square k-space if only the circle inscribed in the square is actually sampled. Since recovering such signal is ill-conditioned, it typically entails an excessive amplification of high-frequency noise, as discussed in Chapter 2. A low-pass filtering is usually applied to $\hat{\mathbf{x}}$ to avoid this.

4.5 Examples

Gridding is illustrated with a 1D and a 2D example in this section. The triangular function shown in Fig. 4.5 serves as object in the 1D example. The Fourier transform of this object is given by

$$s(k_x) = R\,\text{sinc}^2(\pi R k_x), \tag{4.65}$$

where $R = 320$.

Initially, an equidistant sampling of s with k_x in the range $[-\frac{1}{4}, \frac{1}{4})$ and $M = 400$ is assumed. It allows applying an inverse FFT for reference and assessing aliasing or truncation errors introduced by gridding. The object is reconstructed at $r_x = -\frac{N}{2}, -\frac{N}{2}+1, \ldots, \frac{N}{2}-1$ with $N = 800$. Thus, it is enlarged by Fourier interpolation by a factor of 2, involving zero padding in k-space to the range $[-\frac{1}{2}, \frac{1}{2})$. Resulting errors are plotted in Fig. 4.6 and dominated by ringing artifacts at $r_x = -320, 0, 320$. The triangular function is continuous but not differentiable at these points. Therefore, the classical Gibbs phenomenon is not observed, but a poorer approximation by a truncated Fourier series is expected due to the slower decay of the Fourier coefficients with frequency. Compared to the inverse FFT, gridding

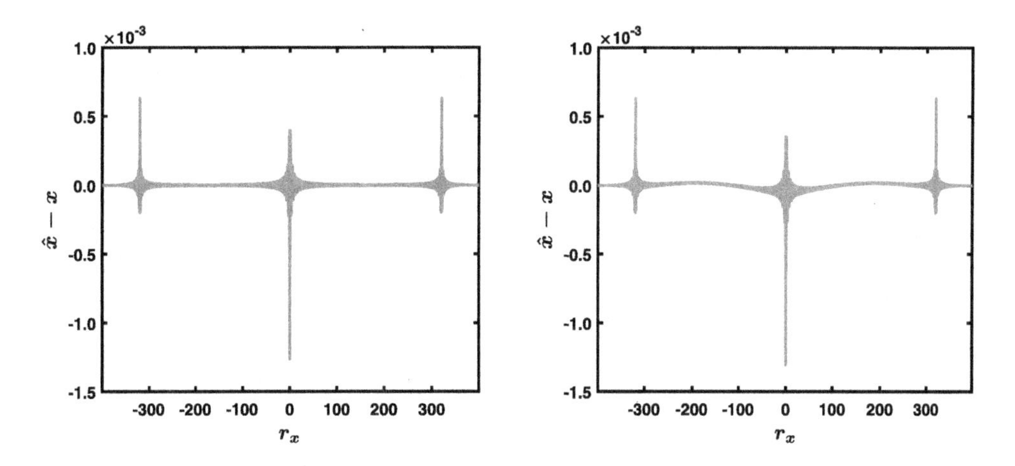

FIGURE 4.6 1D Cartesian Sampling — Inverse FFT versus Gridding

From an equidistant sampling of the frequency spectrum s in Fig. 4.5, 1D objects $\hat{x}$ are reconstructed with twice the spatial resolution using zero padding, once with an inverse FFT (left) and once with gridding (right). Errors are calculated with respect to the continuous triangular function x.

with $f = 2$ and $K = 5$ adds only negligible errors that originate from the truncation of the convolution kernel in this case. In the following, these settings of the oversampling factor and the kernel size are kept.

To simulate a nonequidistant sampling of s, the magnetic field gradient in the frequency-encoding direction, also known as readout gradient, is assumed to ramp up 100 µs after the excitation with a gradient slew rate of $200\,\mathrm{Tm^{-1}s^{-1}}$, at which point in time the data acquisition is also triggered, and to level off at a gradient strength of $40\,\mathrm{Tm^{-1}}$, as illustrated in Fig. 4.7. The gradient system is modeled as a linear time-invariant system, once with an ideal frequency response

$$H(\mathrm{i}\omega) = 1 \tag{4.66}$$

and once with a more realistic frequency response

$$H(\mathrm{i}\omega) = \frac{\mathrm{e}^{\mathrm{i}\omega T}}{1 + \mathrm{i}\omega T}, \tag{4.67}$$

which corresponds to a mono-exponential decay with a time constant T, chosen to be 40 µs in this example. The amplitude and phase of $H(\mathrm{i}\omega)$ are plotted in Fig. 4.8. In the ideal case, the actual gradient waveform is identical to the demand gradient waveform. In the more realistic case, however, the actual gradient waveform is smoothed. This is seen in Fig. 4.7, as well as the resulting k-space sampling positions as a function of time. The data acquisition is performed twice, with positive and negative polarity of the readout gradient, to traverse the 1D k-space from the center, or origin, to the periphery, in positive and negative direction, respectively. Thus, the same range of k_x as before is covered. The sampling positions are located in Fig. 4.9 at the points of intersection of the horizontal lines ($k_y = 0$) with the circles. While the sampling density in principle increases towards the k-space center, $|k_x| < \frac{0.36}{N}$ remains

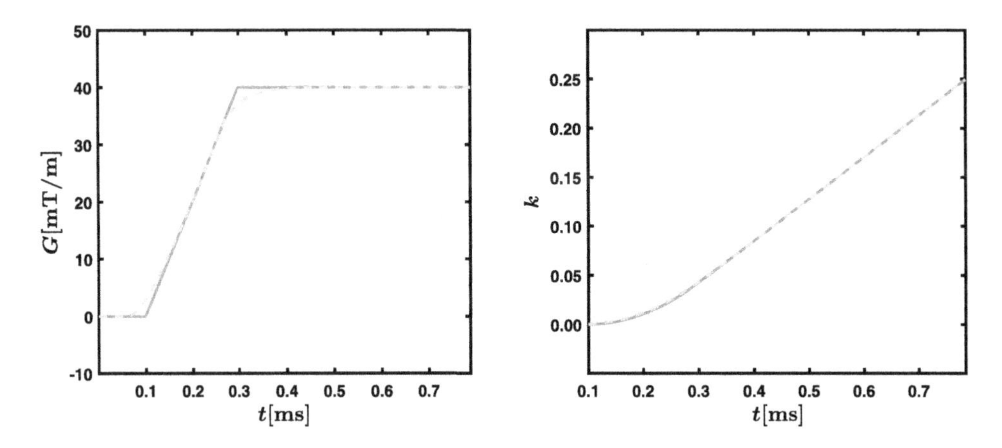

FIGURE 4.7 Gradient Waveform and k-Space Trajectory

The demand gradient waveform (solid red (mid gray in print version), left) linearly increases with the maximum gradient slew rate until the maximum gradient strength is reached. The real gradient system from Fig. 4.8 low-pass filters the gradient waveform (dashed green (gray in print version), left), leading to a distortion of the k-space trajectory (right), which is hardly perceptible at this scale.

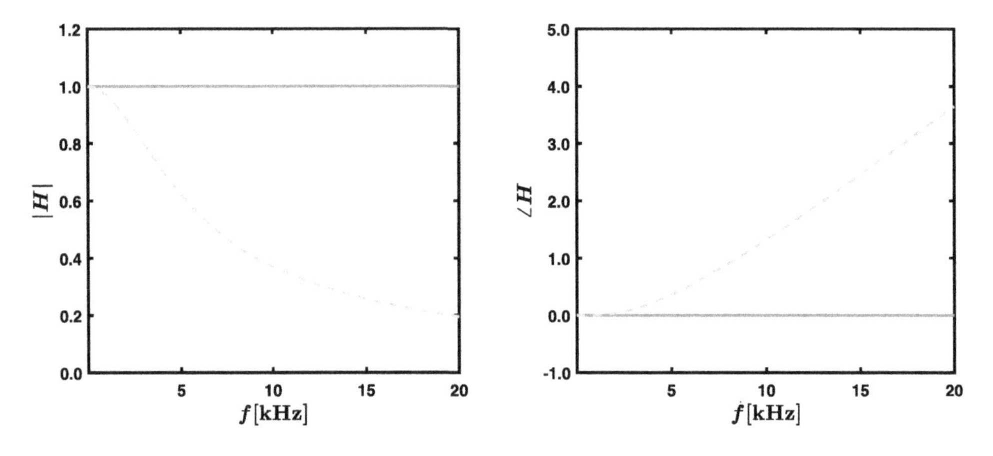

FIGURE 4.8 Gradient System Characterization

An ideal gradient system (solid red (mid gray in print version)) has an invariant amplitude (left) and phase (right) response, whereas a real gradient system (dashed green (gray in print version)) typically exhibits an amplitude response decaying with frequency and a phase response growing with frequency.

uncovered in the more realistic case. It is worth noting that this does not constitute an undersampling per se because the critical sampling distance of $\frac{1}{N}$ is larger than the diameter of the gap.

Applying gridding to the nonequidistant sampling of s produces the errors plotted in Fig. 4.10. With the ideal gradient system, a moderate offset is observed, whereas, with the more realistic gradient system, major errors apparently arise from the gap.

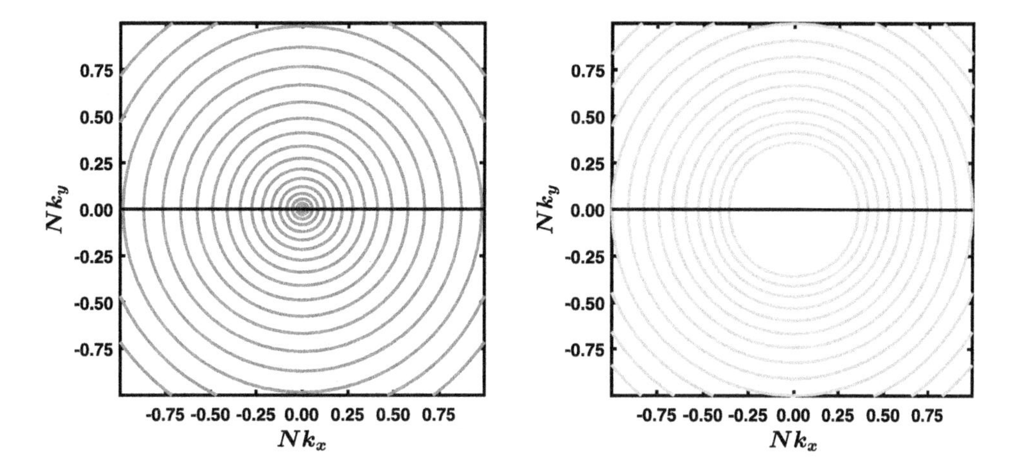

FIGURE 4.9 1D and 2D Sampling Patterns

From the k-space trajectories in Fig. 4.7, 1D (k_x) and 2D (k_x, k_y) sampling patterns are generated by equiangular rotation around the origin. Unlike with the ideal gradient system (left), the sampling pattern obtained with the more realistic gradient system (right) leaves a gap in the k-space center, which is enlarged in both plots.

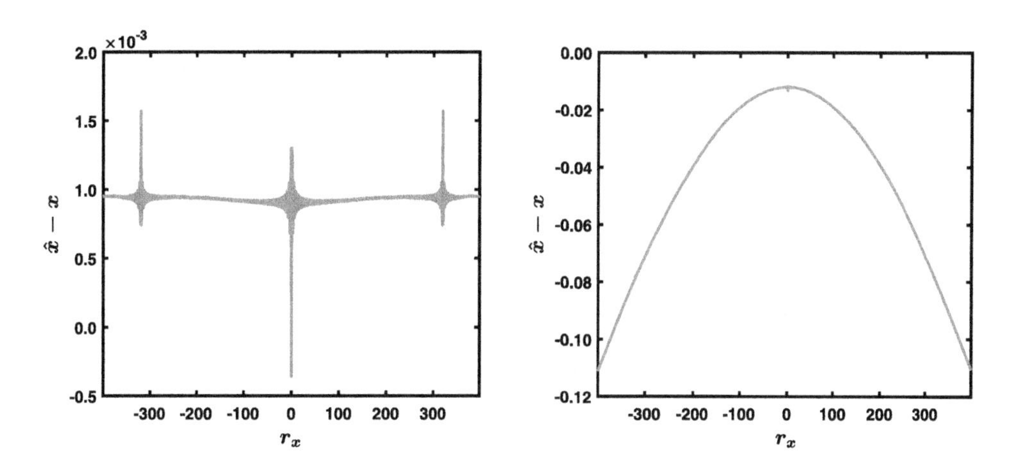

FIGURE 4.10 1D Non-Cartesian Sampling — Gridding

Using the two nonequidistant 1D sampling patterns from Fig. 4.9, 1D objects $\hat{x}$ are reconstructed with twice the spatial resolution with gridding. Errors are calculated with respect to the continuous triangular function x.

So far, a sampling density compensation based on the distance between adjacent k-space sampling positions has been used. Weights obtained with this geometrical approach and with some of the other approaches described in the previous section are juxtaposed in Fig. 4.11, along with corresponding errors, for the problematic case of the more realistic gradient system. The fix-point iteration according to

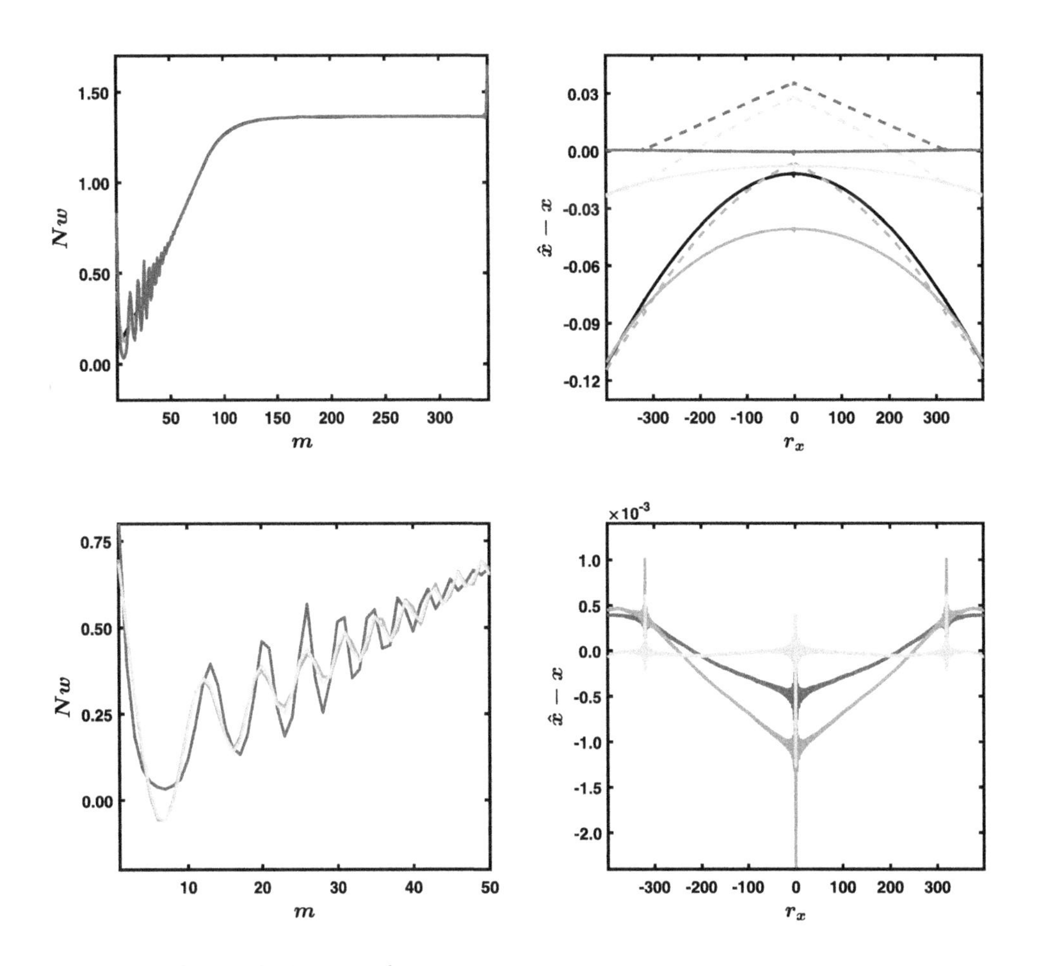

FIGURE 4.11 Sampling Density Compensation

From the second nonequidistant 1D sampling pattern in Fig. 4.9, weights for sampling density compensation are derived (left), geometrically (black, top), numerically based on repeated gridding with 5 (red (mid gray in print version), top), 25 (green (gray in print version), top), and 125 (blue (dark gray in print version), top and bottom) iterations, once with (solid) and once without (dashed) normalization, and algebraically based on Eq. (4.55) (red (mid gray in print version), bottom) and Eq. (4.52) (green (gray in print version), bottom). 1D objects $\hat{x}$ are reconstructed with gridding, and errors are calculated with respect to the continuous triangular function x (right).

Eq. (4.56) slowly converges from $\mathbf{w}^{[0]} = \frac{1}{N}\mathbf{1}$, for which the weights always remain positive. A sinc^2 convolution kernel, truncated to two side lobes, was employed in this example to map nonequidistant samples to nonequidistant samples directly [27]. Errors decrease at first, but soon they are dominated by a systematic overestimation of the weights due to the omitted side lobes. By contrast, enforcing $\|\mathbf{w}\|_1 = 1$ allows reducing errors by two orders of magnitude overall, yet results deteriorate again when proceeding with the iteration. Solving Eq. (4.55) and Eq. (4.52) leads to weights that still show the

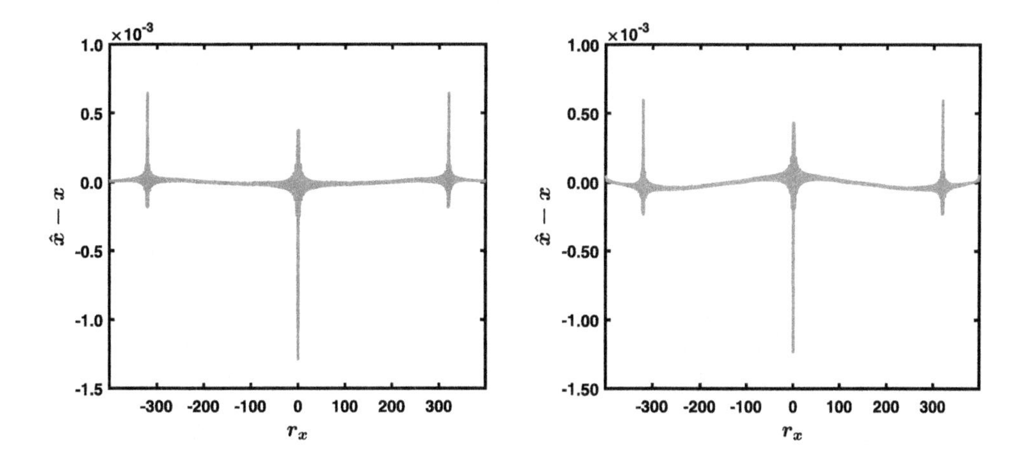

FIGURE 4.12 1D Non-Cartesian Sampling — Iterative Reconstruction

Using the two nonequidistant 1D sampling patterns from Fig. 4.9, 1D objects $\hat{x}$ are iteratively reconstructed with twice the spatial resolution. Errors are calculated with respect to the continuous triangular function x.

striking, unexpected oscillation for the sampling positions in the k-space center, but that are not necessarily positive anymore. The same convolution kernel as before was chosen in the first case, while the Dirichlet kernel from Eq. (4.15) was selected in the second case. A CG method without explicit regularization reliably converges only in the second case, in which also the lowest errors are attained.

Finally, results obtained with an iterative reconstruction are provided for reference in Fig. 4.12. A CG method, initialized with $\hat{\mathbf{x}}^{[0]} = \mathbf{0}$, leads to similar errors as gridding in the case of equidistant sampling and, with favorable sampling density compensation, in the case of nonequidistant sampling. An NFFT and an adjoint NFFT were used in the iterative reconstruction, but no sampling density compensation or explicit regularization.

The conic function shown in Fig. 4.13 is chosen as object to illustrate gridding with a 2D example. Its Fourier transform is given by

$$s(k_r) = \begin{cases} \pi^2 \frac{J_1(Rk_r)H_0(Rk_r) - J_0(Rk_r)H_1(Rk_r)}{k_r^2}, & k_r \neq 0 \\ \frac{1}{3}\pi R^2, & \text{otherwise} \end{cases}, \tag{4.68}$$

where J_0, J_1, and H denote the zeroth and first order Bessel functions of the first kind and the Struve function, respectively [28].

Applying an inverse FFT for reference and gridding to an equidistant sampling of s with k_x and k_y in the range $[-\frac{1}{4}, \frac{1}{4})$ and $M = 400^2$ results in the errors plotted in Fig. 4.14, which are dominated by ringing artifacts at $r_x^2 + r_y^2 = 320^2$ and at the origin, where the conic function is not differentiable. Additional errors introduced by gridding remain negligible, even though they are perceptibly higher than in the 1D example. Gridding requires about 2.75 times the number of floating point operations of the inverse FFT in this case, of which more than 80% are involved in the Fourier transform due to the oversampling.

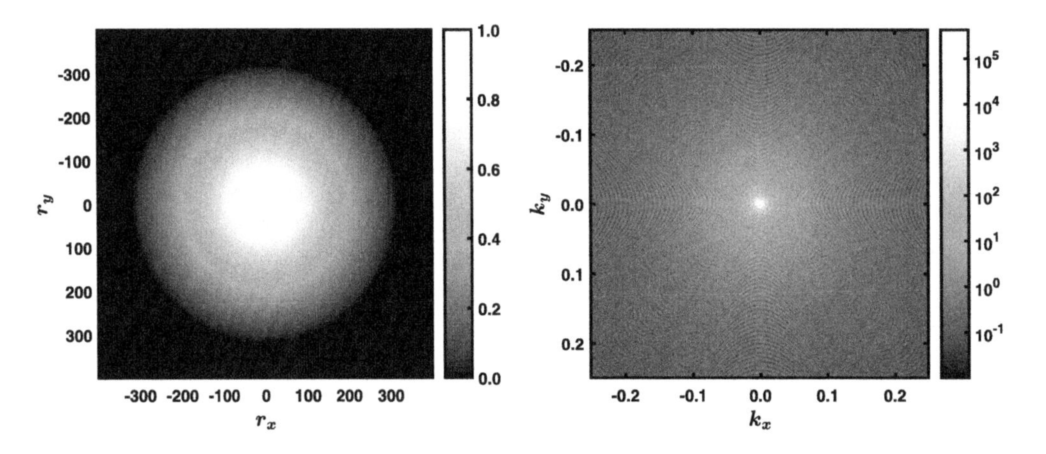

FIGURE 4.13 2D Object

A continuous conic function, covering 50% of the square FOV and 64% of the inscribed circular FOV, serves as 2D object (left). The magnitude of its frequency spectrum is displayed on a logarithmic scale (right).

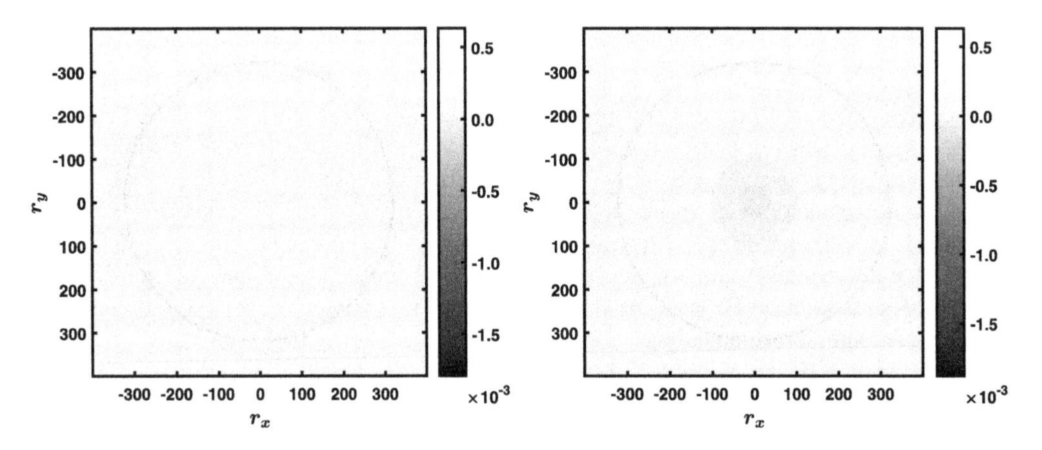

FIGURE 4.14 2D Cartesian Sampling — Inverse FFT versus Gridding

From an equidistant sampling of the frequency spectrum in Fig. 4.13, 2D objects are reconstructed with twice the spatial resolution using zero padding, once with an inverse FFT (left) and once with gridding (right). Errors are calculated with respect to the continuous conic function.

The same readout gradient as in the 1D example is assumed to simulate a nonequidistant sampling of s, but the data acquisition is repeated $\lceil \pi \sqrt{M} \rceil = 1257$ times. The direction of the readout gradient is rotated about the k-space origin by a constant angle after each instance, and with it the sampling positions in the 2D k-space. The number of instances is selected such that the distance between adjacent sampling positions in the azimuthal direction does not exceed $\frac{1}{2\sqrt{M}}$. Thus, undersampling is avoided. The resulting sampling patterns, which cover a circular k-space area, are shown in Fig. 4.9, where the

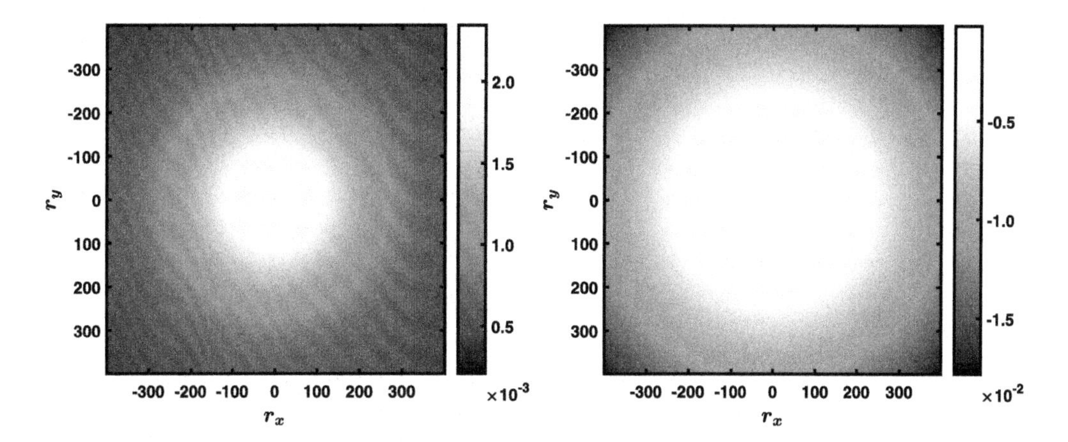

FIGURE 4.15 2D Non-Cartesian Sampling — Gridding

Using the two nonequidistant 2D sampling patterns from Fig. 4.9, 2D objects are reconstructed with twice the spatial resolution with gridding. Errors are calculated with respect to the continuous conic function.

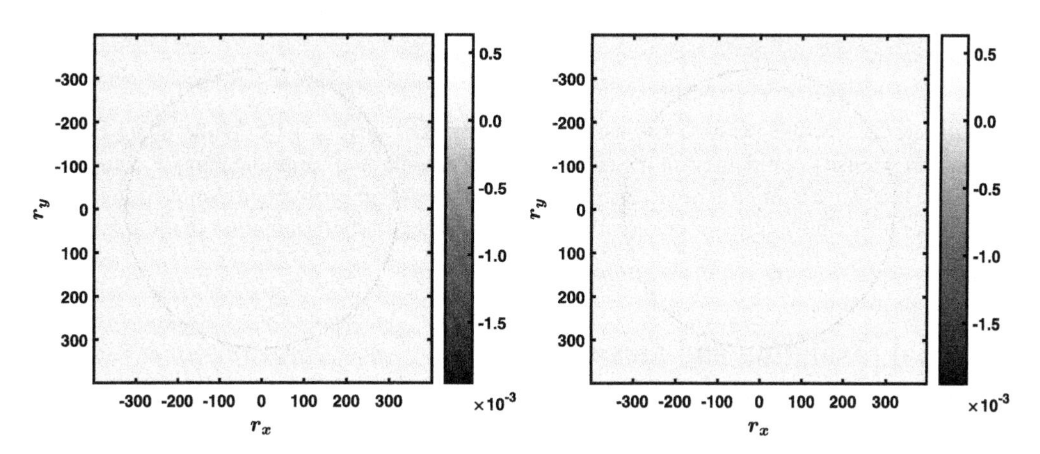

FIGURE 4.16 2D Non-Cartesian Sampling — Iterative Reconstruction

From the two nonequidistant 2D sampling patterns in Fig. 4.9, 2D objects are iteratively reconstructed with twice the spatial resolution. Errors are calculated with respect to the continuous conic function.

seemingly continuous sampling in the azimuthal direction is due to the high sampling density in the k-space center.

Applying gridding and an iterative reconstruction to the nonequidistant sampling of s produces the errors plotted in Fig. 4.15 and Fig. 4.16, respectively. Weights for sampling density compensation were derived with a geometrical approach and only employed in gridding. Most notably, the errors obtained with gridding in the case of the more realistic gradient system are reduced by one order of magnitude relative to the 1D example.

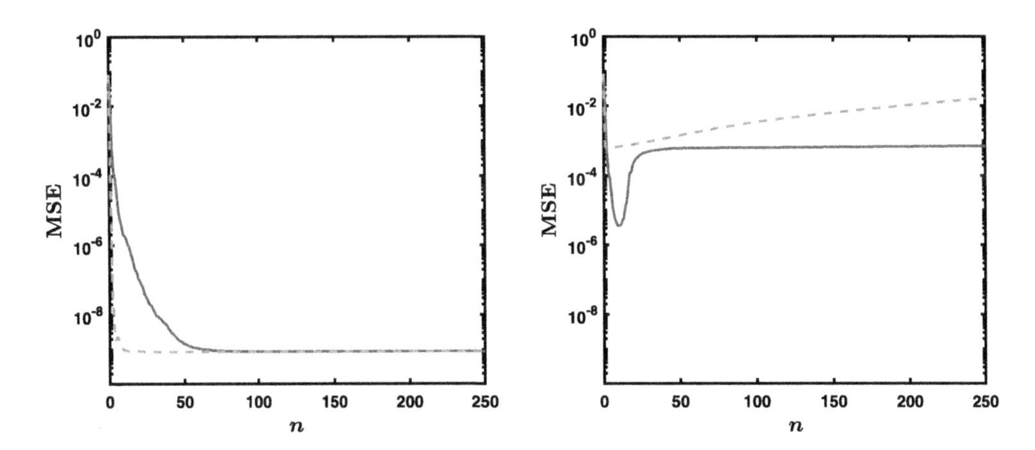

FIGURE 4.17 2D Non-Cartesian Sampling — Convergence

2D objects are iteratively reconstructed without (solid red (mid gray in print version)) and with (dashed green (gray in print version)) sampling density compensation, once without (left) and once with (right) added noise emulating a nominal SNR of 50. Mean squared errors with respect to the continuous conic function are plotted as function of the number of iterations.

The convergence of the iterative reconstruction is analyzed in Fig. 4.17. While using Eq. (4.61) with a sampling density compensation instead of Eq. (4.59) without a sampling density compensation leads to considerable acceleration, it adversely affects the convergence in the presence of noise, which demands either a suitable explicit regularization or a sufficiently early termination of the iteration.

4.6 Spatial resolution and noise

Especially when comparing non-Cartesian imaging with Cartesian imaging, awareness of potential differences in spatial resolution and noise is essential and, therefore, raised in this section.

A common measure of spatial resolution is the full width at half maximum (FWHM) of the main lobe of the PSF. In 1D Cartesian imaging, the PSF is given by Eq. (4.37) and approximated by

$$\mathrm{PSF}(r) \approx \mathrm{sinc}(\pi r) \tag{4.69}$$

for $M = N$ and small $\frac{r}{N}$. The FWHM equals 1.21 since $\mathrm{sinc}(0.603\pi) = 0.5$. In 2D Cartesian imaging, the PSF remains the same along the axes r_x and r_y, but changes to

$$\mathrm{PSF}(r_d) \approx \mathrm{sinc}^2\left(\frac{\pi}{\sqrt{2}} r_d\right) \tag{4.70}$$

along the diagonals r_d, with $r_d = \sqrt{2}\, r_x$ and $r_y = \pm r_x$, assuming a square k-space area to be covered by the acquisition. The FWHM increases to 1.25, which indicates a slightly anisotropic spatial resolution.

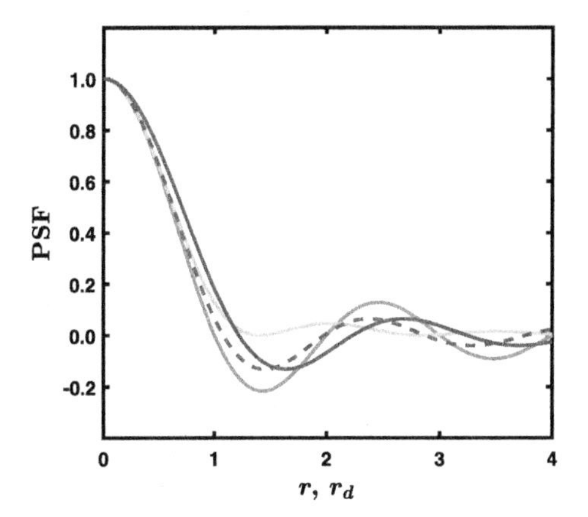

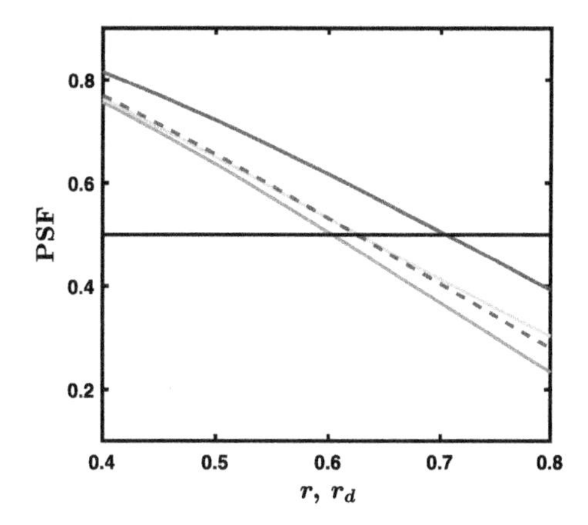

FIGURE 4.18 Spatial Resolution

PSFs are approximated for acquisitions covering a square FOV, along the axes (solid red (mid gray in print version)) and the diagonals (solid green (gray in print version)), and circular FOVs, with the circle inscribed in the square (solid blue (dark gray in print version)) and the area of the circle and the square equalized (dashed blue (dark gray in print version)). Crossings of the main lobes with the horizontal line at half the maximum amplitude are determined.

In 2D non-Cartesian imaging, only a smaller circular k-space area is usually covered, as seen in Fig. 4.1. The PSF is approximated by

$$\mathrm{PSF}(r) \approx 2\,\mathrm{jinc}(\pi r) \tag{4.71}$$

using

$$\mathrm{jinc}(x) = \begin{cases} \frac{J_1(x)}{x}, & x \neq 0 \\ \frac{1}{2}, & \text{otherwise} \end{cases}, \tag{4.72}$$

and the FWHM amounts to 1.41 regardless of direction. Eq. (4.71), as Eq. (4.69) and Eq. (4.70), can be derived by assuming a continuous sampling of the covered k-space area [29].

To match the spatial resolution in 2D Cartesian imaging, the circular k-space area has to be enlarged relative to the square k-space area. Scaling the radius of the former by $\frac{2}{\sqrt{\pi}} = 1.13$ equalizes the area of both [30]. Applied to Eq. (4.71), it results in

$$\mathrm{PSF}(r) \approx 2\,\mathrm{jinc}(2\sqrt{\pi}\,r) \tag{4.73}$$

and an FWHM of 1.25. This is illustrated in Fig. 4.18, in which the PSFs according to Eq. (4.69), Eq. (4.70), Eq. (4.71), and Eq. (4.73) are plotted. It is worth noting that the amplitudes of the first side lobes of the PSF along the axes in 2D Cartesian imaging are the highest. In general, this leads to more pronounced ringing artifacts in 2D Cartesian imaging than in 2D non-Cartesian imaging.

The validity of the continuous sampling approximation is substantiated in Fig. 4.19. Despite the rather small number of samples, the calculated PSFs for PROPELLER, radial, and spiral imaging using

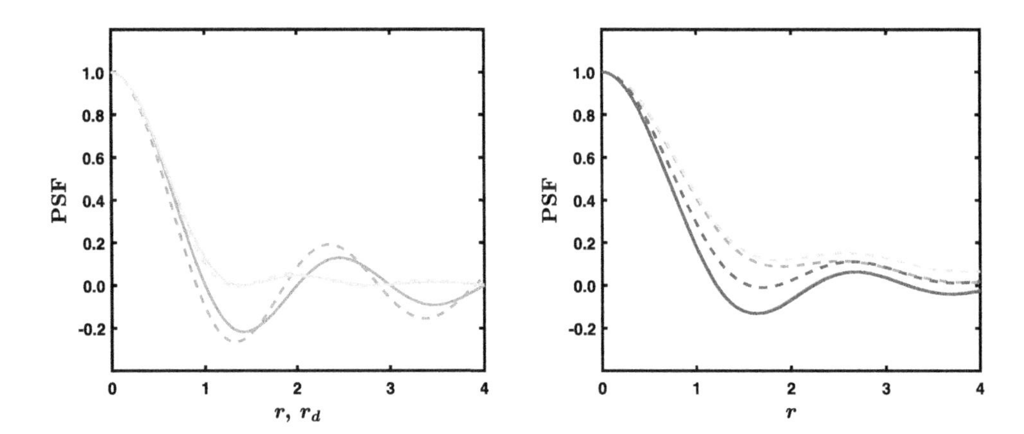

FIGURE 4.19 Point Spread Functions

PSFs are calculated for the sampling patterns in Fig. 4.1, namely for EPI along the axes (red (mid gray in print version), left) and the diagonals (green (gray in print version), left), as well as for PROPELLER (red (mid gray in print version), right), radial (green (gray in print version), right), and spiral (blue (dark gray in print version), right) imaging, and gridding without (dashed) and with (solid) sampling density compensation.

gridding with sampling density compensation closely agree with each other, to the extent that they are not discernible at this scale, and with the approximated PSF in Fig. 4.18. The calculated and approximated PSFs for EPI along the axes and the diagonals are similarly consistent. Naturally, this only holds for the main lobe and the first side lobes.

While the sampling density compensation is crucial to shape the PSF in non-Cartesian imaging using gridding, it entails a reduction in SNR by a factor of

$$\frac{\sum_{m=1}^{M} [\mathbf{w}]_m}{\sqrt{M}\,\|\mathbf{w}\|_2}, \tag{4.74}$$

assuming additive white Gaussian noise [31]. This factor equals 1.0 only if the weights are uniform. For the 2D sampling patterns in Fig. 4.1, it decreases to 0.98, 0.85, 0.87, and 0.96 in the case of echo-planar, PROPELLER, radial, and spiral imaging, respectively. It drops further from 0.87 to 0.80 for the 2D sampling patterns in Fig. 4.9 due to the additional nonequidistant sampling in radial direction.

The same reduction in SNR is observed in the right plot in Fig. 4.17 using an iterative reconstruction without sampling density compensation. The nominal SNR of 50 accounts for a mean squared error of $4 \cdot 10^{-4}$, given that the maximum signal is normalized to a value of 1.0 and that other errors are negligible. The actual mean squared error of $6.3 \cdot 10^{-4}$ after 100 iterations then implies an amplification of the noise variance by a factor of 1.58 and a loss in SNR of 20%. However, a CG method provides a diminishing intrinsic regularization in the course of the iterations [32], which is particularly evident using an iterative reconstruction with sampling density compensation. The amplification of the noise variance is ultimately given by the diagonal elements of the matrix

$$\mathbf{X} = \left(\mathbf{E}^{\mathrm{H}}\,\mathbf{E}\right)^{-1} \tag{4.75}$$

or

$$\mathbf{X} = \left(\mathbf{E}^{\mathrm{H}}\mathbf{W}\,\mathbf{E}\right)^{-1}\left(\mathbf{E}^{\mathrm{H}}\mathbf{W}^{2}\,\mathbf{E}\right)\left(\mathbf{E}^{\mathrm{H}}\mathbf{W}\,\mathbf{E}\right)^{-1} \tag{4.76}$$

and is often unacceptably high without explicit regularization because the system matrix is typically ill-conditioned in non-Cartesian imaging [26].

Variation in sampling density also leads to colored noise, which is characterized by a nonconstant power spectral density. Colored noise notably impairs the perceived spatial resolution of images [33]. This can be mitigated by adding noise or by considering other criteria, for instance contrast, in the sampling density compensation [34].

4.7 Extensions

An NFFT for an efficient conversion of equidistant samples in space into nonequidistant samples in k-space was introduced in a previous section. Moreover, an adjoint NFFT was shown to be an integral part of gridding and, together with an NFFT, of an iterative reconstruction for non-Cartesian imaging. Other NFFTs exist for related conversions, which are of relevance to both Cartesian and non-Cartesian imaging. They are outlined in this section, as well as their applications.

For a mapping of nonequidistant samples in space to equidistant samples in k-space, the approximation

$$\mathrm{e}^{-\mathrm{i}2\pi kr} \approx \frac{1}{f\tilde{c}_{-Nk}} \sum_{n'=-fN/2}^{fN/2-1} \tilde{c}\left(r - \tfrac{n'}{f}\right) \mathrm{e}^{-\mathrm{i}2\pi k \frac{n'}{f}} \tag{4.77}$$

is obtained similarly to Eq. (4.11), taking the different range of r and k and the necessary stretching and scaling of $\tilde{c}$ into account. It leads to the forward model

$$\tilde{s}_n \approx \frac{1}{f\tilde{c}_{-Nk_n}} \sum_{n'=-fN/2}^{fN/2-1} \left(\sum_{m=1}^{M} x_m\, \tilde{c}\left(r_m - \tfrac{n'}{f}\right)\right) \mathrm{e}^{-\mathrm{i}2\pi k_n \frac{n'}{f}}, \tag{4.78}$$

where $k_n = -\frac{1}{2}, -\frac{1}{2} + \frac{1}{N}, \ldots, \frac{1}{2} - \frac{1}{N}$. The inner sum converts M nonequidistant samples in space by a convolution into fN equidistant samples in space. The outer sum amounts to an FFT of length fN, transforming the equidistant samples in space into equidistant samples in k-space. The deapodization is applied last.

Nonequidistant samples in space occur in the presence of an imperfect, namely a spatially nonconstant, magnetic field gradient, which affects the spatial encoding and leads to distortions in images if ignored in the reconstruction. The signal measured in k-space is modeled under these circumstances by

$$s(k) = \int_{-N/2}^{N/2} x(r)\, \mathrm{e}^{-\mathrm{i}2\pi k g(r)} \mathrm{d}r \tag{4.79}$$

with a usually nonlinear function g mapping the true position r to the apparent position r' in space. Provided that g is invertible, a change in variables results in

$$s(k) = \int_{g(-N/2)}^{g(N/2)} x(g^{-1}(r')) \frac{\mathrm{d}g^{-1}(r')}{\mathrm{d}r'} \mathrm{e}^{-\mathrm{i}2\pi k r'} \mathrm{d}r' \tag{4.80}$$

and a spatial discretization in nonequidistant samples of x. Currently, a distortion correction is routinely performed after the reconstruction in space, using interpolation and scaling. However, integrating it into the reconstruction promises advantages, particularly with regard to spatial resolution [35]. While this demands an efficient evaluation of Eq. (4.78) in Cartesian imaging, the forward model has to be generalized to nonequidistant samples in both domains in non-Cartesian imaging.

Nonequidistant samples in space also arise from an imperfect main magnetic field. The signal is described by

$$s(t) = \int_{-N/2}^{N/2} x(r)\, \mathrm{e}^{-\mathrm{i}2\pi k(t) r}\, \mathrm{e}^{-\mathrm{i}2\pi t f(r)} \mathrm{d}r \tag{4.81}$$

in this case. Here, s and k are explicit functions of t, the time after the excitation, and f represents the offset in the frequency of the precession of the nuclear spins induced by the spatial variation of the main magnetic field. In Cartesian imaging, k is a linear function because the readout gradient remains unchanged during the data acquisition. Eq. (4.81) can then be written as Eq. (4.79). In EPI, this still holds in the phase-encoding direction, in which the dominant distortions occur [36].

The consideration of an inhomogeneous main magnetic field in non-Cartesian imaging is treated in Chapter 12. Only an NFFT for an efficient conversion between nonequidistant samples in both domains is derived here, which was shown to be applicable to such an off-resonance correction [37,38]. The inverse Fourier transform of the window function c

$$(\mathcal{F}^{-1}c)(r) = \int_{-\infty}^{\infty} c(k)\, \mathrm{e}^{\mathrm{i}2\pi k r} \mathrm{d}k \tag{4.82}$$

is rewritten as

$$(\mathcal{F}^{-1}c)(r) = \int_{-1/2}^{1/2} \sum_{p=-\infty}^{\infty} c(k+p)\, \mathrm{e}^{\mathrm{i}2\pi (k+p) r} \mathrm{d}k \tag{4.83}$$

and approximated by

$$(\mathcal{F}^{-1}c)(r) \approx \frac{1}{fN} \sum_{n'=-fN/2}^{fN/2-1} \sum_{p=-\infty}^{\infty} c\left(k - \tfrac{n'}{fN} + p\right) \mathrm{e}^{\mathrm{i}2\pi (k - \frac{n'}{fN} + p) r} \tag{4.84}$$

to obtain

$$\mathrm{e}^{-\mathrm{i}2\pi k r} \approx \frac{1}{fN(\mathcal{F}^{-1}c)(r)} \sum_{n'=-fN/2}^{fN/2-1} \sum_{p=-\infty}^{\infty} c\left(k - \tfrac{n'}{fN} + p\right) \mathrm{e}^{-\mathrm{i}2\pi (\frac{n'}{fN} - p) r}. \tag{4.85}$$

The inner sum is then eliminated, for example by simply limiting the range of c to $[-\frac{K}{2fN}, \frac{K}{2fN}]$ and of k to $[-\frac{1}{2} + \frac{K}{2fN}, \frac{1}{2} - \frac{K}{2fN})$, leading to

$$e^{-i2\pi kr} \approx \frac{1}{fN(\mathcal{F}^{-1}c)(r)} \sum_{n'=-fN/2}^{fN/2-1} c'\left(k - \frac{n'}{fN}\right) e^{-i2\pi \frac{n'}{fN} r}. \tag{4.86}$$

Such an NFFT also permits including higher order magnetic field perturbations in the reconstruction, which can be modeled by

$$s(t) = \int_{-N/2}^{N/2} x(r)\, e^{-i2\pi \sum_{l=1}^{L} k_l(t) b_l(r)} dr \tag{4.87}$$

with spatial basis functions b and corresponding time-variant coefficients k [39]. While spherical harmonics are commonly chosen for b, separate calibration measurements, such as gradient impulse response measurements, or concurrent magnetic field monitoring are usually required to determine k [40,41]. The first-order spherical harmonics are equivalent to spatially constant magnetic field gradients and thus describe the actual sampling positions in k-space. Accurate knowledge of the corresponding time-variant coefficients is often necessary but also sufficient to achieve adequate image quality in non-Cartesian imaging.

Finally, exponential signal decay during the data acquisition can be incorporated into Eq. (4.81) by adding a suitably scaled decay rate as imaginary part to f [42].

4.8 Summary

NFFTs are an indispensable part of non-Cartesian MRI reconstruction. They provide a generalization of FFTs to nonequidistant samples and rely on an approximation with controllable accuracy, which is, even with small oversampling factors and kernel sizes, sufficient for MRI. Several optimized implementations are nowadays available for various computing platforms, also under other acronyms, such as NUFFTs and FINUFFTs. They considerably facilitate the development and application of general algorithms coping with various sampling patterns and further alleviate concerns about the running time of NFFTs.

Two such algorithms were described in this chapter, gridding and an iterative reconstruction. Gridding involves, in addition to an adjoint NFFT, a sampling density compensation, with which reasonable accuracy is attained for most popular sampling patterns. The iterative reconstruction comprises both an NFFT and an adjoint NFFT and allows achieving better accuracy at the expense of computational complexity. Both algorithms form the basis for other algorithms discussed in following chapters, which notably support an acceleration of non-Cartesian imaging by undersampling.

NFFTs were also demonstrated to be a valuable part of corrections for an imperfect main magnetic field and magnetic field gradient. Rendering these corrections more robust, in particular without accurate knowledge of nuisance parameters, still constitutes a considerable challenge.

While many non-Cartesian imaging methods have been proposed and investigated to date, only a few have been adopted in clinical routine on a larger scale so far. Whether the use of non-Cartesian imaging methods will further expand and ever rival the use of Cartesian imaging methods remains to be seen.

References

[1] Lauterbur PC. Image formation by induced local interactions: examples employing nuclear magnetic resonance. Nature 1973;242:190–1.

[2] Kumar A, Welti D, Ernst RR. NMR Fourier zeugmatography. J Magn Reson 1975;18:69–83.

[3] Ernst RR. NMR Fourier zeugmatography. J Magn Reson 2011;213:510–2.

[4] Mansfield P. Multi-planar image formation using NMR spin echoes. J Phys C 1977;10:L55–8.

[5] Pipe JG. Motion correction with PROPELLER MRI: application to head motion and free-breathing cardiac imaging. Magn Reson Med 1999;42:963–9.

[6] Ahn CB, Kim JH, Cho ZH. High-speed spiral-scan echo planar NMR imaging - I. IEEE Trans Med Imaging 1986;5:2–7.

[7] Dutt A, Rokhlin V. Fast Fourier transforms for nonequispaced data. SIAM J Sci Comput 1993;14:1368–93.

[8] Plonka G, Potts D, Steidl G, Tasche M. Numerical Fourier analysis. Basel: Birkhäuser; 2019.

[9] O'Sullivan JD. A fast sinc function gridding algorithm for Fourier inversion in computer tomography. IEEE Trans Med Imaging 1985;4:200–7.

[10] Jackson JI, Meyer CH, Nishimura DG, Macovski A. Selection of a convolution function for Fourier inversion using gridding. IEEE Trans Med Imaging 1991;10:473–8.

[11] Slepian D, Pollak HO. Prolate spheroidal wave functions, Fourier analysis and uncertainty - I. Bell Syst Tech J 1961;40:43–63.

[12] Kaiser JF. Digital filters. In: Kuo FF, Kaiser JF, editors. System analysis by digital computer. New York: Wiley; 1966. p. 218–85.

[13] Fessler JA, Sutton BP. Nonuniform fast Fourier transforms using min-max interpolation. IEEE Trans Signal Process 2003;51:560–74.

[14] Wajer FTAW, Woudenberg E, de Beer R, Fuderer M, Mehlkopf AF, van Ormondt D. Simple equation for optimal gridding parameters. In: Proceedings of the 7th annual meeting of the ISMRM; 1999. p. 663.

[15] Beatty PJ, Nishimura DG, Pauly JM. Rapid gridding reconstruction with a minimal oversampling ratio. IEEE Trans Med Imaging 2005;24:799–808.

[16] Dale B, Wendt M, Duerk JL. A rapid look-up table method for reconstructing MR images from arbitrary k-space trajectories. IEEE Trans Med Imaging 2001;20:207–17.

[17] Boada FE, Hancu I, Shen GX. Spherically symmetric kernels for improved convolution gridding. In: Proceedings of the 7th annual meeting of the ISMRM; 1999. p. 1654.

[18] Brouw WN. Aperture synthesis. In: Alder B, Fernbach S, Rotenberg M, editors. Radio astronomy. New York: Academic; 1975. p. 131–75.

[19] Hoge RD, Kwan RKS, Pike GB. Density compensation functions for spiral MRI. Magn Reson Med 1997;38:117–28.

[20] Rasche V, Proksa R, Sinkus R, Börnert P, Eggers H. Resampling of data between arbitrary grids using convolution interpolation. IEEE Trans Med Imaging 1999;18:385–92.

[21] Rosenfeld D. An optimal and efficient new gridding algorithm using singular value decomposition. Magn Reson Med 1998;40:14–23.

[22] Sedarat H, Nishimura DG. On the optimality of the gridding reconstruction algorithm. IEEE Trans Med Imaging 2000;19:306–17.

[23] Pipe JG, Menon P. Sampling density compensation in MRI: rationale and an iterative numerical solution. Magn Reson Med 1999;41:179–86.

[24] Wajer FTAW, Pruessmann KP. Major speedup of reconstruction for sensitivity encoding with arbitrary trajectories. In: Proceedings of the 9th annual meeting of the ISMRM; 2001. p. 767.
[25] Eggers H, Boernert P, Boesiger P. Comparison of gridding- and convolution-based iterative reconstruction algorithms for sensitivity-encoded non-Cartesian acquisitions. In: Proceedings of the 10th annual meeting of the ISMRM; 2002. p. 743.
[26] Pruessmann KP, Weiger M, Börnert P, Boesiger P. Advances in sensitivity encoding with arbitrary k-space trajectories. Magn Reson Med 2001;46:638–51.
[27] Johnson KO, Pipe JG. Convolution kernel design and efficient algorithm for sampling density correction. Magn Reson Med 2009;61:439–47.
[28] Abramowitz M, Stegun IA, editors. Handbook of mathematical functions with formulas, graphs, and mathematical tables. New York: Dover; 1972.
[29] Lauzon ML, Rutt BK. Effects of polar sampling in k-space. Magn Reson Med 1996;36:940–9.
[30] van Gelderen P. Comparing true resolution in square versus circular k-space sampling. In: Proceedings of the 6th annual meeting of the ISMRM; 1998. p. 424.
[31] Pipe JG, Duerk JL. Analytical resolution and noise characteristics of linearly reconstructed magnetic resonance data with arbitrary k-space sampling. Magn Reson Med 1995;34:170–8.
[32] Qu P, Zhong K, Zhang B, Wang J, Shen GX. Convergence behavior of iterative SENSE reconstruction with non-Cartesian trajectories. Magn Reson Med 2005;54:1040–5.
[33] Newbould R, Liu C, Bammer R. Colored noise and effective resolution: data considerations for non-uniform k-space sampling reconstructions. In: Proceedings of the 14th annual meeting of the ISMRM; 2006. p. 2939.
[34] Eggers H, van Yperen GH, Nehrke K. Contrast optimization by data weighting in PROPELLER imaging. In: Proceedings of the 15th annual meeting of the ISMRM; 2007. p. 1737.
[35] Tao S, Trzasko JD, Shu Y, Huston III J, Bernstein MA. Integrated image reconstruction and gradient nonlinearity correction. Magn Reson Med 2015;74:1019–31.
[36] Jezzard P, Balaban RS. Correction for geometric distortion in echo planar images from B_0 field variations. Magn Reson Med 1995;34:65–73.
[37] Sutton BP, Noll DC, Fessler JA. Fast, iterative image reconstruction for MRI in the presence of field inhomogeneities. IEEE Trans Med Imaging 2003;22:178–88.
[38] Eggers H, Knopp T, Potts D. Field inhomogeneity correction based on gridding reconstruction for magnetic resonance imaging. IEEE Trans Med Imaging 2007;26:374–84.
[39] Wilm BJ, Barmet C, Pavan M, Pruessmann KP. Higher order reconstruction for MRI in the presence of spatiotemporal field perturbations. Magn Reson Med 2011;65:1690–701.
[40] Alley MT, Glover GH, Pelc NJ. Gradient characterization using a Fourier-transform technique. Magn Reson Med 1998;39:581–7.
[41] de Zanche N, Barmet C, Nordmeyer-Massner JA, Pruessmann KP. NMR probes for measuring magnetic fields and field dynamics in MR systems. Magn Reson Med 2008;65:176–86.
[42] Knopp T, Eggers H, Dahnke H, Prestin J, Sénégas J. Iterative off-resonance and signal decay estimation and correction for multi-echo MRI. IEEE Trans Med Imaging 2009;28:394–404.

CHAPTER

5 "Early" Constrained Reconstruction Methods

Justin P. Haldar[a] and Zhi-Pei Liang[b]

[a]*University of Southern California, Signal and Image Processing Institute, Ming Hsieh Department of Electrical and Computer Engineering, Los Angeles, CA, United States*

[b]*University of Illinois at Urbana-Champaign, Department of Electrical and Computer Engineering, Urbana, IL, United States*

5.1 Introduction

Constrained image reconstruction has a long history in MRI and an even longer history outside of MRI. For example, there was more than enough material to write a longer than 100-page review article on constrained MRI nearly 30 years ago [79], and the field has continued to see vigorous development over the intervening decades. Although many of the implementation details have evolved over time, the basic concepts and principles from the early literature are still relevant and are frequently embedded in the modern constrained image reconstruction methods. In this chapter, we will discuss some of the key constrained imaging concepts that emerged in the early literature, including those based on support constraints, phase constraints, smoothness constraints, sparsity constraints, linear predictability constraints, and constraints derived from reference scans. To provide notation and context for the discussion, we will briefly review conventional Fourier reconstruction and the relevant literature on early constrained image reconstructions in the remainder of this section.

5.1.1 Basic Fourier reconstruction

For simplicity, our description will focus on 1D Fourier imaging, although all the results extend easily to higher-dimensional imaging scenarios. As discussed in Chapter 2, for an experiment that acquires M k-space measurements at sampling positions k_m for $m = 0, 1, \ldots, M-1$, the measured data $\hat{s}(k_m)$ is related to the original image function $\rho(r)$ by the noisy Fourier transform model:

$$\hat{s}(k_m) = s(k_m) + \eta_m = \int_{-\infty}^{\infty} \rho(r) e^{-i2\pi k_m r} dr + \eta_m, \tag{5.1}$$

where η_m represents measurement noise and $s(k)$ is the ideal noiseless Fourier data.

Without additional assumptions, the image reconstruction problem corresponding to the imaging model from Eq. (5.1) is inherently ill-posed. In particular, there are infinitely many possible images $\rho(r)$ that will be perfectly consistent with the measured data samples $\hat{d}(k_m)$, and when accounting for noise, there is an even larger set of images that will be nearly data-consistent. To obtain a unique reconstruction, it is thus necessary to make additional assumptions about the image $\rho(r)$, which is a cornerstone of *constrained image reconstruction*.

Advances in Magnetic Resonance Technology and Applications, Volume 7, ISSN 2666-9099. https://doi.org/10.1016/B978-0-12-822726-8.00014-2

The classical Fourier reconstruction method [77] is based on the finite sum

$$\hat{\rho}(r) = \mathbb{1}_{FOV}(r) \sum_{m=0}^{M-1} w_m \hat{s}(k_m) e^{i2\pi k_m r}, \tag{5.2}$$

where the weights w_m can all be set to the same value when data is sampled uniformly [77] or can be set to different values to account for nonuniform sampling density [55] (further details can be found in Chapter 4), and $\mathbb{1}_{FOV}(r)$ denotes the indicator function for the field-of-view (FOV). One way to justify Eq. (5.2) is to view it as a simple Riemann sum approximation of the continuous inverse Fourier transform integral, which would have been an analytic perfect-reconstruction formula if we had access to noiseless Fourier data $s(k)$ at all possible (continuous) k-space locations. In addition, assuming that the support of $\rho(r)$ is wholly within the FOV (i.e., $\rho(r) = 0$ for $r \notin [-W/2, W/2]$, where W defines the size of the FOV) and k-space is sampled uniformly at the Nyquist rate (i.e., $\Delta k = 1/W$), then the reconstruction given by Eq. (5.2) with uniform weights corresponds to the unique "minimum-norm reconstruction" [77,131], which has the smallest $\mathcal{L}_2$ norm (i.e., the smallest signal energy) among all data-consistent reconstructions supported within the FOV.

Several notable properties of the Fourier reconstruction can be obtained from the following equation that relates $\hat{\rho}(r)$ obtained from Eq. (5.2) to the true image $\rho(r)$:

$$\hat{\rho}(r) = \int_{-\infty}^{\infty} \rho(\tau) h(r - \tau) d\tau + \sum_{m=0}^{M-1} w_m \eta_m, \tag{5.3}$$

where the point-spread function (PSF) is defined by

$$h(r) \triangleq \sum_{m=0}^{M-1} w_m e^{i2\pi k_m r}. \tag{5.4}$$

- *Resolution*: With uniform weights and uniform Nyquist-rate sampling with symmetric k-space coverage, it can be shown that the PSF for Eq. (5.2) satisfies $h(r) \propto \frac{\sin(\pi M \Delta k r)}{\sin(\pi \Delta k r)}$. Convolution with $h(r)$ leads to spatial blurring. While there are multiple ways to quantify spatial resolution, a common approach based on the effective width of $h(r)$ results in a nominal spatial resolution of $1/(M\Delta k)$ for this case. *Reconstructing images with spatial resolution that is better than this is a "super-resolution" reconstruction problem.*
- *Undersampling*: When k-space sampling violates the Nyquist sampling criterion, i.e., $\Delta k > 1/W$, the PSF will have undesirable characteristics that lead to various kinds of aliasing artifacts. *Reconstructing aliasing-free images from sub-Nyquist data is a "sparse data" reconstruction problem.*
- *Noise*: Let η_m be independent identically distributed noise with zero mean and standard deviation σ. With uniform weights in Eq. (5.2), the image-domain noise standard deviation will be the same at every spatial location and proportional to $\sqrt{M}\sigma$. Thus, if one attempts to improve image resolution by extending k-space coverage from $M\Delta k$ to $2M\Delta k$ by doubling the data acquisition time (already a major sacrifice!), the image signal-to-noise ratio (SNR) will also be $\sqrt{2}\times$ worse. *Reconstructing high-quality images from noisy data is a "denoising" reconstruction problem.*

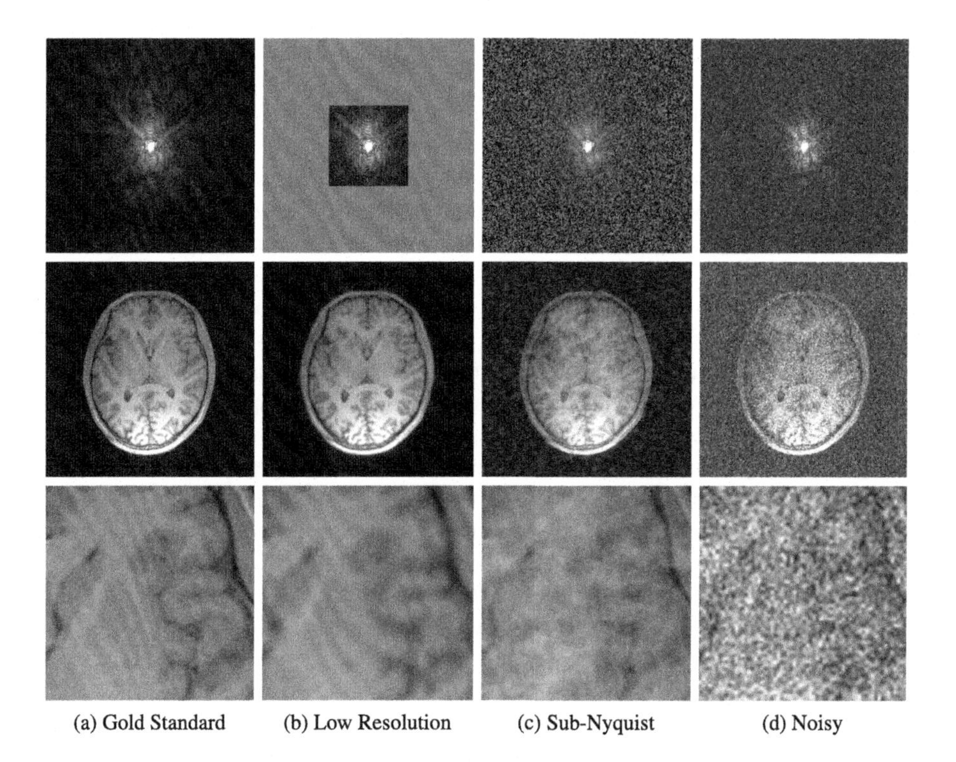

FIGURE 5.1

Illustration of the degradations in image quality for traditional Fourier reconstruction corresponding to (b) low-resolution acquisition, (c) sub-Nyquist acquisition, and (d) noisy data. The top row shows k-space data (with missing samples indicated in red (gray in print version)), while the middle row shows the corresponding reconstructed MRI images, and the bottom row shows image zoom-ins for enhanced visualization.

Illustrations of typical image quality degradations resulting from resolution, undersampling, and noise are shown in Fig. 5.1. The early literature on constrained image reconstruction proposed various techniques to mitigate these degradations, as will be reviewed in the next subsection.

5.1.2 Constrained reconstruction: historical perspective

Constrained image reconstruction has a long history and wide applications that go well beyond MRI. This subsection provides a brief literature review. Because of the large amount of literature that exists, this review is by no means comprehensive. The goal is to provide context for discussing the methods described in Sections 5.2–5.7.

For many years before the invention of MRI, it had been believed that the resolution of a Fourier imaging system was fundamentally limited by k-space coverage. This led to concepts like the Rayleigh limit to spatial resolution, which was based on the characteristics of the PSF [66,123]. However, it later became recognized that missing or degraded frequency information could actually be recovered through the process of constrained reconstruction by utilizing additional prior information [66,103,123].

The concept of constrained reconstruction was used extensively in other modalities before it was first applied to MRI. Some of these early approaches were focused on achieving superresolution by utilizing detailed prior knowledge about the exact spatial support of the image within the FOV, together with the fact that support-limited images will have analytic k-space data (such that the Fourier signal must be smooth). Slepian, Landau, and Pollak's extensive work on support-limited functions and prolate spheroidal wave functions [69,70,115–117] was of significant theoretical value, and the Gerchberg–Papoulis algorithm was perhaps the first practical method to perform support-limited extrapolation for both continuous and discrete data in the presence of measurement noise [26,99]. Other approaches were based on the idea that, when presented with ambiguity and uncertainty about the true solution to an inverse problem, we should avoid choosing solutions that appear to be more complicated or definitive than might be supported by the measured data, and should instead have a preference for solutions that are as simple and noncommittal as possible. This is a general philosophical principle that is sometimes known as Occam's razor. Although measuring the "simplicity" of a reconstruction is subjective, some of the earliest common approaches include a preference towards solutions that have minimal signal energy or maximal spatial smoothness [3,8,24,25,47,122], maximum entropy [31,57], sparsity [3,8,9,24,25,47,48], or nonnegativity [83,108].

In the early days of MRI, constrained reconstruction research was mainly focused on solving the super-resolution problem. Some notable constraints that were used for this purpose include support constraints [79,105,131], phase constraints [17,79,89,94], linear predictability constraints [32,38,78, 79,118], and constraints derived from reference images [12,52,60,74,76,132]. However, it was later recognized (particularly after the emergence of parallel imaging and compressed sensing) that many of these constraints used for superresolution reconstruction could be used much more effectively when applied to sub-Nyquist reconstruction [12,19,30,86,90,106,107] or denoising reconstruction [33,37, 41] problems. As a result, recent research in constrained image reconstruction has shifted away from addressing the superresolution problem per se.

Although sub-Nyquist reconstruction rose to popularity more recently within MRI, there were substantial early theoretical foundations for the recovery of general signals from sub-Nyquist measurements, including Landau's interpolation theory [68] and Papoulis' multichannel sampling theory [100]. In the signal processing community, much work was done by Bresler's group on the development of practical sampling schemes and reconstruction algorithms to achieve the Landau bound [133].

For static MRI applications, Cao and Levin's early Feature-Recognizing MRI approach from 1993 [12] not only utilized sub-Nyquist sampling, but also represents an important precursor to modern data-driven machine learning constrained MRI reconstruction methods [64,81,111]. Feature-Recognizing MRI took a database of training images of the same body part from multiple subjects, and used standard low-rank modeling/principal component analysis [59] to identify a low-dimensional set of basis functions that optimally captured most of the energy of these images. This low-dimensional representation could then be used to reconstruct unseen data of the same body part from a new subject, and could also be used to derive optimal k-space sampling locations (that turned out to be nonuniform). Reeves [107] observed that uniform Nyquist sampling was suboptimal for support-constrained reconstruction and optimized a set of irregularly-spaced k-space sampling locations using estimation-theoretic (Cramér–Rao bound) concepts. Similarly, von Kienlin and Mejia [134] optimized k-space sampling locations under the modeling assumptions of the earlier reference-based SLIM constrained imaging approach [52]. Marseille et al. [90] used Cramér–Rao bounds to discover that, if images were modeled as smooth regions separated by edge structures occurring at unknown locations, then k-space interpolation (i.e., reconstruction of sub-Nyquist data) would be theoretically more reliable than k-space extrapolation (i.e., the super-resolution problem). This can be seen to be conceptually quite similar to more recent compressed sensing approaches that impose sparsity constraints on image edges [7,42,86,124]. Dologlou et al. [19] later combined this nonuniform sampling concept with structured low-rank Hankel matrix concepts, resulting in what was possibly the first linear-predictive structured low-rank matrix completion method for sub-Nyquist constrained MRI reconstruction and appeared long before the recent reinvigoration of interest in this kind of approach [5,34,38,39,58,96,113].

For dynamic MRI applications, Xiang and Henkelman introduced the (k,t)-space formulation in 1993 [137]. Based on the (k,t)-space formulation, it was shown by Liang and Lauterbur in 1997 that motion artifacts were due to temporal undersampling and high-quality image reconstruction from undersampled (k, t)-space data was possible using special temporal basis functions [80]. Many methods were developed for accelerated dynamic imaging using sparse sampling of (k, t)-space: the UNFOLD method [87,128], the (k,t)-BLAST method [129], and the partial separability model-based method [14,15,36,73,138] are just three examples of many methods that were developed to provide practical means to undersample (k,t)-space and to reconstruct high-quality images from sub-Nyquist (k,t)-space data. Some of these methods also serve as the basis for the more recent dynamic low-rank (Casorati) matrix and tensor recovery methods [36,45,82,97,126,127,138]. There were also a number of early methods that accelerated dynamic imaging using sub-Nyquist sampling without invoking the (k,t)-space formulation. One example is the use of spatial-support constraints to achieve sub-Nyquist sampling [49,93], while others include the TRICKS and HYPR methods developed by Mistretta's group, which have been widely used for high-speed angiographic imaging [91,92].

Compared to super-resolution and sub-Nyquist reconstruction, there was relatively little work on the denoising reconstruction problem in the early days of MRI, even though limited sensitivity and low SNR have long been major limitations of MRI signal detection [77]. Although there was a substantial amount of early work on the denoising reconstruction problem for other imaging modalities [8,13,23,25,28,47,71,119], the early MRI work generally treated reconstruction and denoising as two distinct steps instead of combining them together. As a result, the early MRI denoising work often relied on standard image processing tools like wavelet denoising [46,95,104,135,136] and edge-preserving smoothing using anisotropic diffusion methods [27,109,114] or Markov random fields [53], low-rank matrix modeling/principal component analysis methods for denoising time series data [121], or special data acquisition schemes to improve SNR [10,20,88,101,102]. More recently, it has been shown that it is beneficial to acquire noisy high-frequency k-space data and perform denoising and image reconstruction jointly [33,37,41,43]. Such an approach has proved effective for a number of SNR-limited imaging applications, such as high-resolution MR spectroscopic imaging [67] and diffusion imaging [44].

5.2 Support-constrained reconstruction

The support of an image is the set of spatial locations r for which $\rho(r) \neq 0$. Most MRI reconstruction methods employ a weak form of support constraint, assuming that $\rho(r) = 0$ outside of the FOV. However, when we refer to "support-constrained" methods, we specifically mean methods that make the stronger assumption that the image $\rho(r)$ does not actually occupy the entire FOV. Instead, it is assumed that there are regions within the FOV for which $\rho(r) \approx 0$. Mathematically, this can be described as

$$\rho(r) \approx 0 \text{ for } r \notin \Omega, \tag{5.5}$$

where the set Ω denotes the support of the image and is assumed to satisfy $\Omega \subset [-W/2, W/2]$.

In the early literature, it was particularly common for the image support to be known *a priori* from reference images. Given this knowledge, constrained reconstruction could be performed using, e.g., linear algebraic formulations that only allowed voxels within the support to be nonzero or iterative algorithms that alternatingly projected the current image estimate onto (i) the set of data consistent images and (ii) the set of support-limited images. This latter approach can be viewed as a form of the Projection-Onto-Convex Sets (POCS) algorithm [79], which can have nice theoretical convergence characteristics [2].

Although the details are somewhat technical, these approaches can be understood at a high level because of the Fourier transform property that support-limited images $\rho(r)$ will always have Fourier transforms $s(k)$ that are analytic (i.e., smooth and infinitely differentiable). This smoothness characteristic implies that it can be possible to extrapolate or interpolate unmeasured k-space data from incomplete or noisy k-space data.

While support constraints have been applied to extrapolate missing high-frequency information in MRI scenarios [79,105,131], this super-resolution approach is limited by the fact that it tends to be ill-posed and very sensitive to noise, leading to practical gains in image spatial resolution that are not too significant. However, it should be noted that support constraints are still widely used in modern constrained MRI methods, although, for sub-Nyquist reconstruction applications rather than the super-resolution scenarios they were originally developed for. In some cases, it can be hard to see that support constraints are being used because they are implicit rather than explicit. For example, in multichannel parallel imaging methods (see Chapter 6 for more information) like SENSE [106], image support constraints are often an implicitly incorporated into the definition of coil sensitivity maps [130]. It has also been recently shown [34,38] that virtually all linear predictive methods (including methods like GRAPPA [30], SPIRiT [84], and structured low-rank approaches [5,34,38,39,58,96,113]) will implicitly take advantage of image support constraints when reconstructing support-limited images.

5.3 Phase-constrained reconstruction

Phase-constrained reconstruction was one of the earliest successful approaches to constrained reconstruction in MRI. These methods were originally developed to solve the data extrapolation problem associated with half-Fourier (or partial-Fourier) imaging. As the "half-Fourier" name implies, here $s(k)$ is measured only for half k-space, say, $\forall k \geq 0$.

To understand how phase-constrained extrapolation works, consider the simple scenario in which the image function $\rho(r)$ is purely real-valued, with zero imaginary component. In this case, $s(k)$ has perfect Hermitian symmetry, i.e., $s(-k) = s^*(k)$, where * denotes complex conjugation. This symmetry relationship suggests a simple data extrapolation scheme to generate $s(-k)$. Indeed, this approach corresponds to the earliest form of half-Fourier imaging [22]. In practice, $\rho(r)$ is not real-valued due to field inhomogeneity, coil sensitivity variations, flow effects, etc., and more practical data extrapolation methods were developed to cope with the problem, using a pre-estimated phase $\hat{\phi}(r)$ as a constraint.

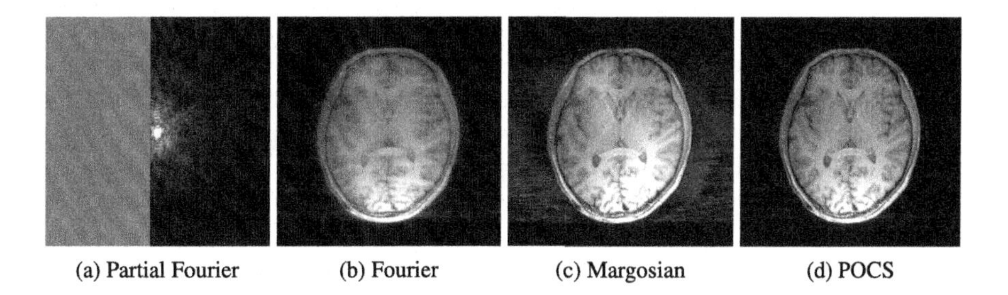

(a) Partial Fourier (b) Fourier (c) Margosian (d) POCS

FIGURE 5.2

Illustration of single-channel phase-constrained partial Fourier reconstruction using the Margosian and POCS algorithms. In this case, (a) a half-Fourier sampling scheme was augmented with an additional set of eight central k-space lines to enable phase estimation. While (b) classical zero-filled Fourier reconstruction of this data is heavily blurred, the (c) Margosian and (d) POCS reconstructions of this data are quite successful at recovering crisp high-resolution image features, although the Margosian method especially demonstrates visible artifacts stemming largely from imperfect phase estimation.

For example, the imaging model from Eq. (5.1) could be replaced with

$$\hat{s}(k_m) = \int_{-\infty}^{\infty} m(r) e^{i\hat{\phi}(r)} e^{-i2\pi k_m r} dr + \eta_m \tag{5.6}$$

where $m(r)$ represents the unknown magnitude image. The fact that $m(r)$ is real-valued means that it has roughly half the degrees of freedom as the complex-valued image $\rho(r)$, resulting in a substantially easier inverse problem.

Assuming that the phase is relatively smooth (which is a good assumption in many cases, though is not always appropriate), such a phase constraint could, e.g., be obtained by collecting a few additional samples across the center of k-space to provide a symmetric data set to estimate $\hat{\phi}(r)$. Imposing this phase constraint along with the data-consistency constraint leads to one version of the phase constrained reconstruction problem. Several methods were proposed for solving this problem including the Margosian method [89], the homodyne detection method by Noll and Macovski [94], the Cuppen method [17], and the POCS method [79]. We have already mentioned the POCS algorithm in the context of support-constrained reconstruction. While the POCS algorithm for phase-constrained reconstruction is different, it still follows the basic principle of iteratively projecting the current estimate of the image onto two different constraint sets. In the context of phase-constrained reconstruction, the image is iteratively projected onto (i) the set of data consistent images and (ii) the set of images that are consistent with the estimated phase $\hat{\phi}(r)$. An illustration of the performance of the Margosian and POCS methods is shown in Fig. 5.2.

Although the phase-constrained extrapolation problem was actively addressed 30+ years ago, the basic concepts of using prior phase information and leveraging Fourier symmetry for constrained reconstruction are still relevant today. While some of the more recent approaches bear relatively close resemblance to some of the earliest methods by relying explicitly on prior phase estimates [11,110], recent years have also seen the introduction of very different kinds of phase constraints that have very different structure. For example, recent years have witnessed the emergence of phase-constrained reconstruction methods that do not actually require any explicit phase-estimation procedure, which eliminates one of the major pitfalls for classical methods. Instead, these newer methods are based on concepts of linear-predictability across different sides of k-space [34,38,54], virtual conjugate coils [6], and structured low-rank matrix recovery [34,38,39,62]. Compared to earlier approaches, these new types of approaches have substantially more flexibility than previous approaches and, for example, are capable of exploiting Fourier symmetry relationships with very different sampling patterns than before. This flexibility allows them to not only be used for super-resolution reconstruction of Nyquist-sampled data (as was the case for the classical methods), but also allows them to be used for reconstruction of data sampled below the Nyquist rate with samples missing from both sides of k-space and without dense sampling of the k-space center [34,38,39,62].

5.4 Linear predictive reconstruction

The autoregressive moving average (ARMA) model was one of the first methods that used linear prediction to extrapolate k-space data [118]. In MRI, a (P, Q)-order ARMA model of an image $\rho(r)$ will often take the form

$$\rho(r) = \frac{\sum_{q=0}^{Q} \beta_q e^{i2\pi \Delta kqr}}{1 - \sum_{p=1}^{P} \alpha_p e^{i2\pi \Delta kpr}}. \tag{5.7}$$

In k-space, this is equivalent to

$$s(n\Delta k) = \sum_{p=1}^{P} \alpha_p s([n-p]\Delta k) + \frac{1}{\Delta k} \sum_{q=0}^{Q} \beta_q \delta[n-q], \tag{5.8}$$

where $\delta[n]$ is the Kronecker delta function. One may notice that Eq. (5.8) is a generalized version of linear prediction that allows one sample in k-space to be predicted from previous neighboring values, which is a perfect fit for extrapolating higher-frequency k-space data from a set of Nyquist-sampled low-frequency k-space measurements. It should also be noted that the ARMA model from Eq. (5.7) reduces to a conventional Fourier model in the form of Eq. (5.2) when $P = 0$ and to a strict autoregressive model when $Q = 0$. Combining both parts of this model together enables the representation of both smooth features (via the coefficients β_q) and spiky/support-limited features (which autoregressive models are excellent at capturing [38]).

There were a number of extensions on the ARMA model for linear predictive extrapolation. One notable approach, proposed by Liang and Haacke [32,78], applies the ARMA model to a weighted version of the k-space data, i.e., $\tilde{s}(k) = (i2\pi k)^p s(k)$ with $p \geq 1$. In the image domain, this weighting of k-space is equivalent to applying a spatial differentiation operator, which is a common sparsifying transform in the modern literature. It was shown that, if the image function is piecewise constant or piecewise polynomial, the ARMA model is precise after such a sparsifying transform; in other words, the k-space data is perfectly linear predictable after this sparsifying transform [72,78]. Interestingly, this work may be viewed as some of the earliest attempts to use sparsity to constrain image reconstruction in MRI and also represents some of the first applications of structured low-rank Hankel matrix modeling in this field [78] (see also Chapters 8 and 10).

All of the early linear predictive models were originally formulated for 1D extrapolation of k-space in order to achieve super-resolution. Unfortunately, while these types of approaches were reasonably effective with noiseless data, 1D extrapolation over long distances can be quite challenging and very sensitive to small amounts of noise. Modern linear predictive methods can not only handle multidimensonal k-space data but also sub-Nyquist data [38], significantly enhancing its practical utility. For example, the popular multichannel parallel imaging (see also Chapter 6) methods GRAPPA [30] and SPIRiT [84] are based on multidimensional linear predictive relationships, and the constrained imaging field is currently enjoying somewhat of a renaissance of linear predictive approaches [38] due to the reemergence of structured low-rank matrix recovery approaches. These latter approaches are quite powerful in that they enable the automatic data-adaptive identification and imposition of a large number of simultaneous imaging constraints, including support constraints, phase constraints, sparsity constraints, parallel imaging constraints, and multicontrast constraints, often without requiring any kind of calibration data [5,34,38,39,58,96,113].

5.5 Rank-constrained reconstruction

MR signals often possess low-rank structures when put in the form of Hankel matrices or Casorati matrices (or tensors) due to linear predictability [38] and partial separability [36,73,112]. These low-rank structures have been exploited in a number of early constrained reconstruction methods for addressing all three of the main constrained reconstruction problems: super-resolution reconstruction, sub-Nyquist data reconstruction, and denoising. This section will describe these low-rank structures and their use for constrained image reconstruction (see also Chapters 9 and 10).

Consider a sequence $s[n]$ of uniformly-spaced k-space samples such that $s[n] \triangleq s(n\Delta k)$, and assume that this sequence satisfies the following shift-invariant autoregressive relationship

$$s[n] = \sum_{p=1}^{P} \alpha_p s[n-p] \tag{5.9}$$

for all possible integer values of n. Note that this corresponds to a form of linear predictive model as described in the previous section. In addition, either general support constraints [34,38] or a specific kind of sparsity constraint [72,78] can be used to guarantee the accuracy of this autoregressive model. If a Hankel or Toeplitz matrix is formed from this kind of data, it can be shown that the matrix will have low-rank characteristics. For example, for any chosen integers N and L with $N > L$, the Hankel matrix $\mathbf{H} \in \mathbb{C}^{(N-L)\times L}$ constructed as

$$\mathbf{H} \triangleq \begin{bmatrix} s[0] & s[1] & s[2] & \dots & s[L-1] \\ s[1] & s[2] & s[3] & \dots & s[L] \\ \vdots & \vdots & \vdots & \ddots & \vdots \\ s[N-L-1] & s[N-L] & s[N-L+1] & \dots & s[N] \end{bmatrix} \tag{5.10}$$

must have $\text{rank}(\mathbf{H}) \leq \min\{P, L, N-L\}$. If N and L are chosen such that $L \gg P$ and $N - L \gg P$, then $\mathbf{H}$ will be heavily rank-deficient.

High-dimensional MR signals are also often partially separable. For example, in dynamic imaging, the ideal k-t space data $s(k,t)$ can often be modeled as being Qth-order partially separable:

$$s(k,t) = \sum_{q=1}^{Q} g_q(k) h_q(t). \tag{5.11}$$

If a Casorati matrix is formed from this kind of data, it can be shown that the matrix will also have low-rank characteristics. In this case, for any chosen integers M and N, the Casorati matrix $\mathbf{C} \in \mathbb{C}^{M \times N}$ constructed as

$$\mathbf{C} \triangleq \begin{bmatrix} s(k_1,t_1) & s(k_1,t_2) & \dots & s(k_1,t_N) \\ s(k_2,t_1) & s(k_2,t_2) & \dots & s(k_2,t_N) \\ \vdots & \vdots & \ddots & \vdots \\ s(k_M,t_1) & s(k_M,t_2) & \dots & s(k_M,t_N) \end{bmatrix} \tag{5.12}$$

must have $\text{rank}(\mathbf{C}) \leq \min\{Q, M, N\}$. If M and N are chosen such that $M \gg Q$ and $N \gg Q$, then $\mathbf{C}$ will be heavily rank-deficient.

The rank constraints can be enforced in a number of ways in constrained reconstruction, both in the early literature and in modern efforts. One approach that has become increasingly common in the modern literature is the use of regularization penalties that prefer lower-rank matrices, although there remain many viable alternatives that may be more or less useful depending on the specific application. Although the concepts of low-rank matrix modeling have been used in MRI for decades, recent years have seen a particular resurgence of interest in structured low-rank matrix recovery methods that use the characteristics of Hankel/Toeplitz matrices to enforce support, phase, sparsity, parallel imaging, and multicontrast constraints [5,34,38,39,58,96,113] without requiring the kinds of calibration data that would often be employed by early methods when using such constraints. In addition, methods based on the low-rank structure of Casorati matrices/tensors [36,45,82,97,126,127,138] are also quite common in the modern literature, particularly in dynamic and multicontrast imaging contexts. Similar to the case of Hankel/Toeplitz low-rank modeling, one of the important differences between the modern literature and the early literature is that modern techniques are frequently capable of identifying the low-rank structure directly from sparsely sampled or low-quality measurements, without requiring reference information or calibration data to pre-estimate the low-rank structure.

5.6 Sparsity-constrained reconstruction

Sparsity-constrained reconstruction is very visible within the modern constrained MRI reconstruction literature due to the rise of compressed sensing concepts more than a decade ago [7,42,61,85,86,124]. However, there had already been substantial work in sparse modeling for constrained MRI before that time. While modern compressed sensing approaches are frequently defined by their use of sparsity together with nonlinear reconstruction and some form of quasi-random/incoherent sub-Nyquist sampling scheme, the early MRI methods frequently did not share these same features, oftentimes relying on sampling patterns that were more uniform and reconstruction methods that were much more linear than modern approaches.

In order to define a simple notion of sparsity, we will assume that the set of functions $\{\phi_n(r)\}$ forms a basis or frame for the space of possible images. In other words, for any possible image $\rho(r)$, there exists a sequence of coefficients $\{\alpha_n\}$ such that

$$\rho(r) = \sum_n \alpha_n \phi_n(r). \tag{5.13}$$

An image $\rho(r)$ is said to have a sparse representation in the domain specified by $\{\phi_n(r)\}$ if most of the α_n coefficients in this representation are equal to zero, with only a small number of nonzero values.

Different approaches were used in the early literature to impose sparsity constraints. One common approach was to pre-estimate the index values where the coefficients α_n would take on nonzero values. This could be done using the low-rank characteristics of Hankel matrices for certain types of sparse signals as described previously [72,78,80], or by using reference images to pre-identify the basis functions that would capture the important structural characteristics of the image [12,52,98]. Once the small number of relevant basis functions has been identified, it is straightforward to use simple linear algebra to estimate the corresponding α_n values. This can be viewed as a form of support-limited reconstruction, where the support constraints are defined with respect to the representation coefficients α_n, rather than original image $\rho(r)$. Similar ideas used in these early methods can be found in modern multicontrast reconstruction methods that utilize the fact that different images of the same anatomy oftentimes share similar sparsity characteristics [4,18,21,29,35,41,43,56,63,65,125]. Similar information-sharing methods were also developed in the early literature for other imaging modalities [13,23,28,71,119].

Another approach for imposing sparsity was based on the use of edge-preserving Markov random fields or edge-preserving anisotropic diffusion filtering schemes that were intentionally designed to preserve sparse image features such as edges. Although these approaches were utilized to some extent in the early MRI literature [50], they were much more commonly found in the early literature for other imaging modalities [3,8,24,25,47]. Widespread modern reconstruction methods like compressed sensing that rely on sparsity-promoting regularization of image edges using penalties like the convex ℓ_1-norm or the nonconvex ℓ_p-norm with $p < 1$ can all be viewed as Markov random field methods. Of course, as said before, a major difference between the early Markov random field methods and modern compressed sensing is the difference in sampling styles—the early work largely used Markov random field models to address super-resolution and/or denoising problems, rather than the sub-Nyquist reconstruction problems that are the primary emphasis of modern compressed sensing methods (which, as described previously, often leads to a much easier inverse problem, yielding higher-quality reconstruction results).

5.7 Reconstruction using side information

In many imaging applications, such as dynamic imaging and spectroscopic imaging, side information, in the form of reference images, is often available. Keyhole imaging [60,132] represents an early attempt in MRI to use reference data for super-resolution. To illustrate how this is done, let's consider the basic keyhole method as applied in dynamic imaging in which a high-resolution reference data set $s_r(k)$ is obtained for $k = -N\Delta k, \cdots, N\Delta k$, followed by a sequence of dynamic data sets $s_t(k)$ for $k = -M\Delta k, \cdots, M\Delta k$, where $M \ll N$. A straightforward method to perform data extrapolation on $s_t(k)$ is to replace the unmeasured high-frequency data with those from $s_r(k)$. In other words, we obtain extrapolated data as follows:

$$\hat{s}_t(k) = \begin{cases} s_t(k), & \text{for } k \in \{-M\Delta k, \cdots, M\Delta k\} \\ s_r(k), & \text{otherwise.} \end{cases} \tag{5.14}$$

This method can perform perfect or high-quality data extrapolation if $s_r(k)$ is equal to or approximately equal to $s_t(k)$ for $|k| > M\Delta k$. This is the case in some dynamic contrast-enhanced imaging experiments where dynamic signal changes take place mainly in the low-frequency k-space data. However, when this assumption is not valid, this simple keyhole data extrapolation scheme would not be effective. This problem can be addressed using a more sophisticated data extrapolation method that uses $s_r(k)$ to construct an optimal set of basis functions for $s_t(k)$ [74,76].

As an example, the generalized-series approach [74,76] represents $\rho_t(r)$ using a series of basis functions $\{\phi_\ell(r)\}$ as

$$\rho_t(r) = \sum_{\ell=1}^{L} \alpha_{t,\ell}\phi_\ell(r). \tag{5.15}$$

An optimal set of the basis functions $\phi_\ell(r)$ according to the minimum cross-entropy model selection principle is, to a first-order approximation, given by [74,76]

$$\phi_\ell(r) = q(r)e^{i2\pi\ell\Delta kr}, \tag{5.16}$$

where $q(r)$ can be chosen to equal the reference image $\rho_r(r)$ or some appropriate constraint function derived from $\rho_r(r)$.

With the basis functions in Eq. (5.16) and taking $q(r) = \rho_r(r)$, Eq. (5.15) can be rewritten as

$$s_t(k) = \sum_{\ell=1}^{L} \alpha_{t,\ell}s_r(k - \ell\Delta k). \tag{5.17}$$

If L is smaller than the number of k-space measurements, the coefficients $\alpha_{t,\ell}$ can easily be estimated using linear algebraic methods (similar to the case described for sparsity-constrained reconstruction with known support in the previous section) from the measured data $s_t(k)$. After the $\alpha_{t,\ell}$ are determined, Eqs. (5.15) and (5.17) can be used to synthesize the high-resolution dynamic image and/or the corresponding extrapolated dynamic k-space data. A quick illustration of the keyhole and generalized series approaches is given for multicontrast MRI data in Fig. 5.3.

High-resolution reference images have also been used in the early constrained MRI literature to construct other specialized image models. For example, a common situation in MR spectroscopic imaging is that a user may desire to measure accurate information about one specific region of interest (ROI) within the FOV, but the ability to make such a measurement is confounded because the data was acquired with low spatial resolution and there is a substantial amount of signal from undesired regions that leaks into the ROI because of PSF blurring effects for conventional Fourier reconstruction. The Spectral Localization by IMaging (SLIM) [52] technique, which was originally formulated for MR

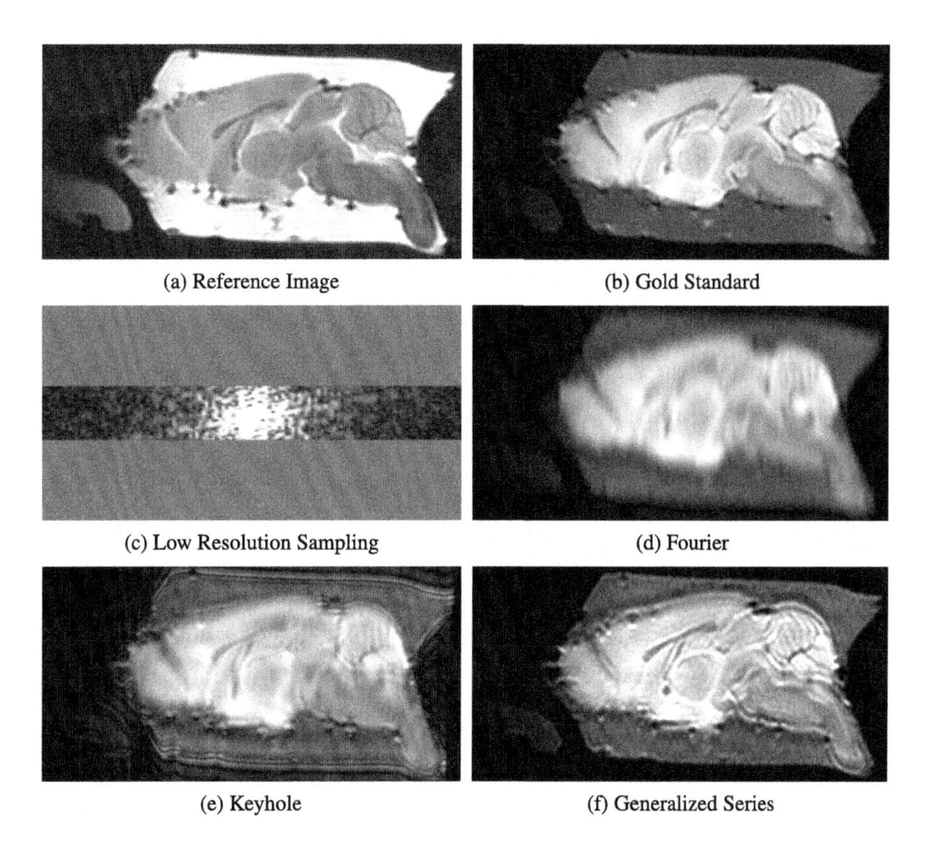

(a) Reference Image

(b) Gold Standard

(c) Low Resolution Sampling

(d) Fourier

(e) Keyhole

(f) Generalized Series

FIGURE 5.3

Illustration of reference-based image reconstruction using multicontrast data of an ex vivo mouse brain. Given a (a) high-resolution reference image, the goal is to reconstruct (b) an image acquired with different contrast parameters from (c) low-resolution k-space data. (d) Standard Fourier reconstruction has the expected blurring artifacts. (e) Keyhole reconstruction has some potential resolution recovery, though also has substantial artifacts and blurring due to a mismatch between the high-frequency k-space signal between the reference image and the target image. On the other hand, (f) generalized series reconstruction is more tolerant of data mismatches, resulting in artifacts that are less severe. Note that, due to phase mismatches between the reference image and the target image, a low-resolution phase estimate from the target image was used to construct a phase-matched reference image that was used for both keyhole and generalized series reconstruction.

spectroscopic imaging but can also be applied more broadly, is perhaps the earliest attempt to overcome this problem by using information from a high-resolution reference image. In particular, SLIM uses the high-resolution reference image to define a small set of generalized voxels (or compartments) defined by basis functions $\psi_\ell(r)$, where each compartment is designed to exactly match (with high resolution) the edge shapes, sizes, and locations of various image features and regions-of-interest within of the image. The image is then modeled simply by

$$\rho(r) = \sum_{\ell=1}^{L} \alpha_\ell \phi_\ell(r), \tag{5.18}$$

and if the number of voxels is small with the respect to the number of measured k-space samples, the coefficients α_ℓ can be estimated using simple linear algebraic methods. Importantly, rather than directly estimating a high-resolution image, this approach can be interpreted as instead measuring the average value of the image over the spatial region spanned by each compartment [51,75]. Further extensions of this general approach were also proposed in the early constrained MRI reconstruction literature [16,50,74,120,134], and the idea of using prior-edge information derived from high-quality reference images still lives on in more recent approaches that impose constraints in a softer way using regularization [18,21,41,56,65] or that directly attempt to estimate the average signal from a prescribed image compartment using less-stringent modeling assumptions [1,40].

5.8 Discussion

We have reviewed several constrained reconstruction methods, each utilizing one of the key constraints that appeared in the early literature, which include support constraints, phase constraints, linear predictability constraints, rank constraints, sparsity constraints, and constraints derived from reference images. Although more recent methods oftentimes rely on similar constraints to those used in early methods, there are a few distinctions. For one, computational resources were substantially more limited in the early days than they are now. As a result, the early methods tended to focus on simpler and less computationally demanding approaches than the methods that are common today. For example, it was quite common for early methods to focus on 1D reconstruction (e.g., in a 2D acquisition, reconstructing the fully-sampled frequency-encoding dimension using standard Fourier methods, leaving only the phase-encoding dimension to be reconstructed using constraints). Modern approaches that use multidimensional formulations with multidimensional constraints tend to be much more effective than the earlier 1D approaches. Another major difference is that early constrained imaging methods tended to use prior information that was much stronger and more explicit (and subsequently, more prone to failure) than modern methods. For example, early methods that used support constraints, phase constraints, or sparsity constraints often required precise knowledge of the spatial image support, the image phase, and the support of the image in the sparsity domain, which was often obtained from auxiliary reference scans of the subject. This level of strong prior information often enabled the use of computationally inexpensive non-iterative reconstruction methods. In contrast, the use of modern nonlinear/iterative problem formulations allows users to know only that the image possesses some form of limited support, smooth phase, or sparsity in a transform domain —the user no longer needs auxiliary scans to obtain appropriate constraints, since modern approaches are powerful enough to actually extract that kind of information directly from the low-resolution, sub-Nyquist, and/or noisy k-space data. This enables the use of less data and more robustness to inaccurate constraints.

It is also worth noting that, in many MRI applications, one often has flexibility to implement various data acquisition schemes while satisfying the experimental constraints. Different acquisition schemes may lead to a super-resolution problem or a denoising problem or sparse data-recovery problem or their combination. In the old days, the super-resolution problem was the most heavily studied, as researchers often designed data acquisition to sample at the Nyquist rate while protecting SNR. However, from

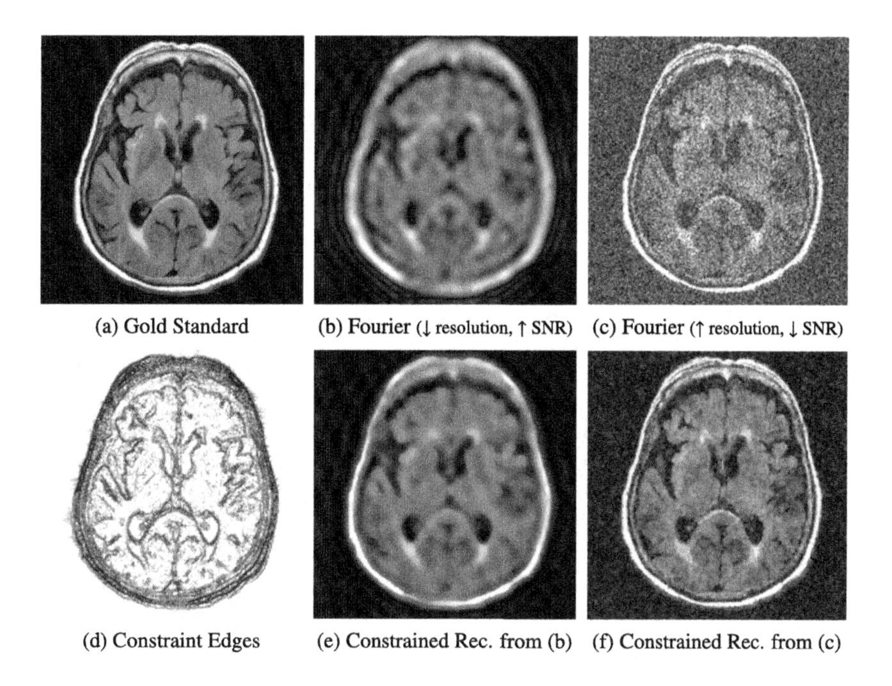

FIGURE 5.4

Illustration of the difference in effectiveness when using the same constraints for different data-acquisition schemes. (a) Gold standard image. (b) Standard Fourier reconstruction of low-resolution high-SNR data. (c) Standard Fourier reconstruction of high-resolution low-SNR data. (d) Image edges (derived from a reference image) that will be used as constraints using the method from Ref. [41]. (e) Constrained reconstruction of low-resolution high-SNR data. (f) Constrained reconstruction of high-resolution low-SNR data.

an image reconstruction standpoint, some of the reconstruction constraints are more effective when combined with some data-acquisition strategies than they are when combined with others. For example, Fig. 5.4 shows a set of results to illustrate the difference in effectiveness when utilizing anatomical boundary constraints (derived from a reference image) for solving a denoising problem (where the data has high spatial resolution but low SNR) compared to solving a super-resolution problem (where the data has low spatial resolution and the extra time was used for averaging to improve SNR). As can be seen, the denoising reconstruction has much better spatial resolution than the super-resolution result. Thus, an important observation from the more recent literature is that, in order to make maximal use of constrained reconstruction ideas, one should formulate and solve the data acquisition problem and the resulting reconstruction problem jointly.

5.9 Summary

Image reconstruction from limited and noisy data is a classical problem that arises in MRI and other imaging applications. A number of constrained reconstruction methods have been developed over the decades to utilize prior information to compensate for the lack of sufficient experimental data and to remove measurement noise. These methods have helped improve the accuracy and efficiency of many MRI techniques and laid a solid technical and conceptual foundation for the development and application of modern image reconstruction methods to be described in later chapters.

References

[1] Bakir T, Reeves SJ. A filter design method for minimizing ringing in a region of interest in MR spectroscopic images. IEEE Trans Med Imaging 2000;19:585–600.
[2] Bauschke HH, Borwein JM. On the convergence of von Neumann's alternating projection algorithm for two sets. Set-Valued Anal 1993;1:185–212.
[3] Besag J. On the statistical analysis of dirty pictures. J R Stat Soc Ser B 1986;48:259–302.
[4] Bilgic B, Goyal VK, Adalsteinsson E. Multi-contrast reconstruction with Bayesian compressed sensing. Magn Reson Med 2011;66:1601–15.
[5] Bilgic B, Kim TH, Liao C, Manhard MK, Wald LL, Haldar JP, et al. Improving parallel imaging by jointly reconstructing multi-contrast data. Magn Reson Med 2018;80:619–32.
[6] Blaimer M, Gutberlet M, Kellman P, Breuer FA, Köstler H, Griswold MA. Virtual coil concept for improved parallel MRI employing conjugate symmetric signals. Magn Reson Med 2009;61:93–102.
[7] Block KT, Uecker M, Frahm J. Undersampled radial MRI with multiple coils. Iterative image reconstruction using a total variation constraint. Magn Reson Med 2007;57:1086–98.
[8] Bouman C, Sauer K. A generalized Gaussian image model for edge-preserving MAP estimation. IEEE Trans Image Process 1993;2:296–310.
[9] Bresler Y, Macovski A. Exact maximum likelihood parameter estimation of superimposed exponential signals in noise. IEEE Trans Acoust Speech Signal Process 1986;34:1081–9.
[10] Brooker HR, Mareci TH, Mao J. Selective Fourier transform localization. Magn Reson Med 1987;5:417–33.
[11] Bydder M, Robson MD. Partial Fourier partially parallel imaging. Magn Reson Med 2005;53:1393–401.
[12] Cao Y, Levin DN. Feature-recognizing MRI. Magn Reson Med 1993;30:305–17.
[13] Chen CT, Ouyang X, Wong WH, Hu X, Johnson VE, Ordonez C, et al. Sensor fusion in image reconstruction. IEEE Trans Nucl Sci 1991;38:687–92.
[14] Christodoulou AG, Brinegar C, Haldar JP, Zhang H, Wu YJL, Foley LM, et al. High-resolution cardiac MRI using partially separable functions and weighted spatial smoothness regularization. In: Proc. IEEE Eng. Med. Bio. Conf.; 2010. p. 883–6.
[15] Christodoulou AG, Babacan SD, Liang ZP. Accelerating cardiovascular imaging by exploiting regional low-rank structure via group sparsity. In: Proc. IEEE Int. Symp. Biomed. Imag.; 2012. p. 330–3.
[16] Constable RT, Henkelman RM. Data extrapolation for truncation artifact removal. Magn Reson Med 1991;17:108–18.
[17] Cuppen JJ, Van Est A. Reducing MR imaging time by one-sided reconstruction. In: Topical Conf. fast MRI techniques; 1987.
[18] Denney TS, Reeves SJ. Bayesian image reconstruction from Fourier-domain samples using prior edge information. J Electron Imaging 2005;14:043009.
[19] Dologlou I, van Ormondt D, Carayannis G. MRI scan time reduction through non-uniform sampling and SVD-based estimation. Signal Process 1996;55:207–19.
[20] Edelstein WA, Glover GH, Hardy CJ, Redington RW. The intrinsic signal-to-noise ratio in NMR imaging. Magn Reson Med 1986;3:604–18.
[21] Eslami R, Jacob M. Robust reconstruction of MRSI data using a sparse spectral model and high resolution MRI priors. IEEE Trans Med Imaging 2010;29:1297–309.

[22] Feinberg DA, Hale JD, Watts JC, Kaufman L, Marks A. Halving MR imaging time by conjugation: demonstration at 3.5 kG. Radiology 1986;161:527–31.
[23] Fessler JA, Clinthorne NH, Rogers WL. Regularized emission image reconstruction using imperfect side information. IEEE Trans Nucl Sci 1992;39:1464–71.
[24] Geman S, Geman D. Stochastic relaxation, Gibbs distributions, and the Bayesian restoration of images. IEEE Trans Pattern Anal Mach Intell 1984;6:721–41.
[25] Geman D, Reynolds G. Constrained restoration and the recovery of discontinuities. IEEE Trans Pattern Anal Mach Intell 1992;14:367–83.
[26] Gerchberg RW. Super-resolution through error energy reduction. Opt Acta 1974;21:709–20.
[27] Gerig G, Kübler O, Kikinis R, Jolesz FA. Nonlinear anisotropic filtering of MRI data. IEEE Trans Med Imaging 1992;11:221–32.
[28] Gindi G, Lee M, Rangarajan A, Zubal IG. Bayesian reconstruction of functional images using anatomical information as priors. IEEE Trans Med Imaging 1993;12:670–80.
[29] Gong E, Huang F, Ying K, Wu W, Wang S, Yuan C. PROMISE: parallel-imaging and compressed-sensing reconstruction of multicontrast imaging using sharable information. Magn Reson Med 2015;73:523–35.
[30] Griswold MA, Jakob PM, Heidemann RM, Nittka M, Jellus V, Wang J, et al. Generalized autocalibrating partially parallel acquisitions (GRAPPA). Magn Reson Med 2002;47:1202–10.
[31] Gull SF, Skilling J. Maximum entropy method in image processing. IEE Proc 1984;131:646–59.
[32] Haacke EM, Liang ZP, Izen SH. Superresolution reconstruction through object modeling and parameter estimation. IEEE Trans Acoust Speech Signal Process 1989;37:592–5.
[33] Haldar JP. Constrained imaging: denoising and sparse sampling. Ph.D. thesis. Urbana, IL, USA: University of Illinois at Urbana-Champaign; 2011. Available from: http://hdl.handle.net/2142/24286.
[34] Haldar JP. Low-rank modeling of local k-space neighborhoods (LORAKS) for constrained MRI. IEEE Trans Med Imaging 2014;33:668–81.
[35] Haldar JP, Liang ZP. Joint reconstruction of noisy high-resolution MR image sequences. In: Proc. IEEE Int. Symp. Biomed. Imag.; 2008. p. 752–5.
[36] Haldar JP, Liang ZP. Spatiotemporal imaging with partially separable functions: a matrix recovery approach. In: Proc. IEEE Int. Symp. Biomed. Imag.; 2010. p. 716–9.
[37] Haldar JP, Liang ZP. On MR experiment design with quadratic regularization. In: Proc. IEEE Int. Symp. Biomed. Imag.; 2011. p. 1676–9.
[38] Haldar JP, Setsompop K. Linear predictability in magnetic resonance imaging reconstruction: leveraging shift-invariant Fourier structure for faster and better imaging. IEEE Signal Process Mag 2020;37:69–82.
[39] Haldar JP, Zhuo J. P-LORAKS: low-rank modeling of local k-space neighborhoods with parallel imaging data. Magn Reson Med 2016;75:1499–514.
[40] Haldar JP, Hernando D, Liang ZP. Shaping spatial response functions for optimal estimation of compartmental signals from limited Fourier data. In: Proc. IEEE Int. Symp. Biomed. Imag.; 2007. p. 1364–7.
[41] Haldar JP, Hernando D, Song SK, Liang ZP. Anatomically constrained reconstruction from noisy data. Magn Reson Med 2008;59:810–8.
[42] Haldar JP, Hernando D, Liang ZP. Compressed-sensing MRI with random encoding. IEEE Trans Med Imaging 2011;30:893–903.
[43] Haldar JP, Wedeen VJ, Nezamzadeh M, Dai G, Weiner MW, Schuff N, et al. Improved diffusion imaging through SNR-enhancing joint reconstruction. Magn Reson Med 2013;69:277–89.
[44] Haldar JP, Liu Y, Liao C, Fan Q, Setsompop K. Fast submillimeter diffusion MRI using gSlider-SMS and SNR-enhancing joint reconstruction. Magn Reson Med 2020;84:762–76.
[45] He J, Liu Q, Christodoulou AG, Ma C, Lam F, Liang ZP. Accelerated high-dimensional MR imaging with sparse sampling using low-rank tensors. IEEE Trans Med Imaging 2016;35:2119–29.
[46] Healy DM, Weaver JB. Two applications of wavelet transforms in magnetic resonance imaging. IEEE Trans Inf Theory 1992;38:840–60.

[47] Hebert T, Leahy R. A generalized EM algorithm for 3-D Bayesian reconstruction from Poisson data using Gibbs priors. IEEE Trans Med Imaging 1989;8:194–202.
[48] Hogbom JA. Aperture synthesis with a non-regular distribution of interferometer baselines. Astron Astrophys Suppl 1974;15:417–26.
[49] Hu X, Parrish T. Reduction of field of view for dynamic imaging. Magn Reson Med 1994;31:691–4.
[50] Hu X, Stillman AE. Technique for reduction of truncation artifact in chemical shift images. IEEE Trans Med Imaging 1991;10:290–4.
[51] Hu X, Wu Z. SLIM revisited. IEEE Trans Med Imaging 1993;12:583–7.
[52] Hu X, Levin DN, Lauterbur PC, Spraggins T. SLIM: spectral localization by imaging. Magn Reson Med 1988;8:314–22.
[53] Hu X, Johnson V, Wong WH, Chen CT. Bayesian image processing in magnetic resonance imaging. Magn Reson Imaging 1991;9:611–20.
[54] Huang F, Lin W, Li Y. Partial Fourier reconstruction through data fitting and convolution in k-space. Magn Reson Med 2009;62:1261–9.
[55] Jackson JI, Meyer CH, Nishimura DG, Macovski A. Selection of a convolution function for Fourier inversion using gridding. IEEE Trans Med Imaging 1991;10:473–8.
[56] Jacob M, Zhu X, Ebel A, Schuff N, Liang ZP. Improved model-based magnetic resonance spectroscopic imaging. IEEE Trans Med Imaging 2007;26:1305–18.
[57] Jaynes ET. Where do we stand on maximum entropy? In: Proc. maximum entropy formalism conference; 1978.
[58] Jin KH, Lee D, Ye JC. A general framework for compressed sensing and parallel MRI using annihilating filter based low-rank Hankel matrix. IEEE Trans Comput Imaging 2016;2:480–95.
[59] Jolliffe IT. Principal component analysis. second ed. New York: Springer-Verlag; 2002.
[60] Jones RA, Haraldseth O, Muller TB, Rinck PA, Oksendal AN. K-space substitution: a novel dynamic imaging technique. Magn Reson Med 1993;29:830–4.
[61] Jung H, Sung K, Nayak KS, Kim EY, Ye JC. k-t FOCUSS: a general compressed sensing framework for high resolution dynamic MRI. Magn Reson Med 2009;61:103–16.
[62] Kim TH, Setsompop K, Haldar JP. LORAKS makes better SENSE: phase-constrained partial Fourier SENSE reconstruction without phase calibration. Magn Reson Med 2017;77:2236–49.
[63] Knoll F, Holler M, Koesters T, Otazo R, Bredies K, Sodickson DK. Joint MR-PET reconstruction using a multi-channel image regularizer. IEEE Trans Med Imaging 2017;36:1–16.
[64] Knoll F, Hammernik K, Zhang C, Moeller S, Pock T, Sodickson DK, et al. Deep-learning methods for parallel magnetic resonance imaging reconstruction: a survey of the current approaches, trends, and issues. IEEE Signal Process Mag 2020;37:128–40.
[65] Kornak J, Young K, Soher BJ, Maudsley AA. Bayesian k-space-time reconstruction of MR spectroscopic imaging for enhanced resolution. IEEE Trans Med Imaging 2010;29:1333–50.
[66] Kosarev EL. Shannon's superresolution limit for signal recovery. Inverse Probl 1990;6:55.
[67] Lam F, Ma C, Clifford B, Johnson CL, Liang ZP. High-resolution ^{1}H-MRSI of the brain using SPICE: data acquisition and image reconstruction. Magn Reson Med 2016;76:1059–70.
[68] Landau HJ. Necessary density conditions for sampling and interpolation of certain entire functions. Acta Math 1967;117:37–52.
[69] Landau HJ, Pollak HO. Prolate spheroidal wave functions, Fourier analysis and uncertainty – II. Bell Syst Tech J 1961;40:65–84.
[70] Landau HJ, Pollak HO. Prolate spheroidal wave functions, Fourier analysis and uncertainty – III: the dimension of the space of essentially time- and band-limited signals. Bell Syst Tech J 1962;41:1295–336.
[71] Leahy R, Yan X. Incorporation of anatomical MR data for improved functional imaging with PET. Inf Process Med Imag 1991;511:105–20.
[72] Liang ZP. Constrained image reconstruction from incomplete and noisy data: a new parametric approach. Ph.D. thesis. Cleveland, OH, USA: Case Western Reserve University; 1989.

[73] Liang ZP. Spatiotemporal imaging with partially separable functions. In: Proc. IEEE Int. Symp. Biomed. Imag.; 2007. p. 988–91.

[74] Liang ZP, Lauterbur PC. A generalized series approach to MR spectroscopic imaging. IEEE Trans Med Imaging 1991;10:132–7.

[75] Liang ZP, Lauterbur PC. A theoretical analysis of the SLIM technique. J Magn Reson, Ser B 1993;102:54–60.

[76] Liang ZP, Lauterbur PC. An efficient method for dynamic magnetic resonance imaging. IEEE Trans Med Imaging 1994;13:677–86.

[77] Liang ZP, Lauterbur PC. Principles of magnetic resonance imaging: a signal processing perspective. New York: IEEE Press; 2000.

[78] Liang ZP, Haacke EM, Thomas CW. High-resolution inversion of finite Fourier transform data through a localised polynomial approximation. Inverse Probl 1989;5:831–47.

[79] Liang ZP, Boada F, Constable T, Haacke EM, Lauterbur PC, Smith MR. Constrained reconstruction methods in MR imaging. Rev Magn Reson Med 1992;4:67–185.

[80] Liang ZP, Jiang H, Hess CP, Lauterbur PC. Dynamic imaging by model estimation. Int J Imaging Syst Technol 1997;8:551–7.

[81] Liang D, Cheng J, Ke Z, Ying L. Deep magnetic resonance imaging reconstruction: inverse problems meet neural networks. IEEE Signal Process Mag 2020;37:141–51.

[82] Lingala SG, Hu Y, DiBella E, Jacob M. Accelerated dynamic MRI exploiting sparsity and low-rank structure: k-t SLR. IEEE Trans Med Imaging 2011;30:1042–54.

[83] Lucy LB. An iterative technique for the rectification of observed distributions. Astron J 1974;79:745–54.

[84] Lustig M, Pauly JM. SPIRiT: iterative self-consistent parallel imaging reconstruction from arbitrary k-space. Magn Reson Med 2010;65:457–71.

[85] Lustig M, Santos JM, Donoho DL, Pauly JM. k-t SPARSE: high frame rate dynamic MRI exploiting spatiotemporal sparsity. In: Proc. Int. Soc. Magn. Reson. Med.; 2006. p. 2420.

[86] Lustig M, Donoho D, Pauly JM. Sparse MRI: the application of compressed sensing for rapid MR imaging. Magn Reson Med 2007;58:1182–95.

[87] Madore B, Glover GH, Pelc NJ. Unaliasing by Fourier-encoding the overlaps using the temporal dimension (UNFOLD), applied to cardiac imaging and fMRI. Magn Reson Med 1999;42:813–28.

[88] Mareci TH, Brooker HR. High-resolution magnetic resonance spectra from a sensitive region defined with pulsed field gradients. J Magn Reson 1984;57:157–63.

[89] Margosian P, Schmitt F, Purdy D. Faster MR imaging: imaging with half the data. Health Care Instrum 1986;1:195–7.

[90] Marseille GJ, Fuderer M, de Beer R, Mehlkopf AF, van Ormondt D. Reduction of MRI scan time through nonuniform sampling and edge-distribution modeling. J Magn Reson, Ser B 1994;103:292–5.

[91] Mistretta CA. Undersampled radial MR acquisition and highly constrained back projection (HYPR) reconstruction: potential medical imaging applications in the post-Nyquist era. J Magn Reson Imaging 2009;29:501–16.

[92] Mistretta CA, Wieben O, Velikina J, Block W, Perry J, Wu Y, et al. Highly constrained backprojection for time-resolved MRI. Magn Reson Med 2006;55:30–40.

[93] Nagle SK, Levin DN. Multiple region MRI. Magn Reson Med 1999;41:774–86.

[94] Noll DC, Nishimura DG, Macovski A. Homodyne detection in magnetic resonance imaging. IEEE Trans Med Imaging 1991;10:154–63.

[95] Nowak RD. Wavelet-based Rician noise removal for magnetic resonance imaging. IEEE Trans Image Process 1999;8:1408–19.

[96] Ongie G, Jacob M. Off-the-grid recovery of piecewise constant images from few Fourier samples. SIAM J Imaging Sci 2016;9:1004–41.

[97] Otazo R, Candès E, Sodickson DK. Low-rank plus sparse matrix decomposition for accelerated dynamic MRI with separation of background and dynamic components. Magn Reson Med 2015;73:1125–36.

[98] Panych LP, Jolesz FA. A dynamically adaptive imaging algorithm for wavelet-encoded MRI. Magn Reson Med 1994;32:738–48.
[99] Papoulis A. A new algorithm in spectral analysis and band-limited extrapolation. IEEE Trans Circuits Syst 1975;CAS-22:735–42.
[100] Papoulis A. Generalized sampling expansion. IEEE Trans Circuits Syst 1977;CAS-24:652–4.
[101] Parker DL, Gullberg GT. Signal-to-noise efficiency in magnetic resonance imaging. Med Phys 1990;17:250–7.
[102] Parker DL, Gullberg GT, Frederick PR. Gibbs artifact removal in magnetic resonance imaging. Med Phys 1987;14:640–5.
[103] Pike ER, McWhirter JG, Bertero M, de Mol C. Generalised information theory for inverse problems in signal processing. IEE Proc 1984;131:660–7.
[104] Pižurica A, Philips W, Lemahieu I, Acheroy M. A versatile wavelet domain noise filtration technique for medical imaging. IEEE Trans Med Imaging 2003;22:323–31.
[105] Plevritis SK, Macovski A. Spectral extrapolation of spatially bounded images. IEEE Trans Med Imaging 1995;14:487–97.
[106] Pruessmann KP, Weiger M, Scheidegger MB, Boesiger P. SENSE: sensitivity encoding for fast MRI. Magn Reson Med 1999;42:952–62.
[107] Reeves SJ. Selection of k-space samples in localized MR spectroscopy of arbitrary volumes of interest. J Magn Reson Imaging 1995;5:245–7.
[108] Richardson WH. Bayesian-based iterative method of image restoration. J Opt Soc Am 1972;62:55–9.
[109] Samsonov AA, Johnson CR. Noise-adaptive nonlinear diffusion filtering of MR images with spatially varying noise levels. Magn Reson Med 2004;52:798–806.
[110] Samsonov AA, Kholmovski EG, Parker DL, Johnson CR. POCSENSE: POCS-based reconstruction for sensitivity encoded magnetic resonance imaging. Magn Reson Med 2004;52:1397–406.
[111] Sandino CM, Cheng JY, Chen F, Mardani M, Pauly JM, Vasanawala SS. Compressed sensing: from research to clinical practice with deep neural networks. IEEE Signal Process Mag 2020;37:117–27.
[112] Sen Gupta A, Liang ZP. Dynamic imaging by temporal modeling with principal component analysis. In: Proc. Int. Soc. Magn. Reson. Med.; 2001. p. 10.
[113] Shin PJ, Larson PEZ, Ohliger MA, Elad M, Pauly JM, Vigneron DB, et al. Calibrationless parallel imaging reconstruction based on structured low-rank matrix completion. Magn Reson Med 2014;72:959–70.
[114] Sijbers J, den Dekker AJ, Van der Linden A, Verhoye M, Van Dyck D. Adaptive anisotropic noise filtering for magnitude MR data. Magn Reson Imaging 1999;17:1533–9.
[115] Slepian D. Prolate spheroidal wave functions, Fourier analysis and uncertainty — IV: extensions to many dimensions; generalized prolate spheroidal functions. Bell Syst Tech J 1964;43:3009–57.
[116] Slepian D. Prolate spheroidal wave functions, Fourier analysis, and uncertainty. V – the discrete case. Bell Syst Tech J 1978;57:1371–430.
[117] Slepian D, Pollak HO. Prolate spheroidal wave functions, Fourier analysis and uncertainty – I. Bell Syst Tech J 1961;40:43–63.
[118] Smith MR, Nichols ST, Henkelman RM, Wood ML. Application of autoregressive moving average parametric modeling in magnetic resonance image reconstruction. IEEE Trans Med Imaging 1986;MI-5:132–9.
[119] Snyder DL. Utilizing side information in emission tomography. IEEE Trans Nucl Sci 1984;NS-31:533–7.
[120] Stokely EM, Twieg DB. Reconstructing magnetic resonance spectroscopic images using spatial domain priors. In: Proc. IEEE Int. Conf. Image Process.; 1994. p. 6–10.
[121] Thomas CG, Harshman RA, Menon RS. Noise reduction in BOLD-based fMRI using component analysis. NeuroImage 2002;17:1521–37.
[122] Tikhonov AN. Solution of incorrectly formulated problems and the regularization method. Sov Math Dokl 1963;5:1035–8.
[123] Toraldo di Francia G. Resolving power and information. J Opt Soc Am 1955;45:497–501.

[124] Trzasko J, Manduca A. Highly undersampled magnetic resonance image reconstruction via homotopic ℓ_0-minimization. IEEE Trans Med Imaging 2009;28:106–21.

[125] Trzasko J, Manduca A. Group sparse reconstruction of vector-valued images. In: Proc. Int. Soc. Magn. Reson. Med.; 2011. p. 2839.

[126] Trzasko J, Manduca A. Local versus global low-rank promotion in dynamic MRI series reconstruction. In: Proc. Int. Soc. Magn. Reson. Med.; 2011. p. 4371.

[127] Trzasko JD, Manduca A. A unified tensor regression framework for calibrationless dynamic, multi-channel MRI reconstruction. In: Proc. Int. Soc. Magn. Reson. Med.; 2013. p. 603.

[128] Tsao J. On the UNFOLD method. Magn Reson Med 2002;47:202–7.

[129] Tsao J, Boesiger P, Pruessmann KP. *k-t* BLAST and *k-t* SENSE: dynamic MRI with high frame rate exploiting spatiotemporal correlations. Magn Reson Med 2003;50:1031–42.

[130] Uecker M, Lai P, Murphy MJ, Virtue P, Elad M, Pauly JM, et al. ESPIRiT – an eigenvalue approach to autocalibrating parallel MRI: where SENSE meets GRAPPA. Magn Reson Med 2014;71:990–1001.

[131] Van de Walle R, Barrett HH, Myers KJ, Altbach MI, Desplanques B, Gmitro AF, et al. Reconstruction of MR images from data acquired on a general nonregular grid by pseudoinverse calculation. IEEE Trans Med Imaging 2000;19:1160–7.

[132] van Vaals JJ, Brummer ME, Dixon WT, Tuithof HH, Engels H, Nelson RC, et al. "Keyhole" method for accelerating imaging of contrast agent uptake. J Magn Reson Imaging 1993;3:671–5.

[133] Venkataramani R, Bresler Y. Perfect reconstruction formulas and bounds on aliasing error in sub-Nyquist nonuniform sampling of multiband signals. IEEE Trans Inf Theory 2000;46:2173–83.

[134] von Kienlin M, Mejia R. Spectral localization with optimal pointspread function. J Magn Reson 1991;94:268–87.

[135] Weaver JB, Xu Y, Healy DM, Cromwell LD. Filtering noise from images with wavelet transforms. Magn Reson Med 1991;21:288–95.

[136] Wood JC, Johnson KM. Wavelet packet denoising of magnetic resonance images: importance of Rician noise at low SNR. Magn Reson Med 1999;41:631–5.

[137] Xiang QS, Henkelman RM. *K*-space description for MR imaging of dynamic objects. Magn Reson Med 1993;29:422–8.

[138] Zhao B, Haldar JP, Christodoulou AG, Liang ZP. Image reconstruction from highly undersampled $(\mathbf{k}, t)$-space data with joint partial separability and sparsity constraints. IEEE Trans Med Imaging 2012;31:1809–20.

PART

2

Reconstruction of Undersampled MRI Data

CHAPTER

Parallel Imaging

6

Evan Cummings[a], Jacob A. Macdonald[b,c], and Nicole Seiberlich[a,c]

[a]*Department of Biomedical Engineering, University of Michigan, Ann Arbor, MI, United States*

[b]*Department of Radiology, Duke University, Durham, NC, United States*

[c]*Department of Radiology, University of Michigan, Ann Arbor, MI, United States*

6.1 Introduction

Parallel imaging is an image reconstruction technique that relies on spatial information from an array of receiver coils to reduce aliasing artifacts arising from undersampled k-space data. Parallel imaging techniques typically exploit regular undersampling of k-space data, such that aliasing artifacts can be predicted and resolved using knowledge of the sensitivity profiles of the receiver coils. However, as we will see in later in this chapter, these techniques can also be extended to other more general acquisition schemes.

Parallel imaging is typically deployed to increase the speed of MRI scans without compromising spatial resolution. As discussed in previous chapters, there are many advantages of moving to accelerated data collection in MRI. Because many images with varied contrasts and orientations must be collected, clinical MRI protocols often take between 30 min to an hour, which can be excessive for sick or uncooperative patients. In addition to patient-comfort concerns, longer scan times for individual images can also negatively impact image quality. A longer scan increases the probability of patient motion during data collection, which can lead to motion artifacts. Furthermore, reducing scan times allows MRI to be used in applications that require high temporal resolution, such as cardiac, perfusion, and interventional imaging. In applications that are not time-sensitive, parallel imaging can be used to increase image resolution within a given scan duration. Parallel imaging can also offer improvements in image quality in sequences such as TSE and EPI by reducing the amount of data that must be collected in a single TR [1].

While there are many different parallel imaging-reconstruction algorithms, four main methods are frequently referenced in modern literature: SENSitivity Encoding (SENSE) [2], GeneRalized Autocalibrating Partially Parallel Acquisitions (GRAPPA) [3], iTerative Self-consistent Parallel Imaging Reconstruction (SPIRiT) [4], and Eigenvalue SPIRiT (ESPIRiT) [5]. Each of these methods will be discussed in this chapter, as well as variants for 3D [6] [7] [8], non-Cartesian [9] [10] [11], and dynamic imaging [12] [13] [14] [15] [16]. By the end of this chapter, the reader will ideally have attained a good understanding of the basic principles of parallel imaging reconstruction and will be able to describe the theory, advantages, and limitations of several state-of-the-art parallel imaging-reconstruction algorithms.

Advances in Magnetic Resonance Technology and Applications, Volume 7, ISSN 2666-9099. https://doi.org/10.1016/B978-0-12-822726-8.00016-6

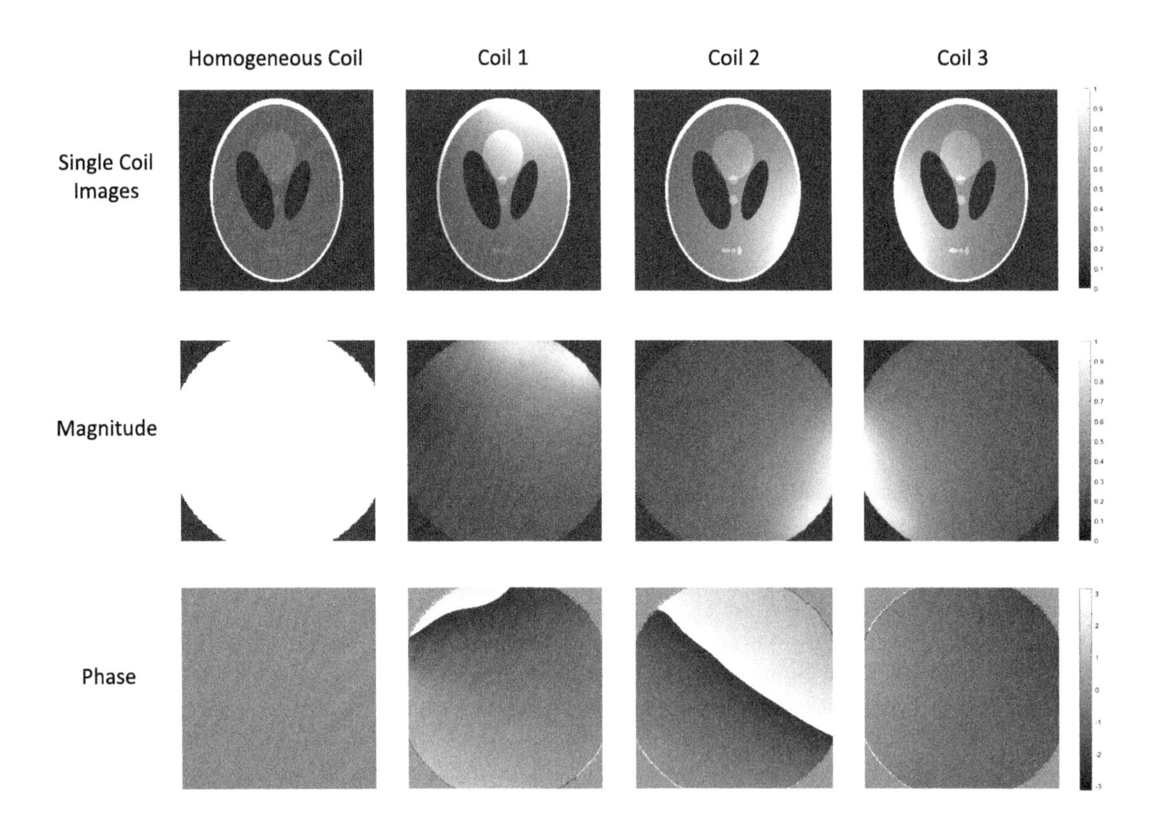

FIGURE 6.1

(*Top*) An image collected with various receiver coils, namely a coil with a homogeneous receive profile and no spatial variation in sensitivity (left), and three different elements of a receiver coil array placed at various locations around the object (center left, center right, and right). (Middle and bottom rows) The coil sensitivity profiles associated with these coils. When acquiring images with these coils, the localized sensitivity information provides an additional element of spatial localization that can complement gradient encoding. This effect can be seen in single coil images (top), where areas near to the coil appear brighter compared to those further away. Note that each coil sensitivity map is complex-valued and has a magnitude (center) and phase component (bottom). In the top row, lighter colors indicate that the coil is more sensitive to that portion of the object, and darker colors indicate both that there is lower sensitivity or that the object is not present (and thus there is no signal for the coil to be sensitive to).

Basic principles of parallel imaging

In MRI, data are collected in k-space using receiver coils. An array consisting of multiple receiver coils is often used to increase SNR, and an additional consequence of the use of such an array is that each individual coil is only sensitive to the portion of the body close to the coil. Fig. 6.1 shows an example of an image collected with a single homogeneous receiver coil, as well as images collected with several individual coils in a receiver array. Due to the local sensitivity patterns exhibited by the individual elements in the array, array coils can also provide some information for determining the spatial location of signals. In parallel imaging, this localized coil information is used to resolve aliasing artifacts present in undersampled images.

The sensitivity profile of a receiver coil can be described using a so-called "coil sensitivity map," as shown for the homogeneous receive coil and the individual elements of the array coil in Fig. 6.1. Coil sensitivity maps are complex-valued and are dependent on a variety of factors such as coil shape, placement, design, and even magnetic loading from the patient in the scanner. As a result, coil sensitivities are not constant and, in fact, can vary substantially from patient to patient. Parallel imaging techniques use a variety of methods for both estimating coil sensitivity information and applying it to reconstruct images.

The MR signal from a single receiver coil is given by the following equation:

$$S_j(\boldsymbol{k}) = \int_V C_j(\boldsymbol{r}) M_{xy}(\boldsymbol{r}) e^{(-2\pi i \boldsymbol{k}\cdot\boldsymbol{r})} d\boldsymbol{r} = \mathcal{F}\{C_j(\boldsymbol{r}) M_{xy}(\boldsymbol{r})\} \tag{6.1}$$

where $\boldsymbol{k}$ is the k-space position vector, $\boldsymbol{r}$ is the spatial position vector, $S_j(\boldsymbol{k})$ is the MRI signal acquired from coil j, $C_j(\boldsymbol{r})$ is the spatially varying coil sensitivity for coil j, and $M_{xy}(\boldsymbol{r})$ is the spatially varying magnetization. This equation can also be expressed in a discretized matrix form:

$$\boldsymbol{s} = \boldsymbol{BFCx} = \boldsymbol{Ex} \tag{6.2}$$

where $\boldsymbol{s}$ represents the MR signal, $\boldsymbol{B}$ represents the sampling operation, $\boldsymbol{F}$ represents the Fourier transform operation, $\boldsymbol{C}$ represents a matrix of discretized coil sensitivities, and $\boldsymbol{x}$ represents the magnetization at each point in the image. Frequently, the $\boldsymbol{BFC}$ operations are combined into the general MRI encoding matrix $\boldsymbol{E}$, which is used in many parallel imaging reconstruction methods.

When data are collected across multiple coils, the resulting single coil images each only contain part of the image, and they need to be combined in order to reconstruct the full image. There are a number of different methods for combining single coil images. Roemer, et al. [17] present an SNR-optimal recombination method in their original paper on multicoil acquisitions. Walsh, et al. [18] created a stochastic method for coil recombination for cases where coil sensitivity values are not exactly known. In cases where the application of these techniques is infeasible, the square root of the sum of squares of the single coil image values can be calculated pixel-wise to obtain an estimate for the true image.

There are a few terms and concepts that are common among all parallel imaging methods (and indeed most MRI reconstruction algorithms). Some of these concepts were already introduced in Chapter 2 and are briefly reviewed here for simplicity. The acceleration factor (Acc, also referred to here as R) describes the degree of undersampling for the acquired k-space data with respect to a fully-sampled k-space (as defined by the Nyquist–Shannon sampling theorem). In Cartesian acquisitions, undersampling is typically performed in the phase encoding direction since readout undersampling has minimal effect on image acquisition time. Therefore, when data are collected with an acceleration factor of R = 4, this usually means that three out of every four lines of k-space are skipped in the acquisition. A diagram illustrating how k-space data are typically collected for a few different acceleration factors is shown in Fig. 6.2. The result of increasing the distance between collected k-space lines is that the image field-of-view is decreased; when parts of the object are outside of the smaller field-of-view, these parts appear in the smaller field-of-view on top of the parts which are still within the field-of-view, a phenomenon known as "aliasing." For example, when k-space data are accelerated by R = 2, the field-of-view is decreased by a factor of two, and pixels which were originally half a field-of-view apart in

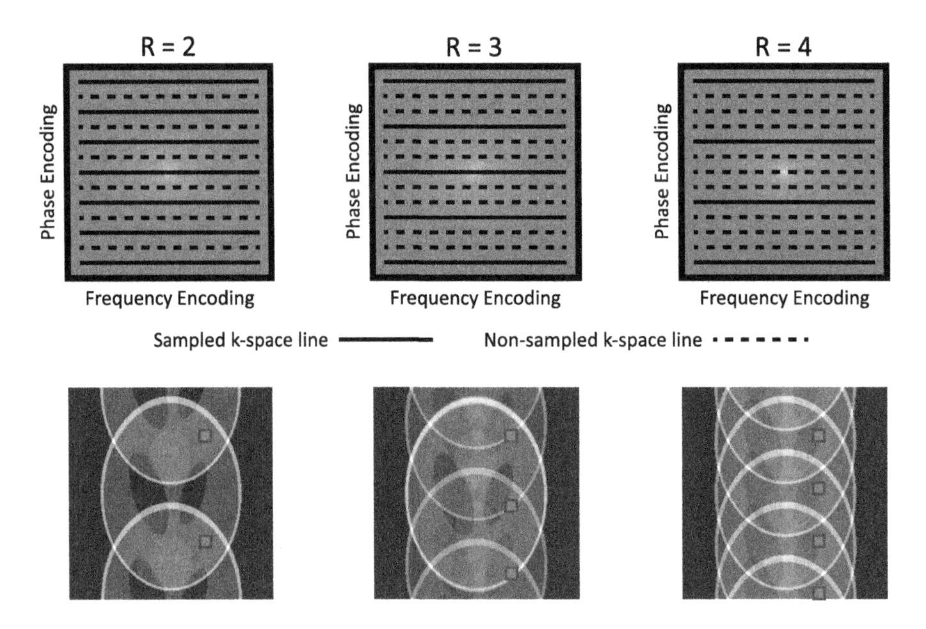

FIGURE 6.2

Sampling and aliasing patterns for various acceleration factors. **(top)** Uniform Cartesian sampling schemes for various acceleration factors. Note that the acceleration factor denotes the scan time reduction relative to a fully-sampled scan. **(bottom)** Aliasing patterns resulting from these undersampling schemes; the red (gray in print version) pixels in each image are examples of aliased points.

the unaccelerated image fall on top of one another (see Fig. 6.2, bottom left). The higher the acceleration factor, the larger number of pixels that fold together in the image, and the more challenging it can be to use the image for clinical or research purposes.

As a theoretical limit, the acceleration factor which can be resolved using parallel imaging is constrained by the number of receiver coils used to collect the data and their geometry with respect to the direction of undersampling. A coil array with completely independent coil sensitivity profiles with respect to the undersampling direction should in theory be able to reconstruct data with an acceleration factor equal to the coil count. However, it is rarely possible to resolve aliasing artifacts using parallel imaging when the acceleration factor is as high as the number of receiver coils in the array. The practical limits on the acceleration factor depend on the type of scan being performed and what tradeoffs can be made between image quality and temporal resolution. If there is sufficient SNR, parallel imaging can be used to accelerate any kind of clinical scan to improve patient compliance and comfort, even if there is no anticipated motion. In such a setting, modest acceleration factors of $R = 2$ are used to ensure high quality images. For specific applications, such as cardiac or perfusion scans, where both information concerning the motion of anatomical structures as well as temporal changes are needed clinically, acceleration factors of three or four are common. Some research scans can have acceleration factors as high as $R = 16$, depending on the size of coil arrays used in the scan. However, such high acceleration rates are only possible with special receiver coil arrays and with advanced reconstruction techniques that are not yet appropriate for widespread clinical adoption.

Another important concept in parallel imaging is how robustly the parallel imaging system can be solved, and thus how well the reconstructed image recapitulates the desired image. As explored later in this chapter, parallel imaging methods estimate missing data by setting up systems of equations and then solve these equations for the values of the unknown nonsampled points. While parallel imaging reconstruction algorithms vary in how these equations are set up and what they represent, all algorithms rely on additional calibration information, which is used to solve these systems of equations. The quality of the reconstructed images is highly dependent on the factor by which the system is overdetermined, which in turn depends on the quality and amount of calibration information available. Reconstructions with equations that are more overdetermined tend to be more robust and less sensitive to noise, but require more computational power to solve and a longer acquisition time to collect this information.

Finally, parallel imaging impacts the SNR of the resultant image [19]. The decrease in SNR can be calculated based on the following equation:

$$SNR_{PI} = \frac{SNR_0}{g\sqrt{R}}, \tag{6.3}$$

where SNR_{PI} is the SNR of the parallel imaging reconstruction, SNR_0 is the SNR of an equivalent fully-sampled image, R is the acceleration factor, and g is the coil geometry factor, or g-factor. The g-factor is dependent on both the aliasing patterns in the accelerated images and the coil sensitivity profiles. The g-factor is (almost) always greater than or equal to one as parallel imaging does not usually improve the SNR of an image. The g-factor is spatially dependent and can be represented as a g-factor map. Explicit knowledge of the g-factor is not required for parallel imaging reconstructions, but it is frequently used to analyze the performance of a parallel imaging technique.

6.2 Fundamental techniques

SENSE and GRAPPA are the most commonly used parallel imaging techniques due to their robustness in a wide range of imaging scenarios. SENSE is an image-domain technique that uses explicit knowledge of the coil sensitivity maps to solve for unaliased pixel values. GRAPPA is a k-space based technique that extracts coil information from a fully-sampled calibration space and reconstructs missing points in k-space. They also form the theoretical framework for other approaches such as CG-SENSE, SPIRiT, and ESPIRiT, and are the basis for a variety of techniques for dynamic data sets, such as k-t SENSE, TGRAPPA, k-t GRAPPA, and PEAK-GRAPPA, all of which are discussed later in this chapter.

Sensitivity encoding (SENSE)

Siemens: mSENSE; GE: ASSET; Phillips: SENSE

SENSE [2] is one of the earliest proposed and most intuitive parallel imaging techniques. SENSE operates in the image domain and uses the principle that each point in an aliased image is the sum of several values in the unaliased image weighted by the local coil sensitivity at each point. This principle is shown in Fig. 6.3. If the coil sensitivity maps are known and enough independent coils are used for data collection, a system of equations can be solved to find the unknown image values.

A simple implementation of the SENSE algorithm for uniformly undersampled Cartesian data considers a set of equations for each point in the aliased image and for each receiver coil used for acquisition. For example, for an acceleration factor of $R = 3$, and data collected with three coils ($N_c = 3$),

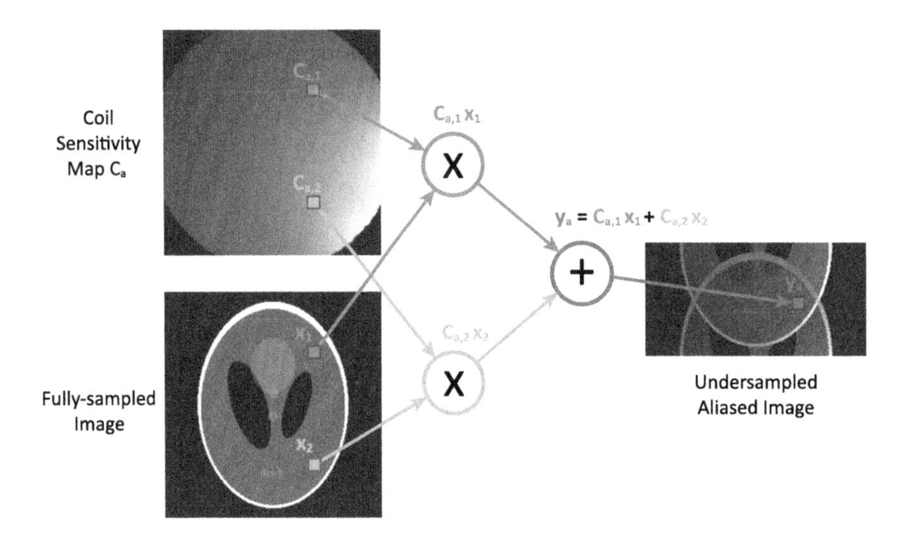

FIGURE 6.3

Visual representation of the SENSE equation for R = 2. The aliased image is the sum of two image pixels, weighted by the appropriate coil sensitivity at each point. This equation is written for each aliased single coil image to define a system of equations that is solved to reconstruct both aliased pixels.

the matrix equations for each pixel would be:

$$\begin{bmatrix} y_a \\ y_b \\ y_c \end{bmatrix} = \begin{bmatrix} c_{a1} & c_{a2} & c_{a3} \\ c_{b1} & c_{b2} & c_{b3} \\ c_{c1} & c_{c2} & c_{c3} \end{bmatrix} \begin{bmatrix} x_1 \\ x_2 \\ x_3 \end{bmatrix} \tag{6.4}$$

Here, y_a denotes the aliased pixel in coil a; y_b and y_c the aliased pixels in coils b and c; x_1, x_2, x_3 are the pixels in the unaliased image to be recovered; and the c elements represent the coil sensitivity value for the coil indicated by the letter subscript at the location indicated with the numerical subscript. This implementation is shown visually in Fig. 6.4. This equation can be written more compactly in matrix form:

$$\boldsymbol{y} = \boldsymbol{C}_p \boldsymbol{x} \tag{6.5}$$

where $\boldsymbol{y}$ is a vector of length N_c containing the aliased pixels for each coil, $\boldsymbol{C}_p$ is an $N_c \times R$ matrix containing the coil sensitivity values for this set of pixels, and $\boldsymbol{x}$ is the unknown but desired vector of R unaliased image pixels. As the values of the aliased pixels are known and the values of the coil sensitivity maps can be determined as subsequently described, the goal is to solve for the R unknown image pixels in $\boldsymbol{x}$. This matrix equation can be solved by applying the inverse of the coil sensitivity encoding matrix $\boldsymbol{C}$ to both sides of the equation:

$$\hat{\boldsymbol{x}} = \boldsymbol{C}_p^{\dagger} \boldsymbol{y}, \tag{6.6}$$

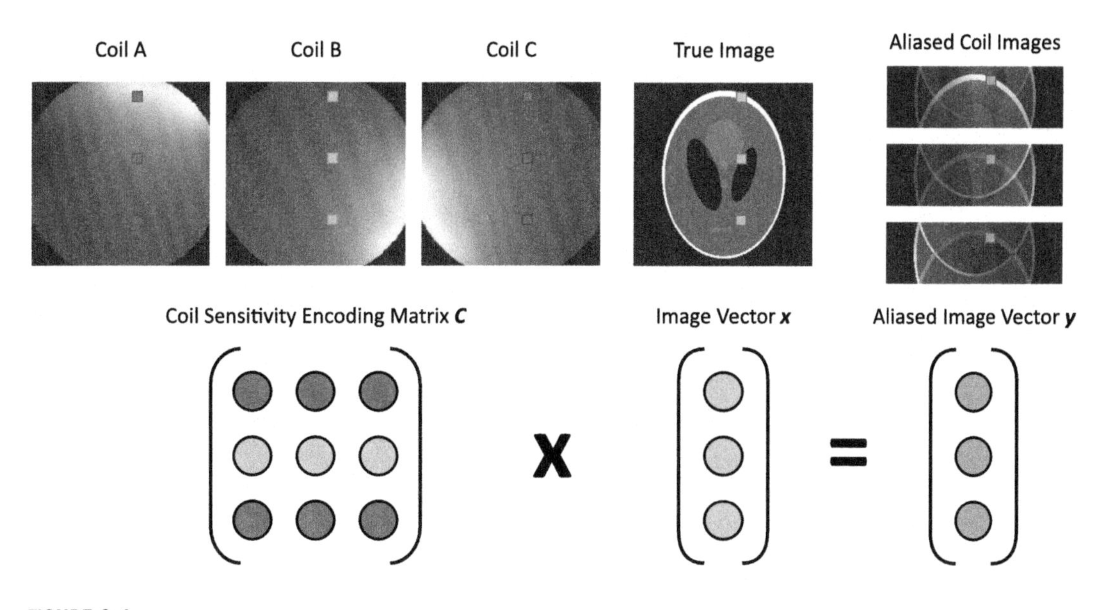

FIGURE 6.4

Matrix representation of the SENSE reconstruction. Coil sensitivity values are placed as row vectors in the coil sensitivity encoding matrix $\boldsymbol{C}$. The true image vector $\boldsymbol{x}$ is related to the pixels in the aliased images $\mathbf{y}$ through the application of this sensitivity encoding matrix, as can be seen in Eq. (6.2).

where the † symbol indicates the matrix pseudoinverse. By solving this equation, a single set of aliased points can be unfolded and placed in the appropriate locations (1, 2, and 3, as previously noted). A similar system of equations must be solved for each set of aliased pixels in order to reconstruct the full image.

At higher acceleration factors, more pixels are aliased together, and thus the length of the vector of the unknown original pixel values, $\boldsymbol{x}$, increases, and additional columns are added to the coil sensitivity matrix. Theoretically, for the SENSE equations to be solvable, the acceleration factor can be at most equal to the number of coils. In practice, it is usually much smaller given the fact that coils are not perfectly independent (increasing the condition number of $\boldsymbol{C}_p$ and thus leading to noise amplification when solving the inversion), and image SNR is reduced for higher acceleration factors. The need to solve for more unknown quantities in each system of equations makes each system less overdetermined, which can negatively impact the quality of the reconstruction. If more (independent) coil elements are used to collect data, more rows are added to the system, making it more overdetermined and robust to noise in coil sensitivity measurements.

As just described, the crucial information required for SENSE reconstruction are maps of the coil sensitivity profiles. A simple method for estimating coil maps is described in the original SENSE paper [2]. Images of the same object can be collected using both the body coil (with a homogeneous sensitivity profile) and the receiver array to be used for parallel imaging. The images collected with the array coils are then divided by the body coil image, yielding a rough estimate for the coil sensitivities at each pixel. These coil sensitivity maps can be thresholded to focus on the area of interest and denoised to reduce

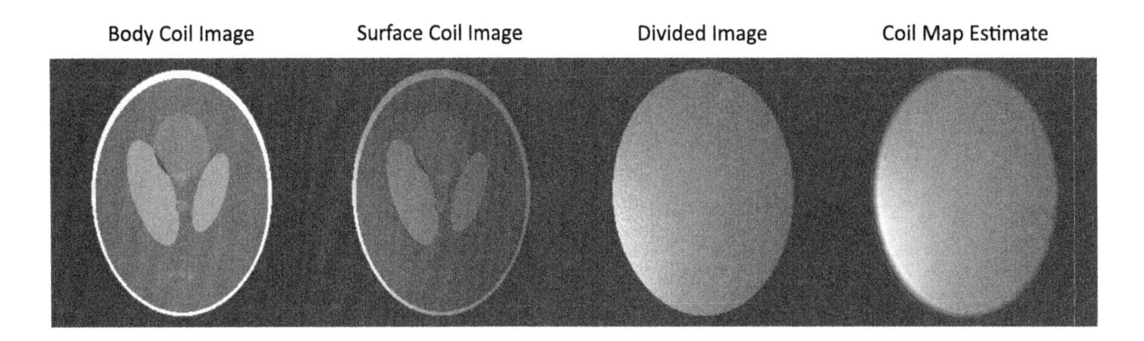

FIGURE 6.5

Example process for calculating coil sensitivity maps. An image collected using an element of the array coil is divided pixel-wise by an image collected using the homogeneous body coil. This produces a noisy estimate of the coil sensitivity map for that array coil element. This initial estimate can be filtered with a 2D polynomial smoothing function to yield a smoother estimate for the coil sensitivity.

noise or other spurious signal that could negatively impact the SENSE reconstruction. This method for estimating coil maps is shown in Fig. 6.5. Other methods for estimating coil maps include adaptive reconstruction [18] and ESPIRiT [5].

As described in Eq. (6.1), the amount of noise present in the reconstruction is related to the g-factor. For SENSE reconstructions, the g-factor can be calculated pixel-wise over the aliased image using the following equation:

$$g_i = \sqrt{[\left(C_p^{\mathrm{H}} C_p\right)^{-1}]_{i,i}[\left(C_p^{\mathrm{H}} C_p\right)]_{i,i}}, \tag{6.7}$$

where ^{H} indicates the matrix conjugate transpose, $^{-1}$ indicates the matrix inverse, and i is the index of the pixel of the original image that aliases onto the point in the aliased image. Note that the i, i subscripts indicate the ith diagonal element of the $C_p^{\mathrm{H}} C_p$ or $(C_p^{\mathrm{H}} C_p)^{-1}$ matrices. A sample g-factor map for a SENSE reconstruction is shown in Fig. 6.6.

The original SENSE implementation also includes a receiver noise covariance matrix term that describes noise levels for a given pixel and how the noise is correlated across coils. The reconstruction equation (Eq. (6.6)) can be restated to include these terms as well:

$$\hat{x} = (C_p^{\mathrm{H}} \Psi^{-1} C_p)^{-1} C_p^{\mathrm{H}} \Psi^{-1} y, \tag{6.8}$$

where Ψ is the receiver noise covariance matrix. In the case where the noise is independent and constant across coils, this matrix is the identity matrix, and this equation simplifies to Eq. (6.4). Including the Ψ term effectively weights the reconstruction matrix so that coils with a higher SNR have more influence on the reconstructed image vector [20], which can reduce errors due to low signal coils.

The SENSE reconstruction method has a few drawbacks. Most notably, it requires accurate coil sensitivity maps to work properly. Inaccuracies, misregistration, or noise in the coil maps can propagate into the reconstructed image, causing errors in reconstruction. In addition, subject motion between

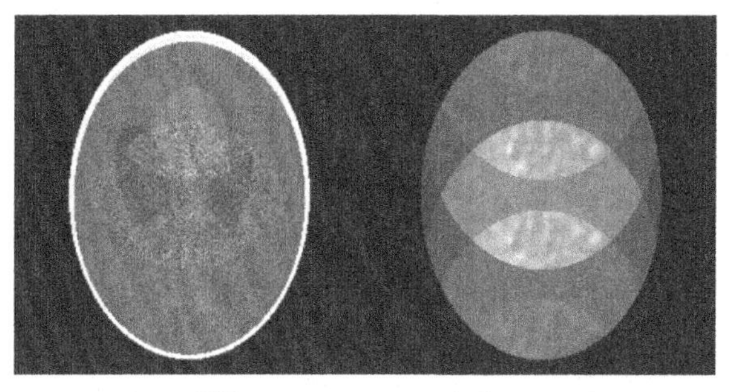

FIGURE 6.6

A sample reconstructed image (left) and g-factor map (right) from a SENSE reconstruction with R = 4. Note the noisy regions in the reconstructed image that correspond to regions with high g-factor values.

coil sensitivity estimation and data acquisition can further corrupt reconstruction quality. The straightforward SENSE implementation described here is also limited to uniformly sampled Cartesian data. The more generalized CG-SENSE algorithm (described later in this chapter) is required to reconstruct images with more complex sampling schemes.

SENSE and related techniques have been used for a wide variety of clinical applications. For example, in the heart, it has been used for cine imaging [21] [22] and myocardial perfusion [23]. In the brain, it has been used to accelerate fMRI scans [24] [25]. SENSE has also been applied in abdominal imaging for renal perfusion scans [26] and abdominal angiography [27], among many other applications.

Generalized autocalibrating partially parallel acquisitions (GRAPPA)

Siemens: GRAPPA; GE: ARC

Unlike SENSE, which unfolds aliased pixels in the image domain, the GRAPPA algorithm is a k-space-based technique. In GRAPPA, the unsampled k-space points are estimated by exploiting relationships between neighboring points [3]. GRAPPA requires a small fully sampled portion of the dataset for calibration (often referred to as the Autocalibration Signal, or ACS). From the ACS data, a GRAPPA weight set can be calculated to capture the relationship between neighboring k-space points. This weight set is then applied to the remaining undersampled data in order to estimate the missing k-space points. Finally, a Fourier transform is used to generate the unaliased reconstructed image from the synthesized k-space data.

The theory behind GRAPPA can be best understood by reexamining the idea of SENSE in k-space. As just described, the individual single coil images acquired with MRI can be imagined as the true image multiplied element-wise by a coil sensitivity map, as shown in Eq. (6.2). In k-space, this process corresponds mathematically to convolving the Fourier transform of the coil sensitivities with the Fourier transform of the image:

$$\mathcal{F}\left\{C(\boldsymbol{r})\,M_{xy}(\boldsymbol{r})\right\} = \mathcal{F}\{C(\boldsymbol{r})\} * \mathcal{F}\left\{M_{xy}(\boldsymbol{r})\right\}. \tag{6.9}$$

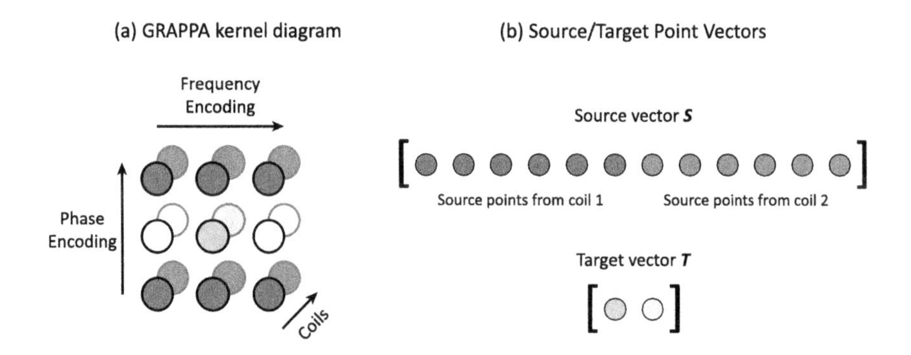

FIGURE 6.7

(a) A sample GRAPPA kernel, with source points (blue (dark gray in print version)) and target points (orange (light gray in print version)), on a data set with an acceleration factor of R = 2. This is a 3 × 2 kernel, referring to the number of source points in the kernel that extend over three readout points and two phase-encoding lines. **(b)** Each set of source and target points from a kernel forms one row vector of data. Row vectors from multiple kernels can be concatenated vertically to form the source and target matrices for reconstruction.

The result of this convolution is that each point of the measured k-space signal contains some information about the missing k-space points; in other words, there is a mathematical relationship between neighboring points in k-space due to the inhomogeneous coil sensitivities. In any small neighborhood of points, these relationships contain mostly structural information about the image. However, when looking at many groupings of local points, the local structural information averages out, and the convolution pattern due to the coil sensitivity can be elucidated.

At the core of the GRAPPA implementation is the GRAPPA kernel, a recurring pattern of neighboring points in k-space. The kernel is made up of source and target points; in an undersampled dataset, the source points are collected, but the target points are unknown and must be reconstructed. A diagram of a typical kernel made up of several source points for each target point is shown in Fig. 6.7; datapoints along the fully sampled phase-encoding lines above and below the target point are often used as the source points. GRAPPA kernels also extend in the coil dimension because data from every coil is used to derive GRAPPA weights and reconstruct target points across all coils.

The mathematical relationship between points in a kernel is captured in the GRAPPA weight set. For kernels in the fully sampled ACS regions, the source points and target points are known, allowing the relationship between these points to be derived. When data are collected along a standard uniform Cartesian trajectory, the ACS data typically take the form of a few readout lines near the center of k-space, and the rest of k-space is undersampled. A diagram of this acquisition scheme with an acceleration factor of R = 2 is shown in Fig. 6.8. The center of k-space is often used as the ACS since this region contains high signal data, reducing the influence of noise on the GRAPPA weights. To determine the GRAPPA weights, source points and target points are collected for each occurrence of the kernel in the ACS data. The number of kernel repetitions in an ACS is given by the following equation:

$$(N_R - s_R + 1)(N_{PE} - R(s_{PE} - 1)), \tag{6.10}$$

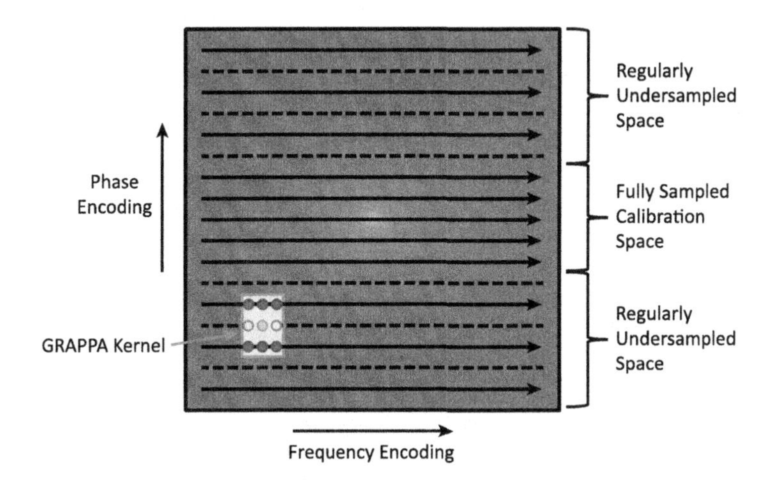

FIGURE 6.8

Sample k-space acquisition scheme and GRAPPA kernel for Cartesian GRAPPA.

where N_R and N_{PE} are the size of the ACS in frequency- and phase-encoding directions, and s_R and s_{PE} are the size of the GRAPPA kernel. For an ACS that is made up of 256 readout points and 9 phase-encoding lines, an acceleration factor of R = 2, and a kernel of size 3 × 2, there are 1778 repetitions of the kernel. These kernel repetitions are rearranged into source and target matrices of size (# of kernel repetitions in ACS) × (# of source or target points in kernel × number of coils). The relationship between the source and target matrices is expressed by the following equation:

$$\boldsymbol{T}_{ACS} = \boldsymbol{S}_{ACS}\boldsymbol{W}, \tag{6.11}$$

where $\boldsymbol{T}$ is the target point matrix, $\boldsymbol{S}$ is the source point matrix, and $\boldsymbol{W}$ is the GRAPPA weight set. In order to solve for the weight set, the pseudoinverse of the source point matrix is applied to both sides of the equation, yielding:

$$\boldsymbol{W} = \boldsymbol{S}^{\dagger}_{ACS}\boldsymbol{T}_{ACS}. \tag{6.12}$$

Once the GRAPPA weights have been determined, the target points in the undersampled k-space data can be recovered by applying these weights to source point matrices constructed using the same kernel structure:

$$\boldsymbol{T} = \boldsymbol{S}\boldsymbol{W}. \tag{6.13}$$

The amount of ACS data required for the GRAPPA reconstruction is related to the size of the GRAPPA kernel and number of coils. The system in Eq. (6.8) must be overdetermined for an accurate reconstruction, and so the number of kernel repetitions in the ACS must be greater than the number of source points in a kernel across all coils.

In addition, there are a few different options for collecting ACS data. First, the contrast of the ACS data does not have to match the rest of the scan, which can be beneficial for sequences with long repetition times. In addition, ACS data can be collected either during the regular scan, or as a separate scan either before or after undersampled data acquisition.

The main advantage of GRAPPA is that it is relatively robust in a variety of clinical imaging scenarios. One of the main reasons for this is that the GRAPPA weights are determined using many ACS kernels, making the system more robust against poor data in any specific kernel and ultimately allowing for high-quality reconstructions in nonideal conditions [1] [28]. GRAPPA also does not require coil sensitivity maps to be explicitly formulated, which can also be a source of reconstruction error [29].

GRAPPA has been used clinically and in a wide variety of research applications, including cardiac cine imaging [22], myocardial perfusion [30], interventional catheter placement [31], spinal TSE imaging [32], and MR angiography [33].

6.3 Advanced techniques

In addition to SENSE and GRAPPA, there are many more advanced parallel imaging methods which work along the same basic principles but which can improve the robustness of the reconstructions or be used to accelerate Cartesian and non-Cartesian data acquisitions with more general sampling patterns. These techniques pose the reconstruction problem within an optimization framework, which can be solved using iterative methods, such as the conjugate gradient algorithm described in Chapter 3, and can also incorporate other regularization options as desired to improve image quality.

Conjugate gradient SENSE (CG-SENSE)

Conjugate Gradient SENSE [9] is a generalization of the SENSE algorithm designed to work with data collected along nonuniform Cartesian and non-Cartesian trajectories. When Cartesian data are undersampled with uniform or regular sampling (e.g., every two lines in k-space), the aliasing artifacts are predictable and can be easily modeled, a feature exploited by SENSE to unfold small groups of aliased pixels. With nonuniform Cartesian (e.g., random Cartesian undersampling) and non-Cartesian (e.g., radial/spiral) sampling, the aliasing patterns are more complex; each pixel can affect every other pixel in the resulting aliased image. Solving this problem using the SENSE approach described before would require a very large coil sensitivity encoding matrix of size (total number of k-space signals) × (image matrix size). Due to its size (and typically ill-conditioned nature), the pseudoinverse of this matrix cannot be easily calculated.

To solve this problem and enable SENSE to be used to reconstruct images collected with undersampled nonuniform Cartesian and non-Cartesian data, we begin with the discretized signal model from Eq. (6.2):

$$\boldsymbol{E}\boldsymbol{x} = \boldsymbol{s}. \tag{6.14}$$

The generalized encoding matrix $\boldsymbol{E}$ corresponds to a multichannel version of the encoding matrix described for a single coil in Chapter 2. For the specific case of non-Cartesian data, $\boldsymbol{E}$ corresponds to a multichannel version of the nonuniform fast Fourier transform (NUFFT) operations: coil multiplication, Fourier transform, and nonuniform sampling. To reconstruct the image, the pseudoinverse of the encoding matrix is applied to the k-space data:

$$\hat{\boldsymbol{x}} = (\boldsymbol{E}^{\mathrm{H}}\boldsymbol{E})^{-1}\boldsymbol{E}^{\mathrm{H}}\boldsymbol{s}, \tag{6.15}$$

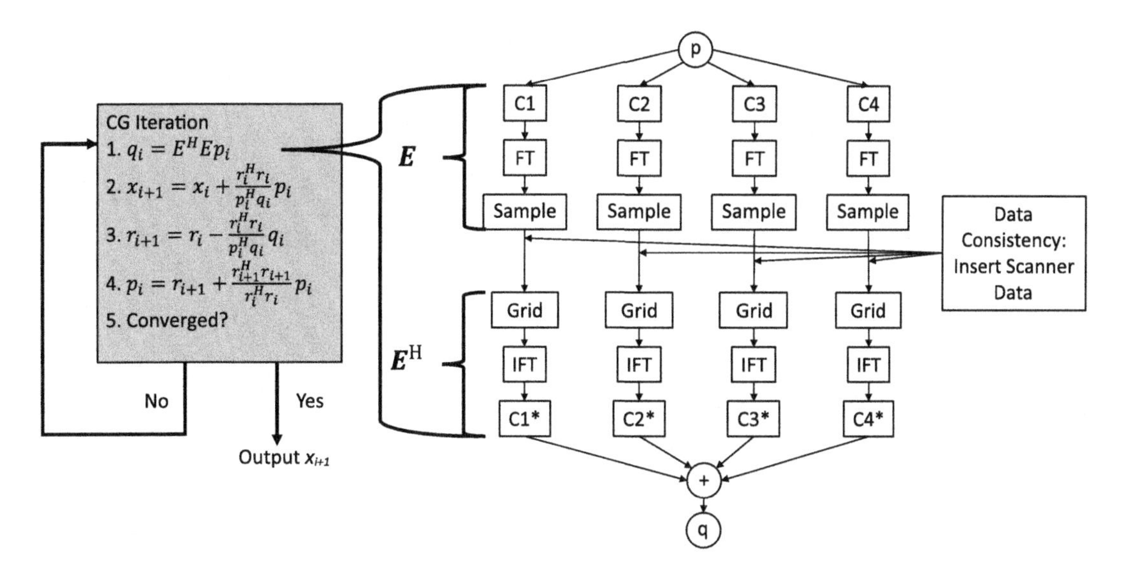

FIGURE 6.9

Diagram of the CG-SENSE algorithm. At the core of this algorithm is the conjugate gradient (CG) algorithm (blue (mid gray in print version) box). The $\boldsymbol{E}^H\boldsymbol{E}$ operation is shown in detail and is broken down into coil multiplication, Fourier transform, re-sampling, data insertion, gridding (in the case of non-Cartesian sampling), inverse Fourier transformation, and coil combination steps.

where $\hat{x}$ is the reconstructed image, $\boldsymbol{s}$ is the k-space signal data, ^{H} indicates the matrix conjugate transpose, and $^{-1}$ indicates the matrix inverse. Since $\mathbf{E}$ is a very large matrix, inverting the matrix $\boldsymbol{E}^H\boldsymbol{E}$ is computationally intensive. However, this matrix inversion can be eliminated by multiplying both sides of the equation by $\boldsymbol{E}^H\boldsymbol{E}$:

$$\boldsymbol{E}^{\mathrm{H}}\boldsymbol{E}\hat{x} = \boldsymbol{E}^{\mathrm{H}}\boldsymbol{s}. \tag{6.16}$$

CG-SENSE uses the conjugate gradient algorithm to iteratively solve this equation. The implementation of the conjugate gradient algorithm for CG-SENSE in shown in Fig. 6.9. CG-SENSE slightly deviates from the standard CG implementation discussed in Chapter 3 by including a data consistency step, where data collected at the MRI scanner are inserted into the points that were originally sampled to ensure that the final solution image is consistent with the non-Cartesian data acquired.

While CG-SENSE is more computationally efficient than inverting the generalized sensitivity-encoding matrix, it still requires multiple NUFFT operations in the case of non-Cartesian trajectories that are relatively computationally intense. In addition, like SENSE, CG-SENSE requires accurate coil sensitivity maps. Errors in these coil sensitivity maps, density compensation required for the NUFFT, or intensity weighting can become exaggerated after iterating through these operations multiple times.

Another benefit of the CG-SENSE algorithm is that regularization terms can easily be incorporated within the CG iterations, thereby enforcing desirable qualities over the entire reconstructed image or k-space signal. Some examples include using Tikhonov regularization to reduce noise [34] or using L1-regularization of the wavelet transform of the image for simultaneous compressed sensing recon-

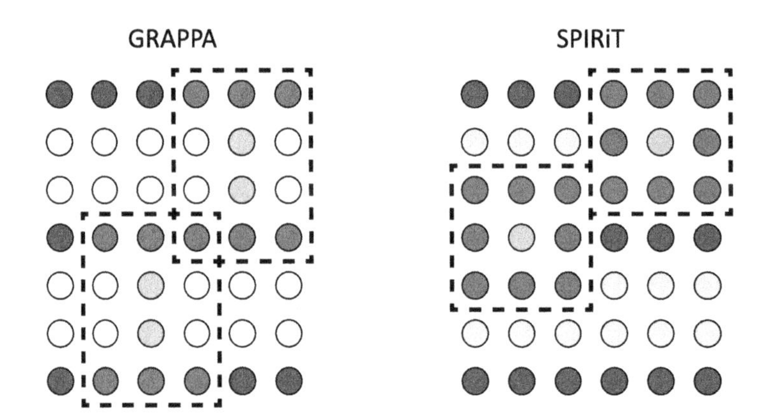

FIGURE 6.10

Differences between GRAPPA kernels and weight applications between GRAPPA and SPIRiT, depicted for a single coil. Source points are shown in blue (mid gray in print version), target points in orange (gray in print version), sampled points in dark gray, missing points in white, and estimated points in light gray. **(left)** In GRAPPA, weight sets are enforced only from sampled to nonsampled lines. **(right)** In SPIRiT, these weights are enforced over every patch of k-space. This allows for iterative improvement in the estimated values for the nonsampled points.

struction [35]. This property also makes CG-SENSE compatible with other iterative reconstruction techniques including compressed sensing, which will be introduced in Chapter 8.

As randomly/nonuniformly sampled Cartesian and non-Cartesian scans are generally not used clinically, CG-SENSE is mostly limited to research applications. However, it was the first major non-Cartesian parallel imaging technique, and it has been studied in cardiac [36] and fMRI [37] applications.

Iterative self-consistent parallel imaging reconstruction (SPIRiT)

In a mathematical sense, SPIRiT [4] is a generalization of the GRAPPA algorithm that considers GRAPPA as an iterative optimization problem. Much like GRAPPA, SPIRiT requires a calibration data set to determine reconstruction weights for a specific GRAPPA kernel that are then applied to recover uncollected k-space points in an undersampled data set.

The main idea behind SPIRiT is that the numerical relationships within a kernel should be satisfied over all kernels found in the fully sampled k-space data. This contrasts with the original GRAPPA algorithm, which only enforces the relationships captured in the GRAPPA weights to be satisfied between the originally sampled and non-sampled points. This principle is shown visually in Fig. 6.10. In SPIRiT, the GRAPPA weight operator, $\boldsymbol{G}$, is a convolution operation that is applied over the entire k-space data to enforce the relationships expected given the coil sensitivities. This operation can be expressed as

$$\boldsymbol{x} = \boldsymbol{G}\boldsymbol{x}. \tag{6.17}$$

Intuitively, this equation can be understood as stating that any point in k-space can be described as some combination of nearby k-space points using the GRAPPA operation. To reach this ideal solution and approximate the missing k-space values from an undersampled data set, this equation is posed as an optimization problem:

$$\operatorname{argmin}_{x} \|(\boldsymbol{G}-\boldsymbol{I})\boldsymbol{x}\|_2^2, \tag{6.18}$$

where $\boldsymbol{I}$ is the identity matrix, and $\|\cdot\|_2$ indicates the L2-norm. This formulation ensures that the GRAPPA relationships are enforced over each local area of k-space.

While this approach could be applied to reconstruct accelerated data, it lacks a term to enforce data consistency and could converge to a solution that does not reflect the original data. For instance, $\boldsymbol{x}=0$ would satisfy the equation regardless of the GRAPPA operator deployed. A regularization term based on data consistency is thus included within the optimization structure

$$\operatorname{argmin}_{x} \|(\boldsymbol{G}-\boldsymbol{I})\boldsymbol{x}\|_2^2 + \lambda \|\boldsymbol{B}\boldsymbol{x}-\boldsymbol{y}\|_2^2, \tag{6.19}$$

where λ is a regularization parameter, $\boldsymbol{B}$ is a linear sampling operator that represents selection of the sampled k-space points, and $\boldsymbol{y}$ is the originally sampled data. In Cartesian sampling, $\boldsymbol{B}$ is a sampling mask matrix. SPIRiT can be extended to non-Cartesian datasets simply by using a gridding operation as the $\boldsymbol{B}$ matrix. Note that $\boldsymbol{x}$ is a Cartesian k-space data matrix (even for non-Cartesian datasets), and $\boldsymbol{B}$ does not typically include any Fourier transform operation. Additional regularization terms can be added to this objective function as desired. This problem can then be solved using any appropriate optimization algorithm, such as the conjugate gradient (CG) algorithm. More information on optimization algorithms can be found in Chapter 3.

The iterative nature of SPIRiT confers both benefits and drawbacks. Like CG-SENSE, regularization can easily be applied to SPIRiT reconstructions. For example, by incorporating L1-regularization to the wavelet transform of the image, the problem can be converted to a hybrid parallel imaging/compressed sensing reconstruction approach [38]. While iterative techniques tend to require more processing power compared to direct techniques such as SENSE and GRAPPA, the SPIRiT algorithm has been implemented with parallel CPU and GPU resources to enable rapid reconstruction on modern computers [38].

SPIRiT is not broadly used in clinical applications, but it has been applied in many types of research scans. For example, SPIRiT has been investigated for acceleration of pediatric scans, including cardiac, abdominal, and knee scans [39]. SPIRiT has also been investigated for potential applications in cardiac cine imaging [40].

Eigenvalue iterative self-consistent parallel imaging reconstruction (ESPIRiT)

The driving motivation behind the ESPIRiT algorithm is to balance the fundamental tradeoffs between k-space-based parallel imaging techniques such as GRAPPA or SPIRiT and coil sensitivity-based techniques such as SENSE. Autocalibrating techniques (like GRAPPA and SPIRiT) are typically more robust to noise and errors in sensitivity maps. However, sensitivity-based methods can have much better results in ideal circumstances and have other advantages, such as straightforward incorporation of regularization. ESPIRiT aims to combine the strengths of both approaches by using autocalibration data to robustly estimate coil maps for a coil-sensitivity-based approach, with improved reconstructions even in the presence of noise.

To achieve this goal, two new techniques are introduced within the SPIRiT framework as ESPIRiT [5]: (1) a novel method of robustly estimating coil sensitivity maps based on SPIRiT calibration data and (2) a variation of the SENSE algorithm that can utilize these eigenvector coil maps.

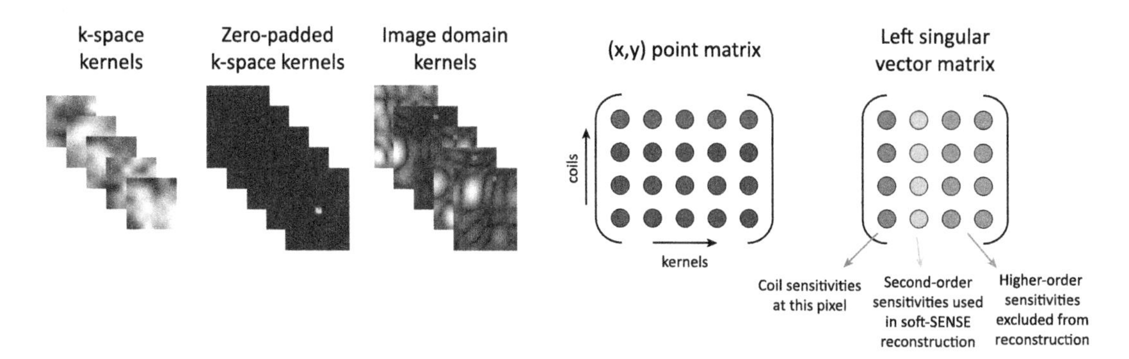

FIGURE 6.11

Diagram of coil estimation process using the ESPIRiT algorithm. First, a set of SPIRiT kernels are extracted from the k-space calibration data. These kernels are zero-padded to the desired coil map resolution and transformed into the image domain. Each image pixel has an associated matrix of size $N_c \times N_k$. The SVD of this matrix is taken, and the left singular vector matrix contains information about coil sensitivities. The first singular vector contains the primary coil sensitivity map values for that pixel, and the other singular vectors can contain higher-order coil sensitivity map values.

Like SPIRiT, ESPIRiT requires a fully sampled calibration dataset, from which a set of kernels is extracted in order to calculate robust coil sensitivity maps. ESPIRiT kernels are different from GRAPPA or SPIRiT kernels in that source and target points are not differentiated from one another. The entire k-space kernel is extracted from the calibration data and placed into a 4D calibration matrix of size $h_k \times w_k \times N_c \times N_k$, where h_k is the kernel height, w_k is the kernel width, N_c is the number of coils, and N_k is the number of kernels. Next, the calibration matrix is normalized, and the kernels are compressed using Principal Component Analysis (PCA) to reduce the number of kernels that must be stored in memory. The kernels are then zero-padded to the size of the image, and a 2D Fourier Transform is performed along the first two dimensions of the calibration matrix, resulting in a set of low-resolution kernel images. For each spatial coordinate (x, y), a corresponding matrix of size $N_c \times N_k$ with the information for each kernel and coil at that point is extracted. The singular value decomposition (SVD) of this matrix is calculated, resulting in a $N_c \times N_c$ left singular vector matrix and a set of N_c singular values. This process is shown in Fig. 6.11. This SVD can also be thought of as an eigendecomposition of a matrix associated with each pixel multiplied by its transpose, which is how the ESPIRiT algorithm derives its name. In this case, the eigenvector matrix is equivalent to the left singular vector matrix, and the eigenvalues are equivalent to the squared singular vectors. After PCA, the greatest variation in the signal along the coil dimension should be due to differences in coil sensitivities, and the most significant singular vectors should consequently contain the coil sensitivity values for that pixel. The most significant singular vectors have a corresponding singular value of 1 since all singular values are constrained to be less than or equal to 1. To create coil sensitivity maps, the first singular vectors are extracted for each pixel position and combined into a sensitivity map matrix of size $N_x \times N_y \times N_c$, where N_x and N_y are the image matrix dimensions. If a pixel's first eigenvalue is less than one, the zero vector is used instead of the first singular vector as it is assumed that the first singular vector is not related to coil sensitivities for this pixel. The resulting coil sensitivity maps can then be used in SENSE or CG-SENSE to reconstruct unaliased images.

In some cases, pixels can have multiple singular values equal to (or nearly equal to) one and multiple sets of associated coil sensitivity values. A second set of coil maps can be calculated in a process similar to that used for the first coil maps: The second singular vectors are extracted and combined into a map matrix, and pixels with a second eigenvalue less than one are set as the zero vector. This process can be repeated up to N_c times. There is value in including higher-order coil sensitivities during the reconstruction step as they can contain useful information missing in the first-order sensitivities. For example, if part of the imaged object lies partially outside the image field of view, the second-order map contains sensitivities for the aliased region. Neglecting to include these values can cause errors in the SENSE reconstruction. Two sets of coil maps are typically used, allowing for better reconstructions over traditional SENSE without requiring significantly additional processing time.

Like SPIRiT, ESPIRiT considers the reconstruction as an optimization problem. To perform an ESPIRiT reconstruction, an objective function incorporating multiple sets of coil maps is created:

$$\operatorname{argmin}_{\boldsymbol{x}} \sum_{i=1}^{N_c} \left\| \boldsymbol{B}\boldsymbol{F} \sum_{j=1}^{N_m} \boldsymbol{C}_{i,j}\boldsymbol{x} - \boldsymbol{s}_i \right\|_2^2, \tag{6.20}$$

where $\boldsymbol{s}_i$ is the sampled signal at the ith coil, $\boldsymbol{B}$ is the sampling operator, $\boldsymbol{F}$ is the Fourier transform operation, N_m is the number of coil sensitivity map sets used for reconstruction, $\boldsymbol{C}_{i,j}$ is the ith coil from the jth coil set, and $\boldsymbol{x}$ is the image. In the case in which there is only one set of coil sensitivity maps ($N_m = 1$), this becomes equivalent to the CG-SENSE algorithm or the traditional SENSE algorithm in the Cartesian case.

Like SPIRiT, ESPIRiT is not yet used routinely for clinical scans. The ESPIRiT coil map-estimation process [41] has been adopted for some research applications, and recent investigations have also used the ESPIRiT algorithm for cine imaging [42].

6.4 3D volumetric parallel imaging

While the techniques discussed so far have been focused on the reconstruction of 2D images, parallel imaging can also be used to effectively accelerate the acquisition of 3D volumetric scans. For such scans, the typical convention is to accelerate in both the phase-encoding and partition-encoding (second phase encoding) directions. By accelerating the acquisition along two dimensions, higher acceleration factors can be deployed due to the orthogonality of the coil-encoding capabilities in various directions. As with 2D applications, undersampling in the frequency encoding (readout) direction yields minimal time savings.

In the volumetric extension of SENSE (called 2D SENSE since undersampling occurs in two directions) [6], the SENSE equation (Eq. (6.6)) is still applicable. For example, a dataset undersampled by $R = 2$ in both the phase-encoding and partition-encoding directions has a total acceleration factor of $R = 4$. As a result of this sampling scheme, four pixels are aliased into one single pixel in the aliased image; two replicas are seen in the phase-encoding direction and two in the partition-encoding direction. The effects of this undersampling scheme are shown in Fig. 6.12. As before, the SENSE equations

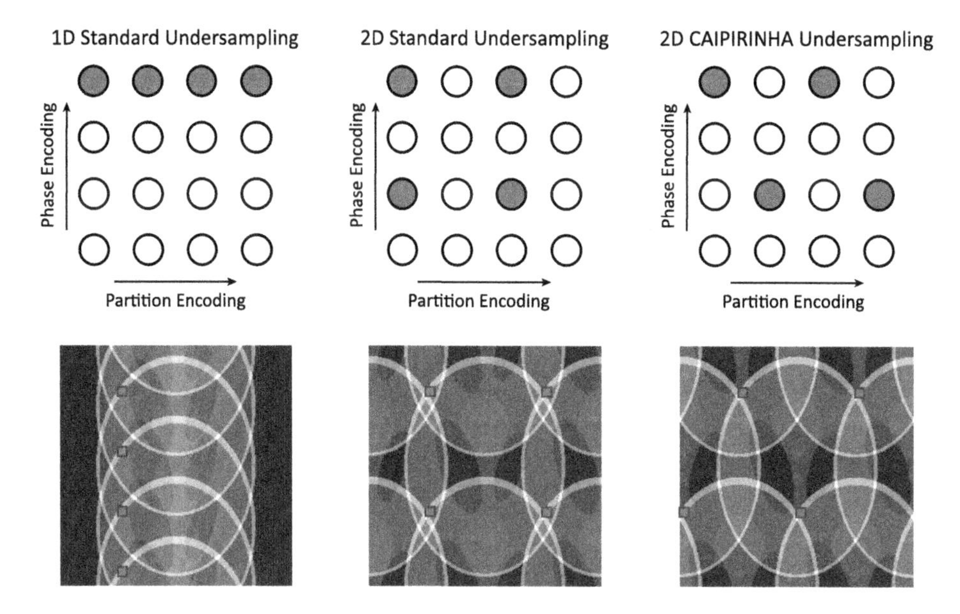

FIGURE 6.12

An example of using different sampling schemes with a total acceleration factor of R = 4. **(left)** Sampling and aliasing patterns for 1D undersampling along the phase-encoding direction. In this image, many areas near the center of the image have three or four overlapping signal-bearing pixels. **(center)** Using an undersampling pattern of R = 2 in both the phase-encoding and partition-encoding directions lowers the amount of overlapping signal in many parts of the image. However, there are still areas in this image where four signal-bearing pixels overlap. **(right)** In the CAIPIRINHA scheme, the maximum number of overlapping signal-bearing pixels is only three, and the total overlapped area is reduced in comparison to the standard 2D undersampling scheme. The reduction in overlapping information leads to a lower g-factor in the reconstructed images and thus less noise and fewer residual aliasing artifacts.

are structured so that the aliased points are the sum of the unaliased pixels multiplied by the coil sensitivity at that pixel location. The coil sensitivity coefficient matrix can be inverted, and the unaliased pixel values can be found.

In 2D GRAPPA [7] (also named for the number of undersampled dimensions, not the number of spatial dimensions), kernels can be defined across three spatial dimensions. An example of a 3D GRAPPA kernel is shown in Fig. 6.13. The remainder of the 2D GRAPPA reconstruction is largely equivalent to standard GRAPPA. The source and target matrices will have more columns since 3D kernels typically have more points than 2D kernels. Once calculated, the GRAPPA weights can be applied over all kernels in the 3D volume to reconstruct the missing data.

Highly accelerated 3D imaging is often facilitated by the application of Controlled Aliasing in Parallel Imaging Results in Higher Acceleration (CAIPIRINHA) [8], a sampling technique in which sampled lines are offset in the phase-encoding and partition-encoding directions, as shown in Fig. 6.12. When using such a sampling scheme, aliasing artifacts are shifted in the phase/partition directions, so there is less overlap in the aliased images. The reduction in signal aliasing leads to a lower g-factor in

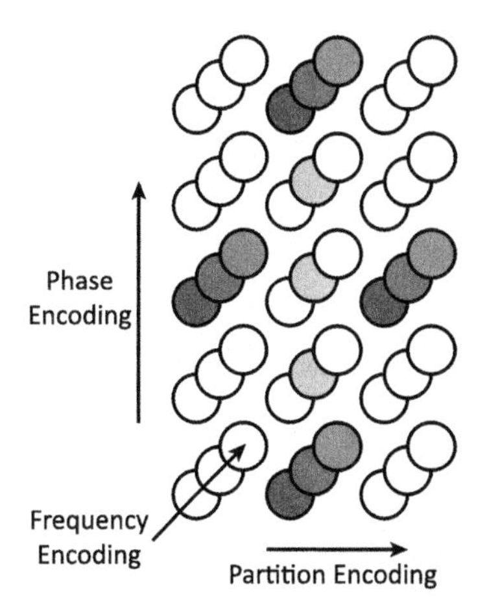

FIGURE 6.13

Example of a GRAPPA kernel defined in three dimensions, with an acceleration factor of R = 4 and a CAIPIRINHA sampling scheme. The source points are shown in blue (dark gray in print version) and target points in orange (light gray in print version). Note that unlike Fig. 6.7, only a single coil is shown here for ease of presentation, but the GRAPPA kernel would actually be made up of source points across multiple coils.

parts of the image and improves image quality. This sampling scheme can be used in conjunction with any parallel imaging algorithm designed to work on 3D datasets.

3D parallel imaging techniques are used clinically alongside traditional GRAPPA and SENSE. Some common applications for 3D imaging techniques include MR angiography [27] and musculoskeletal MRI [43], among many others.

6.5 Dynamic parallel imaging

The previous sections describe parallel imaging approaches that can be applied to a single static image. Dynamic applications, such as cardiac cine imaging, can also greatly benefit from parallel imaging to improve temporal resolution without significant compromises in image quality. While previously discussed approaches can be applied frame-wise in dynamic imaging, it is often beneficial to incorporate temporal information in the reconstruction. The following sections describe a few of these notable dynamic parallel imaging approaches: TSENSE and k-t SENSE, TGRAPPA, PEAK-GRAPPA, and through-time GRAPPA.

TSENSE and k-t SENSE

Similar to the CAIPIRINHA sampling scheme, in which data are collected in an offset pattern, many temporal parallel imaging algorithms require data sampled in an offset pattern in the temporal dimension. For example, with an acceleration factor of R = 2, odd phase-encoding lines would be sampled for the first temporal frame, even phase encoding lines in the second frame, and so on. This sampling scheme introduces aliasing along one spatial dimension and the temporal frequency dimension, which

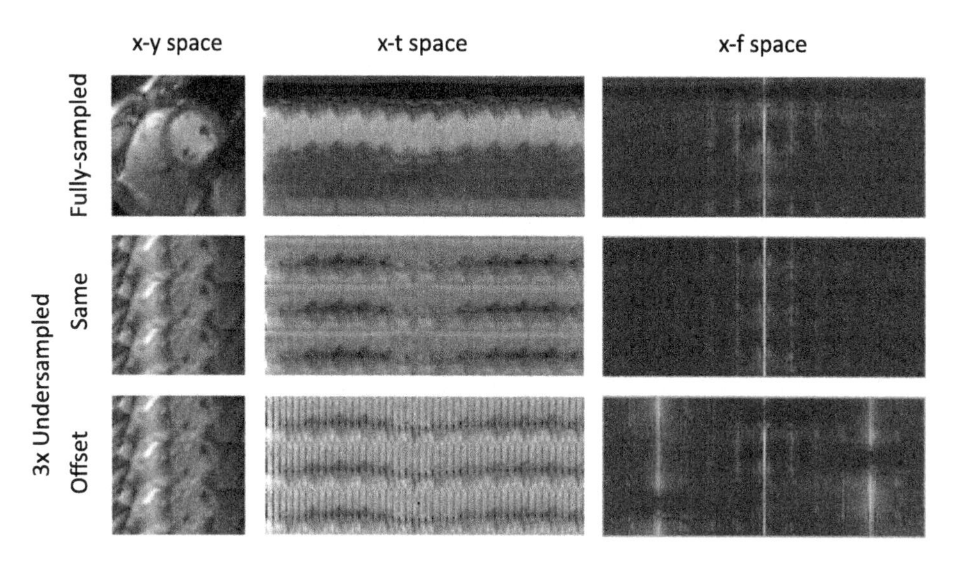

FIGURE 6.14

Example of cardiac MRI data, shown in different domains. **(top)** Representations of fully sampled data, without any aliasing artifacts, shown for reference. **(center)** R = 3 undersampled data, where the same lines of k-space data are collected for each frame. **(bottom)** R = 3 undersampled data, where the sampled lines are alternated in each frame as described in the text. **(left)** Image domain representations of cardiac data. **(center)** x-t space representation, showing action along a single line through time. Note the "flickering" in the offset undersampled x-t space that arises from the interleaved sampling pattern. **(right)** x-f space representations generated by calculating the temporal Fourier transform of the x-t representations. In the R = 3 undersampled x-f space, multiple replicates of the x-f space signal can be seen, where each replica is slightly shifted, again due to the sampling pattern. When the same k-space line is sampled, these replicates are only shifted along the x dimension and are aliased over the original spectrum. When sampled in the offset pattern, these replicates are shifted in both spatial (x) and frequency (f) dimensions, resulting in significantly less overlap between replicates in the x-f domain.

reduces the amount of signal overlap occurring in the spatial position/temporal frequency domain, also known as x-f space. An example of information in the x-f space is shown in Fig. 6.14. Since the temporal frequency domain is relatively sparse, higher acceleration factors can be deployed before significant aliasing occurs.

TSENSE [12] reconstructs data in a two-step process, combining the SENSE and UNFOLD [44] algorithms. The SENSE algorithm is first applied to each undersampled frame in the image domain, reducing spatial aliasing in the data set. A temporal low-pass filter is then applied to the data, reducing temporal aliasing effects. Between these two methods for reducing aliasing artifacts, higher acceleration factors can be achieved.

k-t SENSE [13] works in a similar manner and is an extension of the k-t BLAST algorithm described in Chapter 5. The k-t BLAST algorithm uses initial estimates of signal strength to determine the most effective filter to remove aliasing in the x-f space. At each point in the aliased image, an estimate is calculated for the fraction of the signal attributed to each signal source point. The aliased pixel signal is then redistributed to reconstruct pixels based on these estimated signal fractions.

k-t SENSE improves upon this algorithm by incorporating coil sensitivity profiles into the estimated signal strength matrix. Much like traditional SENSE, k-t SENSE is performed pixel-wise on the aliased image. This can be achieved by inserting the coil sensitivity matrix into the k-t BLAST equations

$$\hat{\boldsymbol{x}} = \hat{\boldsymbol{x}}_0 + \boldsymbol{M}^2 \boldsymbol{C}_p^{\mathrm{H}} \left(\boldsymbol{C}_p \boldsymbol{M}^2 \boldsymbol{C}_p^{\mathrm{H}} \right)^{-1} \left(\boldsymbol{x}_{alias} - \boldsymbol{C}_p \hat{\boldsymbol{x}}_0 \right) \tag{6.21}$$

In this equation, $\boldsymbol{x}_{alias}$ is the aliased image space signal vector of length N_c (# of coils), $\hat{\boldsymbol{x}}$ is the resulting estimated signal vector of length R (the acceleration factor), $\hat{\boldsymbol{x}}_0$ is an initial estimate vector of length R, $\boldsymbol{M}^2$ represents the $R \times R$ covariance matrix of signal deviation from $\hat{\boldsymbol{x}}_0$, and $\boldsymbol{C}_p$ is the $N_c \times R$ encoding matrix of coil sensitivity values used in the traditional SENSE reconstruction. This equation can be solved for each point in the aliased image to reconstruct the dynamic data.

In addition to k-t SENSE, the CG-SENSE algorithm can also be adapted for use in non-Cartesian dynamic imaging applications. For dynamic CG-SENSE, regularization terms can be added to the CG-SENSE cost function to enforce certain image properties over time. For example, Otazo, et al. [45] used spectral sparsity (i.e., L1-norm of the temporal Fourier transform) as a regularizer for cardiac perfusion data. More information on this type of technique can be found in Chapter 8.

One of the greatest benefits of using temporal methods for dynamic image reconstructions is that they exploit redundancies in the time domain. In this case, the temporal frequency domain is relatively sparse and most information is contained near the $f = 0$ point. This sparsity allows for a high degree of undersampling before temporal signals alias in the frequency domain, resulting in improved reconstructions at higher acceleration factors than SENSE or k-t BLAST alone. k-t SENSE is commonly used in place of SENSE for applications in dynamic imaging, such as cardiac cine [46] or myocardial perfusion [47].

TGRAPPA, k-t GRAPPA, & PEAK-GRAPPA

A few different variants of the GRAPPA algorithm have been created specifically for dynamic imaging. Three of these techniques are TGRAPPA, k-t GRAPPA, and Parallel MRI with Extended and Averaged Kernels GRAPPA (PEAK-GRAPPA).

TGRAPPA [14] is the most straightforward of the three as it works with the same kernel design as traditional GRAPPA. Much like k-t BLAST/SENSE, the phase-encoding lines that are collected are interleaved and alternated for each frame. After collecting a series of R frames, the data can then be combined to create a fully sampled k-space, which is used as the ACS data set, as shown in Fig. 6.15a. GRAPPA weights are determined using this temporally combined data and then applied to the undersampled frames to reconstruct the missing data points for each frame. This approach enables imaging with a higher frame rate without the need to collect dedicated ACS data.

k-t GRAPPA [15] uses a slightly different approach for reconstructing temporal data. k-t GRAPPA uses a similar offset phase-encoding sampling pattern as TGRAPPA, but uses a different type of GRAPPA kernel. Rather than just using spatial k-space neighbors, k-t GRAPPA also includes the two nearest temporal neighbors in the GRAPPA kernel as well. A diagram of this sampling scheme is shown in Fig. 6.15b. k-t GRAPPA can be calibrated using either a dedicated ACS dataset, in the form of several central k-space lines collected along with each frame, or using a TGRAPPA calibration scheme where the ACS is generated from a series of combined frames from an offset sampling pattern.

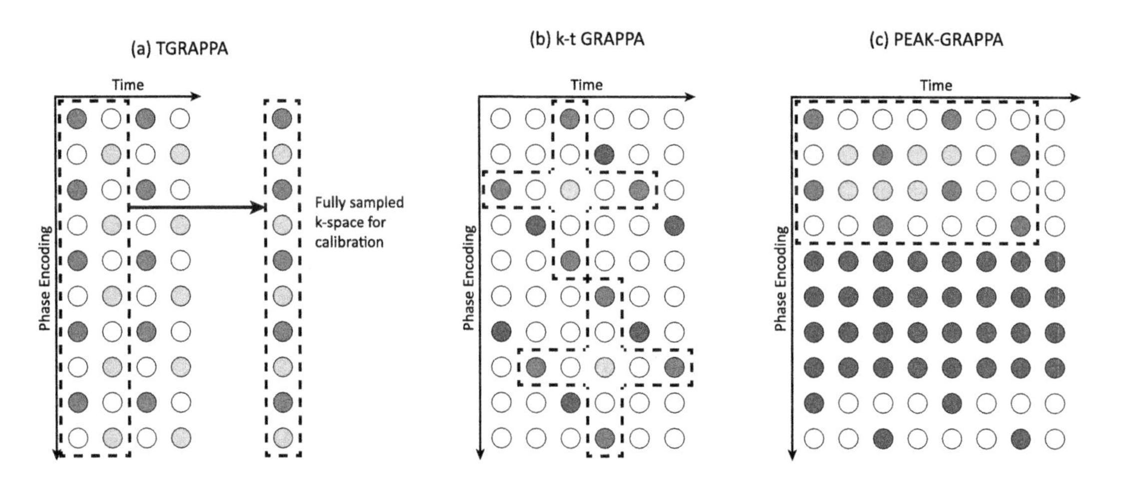

FIGURE 6.15

(a) TGRAPPA sampling. Different interleaves of undersampled data are collected at each time frame. If data are combined over R of these time frames, the result is a fully sampled dataset that can be used for calibration. **(b)** k-t GRAPPA sampling and kernel structure. In k-t GRAPPA, kernels include source points in k-space from the same time frame, as well as adjacent time frames. Kernels contain a single target point (orange (light gray in print version)), and the closest neighbors both in time and in k-space are used as source points (blue (mid gray in print version)) **(c)** PEAK-GRAPPA sampling and kernel definition. A sample kernel is shown near the top of the figure, with source (blue (mid gray in print version)) and target (orange (light gray in print version)) points. Sampled points outside this kernel are shown in dark gray. Unlike TGRAPPA, central calibration lines are still required since temporal information is needed to properly calibrate the temporal kernels.

PEAK-GRAPPA [16] is a more generalized form of k-t GRAPPA. Rather than only using a single target point and its four nearest spatiotemporal neighbors as source points, PEAK-GRAPPA uses a larger kernel. A diagram of a PEAK-GRAPPA kernel and sampling scheme is shown in Fig. 6.15c. PEAK-GRAPPA also includes the collection of a dedicated ACS dataset to enable accurate characterization of kernel time dynamics.

Temporal GRAPPA approaches can be used in place of GRAPPA for dynamic applications since temporal redundancies can allow for higher acceleration factors. Like k-t SENSE, this includes applications such as cardiac cine [48] [49] or perfusion imaging [50].

Through-time non-Cartesian GRAPPA

Through-time non-Cartesian GRAPPA [51] is a method for applying dynamic parallel imaging to non-Cartesian data sets. Through-time non-Cartesian GRAPPA uses a different approach compared to other dynamic parallel imaging techniques: Rather than using temporal data to reduce x-f space aliasing, through-time non-Cartesian GRAPPA uses temporal information to increase the amount of data available for the calibration step.

Several variations of the GRAPPA algorithm exist for non-Cartesian sampling schemes [10] [11]. When working with non-Cartesian data, there are some additional considerations that must be made due to the sampling geometry [52]. In Cartesian GRAPPA, the shape of the GRAPPA kernel is constant

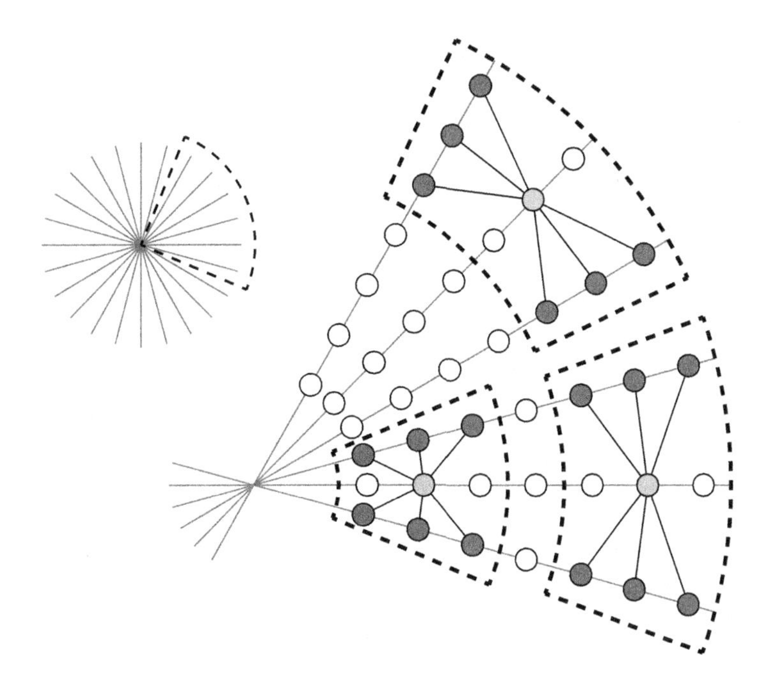

FIGURE 6.16

GRAPPA kernel definition in radial k-space data. In radial GRAPPA, kernel geometries are not consistent across all of k-space. Distances between the points in kernels near the center of k-space are much smaller than distances between points in kernels near the edge of k-space, and so there is no longer a single weight set that can be applied to both kernels to accurately reconstruct the missing points. Rotations in k-space also distort the positioning between neighboring points, and so separate weight sets must be used for each readout line.

throughout k-space, which allows weights to be generated from a large ACS dataset for the single kernel geometry and applied uniformly across all of k-space. In non-Cartesian GRAPPA, the shape of the GRAPPA kernel is not constant and shifts depending on the positions of the sampled k-space points. Thus, a single kernel shape cannot be used to describe the relationship between all subsets of collected and missing k-space points, as shown in Fig. 6.16. Despite this challenge, there are a few ways to deploy GRAPPA for non-Cartesian data. In general, formulations of non-Cartesian GRAPPA assume that kernel geometries are constant within small segments of non-Cartesian k-space. Weights can then be calculated for each segment of k-space from a fully-sampled non-Cartesian dataset and then applied locally to the undersampled data.

While GRAPPA can be used to generate an image from undersampled non-Cartesian data, the need for a fully sampled dataset for calibration reduces the efficiency of the acquisition. However, in the case of dynamic imaging, where the calibration time is less important than the temporal resolution of the accelerated images, non-Cartesian GRAPPA can provide high frame rates with low levels of residual aliasing artifacts. In through-time non-Cartesian GRAPPA [10] [11], several frames of fully-sampled k-space are collected prior to a dynamic MRI scan. GRAPPA weights are then calculated over very small segments of k-space, primarily relying on multiple kernel repetitions through time to generate stable GRAPPA weights. Once GRAPPA weights are generated for these local segments, they can be applied to a dynamic series of highly-accelerated non-Cartesian frames to achieve high temporal resolution.

Through-time non-Cartesian GRAPPA has been applied to accelerate dynamic imaging with high frame rate requirements, such as cardiac [51] and perfusion [53] imaging, where higher temporal resolution can capture rapid motion or contrast changes. In addition, through-time non-Cartesian GRAPPA reconstructions can be performed rapidly for real-time imaging applications [54].

6.6 Artifacts in parallel imaging

While parallel imaging is used in nearly all clinical scans to accelerate data collection, there are a number of artifacts that can arise when the parallel imaging algorithm fails to completely resolve the aliasing due to undersampling. Noise enhancement is one common artifact seen in parallel imaging. While some decrease in SNR is expected with parallel imaging, this artifact refers to a dramatic decrease in SNR in a specific region of an image. This artifact is often seen in image areas where many pixels alias together, and the coil sensitivity profiles contain less distinct values (resulting in ill-conditioning of the matrix and thus noise amplification during inversion) [29]. This situation typically occurs near the center of images, where coil sensitivity maps can have relatively similar values. This can cause the SENSE coil sensitivity encoding matrix or GRAPPA ACS source point matrix to be nearly singular, thereby increasing the g-factor of the region. Some examples of this artifact are shown in Figs. 6.17 and 6.19. Noise enhancement artifacts can typically be resolved by reducing the acceleration factor [28].

Residual aliasing occurs as a result of poor estimation of SENSE coil sensitivity maps or GRAPPA weights [29]. In this case, errors in the sensitivity information can cause some aliasing artifacts to remain in the image even after reconstruction. This situation can occur in cases where the patient moves between the coil sensitivity estimation or ACS collection and the actual scan. This motion causes the coils to be loaded differently, changing their sensitivity profiles [1]. In GRAPPA, this can also occur if the ACS is chosen to be too small, which can cause inaccuracies in the GRAPPA weights [28]. Examples of residual aliasing artifacts are shown in Figs. 6.18 and 6.19. Residual aliasing can also occur alongside noise enhancement because singularity in the coil sensitivity matrix or source point matrix introduces errors in the reconstruction, which can manifest as residual aliasing. Residual aliasing artifacts can typically be resolved by recollecting or increasing the amount of calibration data [28].

6.7 Summary

Parallel Imaging is a class of reconstruction algorithms in which multiple receiver coils are used to resolve aliasing artifacts intentionally generated by undersampling k-space. These techniques can be used to greatly reduce the amount of time required to collect many different types of MRI scans or improve spatial resolution with no additional required imaging time. The most commonly used techniques include SENSE, GRAPPA, SPIRiT, and ESPIRiT, each of which have been used to accelerate MRI scans in the clinic. Variations of these techniques have been developed to efficiently reconstruct accelerated nonuniform Cartesian, non-Cartesian, volumetric, and dynamic data sets. Moreover, in recent years, parallel imaging has been successfully combined with methods that exploit other sources of

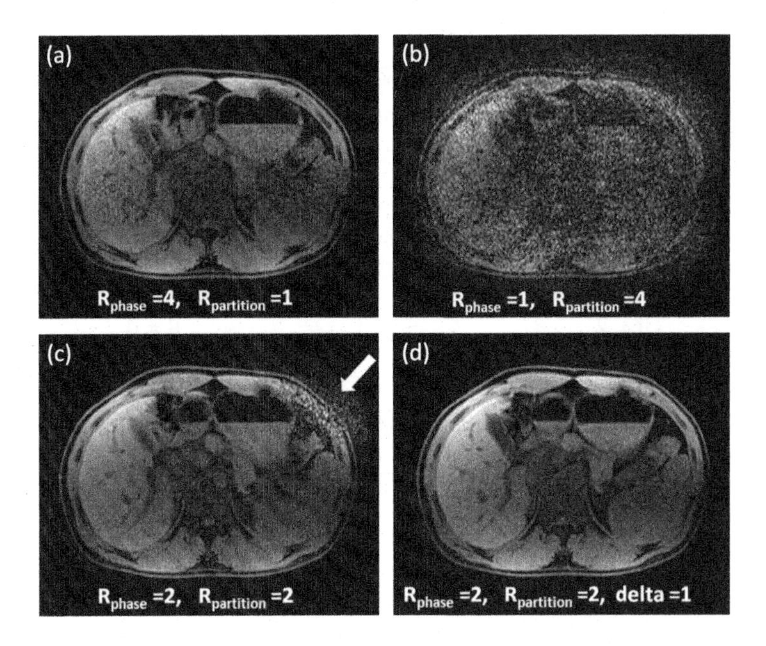

FIGURE 6.17

Example of noise enhancement in a 3D abdominal scan reconstructed with 2D GRAPPA, originally published in [28]. **(a)** Image showing noise enhancement artifacts due to excessive oversampling in the phase-encoding direction. **(b)** Image showing noise enhancement artifacts due to oversampling in the partition-encoding direction. Partition-encoding artifacts are worse than the phase-encoding artifacts because the coil sensitivities were similar along the partition-encoding direction compared to the phase-encoding direction. **(c)** Image showing a reconstruction that used undersampling along both phase-encoding and partition-encoding directions. This sampling scheme resolves most of the noise-enhancement artifacts, but there is still a small region where the coils do not support reconstruction for this undersampling scheme. **(d)** Image reconstructed using a CAIPIRINHA sampling scheme that effectively removes the artifacts and reduces noise enhancement.

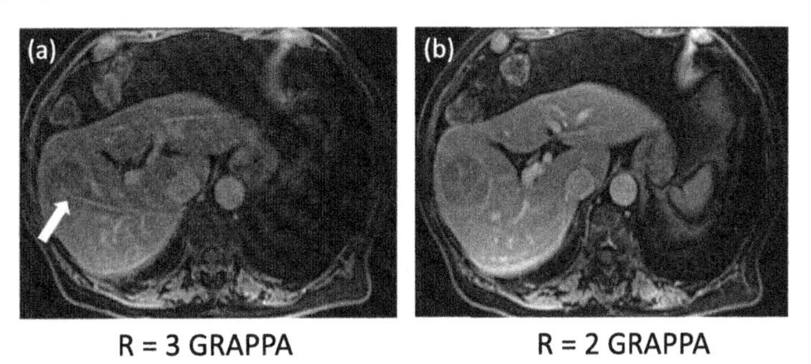

FIGURE 6.18

Example of residual aliasing artifact in an abdominal scan reconstructed with GRAPPA, originally published in [28]. **(a)** Image with an acceleration factor of R = 3. In this case, residual aliasing can be seen near the center of the image. **(b)** Image with an acceleration factor of R = 2. In this case, the residual aliasing artifact is cleared by reducing the acceleration factor to a level that the coil geometry can support.

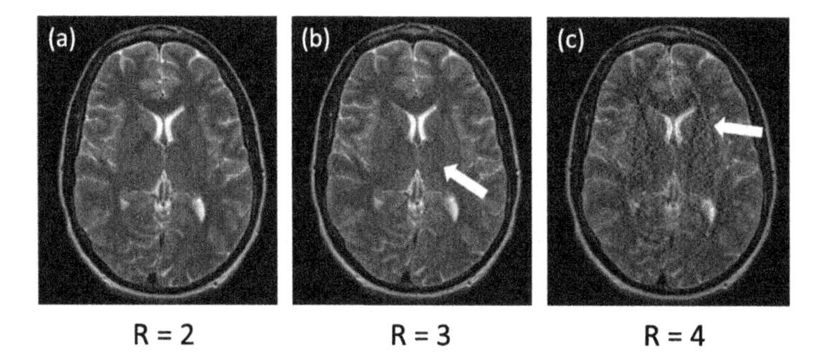

FIGURE 6.19

Example of residual aliasing co-occurring with noise enhancement in a brain scan reconstructed with GRAPPA, originally published in [28]. As the acceleration factor is increased, noise enhancement and residual aliasing artifacts appear since the coil geometry is no longer able to support an accurate parallel imaging reconstruction.

prior information and regularization (for example, low-rank reconstructions and compressed sensing); these combined techniques will be described in the following chapters.

References

[1] Hamilton J, Franson D, Seiberlich N. Recent advances in parallel imaging for MRI. Prog Nucl Magn Reson Spectrosc 2017;101:71–95.

[2] Pruessman KP, Weiger M, Scheidigger MB, Boesiger P. SENSE: sensitivity encoding for fast MRI. Magn Reson Med 1999;42:952–62.

[3] Griswold MA, Jakob PM, Heidemann RM, Nittka M, Jellus V, Wang J, et al. Generalized autocalibrating partially parallel acquisitions (GRAPPA). Magn Reson Med 2002;47:1202–10.

[4] Lustig M, Pauly JM. SPIRiT: iterative self-consistent parallel imaging reconstruction from arbitrary k-space. Magn Reson Med 2010;64(2):457–71.

[5] Uecker M, Lai P, Murphy MJ, Virtue P, Elad M, Pauly JM, et al. ESPIRiT- an eigenvalue approach to autocalibrating parallel MRI: where SENSE meets GRAPPA. Magn Reson Med 2014;71:990–1001.

[6] Weiger M, Pruessman KP, Boesiger P. 2D SENSE for faster 3D MRI. Magn Reson Mater Phys 2002;14:10–9.

[7] Blaimer M, Breuer FA, Mueller M, Seiberlich N, Ebel D, Heidemann RM, et al. 2D-GRAPPA-operator for faster 3D parallel MRI. Magn Reson Med 2006;56:1359–64.

[8] Breuer FA, Blaimer M, Mueller MF, Seiberlich N, Heidemann RM, Griswold MA, et al. Controlled aliasing in volumetric parallel imaging (2D CAIPIRINHA). Magn Reson Med 2006;55:549–56.

[9] Pruessman KP, Weiger M, Bornet P, Boesiger P. Advances in sensitivity encoding with arbitrary k-space trajectories. Magn Reson Med 2001;46:638–51.

[10] Griswold MA, Heidemann RM, Jakob PM. Direct parallel imaging reconstruction of radially sampled data using GRAPPA with relative shifts. In: 11th annual meeting of the ISMRM; 2003.

[11] Heidemann RM, Griswold MA, Seiberlich N, Kruger G, Kannengiesser SA, Kiefer B, et al. Direct parallel image reconstructions for spiral trajectories using GRAPPA. Magn Reson Med 2006;56:317–26.

[12] Kellman P, Epstein FH, McVeigh ER. Adaptive sensitivity encoding incorporating temporal filtering (TSENSE). Magn Reson Med 2001;45:846–52.

[13] Tsao J, Boesiger P, Pruessman KP. k-t BLAST and k-t SENSE: dynamic MRI with high frame rate exploiting spatiotemporal correlations. Magn Reson Med 2003;50:1031–42.
[14] Breuer FA, Kellman P, Griswold MA, Jakob PM. Dynamic autocalibrated parallel imaging using temporal GRAPPA. Magn Reson Med 2005;53:981–5.
[15] Huang F, Akao J, Vijayakumar S, Duensing GR, Limkeman M. k-t GRAPPA: a k-space implementation for dynamic MRI with high reduction factor. Magn Reson Med 2005;54:1172–84.
[16] Jung B, Ullmann P, Honal M, Bauer S, Hennig J, Markl M. Parallel MRI with extended and averaged GRAPPA kernels (PEAK-GRAPPA): optimized spatiotemporal dynamic imaging. J Magn Reson Imaging 2008;28:1226–32.
[17] Roemer PB, Edelstein WA, Hayes CE, Souza SP, Mueller OM. The NMR phased array. Magn Reson Med 1990;16:192–225.
[18] Walsh DO, Gmitro AF, Marcellin MW. Adaptive reconstruction of phased array MR imagery. Magn Reson Med 2000;43:682–90.
[19] Robson PM, Grant AK, Madhuranthakam AJ, Lattanzi R, Sodickson DK, McKenzie CA. Comprehensive quantification of signal-to-noise ratio and g-factor for image-based and k-space-based parallel imaging reconstructions. Magn Reson Med 2008;60:895–907.
[20] Aja-Fernández S, Vegas-Sánchez-Ferero G, Tristán-Vega A. Noise estimation in parallel MRI: GRAPPA and SENSE. Magn Reson Imaging 2014;32:281–90.
[21] Wintersperger BJ, Reeder SB, Nikolaou K, Dietrich O, Huber A, Greiser A, et al. Cardiac CINE MR imaging with a 32-channel cardiac coil and parallel imaging: impact of acceleration factors on image quality and volumetric accuracy. J Magn Reson Imaging 2006;23:222–7.
[22] Hunold P, Maderwald S, Ladd ME, Jellus V, Barkhausen J. Parallel acquisition techniques in cardiac cine magnetic resonance imaging using TrueFISP sequences: comparison of image quality and artifacts. J Magn Reson Imaging 2004;20:506–11.
[23] Irwan R, Lubbers DD, van der Vleuten PA, Kappert P, Gotte MJ, Siejens PE. Parallel imaging for first-pass myocardial perfusion. Magn Reson Imaging 2006;25:678–83.
[24] Golay X, Pruessman KP, Weiger M, Crelier GR, Folkers PJ, Kollias SS, et al. PRESTO-SENSE: an ultrafast whole-brain fMRI technique. Magn Reson Med 2000;43(6):779–86.
[25] Preibisch C, Pilatus U, Bunke J, Hoogenraad F, Zanella F, Lanfermann H. Functional MRI using sensitivity-encoded echo planar imaging (SENSE-EPI). NeuroImage 2003;19(2):412–21.
[26] Gardener AG, Francis ST. Multislice perfusion of the kidneys using parallel imaging: image acquisition and analysis strategies. Magn Reson Med 2010;63:1627–36.
[27] Muthupillai R, Douglas E, Huber S, Lambert B, Pereyra M, Wilson GJ, et al. Direct comparison of sensitivity encoding (SENSE) accelerated and conventional 3D contrast enhanced magnetic resonance angiography (CE-MRA) of renal arteries: effect of increasing spatial resolution. J Magn Reson Imaging 2010;31:149–59.
[28] Deshmane A, Gulani V, Griswold MA, Seiberlich N. Parallel MR imaging. J Magn Reson Imaging 2012;36:55–72.
[29] Blaimer M, Breuer F, Mueller M, Heidemann RM, Griswold MA, Jakob PM. SMASH, SENSE, PILS, GRAPPA: how to choose the optimal method. Top Magn Reson Imaging 2004;15(4):223–36.
[30] Jao TR, Do HP, Nayak KS. Myocardial ASL-CMR perfusion imaging with improved sensitivity using GRAPPA. In: 19th annual SCMR scientific sessions; 2016.
[31] Bock M, Muller S, Zuehlsdorff S, Speier P, Fink C, Hallscheidt P, et al. Active catheter tracking using parallel MRI and real-time image reconstruction. Magn Reson Med 2006;55:1454–9.
[32] Nölte I, Gerigk L, Brockmann MA, Kemmling A, Groden C. MRI of degenerative lumbar spine disease: comparison of non-accelerated and parallel imaging. Neuroradiology 2008;50:403–9.
[33] Zengo MO, Vogt FM, Brauk K, Jokel M, Barkhausen J, Kannengiesser S, et al. High-resolution continuously acquired MR angiography featuring partial parallel imaging GRAPPA. Magn Reson Med 2006;56(4):859–65.
[34] Lin F-H, Kwong KK, Belliveau JW, Wald LL. Parallel imaging reconstruction using automatic regularization. Magn Reson Med 2004;51:559–67.

[35] Liang D, Liu B, Wang J, Ying L. Accelerating SENSE using compressed sensing. Magn Reson Med 2009;62:1574–84.
[36] Boubertakh R, Prieto C, Batchelor PG, Uribe S, Atkinson D, Eggers H, et al. Whole-heart imaging using undersampled radial phase encoding (RPE) and iterative sensitivity encoding (SENSE) reconstruction. Magn Reson Med 2009;62(5):1331–7.
[37] Weiger M, Pruessman KP, Osterbauer R, Börnert P, Boesiger P, Jezzard P. Sensitivity-encoded single-shot imaging for reduced susceptibility artifacts in BOLD MRI. Magn Reson Med 2002;48(5):860–6.
[38] Murphy M, Alley M, Demmel J, Keutzer K, Vasanawala S, Lustig M. Fast L1-SPIRiT compressed sensing parallel imaging MRI: scalable parallel implementation and clinically feasable runtime. IEEE Trans Med Imaging 2012;31(6):1250–62.
[39] Vasanawala SS, Alley MT, Hargreaves BA, Barth RA, Pauly JM, Lustig M. Improved pediatric MR imaging with compressed sensing. Radiology 2010;256(2):607–16.
[40] Xue H, Kellman P, LaRocca G, Arai AE, Hansen MS. High spatial and temporal resolution retrospective cine cardiovascular magnetic resonance from shortened free breathing real-time acquisitions. J Cardiovasc Magn Reson 2013;15:102.
[41] Uecker M, Lustig M. Estimating absolute-phase maps using ESPIRiT and virtual conjugate coils. Magn Reson Med 2017;77(3):1201–7.
[42] Santelli C, Kozerke S. L1 k-t ESPIRiT: accelerating dynamic MRI using efficient auto-calibrated parallel imaging and compressed sensing reconstruction. In: 19th annual SCMR scientific sessions; 2016.
[43] Li CQ, Chen W, Rosenberg JK, Beatty PJ, Kijowski R, Hargreaves BA, et al. Optimizing isotropic 3D FSE methods for imaging the knee. J Magn Reson Imaging 2014;39(6):1417–25.
[44] Madore B, Glover GH, Pelc NJ. Unaliasing by Fourier-encoding the overlaps using the temporal dimension (UNFOLD), applied to cardiac imaging and fMRI. Magn Reson Med 1999;42:813–28.
[45] Otazo R, Kim D, Axel L, Sodickson DK. Combination of compressed sensing and parallel imaging for highly accelerated first-pass cardiac perfusion MRI. Magn Reson Med 2010;64:767–76.
[46] Maredia N, Kozerke S, Larghat A, Abidin N, Greenwood JP, Boesiger P, et al. Measurement of left ventricular dimensions with contrast-enhanced three-dimensional cine imaging facilitated by k-t SENSE. J Cardiovasc Magn Reson 2008;10:27.
[47] Plein S, Ryf S, Schwitter J, Radjenovic A, Boesiger P, Kozerke S. Dynamic contrast-enhanced myocardial perfusion MRI accelerated with k-t SENSE. Magn Reson Med 2007;58(4):777–85.
[48] Theisen D, Sandner TA, Bamberg F, Bauner KU, Schwab F, Schwarz F, et al. High-resolution cine MRI with TGRAPPA for fast assessment of left ventricular function at 3 Tesla. Eur J Radiol 2013;82(5):209–24.
[49] Jung B, Honal M, Ullman P, Hennig J, Markl M. Highly k-t-space-accelerated phase-contrast MRI. Magn Reson Med 2008;60:1169–77.
[50] Jung B, Honal M, Hennig J, Markl M. k-t-space accelerated myocardial perfusion. J Magn Reson Imaging 2008;28:1080–5.
[51] Seiberlich N, Ehses P, Duerk J, Gilkeson R, Griswold M. Improved radial GRAPPA calibration for real-time free-breathing cardiac imaging. Magn Reson Med 2011;65:492–505.
[52] Wright KL, Hamilton JI, Griswold MA, Gulani V, Seiberlich N. Non-Cartesian parallel imaging reconstruction. J Magn Reson Imaging 2014;40:1022–40.
[53] Wright KL, Chen Y, Saybasili H, Griswold MA, Seiberlich N, Gulani V. Quantitative high resolution renal perfusion imaging using 3D through-time radial GRAPPA. Invest Radiol 2014;49(10):666–74.
[54] Saybasili H, Herzka DA, Seiberlich N, Griswold MA. Real-time imaging with radial GRAPPA: implementation on a heterogeneous architecture for low-latency reconstructions. Magn Reson Med 2014;32(6):747–58.

Suggested readings

For additional information on parallel imaging in general:

[55] Schönberg SO, Dietrich O, Reiser MF. Parallel imaging in clinical MR applications. Berlin/Heidelberg, Germany: Springer; 2006.

[56] Larkman DJ, Nunes RG. Parallel magnetic resonance imaging. Phys Med Biol 2007;52(7):15–55.

[57] Hamilton J, Franson D, Seiberlich N. Recent advances in parallel imaging for MRI. Prog Nucl Magn Reson Spectrosc 2017;101:71–95.

[58] Deshmane A, Gulani V, Griswold MA, Seiberlich N. Parallel MR imaging. J Magn Reson Imaging 2012;36:55–72.

For additional information on non-Cartesian parallel imaging:

[59] Wright KL, Hamilton JI, Griswold MA, Gulani V, Seiberlich N. Non-Cartesian parallel imaging reconstruction. J Magn Reson Imaging 2014;40:1022–40.

CHAPTER

7 Simultaneous Multislice Reconstruction

Steen Moeller[a] **and Suchandrima Banerjee**[b]
[a]*Center for Magnetic Resonance Research, University of Minnesota, Minneapolis, MN, United States*
[b]*GE Healthcare, Menlo Park, CA, United States*

7.1 Introduction

Shorter MR scan times are desirable from considerations of scan throughput, patient comfort, and reducing incidences of motion. Parallel imaging accelerates an MR scan by acquiring fewer number of phase-encode steps, at less dense spacing than needed to meet the Nyquist criterion and avoid aliasing, while keeping the spatial frequency extent unchanged. The missing information is derived by exploiting the additional spatial encoding provided by signal reception with multiple coil elements [23,56,63]. Other sparse data sampling approaches have also been introduced to shorten the scan time [35]. These techniques have been successfully applied to speed up two dimensional (2D) and three dimensional (3D) Cartesian imaging with some extensions to non-Cartesian trajectories [59].

While 3D acquisitions have gained traction in some applications, the majority of MRI acquisitions rely on 2D multislice encoding and acquisition. In the case of sequences such as FLASH/Gradient Recalled Echo, where a single slice is excited and acquired per repetition time (TR), the volume acquisition time (VAT) is directly proportional to the number of slices and the number of phase-encoding lines. Shortening the volume acquisition time is desirable not only in the interest of shorter scan time but also for applications such as body and cardiac and to reduce the temporal footprint of the acquired volume against physiological motion cycles. By exciting and acquiring N slice locations simultaneously, the acquisition time could be shortened by N folds, and is an independent encoding to phase-encoding undersampling and additionally can for sequences employing echo-trains or optimized k-space sampling be used without impacting the TE or sampling pattern. This is the underlying motivation for Simultaneous MultiSlice (SMS) Imaging. For sequences employing extended echo-trains the scan-time reduction with SMS is less straightforward and would result from either the reduction in number of slice packets (interleaves) or the duration of TR.

Let us consider the specific case of single shot (SS) MRI pulse sequences that employ a train of excitation pulses to acquire all the locations in the volume being imaged within a single repetition time (TR), and acquire the complete k-space data for the excited slice location after each RF excitation. The most widely used single-shot techniques include single-shot echo-planar imaging (EPI) and single-shot fast spin-echo (FSE) sequences. In a multishot version, a fraction of the k-space in each excited location is sampled in each TR or shot, and the data from the multiple shots are then combined during reconstruction to get the complete k-space coverage at each location. The minimum TR for such sequences would be determined by the time needed to image all the slices, in addition to a fixed block of time that might be allocated for image-contrast preparation.

Advances in Magnetic Resonance Technology and Applications, Volume 7, ISSN 2666-9099. https://doi.org/10.1016/B978-0-12-822726-8.00017-8

The ability to excite multiple slice locations simultaneously would reduce the number of independent excitations that need to be carried out. If we consider the application of functional MRI where usually an ssEPI sequence is employed, and the contrast preparation time is minimal, simultaneously exciting and acquiring multiple slices would shorten TR (improve the temporal resolution) and allow for many more volumes or higher resolution to be acquired during a particular scan-time session. Single or multishot echo-planar diffusion-tensor imaging with many diffusion directions had previously been unfeasible in clinical settings because of the long scan times. Even though a significant block of time in diffusion imaging is allocated for contrast preparation, SMS has proved still beneficial in significantly shortening TR and scan time, improving SNR efficiency. These are a few example applications where simultaneous multislice excitation would be very beneficial.

7.2 Basics of SMS encoding

The idea of simultaneous multislice (SMS) excitation was first introduced long before the advent of parallel imaging [39,46] with the goal of improving sensitivity. With single-channel receiver coils being the norm at that time, researchers relied on phase manipulation to resolve the signal from simultaneously acquired slices. The simplest multiband (MB) excitation RF pulse can be written in the form of

$$RF_{MB}(t) = A(t) \cdot \sum_{n=1}^{N} e^{i\,\Delta\omega_n t + \varphi_n}, \tag{7.1}$$

where A(t) is the complex RF waveform for a single slice, ω and φ determine the position and phase of each slice excited respectively, and N is the SMS factor. One of the initial approaches proposed Hadamard encoding of the multiband excitation pulses. In this approach the measurements are repeated N times with different phase patterns φ_n for each measurement to separate the signal from the N simultaneously excited slices. The slices are resolved on a voxel-by-voxel basis, by using the N individual images to create appropriate Hadamard-encoded combinations. This would not provide a scan-time benefit unless for certain specific multiaverage acquisition protocol. Also, signal-reconstruction schemes involving addition/subtraction from multiple measurements are generally prone to motion artifacts and additional phase errors. A variation of this approach was proposed with the Phase-Offset multiplanar (POMP) method [21], which added more flexibility over Hadamard encoding by introducing a specific k_y dependent-phase offset to the RF subpulses. By the Fourier shift Theorem, a shift in the time/spatial domain corresponds to a linear phase term in the frequency domain. Let us consider a 1D image $\boldsymbol{x}(y)$ with field of view FOV_y which can be shown to be the inverse discrete Fourier transform of the k-space signal $\boldsymbol{s}(k_y)$. A shift of y_0 in the image domain can be affected by a k_y-dependent phase offset in the frequency/ k-space domain as

$$\begin{aligned} &\boldsymbol{x}(y - y_0) \overset{\mathcal{F}}{\Leftrightarrow} \boldsymbol{s}(k_y) e^{-i2\pi k_y y_0}; \\ &k_y = n\Delta k = \frac{n}{N\Delta y} = \frac{n}{FOVy}. \end{aligned} \tag{7.2}$$

In the case of POMP with two-slice excitation, a half FOV shift in the image from the second slice would be affected by a linear phase modulation of:

$$e^{i2\pi \frac{n}{FOV_y} * \frac{FOV_y}{2}} = e^{i\pi n} = (-1)^n \tag{7.3}$$

This corresponds to a sign alteration in every other measurement, which is also known as "phase chopping". In contrast with the Hadamard approach, POMP allows for a direct reconstruction. To separate out the contributions from two simultaneously excited locations the FOV would roughly need to be extended by a factor of 2, which requires doubling the number of phase encodes during k-space data acquisition. While this method also does not provide a scan-time reduction, unless in specific protocol settings, many of the recent SMS developments draw from the POMP methodology.

The introduction of parallel imaging [23,56,63] provided technology to overcome the inefficiencies with POMP, while also providing new insight into increasing the scan efficiency. Like POMP, multislice Controlled Aliasing in Parallel Imaging Results in Higher Acceleration (CAIPIRINHA) introduced by Breuer et al. [8] uses phase modulation, to shift the individual slices. But unlike POMP, it does not need oversampling by a factor of R because it uses the coil sensitivity information to resolve the signal from superimposed slices. In CAIPIRINHA, the FOV is still shifted between adjacent slices since this reduces the noise enhancement/improves the geometry factor of the parallel imaging reconstruction by increasing the spatial-encoding differences between locations that contribute signal to an aliased pixel. The differences in POMP and MS CAIPIRINHA acquisitions for the case of simultaneous excitation of two slices is shown in Fig. 7.1, where slice 2 has a half-FOV shift relative to slice 1.

Simultaneous MultiSlice (SMS) and MultiBand (MB) have often been used synonymously, whereas multiband terminology is also used in other applications and referring specifically to aspects of the excitations, SMS currently has a singular reference to both the excitation and reconstruction framework. For the rest of this chapter, we will refer to this technology as SMS.

When considering how to reconstruct SMS data using parallel imaging, several different techniques have been considered, similarly to the image and k-space approaches for phase-encoding undersampled parallel imaging (as detailed in Chapter 6). Both the image-space and the k-space approaches are relatively straightforward with parallel-imaging knowledge, and more easily appreciated as undersampled POMP-encoded acquisitions as illustrated in Fig. 7.1. This in particular brings back the need for using the encoding schemes from Eq. (7.3), in which for a half FOV shift sequentially phase-encoding lines are modulated by a phase modulation of the RF subpulse alternating with [0 pi 0 pi 0 pi ...] between TRs.

7.3 Reconstruction of SMS using parallel imaging concepts

The SENSE framework [56] is well-suited for understanding how to reconstruct SMS data. For traditional SMS, the encoding of the individual slices in the simultaneously acquired slices is the same, and the aliasing in the simultaneously acquired slices is a direct summation of the individual slices. The GRAPPA framework [23] has also been adapted to SMS reconstruction through a progression in algorithms (SENSE-GRAPPA, RO-SENSE-GRAPPA and slice-GRAPPA) as will be discussed subsequently.

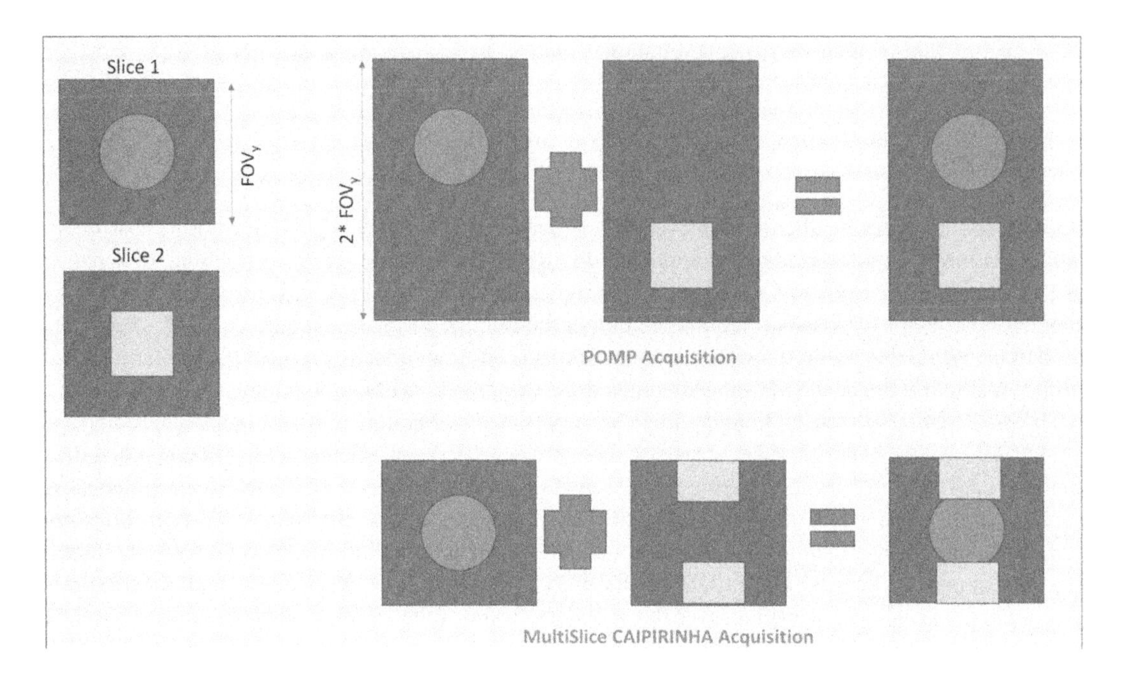

FIGURE 7.1

Schematic Diagram of the acquisition differences between POMP and MS-CAIPIRINHA. Two distinct slices are indicated by the circle and square respectively. Phase-offset excitation results in relative displacement of the simultaneously excited slices in the image. The FOV has been doubled in the POMP acquisition (top row) but not enlarged in the case of the MS-CAIPIRINHA acquisition, where a half-FOV shift is applied to the second slice (bottom row).

7.3.1 SENSE

The SENSE framework was introduced by Larkman et al. [32] and the schematic is shown in Fig. 7.2. If the spatial variation in sensitivity profile is known, then the simultaneous acquired signals can be considered as the solution of Eq. (7.1), where $x_1, \cdots x_m$ are the images in individual channels, $\hat{x}_1, \cdots \hat{x}_n$ the unaliased and sensitivity weighted images for the resolved slices, and C_{ij} the sensitivity profile for location i and channel j:

$$\begin{bmatrix} C_{11} & \cdots & C_{1n} \\ \vdots & \ddots & \vdots \\ C_{m1} & \cdots & C_{mn} \end{bmatrix} \begin{bmatrix} \hat{x}_1 \\ \vdots \\ \hat{x}_n \end{bmatrix} = \begin{bmatrix} x_1 \\ \vdots \\ x_m \end{bmatrix}. \tag{7.4}$$

With known/estimated sensitivity profiles, the least squares solution for Eq. (7.4) produces the images of the individual slice locations.

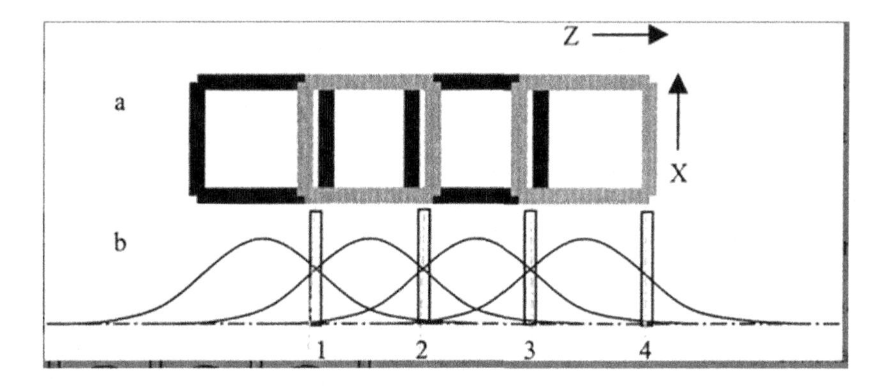

FIGURE 7.2

First demonstration of resolving signals from SMS excitation using array coil sensitivity profiles from Larkman et al. [32]. In **(a)** four overlapping surface coils are shown, and in **(b)** magnitude sensitivity profiles for the coils at their center line and slice locations (1, 2, 3, 4) are shown. For this example the slice positions were chosen such that the magnitudes of the sensitivities of nearest neighbor coils are approximately equal.

7.3.2 Extended FOV methods

The first method, originally introduced as SENSE-GRAPPA, uses the GRAPPA framework for reconstruction, and the aliasing intuition from the SENSE framework for creating an extended FOV—the POMP acquisition framework.

7.3.2.1 *SENSE-GRAPPA*

The data-driven k-space approach for solving Eq. (7.4) is best understood by referring back to the POMP technique as illustrated in Fig. 7.1. In this case the acquired FOV for two slices when concatenated can be thought of as two times as large, and when only every other phase-encoding line is acquired in that concatenated FOV, the resulting image is the same as the collapsed image from two simultaneously excited slices. This approach is the concept underlying SENSE-GRAPPA. The POMP-like images used for calibration of coil-weighting factors in SENSE-GRAPPA were acquired, unlike in POMP, by separately acquiring a regular FOV image for each slice and then concatenating them to synthesize the extended FOV image. Parallel imaging allows for SMS to both have an R-fold undersampling in phase-encoding and N-fold slice-aliasing, and was tackled in SENSE-GRAPPA by solving two conventional GRAPPA problems, each with a 1D undersampling of N and R, respectively.

In the SENSE-GRAPPA method, the most common approach for reconstructing separated images is to apply the Fourier Transform (FT) along the extended/concatenated phase FOV dimension. This is then followed by partitioning the FOV into the two or more appropriate slices. A flow-diagram of the calibration data preparation and reconstruction is shown in Fig. 7.3. As an alternative approach, and again drawing on the POMP reference, after obtaining an extended k-space, the k-space signal for each slice (and then sampled at the Nyquist rate) can be obtained as

$$\mathbf{s}_{sl} = e^{-i\phi_{sl}} \left(\mathbf{s}^{ext.\ FOV} * e^{i\phi_{sl}} \mathrm{Sinc}_{FOV} \right), \tag{7.5}$$

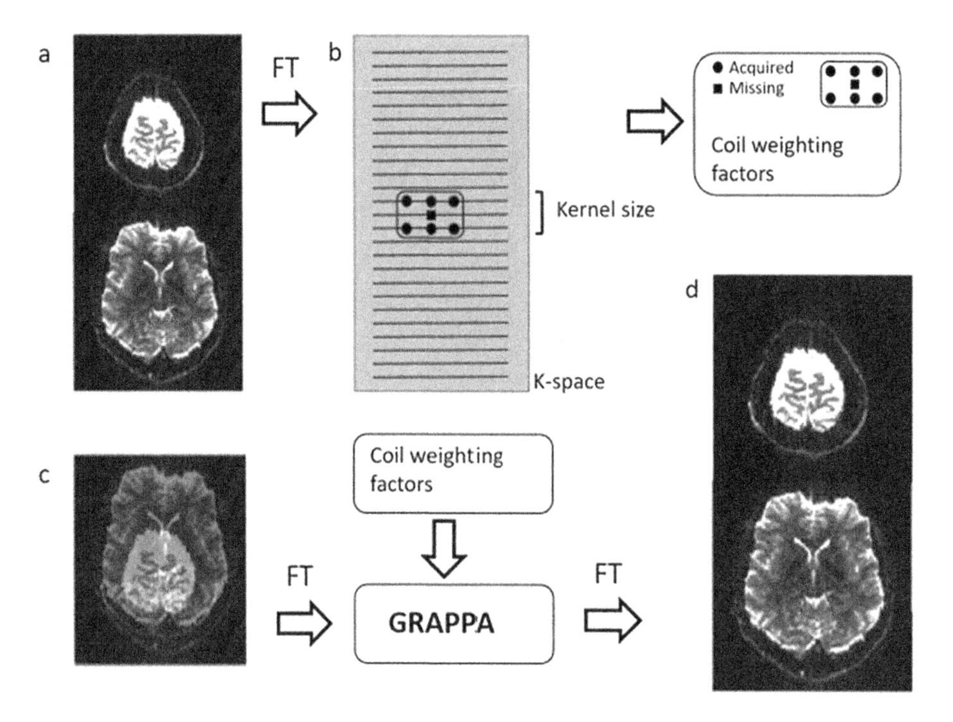

FIGURE 7.3

Flow diagram for the SENSE-GRAPPA calibration and reconstruction. Top row left, A, the individual slice images in calibration data have been concatenated along an extended phase FOV. In the middle, B, the standard GRAPPA kernel calibration is illustrated for the extended k-space. In the bottom row, C, SMS data for the same number of slices is shown. GRAPPA reconstruction of the SMS data in k-space yields an image, D, with the same extended FOV in image space from which the two individual slice images can be segmented.

where Sinc_{FOV} designates the sinc function defined in k-space, which is bandlimited in image space with width FOV, and ϕ_{sl} the phase for locating the bandlimited function within the extended FOV.

7.3.2.2 *RO-SENSE-GRAPPA*

Since the FOV in SENSE-GRAPPA is artificial, the same algorithm can be used with interchanging of the direction of the phase-encoding and the readout. This is then the RO-SENSE-GRAPPA view of the SMS reconstruction, as shown in Fig. 7.4.

This was demonstrated in the first 7T application of SMS EPI with $N \times R = 4 \times 4$ acceleration using a 16 channel coil [44]. The concatenation of the slice from SENSE/GRAPPA was altered from extending along the phase-encoding direction to be in the read-out direction, which then no longer has the same connection to POMP. This was done solely to accommodate the slice-dependent EPI ghost, and happened to be compatible with the subsequently developed blipped-CAIPI that will be described later in this chapter in connection with EPI. Whereas for in-plane parallel imaging with

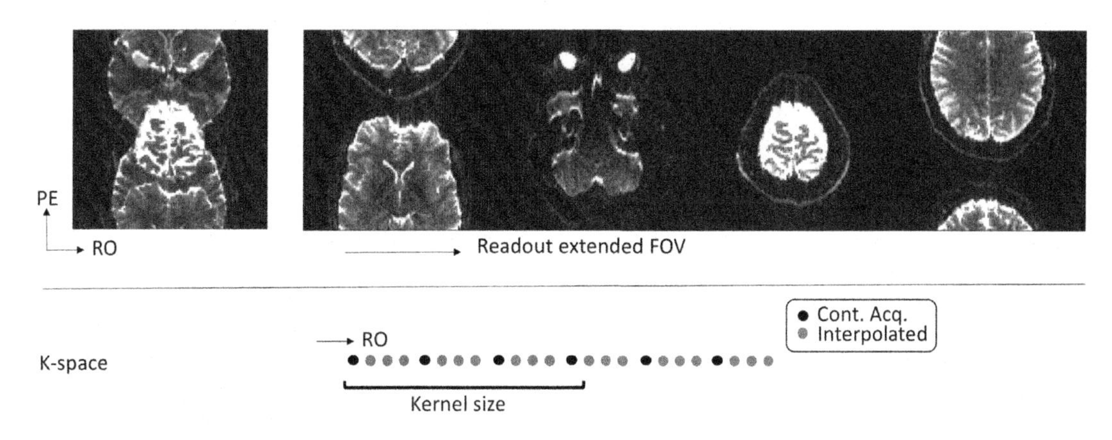

FIGURE 7.4

Reconstruction of SMS with N = 4 and 1/4 FOV shift between slices in the concatenated RO-FOV for SENSE-GRAPPA. The CAIPIRINHA shifts in the PE direction are preserved with the GRAPPA reconstruction. Separating the slices at the edges does not introduce artifacts, and the FOV shift can be corrected afterwards for each slice. In the bottom of the figure, the dimensions of the GRAPPA kernel is illustrated for the extended k-space in the readout direction.

EPI, the navigator correction is applied before phase-encoding parallel-imaging reconstruction, it was applied for RO-SENSE-GRAPPA after slice-separation and before phase-encoding interpolation.

Kernel calibration

The kernel calibration for SENSE-GRAPPA and RO-SENSE-GRAPPA follow the same considerations as for GRAPPA, with the only difference being in the calibration data preparation from acquired ACS data. Following the pictorial guide of SENSE-GRAPPA from Blaimer et al. [5] as shown in Fig. 7.5, for RO-SENSE-GRAPPA, one first constructs extended FOV image from the calibration data. From this extended FOV image, a pseudo-k-space is constructed and used to compute coil weighting factors using the conventional GRAPPA technique. The obtained coil weighting factors are then applied directly to the SMS k-space, and an extended pseudo-k-space is reconstructed. Following the Fourier transform along the readout the signals can be resolved to the specific slices.

7.3.3 Slice-GRAPPA

The k-space slice separation for SENSE-GRAPPA from Eq. (7.4) forms the basis for slice-GRAPPA [61]. But in this case, instead of reconstructing an extended FOV as in SENSE-GRAPPA or RO-SENSE-GRAPPA, the data are reconstructed directly into the separated slices as depicted in Fig. 7.6a. In Fig. 7.6b, the convolution kernels for slice-GRAPPA are illustrated for reconstructing a single slice and channel-specific data-points. For each channel and slice, a unique set of convolution kernels are computed that are used to reconstruct each slice independently.

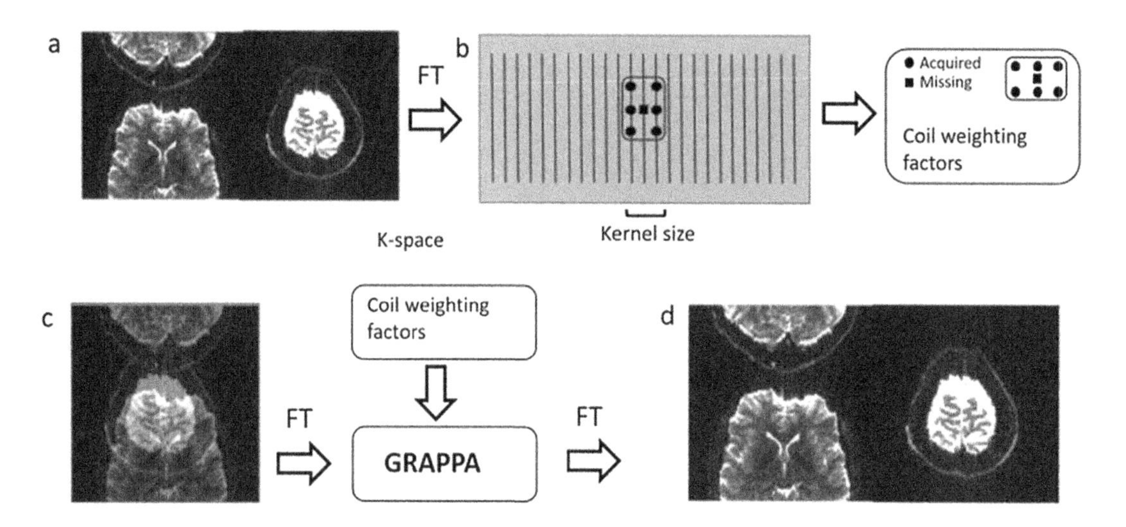

FIGURE 7.5

Flow diagram for the RO-SENSE-GRAPPA calibration and reconstruction. Top row, left, A, shows the calibration data from individual slices concatenated into an extended FOV along the readout direction. In the middle, B, the standard GRAPPA kernel calibration is illustrated for the artificially extended k-space. In the bottom row, C, SMS data for the same number of slices is shown. GRAPPA reconstruction of the SMS data in k-space yields an image with identical extended FOV in image space, D, from which the images of the individual slices can be separated.

Kernel calibration

Slice-GRAPPA is a different reconstruction compared with the other GRAPPA variants. Whereas GRAPPA is used for interpolating missing data using small convolutional kernels, in slice-GRAPPA the convolutional kernels are a projection onto the solution space. For GRAPPA the reconstructed image is a combination of the measured data and the interpolated data, but for slice-GRAPPA the measured data is not used together with the reconstructed data to resolve the N simultaneously acquired slices. The coil weights $\boldsymbol{G}_j$ in slice-GRAPPA, for acquired SMS kspace $\boldsymbol{s}_{MB}$ and targeted slice resolved k-space $\boldsymbol{s}_j$, are constructed such that

$$\boldsymbol{G}_j(\boldsymbol{s}_{MB}(k)) = \boldsymbol{s}_j(k) \quad \forall j\, \forall k \qquad where\ j = 1..N, \tag{7.6}$$

and the coil weights $\boldsymbol{G}_j$ are determined by solving for $\boldsymbol{G}_j$ in Eq. (7.6) by substituting

$$\boldsymbol{s}_{MB} = \sum_j \boldsymbol{s}_j. \tag{6a}$$

Pictorially, the method is illustrated in Fig. 7.7 for a kernel that independently reconstructs a single slice from the superimposed slices (the multichannel aspect is suppressed in the figure). This kernel construction was initially adapted due to its superior g-factor performance [77].

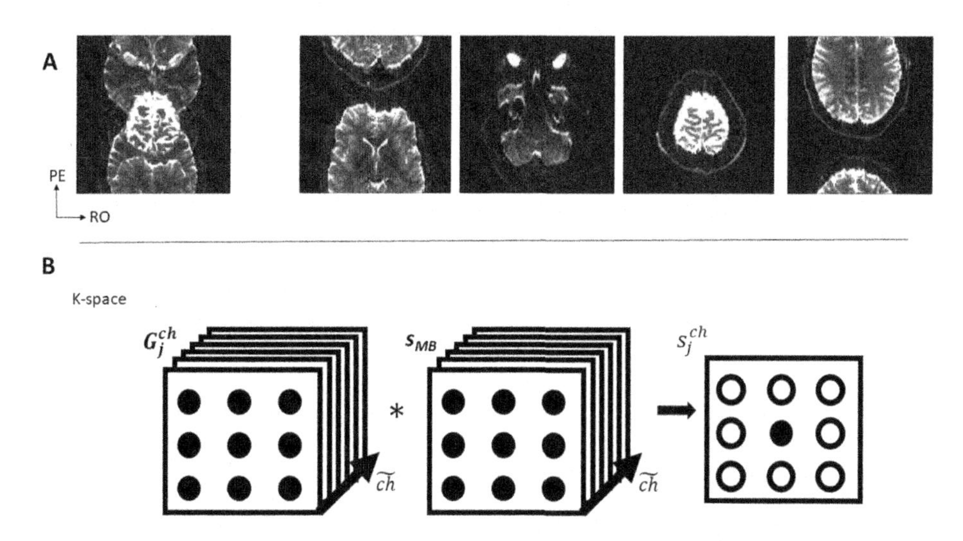

FIGURE 7.6

Reconstruction of SMS with N = 4 and 1/4 FOV shift between slices in Slice-GRAPPA. The CAIPIRINHA shifts in the PE direction is preserved with the slice-GRAPPA reconstruction. Reconstructing each slice with a projection reconstruction does not introduce artifacts, and the FOV shift can be corrected afterwards for each slice. In the bottom line b, the construction of the convolution kernels is illustrated.

A similar SMS extension of the ARC [2,6] method of parallel imaging for reconstruction of individual slices from the superimposed slice data is known as the slice-ARC technique.

7.3.4 Split-Slice-GRAPPA

Slice-GRAPPA was introduced as part of the developments in the human connectome projects and quickly adapted in these studies because of the superior image quality. Eventually, it also revealed short-comings for residual slice aliasing and temporal SNR (tSNR), which could be improved upon with a more complex dynamic calibration techniques [42] and subsequently more elegantly with Split-Slice-GRAPPA. The latter has turned out to be similar to unbiased slice-GRAPPA in minimizing slice leakage and analogous in performance to SENSE-GRAPPA.

It was realized in [11] that the slice-GRAPPA could be trained to minimize the sensitivity to phase variations that varied significantly from the calibration phases by ensuring the MB data mapped to the proper SB data, while not mapping to alternate slices. Formally for Split Slice-GRAPPA (SpSg), the kernels were determined such that

$$\boldsymbol{G}_j\left(\sum_i s_i\right) = s_j \; \forall j \; AND \; \boldsymbol{G}_j(s_i) = 0, \; i \neq j. \tag{7.7}$$

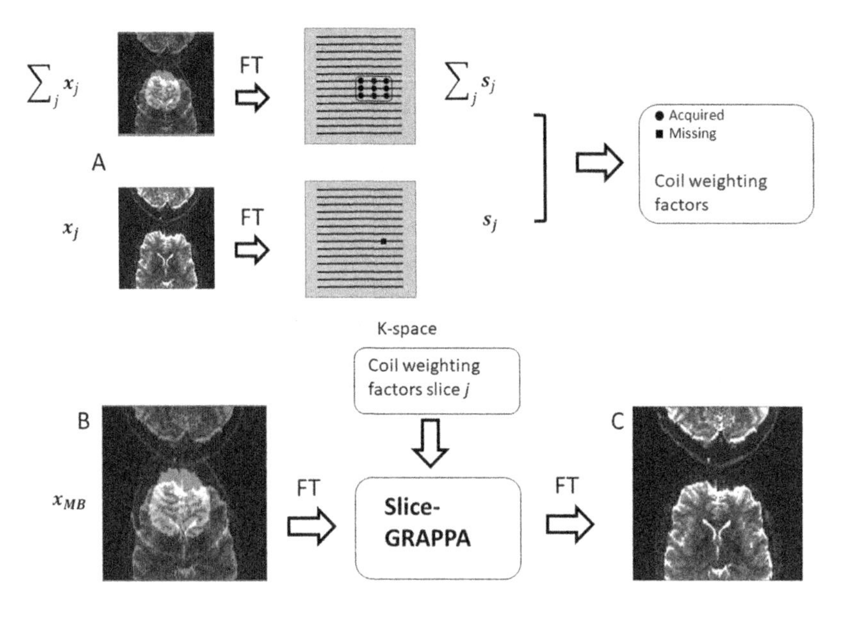

FIGURE 7.7

Flow diagram for the slice-GRAPPA calibration and reconstruction is shown. Top left, A, individual slices acquired with single-band excitation and their summation, which are used for calibration of the coil weighting factors, are shown. In the bottom row, B, SMS data for the same number of slices are shown. After slice-GRAPPA reconstruction in k-space, a resolved slice, C, is reconstructed.

Kernel calibration

To solve for this, the system of equations was adapted to solve the penalized least squares, with $\lambda = 1$, yielding

$$\arg\min_{\substack{G_i \\ \forall i}} \left\| G_i \left(\sum_j^N s_j \right) - s_i \right\|_2^2 + \lambda \sum_{j \neq i} \| (G_i (s_j)) \|_2^2 . \tag{7.8}$$

The construct with suppressing residual aliasing in slice-GRAPPA is an approach that does not have an obvious analogy for conventional parallel imaging. Nonetheless, the importance of the SpSg algorithm was quickly realized for a variety of SMS applications and adapted for large-scale projects, such as the human connectome projects [69].

7.3.5 SMS with phase-encoding undersampling

Combining phase-encoding undersampling, or parallel imaging as it is more commonly described, with SMS can provide significant scan-time reduction or other advantages such as reduction in distortion and blurring due to reduced echo-spacing in the case of single-shot acquisitions. So it is a natural extension to combine SMS with phase-encoding undersampling. In both SENSE-GRAPPA and slice-

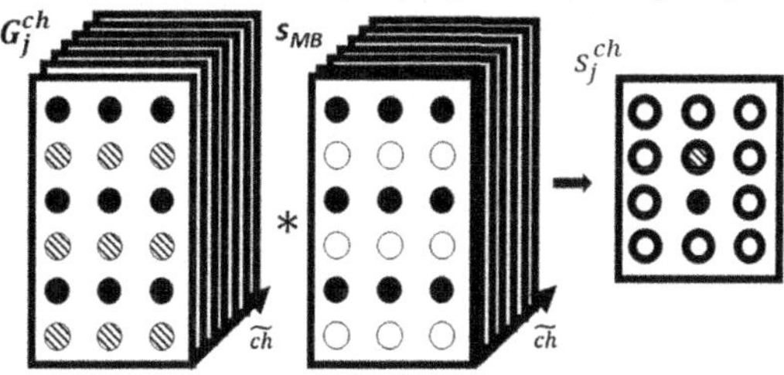

FIGURE 7.8

The slice-GRAPPA convolution kernels for reconstructing SMS with phase-encoding undersampling. From the acquired k-space with R different kernels, the calculated values for R different k-space locations is simultaneously determined.

GRAPPA, this is typically accomplished by sequential application of two kernels. Firstly, the SMS signal is resolved into the channels for each slice, and then conventional GRAPPA is used for interpolating the missing phase-encoding samples. These can also be used jointly, such that the acquired signal is projected onto the entire fully resolved signal space. The slice-GRAPPA kernels calculated from Eq. (7.8) or Eq. (7.10), are constructed such that, for an R-fold phase-encoding undersampling, R slice-GRAPPA kernels are calculated from the same system of equations from a reference acquisition with Nyquist sampled phase-encoding (Fig. 7.8).

7.3.6 Reconstruction of SMS for EPI

Historically, the importance of SMS became apparent with its use in EPI functional MRI (fMRI) and Diffusion Tensor Imaging (DTI) applications. In this section, we will discuss the additional considerations for the extension of SMS acquisition and reconstruction to echo-planar imaging.

7.3.6.1 *Blipped-wideband and blipped-CAIPI encoding*

Since the data in single-shot EPI are collected in a single shot/TR, the RF phase-cycling approach for generating CAIPIRINHA shifts is not feasible. Instead, the slice gradient, a gradient perpendicular to the slice, is used to augment the encoding. This was first proposed by Nunes [57] using a blipped wideband approach. In the blipped wideband approach, a short blipped slice-gradient is added simultaneously with the phase-encoding blip. The gradient moment is calculated to continuously adding a slice-specific phase, such for a slice with 1/2 FOV shift, the accumulating phase is [0, 180°, 360°, 540°, ...]

This was subsequently improved by Setsompop et al. [61] using a balanced blipping approach, such that the added phase for the same 1/2 FOV shift is [0, 180°, 0°, 180°, ...]. Mathematically, there is no difference, but experimentally there is a significant difference since the added phase varies subtly across the slice. As the added phase increases, as it does in blipped wideband, the phase across the slice leads to signal loss. The distinction is illustrated in Fig. 7.9, from Setsompop et al. [61].

Instead of the slice-gradient blipping, the slice-gradient can also be modulated during each readout in the EPI echo-train [49,73,4].

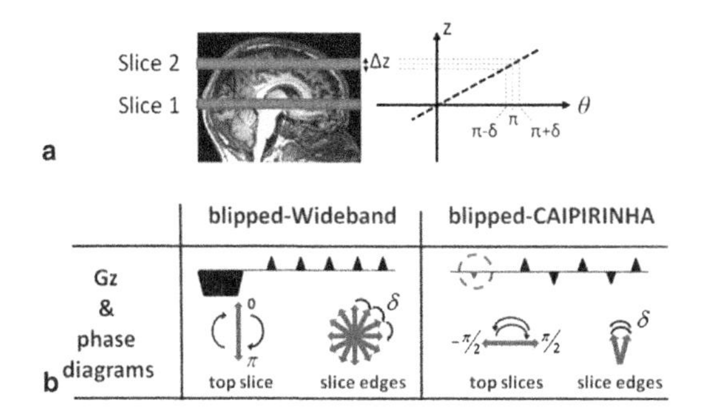

FIGURE 7.9

Illustration from Setsompop et al. [61] of the accumulating phase for blipped-wideband and blipped-CAIPI slice encoding.

Nyquist ghosting, one of the most common artifacts in echo-planar imaging, originates from eddy-current delays in the alternating readout gradients leading to nonalignment of the odd and even k-space lines, B0 inhomogeneity, eddy currents from slice-encoding blips, and other sources. The ghost correction becomes particularly tricky in multiband EPI since ghost correction prior to separation of the simultaneously excited slices would theoretically correct for slice-averaged ghosting only. A "tailored ghost correction" where residual ghosting after separation of slices was corrected using navigator information from the single-band calibration acquisition has been proposed [44].

Just as residual N/2 eddy-current ghosts are hard to separate from aliasing in parallel imaging, they complicate slice unaliasing as well. The eddy currents from opposite's polarities of the readouts create ghosting that can normally be corrected with simple time-correction, with the linear phase estimated from the EPI navigator. While the simplest eddy-current model is a spatially and temporally invariant single timing correction, the eddy currents from gradient switching vary spatially and different slices may need different correction [3, Chapter 16.1]. The common extensions from this simplest model are slice-dependent eddy-current and 2D-eddy-current correction models. The latter is more common for oblique imaging, where two gradient elements are used simultaneously. The 2D eddy-current correction will not be considered here, but references for the more complex models are Nyquist Ghost Corrected SENSE-GRAPPA (NGC-SENSE-GRAPPA) [31] image space [37], k-space [33]. When the eddy-currents have enough spatial variation, the ghosting in the PE direction acts similarly, albeit with more subtlety, to the blipped-CAIPI shift. To avoid the interference of eddy current-induced ghosting with slice and phase unaliasing, the SENSE-GRAPPA approach can be applied with RO as the extended dimension as in RO-SENSE-GRAPPA or the slice-GRAPPA approach that reconstructs individual slices can be used. Once the slices have been resolved, for either approach, the slice-specific eddy-current correction can be applied to obtain an "artifact free" SMS-EPI reconstruction.

7.3.6.2 *Slice-GRAPPA with dual polarity*

In the slice-GRAPPA approach, to account for eddy currents from alternating phase-encoding, the slice-GRAPPA projection was modified to map groupings of signals that are consistent. Essentially, a unique kernel is needed to distinguish between even–odd–even phase-encoding patches and odd–even–odd phase-encoding patches, and these can be constructed as a single kernel. For example, the 3×3 kernels

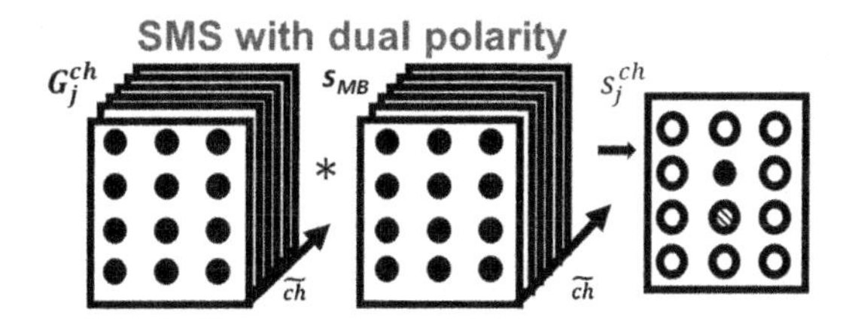

FIGURE 7.10

Illustration of the dual-polarity kernels as a single interpolation kernel.

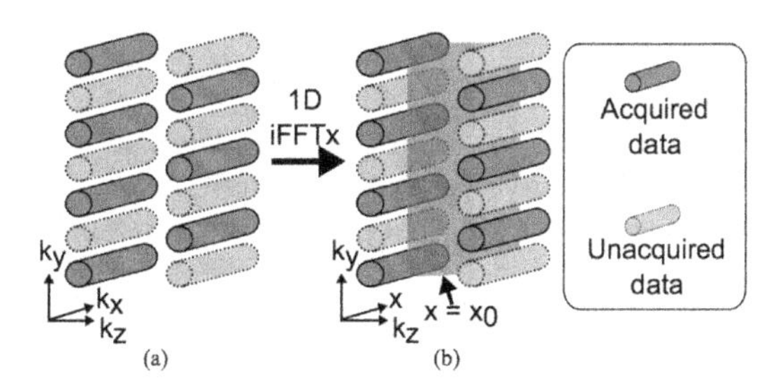

FIGURE 7.11

The hybrid-space SENSE reconstruction adopts a 3D representation of SMS acquisitions. The 3D k-space data, **a**, is transformed into the hybrid space, b, by a 1D inverse FFT. For each point $\boldsymbol{x} = x_o$, the unknown magnetization on the $\boldsymbol{y}$-$\mathbf{z}$ plane is solved by $\boldsymbol{m}_{x_o} = \boldsymbol{E}_{x_o}^{-1}\boldsymbol{s}_{x0}$ where $\boldsymbol{E}_{x_o}$ is the encoding matrix consisting of sensitivity and $\boldsymbol{k}_y - \boldsymbol{k}_z$ phase encodings, and $\boldsymbol{s}_{x0}$ is a vector containing the measured signals on the $\boldsymbol{k}_y - \boldsymbol{k}_z$ plane.

were effectively extended to a 4 × 3 kernel as illustrated in Fig. 7.10. These extended slice-GRAPPA kernels are not shift-invariant as the k-space is normally organized, but shift-invariant for double shifts in the phase-encoding direction. The similar reasoning has subsequently been used in in-plane parallel imaging as dual-polarity GRAPPA [27,28].

7.3.6.3 SENSE-model for EPI

For parallel imaging reconstruction, there is clear link between SENSE- and GRAPPA-type reconstructions, and the same constructs holds for SMS. As discussed earlier, the SENSE model might indeed be the simplest way to consider how to separate fully sampled SMS, and it is conceptually equivalent to the Hadamard construction, where the multiple phase-modulated acquisitions have been replaced with single acquisition with multiple receivers. Both phase-encoding undersampling and the Nyquist ghost from the EPI echo train can be incorporated in the SENSE framework using a more complex encoding model. Zhu et al. [81] considered the same eddy-current model as illustrated in Fig. 7.11.

For SENSE with spatially varying eddy currents, the authors introduced the matrix-decoding for SENSE, represented as

$$s_j\left(x_0, k_y(n), k_z(n)\right) = \sum_{y=0}^{N_y-1}\sum_{z=0}^{N_s-1} x(x_0, y, z)\cdot C_j(x_0, y, z)\cdot e^{-i[k_y(n)\cdot y + k_z(n)\cdot z]} e^{i\theta(x_0, n, z)} \tag{7.9}$$

for $n = 1, 2, \cdots, N_p$ and $j = 1, 2, \cdots, N_c$ with N_p the number of slices and N_c the number of channels, and x_0 a fixed readout frequency. Using this formulation and relying on well-defined sensitivity profiles provided equivalent results to SpSg and was reformulated in computational efficient form by Hennel et al. [25].

7.4 Calibration and reference scans

As with in-plane parallel imaging, how to acquire calibration data, be it for deriving of coil sensitivity information in the SENSE framework or for computation of calibration weights in GRAPPA based approaches, is a key design consideration in SMS imaging. While a prescan-based coil sensitivity estimation was originally proposed for SENSE by Pruessmann et al. [56], autocalibration scan (ACS) has the advantage of sharing the same image contrast as the undersampled data [23]. The pros and cons and the applicability of various calibration schemes is discussed in detail by Griswold et al. [22], and these considerations in general hold for SMS as well.

The main difference in calibration need for SMS compared to in-plane parallel imaging is that calibration information is needed from the multiple individual locations that are excited simultaneously in the accelerated acquisition. In case of SMS structural imaging, for expediency, calibration data is usually extracted from a rapid gradient-echo sequence with a conventional single-band excitation since the reconstruction methods are largely robust against differences in contrast between the imaging and calibration data. An exception to this is echo-planar sequences.

7.4.1 Calibration and reference scans for EPI

In EPI it is preferred that the calibration data has similar presence of Nyquist ghosts as the accelerated data so that these can be accounted for in the reconstruction either explicitly or implicitly. This can be achieved with a single-band EPI calibration using otherwise identical scan parameters including matched in-plane acceleration as the SMS scan, and additionally an ACS dataset that is used to compute reconstruction weights for in-plane undersampled data once the SMS data is resolved into individual slices [44] (Fig. 7.12).

Since fMRI and DTI scans include repeated volume measurements, it might be possible to spend a little more additional time on the calibration. One such approach is to acquire a fully distortion-matched and Nyquist ghosting-matched echo-planar calibration dataset by acquiring N phase cycled repetitions with the multiband excitation and R undersampled shots, where N and R are the SMS and in-plane parallel imaging-acceleration factor, respectively. In each of the phase cycled repetitions p, the phase offset between the sub-bands is varied so that they are uniformly distributed over a unit circle.

$$RF_{MB,p}(t) = A(t)\,.\sum_{n=1}^{N} e^{i\,\Delta\omega_n(t)+\varphi_{n,p}}\,. \tag{7.10}$$

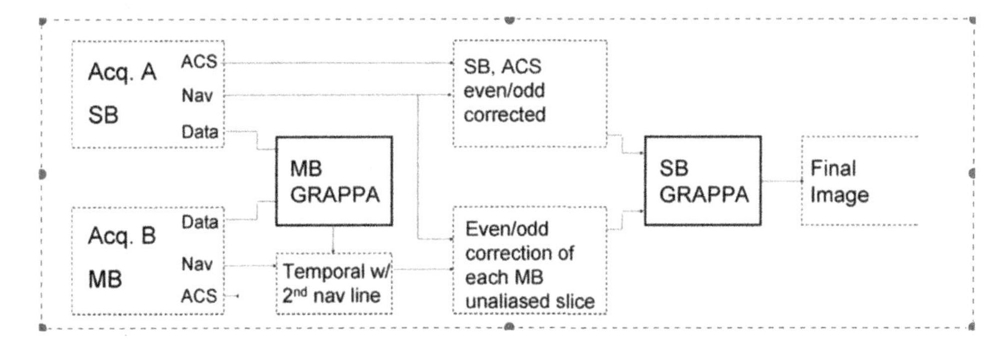

FIGURE 7.12

The flow diagram for RO-SENSE-GRAPPA for EPI with phase-encoding undersampling from Moeller et al. [44]. Each of the single-band data have an ACS acquisition and a phase-encoding undersampled acquisition in addition to a the-line navigator. The single-band phase-encoding undersampled acquisitions are used in an MB-GRAPPA (RO-SENSE-GRAPPA) reconstruction to resolved the SMS (MB) acquisition. After resolving the slice, the resolved slices are corrected for slice-specific eddy currents determined from the single-band acquisition. Lastly, a conventional GRAPPA reconstruction is used to reconstruct the full-space for each channel and each slice.

For an SMS factor of 3, the phase offset between slices would be 0, $\pi/3$, $2\pi/3$. The calibration data for each individual slice location can be extracted by application of the Fourier theorem to the phase-cycled multiband data, and the fully sampled data for each slice location could be collated from the individual shots. This has been demonstrated to provide a robust calibration and low ghosting levels, especially at 7T [30,64]. However, the NxR additional TRs needed for calibration acquisition would undercut the scan-time savings and would not be suited to clinical DWI and anatomical SMS imaging.

The FLEET technique (fast low-angle excitation echo-planar technique) [51,76] is a popular single-band calibration technique, which acquires all segments belonging to a slice in immediate succession rather than further apart in time, alleviating the concern of motion or breathing between segmented acquisitions. It has also been shown to exhibit low residual aliasing in SMS reconstructed images. To avoid complexities of echo-planar data altogether, fast gradient-echo sequences with low flip angles integrated in prescan are also a practical choice for calibration [66]. The optimal choice of calibration acquisition ultimately depends on the specific needs of the imaging application.

The g-factor for the k-space based methods differ based on the SNR of the ACS data. The noise in the ACS data acts a regularization term, as discussed in Ding et al. [16], and this is similar across SMS reconstruction methods. Some of these considerations are taken into account in an optimization of the hyperparameters of SMS reconstruction by Muftuler et al. [45].

7.5 Reconstruction metrics

The Signal-to-Noise (SNR) ratio is the primary measure for data quality and where SMS has an advantage over parallel imaging. With SMS, the thermal noise is shared between the N slice locations

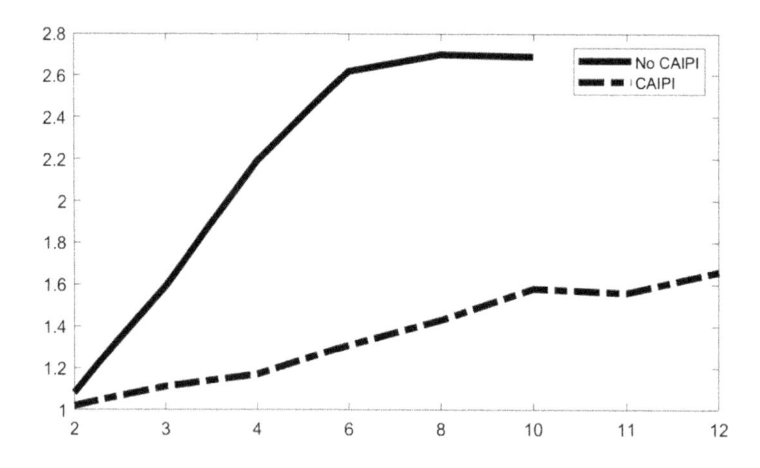

FIGURE 7.13

Plot of mean g-factors from Xu et al. [77] for SMS for whole brain imaging without CAIPI and for CAIPI with 1/4 the FOV shift between slices. Without CAIPI the g-factor increases rapidly and plateaus around N = 6. For CAIPI the g-factor increases more gradually.

that are excited and acquired simultaneously, and there is some SNR loss associated with resolving these N slices, but there is not a sqrt(N) SNR penalty due to reduced sampling—as is the case for phase-encoding undersampling.

For completeness, we should point out that SMS will have lower SNR by a factor of sqrt(N) when compared with a fully sampled acquisition such as the Hadamard acquisition discussed earlier in the chapter, which would be composed of N individual SMS acquisitions. This sqrt(N) SNR penalty is analogous to the phase-encoding undersampling in 3D acquisitions with N-fold undersampling.

7.5.1 Noise amplification

Noise amplification in reconstruction beyond that due to reduced data measurement in k-space is quantified by the geometry factor, better known as the g-factor, in conventional parallel imaging. An analogous g-factor can be calculated for the SMS reconstruction methods previously described since it can be for any linear reconstruction. The calculation can be done with the encoding matrix for SENSE [56], Monte Carlo simulation method for linear k-space methods [58], or directly from the GRAPPA-kernels [9].

An initial evaluation for g-factors in SMS brain imaging was performed with a GRE acquisition for SMS = 2, …, 12, [77] using various CAIPIRINHA shifts. In Fig. 7.13, g-factors for adjacent slices using slice-GRAPPA without CAIPIRINHA and with a CAIPIRINHA shift of 1/4 the FOV are shown. From this figure the mean g-factor without CAIPIRINHA increases linearly up to SMS = 6 and then plateaus. Likewise, the g-factor with CAIPIRINHA increases linearly with increases in SMS factors, but at a slower rate. The plateau seen without CAIPIRINHA is a consequence of a few observations. For the same amount of ACS data, the linear system has more unknowns as the SMS factor increases. Furthermore as the SMS factor increases, the kernel sizes also have to increase to encode the systems better, further increasing the number of unknowns, and making the system less overdetermined. Since the system is less overdetermined, it is easier to obtain a solution that fits well to the calibration data, but which might not be as reliable when applied to new data—typically reflected as residual aliasing. For CAIPIRINHA the same challenges of increasing number of unknowns is inherent, but the encoding model is better-conditioned, and as such it is easier to find the "right" solution. The dip for SMS = 11 and SMS = 12 leads to the question of what is being lost at the higher acceleration factors since it does not appear to be the noise amplification that is the limitation.

7.5.2 Residual aliasing

These observations are what led to considering residual aliasing as a metric, and also led to the proposition and application of SpSg. Considering the residual leakage provided the desired complementary insight. Two different approaches for this have been pursued: The simplest is to consider $\widetilde{MB}(\cdot) = [\cdots MB_j(\cdot)\cdots]$ since the combination of individual MB_j projection operators for recovering signal from slice j such that

$$\widetilde{MB}\left(\sum_{j}^{N} s_j\right) = \widetilde{MB}(s_1) + \cdots + \widetilde{MB}(s_N) = \sum_{i,j} \tilde{s}_{ij} = \sum_{\substack{i,j \\ i \neq j}} \tilde{s}_{ij} + \sum_{i=j} \tilde{s}_{ij} + \epsilon, \tag{7.11}$$

where $\tilde{s}_{ij} = MB_j(s_i)$ is the reconstruction across all channels for slice i of the input signal from slice j only. The first term is the sum of all aliased signal components, and ϵ denotes the inherent noise amplification. The original approach used a spectral analysis of the reconstructions for a slice $\hat{s_j}(t)$ determined from

$$MB_j\left(\sum_{j}^{N} \mu_n(t)\, s_j(\vec{k})\right) = \hat{s_j}\left(\vec{k}, t\right) = E^H(\hat{\boldsymbol{x}}_j(t)), \tag{7.12}$$

where E is the sensitivity weighted-channel combination and $\mu_n(t) = (1 + 0.1 \cdot \cos(4\pi \cdot n \cdot t))$ is chosen as a modulation function that is nonzero and resolvable with the Fourier transform. For both Eq. (7.11) and Eq. (7.12), the reconstructions have three components, the "true" resolved signal, the residual aliased signal and the noise amplification. For either method, comparing the "true" resolved signal to the residual aliased signal is elevated by the noise. One approach taken to quantitatively determine the signal leakage was to compute the "temporal" spectrum

$$\hat{x_j}(f) = F_t\left(\hat{x_j}(t)\right), \tag{7.13}$$

where μ_n defines, through construction, the frequencies f_n unique relative to the input of each slice. For a given frequency f_n, the existence/contribution of the signal in unaliased slices that are different from the input slice is then the amount of residual signal (slice leakage signal). The amount of signal leakage is quantified by the L-factor and determined from Eq. (7.13) as the largest signal in a slice distinct from the source slice. To this end, signal contributions to a slice, with a uniquely probed frequency f_n relative to other probed frequencies $f_{n'}$ are identified, masked ($\Omega = \{x_j \gg 0\}$), and summed

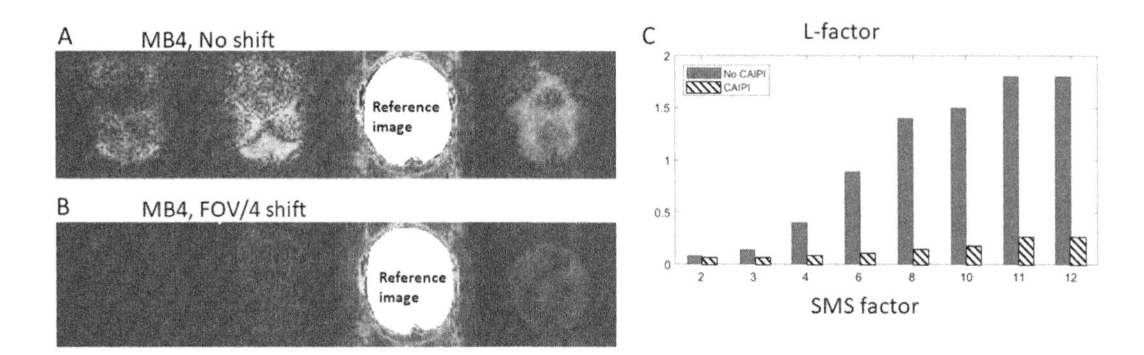

FIGURE 7.14

Illustration from Xu et al. [77] and adapted to grayscale for slice-leakage without, **A**, and with, **B**, 1/4 the FOV shift between slices for SMS with N = 4. The right figure, **C**, is a barplot of the numbers from Xu et al. [77] of the slice-leakage, and where the g-factor plateaued for N = 6 without CAIPI in Fig. 7.12, is when the slice-leakage effectively reached 100%.

$$L(n) = \sum_{r \in \Omega} \sum_{n' \neq n} \frac{\hat{x}_n(r, f_{n'})}{\hat{x}_n(r, f_n)} \tag{7.14}$$

and the maximum and average of $L(n)$ respectively quantified.

7.5.3 Qualitative effect of slice leakage

The qualitative effect of slice leakage, as introduced in Xu et al. [77], is shown for SMS with SMS = 4 in Fig. 7.14A, and quantitatively in Fig. 7.14B.

Determining how much slice-leakage is tolerable, and correspondingly which algorithm should be used for balancing noise-amplification against aliasing, can be guided by the application. Using a functional task, it was shown in Todd et al. [68] that while SpSg reduced false-positives due to residual aliasing in the SMS, it also reduced the temporal SNR as shown in Fig. 7.15. This decrease in temporal SNR correlates with an increase in g-factor for the SpSg kernels.

For rs-fMRI a detailed evaluation of SMS acquisitions with SMS = 3, and no CAIPI shift did not indicate any detrimental effects from residual aliasing. The level of residual aliasing as measured with the L-factor characterized these levels as 0.03 for the mean L-factor, 0.14 for the 99 percentile, and 0.44 for the maximum (e.g., in the order of 3, 14 and 44%, respectively). These levels were further increased in Xu et al. [77], to 0.05, 0.3 and 0.9, respectively, and were then used to justify the use of SMS = 8 with blipped CAIPI, although they indicated that higher SMS factors could be used without excessive increases in residual aliasing. In this case the Leakage factor provided a better discriminator compared with the g-factor in establishing feasible ranges for SMS. In addition to the application specific threshold in [77], the signal leakage is further reduced by using any of the unbiased methods, with an increase in the g-factor, which then can be evaluated in the classical SNR-efficiency sense.

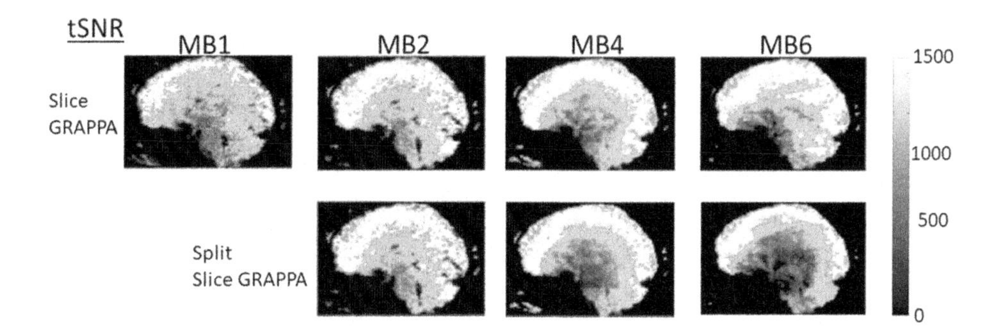

FIGURE 7.15

The effect of SMS reconstruction strategy on temporal SNR (tSNR) in fMRI adapted to gray-scale from Todd et al. [68]. The temporal SNR (tSNR) between slice-GRAPPA and split sliceGRAPPA is negligible for MB2 and small for MB4. For MB6 the difference in tSNR is larger both medially and peripherally. For fMRI the reduction in false-positives for MB6 with split slice-GRAPPA was superior to the better tSNR with slice-GRAPPA.

7.6 Extensions of SMS

7.6.1 SMS and 3D imaging

SMS can be viewed as a 3D imaging method in that it handles multislice acquisition, but it differs from most common 3D MRI sequences in the sense that it does not acquire a single contiguous volume. For 3D encoding, CAIPIRINHA uses either RF or gradient-based phase modulation to obtain (undersampled) data with reduced effective aliasing. By considering the SMS encoding in the context of 3D encoding, the link between the first introduction of Hadamard encoding and the blipped CAIPI is more transparent, and it even makes it clearer how to extend SMS for specific applications.

Considering first Hadamard encoding for two slices, A and B, as illustrated in Fig. 7.16a, with a 180° phase modulation for the second slice. Each Hadamard acquisition is a specific k_z plane in the SMS k-space. The CAIPIRINHA sampling alternates between the two Hadamard acquisition conditions, and for each phase-encoding it only samples one of the Hadamard encoding conditions. In CAIPIRINHA both Hadamard conditions are acquired, both Hadamard conditions are undersampled, and the undersampling pattern is shifted relative to one another. When fully sampled, the FFT can be applied through k_z, which is the same as the Hadamard combination of slice A + B and slice A–B. As a second example, in Fig. 7.16b, for $N \times R = 8 \times 2$ with a regular CAIPIRINHA shift of FOV/4, a similar seesaw sampling pattern is encoded. The pattern for which k_z phase-encodes are acquired in a SMS acquisition is reflected in how the CAIPIRINHA shifts are implemented with slice gradient variations, and, since the slices are getting closer, a larger gradient blip is necessary to impart the slice-specific phases.

The blipped-wideband and blipped-CAIPI approaches can also be considered in this construct so that the monotonic increasing phase in blipped-wideband is equivalent to acquiring on a diagonal in the SMS k-space, which is equivalent to the voxel-tilting effect [61].

Previously, a multislab approach that divided the whole acquisition into several sub-volumes [19] and conducted a 3D encoding for each subvolume has been demonstrated for whole brain diffusion and fMRI studies [53,54]. High SNR efficiency and the achievable high resolution make these approaches appealing, though there are technical challenges from intershot phase errors varying in all three direc-

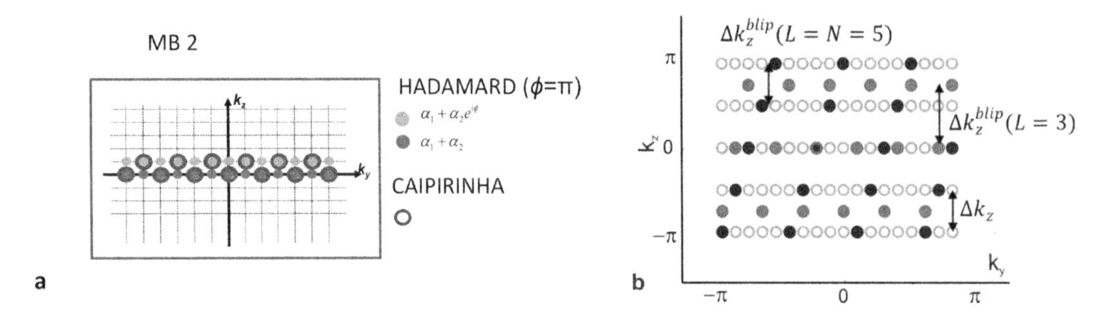

FIGURE 7.16

3D k-space and kz phase-encoding in SMS. **a)** The phase in Hadamard encoding, and the corresponding CAIPIRINHA samples for two slices. *For CAIPIRINHA, when the FFT is applied across all the phase-encoding samples, slice* a_1 *and* a_2 *are fully sampled, and the FOV of* a_2 *is shifted by 1/2 the FOV when* $\phi = \pi$. For **b)**, an example from Zahneisen et al. [80] of the sampling with five slices, and a FOV/3 shift between adjacent slices. Each row is the phase combination that would be obtained in one of the five acquisitions in a Hadamard acquisition. The blue (dark gray in print version) dots is the sampling for a 1/5 FOV shift, and the red (mid gray in print version) dots are the sampling acquired when using a 1/3 FOV shift [80].

tions and slab-boundary artifacts [17,70,75]. A combined multislab SMS approach was demonstrated, where high resolution is achieved by simultaneous reconstruction of overlapping slabs [62,72]. In the gSLIDER-SMS approach, slabs were RF encoded with a scheme similar to Hadamard to further improve orthogonality of the reconstruction bases [60]. SMSlab, a novel 3D k-space Fourier encoding and reconstruction framework proposed by Dai et al., applies joint RF and gradient encoding to a more traditional 3D multislab acquisition where interslab gaps exist [14]. Bruce et al. [10] formulated multislab multishot SMS data reconstruction in an expanded SENSE formalism that addressed both aliasing and motion correction.

7.6.2 Non-Cartesian SMS

The primary use of SMS is for Cartesian applications, but as with most encodings, it can also be used for non-Cartesian acquisitions. For most non-Cartesian parallel imaging reconstructions, the CG-SENSE framework is used [55]. For CG-SENSE any known MR specific encoding effects can be included into the reconstruction and combined with undersampled encoding when used with multichannel systems.

For SMS in non-Cartesian applications, two different classes of encoding are used: the equivalent of CAIPIRINHA and blipped-CAIPI. To consider the differences in the FOV shift for non-Cartesian encoding, we briefly discuss both radial and spiral SMS.

For radial acquisitions [79], the CAIPIRINHA (or Hadamard) encoding is used for alternating spokes since each SMS spoke can be encoded with different phases. Especially for the case of two simultaneous slices, [A, B], the application of alternating pairs $[0°, 0°]$ and $[0°, 180°]$ has the unique property that the signal from slice B is essentially cancelled when an NUFFT is used on the series of SMS spokes. To recover the signal from slice B, the SMS spokes are modulated by an alternating 0° and 180° phase, such that the NUFFT of the SMS signal instead cancels slice A and recovers slice B.

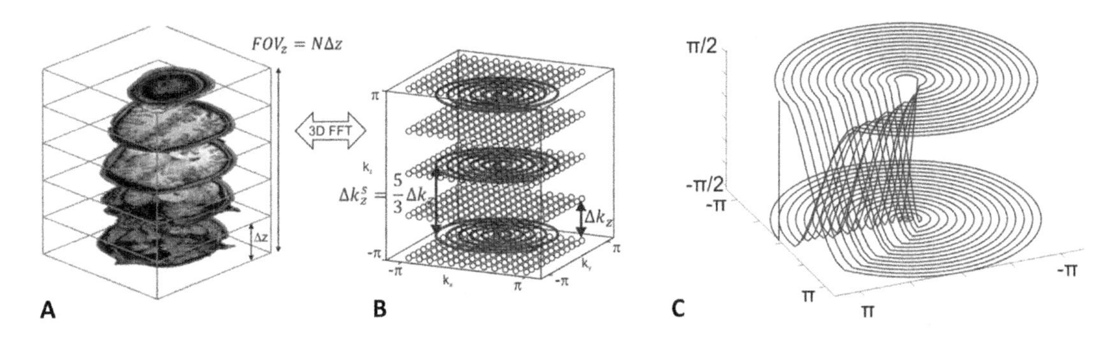

FIGURE 7.17

Illustration of the connection between SMS encoding in image space **(a)**, and sampling in k-space **(b)** for an SMS acquisition with $N = 5$, and 1/3 FOV between slices. Each k_z-plane can be sampled either with Cartesian or non-Cartesian sampling. **(c)** illustrates how to add slice-encoding blips into a single-shot spiral readout. Adapted from Zahneisen et al. [80].

This is then independent of multichannel systems, but, for more than two slices, they found that a more complete reconstruction is necessary analogous to the previous two-slice Hadamard example.

For spiral acquisitions, the blipped-CAIPI with slice-encoding blips is more amendable, although for segmented spiral acquisitions (and notably similarly for segmented Cartesian EPI), both blipped-CAIPI and CAIPIRINHA can be used. For EPI, the blipped-CAIPI slice-encoding are designed during the acquisitions dead time where the phase-encoding blips is used. For spiral acquisitions there is no inherent dead time where slice-encoding blips can be inserted. Instead, acquisition dead times have to be "created" where slice-encoding blips can be inserted. The design of the blips are chosen such that the inherent 3D-SMS k-space for Fig. 7.16b is approximately uniformly undersampled. For SMS spiral blipped-CAIPI, the k-space and sampling is illustrated in Fig. 7.17b.

For a technique such as spiral blipped-CAIPI, improvements from field monitoring have been shown to be advantageous [71]. When segmented acquisitions are used, it has been demonstrated in whole-heart myocardial perfusion imaging to be more advantageous to use a Hadamard cycling scheme [78], instead of spiral blipped-CAIPI additionally reducing the demands on the gradient performance.

Similar to the use of iterative reconstruction with constraints for calibration consistency and acquired data consistency that were previously applied to improve non-Cartesian in-plane undersampled data [36], slice-SPIRiT model has been developed to improve MB spiral cardiac MR data [65,78] and Cartesian cardiac data [15].

7.7 Applications of SMS

SMS imaging has proved to be a useful speedup tool, in addition to parallel imaging, in several clinical applications. Reports on SMS-accelerated clinical knee and spine imaging were presented by Fritz et al. [18] and Longo et al. [34], respectively. Norbeck et al. [47] applied SMS to PROPELLER [50] where they acquired the calibration data with single-band excitation from a single blade in central k-space.

From imaging contrast considerations, short TRs enabled by SMS might help improve T1 contrast. In T2 and T2 FLAIR-weighted imaging, longer TRs are needed to achieve the desired image-contrast weighting. In these applications, SMS is better utilized by extending slice coverage or acquiring thinner slices for the same slice coverage while keeping the TR same as in the original protocol.

One factor limiting the adoption of SMS technique to Rapid Acquisition with Refocusing Echoes (RARE), also known as turbo-spin echo (TSE) or fast-spin echo (FSE), which is the clinical workhorse for anatomical brain, spine and musculoskeletal imaging has been the higher peak B1 amplitude and energy of multiband pulses in a sequence that is already high in energy deposition. The use of PTx can help with reducing the increases in B1 [43]. Increasing the duration of the RF pulse would reduce peak B1 but would increase the echo spacing and have detrimental effect on image quality. Other mitigation strategies such as variable rate excitation [13] and Power Independent of Number of Slices (PINS) radiofrequency pulses [48] have associated challenges such as sensitivity to off-resonance and periodic slice profiles. Massive acceleration in brain imaging was demonstrated by Gagoski et al. [20] with SMS Wave-CAIPI and multiPINS pulses.

Comparatively, SMS has been widely adopted with EPI to enable higher direction diffusion tensor imaging in clinically feasible scan times and to enable higher temporal/spatial resolution for functional MRI.

The benefit of SMS in functional MRI is quite obvious—acquisition of more time points within a given scan time increases the statistical power. In the human connectome protocol, an SMS factor of 8 is employed for the GRE EPI fMRI protocol at 3 Tesla.

SMS can also help achieve a high enough spatial and temporal resolution in GRE EPI based Dynamic Susceptibility Contrast (DSC) perfusion techniques. An SMS multi-echo EPI sequence was used for DSC perfusion imaging with leakage correction and vascular permeability measurements [74]. SMS was demonstrated with Spin and Gradient Echo (SAGE) EPI to provide more extensive slice coverage and detailed information about tumor vasculature compared with conventional GRE DSC perfusion imaging [24].

For SE EPI Diffusion Imaging, both the excitation and the refocusing pulses need to be multiband, so typically SMS factors are limited to a factor of 3. Application of SMS has now enabled diffusion tensor imaging with a high number of directions and multiple shells in clinically feasible scan times. Furthermore, in diffusion imaging, shortening the TR below the 3000–3200 ms range at 3T for multislice imaging is not recommended because of T1 relaxation considerations in gray matter [43]. It needs to be noted that reduction in TR by using higher SMS factors can also lead to a reduction in the SNR of the SMS data.

Since motion poses serious challenges to SMS reconstruction, adoption of SMS in diffusion outside the brain has been limited to anatomies less affected by physiological motion, such as the breast and prostate. Segmented diffusion techniques, such as readout-segmented echo-planar imaging (rsEPI) [52] and multiplexed sensitivity-encoding (MUSE) [12], enable high-resolution diffusion, but the scan time increases proportionally to the number of shots. When SMS is extended to such techniques to reduce the TR and scan time [26,29,38], additional steps need to be considered for phase reconciliation between the segments. Multislab SMS acquisitions are generally used only in applications where submillimeter resolution is desired. McKay et al. [40] did a study in breast diffusion in which they used SMS to obtain the left-right coverage with thin-slice sagittal acquisitions and compared the axially reformatted slices against axially acquired rsEPI with five shots with comparable scan time and a faster regular ssEPI DWI acquisition. Image quality for the axially reformatted SMS acquisition was found to be the highest. Some innovative extensions of SMS in DWI include multiband refocusing with 2D RF excitation [1] and in-plane multiband (multiplexed phase FOV) with 2D RF excitation pulse for high-resolution breast DWI [67].

SMS remains an active field, and new clinical applications employing this technology continue to emerge. In this section we have attempted to provide a few examples of its clinical utility rather than an exhaustive list.

7.8 Summary

In summary, this chapter describes the evolution of SMS from multislice Hadamard and POMP techniques that were proposed before the advent of either multicoil receiver arrays or parallel imaging. It then explains the various SMS reconstruction algorithms with special emphasis on adaptations of these methods to echo-planar imaging, lays the framework for comparison of the different reconstruction methods based on reconstruction metrics such as noise amplification and residual aliasing, and discusses a few practical design considerations and trade-offs, such as choice of calibration scan and protocol optimization in the context of imaging applications that have widely adopted SMS technology.

Similar to slice-encoding undersampling in parallel imaging (Chapter 6), SMS methods use the spatial encoding of coils to resolve aliasing between simultaneously excited slices and hence can only be supported by a receiver coil with appropriate coil geometry having separation of elements in the slice direction.

FOV shifts, commonly known as CAIPIRINHA shifts, effected by RF or gradient-phase modulation, are ubiquitously used with SMS to alter the location of aliased to improve the geometry factor and the conditioning of the reconstruction matrix. The encoding with SMS and CAIPIRINHA reduces dependency on the coil geometry and accommodates the use of higher acceleration with relatively few channels, as opposed to parallel imaging which, due to higher g-factors, generally is not used for accelerations larger than two with eight or less channel elements.

To conclude, SMS technology is a powerful speedup tool that can be used independently or in conjunction with parallel imaging to shorten scans or to accommodate more flexibility in the imaging protocol such as increased resolution or coverage.

7.9 Exercise

Workshop Tutorial from the ISMRM Workshop on Simultaneous MultiSlice Imaging: Neuroscience & Clinical Applications, Pacific Grove, CA, USA, 19–22 July 2015.

Material is created primarily by Felix Breuer, with contributions from Kawin Setsompop and Steen Moeller for the Split slice-GRAPPA and unbiased slice-GRAPPA reconstructions (https://dx.doi.org/10.5281/zenodo.5149780) [7].

7.9.1 Content of tutorial

Part I: SENSE
Part IIa: SENSE/GRAPPA Hyprid with ACS arranged according two cmaps in SENSE

Part IIb: SENSE/GRAPPA hybrid with extended FOVs
Part III: Slice-GRAPPA

7.9.2 Questions

The default setup is SMS with N = 4, and CAIPIRINHA with a 1/4 FOV shift between adjacent slices.

a) Construct CAIPIRINHA shifts for 1/2, 1/3, 1/4 and 1/5 FOV shifts.

- If the FOV shifts are larger than 1, what does it mean for gradient and RF encoding respectively?

b) For reconstruction with CAIPIRINHA encoding using either SENSE-GRAPPA or slice-GRAPPA, implement the FOV shift in k-space to correct for the CAIPIRINHA encoding.
c) Determine how the mean and maximum g-factor vary for the SENSE-GRAPPA, slice-GRAPPA, Split slice-GRAPPA and unbiased slice-GRAPPA methods for various multiband factors and various FOV shifts.
d) Generate and calculate the leakage factor for SENSE-GRAPPA, slice-GRAPPA, Split slice-GRAPPA and unbiased slice-GRAPPA.

Appendix 7.A Extended FOV methods for SMS

7.A.1 PE-SENSE-GRAPPA

Let us now revisit the SENSE-GRAPPA technique and pictorially consider the differences between the concepts of the different k-space techniques in the context of the CAIPIRINHA shift (the FOV shift in the PE direction for different slices). For SENSE-GRAPPA as described by Breuer et al. [8] (which we will term PE-SENSE-GRAPPA), the reconstruction aims to resolve multiple slices as shown in Fig. 7.18.

As discussed in the chapter on parallel imaging, small convolution k-space kernels can also be used as image multiplication kernels. These kernels will be smooth in the image domain since they are small in k-space they have a broader support in image-space. Likewise, since they only have a small number of Fourier coefficients with a large zero-filling component, they will also be smooth relative to the large FOV. These kernels are not well-suited to resolving the slices, when there is signal right up to the edge of the slice, such as the case when the FOV is shifted, or when it is aliased. The original introduction of CAIPIRINHA with a SENSE-GRAPPA reconstruction addressed this by adding artificial blank slices to the calibration volume, but, since this has not received much attention, that "trick" will not be further belabored in this chapter.

7.A.2 Unbiased slice-GRAPPA

The construct, with suppressed residual aliasing in SpSg, can also be achieved in a way that is more analogous to parallel imaging, using the SENSE-GRAPPA framework and a link between POMP and Hadamard encoding. The MB (or equivalently the $\sum_j s_j$ in slice-GRAPPA) is a specific subset of the extended kspace for SENSE-GRAPPA. This subset is the 1/N regular undersampled data from SENSE-GRAPPA. The other (N-1) subsets in the extended FOV are constructions of various combinations of

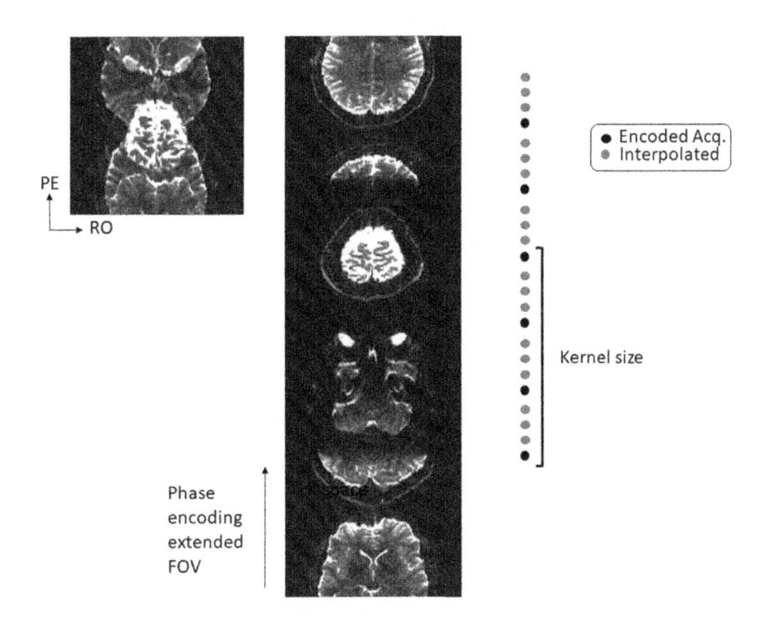

FIGURE 7.18

Reconstruction of SMS with N = 4 and 1/4 FOV shift between slices in the concatenated PE-FOV for SENSE-GRAPPA. The CAIPIRINHA shifted in the PE direction is preserved with the GRAPPA reconstruction, and separating the slices introduces artifacts at the edges of the FOV. In the bottom of the figure, the dimensions of the GRAPPA kernel is illustrated for the extended k-space in the phase-encoding direction.

the single-band in addition to $\sum_j \boldsymbol{s}_j$. To extend slice-GRAPPA to include the additional information, the unbiased slice-GRAPPA is framed as solving

$$\boldsymbol{G}_j\left(\sum_i \alpha_i \boldsymbol{s}_i\right) = \boldsymbol{s}_j \quad \forall j \quad \forall \alpha_i \text{ with } |\alpha_i| = 1 \quad \text{Unbiased Slice-GRAPPA} \tag{7.A.1}$$

Practically, and since $\boldsymbol{G}_j$ is a linear operator, only a finite set of different combinations is necessary, and these can be created by the using α_i as the basis-vector in the Fourier matrix (or are possible, in general, in any complex Hadamard matrix). The solution is then

$$\arg\min_{\substack{G_i \\ \forall i}} \left\| \boldsymbol{G}_i \left(\sum_j^N \tilde{s}_j^k\right) - \tilde{s}_j^k \right\|_2^2 \tag{7.A.2}$$

where $\tilde{s}_j^k = H(i,k)\, s_i$ with H the complex Fourier matrix. This is equivalent to increasing the number of equations an additional (N-1)-fold.

With this formulation, Sp-Sg, Unbiased slice-GRAPPA and RO-SENSE-GRAPPA, are equivalent in performance [41] with respect to both g-factor noise and interslice leakage, and all superior to the original slice-GRAPPA in terms and accuracy and inferior in terms of noise-amplification.

References

[1] Banerjee S, Saritas EU, Connett R, Shankaranarayanan A. Reduced field-of-view diffusion with 2D echo-planar RF excitation and multiband refocusing for extended slice coverage and robust fat suppression. Int Soc Magn Reson Med 2014;4437.

[2] Beatty Philip James, Brau Anja C, Chang Shaorong, Joshi Sanjay M, Michelich Charles R, Bayram Ersin, et al. A method for autocalibrating 2-D accelerated volumetric parallel imaging with clinically practical reconstruction times. Int Soc Magn Reson Med 2007;1749.

[3] Bernstein MA, King KF, Zhou ZJ. Handbook of MRI pulse sequences. Amsterdam, Boston: Academic Press; 2004.

[4] Bilgic B, Gagoski BA, Cauley SF, Fan AP, Polimeni JR, Grant PE, et al. Wave-CAIPI for highly accelerated 3D imaging. Magn Reson Med 2015;73:2152–62.

[5] Blaimer M, Breuer FA, Seiberlich N, Mueller MF, Heidemann RM, Jellus V, et al. Accelerated volumetric MRI with a SENSE/GRAPPA combination. J Magn Reson Imaging 2006;24:444–50.

[6] Brau AC, Beatty PJ, Skare S, Bammer R. Comparison of reconstruction accuracy and efficiency among autocalibrating data-driven parallel imaging methods. Magn Reson Med 2008;59:382–95.

[7] Breuer F, Setsompop K, Moeller S. Tutorial for performing simultaneous multi-slice reconstructions in MRI. In: Magnetic resonance image reconstruction. Elsevier; 2021.

[8] Breuer FA, Blaimer M, Heidemann RM, Mueller MF, Griswold MA, Jakob PM. Controlled aliasing in parallel imaging results in higher acceleration (CAIPIRINHA) for multi-slice imaging. Magn Reson Med 2005;53:684–91.

[9] Breuer FA, Kannengiesser SA, Blaimer M, Seiberlich N, Jakob PM, Griswold MA. General formulation for quantitative G-factor calculation in GRAPPA reconstructions. Magn Reson Med 2009;62:739–46.

[10] Bruce IP, Chang HC, Petty C, Chen NK, Song AW. 3D-MB-MUSE: a robust 3D multi-slab, multi-band and multi-shot reconstruction approach for ultrahigh resolution diffusion MRI. NeuroImage 2017;159:46–56.

[11] Cauley SF, Polimeni JR, Bhat H, Wald LL, Setsompop K. Interslice leakage artifact reduction technique for simultaneous multislice acquisitions. Magn Reson Med 2014;72:93–102.

[12] Chen NK, Guidon A, Chang HC, Song AW. A robust multi-shot scan strategy for high-resolution diffusion weighted MRI enabled by multiplexed sensitivity-encoding (MUSE). NeuroImage 2013;72:41–7.

[13] Conolly S, Nishimura D, Macovski A, Glover G. Variable-rate selective excitation. J Magn Res 1988;78:440–58.

[14] Dai E, Wu Y, Wu W, Guo R, Liu S, Miller KL, et al. A 3D k-space Fourier encoding and reconstruction framework for simultaneous multi-slab acquisition. Magn Reson Med 2019;82:1012–24.

[15] Demirel OB, Weingartner S, Moeller S, Akcakaya M. Improved simultaneous multislice cardiac MRI using readout concatenated k-space SPIRiT (ROCK-SPIRiT). Magn Reson Med 2021;85:3036–48.

[16] Ding Y, Xue H, Ahmad R, Chang TC, Ting ST, Simonetti OP. Paradoxical effect of the signal-to-noise ratio of GRAPPA calibration lines: a quantitative study. Magn Reson Med 2015;74:231–9.

[17] Engstrom M, Martensson M, Avventi E, Skare S. On the signal-to-noise ratio efficiency and slab-banding artifacts in three-dimensional multislab diffusion-weighted echo-planar imaging. Magn Reson Med 2015;73:718–25.

[18] Fritz J, Fritz B, Zhang J, Thawait GK, Joshi DH, Pan L, et al. Simultaneous multislice accelerated turbo spin echo magnetic resonance imaging: comparison and combination with in-plane parallel imaging acceleration for high-resolution magnetic resonance imaging of the knee. Invest Radiol 2017;52:529–37.

[19] Frost R, Miller KL, Tijssen RH, Porter DA, Jezzard P. 3D multi-slab diffusion-weighted readout-segmented EPI with real-time cardiac-reordered K-space acquisition. Magn Reson Med 2014;72:1565–79.

[20] Gagoski BA, Bilgic B, Eichner C, Bhat H, Grant PE, Wald LL, et al. RARE/turbo spin echo imaging with Simultaneous Multislice Wave-CAIPI. Magn Reson Med 2015;73:929–38.

[21] Glover GH. Phase-offset multiplanar (POMP) volume imaging: a new technique. J Magn Reson Imaging 1991;1:457–61.

[22] Griswold MA, Breuer F, Blaimer M, Kannengiesser S, Heidemann RM, Mueller M, et al. Autocalibrated coil sensitivity estimation for parallel imaging. NMR Biomed 2006;19:316–24.

[23] Griswold MA, Jakob PM, Heidemann RM, Nittka M, Jellus V, Wang J, et al. Generalized autocalibrating partially parallel acquisitions (GRAPPA). Magn Reson Med 2002;47:1202–10.

[24] Han M, Yang B, Fernandez B, Lafontaine M, Alcaide-Leon P, Jakary A, et al. Simultaneous multi-slice spin- and gradient-echo dynamic susceptibility-contrast perfusion-weighted MRI of gliomas. NMR Biomed 2021;34:e4399.

[25] Hennel F, Buehrer M, von Deuster C, Seuven A, Pruessmann KP. SENSE reconstruction for multiband EPI including slice-dependent N/2 ghost correction. Magn Reson Med 2016;76:873–9.

[26] Herbst M, Deng W, Ernst T, Stenger VA. Segmented simultaneous multi-slice diffusion weighted imaging with generalized trajectories. Magn Reson Med 2017;78:1476–81.

[27] Hoge WS, Polimeni JR. Dual-polarity GRAPPA for simultaneous reconstruction and ghost correction of echo planar imaging data. Magn Reson Med 2016;76:32–44.

[28] Hoge WS, Setsompop K, Polimeni JR. Dual-polarity slice-GRAPPA for concurrent ghost correction and slice separation in simultaneous multi-slice EPI. Magn Reson Med 2018;80:1364–75.

[29] Jiang JS, Zhu LN, Wu Q, Sun Y, Liu W, Xu XQ, et al. Feasibility study of using simultaneous multi-slice RESOLVE diffusion weighted imaging to assess parotid gland tumors: comparison with conventional RESOLVE diffusion weighted imaging. BMC Med Imaging 2020;20:93.

[30] Kelley DAC, Banerjee S, Bian W, Owen JP, Hess CP, Nelson SJ. Improving SNR and spatial coverage for 7T DTI of human brain tumor using B1 mapping and multiband acquisition. Int Soc Magn Reson Med 2013;3642.

[31] Koopmans PJ. Two-dimensional-NGC-SENSE-GRAPPA for fast, ghosting-robust reconstruction of in-plane and slice-accelerated blipped-CAIPI echo planar imaging. Magn Reson Med 2017;77:998–1009.

[32] Larkman DJ, Hajnal JV, Herlihy AH, Coutts GA, Young IR, Ehnholm G. Use of multicoil arrays for separation of signal from multiple slices simultaneously excited. J Magn Reson Imaging 2001;13:313–7.

[33] Liu Y, Lyu M, Barth M, Yi Z, Leong ATL, Chen F, et al. PEC-GRAPPA reconstruction of simultaneous multislice EPI with slice-dependent 2D Nyquist ghost correction. Magn Reson Med 2019;81:1924–34.

[34] Longo MG, Fagundes J, Huang S, Mehan W, Witzel T, Bhat H, et al. Simultaneous multislice-based 5-minute lumbar spine MRI protocol: initial experience in a clinical setting. J Neuroimaging 2017;27:442–6.

[35] Lustig M, Donoho D, Pauly JM. Sparse MRI: the application of compressed sensing for rapid MR imaging. Magn Reson Med 2007;58:1182–95.

[36] Lustig M, Pauly JM. SPIRiT: iterative self-consistent parallel imaging reconstruction from arbitrary k-space. Magn Reson Med 2010;64:457–71.

[37] Lyu M, Barth M, Xie VB, Liu Y, Ma X, Feng Y, et al. Robust SENSE reconstruction of simultaneous multislice EPI with low-rank enhanced coil sensitivity calibration and slice-dependent 2D Nyquist ghost correction. Magn Reson Med 2018;80:1376–90.

[38] Mani M, Jacob M, McKinnon G, Yang B, Rutt B, Kerr A, et al. SMS MUSSELS: a navigator-free reconstruction for simultaneous multi-slice-accelerated multi-shot diffusion weighted imaging. Magn Reson Med 2020;83:154–69.

[39] Maudsley AA. Multiple-line-scanning spin density imaging. J Magn Res 1980;41:112–26.

[40] McKay JA, Church AL, Rubin N, Emory TH, Hoven NF, Kuehn-Hajder JE, et al. A comparison of methods for high-spatial-resolution diffusion-weighted imaging in breast MRI. Radiology 2020;297:303–12.

[41] Moeller S. Simultaneous multi-slice methods. ISMRM 2017:8111.

[42] Moeller S, Auerbach E, Xu J, Lenglet C, Ugurbil K, Yacoub E. Dynamic multiband calibration for improved signal fidelity. In: ISMRM scientific workshop, data sampling & image reconstruction, vol. 19, 2013.

[43] Moeller S, Pisharady Kumar P, Andersson J, Akcakaya M, Harel N, Ma RE, et al. Diffusion imaging in the post HCP era. J Magn Reson Imaging 2021;54:36–57.

[44] Moeller S, Yacoub E, Olman CA, Auerbach E, Strupp J, Harel N, et al. Multiband multislice GE-EPI at 7 tesla, with 16-fold acceleration using partial parallel imaging with application to high spatial and temporal whole-brain fMRI. Magn Reson Med 2010;63:1144–53.
[45] Muftuler LT, Arpinar VE, Koch K, Bhave S, Yang B, Kaushik S, et al. Optimization of hyperparameters for SMS reconstruction. Magn Reson Imaging 2020;73:91–103.
[46] Muller S. Multifrequency selective rf pulses for multislice MR imaging. Magn Reson Med 1988;6:364–71.
[47] Norbeck O, Avventi E, Engstrom M, Ryden H, Skare S. Simultaneous multi-slice combined with PROPELLER. Magn Reson Med 2018;80:496–506.
[48] Norris DG, Boyacioglu R, Schulz J, Barth M, Koopmans PJ. Application of PINS radiofrequency pulses to reduce power deposition in RARE/turbo spin echo imaging of the human head. Magn Reson Med 2014;71:44–9.
[49] Paley MN, Lee KJ, Wild JM, Griffiths PD, Whitby EH. Simultaneous parallel inclined readout image technique. Magn Reson Imaging 2006;24:557–62.
[50] Pipe JG. Motion correction with PROPELLER MRI: application to head motion and free-breathing cardiac imaging. Magn Reson Med 1999;42:963–9.
[51] Polimeni JR, Bhat H, Witzel T, Benner T, Feiweier T, Inati SJ, et al. Reducing sensitivity losses due to respiration and motion in accelerated echo planar imaging by reordering the autocalibration data acquisition. Magn Reson Med 2016;75:665–79.
[52] Porter DA, Heidemann RM. High resolution diffusion-weighted imaging using readout-segmented echo-planar imaging, parallel imaging and a two-dimensional navigator-based reacquisition. Magn Reson Med 2009;62:468–75.
[53] Posse S, Ackley E, Mutihac R, Rick J, Shane M, Murray-Krezan C, et al. Enhancement of temporal resolution and BOLD sensitivity in real-time fMRI using multi-slab echo-volumar imaging. NeuroImage 2012;61:115–30.
[54] Posse S, Ackley E, Mutihac R, Zhang T, Hummatov R, Akhtari M, et al. High-speed real-time resting-state FMRI using multi-slab echo-volumar imaging. Front Human Neurosci 2013;7:479.
[55] Pruessmann KP, Weiger M, Bornert P, Boesiger P. Advances in sensitivity encoding with arbitrary k-space trajectories. Magn Reson Med 2001;46:638–51.
[56] Pruessmann KP, Weiger M, Scheidegger MB, Boesiger P. SENSE: sensitivity encoding for fast MRI. Magn Reson Med 1999;42:952–62.
[57] Nunes RG, Hajnal JH, Golay X, Larkman DJ. Simultaneous slice excitation and reconstruction for single shot EPI. ISMRM 2006;2006:293.
[58] Robson PM, Grant AK, Madhuranthakam AJ, Lattanzi R, Sodickson DK, McKenzie CA. Comprehensive quantification of signal-to-noise ratio and g-factor for image-based and k-space-based parallel imaging reconstructions. Magn Reson Med 2008;60:895–907.
[59] Seiberlich N, Breuer F, Heidemann R, Blaimer M, Griswold M, Jakob P. Reconstruction of undersampled non-Cartesian data sets using pseudo-Cartesian GRAPPA in conjunction with GROG. Magn Reson Med 2008;59:1127–37.
[60] Setsompop K, Fan Q, Stockmann J, Bilgic B, Huang S, Cauley SF, et al. High-resolution in vivo diffusion imaging of the human brain with generalized slice dithered enhanced resolution: simultaneous multislice (gSlider-SMS). Magn Reson Med 2018;79:141–51.
[61] Setsompop K, Gagoski BA, Polimeni JR, Witzel T, Wedeen VJ, Wald LL. Blipped-controlled aliasing in parallel imaging for simultaneous multislice echo planar imaging with reduced g-factor penalty. Magn Reson Med 2012;67:1210–24.
[62] Setsompop K, Bilgic B, Nummenmaa A, et al. Slice dithered enhanced resolution simultaneous multislice (SLIDER-SMS) for high resolution (700 um) diffusion imaging of the human brain. In: Proceedings of the 23rd annual meeting of ISMRM; 2015. p. 0339.
[63] Sodickson DK, Manning WJ. Simultaneous acquisition of spatial harmonics (SMASH): fast imaging with radiofrequency coil arrays. Magn Reson Med 1997;38:591–603.

[64] Banerjee S, Takahashi A, Zur Y, Shankaranarayanan A, Kelly DAC. Robust calibration strategy for multiband EPI at 7 Tesla. Int Soc Magn Reson Med 2012;2298.
[65] Sun C, Yang Y, Cai X, Salerno M, Meyer CH, Weller D, et al. Non-Cartesian slice-GRAPPA and slice-SPIRiT reconstruction methods for multiband spiral cardiac MRI. Magn Reson Med 2020;83:1235–49.
[66] Talagala SL, Sarlls JE, Liu S, Inati SJ. Improvement of temporal signal-to-noise ratio of GRAPPA accelerated echo planar imaging using a FLASH based calibration scan. Magn Reson Med 2016;75:2362–71.
[67] Taviani V, Alley MT, Banerjee S, Nishimura DG, Daniel BL, Vasanawala SS, et al. High-resolution diffusion-weighted imaging of the breast with multiband 2D radiofrequency pulses and a generalized parallel imaging reconstruction. Magn Reson Med 2017;77:209–20.
[68] Todd N, Moeller S, Auerbach EJ, Yacoub E, Flandin G, Weiskopf N. Evaluation of 2D multiband EPI imaging for high-resolution, whole-brain, task-based fMRI studies at 3T: sensitivity and slice leakage artifacts. NeuroImage 2016;124:32–42.
[69] Ugurbil K, Xu J, Auerbach EJ, Moeller S, Vu AT, Duarte-Carvajalino JM, et al. Pushing spatial and temporal resolution for functional and diffusion MRI in the Human Connectome Project. NeuroImage 2013;80:80–104.
[70] Van AT, Aksoy M, Holdsworth SJ, Kopeinigg D, Vos SB, Bammer R. Slab profile encoding (PEN) for minimizing slab boundary artifact in three-dimensional diffusion-weighted multislab acquisition. Magn Reson Med 2015;73:605–13.
[71] Vionnet L, Aranovitch A, Duerst Y, Haeberlin M, Dietrich BE, Gross S, et al. Simultaneous feedback control for joint field and motion correction in brain MRI. NeuroImage 2021;226:117286.
[72] Vu AT, Beckett A, Setsompop K, Feinberg DA. Evaluation of SLIce dithered enhanced resolution simultaneous MultiSlice (SLIDER-SMS) for human fMRI. NeuroImage 2018;164:164–71.
[73] Wu EL, Chiueh TD, Chen JH. Multiple-frequency excitation wideband MRI (ME-WMRI). Med Phys 2014;41:092304.
[74] Wu J, Saindane AM, Zhong X, Qiu D. Simultaneous perfusion and permeability assessments using multiband multi-echo EPI (M2-EPI) in brain tumors. Magn Reson Med 2019;81:1755–68.
[75] Wu W, Koopmans PJ, Frost R, Miller KL. Reducing slab boundary artifacts in three-dimensional multislab diffusion MRI using nonlinear inversion for slab profile encoding (NPEN). Magn Reson Med 2016;76:1183–95.
[76] Wu W, Poser BA, Douaud G, Frost R, In MH, Speck O, et al. High-resolution diffusion MRI at 7T using a three-dimensional multi-slab acquisition. NeuroImage 2016;143:1–14.
[77] Xu J, Moeller S, Auerbach EJ, Strupp J, Smith SM, Feinberg DA, et al. Evaluation of slice accelerations using multiband echo planar imaging at 3 T. NeuroImage 2013;83:991–1001.
[78] Yang Y, Meyer CH, Epstein FH, Kramer CM, Salerno M. Whole-heart spiral simultaneous multi-slice first-pass myocardial perfusion imaging. Magn Reson Med 2019;81:852–62.
[79] Yutzy SR, Seiberlich N, Duerk JL, Griswold MA. Improvements in multislice parallel imaging using radial CAIPIRINHA. Magn Reson Med 2011;65:1630–7.
[80] Zahneisen B, Poser BA, Ernst T, Stenger VA. Three-dimensional Fourier encoding of simultaneously excited slices: generalized acquisition and reconstruction framework. Magn Reson Med 2014;71:2071–81.
[81] Zhu K, Dougherty RF, Wu H, Middione MJ, Takahashi AM, Zhang T, et al. Hybrid-space SENSE reconstruction for simultaneous multi-slice MRI. IEEE Trans Med Imaging 2016;35:1824–36.

CHAPTER

Sparse Reconstruction

Li Feng
Department of Radiology, Icahn School of Medicine at Mount Sinai, New York, NY, United States

8.1 Introduction

The quest for fast imaging speed started early in the history of MRI. Fast MRI acquisition helps shorten exam times and thus reduces patient discomfort during MRI exams. Moreover, increased imaging speed can be leveraged to improve spatiotemporal resolution and/or volumetric coverage, all of which play an important role in routine clinical applications. Parallel imaging (or parallel MRI, see Chapter 6), introduced in the 1990s, has proven to be one of the most disruptive fast MRI techniques, serving to robustly generate an MR image from undersampled measurements using coil arrays [1–3]. Parallel imaging was later extended via many different variants [4–7], and it is now available in most MRI scanners for routine clinical use. However, the acceleration capability and reconstruction performance of parallel imaging are limited by the design of coil arrays and the number of coil elements, and it is ultimately restricted by fundamental electrodynamic principles [8,9]. In routine clinical applications, parallel imaging typically allows for a moderate acceleration rate, and excessive acceleration often results in noise amplification that degrades image quality [10].

The rise of compressed sensing in the 2000s [11,12] has generated great excitement, and its application to MRI (known as compressed sensing MRI or sparse MRI) has introduced a new and powerful approach to improve imaging speed by exploiting image sparsity/compressibility [13]. Here, image sparsity/compressibility can be understood in a way that most images can be represented with much less data than the number of pixels, due to correlations between pixels [14]. Fig. 8.1 shows a specific example, where a brain MR image can be represented by only 10% of its largest wavelet coefficients without perceptual loss of important information. This implies a so-called transform coding framework that plays an essential role behind the JPEG, JPEG2000, and MPEG image/video compression standards and is widely used in daily life. The ability to compress images so effectively raises a question: Instead of spending so much effort to acquire an image and then discarding most of its coefficients for compression, why not directly acquire the image in a compressed form? In other words, knowing that images are compressible, perhaps we can build the compression process directly into the sampling/sensing step so that one only acquires the information that is needed to represent the image (e.g., acquisition based on the information rate instead of the Nyquist–Shannon rate). This hypothesis was proved by Candes et al., and Donoho (2006), who established a theory that is well known as compressed sensing or compressive sampling [11,12]. Soon after that, Lustig et al. demonstrated that an MR image, under certain conditions, can be successfully recovered from a number of measurements that are far below the Nyquist–Shannon limit [13,14]. About the same time, Block et al. also demonstrated that an MR image can be reconstructed from undersampled multicoil radial measurements by exploiting im-

Advances in Magnetic Resonance Technology and Applications, Volume 7, ISSN 2666-9099. https://doi.org/10.1016/B978-0-12-822726-8.00018-X

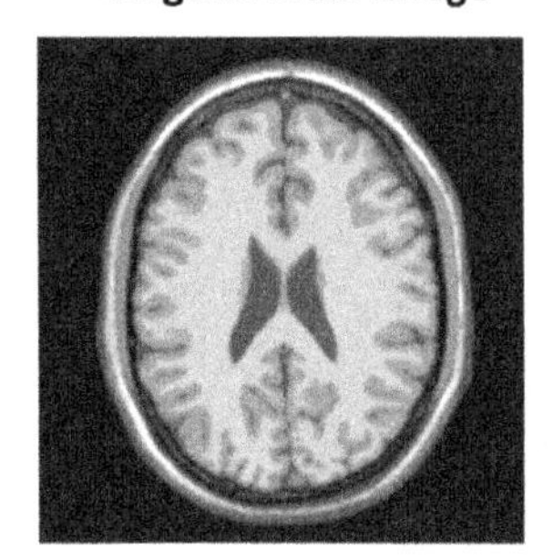

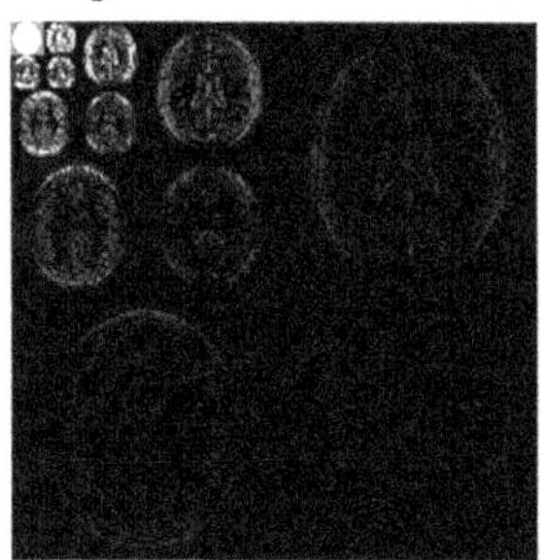

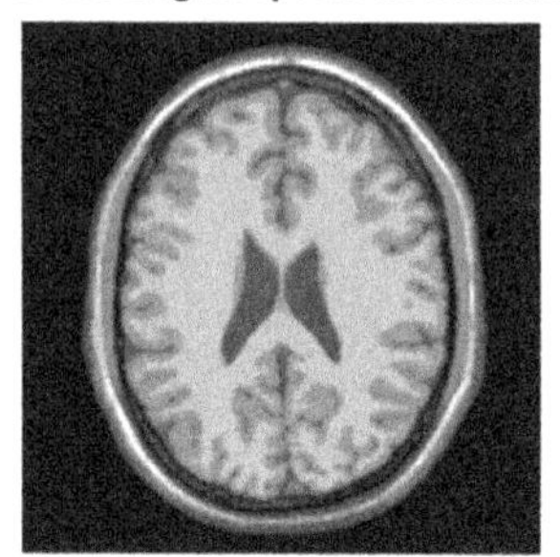

FIGURE 8.1

Demonstration of the compressibility of a brain MR image. Although the image is not sparse by itself, it has a sparse representation in the wavelet domain after applying the wavelet transform to the image (e.g., most of coefficients have a value close to zero in the wavelet domain). With an inverse wavelet transform using only 10% of the largest wavelet coefficients, the image can be recovered without perceptual loss of information. This example demonstrates the compressibility of a brain MR image.

age sparsity [15]. Since then, there has been an explosive growth of various sparsity-based fast imaging techniques [16–18], and various studies have also shown that appropriate combinations of compressed sensing MRI with parallel imaging can enable higher acceleration and improved reconstruction quality compared to either of them alone [19–23]. These advances have all synergistically led to remarkably increased imaging speed with previously unachieveable imaging performance [18,24].

Sparse image reconstruction has a long history and goes far beyond standard compressed sensing MRI. Although the role of image sparsity in image reconstruction was not formally established until the advent of compressed sensing, many early constrained reconstruction methods, such as the keyhole [25], TRICKS/HYPR [26,27], and k-t BLAST [6] methods (see Chapter 5), took advantage of spatiotemporal correlations in dynamic MR images, which represent a different notion of image sparsity. In the meantime, most post-compressed sensing image-reconstruction techniques, such as low-rank or dictionary learning-based reconstruction methods, which will be described in Chapters 9–10, also rely on image sparsity. Artificial intelligence-enabled image reconstruction, as the most recent trend in rapid MRI [28–30] that will be reviewed in Chapter 11, also finds its root in the use of image sparsity/correlations and shares certain similarities with the compressed sensing framework. These advances all indicate that we have entered an era of sparsity for rapid MRI, and understanding sparse reconstruction will help pave a path to further develop more powerful and more advanced image reconstruction techniques to tackle yet more clinical problems.

This chapter will focus on compressed sensing and its applications in rapid MRI. It is organized as follows: Section 8.2 briefly reviews the compressed sensing theory; Section 8.3 presents the translation of compressed sensing to MRI and the specific requirements for data acquisition and image reconstruction; Section 8.4 describes representative frameworks to combine compressed sensing MRI with parallel imaging and shows how they can synergistically result in improved reconstruction quality; Section 8.5 describes representative clinical applications of compressed sensing MRI; Sections 8.6 and 8.7 then conclude the chapter with discussion of the existing challenges of compressed sensing

MRI and potential solutions, along with a summary of the chapter. Finally, a MATLAB® tutorial that demonstrates compressed sensing reconstruction for an undersampled brain MRI data is presented in Section 8.8. With knowledge of sparsity and incoherence, the readers will see how they can be further used to develop the more advanced sparse reconstruction methods that are covered in the following chapters.

8.2 Compressed sensing theory: a brief overview

Compressed sensing involves acquiring a sparse/compressible signal in an efficient manner using an incoherent measurement basis and then recovering the signal without loss of important information [31]. Here, efficient acquisition is achieved by sampling the signal at a rate that is far below the Nyquist–Shannon requirement. As a result, there are three immediate questions related to compressed sensing: (i) why sparsity is needed and useful; (ii) what incoherent sampling is and why we need it; and (iii) how the sparse signal is recovered. This section briefly reviews the answers to these questions and helps readers better understand the principles behind compressed sensing MRI.

8.2.1 Sparsity and incoherence: a first look

We first define a one-dimensional (1D) sensing problem that is typically implemented for signal acquisition and is related to that described in Chapter 2. The sampling of a signal vector $\mathbf{x}$ (size $= N \times 1$) can be interpreted as a projection of the signal onto different sampling waveforms $\mathbf{e}_i$ with the same size:

$$s_i = \langle \mathbf{x}, \mathbf{e}_i \rangle \quad i = 1, 2, 3, ..., M. \tag{8.1}$$

Here, M denotes the total number of measurements and s_i denotes the ith measurement sampled with the ith waveform. In matrix notation, this becomes

$$\mathbf{s} = \mathbf{E}\mathbf{x}, \tag{8.2}$$

where $\mathbf{E}$ denotes a measurement matrix consisting of a total of M sampling waveforms (size $= M \times N$), which is also referred to as an encoding operator, and $\mathbf{s}$ denotes the measurements (size $= M \times 1$). Connecting this to MRI reconstruction, $\mathbf{s}$ can be treated as the k-space measurements that are concatenated into a vector; $\mathbf{x}$ can be treated as the digitalized MR image to be acquired in the form of a vector; and $\mathbf{E}$ can treated as a transform that operates between image and k-space without loss of generality. As seen in Chapter 2, when $M = N$ and the encoding operator $\mathbf{E}$ has a full rank, Eq. (8.2) is a well-defined linear equation with a unique solution (Fig. 8.2a). However, when the number of measurements is smaller than the length of the signal ($M < N$), Eq. (8.2) is an under-determined or undersampled linear equation with an infinite number of possible solutions (Fig. 8.2b). Here comes the role of sparsity: if $\mathbf{x}$ is a K-sparse signal (i.e., $\mathbf{x}$ has at most K nonzero coefficients and $K \ll N$) and the locations of the nonzero components are also known (Fig. 8.2c), the signal $\mathbf{x}$ then becomes $\mathbf{x}_K$ with reduced dimensionality (size $= K \times 1$). In a similar way, $\mathbf{E}$ can be reduced to $\mathbf{E}_K$ with size $M \times K$ by keeping only the K columns corresponding to the nonzero locations in $\mathbf{x}$ (see Fig. 8.2c). With the knowledge of signal sparsity and sparse locations, Eq. (8.2) becomes Eq. (8.3), and, assuming that $\mathbf{E}_K$ has a full

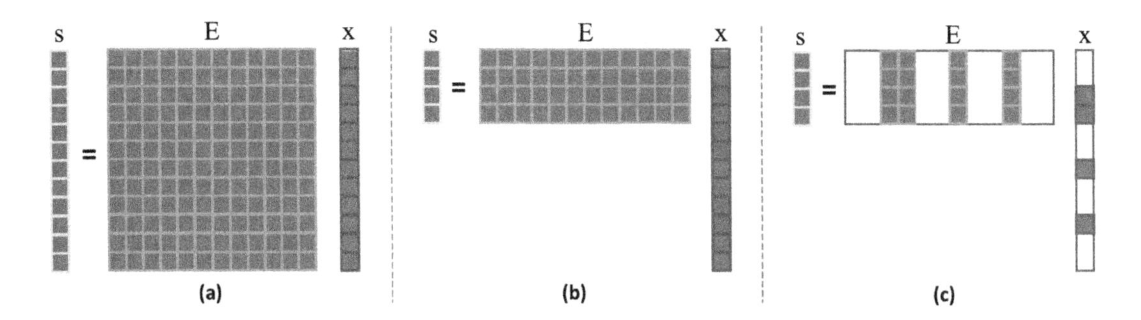

FIGURE 8.2

(a) A linear equation is well-determined when the number of measurements ($\mathbf{s}$, size $= 12 \times 1$) equals to the length of the signal ($\mathbf{x}$, size $= 12 \times 1$) to be sampled, which ensures a unique solution if $\mathbf{E}$ has a full rank. **(b)** The linear equation becomes under-determined when the number of measurements ($\mathbf{s}$, size $= 4 \times 1$) is smaller than the length of the signal ($\mathbf{x}$, size $= 12 \times 1$) to be sampled, which results in more than one possible solution. **(c)** if $\mathbf{x}$ is a K-sparse with $K \ll 12$ (e.g., $\mathbf{x}$ has at most $K = 4$ nonzero coefficients in this example) and the locations of the nonzero components are known, the signal $\mathbf{x}$ then becomes $\mathbf{x}_K$ with reduced dimensionality (size $= 4 \times 1$), and $\mathbf{E}$ also reduces to $\mathbf{E}_K$ with a reduced size (4×4) by keeping only the four columns corresponding to the nonzero locations in $\mathbf{x}$. In this way, a unique solution can be ensured with only four measurements assuming that $\mathbf{E}_K$ has a full column rank.

column rank, one would only need K ($K \ll N$) measurements (instead of N in traditional acquisition) to ensure a unique solution

$$\mathbf{s} = \mathbf{E}_K \mathbf{x}_K. \tag{8.3}$$

This simple example tells us how sparsity plays a role in reducing the number of measurements in a linear equation. However, it assumes that the locations of the sparse coefficients are known in advance, which is impossible in practice. Thus, more measurements are typically needed in actuality. In general, if we have $2K$ measurements for a K-sparse signal (the size of $\mathbf{s} = 2K \times 1$ and the size of $\mathbf{E} = K \times N$, $2K < N$), one can guarantee a unique solution to Eq. (8.2) if every $2K$ columns in the encoding operator $\mathbf{E}$ are linearly independent (see Appendix 8.A for more details) [32]. Here, in addition to signal sparsity, an additional condition is also enforced to the encoding operator $\mathbf{E}$, and this requirement (every $2K$ columns in $\mathbf{E}$ are linearly independent) is essentially what incoherence implies in compressed sensing. This also suggests that the level of sparsity (i.e., the number of nonzero coefficients) determines the minimum measurements required to solve an under-determined linear equation.

With this simple demonstration, we get to know how sparsity plays a role in signal acquisition and the basic understanding of what incoherence is. Now we will discuss how a sparse signal can be recovered in practice and how we can meet the requirement of incoherence.

8.2.2 Compressed sensing reconstruction

The aim of compressed sensing reconstruction is to recover a signal from a reduced number of measurements. By definition, the sparsity of a signal $\mathbf{x}$ is calculated by its zero-norm (L0-norm), which is given by counting the number of nonzero coefficients in the signal. As a result, the recovery of a sparse signal in an undersampled linear equation can be formulated as the following L0-minimized optimization problem:

$$\tilde{\mathbf{x}} = \underset{\mathbf{x}}{\arg\min} \|\mathbf{x}\|_0 \quad s.t. \quad \mathbf{s} = \mathbf{E}\mathbf{x}. \tag{8.4}$$

From Section 8.2.1, we know that Eq. (8.2) has infinite solutions if $M < N$ in the encoding operator $\mathbf{E}$. In this case, Eq. (8.4) aims to find a solution that has the smallest number of nonzero coefficients from all possible solutions. When the requirement of incoherence (as described in 8.2.1) is satisfied, this solution is unique.

In practice, we will face two problems when solving Eq. (8.4). First, few signals are truly sparse, and most signals are not sparse by themselves. For example, the brain MR image shown in Fig. 8.1 is not sparse, but it has a representation in the wavelet domain in which its coefficients decay rapidly with most of its coefficients close to zero, and the wavelet image can be well-approximated as a K-sparse signal. In this case, this signal is said to be compressible. We can expand the sensing problem (Eq. (8.2)), so that a compressible signal $\mathbf{x}$ can be expressed with a sparse basis and associated sparse representation as follows:

$$\mathbf{d} = \Phi\mathbf{x} \tag{8.5}$$

Here, Φ denotes the basis under which $\mathbf{d}$ can sparsely represent the original signal $\mathbf{x}$, and it is assumed to be an orthonormal basis for simplicity. It is also known as a sparsifying transform. With this definition, Eqs. (8.2) and (8.5) can now be combined as

$$\mathbf{s} = \mathbf{E}\mathbf{x} = \mathbf{E}\Phi^{\mathrm{H}}\mathbf{d}, \tag{8.6}$$

where $\mathbf{E}\Phi^{\mathrm{H}}$ combines the encoding operator with the sparsity basis and $\mathbf{H}$ denotes the Hermitian transpose. Second, signal measurements are also contaminated by noise, which needs to be taken into consideration in signal recovery. With these two conditions, Eq. (8.4) can be modified to

$$\tilde{\mathbf{d}} = \underset{\mathbf{d}}{\arg\min} \|\mathbf{d}\|_0 \quad s.t. \quad \left\|\mathbf{s} - \mathbf{E}\Phi^{\mathrm{H}}\mathbf{d}\right\|_2^2 < \varepsilon. \tag{8.7}$$

Here, the L2-norm is employed to quantify the difference between the measurements and the estimated solution and ε is related to the noise level in signal measurement.

Challenges to solving Eq. (8.7) are that L0-norm minimization is a computationally intensive problem and real signals are generally compressible but not truly sparse as previously described. Therefore, a relaxed strategy has been proposed to make the optimization problem tractable. Commonly used sparse-recovery algorithms utilize the L1-norm as a surrogate measure of signal sparsity [11,32,33], and the corresponding optimization problem minimizing the L1-norm is given by

$$\begin{aligned} \tilde{\mathbf{d}} &= \arg\min_{\mathbf{d}} \|\mathbf{d}\|_1 \\ s.t. &\quad \left\|\mathbf{s} - \mathbf{E}\Phi^{\mathrm{H}}\mathbf{d}\right\|_2^2 < \varepsilon. \end{aligned} \tag{8.8}$$

Here, the L1-norm is defined as the summation of the absolute values of all coefficients in a signal

$$\|\mathbf{d}\|_1 = \sum_i |d_i|. \tag{8.9}$$

L1-norm minimization is an effective alternative because it promotes sparsity, and it is a convex problem guaranteeing a global minimum, as shows in Fig. 8.3. L1-norm minimization has also been shown to have the same solution as L0-norm minimization under appropriate conditions [11,32,33]. Instead of L1-norm minimization, different approaches have also been proposed to solve the compressed sensing problem based on Lp-norm minimization with $0 < p < 1$ [34–36]. This likely leads to a sparse solution, as seen in Fig. 8.3c, but it results in a nonconvex optimization problem and requires more complex algorithms to find the right solution while avoiding local minima.

Alternatively, we can also reformat Eq. (8.9) as

$$\begin{aligned} \tilde{\mathbf{x}} &= \arg\min_{\mathbf{x}} \|\Phi\mathbf{x}\|_1 \\ s.t. &\quad \|\mathbf{s} - \mathbf{E}\mathbf{x}\|_2^2 < \varepsilon. \end{aligned} \tag{8.10}$$

The constrained optimization problem in Eqs. (8.8) and (8.10) can be further formulated as an unconstrained problem using Lagrange multipliers as shown below, and they can be efficiently solved using different optimization algorithms previously described in Chapter 3:

$$\tilde{\mathbf{d}} = \arg\min \left\|\mathbf{s} - \mathbf{E}\Phi^{\mathrm{H}}\mathbf{d}\right\|_2^2 + \lambda \|\mathbf{d}\|_1 \tag{8.11}$$

$$\tilde{\mathbf{x}} = \arg\min \|\mathbf{s} - \mathbf{E}\mathbf{x}\|_2^2 + \lambda \|\Phi\mathbf{x}\|_1 \tag{8.12}$$

Eq. (8.11) is known as the synthesis formulation of compressed sensing reconstruction since it aims to reconstruct the sparse representation of the signal under the basis Φ, and the final signal to be reconstructed needs to be generated (or synthesized) as $\Phi^{\mathrm{H}}\mathbf{d}$. Eq. (8.12) is called the analysis formulation that directly reconstructs the signal $\mathbf{x}$. When the sparsifying transform Φ is orthonormal, these two formulations are equivalent.

Note that both Eqs. (8.11) and (8.12) are nonlinear, thus requiring a nonlinear optimization algorithm. A regularization parameter λ is required to control the balance between the sparsity term (term on the right) and the consistency with acquired measurements (on the left). With L1-norm minimization, more measurements are expected to ensure successful signal recovery. It has been shown that under proper conditions, a K-sparse signal can be reconstructed with high probability if the number of measurements $M > \mu^2 K \log^4 N$ [11,33], where μ indicates the coherence level. In practice, a good rule of thumb is to have ~3–5 K measurements to recover a K-sparse signal based on L1-norm minimization [37,38].

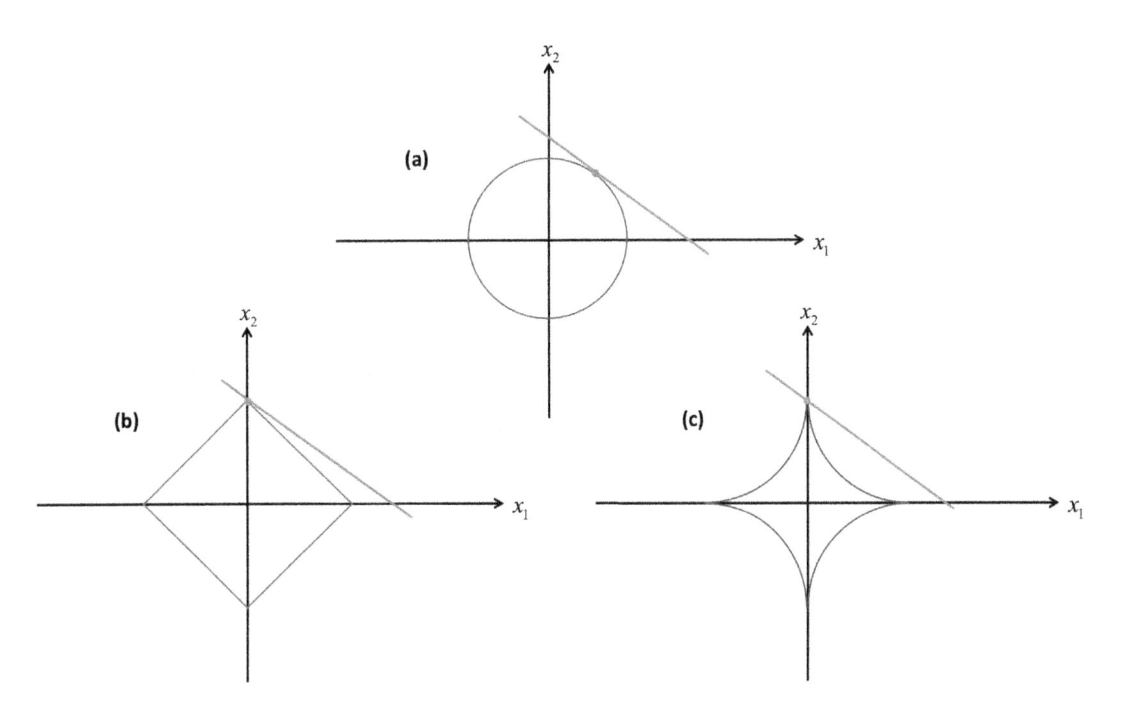

FIGURE 8.3

Plots of unit circles for **(a)** the L2-norm (purple), **(b)** the L1-norm (blue) and **(c)** the Lp-norm $(0 < p < 1)$ (black) calculated for a two-element vector $[x_1, x_2]$. The red line represents a linear equation. The plots suggest that (i) a sparse solution (i.e., $x_1 = 0$ in the blue dot) can be obtained based on L1-norm optimization and Lp-norm $(0 < p < 1)$, while the use of L2-norm fail to promote sparsity (green dot); (ii) although both L1-norm and Lp-norm $(0 < p < 1)$ promote sparsity, L1-norm is convex to guarantee a global minimum while Lp-norm $(0 < p < 1)$ is nonconvex (color figure is only available on the web).

8.2.3 Conditions for compressed sensing reconstruction

Section 8.2.1 describes the role of sparsity in sparse signal recovery and what incoherent sampling is. Section 8.2.2 describes how a sparse or compressible signal can be recovered. This section presents how to choose a sparsifying transform and how to design an incoherent sampling scheme.

Compressed sensing involves three domains, including: the original signal domain, the sampling domain in which a signal is measured, and the sparse domain in which a signal can have a sparse representation, as shown in Fig. 8.4. The first requirement for successful compressed sensing reconstruction is that a compressible signal should have a sparse representation in the sparse domain and should not be sparse in the sampling domain. In other words, a signal with a sparse representation in the sparse domain must have a dense representation in the sampling domain. This ensures that sparse signal recovery is possible from only a small subset of measurements. As a specific example, considering the worst scenario with $\mathbf{E} = \Phi$, the sampling domain becomes the same as the sparse domain. In this situation, when samples are not acquired, they are permanently lost and cannot be recovered.

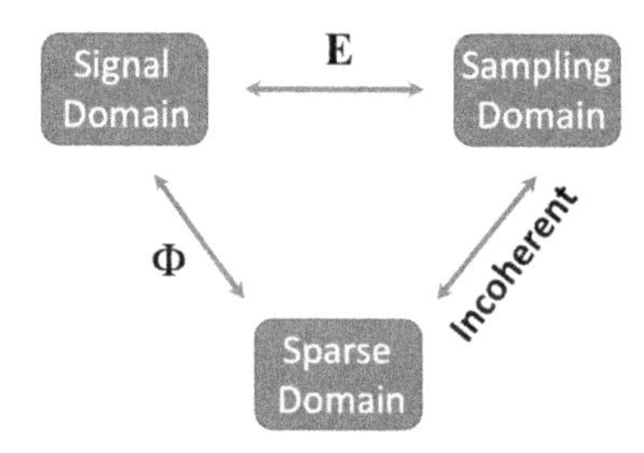

FIGURE 8.4

Compressed sensing reconstruction involves operation in three domains, including the original signal domain, the sampling domain in which a signal is measured, and the sparse domain in which a signal can have a sparse representation. Compressed sensing requires that a compressible signal should have a sparse representation in the sparse domain and should not be sparse in the sampling domain. That means a signal with a sparse representation in the sparse domain must have a dense representation in the sampling domain.

Another component that is important to ensure successful signal recovery in compressed sensing is the design of an undersampling measurement scheme, with which the columns of the encoding matrix can be largely independent in an undersampled linear equation. This is generally referred to as incoherent undersampling in compressed sensing, and the requirement needs to be fulfilled when we design and build the encoding operator **E** in Eqs. (8.11) and (8.12). It has been suggested that the mutual coherence of a matrix can be used to easily evaluate the incoherence level of **E** [39,40], which is given by the largest absolute inner product between any two columns of **E**

$$\mu(\mathbf{E}) = \max_{i \neq j} \left| \langle e_i, e_j \rangle \right| \tag{8.13}$$

Eq. (8.13) evaluates the off-diagonal entries of the Gram matrix $\mathbf{E}^H\mathbf{E}$. When the off-diagonal entries of matrix $\mathbf{E}^H\mathbf{E}$ are small (e.g., $\mathbf{E}^H\mathbf{E}$ is close to the identity matrix), **E** exhibits low mutual coherence and thus high incoherence. On the other hand, when the off-diagonal entries are large, then **E** exhibits high mutual coherence, which is not desired in compressed sensing [39]. The readers will see how this property can be evaluated in compressed sensing MRI in the next section.

8.3 Compressed sensing MRI

The compressed sensing theory described in the previous section has three key components, including: 1) sparsity, 2) incoherence, and 3) nonlinear reconstruction promoting sparsity. Accordingly, applying compressed sensing to MRI applications involves three objectives, including: (i) finding the right sparsifying transform to represent an MR image; (ii) constructing a good sensing matrix and sampling pattern for incoherent undersampling (thus acceleration of data acquisition); and (iii) building an iterative nonlinear optimization algorithm to reconstruct the MR image [13,14]. In this section, we will learn how these requirements can be fulfilled in compressed sensing MRI in the context of single-coil acquisition.

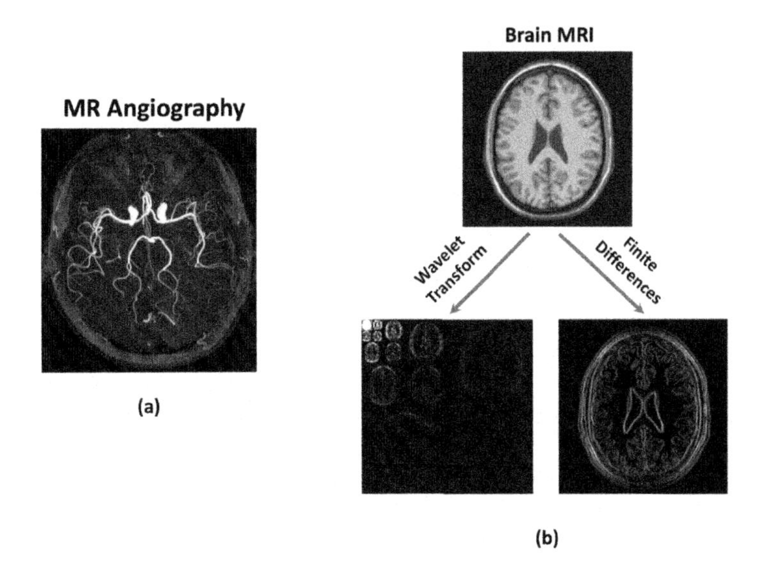

FIGURE 8.5

(a) An MR angiography image is sparse by itself, with the value of most image coefficients close to zero. **(b)** A brain MR image is not sparse by itself, but it can have a sparse representation after applying a wavelet transform or spatial finite differences.

8.3.1 Sparsifying transform and transform sparsity

A signal is sparse if most of its coefficients are zero or approximately sparse if they are close to zero. With this definition, certain MR images (e.g., images from MR angiography) can be considered sparse in the image domain, as shown in Fig. 8.5a. However, most MR images are not sparse, but they typically have a sparse representation in an appropriate transform domain [13]. Popular sparsifying transforms include the wavelet transform and finite differences, as shown in Fig. 8.5b for concrete examples. A combination of two or more sparsifying transforms can also be applied in compressed sensing MRI to improve reconstruction quality [13].

In general, higher image dimensions can result in increased image sparsity. As a result, 3D MRI images can have a sparser representation by employing a 3D sparsifying transform to exploit its pixel correlations in all three spatial dimensions. Dynamic MR images usually have a much higher level of sparsity than static images after applying a temporal sparsifying transform along the dynamic dimension [41], as shown in Fig. 8.6. This is because of the presence of large temporal correlations associated with periodic motion or smooth signal evolution that are often seen in dynamic MRI, similar to temporal correlations in videos leading to higher compression ratio than images. Commonly used generic temporal sparsifying transforms include 1D temporal wavelet transform, temporal Fourier transform, and temporal finite differences. Meanwhile, spatial and temporal sparsifying transforms can be combined to jointly exploit the spatiotemporal correlations in dynamic MR images.

In addition to the generic transforms just mentioned, a sparsifying transform can also be learned from MR images, which could be more effective to obtain a sparser and often more accurate represen-

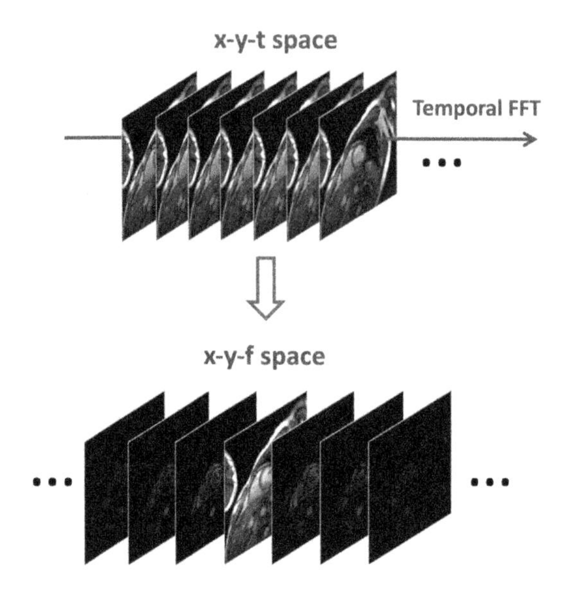

FIGURE 8.6

A dynamic image series (x-y-t space) can have a sparse representation by applying a temporal sparsifying transform such as a temporal Fourier transform (x-y-f space). Dynamic MR images typically have a much higher level of sparsity than static images because of the presence of significant temporal correlations associated with periodic motion or smooth signal evolution that are often seen in dynamic MRI.

tation [42–45]. Specifically, a sparse representation can be learned from an MRI database consisting of similar images in advance or from the to-be-reconstructed undersampled images directly [16]. Low-rank-based MRI reconstruction, as will be seen in more detail in the next chapter, is also related to the concept of compressed sensing. They both aim to reconstruct sparse images from data acquired with an incoherent undersampling scheme using a nonlinear iterative reconstruction algorithm.

8.3.2 Incoherent data acquisition

In addition to image sparsity, successful implementation of compressed sensing MRI requires the conditions described in Section 8.2.3. For the first condition, since MRI acquisition is performed in the Fourier domain (k-space), this naturally ensures that any single measurement in k-space has contribution to all pixels in the image domain. In other words, an MR image that is sparse in the image domain will have a dense representation in the sampling domain (k-space), and thus it is possible to reconstruct the image, even if some measurements (k-space data) are omitted. Therefore, as long as the selected sparsifying transform basis is sufficiently incoherent with the measurement basis (Fourier basis in this case), this condition will be satisfied.

For the second condition, the nature of sequential k-space sampling in MRI offers the flexibility to acquire a user-defined subset of k-space measurements, in most cases. Here, considering standard 2D Cartesian acquisition with 1D undersampling, **E** represents undersampled Fourier basis (size $= M \times N$). According to Eq. (8.13), one can assess whether **E** has low mutual coherence by checking the

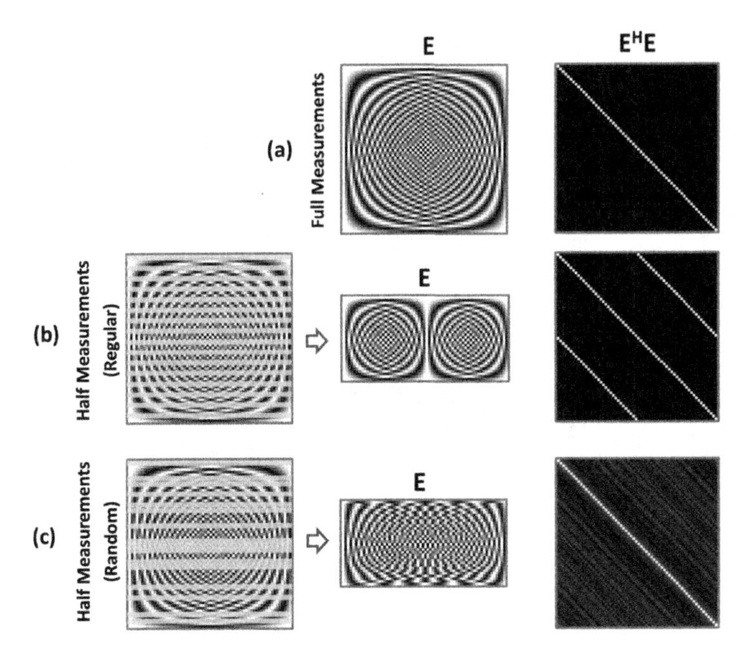

FIGURE 8.7

Demonstration of the effect of undersampling scheme with Fourier encoding. For the fully sampling case **(a)**, $\mathbf{E}$ represents the full Fourier basis to implement fast Fourier transform (FFT). The resulting measurement matrix thus has a low mutual coherence, as reflected by the zero off-diagonal entries in the Gram matrix $\mathbf{E}^H\mathbf{E}$. When only half of the Fourier measurements are sampled regularly **(b)**, the measurement matrix has high mutual coherence, as indicated by the replication of diagonal entries in the Gram matrix $\mathbf{E}^H\mathbf{E}$. When half of the k-space lines are sampled randomly, the measurement matrix has low mutual coherence, as indicated by the low off-diagonal entries in the Gram matrix $\mathbf{E}^H\mathbf{E}$.

off-diagonal entries of $\mathbf{E}^H\mathbf{E}$. Several undersampling schemes are given in Fig. 8.7, including: (a) the full Fourier basis (no undersampling); (b) 2-fold undersampled Fourier basis generated by taking every other row from the full Fourier basis; and (c) 2-fold undersampled Fourier basis generated by randomly taking half of the rows from the full Fourier basis. It can be seen that random undersampling generates lower off-diagonal values than regular undersampling. This simple example shows the connection between the compressed sensing theory and practical sampling scheme design in MRI, suggesting random undersampling as a potential scheme for implementation of compressed sensing MRI [13].

In the original demonstration of compressed sensing MRI, Lustig et al. proposed an easier way to evaluate this condition by calculating the point spread function (PSF) as a measure of incoherence, where low sidelobes in the PSF suggest high incoherence, so that undersampling-induced artifacts should look like added noise and most image content can be preserved [13]. The PSF of 2D Cartesian undersampling patterns can be obtained by preforming 2D FFT on corresponding zero-filled sampling masks. As shown in Fig. 8.8, high incoherence can be obtained using a random undersampling pattern resulting in incoherent artifacts, and high coherence is obtained using a regular undersampling pattern resulting in replication of the image. It should also be noted that, in practice, undersampling is typ-

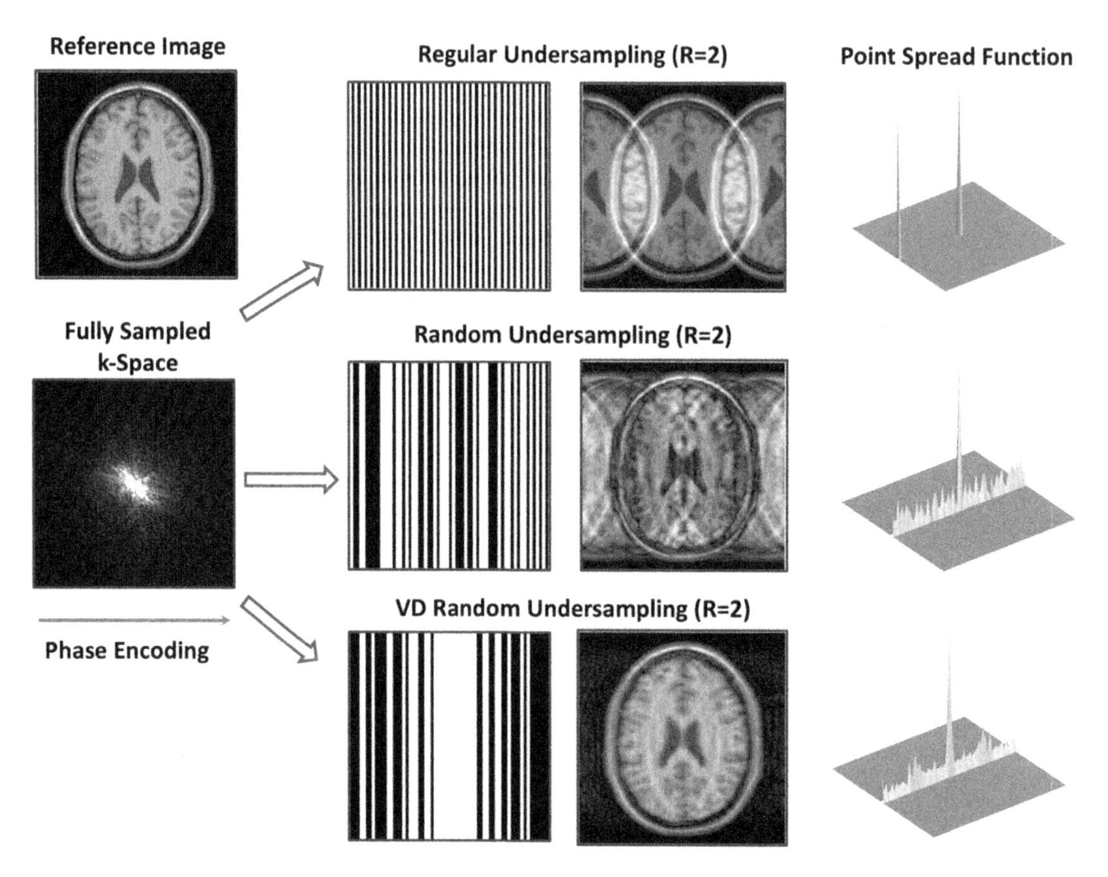

FIGURE 8.8

Undersampling in MRI acquisition is typically performed along the phase-encoding dimension. Regular undersampling results in strong coherent aliasing artifacts (top row), while random undersampling results in less coherent artifacts (middle row). Variable-density undersampling, with which the center of k-space is sampled more than the outer region, can further promote incoherence to reduce undersampling-induced artifacts (bottom row). The point spread function (PSF) can be used to evaluate the incoherence of a sampling scheme, where low sidelobes in the PSF indicate high incoherence. The PSF can be obtained by preforming 2D FFT on corresponding zero-filled sampling masks.

ically not implemented along the frequency-encoding dimension in MRI acquisition because it does not save scan time. So, practical implementation of undersampling should take this fact into consideration. Fig. 8.8 also compares undersampled images using a pure random pattern and a variable-density random pattern. The variable-density random pattern samples the k-space center more frequently than the periphery, which results in a cleaner undersampled image. This is because the energy of k-space is mostly located in its center, which is important to determine the image contrast and major contents. Thus, a variable-density undersampling pattern can achieve favorable reconstruction quality compared to a standard uniform undersampling pattern.

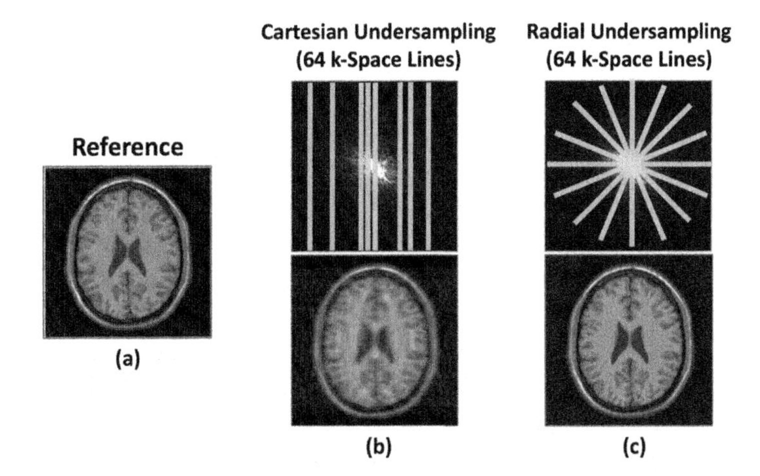

FIGURE 8.9

Comparison of undersampled MRI acquisition based on a Cartesian trajectory and a radial trajectory. Although the frequency-encoding dimension is fully sampled for both trajectories, radial sampling does not have a fixed phase-encoding direction, which effectively generates undersampling along two spatial dimensions by omitting certain phase-encoding measurements. As a results, radial undersampling has higher incoherent undersampling behavior than Cartesian undersampling for the same number of measurements.

In addition to random undersampling based on a Cartesian trajectory, various non-Cartesian sampling schemes, such as radial or spiral trajectories, have attracted substantial attention with the introduction of compressed sensing to MRI [15,46–49]. This is because non-Cartesian sampling intrinsically allows for undersampling along all spatial dimensions, thus creating a high level of incoherence. Fig. 8.9 compares undersampling on a Cartesian trajectory and a radial trajectory. Although the frequency-encoding dimension is fully sampled for both cases, radial sampling does not have a fixed phase-encoding direction, which effectively performs undersampling along two spatial dimensions by omitting certain phase-encoding measurements. As non-Cartesian sampling repeatedly acquires k-space center, it also has a lower sensitivity to motion artifacts [50]. However, compared to standard Cartesian MRI, non-Cartesian MRI suffers from several challenges, including increased sensitivity to system imperfections (e.g., off-resonance and gradient-delay effects) and more complex and prolonged reconstruction requiring gridding reconstruction (see Chapter 4) [51]. Selection of desired sampling schemes depends on specific clinical needs, such as reconstruction time, sequence requirements, and motion sensitivity.

3D acquisition generally enables higher acceleration rates compared to 2D acquisition in compressed sensing MRI. This is because: (a) increased sparsity is generally available in 3D images, as previously mentioned; and (b) undersampling in 3D imaging can be performed along two phase-encoding dimensions (including the slice-encoding dimension), which distributes undersampling-induced artifacts along both dimensions to achieve higher incoherence. Fig. 8.10a-b show two different 3D Cartesian random undersampling patterns based on (a) a variable-density random pattern and (b) a Poisson-disc pattern that has more regular random undersampling [21]. Similarly, dynamic MRI can be treated as 3D (2D spatial + 1D temporal) or 4D (3D spatial + 1D temporal) acquisition, making it a

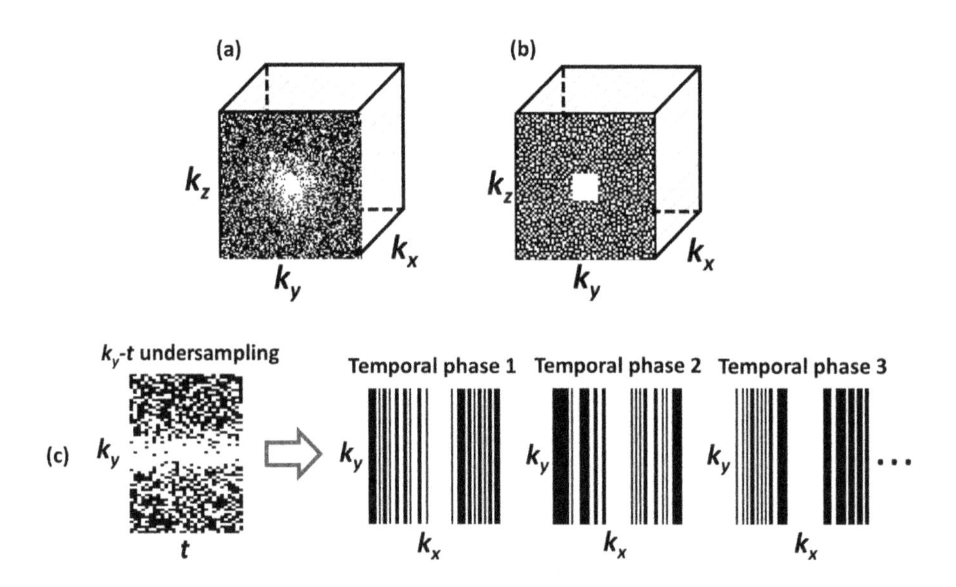

FIGURE 8.10

(a-b) Two different 3D Cartesian random undersampling patterns based on a variable-density pattern and Poisson-disc distributions, where undersampling can be applied along two phase-encoding dimensions ($\boldsymbol{k}_y$ and $\boldsymbol{k}_z$). For the Poisson-disc pattern, a small region in k-space center is fully sampled to form a variable-density distribution. **(c)** A typical $\boldsymbol{k}_y\boldsymbol{-t}$ (phase encoding and time dimensions) undersampling pattern applied in compressed sensing dynamic MRI, where the undersampling scheme can be varied along time to introduce temporal incoherence. For all the sampling patterns, the frequency-encoding dimension ($\boldsymbol{k}_x$) is fully sampled. Figure is adapted from Feng L et al. (J Magn Reson Imaging 2017 Apr;45(4):966–987 (with permission from John Wiley & Sons, Inc.))

good candidate for the application of compressed sensing methods. For dynamic MRI, high acceleration rates can usually be achieved by varying the undersampling pattern along the temporal dimension. This so-called *ky-t* (phase encoding and time dimensions) undersampling scheme [41], as shown in Fig. 8.10c, is widely used in compressed sensing-based dynamic MRI.

8.3.3 Image reconstruction

Most compressed sensing MRI reconstruction methods follow Eqs. (8.11) or (8.12) (for single-coil acquisition). As just mentioned, the reconstruction formulation in Eqs. (8.11) and (8.12) are equivalent if Φ is orthonormal. However, it should be noted that they can give different solutions when Φ is a overcomplete basis (e.g., overcomplete wavelets or redundant dictionaries) [52], and it has been shown that the analysis-based model provides more accurate results with overcomplete regularization [53]. Representative algorithms for compressed sensing reconstruction include the iterative soft thresholding algorithm (ISTA) and its different variants, the nonlinear conjugate gradient method, or the alternating direction method of multipliers (ADMM) method. In this section, a simple example is provided

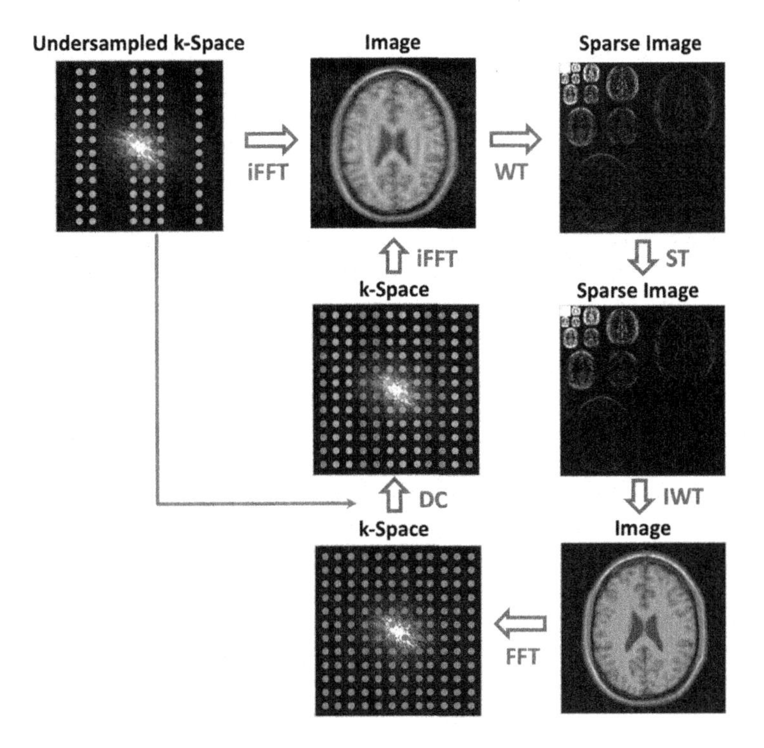

FIGURE 8.11

Compressed sensing MRI reconstruction using an iterative soft-thresholding algorithm. Specifically, image reconstruction iteratively loops among three different domains, including the Fourier domain (k-space), image domain, and sparse domain. The reconstruction process implements a soft thresholding process to remove incoherent aliasing artifacts, while maintaining consistency with the acquired undersampled measurements. WT: wavelet transform; IWT: inverse wavelet transform; ST: soft thresholding; DC: data consistency.

to show how we can perform compressed sensing reconstruction to generate an MR image from 2D undersampled Cartesian k-space.

Fig. 8.11 shows a flow chart to demonstrate the reconstruction workflow using ISTA, which can be formulated using the equation

$$\mathbf{d}_{i+1} = \mathrm{ST}_\lambda(\mathbf{d}_i - \Phi^{\mathrm{H}}\mathbf{E}(\mathbf{E}\Phi^{\mathrm{H}}\mathbf{d}_i - \mathbf{s})), \tag{8.14}$$

where the soft thresholding operator is given as

$$\mathrm{ST}_\lambda(\mathbf{d}) = \begin{cases} (|d_i| - \lambda)\cdot \mathrm{sign}(d_i), & \text{if } |d_i| \geq \lambda \\ 0, & \text{if } |d_i| < \lambda \end{cases}. \tag{8.15}$$

This corresponds to the proximal gradient method (see Chapter 3) with a step size of 1. For Cartesian MRI, the practical implementation of compressed sensing reconstruction starts from a zero-filled undersampling k-space, whose size is typically the same as the image to be reconstructed. The zero-filled reconstruction, given by direct Fourier transform of the zero-filled k-space, serves as the starting point of iterative nonlinear reconstruction, as shown in Fig. 8.11. As a result, the encoding operator $\mathbf{E}$ in Eq. (8.12) can be expressed as $\mathbf{BF}$, where $\mathbf{F}$ is the Fourier transform and $\mathbf{B}$ is a binary sampling mask in which 1 indicates locations acquired and 0 indicates locations that are not acquired in the original undersampled k-space.

Here, the reconstruction iteratively loops among three different domains, including: Fourier domain (k-space); image domain (where we generate MR images); and sparse domain (where we have a sparse representation of MR images under a sparsifying transform). Specifically, for each step, a sparsifying transform (a 2D wavelet transform in this case) is performed on the image to obtain its sparse coefficients, followed by a soft thresholding process in the wavelet domain to remove aliasing artifacts, which is equivalent to L1-norm minimization. Here, we can treat the threshold as the regularization parameter shown in Eq. (8.11), and the threshold can be pre-selected or can be adapted in different iteration steps. After the soft thresholding step, the image can be updated as $\mathbf{d}_i - \Phi^{\mathrm{H}}\mathbf{E}(\mathbf{E}\Phi^{\mathrm{H}}\mathbf{d}_i - \mathbf{s})$, where the index i indicates the ith loop. This step is also known as data consistency. This entire process can be repeated for a number of loops until the reconstruction converges. The ISTA algorithm is invoked to reconstruct an image in the sparse domain ($\mathbf{d}$), and the final image is then given as $\Phi^{\mathrm{H}}\mathbf{d}$.

The example shown in Fig. 8.11 is a simple demonstration for readers to understand the process behind iterative compressed sensing MRI. This example also shows that compressed sensing reconstruction can be intuitively understood as a thresholding process with an additional data-consistency constraint. The source code for this reconstruction example is provided in Section 8.8. A summary of various compressed sensing reconstruction demos that are available online is also given at the end of this chapter.

8.4 Combination of compressed sensing MRI with parallel imaging

In this section, joint frameworks combining compressed sensing with parallel imaging are described. The rationale behind this synergistic combination is first explained, and representative techniques will then be presented and discussed.

8.4.1 Why compressed sensing + parallel imaging?

Compressed sensing MRI can be combined with parallel imaging to form a synergistic framework that enables to fully exploit the coil sensitivity encoding and sparsity jointly for further acceleration of imaging speed [19,21]. Image sparsity and coil sensitivity encoding are complementary sources of information. On one hand, sparsity can serve as a regularizer to prevent noise amplification in parallel imaging; while on the other hand, multicoil acquisition provides additional spatial encoding capabilities to help reduce incoherent aliasing artifacts, thus enabling for higher acceleration rates.

Going back to the compressed sensing theory described in Section 8.2.1, we learned that the minimum requirement for successful recovery of a K-sparse signal using compressed sensing is that every $2K$ columns in the encoding/sensing matrix are linearly independent. Now, suppose we have two completely independent sensors to sample the signal, as shown in Fig. 8.12; then, we could then potentially halve the number of samples in each sensor (K measurements from each sensor) while still maintaining

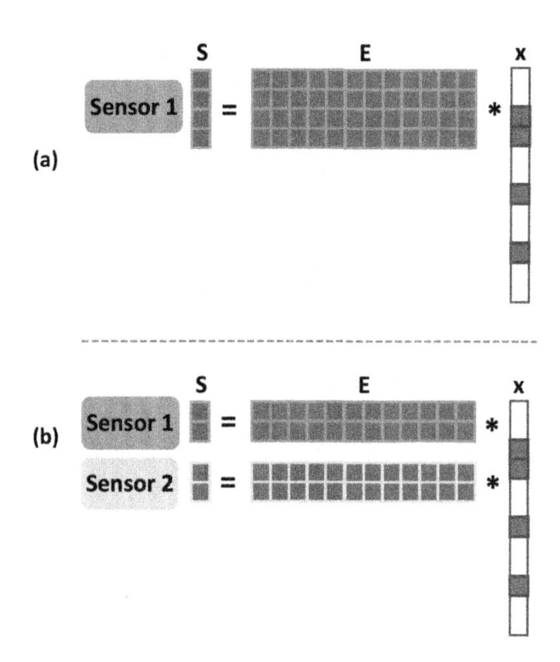

FIGURE 8.12

Demonstration of the use of multiple sensors to reduce the number of measurements in compressed sensing. When two sensors are available and if the sensors are completely independent, the number of samples in each sensor could be reduced to half while maintaining the incoherence requirement for sparse signal recovery.

the requirement for sparse signal recovery (a total of $2K$ measurements that are linearly independent). Another way to understand the improved MRI reconstruction quality from a combination of compressed sensing with parallel imaging is that the total number of measurements in the encoding operator (**E** in Eq. (8.2)) changes from M to $M * C$, where C indicates the number of coil elements used for MRI acquisition and M is the number of measurements in each element. As a result, we can have $M * C \geq N$ when the number of coil elements is larger than or equal to the acceleration rate, making Eq. (8.2) no longer an underdetermined linear equation and thus more stable.

Translating this multisensor framework to MRI, we can treat multiple sensors as multicoil elements with varying coil-sensitivity profiles. Of course, this simple example does not consider the real situation in MRI that coil elements cannot be fully separated, but it intuitively shows why compressed sensing can benefit from the use of multiple sensors (e.g., multicoil arrays in MRI). Fig. 8.13 provides two specific examples comparing 4-fold and 8-fold undersampled images in both single-coil and eight-coil scenarios, where the multicoil images have a lower level of incoherent artifacts.

8.4.2 Representative compressed sensing + parallel imaging methods

There are various ways to combine compressed sensing with parallel imaging. They can be combined in a sequential manner [20] or as a joint reconstruction framework [19,21]. The current consensus is that joint reconstruction is more efficient, and it has been implemented in most of the state-of-art compressed

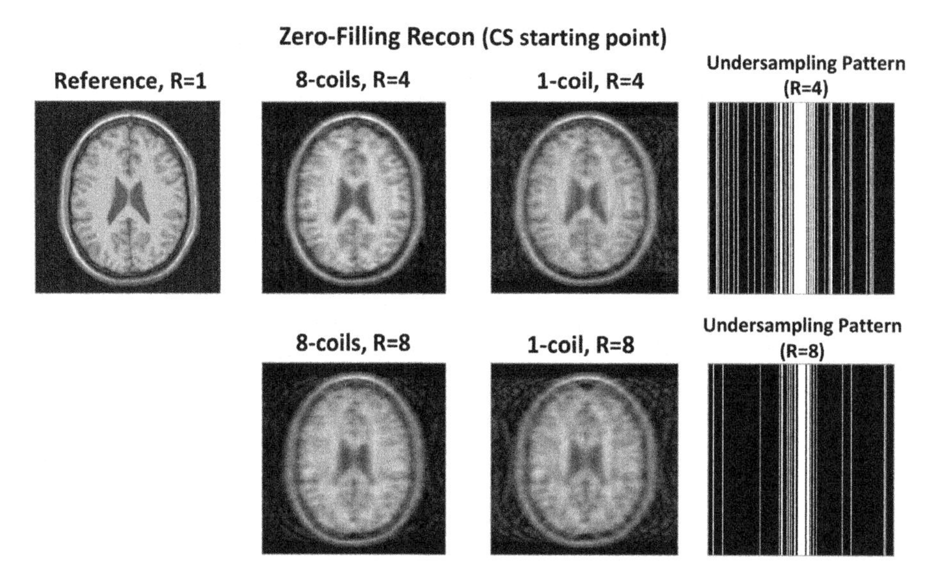

FIGURE 8.13

Comparison of 4-fold and 8-fold undersampled images using single-coil and eight-coil acquisitions. The resulting undersampled image from multicoil acquisition shows a lower level of incoherence artifacts compared to single-coil acquisition.

sensing MRI methods. Since parallel imaging can be implemented either in image domain or in k-space, its combination with compressed sensing can also be performed in both domains. Specifically, there are two well-accepted approaches, one combining compressed sensing with parallel imaging using an iterative SENSE-type framework (known as SparseSENSE [19] or its variants) and the other one combining them using a SPIRiT-type framework (known as L1-SPIRiT [21]). Before going to details, readers should refer to Chapter 6 for the principles of both SENSE and SPIRiT.

The SparseSENSE reconstruction framework can be expressed in a same way as in Eq. (8.12). For multicoil Cartesian acquisition, **s** becomes multicoil k-space (size $= C * M \times 1$ with C indicating the number of coil elements after concatenating into a vector), **x** denotes the image to be reconstructed (size $= N \times 1$), and the encoding operator $\mathbf{E} = \mathbf{BFC}$ combining Fourier transform (**F**), coil sensitivity encoding with C coil elements (or coil sensitivity maps, indicated as **C**), and a binary sampling mask (**B**) as in the single-coil case (see Section 8.3.3). As mentioned, the size of **E** is now $M * C \times N$. Compared to single-coil reconstruction, the L1-norm term in SparseSENSE enforces joint multicoil sparsity from coil-combined images. As in standard SENSE reconstruction, the coil sensitivity maps typically need to be pre-estimated from an additional calibration scan. The SparseSENSE framework can be extended to dynamic MRI reconstruction by employing a temporal sparsity transform, in which case the framework is referred as k-t SPARSE-SENSE [22]. This technique is also called L1-regularized SENSE reconstruction in some studies [54] since it uses a L1-norm constraint to suppress noise amplification in iterative SENSE reconstruction. The SparseSENSE framework can also be implemented with non-Cartesian acquisition by replacing the Fourier transform operation with a nonuniform FFT

(NUFFT) operation in the encoding operator **E** [15,46]. Please refer to Chapter 4 for more details about non-Cartesian reconstruction.

The L1-SPIRiT reconstruction can be extended from the SPIRiT reconstruction by adding an additional regularization term promoting sparsity as follows:

$$\tilde{\mathbf{s}}_{\mathbf{r}} = \arg\min_{\mathbf{s}_{\mathbf{r}}} \|\mathbf{s} - \mathbf{B}\mathbf{s}_{\mathbf{r}}\|_2^2 + \lambda_1 \|(\mathbf{G} - \mathbf{I})\mathbf{s}_{\mathbf{r}}\|_2^2 + \lambda_2 R(\mathbf{F}^H \mathbf{s}_{\mathbf{r}}). \tag{8.16}$$

Here, $\mathbf{s}_{\mathbf{r}}$ becomes the to-be-reconstructed multicoil k-space in a Cartesian grid (size $= N * C \times 1$). **B** is a linear operator that relates $\mathbf{s}_{\mathbf{r}}$ to the acquired undersampled k-space **s**. In the case of Cartesian acquisition, **B** simply selects acquired k-space locations from $\mathbf{s}_{\mathbf{r}}$, while in the case of non-Cartesian acquisition, **B** becomes an interpolation operator to transfer Cartesian k-space ($\mathbf{s}_{\mathbf{r}}$) onto a non-Cartesian grid in which data points in **s** are sampled. The differences inside the L2-norm are concatenated as column vectors to enforce data consistency (the left term) and kernel consistency (the middle term—see Chapter 6 for detail), respectively. A regularization can be placed on the image (the right term) to enforce sparsity. After image reconstruction, an image **x** can be generated from $\mathbf{s}_{\mathbf{r}}$ with a Fourier transform followed by multicoil combination (e.g., sum of square). In the original implementation of L1-SPIRiT, a L1,2-norm was proposed to exploit the joint sparsity of multicoil phased array images as

$$\left\| \begin{matrix} \mathbf{x}_1 \\ \vdots \\ \mathbf{x}_C \end{matrix} \right\|_{1,2} = \sum_{j=1}^{N} \left(\sqrt{\sum_{i=1}^{C} x_{ij}^2} \right), \tag{8.17}$$

where C and N indicate the number of coil elements and the number of pixels in the image, respectively, and i and j represent the coil and pixel indexes, respectively. The inner L2-norm is used to exploit correlations between different coil elements, and the out L1-norm is employed to enforce sparsity. Alternatively, as demonstrated in the original paper [21], L1-SPIRiT can also be implemented in image domain considering that the SPIRiT operations in k-space are linear shift-invariant convolutions.

The performance of SparseSENSE and L1-SPIRiT is expected to be similar, although coil sensitivities are used in different ways (e.g., explicit estimation of coil sensitivity maps in SparseSENSE and calibration of SPIRiT weights in L1-SPIRiT). The introduction of L1-ESPIRiT (Eigenvector-based SPIRiT) later connects these two types of techniques, which enables a SparseSENSE-type reconstruction formulation that can be implemented with reduced computation complexity compared to L1-SPIRiT. More details of ESPIRiT can be found in Chapter 6.

8.5 Clinical applications of compressed sensing MRI

Since the initial introduction of compressed sensing to MRI, there has been an explosive growth of various sparse MRI techniques and clinical applications spanning the body from the head to the foot [17, 18,24,55–57]. In particular, the combination of compressed sensing with parallel imaging has enabled highly accelerated data acquisition in various MRI studies, and these improvements can be leveraged to further increase spatial resolution and/or temporal resolution. Meanwhile, various vendors have also developed their own compressed sensing-related fast imaging techniques, which now enable routine clinical applications in everyday MRI exams [58–64].

In principle, compressed sensing can be applied to accelerate all MRI scans, spanning from simple 2D imaging to multidimensional dynamic imaging, and these accelerated MRI scans can help reduce the overall MRI exam times and improve patient throughput. However, the achievable acceleration using compressed sensing depends on many factors, including image sparsity/compressibility (e.g., image content and/or the dimensionality of images), and the sequence used for data acquisition that can lead to different contrast-to-noise (CNR) and single-to-noise (SNR). Meanwhile, injection of a gadolinium-based contrast agent can improve the SNR level, which could result in better reconstruction quality.

In general, moderate acceleration rates can be achieved for the application of compressed sensing in 2D MRI. This is mainly because of relatively limited image correlations in a 2D image and the restriction of undersampling to phase-encoding only. Higher acceleration rates can be achieved when applying compressed sensing to 3D MRI due to higher image compressibility/correlations than 2D and the flexibility of undersampling along the two phase-encoding dimensions to generate increased incoherence. For example, compressed sensing has been applied to highly accelerated 3D MR angiography (MRA) [60,89–99] and 3D late gadolinium enhancement (LGE) imaging [65–71].

In addition to accelerated static MRI, dynamic MRI can greatly benefit from compressed sensing, and highly-accelerated dynamic MRI studies have been demonstrated in various clinical applications. This is due to the extensive spatiotemporal correlations in dynamic MR images and the increased flexibility to design a time-varying undersampling scheme for improving incoherence [72]. For example, many studies have demonstrated highly accelerated cardiac cine MRI [73–80], phase-contrast cine MRI [81–84], and myocardial perfusion MRI [22,85–96]. Some of these techniques have already seen increased routine clinical use [59,79]. Dynamic contrast-enhanced MRI (DCE-MRI) is another application that has been increasingly combined with compressed sensing. The fast imaging speed offered by compressed sensing enables better capture of contrast changes for various oncological applications in the brain, neck, chest, and various body organs [46,97–103]. Compressed sensing has also been combined with non-Cartesian sampling schemes to enable free-breathing DCE-MRI in the chest and abdomen, where it is important to minimize motion-induced artifacts [18]. More application of compressed sensing to high-dimensional MRI application can be found in Chapter 9.

The use of compressed sensing has also been applied to many other MRI techniques and acquisitions on the research level, which could potentially promote the adoption of these techniques for clinical applications. For example, MR parameter mapping, such as T1/T2/R2*/T1rho mapping, can substantially benefit from the use of compressed sensing for accelerated data acquisition [44,104–114]. Other applications of compressed sensing in MRI can be seen in time-resolved MR angiography [115–117], arterial spin labeling (ASL) MRI [117], dynamic musculoskeletal imaging for metal artifact reduction [118], non-proton MRI [119,120], spectroscopic imaging [121–123], and hyperpolarized MRI [124,125]. Due to limited space, these interesting applications cannot be fully covered in this chapter, but readers can refer to the cited references for more details.

8.6 Challenges of compressed sensing MRI

As a powerful rapid imaging technique, compressed sensing has achieved a profound impact in MRI since its initial demonstration. However, to date, compressed sensing MRI still needs to deal with several important challenges despite more than a decade of development and optimization. In this final section, we summarize these existing challenges and discuss potential solutions to address them.

The first challenge of compressed sensing MRI is the relatively long reconstruction time compared to non-iterative reconstruction (e.g., parallel MRI reconstruction). Currently, compressed sensing MRI has been implemented by multiple vendors, and it has been broadly adopted clinically for which the reconstruction time does not impact the clinical workflow. However, the slow reconstruction speed still remains a challenge for reconstructing large datasets (e.g., dynamic 4D MRI reconstruction) and non-Cartesian datasets. To overcome this problem, researchers have proposed various off-line reconstruction platforms to bring compressed sensing into the clinic. For example, Block et al. have developed a software tool that can automatically transfer MRI raw data to an external server [126]. Iterative image reconstruction can be performed in the server, and reconstructed images are automatically transferred back to the PACS system. Other similar software tools have also been proposed during recent years [65,127]. However, even with these software tools, radiologists may still experience substantial delays between data acquisition and image generation, making its dissemination challenging. Recent adoption of deep learning for MRI reconstruction has provide a nice solution to address this problem for accelerated 2D or even 3D MRI [28,29,128], as will be more evident in Chapter 11, but it is still challenging to apply such methods to high-dimensional MRI reconstruction due to the need for reference images and memory limitations in training deep neural networks.

Second, compressed sensing MRI requires proper selection of a regularization parameter (or more than one, when more regularizations are used) to control the contribution from the sparsity constraint. As evident in Fig. 8.14, a very large regularization value can lead to overregularization, causing loss of detail, while a too small parameter may result in insufficient suppression of aliasing artifacts. With proper signal normalization and estimation of baseline noise level, the parameter selection may become automatic, and commercially available implementations have extensively demonstrated the robust reconstruction performance of such selection scheme in a large number of clinical scans. However, compared to conventional qualitative MRI, quantitative MRI reconstruction may be more sensitive to the selection of regularization parameters and requires additional care.

Third, compressed sensing has introduced new types of artifacts that are different from artifacts generated from more conventional fast imaging techniques (e.g., parallel imaging). These artifacts include blurring, loss of resolution and contrast, and blocky artifacts. These artifacts could be due to too much compression (excessive acceleration) and/or over regularization. Sometimes the artifacts can be subtle and thus require additional attention from the reading radiologists.

Fourth, since the compressed sensing performance is dependent on the level of sparsity, its acceleration capability is somehow related to the dimensionality of images to be reconstructed. Although compressed sensing still provides moderate advantage in accelerating 2D static compared to parallel imaging, there is a consensus in the field that the achievable acceleration rates using compressed sensing can be higher in 3D MRI and dynamic MRI. However, it should be noted that the computational burden is also increased for high-dimensional MRI reconstruction, posing a trade-off.

Fifth, compressed sensing employs a nonlinear reconstruction process, which changes noise statistics, and the true noise level is hard to estimate. For example, one can simply generate a very clean but completely wrong image by using a very high regularization parameter. As a result, most commonly used metrics for assessing image quality, such as the signal-to-noise ratio (SNR) or contrast-to-noise ratio (CNR), cannot be applied. To date, there is still a lack of reliable metrics for assessment of images reconstructed using compressed sensing. Instead, researchers often perform reader studies, which rely on experienced radiologists to visually evaluate image quality. This, however, is subjective and suffers from observer-dependent variability.

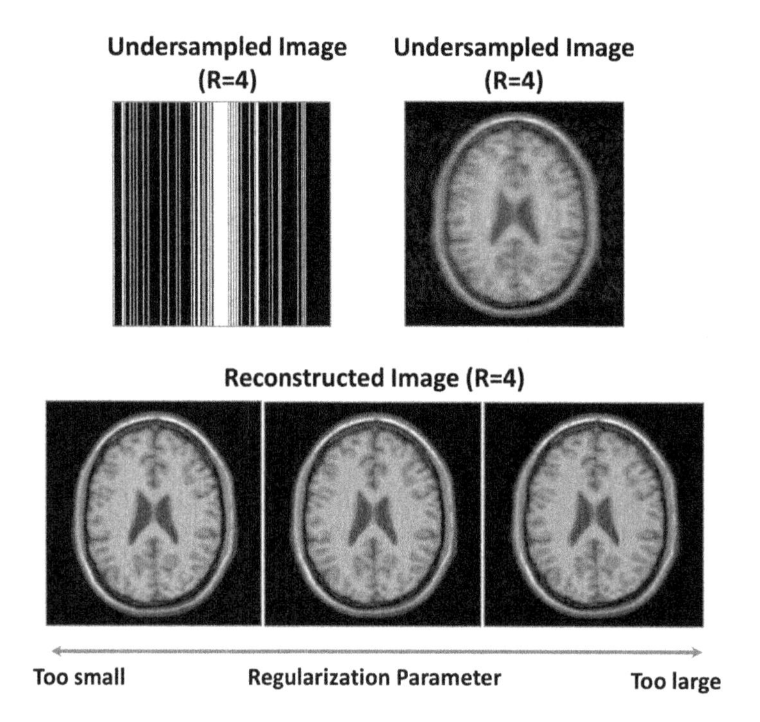

FIGURE 8.14

The influence of regularization parameters on compressed sensing MRI reconstruction. A too large value of parameter can lead to over regularization causing loss of details and/or generation of compression artifacts, while a too small value of parameter may result in insufficient suppression of aliasing artifacts.

8.7 Summary

In this chapter, we learned the three key components of compressed sensing (sparsity, incoherence and nonlinear reconstruction), why we need them, and how they can be implemented in accelerated MRI. Representative frameworks to combine compressed sensing with parallel imaging were presented. Multiple clinical applications of compressed sensing MRI and its existing challenges were summarized. Although image sparsity and incoherent undersampling are usually connected with compressed sensing, it should be realized that both of them play an essential role in more advanced reconstruction techniques, such as the low-rank-based reconstruction, dictionary-based reconstruction, and deep learning-based reconstruction, as will be seen in more details in the following chapters.

8.8 Tutorial

This section provides a demo to reconstruct an undersampled brain image using compressed sensing reconstruction using an iterative soft-thresholding algorithm in MATLAB (Mathworks).

```
clear all;close all;
clc

% Load fully-sampled data
% data: multicoil fully-sampled images
% cmaps: coil sensitivity maps
% mask: 3-fold or 4-fold undersampling pattern
load Brain_8ch.mat;
load('mask_R3.mat') %
cmaps=double(cmaps);
nx,ny,nc=size(data);

% Plot the PSF of the undersampling pattern
PSF=abs(fft2c_mri(mask));
figure,surfl(PSF/max(PSF(:))); colormap(gray),axis square;shading flat,axis
off

% Multicoil combination using sum of square
Img=sum(abs(data.^2),3).^(1/2);
```

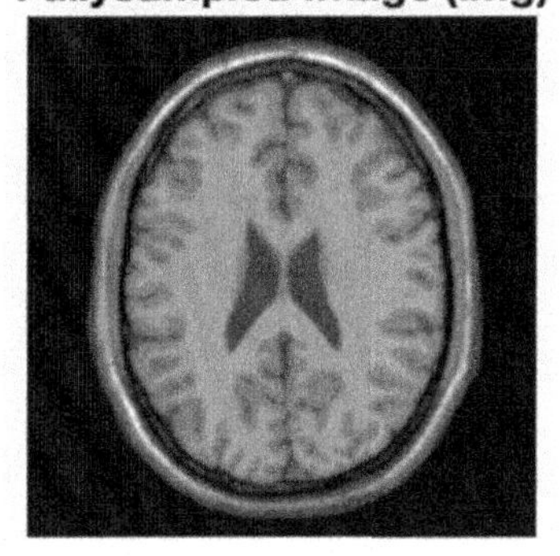


```
% Performing 1D variable-density undersampling
mask=repmat(mask,1 1 8);
kdata=double(fft2c_mri(data).*mask);
```

Point Spread Function

```
% Multicoil Encoding Operator
E=Emat(mask,cmaps);

% Wavelet transform
W=Wavelet('Daubechies',4,4);

% Regularization Parameter
lambda=0.002;

% Number of iterations
nite=50;

% Get undersampled images
Img_u=E'*kdata;
```

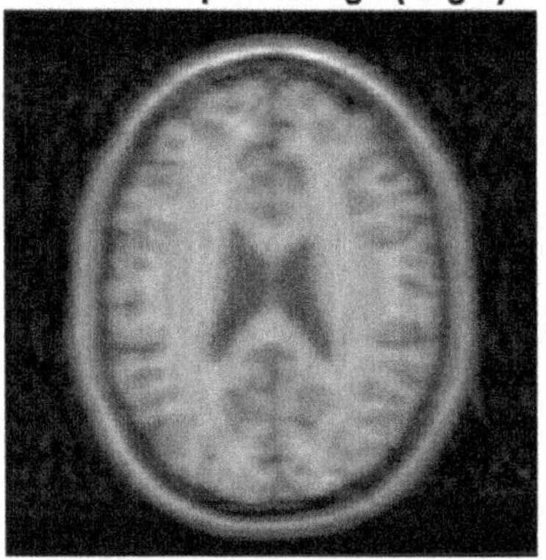


```
% From image domain to sparse domain
Img_sparse=W*Img_u;
```

```
% Iterative reconstruction
for ite=1:nite
    Img_tmp=Img_sparse;

    %Soft-thresholding, as shown in Equation 8.16
    Img_sparse=(abs(Img_sparse)
-lambda).*Img_sparse./abs(Img_sparse).*(abs(Img_sparse)>lambda);
    Img_sparse(isnan(Img_sparse))=0;

    % Image update as shown in Equation 8.15 (data consistency)
    Img_sparse=Img_sparse-W*(E'*(E*(W'*Img_sparse)-kdata));

    %Recon information
    fprintf(' ite: %d, update: %f3\n', ite,norm(Img_sparse(:)-
Img_tmp(:))/norm(Img_tmp(:)));
end

%From sparse domain to image domain
Img_recon=W'*Img_sparse;

% Display images (left to right: Fullysampled Image,Undersampled
Image,Reconstructed Image)
figure,imshow(abs(cat(2,Img,Img_u,Img_recon)),0 1),title('Fullysampled
Image,      Undersampled Image,      Reconstructed Image')
```

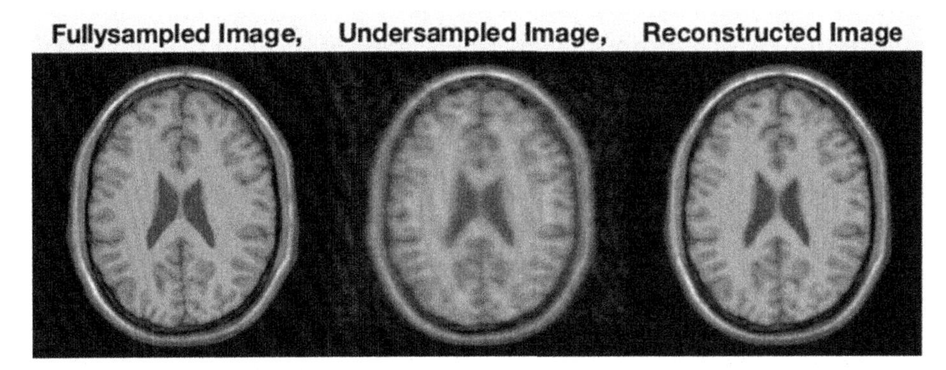

The source code of this demo, together with the sample dataset, can be found along with this chapter. The readers can try the demo with multiple datasets and undersampling patterns. In addition, there are a number of compressed sensing MRI demos that are available online, such those listed here:

Sparse MRI, L1-SPIRiT, L1-ESPIRiT
http://people.eecs.berkeley.edu/~mlustig/Software.html
SparseSENSE, k-t SPARSE-SENSE
https://cai2r.net/resources/k-t-sparse-sense-matlab-code/
Radial compressed sensing MRI
https://cai2r.net/resources/iterative-reconstruction-from-undersampled-radial-data-using-a-total-variation-constraint-matlab-code/
https://cai2r.net/resources/grasp-matlab-code/

Compressed sensing MRI using the split-Bregman algorithm
https://github.com/HGGM-LIM/Split-Bregman-ST-Total-Variation-MRI
Compressed sensing MRI using the alternating direction method of multipliers (ADMM) algorithm
http://jtamir.github.io/t2shuffling-support/
What's more, various implementation compressed sensing algorithms are available in a software tool called BART (Berkeley Advanced Reconstruction Toolbox), which is also available online.
http://wwwuser.gwdg.de/~muecker1/software.html

Acknowledgments

The brain MR image used in this chapter was obtained from the ISMRM Sunrise Course on parallel imaging (http://hansenms.github.io/sunrise/sunrise2013/).

Appendix 8.A Conditions for a unique solution in compressed sensing

This appendix shows why a unique solution can be guaranteed with only $2K$ measurements in an underdetermined linear equation (Eq. (8.2)) if $\mathbf{x}$ is a K-sparse signal (at most K nonzero coefficients) and every $2K$ columns in the measurement matrix $\mathbf{E}$ are linearly independent. First, we state the following definitions:

$\mathbb{N} = \{\hat{\mathbf{x}} : \mathbf{E}\hat{\mathbf{x}} = 0\}$ is the null space of the measurement matrix $\mathbf{E}$;
$\sum K$ is the union of all possible K-sparse signals.
$\sum 2K$ is the union of all possible $2K$-sparse signals.

Now, if we have two solutions to Eq. (8.2), $\mathbf{x}_1$ and $\mathbf{x}_2$, and then we have $\mathbf{x}_1 \in \sum K$, $\mathbf{x}_2 \in \sum K$, $\mathbf{x}_1 + \mathbf{x}_2 \in \sum 2K$, $\mathbf{x}_1 - \mathbf{x}_2 \in \sum 2K$, and $\mathbf{E}(\mathbf{x}_1 - \mathbf{x}_2) = \mathbf{E}\mathbf{x}_1 - \mathbf{E}\mathbf{x}_2 = 0$. To guarantee a unique solution, we need to ensure $\mathbf{x}_1 - \mathbf{x}_2 = 0$. This indicates that we will need to have $\mathbb{N} \cap \sum 2K = \{0\}$ so that $\mathbf{x}_1$ must equal to $\mathbf{x}_2$ to satisfy $\mathbf{E}(\mathbf{x}_1 - \mathbf{x}_2) = 0$. This, in turn, suggests that all $2K$ columns in the measurement matrix $\mathbf{E}$ need to be linearly independent to ensure $\mathbb{N} \cap \sum 2K = \{0\}$.

References

[1] Sodickson DK, Manning WJ. Simultaneous acquisition of spatial harmonics (SMASH): fast imaging with radiofrequency coil arrays. Magn Reson Med 1997;38:591–603. https://doi.org/10.1002/mrm.1910380414.
[2] Pruessmann KP, Weiger M, Scheidegger MB, Boesiger P. SENSE: sensitivity encoding for fast MRI. Magn Reson Med 1999;42:952–62. https://doi.org/10.1002/(SICI)1522-2594(199911)42:5<952::AID-MRM16>3.0.CO;2-S.
[3] Griswold MA, Jakob PM, Heidemann RM, Nittka M, Jellus V, Wang J, et al. Generalized autocalibrating partially parallel acquisitions (GRAPPA). Magn Reson Med 2002;47:1202–10. https://doi.org/10.1002/mrm.10171.
[4] Kellman P, Epstein FH, McVeigh ER. Adaptive sensitivity encoding incorporating temporal filtering (TSENSE). Magn Reson Med 2001;45:846–52. https://doi.org/10.1002/mrm.1113.

[5] Breuer FA, Kellman P, Griswold MA, Jakob PM. Dynamic autocalibrated parallel imaging using temporal GRAPPA (TGRAPPA). Magn Reson Med 2005;53:981–5. https://doi.org/10.1002/mrm.20430.
[6] Tsao J, Boesiger P, Pruessmann KP. k-t BLAST and k-t SENSE: dynamic MRI with high frame rate exploiting spatiotemporal correlations. Magn Reson Med 2003;50:1031–42. https://doi.org/10.1002/mrm.10611.
[7] Huang F, Akao J, Vijayakumar S, Duensing GR, Limkeman M. K-t GRAPPA: a k-space implementation for dynamic MRI with high reduction factor. Magn Reson Med 2005;54:1172–84. https://doi.org/10.1002/mrm.20641.
[8] Ohliger MA, Grant AK, Sodickson DK. Ultimate intrinsic signal-to-noise ratio for parallel MRI: electromagnetic field considerations. Magn Reson Med 2003;50:1018–30. https://doi.org/10.1002/mrm.10597.
[9] Wiesinger F, Van De Moortele PF, Adriany G, De Zanche N, Ugurbil K, Pruessmann KP. Parallel imaging performance as a function of field strength - an experimental investigation using electrodynamic scaling. Magn Reson Med 2004;52:953–64. https://doi.org/10.1002/mrm.20281.
[10] Deshmane A, Gulani V, Griswold MA, Seiberlich N. Parallel MR imaging. J Magn Reson Imaging 2012;36:55–72. https://doi.org/10.1002/jmri.23639.
[11] Candès EJ, Romberg J, Tao T. Robust uncertainty principles: exact signal reconstruction from highly incomplete frequency information. IEEE Trans Inf Theory 2006;52:489–509. https://doi.org/10.1109/TIT.2005.862083.
[12] Donoho DL. Compressed sensing. IEEE Trans Inf Theory 2006;52:1289–306. https://doi.org/10.1109/TIT.2006.871582.
[13] Lustig M, Donoho D, Pauly JM. Sparse MRI: the application of compressed sensing for rapid MR imaging. Magn Reson Med 2007;58:1182–95. https://doi.org/10.1002/mrm.21391.
[14] Lustig M, Donoho DL, Santos JM, Pauly JM. Compressed sensing MRI: a look at how CS can improve on current imaging techniques. IEEE Signal Process Mag 2008;25:72–82. https://doi.org/10.1109/MSP.2007.914728.
[15] Block KT, Uecker M, Frahm J. Undersampled radial MRI with multiple coils. Iterative image reconstruction using a total variation constraint. Magn Reson Med 2007. https://doi.org/10.1002/mrm.21236.
[16] Ye JC. Compressed sensing MRI: a review from signal processing perspective. BMC Biomed Eng 2019;1:1–17. https://doi.org/10.1186/s42490-019-0006-z.
[17] Yang ASC, Kretzler M, Sudarski S, Gulani V, Seiberlich N. Sparse reconstruction techniques in magnetic resonance imaging. Invest Radiol 2016;51:349–64. https://doi.org/10.1097/RLI.0000000000000274.
[18] Feng L, Benkert T, Block KT, Sodickson DK, Otazo R, Chandarana H. Compressed sensing for body MRI. J Magn Reson Imaging 2017;45:966–87. https://doi.org/10.1002/jmri.25547.
[19] Liu B, Zou YM, Ying L. Sparsesense: application of compressed sensing in parallel MRI. In: 5th Int. Conf. Inf. Technol. Appl. Biomed. ITAB 2008 Conjunction with 2nd Int. Symp. Summer Sch. Biomed. Heal. Eng. IS3BHE 2008; 2008. p. 127–30.
[20] Liang D, Liu B, Wang J, Ying L. Accelerating SENSE using compressed sensing. Magn Reson Med 2009;62:1574–84. https://doi.org/10.1002/mrm.22161.
[21] Lustig M, Pauly JM. SPIRiT: iterative self-consistent parallel imaging reconstruction from arbitrary k-space. Magn Reson Med 2010;64:457–71. https://doi.org/10.1002/mrm.22428.
[22] Otazo R, Kim D, Axel L, Sodickson DK. Combination of compressed sensing and parallel imaging for highly accelerated first-pass cardiac perfusion MRI. Magn Reson Med 2010;64:767–76. https://doi.org/10.1002/mrm.22463.
[23] Uecker M, Lai P, Murphy MJ, Virtue P, Elad M, Pauly JM, et al. ESPIRiT - an eigenvalue approach to autocalibrating parallel MRI: where SENSE meets GRAPPA. Magn Reson Med 2014;71:990–1001. https://doi.org/10.1002/mrm.24751.
[24] Jaspan ON, Fleysher R, Lipton ML. Compressed sensing MRI: a review of the clinical literature. Br J Radiol 2015;88. https://doi.org/10.1259/bjr.20150487.
[25] Van Vaals JJ, Brummer ME, Dixon WT, Tuithof HH, Engels H, Nelson RC, et al. 'Keyhole' method for accelerating imaging of contrast agent uptake. J Magn Reson Imaging 1993 Jul-Aug;3(4):671–5.

[26] Korosec FR, Frayne R, Grist TM, Mistretta CA. Time-Resolved Contrast-Enhanced 3D MR Angiography; 1996.

[27] Mistretta CA, Wieben O, Velikina J, Block W, Perry J, Wu Y, et al. Highly constrained backprojection for time-resolved MRI. Magn Reson Med 2006;55:30–40. https://doi.org/10.1002/mrm.20772.

[28] Zhu B, Liu JZ, Cauley SF, Rosen BR, Rosen MS. Image reconstruction by domain-transform manifold learning. Nature 2018;555:487–92. https://doi.org/10.1038/nature25988.

[29] Hammernik K, Klatzer T, Kobler E, Recht MP, Sodickson DK, Pock T, et al. Learning a variational network for reconstruction of accelerated MRI data. Magn Reson Med 2018;79:3055–71. https://doi.org/10.1002/mrm.26977.

[30] Liu F, Samsonov A, Chen L, Kijowski R, Feng L. SANTIS: Sampling-Augmented Neural neTwork with Incoherent Structure for MR image reconstruction. Magn Reson Med 2019;82:1890–904. https://doi.org/10.1002/mrm.27827.

[31] Candes EJ, Wakin MB. An introduction to compressive sampling: a sensing/sampling paradigm that goes against the common knowledge in data acquisition. IEEE Signal Process Mag 2008;25:21–30. https://doi.org/10.1109/MSP.2007.914731.

[32] Donoho DL, Elad M. Optimally sparse representation in general (nonorthogonal) dictionaries via $\ell 1$ minimization. Proc Natl Acad Sci USA 2003;100:2197–202. https://doi.org/10.1073/pnas.0437847100.

[33] Donoho DL. For most large underdetermined systems of linear equations the minimal ℓ 1-norm solution is also the sparsest solution. Commun Pure Appl Math 2006;59:797–829. https://doi.org/10.1002/cpa.20132.

[34] Ge D, Jiang X, Ye Y. A note on the complexity of Lp minimization. Math Program 2011;129:285–99. https://doi.org/10.1007/s10107-011-0470-2.

[35] Mourad N, Reilly JP. Minimizing nonconvex functions for sparse vector reconstruction. IEEE Trans Signal Process 2010;58:3485–96. https://doi.org/10.1109/TSP.2010.2046900.

[36] Chartrand R. Exact reconstruction of sparse signals via nonconvex minimization. IEEE Signal Process Lett 2007;14:707–10. https://doi.org/10.1109/LSP.2007.898300.

[37] Tsaig Y, Donoho DL. Extensions of compressed sensing. Signal Process 2006;86:549–71. https://doi.org/10.1016/j.sigpro.2005.05.029.

[38] Candes EJ, Romberg JK. Signal recovery from random projections. In: Comput. Imaging III, vol. 5674. SPIE; 2005. p. 76.

[39] Davenport MA, Duarte MF, Eldar YC, Kutyniok G. Introduction to compressed sensing. Compress Sens Theory Appl 2012:1–64. https://doi.org/10.1017/CBO9780511794308.002.

[40] Tropp JA, Tropp JA, Gilbert AC. Signal recovery from partial information via orthogonal matching pursuit. In: IEEE TRANS INFORM THEORY; 2005.

[41] Lustig M, Santos JM, Donoho DL, Pauly JM k-t SPARSE: High frame rate dynamic MRI exploiting spatio-temporal sparsity. In: ISMRM 2006 Annu. Meet. Proceedings. page 2420.

[42] Ravishankar S, Bresler Y. Sparsifying transform learning for compressed sensing MRI. In: Proc. - Int. Symp. Biomed. Imaging; 2013. p. 17–20.

[43] Ravishankar S, Bresler Y. Data-Driven learning of a union of sparsifying transforms model for blind compressed sensing. IEEE Trans Comput Imaging 2016;2:294–309. https://doi.org/10.1109/TCI.2016.2567299.

[44] Doneva M, Börnert P, Eggers H, Stehning C, Sénégas J, Mertins A. Compressed sensing reconstruction for magnetic resonance parameter mapping. Magn Reson Med 2010;64:1114–20. https://doi.org/10.1002/mrm.22483.

[45] Akçakaya M, Basha TA, Goddu B, Goepfert LA, Kissinger KV, Tarokh V, et al. Low-dimensional-structure self-learning and thresholding: regularization beyond compressed sensing for MRI Reconstruction. Magn Reson Med 2011;66:756–67. https://doi.org/10.1002/mrm.22841.

[46] Feng L, Grimm R, Block K.T., Chandarana H, Kim S, Xu J, et al. Golden-angle radial sparse parallel MRI: combination of compressed sensing, parallel imaging, and golden-angle radial sampling for fast and flexible dynamic volumetric MRI. Magn Reson Med 2014;72:707–17. https://doi.org/10.1002/mrm.24980.

[47] Tolouee A, Alirezaie J, Babyn P. Compressed sensing reconstruction of cardiac cine MRI using golden angle spiral trajectories. J Magn Reson 2015;260:10–9. https://doi.org/10.1016/j.jmr.2015.09.003.

[48] Valvano G, Martini N, Landini L, Santarelli MF. Variable density randomized stack of spirals (VDR-SoS) for compressive sensing MRI. Magn Reson Med 2016;76:59–69. https://doi.org/10.1002/mrm.25847.

[49] Feng L. Golden-angle radial MRI: basics, advances, and applications. J Magn Reson Imaging 2022. https://doi.org/10.1002/JMRI.28187.

[50] Glover GH, Pauly JM. Projection reconstruction techniques for reduction of motion effects in MRI. Magn Reson Med 1992;28:275–89. https://doi.org/10.1002/mrm.1910280209.

[51] Block KT, Chandarana H, Milla S, Bruno M, Mulholland T, Fatterpekar G, et al. Towards routine clinical use of radial stack-of-stars 3D gradient-echo sequences for reducing motion sensitivity. J Korean Soc Magn Reson Med 2014;18:87. https://doi.org/10.13104/jksmrm.2014.18.2.87.

[52] Elad, M. Milanfar, P. Rubinstein, R. Analysis versus synthesis in signal priors. In: 2006 14th European Signal Processing Conference. Print ISSN: 2219-5491.

[53] Selesnick IW, Figueiredo MAT. Signal restoration with overcomplete wavelet transforms: comparison of analysis and synthesis priors. In: Proc. SPIE 7446, Wavelets XIII, 74460D; 4 September 2009.

[54] Allen BD, Carr M, Botelho MPF, Rahsepar AA, Markl M, Zenge MO, et al. Highly accelerated cardiac MRI using iterative SENSE reconstruction: initial clinical experience. Int J Cardiovasc Imaging 2016;32:955–63. https://doi.org/10.1007/s10554-016-0859-3.

[55] Hollingsworth KG. Reducing acquisition time in clinical MRI by data undersampling and compressed sensing reconstruction. Phys Med Biol 2015;60:R297–322. https://doi.org/10.1088/0031-9155/60/21/R297.

[56] Delattre BMA, Boudabbous S, Hansen C, Neroladaki A, Hachulla AL, Vargas MI. Compressed sensing MRI of different organs: ready for clinical daily practice? Eur Radiol 2020;30:308–19. https://doi.org/10.1007/s00330-019-06319-0.

[57] Yoon JH, Nickel MD, Peeters JM, Lee JM. Rapid imaging: recent advances in abdominal MRI for reducing acquisition time and its clinical applications. Korean J Radiol 2019;20:1597–615. https://doi.org/10.3348/kjr.2018.0931.

[58] Sartoretti E, Sartoretti T, Binkert C, Najafi A, Schwenk Á, Hinnen M, et al. Reduction of procedure times in routine clinical practice with Compressed SENSE magnetic resonance imaging technique. PLoS ONE 2019;14:e0214887. https://doi.org/10.1371/JOURNAL.PONE.0214887.

[59] Ma Y, Hou Y, Ma Q, Wang X, Sui S, Wang B. Compressed SENSE single-breath-hold and free-breathing cine imaging for accelerated clinical evaluation of the left ventricle. Clin Radiol 2019;74:325.e9-e17. https://doi.org/10.1016/J.CRAD.2018.12.012.

[60] Sasi SD, Ramaniharan AK, Bhattacharjee R, Gupta RK, Saha I, Van Cauteren M, et al. Evaluating feasibility of high resolution T1-perfusion MRI with whole brain coverage using compressed SENSE: application to glioma grading. Eur J Radiol 2020;129:109049. https://doi.org/10.1016/J.EJRAD.2020.109049.

[61] Pennig L, Wagner A, Weiss K, Lennartz S, Huntgeburth M, Hickethier T, et al. Comparison of a novel Compressed SENSE accelerated 3D modified relaxation-enhanced angiography without contrast and triggering with CE-MRA in imaging of the thoracic aorta. Int J Cardiovasc Imaging 2021;37:315–29. https://doi.org/10.1007/s10554-020-01979-2.

[62] Boyarko AC, Dillman JR, Tkach JA, Pednekar AS, Trout AT. Comparison of compressed SENSE and SENSE for quantitative liver MRI in children and young adults. Abdom Radiol 2021;46:4567–75. https://doi.org/10.1007/S00261-021-03092-X.

[63] Hur S-J, Choi Y, Yoon J, Jang J, Shin N-Y, Ahn K-J, et al. Intraindividual comparison between the contrast-enhanced golden-angle radial sparse parallel sequence and the conventional fat-suppressed contrast-enhanced T1-weighted spin-echo sequence for head and neck MRI. Am J Neuroradiol 2021. https://doi.org/10.3174/AJNR.A7285.

[64] Tomppert A, Wuest W, Wiesmueller M, Heiss R, Kopp M, Nagel AM, et al. Achieving high spatial and temporal resolution with perfusion MRI in the head and neck region using golden-angle radial sampling. Eur Radiol 2020;31:2263–71. https://doi.org/10.1007/S00330-020-07263-0.

[65] Basha TA, Akçakaya M, Liew C, Tsao CW, Delling FN, Addae G, et al. Clinical performance of high-resolution late gadolinium enhancement imaging with compressed sensing. J Magn Reson Imaging 2017;46:1829–38. https://doi.org/10.1002/jmri.25695.
[66] Akçakaya M, Rayatzadeh H, Basha TA, Hong SN, Chan RH, Kissinger KV, et al. Accelerated late gadolinium enhancement cardiac MR imaging with isotropic spatial resolution using compressed sensing: initial experience. Radiology 2012;264:691–9. https://doi.org/10.1148/radiol.12112489.
[67] Zeilinger MG, Wiesmüller M, Forman C, Schmidt M, Munoz C, Piccini D, et al. 3D dixon water-fat LGE imaging with image navigator and compressed sensing in cardiac MRI. Eur Radiol 2021;31:3951–61. https://doi.org/10.1007/s00330-020-07517-x.
[68] Suekuni H, Kido T, Shiraishi Y, Takimoto Y, Hirai K, Nakamura M, et al. Detecting a subendocardial infarction in a child with coronary anomaly by three-dimensional late gadolinium enhancement MRI using compressed sensing. Radiol Case Rep 2021;16:377–80. https://doi.org/10.1016/j.radcr.2020.11.048.
[69] Kamesh Iyer S, Tasdizen T, Burgon N, Kholmovski E, Marrouche N, Adluru G, et al. Compressed sensing for rapid late gadolinium enhanced imaging of the left atrium: a preliminary study. Magn Reson Imaging 2016;34:846–54. https://doi.org/10.1016/j.mri.2016.03.002.
[70] Pennig L, Lennartz S, Wagner A, Sokolowski M, Gajzler M, Ney S, et al. Clinical application of free-breathing 3D whole heart late gadolinium enhancement cardiovascular magnetic resonance with high isotropic spatial resolution using Compressed SENSE. J Cardiovasc Magn Reson 2020;22. https://doi.org/10.1186/s12968-020-00673-5.
[71] Adluru G, Chen L, Kim SE, Burgon N, Kholmovski EG, Marrouche NF, et al. Three-dimensional late gadolinium enhancement imaging of the left atrium with a hybrid radial acquisition and compressed sensing. J Magn Reson Imaging 2011;34:1465–71. https://doi.org/10.1002/jmri.22808.
[72] Gamper U, Boesiger P, Kozerke S. Compressed sensing in dynamic MRI. Magn Reson Med 2008;59:365–73. https://doi.org/10.1002/MRM.21477.
[73] Usman M, Atkinson D, Odille F, Kolbitsch C, Vaillant G, Schaeffter T, et al. Motion corrected compressed sensing for free-breathing dynamic cardiac MRI. Magn Reson Med 2013;70:504–16. https://doi.org/10.1002/mrm.24463.
[74] Feng L, Srichai MB, Lim RP, Harrison A, King W, Adluru G, et al. Highly accelerated real-time cardiac cine MRI using k-t SPARSE-SENSE. Magn Reson Med 2013;70:64–74. https://doi.org/10.1002/mrm.24440.
[75] Jung H, Park J, Yoo J, Ye JC. Radial k-t FOCUSS for high-resolution cardiac cine MRI. Magn Reson Med 2010;63:68–78. https://doi.org/10.1002/mrm.22172.
[76] Haji-Valizadeh H, Feng L, Ma LE, Shen D, Block KT, Robinson JD, et al. Highly accelerated, real-time phase-contrast MRI using radial k-space sampling and GROG-GRASP reconstruction: a feasibility study in pediatric patients with congenital heart disease. NMR Biomed 2020 May;33(5):e4240. https://doi.org/10.1002/nbm.4240. Epub 2020 Jan 24.
[77] Vincenti G, Monney P, Chaptinel J, Rutz T, Coppo S, Zenge MO, et al. Compressed sensing single-breath-hold CMR for fast quantification of LV function, volumes, and mass. JACC Cardiovasc Imaging 2014;7:882–92. https://doi.org/10.1016/j.jcmg.2014.04.016.
[78] Feng L, Axel L, Latson LA, Xu J, Sodickson DK, Otazo R. Compressed sensing with synchronized cardio-respiratory sparsity for free-breathing cine MRI: initial comparative study on patients with arrhythmias. J Cardiovasc Magn Reson 2014;16. https://doi.org/10.1186/1532-429x-16-s1-o17.
[79] Kido T, Kido T, Nakamura M, Watanabe K, Schmidt M, Forman C, et al. Compressed sensing real-time cine cardiovascular magnetic resonance: accurate assessment of left ventricular function in a single-breath-hold. J Cardiovasc Magn Reson 2016;18:1–11. https://doi.org/10.1186/s12968-016-0271-0.
[80] Sudarski S, Henzler T, Haubenreisser H, Dösch C, Zenge MO, Schmidt M, et al. Free-breathing sparse sampling cine MR imaging with iterative reconstruction for the assessment of left ventricular function and mass at 3.0 T. Radiology 2017;282:74–83. https://doi.org/10.1148/radiol.2016151002.
[81] Kim D, Dyvorne HA, Otazo R, Feng L, Sodickson DK, Lee VS. Accelerated phase-contrast cine MRI using k-t SPARSE-SENSE. Magn Reson Med 2012;67:1054–64. https://doi.org/10.1002/mrm.23088.

[82] Hsiao A, Lustig M, Alley MT, Murphy MJ, Vasanawala SS. Evaluation of valvular insufficiency and shunts with parallel-imaging compressed-sensing 4D phase-contrast MR imaging with stereoscopic 3D velocity-fusion volume-rendered visualization. Radiology 2012;265:87–95. https://doi.org/10.1148/radiol.12120055.

[83] Hsiao A, Lustig M, Alley MT, Murphy M, Chan FP, Herfkens RJ, et al. Rapid pediatric cardiac assessment of flow and ventricular volume with compressed sensing parallel imaging volumetric cine phase-contrast MRI. Am J Roentgenol 2012;198:W250–9. https://doi.org/10.2214/AJR.11.6969.

[84] Vasanawala SS, Alley MT, Hargreaves BA, Barth RA, Pauly JM, Lustig M. Improved pediatric MR imaging with compressed sensing. Radiology 2010;256:607–16. https://doi.org/10.1148/radiol.10091218.

[85] Adluru G, Awate SP, Tasdizen T, Whitaker RT, DiBella EVR. Temporally constrained reconstruction of dynamic cardiac perfusion MRI. Magn Reson Med 2007;57:1027–36. https://doi.org/10.1002/mrm.21248.

[86] Feng L, Axel L, Chandarana H, Block KT, Sodickson DK, Otazo R. XD-GRASP: golden-angle radial MRI with reconstruction of extra motion-state dimensions using compressed sensing. Magn Reson Med 2016;75:775–88. https://doi.org/10.1002/mrm.25665.

[87] Haji-Valizadeh H, Rahsepar AA, Collins JD, Bassett E, Isakova T, Block T, et al. Validation of highly accelerated real-time cardiac cine MRI with radial k-space sampling and compressed sensing in patients at 1.5T and 3T. Magn Reson Med 2018;79:2745–51. https://doi.org/10.1002/mrm.26918.

[88] Roy CW, Seed M, Kingdom JC, Macgowan CK. Motion compensated cine CMR of the fetal heart using radial undersampling and compressed sensing. J Cardiovasc Magn Reson 2017;19:1–14. https://doi.org/10.1186/s12968-017-0346-6.

[89] Adluru G, McGann C, Speier P, Kholmovski EG, Shaaban A, Dibella EVR. Acquisition and reconstruction of undersampled radial data for myocardial perfusion magnetic resonance imaging. J Magn Reson Imaging 2009;29:466–73. https://doi.org/10.1002/jmri.21585.

[90] Yang Y, Kramer CM, Shaw PW, Meyer CH, Salerno M. First-pass myocardial perfusion imaging with whole-heart coverage using L1-SPIRiT accelerated variable density spiral trajectories. Magn Reson Med 2016;76:1375–87. https://doi.org/10.1002/mrm.26014.

[91] Sharif B, Dharmakumar R, Arsanjani R, Thomson L, Bairey Merz CN, Berman DS, et al. Non-ECG-gated myocardial perfusion MRI using continuous magnetization-driven radial sampling. Magn Reson Med 2014;72:1620–8. https://doi.org/10.1002/mrm.25074.

[92] Naresh NK, Haji-Valizadeh H, Aouad PJ, Barrett MJ, Chow K, Ragin AB, et al. Accelerated, first-pass cardiac perfusion pulse sequence with radial k-space sampling, compressed sensing, and k-space weighted image contrast reconstruction tailored for visual analysis and quantification of myocardial blood flow. Magn Reson Med 2019;81:2632–43. https://doi.org/10.1002/mrm.27573.

[93] Chen X, Salerno M, Yang Y, Epstein FH. Motion-compensated compressed sensing for dynamic contrast-enhanced MRI using regional spatiotemporal sparsity and region tracking: block low-rank sparsity with motion-guidance (BLOSM). Magn Reson Med 2014;72:1028–38. https://doi.org/10.1002/mrm.25018.

[94] Pflugi S, Roujol S, Akçakaya M, Kawaji K, Foppa M, Heydari B, et al. Accelerated cardiac MR stress perfusion with radial sampling after physical exercise with an MR-compatible supine bicycle ergometer. Magn Reson Med 2015;74:384–95. https://doi.org/10.1002/mrm.25405.

[95] Akçakaya M, Basha TA, Pflugi S, Foppa M, Kissinger KV, Hauser TH, et al. Localized spatio-temporal constraints for accelerated CMR perfusion. Magn Reson Med 2014;72:629–39. https://doi.org/10.1002/mrm.24963.

[96] Paul J, Wundrak S, Bernhardt P, Rottbauer W, Neumann H, Rasche V. Self-gated tissue phase mapping using golden angle radial sparse SENSE. Magn Reson Med 2016;75:789–800. https://doi.org/10.1002/mrm.25669.

[97] Chandarana H, Feng L, Block TK, Rosenkrantz AB, Lim RP, Babb JS, et al. Free-breathing contrast-enhanced multiphase MRI of the liver using a combination of compressed sensing, parallel imaging, and golden-angle radial sampling. Invest Radiol 2013;48:10–6. https://doi.org/10.1097/RLI.0b013e318271869c.

[98] Chandarana H, Block TK, Ream J, Mikheev A, Sigal SH, Otazo R, et al. Estimating liver perfusion from free–breathing continuously acquired dynamic gadolinium-ethoxybenzyl-diethylenetriamine pentaacetic acid–enhanced acquisition with compressed sensing reconstruction. Invest Radiol 2015;50:88–94. https://doi.org/10.1097/RLI.0000000000000105.

[99] Winkel DJ, Heye TJ, Benz MR, Glessgen CG, Wetterauer C, Bubendorf L, et al. Compressed sensing radial sampling MRI of prostate perfusion: utility for detection of prostate cancer. Radiology 2019;290:702–8. https://doi.org/10.1148/radiol.2018180556.

[100] Guo Y, Lingala SG, Zhu Y, Lebel RM, Nayak KS. Direct estimation of tracer-kinetic parameter maps from highly undersampled brain dynamic contrast enhanced MRI. Magn Reson Med 2017;78:1566–78. https://doi.org/10.1002/mrm.26540.

[101] Zhang T, Chowdhury S, Lustig M, Barth RA, Alley MT, Grafendorfer T, et al. Clinical performance of contrast enhanced abdominal pediatric MRI with fast combined parallel imaging compressed sensing reconstruction. J Magn Reson Imaging 2014;40:13–25. https://doi.org/10.1002/jmri.24333.

[102] Zhang J, Feng L, Otazo R, Kim SG. Rapid dynamic contrast-enhanced MRI for small animals at 7T using 3D ultra-short echo time and golden-angle radial sparse parallel MRI. Magn Reson Med 2019;81:140–52. https://doi.org/10.1002/mrm.27357.

[103] Chen L, Liu D, Zhang J, Xie B, Zhou X, Grimm R, et al. Free-breathing dynamic contrast-enhanced MRI for assessment of pulmonary lesions using golden-angle radial sparse parallel imaging. J Magn Reson Imaging 2018;48:459–68. https://doi.org/10.1002/jmri.25977.

[104] Huang C, Graff CG, Clarkson EW, Bilgin A, Altbach MI. T 2 mapping from highly undersampled data by reconstruction of principal component coefficient maps using compressed sensing. Magn Reson Med 2012;67:1355–66. https://doi.org/10.1002/mrm.23128.

[105] Assländer J, Cloos MA, Knoll F, Sodickson DK, Hennig J, Lattanzi R. Low rank alternating direction method of multipliers reconstruction for MR fingerprinting. Magn Reson Med 2018;79:83–96. https://doi.org/10.1002/mrm.26639.

[106] Zhao B, Setsompop K, Adalsteinsson E, Gagoski B, Ye H, Ma D, et al. Improved magnetic resonance fingerprinting reconstruction with low-rank and subspace modeling. Magn Reson Med 2018;79:933–42. https://doi.org/10.1002/mrm.26701.

[107] Feng L, Otazo R, Jung H, Jensen JH, Ye JC, Sodickson DK, et al. Accelerated cardiac T_2 mapping using breath-hold multiecho fast spin-echo pulse sequence with k-t FOCUSS. Magn Reson Med 2011;65:1661–9. https://doi.org/10.1002/mrm.22756.

[108] Sumpf TJ, Uecker M, Boretius S, Frahm J. Model-based nonlinear inverse reconstruction for T2 mapping using highly undersampled spin-echo MRI. J Magn Reson Imaging 2011;34:420–8. https://doi.org/10.1002/jmri.22634.

[109] Zhang T, Pauly JM, Levesque IR. Accelerating parameter mapping with a locally low rank constraint. Magn Reson Med 2015;73:655–61. https://doi.org/10.1002/mrm.25161.

[110] Zhao B, Lu W, Hitchens TK, Lam F, Ho C, Liang ZP. Accelerated MR parameter mapping with low-rank and sparsity constraints. Magn Reson Med 2015;74:489–98. https://doi.org/10.1002/mrm.25421.

[111] Wang X, Roeloffs V, Klosowski J, Tan Z, Voit D, Uecker M, et al. Model-based T_1 mapping with sparsity constraints using single-shot inversion-recovery radial FLASH. Magn Reson Med 2018;79:730–40. https://doi.org/10.1002/mrm.26726.

[112] Zibetti MVW, Baboli R, Chang G, Otazo R, Regatte RR. Rapid compositional mapping of knee cartilage with compressed sensing MRI. J Magn Reson Imaging 2018;48:1185–98. https://doi.org/10.1002/jmri.26274.

[113] Davies M, Puy G, Vandergheynst P, Wiaux Y. A compressed sensing framework for magnetic resonance fingerprinting. SIAM J Imaging Sci 2014;7:2623–56. https://doi.org/10.1137/130947246.

[114] Cruz G, Jaubert O, Schneider T, Botnar RM, Prieto C. Rigid motion-corrected magnetic resonance fingerprinting. Magn Reson Med 2019;81:947–61. https://doi.org/10.1002/mrm.27448.

[115] Zanardo M, Sardanelli F, Rainford L, Monti CB, Murray JG, Secchi F, et al. Technique and protocols for cardiothoracic time-resolved contrast-enhanced magnetic resonance angiography sequences: a systematic review. Clin Radiol 2020;0. https://doi.org/10.1016/j.crad.2020.08.028.

[116] Rapacchi S, Natsuaki Y, Plotnik A, Gabriel S, Laub G, Finn JP, et al. Reducing view-sharing using compressed sensing in time-resolved contrast-enhanced magnetic resonance angiography. Magn Reson Med 2015;74:474–81. https://doi.org/10.1002/mrm.25414.

[117] Zhou Z, Han F, Yu S, Yu D, Rapacchi S, Song HK, et al. Accelerated noncontrast-enhanced 4-dimensional intracranial MR angiography using golden-angle stack-of-stars trajectory and compressed sensing with magnitude subtraction. Magn Reson Med 2018;79:867–78. https://doi.org/10.1002/mrm.26747.

[118] Otazo R, Nittka M, Bruno M, Raithel E, Geppert C, Gyftopoulos S, et al. Sparse-SEMAC: rapid and improved SEMAC metal implant imaging using SPARSE-SENSE acceleration. Magn Reson Med 2017;78:79–87. https://doi.org/10.1002/mrm.26342.

[119] Madelin G, Chang G, Otazo R, Jerschow A, Regatte RR. Compressed sensing sodium MRI of cartilage at 7T: preliminary study. J Magn Reson 2012;214:360–5. https://doi.org/10.1016/j.jmr.2011.12.005.

[120] Parasoglou P, Feng L, Xia D, Otazo R, Regatte RR. Rapid 3D-imaging of phosphocreatine recovery kinetics in the human lower leg muscles with compressed sensing. Magn Reson Med 2012;68:1738–46. https://doi.org/10.1002/mrm.24484.

[121] Larson PEZ, Hu S, Lustig M, Kerr AB, Nelson SJ, Kurhanewicz J, et al. Fast dynamic 3D MR spectroscopic imaging with compressed sensing and multiband excitation pulses for hyperpolarized 13C studies. Magn Reson Med 2011;65:610–9. https://doi.org/10.1002/mrm.22650.

[122] Furuyama JK, Wilson NE, Burns BL, Nagarajan R, Margolis DJ, Thomas MA. Application of compressed sensing to multidimensional spectroscopic imaging in human prostate. Magn Reson Med 2012;67:1499–505. https://doi.org/10.1002/mrm.24265.

[123] Bogner W, Otazo R, Henning A. Accelerated MR spectroscopic imaging—a review of current and emerging techniques. NMR Biomed 2020. https://doi.org/10.1002/nbm.4314.

[124] Chan H, Stewart NJ, Parra-Robles J, Collier GJ, Wild JM. Whole lung morphometry with 3D multiple b-value hyperpolarized gas MRI and compressed sensing. Magn Reson Med 2017;77:1916–25. https://doi.org/10.1002/mrm.26279.

[125] Ajraoui S, Lee KJ, Deppe MH, Parnell SR, Parra-Robles J, Wild JM. Compressed sensing in hyperpolarized 3He lung MRI. Magn Reson Med 2010;63:1059–69. https://doi.org/10.1002/mrm.22302.

[126] Block KT, Grimm R, Feng L, Otazo R, Chandarana H, Bruno M, et al. Bringing Compressed Sensing to Clinical Reality: Prototypic Setup for Evaluation in Routine Applications. Proc Intl Soc Mag Reson Med. page 3809.

[127] Hansen MS, Sørensen TS. Gadgetron: an open source framework for medical image reconstruction. Magn Reson Med 2013;69:1768–76. https://doi.org/10.1002/mrm.24389.

[128] Bustin A, Fuin N, Botnar RM, Prieto C. From compressed-sensing to artificial intelligence-based cardiac MRI reconstruction. Front Cardiovasc Med 2020;7:17. https://doi.org/10.3389/fcvm.2020.00017.

CHAPTER

9 Low-Rank Matrix and Tensor–Based Reconstruction

Anthony G. Christodoulou
Biomedical Imaging Research Institute, Cedars-Sinai Medical Center, Los Angeles, CA, United States

9.1 Introduction

Low-rank modeling [1] is a powerful approach to reconstruct images from sparsely sampled $(\mathbf{k}, t)$-space data. Like compressed sensing, low-rank methods leverage efficient representations of an image to reduce its degrees of freedom and allow its recovery from sparse samples. A major difference is that, in low-rank reconstructions, the "sparse domain" is adaptive, changing scan to scan based on correlations observed within the acquired data.

Low-rank image reconstruction is used primarily in the context of recovering dynamic images, multicontrast images (e.g., for parameter mapping), multichannel images, and/or image patch arrays. A collection of multiple images or patches naturally lends itself to an interpretation of image sequences and images as arrays, expressed either as matrices (2D arrays) or generalizable to higher-order tensors ($\geq$3D arrays)—especially when imaging multiple simultaneous dynamic processes. Spatiotemporal correlation induces linear dependence between columns, rows, and fibers of the array, leading to low-rankness which is then exploited for accelerated imaging.

A wide range of imaging applications has been shown to admit low-rank representations, including cardiovascular imaging [2–5], quantitative parameter mapping and multicontrast imaging [6–10], MR fingerprinting [11–13]), dynamic contrast enhancement imaging [1,14,15], spectroscopic imaging [16–18], and fMRI [19–21], among others. Furthermore, the use of higher-order tensor models has also allowed simultaneous imaging of various combinations of such dynamic processes [22–26].

Section 9.2 generalizes dynamic imaging and multicontrast imaging problems in particular to a context that will permit high-order tensor interpretations; Section 9.3 introduces several low-rank matrix models; Section 9.4 extends modeling to higher-order tensors; and Section 9.5 concludes the chapter with a brief outlook.

9.2 Problem formulation

This chapter focuses on reconstructing dynamic images and multicontrast images (sequences of multiple image frames changing with, e.g., motion or variable pulse sequence/timing parameters). We generalize the concept of dynamic and multicontrast imaging by first considering an underlying multidimensional image $\dot{x}(\mathbf{r}, \tau_1(t), \tau_2(t), \ldots, \tau_L(t))$, which is a function of voxel location $\mathbf{r} = [x, y, z]^T$ and $L \geq 1$ time-varying independent variables $\{\tau_\ell(t)\}_{\ell=1}^{L}$, each representing a different physical or physiological dynamic process, including but not limited to motion, NMR relaxation, and dynamic contrast

Advances in Magnetic Resonance Technology and Applications, Volume 7, ISSN 2666-9099. https://doi.org/10.1016/B978-0-12-822726-8.00019-1

enhancement. Progression of these dynamic processes changes the image that can be observed as well as the signal that can be measured.

The dynamic MR signal from $\tilde{x}$ is measured in $(\mathbf{k}, t)$-space:

$$s_c(\mathbf{k}, t) = E_c\{x(\mathbf{r}, t)\} + \eta_c(\mathbf{k}, t) \tag{9.1}$$

$$x(\mathbf{r}, t) = \tilde{x}(\mathbf{r}, \tau_1(t), \tau_2(t), \ldots, \tau_L(t)), \tag{9.2}$$

where $E_c\{\cdot\}$ spatially encodes x at coil c and η denotes complex Gaussian noise. The image $x(\mathbf{r}, t)$ describes how $\tilde{x}$ plays out in the scanner as the $\{\tau_\ell(t)\}_{\ell=1}^{L}$ vary.

It is common to target only a single dynamic process by "freezing" all others during acquisition, i.e., the ℓth dynamic process represented by $\tau_\ell(t)$ is often isolated by acquiring data only when $\tau_q(t) = \kappa_q \ \forall \ q \neq \ell$, for some set of constant values $\{\kappa_q\}_{q\neq\ell}$. This is practically achieved by: 1) pausing the actual processes, physically holding the $\tau_q(t)$ constant (e.g., pausing respiration with breath holds); and/or 2) synchronizing acquisition to the process, only collecting or storing $(\mathbf{k}, t)$-space data acquired when $\tau_q(t)$ matches κ_q (e.g., removing the influence of cardiac motion with electrocardiogram (ECG) triggering to a specific cardiac phase). The decision of which dynamic processes to retain and which to remove depends on the specific imaging application.

The goal of low-rank image reconstruction is to recover either x or an $\tilde{x}(\mathbf{r}, \tau_\ell)$ with a single dynamic process (Sections 9.3 and 9.4.2) or the entire $\tilde{x}(\mathbf{r}, \tau_1, \tau_2, \ldots, \tau_L)$ (Section 9.4.3) from undersampled $(\mathbf{k}, t)$-space measurements through the use of low-rank image modeling.

9.3 Matrix-based approaches

This section focuses on reconstructing $x(\mathbf{r}, t)$ using low-rank matrix models.

It is useful to begin with the singular value decomposition (SVD), which decomposes any matrix $\mathbf{X} \in \mathbb{C}^{M\times N}$ into a sum of rank-1 matrices:

$$\mathbf{X} = \mathbf{U}\boldsymbol{\Sigma}\mathbf{V}^{\mathbf{H}} = \sum_{r=1}^{\min(M,N)} \sigma_r \mathbf{u}_r \mathbf{v}_r^{\mathbf{H}}. \tag{9.3}$$

The left and right complex singular vectors $\{\mathbf{u}_r\}_{r=1}^{\min(M,N)}$ and $\{\mathbf{v}_r\}_{r=1}^{\min(M,N)}$ are the columns of $\mathbf{U}$ and $\mathbf{V}$, and are orthonormal bases for the columns of $\mathbf{X}$ and $\mathbf{X}^{\mathbf{H}}$, respectively. The singular values $\{\sigma_r\}_{r=1}^{\min(M,N)}$ lie along the diagonal of $\boldsymbol{\Sigma}$ ($[\boldsymbol{\Sigma}]_{rr} = \sigma_r$), and are real nonnegative values in descending order. The rank of $\mathbf{X}$ is equal to the number of nonzero singular values; when there are only $R < \min(M, N)$ nonzero singular values (i.e., when $\sigma_r = 0 \ \forall \ r > R$), then $\text{rank}(\mathbf{X}) = R < \min(M, N)$, and $\mathbf{X}$ is said to be low-rank. In this case, singular values and vectors corresponding to $r > R$ are unnecessary for signal representation. For a rank-R $\mathbf{X}$, the summation in Eq. (9.3) can therefore be truncated without loss of information, such that $\mathbf{X}$ is the sum of R rank-1 matrices: $\mathbf{X} = \sum_{r=1}^{R} \sigma_r \mathbf{u}_r \mathbf{v}_r^{\mathbf{H}}$. The ability to efficiently represent a low-rank matrix by its truncated SVD implies that low-rank matrices have fewer degrees of freedom than full-rank matrices: Whereas a full-rank $\mathbf{X} \in \mathbb{C}^{M\times N}$ has $2MN$ real degrees of freedom, a rank-R $\mathbf{X}$ has only $2(M + N - R)R$ real degrees of freedom. Note also that truncating the SVD below the true rank of $\mathbf{X}$ ($R < \text{rank}(\mathbf{X})$) produces the least-squares optimal rank-R approximation to $\mathbf{X}$. As a result, the SVD is a critical component of several low-rank reconstruction and approximation algorithms.

A more general decomposition of $\mathbf{X} \in \mathbb{C}^{M \times N}$ is the matrix factorization

$$\mathbf{X} = \mathbf{A}\boldsymbol{\Phi}. \tag{9.4}$$

Here, the left matrix $\mathbf{A} \in \mathbb{C}^{M \times R}$ and right matrix $\boldsymbol{\Phi} \in \mathbb{C}^{R \times N}$ are constrained neither to be orthonormal, nor to have any special ordering of columns and rows, respectively. As a result, the decomposition in Eq. (9.4) is not unique. The SVD can provide forms of this factorization, for example, by calculating $\mathbf{A} = \mathbf{U}\boldsymbol{\Sigma}$, $\boldsymbol{\Phi} = \mathbf{V}^{\mathbf{H}}$, but it is only one example construction. Note that the decomposition in Eq. (9.4) places an upper bound on the rank of $\mathbf{X}$, specifically that $\text{rank}(\mathbf{X}) \leq R$ (ranks lower than R occur if either the columns of $\mathbf{A}$ or the rows of $\boldsymbol{\Phi}$ are linearly dependent). Its generality is useful for low-rank reconstruction algorithms and will be used in this chapter to tie together several low-rank image models.

To reconstruct $x(\mathbf{r}, t)$ using low-rank matrix models, we can express Eq. (9.1) in matrix-vector form as $\mathbf{s} = E(\mathbf{X}) + \boldsymbol{\eta}$, where the matrix $\mathbf{X} \in \mathbb{C}^{M \times N}$ has elements $[\mathbf{X}]_{mn} = x(\mathbf{r}_m, t_n)$, and $E(\cdot)$ models spatial encoding:

$$\mathbf{X} = \begin{bmatrix} x(\mathbf{r}_1, t_1) & x(\mathbf{r}_1, t_2) & \cdots & x(\mathbf{r}_1, t_N) \\ x(\mathbf{r}_2, t_1) & x(\mathbf{r}_2, t_2) & \cdots & x(\mathbf{r}_2, t_N) \\ \vdots & \vdots & \ddots & \vdots \\ x(\mathbf{r}_M, t_1) & x(\mathbf{r}_M, t_2) & \cdots & x(\mathbf{r}_M, t_N) \end{bmatrix}.$$

The matrix $\mathbf{X}$ is sometimes referred to as the Casorati matrix of $x(\mathbf{r}, t)$. Note that a Casorati matrix can also be formed from $\tilde{x}(\mathbf{r}, \tau_\ell)$ with a single dynamic process as $[\mathbf{X}]_{mn} = \tilde{x}(\mathbf{r}_m, \tau_{\ell,n})$. As a result, the approaches discussed here also apply to imaging an $\tilde{x}(\mathbf{r}, \tau_\ell)$, for which $(\mathbf{k}, t)$-space data may have already been reorganized according to, e.g., the cardiac phase.

9.3.1 Global low-rank modeling

The Rth-order partial separability (PS) model [1] expresses $x(\mathbf{r}, t)$ as

$$x(\mathbf{r}, t) = \sum_{r=1}^{R} a_r(\mathbf{r})\varphi_r(t), \tag{9.5}$$

decomposing the image into R spatial coefficient maps $\{a_r(\mathbf{r})\}_{r=1}^{R}$ and R temporal functions $\{\varphi_r(t)\}_{r=1}^{R}$ (Fig. 9.1). This model is consistent with the matrix factorization in Eq. (9.4), where $\mathbf{A}$ has elements $[\mathbf{A}]_{mr} = a_r(\mathbf{r}_m)$ and $\boldsymbol{\Phi}$ has elements $[\boldsymbol{\Phi}]_{rn} = \varphi_r(t_n)$. This image property is termed *partial* separability because it separates groups of variables (e.g., $\mathbf{r}$ as a whole) as opposed to every individual variable (e.g., each x, y, z).

This formulation reveals that strong spatiotemporal correlation in $x(\mathbf{r}, t)$ induces $\mathbf{X}$ to be low-rank: When the set of voxel functions $\{x(\mathbf{r}_m, t)\}_{m=1}^{M}$ (the rows of $\mathbf{X}$) and the set of image frames $\{x(\mathbf{r}, t_n)\}_{n=1}^{N}$ (the columns of $\mathbf{X}$) are each so highly correlated to be linearly dependent, $\mathbf{X}$ will be low-rank such that $\text{rank}(\mathbf{X}) < \min(M, N)$. A low model order $R < \min(M, N)$ provides an avenue for image reconstruction from sparse $(\mathbf{k}, t)$-space samples due to the reduced degrees of freedom in low-rank matrices described previously. See Fig. 9.2.

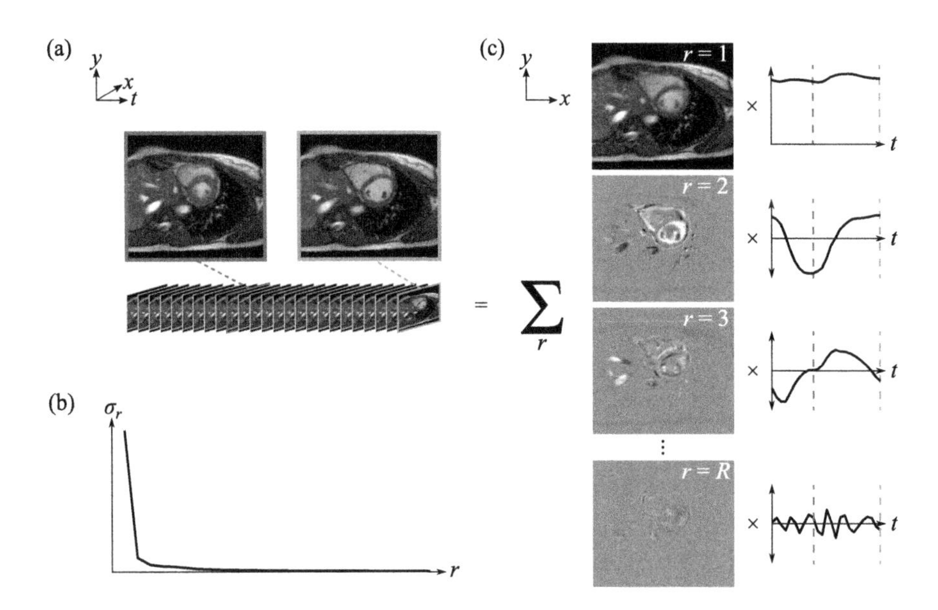

FIGURE 9.1 Global low-rank decomposition

(**a**) A cardiac MR cine image sequence with 25 frames. (**b**) Singular value plot demonstrating the quick decay and low effective rank of the image matrix. (**c**) Decomposition of the image into spatial coefficient maps and their corresponding temporal functions.

Eq. (9.4) implies that $\mathbf{X}$ lies in either of two R-dimensional subspaces: the column space of $\mathbf{A}$ and the row space of $\mathbf{\Phi}$. In the context of Eq. (9.5), the column space of $\mathbf{A}$ becomes a spatial subspace, and the row space of $\mathbf{\Phi}$ becomes a temporal subspace. Similarly, when the columns of $\mathbf{A}$ or the rows of $\mathbf{\Phi}$ are orthonormal, they constitute a spatial basis or a temporal basis for $\mathbf{X}$, respectively. This subspace view can inform data acquisition and image reconstruction (as will be seen in Sections 9.3.1.1 and 9.3.1.2, respectively) and reveals a connection to the synthesis formulation of compressed sensing (CS) with a temporal sparsifying basis. In CS, $\mathbf{\Phi}$ is a pre-selected sparsifying basis with $R = N$, and $\mathbf{A}$ are the sparse coefficients in the transform domain $\mathbf{A} = \mathbf{X}\mathbf{\Phi}^{-1}$. Therefore, where CS seeks a sparse $\mathbf{A}$ with $R = N$ and pre-selected $\mathbf{\Phi}$, low-rank image reconstruction seeks an adaptive "sparsifying basis" $\mathbf{\Phi}$ that depends on the measured signal, and also seeks an $\mathbf{A}$ which is not necessarily sparse. In the context of low-rank image models, "sparsity" is instead achieved by rank-reduction, which induces the sparsity of the singular values: A rank-R matrix has only R nonzero singular values (i.e., $\sigma_r = 0 \; \forall \; r > R$). The fast decay of singular values for an example dynamic MR image is displayed in Fig. 9.1b.

9.3.1.1 Sampling

When low-rank image reconstruction is to be employed, there are two primary considerations for sampling $(\mathbf{k}, t)$-space: 1) incoherence and 2) whether to collect interleaved "training data".

As discussed in Chapter 8, measurement incoherence is dependent on both the sampling pattern and the sparsifying basis used in image reconstruction. For low-rank image reconstruction, there would ideally be incoherence between the sampling operator and the low-rank "sparsifying basis" $\mathbf{\Phi}$ [27].

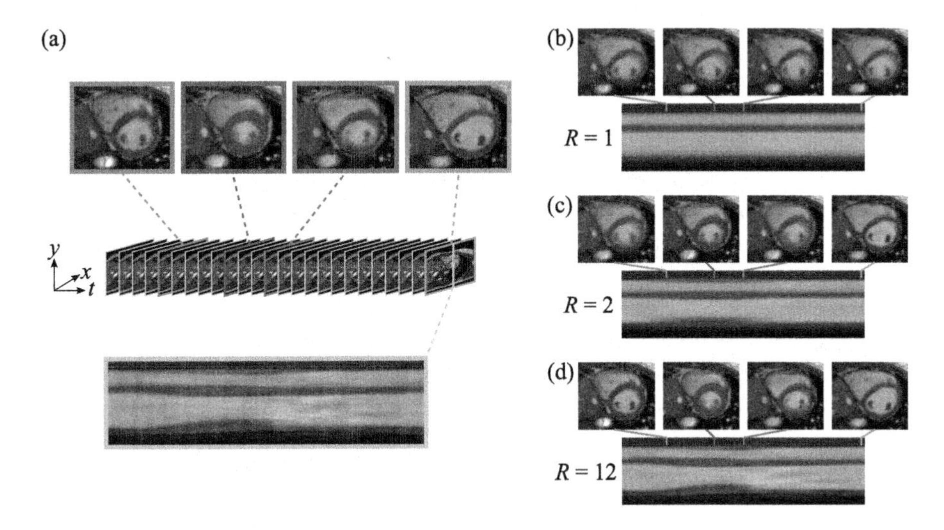

FIGURE 9.2 Effect of rank on image approximation

(**a**) Zoomed-in regions of four frames and a temporal profile of the cardiac cine MR image, and its (**b**) rank-1, (**c**) rank-2, and (**d**) rank-12 SVD approximations. Note that morphology does not change in a rank-1 image since there is only one spatial coefficient map being modulated over time. Additional dynamic detail is present as rank increases, and the $R = 12$ image is virtually indistinguishable from the original 25-frame dynamic image.

Since $\mathbf{\Phi}$ is adaptive in low-rank image modeling, it is generally not known *a priori* exactly how much incoherence a given sampling pattern will achieve, presenting a challenge for optimal sampling design. However, some known-coherent scenarios can be readily avoided. For example, a repetitive sampling pattern synchronized to a periodic dynamic $\tau_\ell(t)$ (e.g., cardiac motion, respiratory motion, magnetization preparation intervals) would result in strong coherence between the sampling operator and $\mathbf{\Phi}$. This coherence can be mitigated by avoiding sampling pattern periodicity within the physiologically expected ranges for cardiac and respiratory motion and synchronization with the known magnetization preparation period. Incoherence can also be practically achieved in most scenarios through randomly ordered, nonrepetitive sampling patterns—a common choice when sampling for low-rank image reconstruction. Real-time motion tracking techniques such as phase ordering with automatic window selection (PAWS) [28] can update sampling patterns on the fly to minimize redundant samples, which can also benefit incoherence. However, all of these are only heuristic approaches for achieving a good sampling patterns; none of them directly measure or optimize incoherence between the sampling operator and $\mathbf{\Phi}$, as it is not known *a priori* in the presence of motion. Real-time feedback approaches have been proposed to estimate $\mathbf{\Phi}$ while data are still being acquired and subsequently update sampling patterns in real-time [29], but have not yet been adopted in practice. In situations where $\mathbf{\Phi}$ can be estimated *a priori* (e.g., for multicontrast imaging/parameter mapping without motion [6,30]), incoherence can be measured and optimized prospectively.

The second consideration is the inclusion of regular frequent sampling of a subset of **k**-space, variably referred to as training data, subspace training data, navigator data, dense data, or auxiliary data.

These sampling patterns interleave training data alongside incoherent sparse sampling of $(\mathbf{k}, t)$-space (sometimes referred to as imaging data or sparse data) at more **k**-space locations. The rationale for this is two-fold: 1) because low-rank models decompose $\mathbf{X}$ into a spatial factor $\mathbf{A}$ and a temporal factor $\mathbf{\Phi}$, the classical trade-off between spatial and temporal resolution is somewhat decoupled, and sampling can be split into subsets—one targeting $\mathbf{A}$ and the other targeting $\mathbf{\Phi}$. The imaging data have the extensive **k**-space coverage required for high spatial resolution and therefore target $\mathbf{A}$; the training data are sampled very frequently and therefore target $\mathbf{\Phi}$. 2) Interleaved acquisition also makes two-step explicit recovery of $\mathbf{A}$ and $\mathbf{\Phi}$ possible since $\mathbf{\Phi}$ can be directly estimated from the training data, after which image reconstruction reduces to recovering $\mathbf{A}$ from the imaging data given a known $\mathbf{\Phi}$. However, collecting training data usually comes at the cost of collecting more imaging data elsewhere in **k**-space; if every Nth readout is a training readout instead of an imaging readout, then the imaging data acquisition efficiency is $\frac{N-1}{N}$ (e.g., 50% efficient when collecting training data every other readout, and 87.5% efficient when collecting training data every eighth readout). Furthermore, frequent training readouts also have the potential to induce eddy-current artifacts from large jumps in **k**-space, especially for balanced SSFP sequences [31]. Although there are self-contained training data schemes that collect training data during slice rephase and read dephase gradients that already occur prior to imaging readouts [32], this strategy extends the minimum achievable echo time and is therefore primarily appropriate for T_2^*-weighted imaging.

9.3.1.2 Image reconstruction

There are two basic approaches to image reconstruction: 1) explicit low-rank reconstruction, which aims to recover a rank-R $\mathbf{X}$ (typically by recovering the factors $\mathbf{A}$ and $\mathbf{\Phi}$); and 2) implicit low-rank reconstruction, which aims to recover an $\mathbf{X}$ which minimizes a rank surrogate.

Explicit low-rank reconstruction

The objective of explicit low-rank reconstruction is to reconstruct a $\mathbf{X}$ with at most rank R. This sets up a rank-constrained optimization problem

$$\hat{\mathbf{X}} = arg\ min_{\mathbf{X}} \|E(\mathbf{X}) - \mathbf{s}\|_2^2 \quad s.t. \quad \text{rank}(\mathbf{X}) \leq R, \tag{9.6}$$

which can be parameterized as

$$\left\{\hat{\mathbf{A}}, \hat{\mathbf{\Phi}}\right\} = arg\ min_{\mathbf{A},\mathbf{\Phi}} \|E(\mathbf{A\Phi}) - \mathbf{s}\|_2^2. \tag{9.7}$$

Eq. (9.7) frames the problem as recovery of the model factors $\mathbf{A}$ and $\mathbf{\Phi}$ from measured data. This permits efficient image reconstruction algorithms that store and operate upon $\mathbf{A}$ and $\mathbf{\Phi}$ instead of storing and operating upon the larger $\mathbf{X}$ matrix; in fact, $\mathbf{X}$ needs never to be explicitly calculated [33], permitting image reconstruction even on systems that cannot hold the full $\mathbf{X}$ in memory.

Fixed-subspace reconstruction

Several imaging methods use a two-step approach to recover $\mathbf{A}$ and $\mathbf{\Phi}$ [1,3]. These methods require collecting training data as previously described in Section 9.3.1.1 that can then be arranged into a multichannel $(\mathbf{k}, t)$-space Casorati matrix $\mathbf{S}_{\mathrm{tr}}$ that lies in the row space of $\mathbf{\Phi}$:

$$\mathbf{S}_{\mathrm{tr}} = \begin{bmatrix} \mathbf{S}_{\mathrm{tr},1} \\ \mathbf{S}_{\mathrm{tr},2} \\ \vdots \\ \mathbf{S}_{\mathrm{tr},C} \end{bmatrix}, \quad \mathbf{S}_{\mathrm{tr},c} = \begin{bmatrix} s_c(\mathbf{k}_1, t_1) & s_c(\mathbf{k}_1, t_2) & \cdots & s_c(\mathbf{k}_1, t_N) \\ s_c(\mathbf{k}_2, t_1) & s_c(\mathbf{k}_2, t_2) & \cdots & s_c(\mathbf{k}_2, t_N) \\ \vdots & \vdots & \ddots & \vdots \\ s_c(\mathbf{k}_M, t_1) & s_c(\mathbf{k}_M, t_2) & \cdots & s_c(\mathbf{k}_M, t_N) \end{bmatrix}.$$

The two image reconstruction steps are then serial recovery of $\mathbf{\Phi}$ and $\mathbf{A}$ from the training data and imaging data, respectively:

$$\hat{\mathbf{\Phi}} = arg\ min_{\mathbf{\Phi}} \left\| \mathbf{S}_{\mathrm{tr}} - \mathbf{S}_{\mathrm{tr}} \left(\mathbf{\Phi}^{\dagger} \mathbf{\Phi} \right) \right\|_F^2, \tag{9.8}$$

$$\hat{\mathbf{A}} = arg\ min_{\mathbf{A}} \left\| E(\mathbf{A}\hat{\mathbf{\Phi}}) - \mathbf{s} \right\|_2^2. \tag{9.9}$$

Eq. (9.8) can be efficiently solved by calculating the SVD of $\mathbf{S}_{\mathrm{tr}}$ and defining $\hat{\mathbf{\Phi}}$ from the R most significant right singular vectors. With a known $\hat{\mathbf{\Phi}}$, image reconstruction then reduces to Eq. (9.9), simple linear least-squares recovery of $\mathbf{A}$. Eq. (9.9) can be viewed as recovering the coordinates of $\mathbf{X}$ within a fixed subspace (the row space of $\hat{\mathbf{\Phi}}$).

Although fixed-subspace reconstruction is computationally efficient, it does of course require collection of training data, which comes at the cost of sparser sampling elsewhere in $\mathbf{k}$-space, as described in Section 9.3.1.1. Two-step reconstruction also potentially introduces model bias through its choice of $\hat{\mathbf{\Phi}}$: Although the temporal basis calculated from the SVD of the training data is least squares optimal for describing $\mathbf{S}_{\mathrm{tr}}$, it is generally not the optimal temporal basis for describing $\mathbf{X}$ as well [32], unless certain conditions on the training $\mathbf{k}$-space trajectory as well as the spark and joint sparsity of the columns of $\mathbf{X}$ are satisfied [34]. Notably, this strategy requires advance selection of the model order R, typically by heuristically fixing R for a given application or by analyzing the singular values of $\mathbf{S}_{\mathrm{tr}}$.

Alternating reconstruction

To address this, schemes for jointly recovering $\mathbf{A}$ and $\mathbf{\Phi}$ from the whole of the collected data have been proposed [35,36]. These methods seek a solution to Eq. (9.7) for both $\mathbf{A}$ and $\mathbf{\Phi}$. This optimization problem is nonconvex, but a satisfactory local minimum can typically be found by alternating least squares (ALS) minimization:

$$\hat{\mathbf{\Phi}}^{[it]} = arg\ min_{\mathbf{\Phi}} \left\| E\left(\hat{\mathbf{A}}^{[it-1]} \mathbf{\Phi} \right) - \mathbf{s} \right\|_2^2 \tag{9.10}$$

$$\hat{\mathbf{A}}^{[it]} = arg\ min_{\mathbf{A}} \left\| E\left(\mathbf{A} \hat{\mathbf{\Phi}}^{[it]} \right) - \mathbf{s} \right\|_2^2, \tag{9.11}$$

each iteration of which updates $\mathbf{\Phi}$ and $\mathbf{A}$ by linear least squares recovery. Fast convergence can be achieved by starting from $R = 1$ and periodically incrementing R during optimization until the desired model order is reached [37]. This approach does not require any training data, but as a nonconvex optimization technique is dependent on initial guesses. Like fixed-subspace reconstruction, this approach also requires selection of the specific rank R.

Implicit low-rank reconstruction

The objective of implicit low-rank reconstruction is to recover a matrix $\mathbf{X}$ without explicitly choosing an R or parameterizing $\mathbf{X}$ as $\mathbf{A\Phi}$. Without a known R, low-rank image reconstruction can be reframed as a rank-minimization problem

$$\hat{\mathbf{X}} = arg\ min_{\mathbf{X}} \text{rank}(\mathbf{X}) \quad s.t. \quad \|E(\mathbf{X}) - \mathbf{s}\|_2^2 \leq \epsilon \tag{9.12}$$

that seeks the lowest-rank $\mathbf{X}$ within a predetermined data consistency tolerance ϵ. Eq. (9.12) is nonconvex and can be NP-hard, so several alternative rank surrogates have been proposed in the framework of regularized reconstruction:

$$\hat{\mathbf{X}} = arg\ min_{\mathbf{X}} \|E(\mathbf{X}) - \mathbf{s}\|_2^2 + \lambda R_{\text{rank}}(\mathbf{X}), \tag{9.13}$$

where $R_{\text{rank}}(\cdot)$ is a rank surrogate and λ is the regularization hyperparameter that controls effective rank.

A natural surrogate for matrix rank uses the Schatten p-norm $\|\cdot\|_p$ [4,38]

$$R_{\text{rank}}(\mathbf{X}) = \|\mathbf{X}\|_p^p = \sum_{r=1}^{\min(M,N)} [\sigma_r(\mathbf{X})]^p, \tag{9.14}$$

that promotes sparsity of the singular values of $\mathbf{X}$ when $p \leq 1$. The Schatten p-norm for $p < 1$ is still nonconvex, so a common practical choice is $p = 1$ [4,20,39], which corresponds to the nuclear norm $\|\cdot\|_*$, the tightest convex relaxation of matrix rank [40,41]:

$$R_{\text{rank}}(\mathbf{X}) = \|\mathbf{X}\|_* = \sum_{r=1}^{\min(M,N)} \sigma_r(\mathbf{X}). \tag{9.15}$$

Because the nuclear norm is analogous to the L1-norm of the singular values, Eq. (9.15) can, for example, be solved by iterative soft-thresholding of the singular values, known as the singular value thresholding algorithm [42].

A major benefit of implicit low-rank matrix-recovery approaches that solve forms of Eq. (9.13) is that they do not require *a priori* knowledge of matrix rank R, but they do still require determination of a hyperparameter λ to control the matrix effective rank. Like explicit ALS approaches, implicit approaches do not require collection of training data, allowing flexibility in sampling. However, these benefits come at the cost of computational efficiency: By recovering a $\mathbf{X}$ that can be factored rather than by recovering the factors $\mathbf{A}$ and $\mathbf{\Phi}$, the memory usage of implicit low-rank reconstruction algorithms typically do not benefit from low-rank matrix compression to the same extent as explicit low-rank reconstruction algorithms. Furthermore, when $\mathbf{X}$ is not parameterized, Fourier encoding and sensitivity encoding must be calculated for all N image frames of $\mathbf{X}$ during each iteration, rather than just for the R frames of $\mathbf{A}$. For small N, these computational considerations are not generally a concern, but can become prohibitive for large-scale dynamic image recovery.

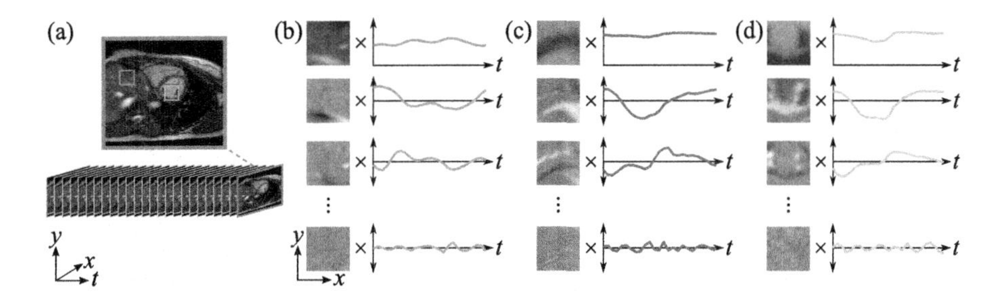

FIGURE 9.3 Local low-rank decomposition

(a) The dynamic image can be split into several spatial regions or patches. (b, c, d) Each patch constitutes its own low-rank matrix and has its own temporal basis that does not have to be shared with other patches. Note that the patch in the liver (*red* (gray in print version)) has quite different temporal basis functions than the two distant patches over the heart (*green, yellow* (mid gray, light gray in print version)), which neighbor each other and have more similar temporal basis functions.

9.3.2 Local low-rank modeling

The global low-rank model in Eq. (9.5) assumes that all voxels are globally correlated, i.e., that the entire set of voxel functions $\{x(\mathbf{r}_m, t)\}_{m=1}^{M}$ lies in the same temporal subspace span $\{\varphi_r(t)\}_{r=1}^{R}$. However, correlation may be higher in neighboring voxels from the same organ than in distant voxels from other organs, in which case it is more efficient to restrict the assumption of spatiotemporal correlation to local voxel neighborhoods (e.g., image patches) via local low-rank (LLR) modeling [43,7,10,44,45].

LLR models split the image into multiple spatial regions $\{\Omega_i\}_{i\geq 1}$, each of which is assigned to its own temporal subspace span $\{\varphi_{i,r}(t)\}_{r=1}^{R_i}$:

$$x(\mathbf{r}, t) = \begin{cases} \sum_{r=1}^{R_1} a_r(\mathbf{r})\varphi_{1,r}(t), & \text{if } \mathbf{r} \in \Omega_1 \\ \sum_{r=2}^{R_2} a_r(\mathbf{r})\varphi_{2,r}(t), & \text{if } \mathbf{r} \in \Omega_2 \\ \vdots & \\ \sum_{r=1}^{R_i} a_r(\mathbf{r})\varphi_{i,r}(t), & \text{if } \mathbf{r} \in \Omega_i \\ \vdots & \end{cases} \tag{9.16}$$

This model allows different patches of $x(\mathbf{r}, t)$ to lie in different temporal subspaces (Fig. 9.3). The image $x(\mathbf{r}, t)$ can be said to lie in a union of subspaces rather than a single shared temporal subspace.

The LLR property can be imposed during image reconstruction by solving

$$\hat{\mathbf{X}} = arg\ min_{\mathbf{X}} \|E(\mathbf{X}) - \mathbf{s}\|_2^2 + \lambda \sum_{i\geq 1} R_{\text{rank}}(\mathbf{P}_i\mathbf{X}), \tag{9.17}$$

where $\mathbf{P}_i$ extracts the ith spatial patch/region (a subset of the rows of $\mathbf{X}$) to form a spatially-localized Casorati matrix. This implicit LLR formulation does not require strict adherence to Eq. (9.16), in the sense that it optionally permits the use of overlapping patches to reduce artifacts at the boundaries of patches. LLR image reconstruction is generally compatible with explicit ALS algorithms. However, it is not always compatible with fixed-subspace modeling since it is difficult to determine a separate $\mathbf{\Phi}_i$ for each spatial region/patch. Still, fixed-subspace reconstruction is sometimes possible when the training scheme allows low spatial resolution reconstruction of the training data [46,47] or when the regions are allowed to share some temporal basis functions but vary in model order [48].

When there is more local correlation than global correlation, LLR reconstruction can produce less temporal blurring from model bias than global low-rank reconstruction, at the expense of ignoring any global correlation that may still be present. LLR reconstruction especially benefits when the average patch model order can be set quite low, in which case the additional degrees of freedom associated with allowing multiple temporal subspaces are counteracted by the lower average model order. LLR approaches require selection of a patch size: Small patch sizes typically have very low rank, but may not be sensitive enough to capture all nearby temporally correlated structures; on the other hand, large patch sizes capture more correlation but are not very specific, and may have higher rank due to the inclusion of more distant, less temporally correlated structures. A patch size that spans the entire image field of view is equivalent to the global low-rank model.

9.3.3 Low-rank and sparse modeling

It is important to note that compressed sensing with fixed sparsifying bases and low-rank image modeling with adaptive bases are not mutually exclusive. For both techniques, incoherence is commonly achieved by the same means (e.g., randomized sampling), so their data acquisition requirements are often compatible. As Chapter 8 has established, MR images are known to be sparse, and as Sections 9.3.1 and 9.3.2 have established, several types of MR images are also known to be globally or locally low-rank. As a result, several methods have been developed to restrict the solution space to those images that are both low-rank and sparse (L&S).

In practice, low-rank image reconstruction can be combined with compressed sensing by introducing a sparse regularization term $\|\Psi(\mathbf{A\Phi})\|_1$ (for explicit low-rank reconstruction) or $\|\Psi(\mathbf{X})\|_1$ (for implicit low-rank reconstruction) to the main objective function, where $\Psi(\cdot)$ is a spatial and/or temporal sparsifying transform. The L&S forms of the previous optimization problems Eqs. (9.7), (9.9), (9.13), and (9.17) become:

$$\left\{\hat{\mathbf{A}}, \hat{\mathbf{\Phi}}\right\} = arg\ min_{\mathbf{A},\mathbf{\Phi}} \|E(\mathbf{A\Phi}) - \mathbf{s}\|_2^2 + \lambda \|\Psi(\mathbf{A\Phi})\|_1 \quad (9.18)$$

$$\hat{\mathbf{A}} = arg\ min_{\mathbf{A}} \left\|E(\mathbf{A}\hat{\mathbf{\Phi}}) - \mathbf{s}\right\|_2^2 + \lambda \left\|\Psi(\mathbf{A}\hat{\mathbf{\Phi}})\right\|_1 \quad (9.19)$$

$$\hat{\mathbf{X}} = arg\ min_{\mathbf{X}} \|E(\mathbf{X}) - \mathbf{s}\|_2^2 + \lambda_1 R_{\text{rank}}(\mathbf{X}) + \lambda_2 \|\Psi(\mathbf{X})\|_1 \quad (9.20)$$

$$\hat{\mathbf{X}} = arg\ min_{\mathbf{X}} \|E(\mathbf{X}) - \mathbf{s}\|_2^2 + \lambda_1 \sum_{i\geq 1} R_{\text{rank}}(\mathbf{P}_i\mathbf{X}) + \lambda_2 \|\Psi(\mathbf{X})\|_1, \quad (9.21)$$

where the choice of Ψ is motivated by the specific MRI application. This can be a fixed transform [4,33] to impose transform sparsity on $\mathbf{X}$ in direct analogy to compressed sensing; in the cases of Eqs. (9.18) and (9.19), Ψ can be an adaptive transform incorporating $\mathbf{\Phi}^\dagger$ (e.g., $\Psi(\mathbf{X}) = \mathbf{\Psi X \Phi}^\dagger = \mathbf{\Psi A}$) to impose transform sparsity on the spatial coefficients $\mathbf{A}$ and to retain the memory benefits of explicit low-rank modeling [36,48,49]. The additional regularization power of compressed sensing allows more flexible

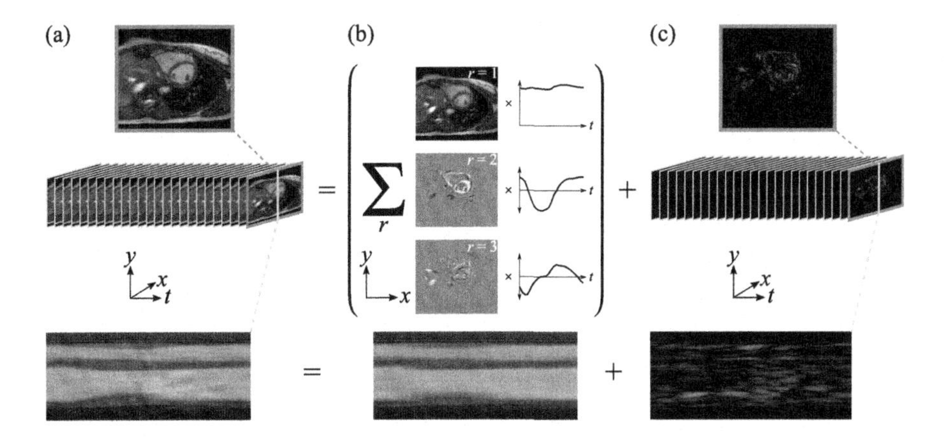

FIGURE 9.4 Low-rank plus sparse decomposition

(**a**) A dynamic image can also be decomposed into the sum of a (**b**) low-rank term and (**c**) a sparse (or transform sparse) term. The low-rank term can be set with a very low model order, leaving additional detail to be absorbed into the sparse term. Note here that the sparse term is being displayed as a magnitude image at 3x brightness for clarity.

selection of low-rank model orders since sparse regularization improves the conditioning of inverse problems with high R or low λ_1 [33].

9.3.4 Low-rank plus sparse modeling

An alternative approach to incorporating sparse modeling into low-rank image reconstruction is to model the image not as low-rank *and* sparse, but rather as a low-rank term *plus* a sparse residual [50–53]. The use of a low-rank plus sparse (L+S) decomposition $\mathbf{X} = \mathbf{L} + \mathbf{S}$ uses a low-rank matrix $\mathbf{L}$ to coarsely capture correlation, absorbing finer detail and transient events into the sparse residual $\mathbf{S}$ (Fig. 9.4). This decomposition is unique as long as $\mathbf{L}$ is not sparse and $\mathbf{S}$ is not low-rank [52].

L+S modeling can be used with either the explicit parameterization $\mathbf{L} = \mathbf{A\Phi}$ [53] or with implicit low-rank modeling penalizing a rank surrogate of $\mathbf{L}$ [50–52]. Explicit low-rank modeling of $\mathbf{L}$ suggests the formulation

$$\left\{\hat{\mathbf{A}}, \hat{\mathbf{\Phi}}, \hat{\mathbf{S}}\right\} = arg\ min_{\mathbf{A},\mathbf{\Phi},\mathbf{S}} \left\| E(\mathbf{A\Phi} + \mathbf{S}) - \mathbf{s} \right\|_2^2 + \lambda \left\| \text{vec}(\mathbf{S}) \right\|_1, \tag{9.22}$$

where $\|\text{vec}(\mathbf{S})\|_1$ promotes sparsity of $\mathbf{S}$. Either the fixed-subspace reconstruction or ALS strategies could be applied to solve Eq. (9.22). Note that, although parameterization of $\mathbf{L} = \mathbf{A\Phi}$ has memory benefits, $\mathbf{S}$ is still stored in full.

Likewise, implicit low-rank modeling of $\mathbf{L}$ suggests the formulation

$$\left\{\hat{\mathbf{L}}, \hat{\mathbf{S}}\right\} = arg\ min_{\mathbf{L},\mathbf{S}} \left\| E(\mathbf{L} + \mathbf{S}) - \mathbf{s} \right\|_2^2 + \lambda_1 R_{\text{rank}}(\mathbf{L}) + \lambda_2 \left\| \text{vec}(\mathbf{S}) \right\|_1, \tag{9.23}$$

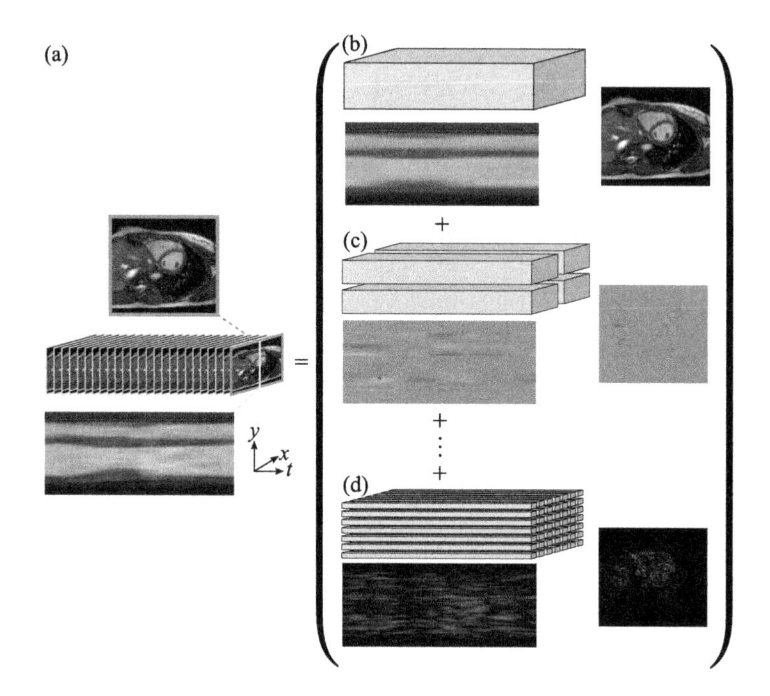

FIGURE 9.5 Multiscale low-rank decomposition

(**a**) A dynamic image can also be represented as the sum of low-rank images at several scales: (**a**) global, (**b**) local, (**c**) voxel-wise. Note that the voxel-wise scale can be treated as a sparse term, displayed here as a magnitude image at 3x brightness.

where $R_{\text{rank}}(\mathbf{L})$ promotes low-rankness of $\mathbf{L}$. This formulation has roots in robust principal component analysis (RPCA) [54] but as applied to matrix recovery rather than image denoising or analysis.

In the event that $\mathbf{S}$ is sparser in a different domain, both Eqs. (9.22) and (9.23) can be modified to impose transform sparsity of $\mathbf{S}$, simply by replacing the regularization term $\|\text{vec}(\mathbf{S})\|_1$ with a transform sparse regularization term $\|\Psi(\mathbf{S})\|_1$.

9.3.5 Multiscale low-rank modeling

Multiscale low-rank modeling generalizes the L+S concept by further decomposing the $\mathbf{L}$ term as the sum of globally low-rank and locally low-rank components, thereby bridging global low-rank modeling, local low-rank modeling, and L+S modeling using a single framework [55,56]. The multiscale low-rank decomposition expresses the image matrix as $\mathbf{X} = \sum_{j=1}^{J} \mathbf{X}_j$, where $\mathbf{X}_j$ is the term at the jth spatial scale: $\mathbf{X}_1$ is globally low rank, $\mathbf{X}_2$ through $\mathbf{X}_{J-1}$ are locally low-rank for increasingly smaller patch sizes, and the voxel-wise term $\mathbf{X}_J$ is analogous to the sparse term in L+S modeling (Fig. 9.5). Hence, this decomposition generalizes global low-rank modeling ($\mathbf{X}_1$ only), local low-rank modeling ($\mathbf{X}_j$ only, $1 < j < J$), and L+S modeling ($\mathbf{X}_1$ and $\mathbf{X}_J$ only).

Multiscale low-rank modeling is compatible with both explicit low-rank factorization [56] and implicit low-rank regularization [55]. The explicit formulation expresses each term as $\mathbf{X}_j = \mathcal{M}_j\left(\mathbf{A}_j, \{\mathbf{\Phi}_{ij}\}_{i\geq 1}\right)$, where $\mathbf{A}_j \in \mathbb{C}^{M\times R_j}$ are spatial coefficients as previously described, where $\mathbf{\Phi}_{ij} \in \mathbb{C}^{R_j\times N}$ is the temporal factor for the ith spatial patch at the jth spatial scale, and where $\mathcal{M}_j$ selectively multiplies each patch of $\mathbf{A}_j$ by the corresponding $\mathbf{\Phi}_{ij}$ to produce an $M \times N$ matrix. Image reconstruction can then be carried out in the basic framework of

$$arg\ min_{\{\mathbf{A}_j\}_{j\geq 1}, \{\mathbf{\Phi}_{ij}\}_{i\geq 1, j\geq 1}} \left\| E\left(\sum_{j=1}^{J} \mathcal{M}_j\left(\mathbf{A}_j, \{\mathbf{\Phi}_{ij}\}_{i\geq 1}\right)\right) - \mathbf{s}\right\|_2^2. \tag{9.24}$$

In practice, Eq. (9.24) incorporates an additional regularization term to further penalize the rank at each spatial scale: $\sum_j \lambda_j \left(\|\mathbf{A}_j\|_F^2 + \sum_i \|\mathbf{\Phi}_{ij}\|_F^2\right)$, which draws upon a relaxation of the nuclear norm for explicitly parameterized low-rank matrices [56].

An implicit multiscale low-rank formulation is

$$arg\ min_{\mathbf{X}} \left\| E\left(\sum_{j=1}^{J} \mathbf{X}_j\right) - \mathbf{s}\right\|_2^2 + \sum_{i\geq 1}\sum_{j=1}^{J} \lambda_j R_{\text{rank}}\left(\mathbf{P}_{ij}\mathbf{X}_j\right), \tag{9.25}$$

where $\mathbf{P}_{ij}$ extracts the ith spatial patch at the jth spatial scale. Note that the storage requirements for the whole set of unfactored $\{\mathbf{X}_j\}_{j=1}^{J}$ can be prohibitive for large-scale problems, motivating the explicit strategy [56] when imaging at ambitious scales.

9.4 Tensor-based approaches

Whereas Section 9.3 interpreted image sequences as matrices (2-dimensional arrays), this section interprets them as higher-order tensors ($\geq$3-dimensional arrays). There are several ways to define additional dimensions that permit a tensor interpretation, e.g., coil dimensions from multichannel receive arrays, preservation of separate spatial dimensions, aggregation of similar patches from different spatial locations, and recovery of the multidynamic $\tilde{x}(\mathbf{r}, \tau_1, \tau_2, \ldots, \tau_L)$ with multiple "time dimensions" $\{\tau_\ell\}_\ell^L$.

9.4.1 Tensor definitions

Tensor notation, operations, and decompositions are not as universally defined as matrix operations so here we introduce some general background before moving on to tensor-based imaging.

We refer to an N-dimensional array $\mathcal{X}$ as an N-way tensor $\mathcal{X} \in \mathbb{C}^{M_1\times M_2\times\cdots\times M_N}$. The collection of values with all indices fixed except along the nth tensor mode (or dimension) is referred to as the mode-n fiber; fibers are simply the higher-order analogues of matrix columns (mode-1 fibers) and rows (mode-2 fibers). When extracted from tensors, fibers are typically represented as column vectors, regardless of original mode. It is sometimes useful to unfold or flatten tensors into matrices. The mode-n unfolding of $\mathcal{X}$ into a matrix is denoted as $\mathbf{X}_{(n)} \in \mathbb{C}^{M_n\times\prod_{i\neq n} M_i}$, which places the nth dimension of $\mathcal{X}$ along the columns of $\mathbf{X}_{(n)}$. Like matrices, tensors can be multiplied. We denote the n-mode matrix product of $\mathcal{X}$ with a matrix $\mathbf{Q} \in \mathbb{C}^{J\times M_n}$ as $\mathcal{Y} = \mathcal{X} \times_n \mathbf{Q}$ with $\mathcal{Y} \in \mathbb{C}^{M_1\times\cdots\times M_{n-1}\times J\times M_{n+1}\times\cdots\times M_N}$, which has

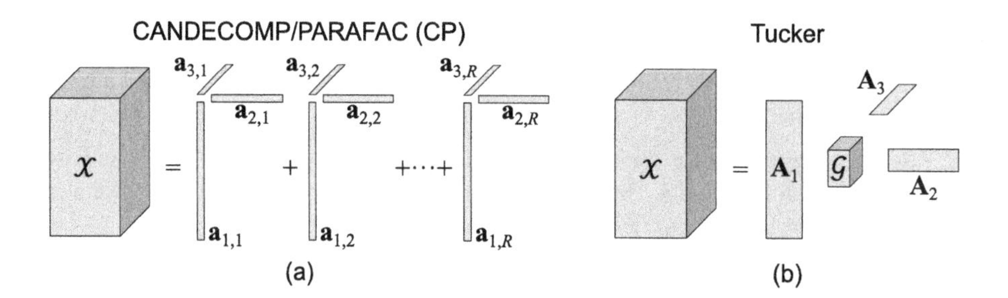

FIGURE 9.6 Conceptual illustration of tensor decompositions

Decomposition of a 3-way tensor $\mathcal{X}$ in (**a**) CP form and (**b**) Tucker form.

elements $[\mathcal{Y}]_{m_1 m_2 \cdots m_{n-1} j m_{n+1} \cdots m_N} = \sum_{m_n=1}^{M_n} [\mathbf{Q}]_{j m_n} [\mathcal{X}]_{m_1 m_2 \cdots m_{n-1} m_n m_{n+1} \cdots m_N}$. This can be compactly represented in mode-n unfolded form as $\mathbf{Y}_{(n)} = \mathbf{Q}\mathbf{X}_{(n)}$.

Similarly to low-rank matrix image reconstruction, low-rank tensor (LRT) image reconstruction can also be divided into explicitly low-rank methods (that parameterize $\mathcal{X}$ by decomposing it into several factors) and implicitly low-rank methods (that minimize a surrogate for the rank of $\mathcal{X}$). A great many tensor decompositions are available for this parameterization and for the definition of tensor rank [57], but we will focus on the two most commonly used for MRI: canonical decomposition [58]/parallel factors [59] (CANDECOMP/PARAFAC, further abbreviated as CP), and the Tucker decomposition [60] (Fig. 9.6). The interested reader is referred to [57] for in-depth discussion of these and other tensor decompositions.

9.4.1.1 CP decomposition

The CP decomposition expresses $\mathcal{X}$ as the sum of rank-1 tensors:

$$\mathcal{X} = \sum_{r=1}^{R} \mathbf{a}_{1,r} \circ \mathbf{a}_{2,r} \circ \cdots \circ \mathbf{a}_{Nr}, \tag{9.26}$$

where $\circ$ denotes the vector outer product and $\mathbf{a}_{n,r} \in \mathbb{C}^{M_n \times 1}$. Element-wise, this becomes

$$[\mathcal{X}]_{m_1 m_2 \cdots m_N} = \sum_{r=1}^{R} [\mathbf{a}_{1,r}]_{m_1} [\mathbf{a}_{2,r}]_{m_2} \ldots [\mathbf{a}_{N,r}]_{m_N} \tag{9.27}$$

$$= \sum_{r=1}^{R} [\mathbf{A}_1]_{m_1 r} [\mathbf{A}_2]_{m_2 r} \ldots [\mathbf{A}_N]_{m_N r}, \tag{9.28}$$

where $\mathbf{A}_n \in \mathbb{C}^{M_n \times R}$ is the factor matrix for the nth dimension. This further admits the mode-1 unfolded representation

$$\mathbf{X}_{(1)} = \mathbf{A}_1 \left(\mathbf{A}_N \odot \mathbf{A}_{N-1} \odot \cdots \odot \mathbf{A}_2 \right)^T, \tag{9.29}$$

where $\odot$ denotes the Khatri–Rao product. Using the CP decomposition, $\mathcal{X}$ is said to be rank-R.

9.4.1.2 Tucker decomposition

The Tucker decomposition decomposes $\mathcal{X}$ with a different model order for each dimension:

$$\mathcal{X} = \sum_{r_1=1}^{R_1} \sum_{r_2=1}^{R_2} \cdots \sum_{r_N=1}^{R_N} g_{r_1 r_2 \cdots r_N} \mathbf{a}_{1,r_1} \circ \mathbf{a}_{2,r_2} \circ \cdots \circ \mathbf{a}_{N r_N}. \tag{9.30}$$

Element-wise, this becomes

$$[\mathcal{X}]_{m_1 m_2 \cdots m_N} = \sum_{r_1=1}^{R_1} \sum_{r_2=1}^{R_2} \cdots \sum_{r_N=1}^{R_N} g_{r_1 r_2 \cdots r_N} [\mathbf{a}_{1,r_1}]_{m_1} [\mathbf{a}_{2,r_2}]_{m_2} \cdots [\mathbf{a}_{N,r_3}]_{m_N} \tag{9.31}$$

$$= \sum_{r_1=1}^{R_1} \sum_{r_2=1}^{R_2} \cdots \sum_{r_N=1}^{R_N} g_{r_1 r_2 \cdots r_N} [\mathbf{A}_1]_{m_1 r_1} [\mathbf{A}_2]_{m_2 r_2} \cdots [\mathbf{A}_N]_{m_N r_N}, \tag{9.32}$$

where $\mathbf{A}_n \in \mathbb{C}^{M_n \times R_n}$ is the factor matrix for the nth dimension. Typically, the columns within each $\mathbf{A}_n$ are chosen to be orthonormal, such that $\mathbf{A}_n$ constitutes a basis for the nth dimension of $\mathcal{X}$. This admits the n-mode product representation

$$\mathcal{X} = \mathcal{G} \times_1 \mathbf{A}_1 \times_2 \mathbf{A}_2 \cdots \times_N \mathbf{A}_N \tag{9.33}$$

that gives rise to a core tensor $\mathcal{G} \in \mathbb{C}^{R_1 \times R_2 \times \cdots \times R_N}$ with elements $[\mathcal{G}]_{r_1 r_2 \cdots r_N} = g_{r_1 r_2 \cdots r_N}$. This core tensor governs the interactions between the factor matrices and is the higher-order analogue to the matrix singular values but does not have to be diagonal [61]. The mode-1 unfolded representation becomes

$$\mathbf{X}_{(1)} = \mathbf{A}_1 \mathbf{G}_{(1)} \left(\mathbf{A}_N \otimes \mathbf{A}_{N-1} \otimes \cdots \otimes \mathbf{A}_2\right)^T, \tag{9.34}$$

where $\otimes$ denotes the Kronecker product. Using the Tucker decomposition, $\mathcal{X}$ is said to be rank-(R_1, $R_2, \ldots, R_N$).

The CP and Tucker decompositions have both been used for a wide range of applications, with CP tending to be used for analysis and Tucker tending to be used for compression (in many scenarios, each R_n for the Tucker decomposition is often much smaller than R from the CP decomposition, leading to smaller factor matrices) [57]. In the context of MRI, the CP decomposition may permit more accurate signal representation [26], at the expense of memory usage. For matrices (i.e., $N = 2$), both the CP and Tucker decompositions lead to the same representation.

9.4.1.3 Tensor rank surrogates

As in low-rank matrix image reconstruction, implicit low-rank tensor reconstruction will seek an $\mathcal{X}$ which minimizes a surrogate for tensor rank. As the comparison of rank definitions for the CP and Tucker decompositions shows, there are several definitions of tensor rank, and the task of determining the rank of a known tensor can be NP-hard [62]. As such, there are also several surrogates $R_{\text{rank}}(\cdot)$ for tensor rank [63–65], the most well-known of which is the sum of nuclear norms of each unfolding of $\mathcal{X}$ [63]:

$$R_{\text{rank}}(\mathcal{X}) = \sum_{n=1}^{N} \left\|\mathbf{X}_{(n)}\right\|_*, \tag{9.35}$$

which serves as a convex relaxation of tensor rank. However, the formulations in this section extend to other definitions of $R_{\text{rank}}(\mathcal{X})$ as well.

9.4.2 Reinterpreting dynamic images as tensors

A dynamic or multicontrast image $x(\mathbf{r}, t)$ or $\tilde{x}(\mathbf{r}, \tau_\ell)$ is not limited to a matrix formation as in Section 9.3; several higher-dimensional formations are possible as well. Additional dimensions can be defined from multiple receive coil channels, by enforcing separability of the individual spatial dimensions, by stacking similar image patches centered at different voxels, and more.

9.4.2.1 *Coil modeling*

Modern MR systems use multiple receive channels for signal detection, for which the image $x_c(\mathbf{r}, t)$ received at channel c varies with the coil sensitivity pattern $x_c(\mathbf{r}, t) = C_c(\mathbf{r})x(\mathbf{r}, t)$. When stacking the images from different receive channels into a tensor $\mathcal{X} \in \mathbb{C}^{M \times N \times C}$ with elements $[X]_{mnc} = x_c(\mathbf{r}_m, t_n)$, coil compressibility and smoothness of the coil sensitivity patterns $\{C_c(\mathbf{r})\}_{c=1}^{C}$ will induce weak global correlation and strong local correlation in $\mathcal{X}$ [66,67]. As a result, LRT modeling can be used to perform calibrationless parallel dynamic imaging without knowledge of the coil sensitivities or an autocalibration signal. The best-known form of this reconstruction problem takes the form of an implicit LLR tensor problem [68]:

$$arg\ min_{\mathbf{X}} \|E(\mathcal{X}) - \mathbf{s}\|_2^2 + \lambda \sum_{i \geq 1} R_{\text{rank}}(\mathcal{X} \times_1 \mathbf{P}_i), \tag{9.36}$$

where as in Eq. (9.17), $\mathbf{P}_i$ extracts the ith spatial patch, but where $E(\cdot)$ no longer includes multiplication by known coil sensitivities. This problem is also compatible with explicit ALS formulations.

9.4.2.2 *Patch similarity modeling*

Additional dimensionality can also be defined by stacking spatially correlated patches from various spatial locations [69–73]. Several sets of patches can be aggregated and then formed into a tensor, either by stacking correlated static patches from individual timepoints (subsets of single columns of $\mathbf{X}$) into a matrix or by stacking correlated dynamic patches (N-column submatrices of $\mathbf{X}$) into a tensor. We will denote the jth collection of patches as $\mathcal{X}_j$. Enforcing low-rankness of each $\mathcal{X}_j$ promotes spatial or spatiotemporal self-similarity throughout $x(\mathbf{r}, t)$.

Low-rankness of the $\{\mathcal{X}_j\}_{j \geq 1}$ can be enforced by solving a problem such as

$$arg\ min_{\mathbf{X}} \|E(\mathbf{X}) - \mathbf{s}\|_2^2 + \lambda \sum_{j \geq 1} R_{\text{rank}}\left(P_j(\mathbf{X})\right), \tag{9.37}$$

where P_j concatenates a set of several similar patches from $\mathbf{X}$ into a matrix or tensor such that $\mathcal{X}_j = P_j(\mathbf{X})$.

There are also several ways to systematically stack shifting, overlapping spatial or **k**-space patches that fall into the class of structured low-rank matrix completion methods, as will be discussed in Chapter 10.

9.4.2.3 Spatial separability

The methods discussed so far have flattened spatial dimensions, conceptualizing an image frame as a vector that is placed as a column of $\mathbf{X}$. Several methods opt to instead preserve spatial dimensionality without vectorization [74]. An $M_x \times M_y \times M_z \times N$ image sequence is therefore taken to be a $\mathcal{X} \in \mathbb{C}^{M_x \times M_y \times M_z \times N}$ with elements $[\mathcal{X}]_{k\ell mn} = x([x_k, y_\ell, z_m]^T, t_n)$.

Whereas Eq. (9.5) assumes low-order partial separability of $\mathbf{X}$ by grouping together all voxels, LRT treatments of this $\mathcal{X}$ assume low-order separability of all variables:

$$x(\mathbf{r}, t) = \sum_{r=1}^{R} u_r(x) v_r(y) w_r(z) \varphi_r(t) \tag{9.38}$$

or

$$x(\mathbf{r}, t) = \sum_{d=1}^{D} \sum_{p=1}^{P} \sum_{q=1}^{Q} \sum_{r=1}^{R} g_{dpqr} u_d(x) v_p(y) w_q(z) \varphi_r(t), \tag{9.39}$$

corresponding to the CP and Tucker decompositions, respectively. These decompositions place stronger assumptions on $x(\mathbf{r}, t)$ than Eq. (9.5) since they assume not only spatiotemporal correlation, but also low-rank spatial structure (linearly-dependent rows/columns/fibers).

Implicit low-rank reconstruction of $\mathcal{X}$ [74,75] can be performed by solving:

$$arg\ min_{\mathbf{X}} \|E(\mathcal{X}) - \mathbf{s}\|_2^2 + \lambda R_{\text{rank}}(\mathcal{X}). \tag{9.40}$$

Separable imaging can also be performed with explicit low-rank modeling by parameterizing $\mathcal{X}$ according to, e.g., Eqs. (9.38) and (9.39) [76–78]. Furthermore, the many variations of low-rank modeling described in Section 9.3 can be readily extended to accommodate spatial separability, including LLR [26,79], L&S, L+S [80], and multiscale LRT image reconstruction.

9.4.3 Multidynamic tensors

Sections 9.3 and 9.4.2 previously discussed reconstructing $x(\mathbf{r}, t)$ or an $\tilde{x}(\mathbf{r}, \tau_\ell)$, which required isolating a single dynamic process. Here we instead discuss reconstructing the underlying multidynamic image $\tilde{x}(\mathbf{r}, \tau_1, \tau_2, \ldots, \tau_L)$, which has L "time dimensions" corresponding to the independent variables $\{\tau_\ell\}_{\ell=1}^{L}$ [81,22,82,24]. This multidimensional structure lends itself well to tensor representations, as $\tilde{x}$ can be readily expressed as an $(L+1)$-way tensor $\mathcal{X} \in \mathbb{C}^{M \times N_1 \times N_2 \times \cdots \times N_L}$ with elements $[\mathcal{X}]_{mn_1n_2\ldots n_L} = \tilde{x}(\mathbf{r}_m, \tau_{1,n_1}, \tau_{2,n_2}, \ldots, \tau_{L,n_L})$.

There are several benefits of multidynamic imaging of $\tilde{x}(\mathbf{r}, \tau_1, \tau_2, \ldots, \tau_L)$ over imaging $x(\mathbf{r}, t)$ or an $\tilde{x}(\mathbf{r}, \tau_\ell)$ with a single time dimension. For one, multidynamic imaging does not require dynamic processes to be frozen by prospectively enforcing $\tau_q(t_n) = \kappa_q\ \forall\, q \neq \ell, n = 1, 2, \ldots, N$. Instead, all L dynamic processes are resolved during imaging, after which an equivalent "frozen" image can be retroactively produced from $\tilde{x}$ by extracting the temporal slice at $\tau_q = \kappa_q\ \forall\, q \neq \ell$: $\tilde{x}(\mathbf{r}, \kappa_1, \kappa_2, \ldots, \tau_\ell, \ldots, \kappa_L)$. This reduces dependence on techniques such as breath holds and ECG triggering since motion can be resolved in $\tilde{x}$ instead of paused or removed. Furthermore, by allowing $\{\kappa_q\}_{q \neq \ell}$ to be chosen retrospectively, imaging decisions such as contrast weighting can be made after imaging rather than scouting for,

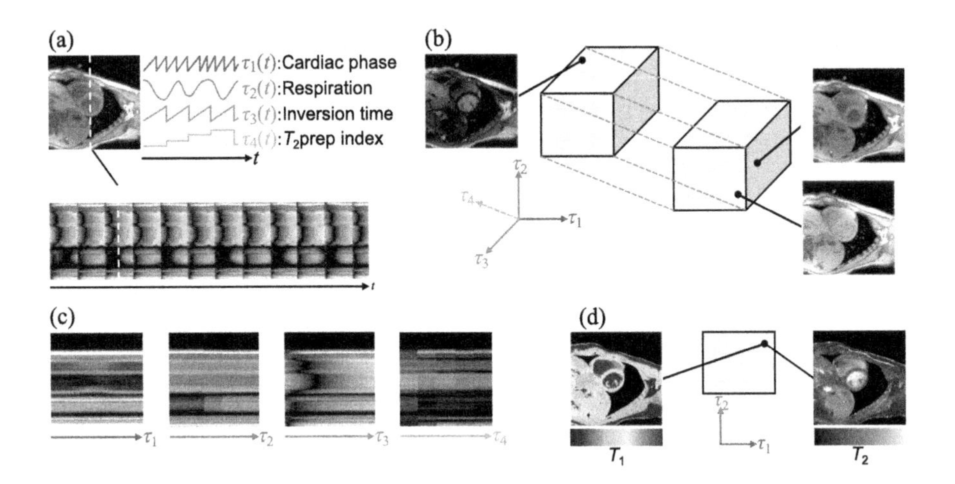

FIGURE 9.7 Illustration of the multidynamic imaging concept

(**a**) Imaging $L=4$ overlapping dynamic processes: cardiac motion, respiratory motion, T1 recovery and T2 decay. (**b**) Four-dimensional hypercube representing all four time dimensions. Tensor reconstruction [24] enables (**c**) visualization of retrospectively isolated dynamic processes, and (**d**) quantitative maps at each motion state combination. *Source: Adapted with permission from Ref. [83].*

e.g., ideal inversion times for nulling tissue. Multidynamic imaging is especially beneficial when multiple dynamic processes are of clinical interest (e.g., different types of motion in addition to multiple contrast weightings for parametric mapping) because recovering the entire $\tilde{x}(\mathbf{r}, \tau_1, \tau_2, \ldots, \tau_L)$ in one scan eliminates the need to perform multiple scans, each resolving a different individual τ_ℓ (Fig. 9.7).

9.4.3.1 Tensor-based compressed sensing

High-dimensional CS can be used to reconstruct $\mathbf{X}$ by solving a problem such as

$$\hat{\mathcal{X}} = arg\ min_{\mathcal{X}} \|E(\mathcal{X}) - \mathbf{s}\|_2^2 + \sum_{i=1}^{L+1} \lambda_i \|\mathrm{vec}(\mathbf{\Psi}_i \mathbf{X}_{(i)})\|_1, \tag{9.41}$$

where the rows of $\mathbf{X}_{(1)} \in \mathbb{C}^{M \times \prod_{\ell=1}^{L} N_\ell}$ index $\mathbf{r}$, the rows of $\mathbf{X}_{(i)} \in \mathbb{C}^{N_{i-1} \times M \prod_{\ell \neq i-1} N_\ell}$ index τ_{i-1} for $i > 1$, and the sparsifying transform $\mathbf{\Psi}_i$ operates along $\mathbf{r}$ for $i=1$ and τ_{i-1} for $i > 1$.

The advantage of multidynamic CS over single-dynamic CS—in addition to the general benefits of multidynamic imaging—is that an image can typically be more efficiently compressed along each individual time dimension τ_ℓ rather than along t. For example, the difference image along τ_ℓ represents the change from a single dynamic process

$$\nabla_{\tau_\ell}\{\tilde{x}\} = \tilde{x}(\mathbf{r}, \tau_1, \tau_2, \cdots, \tau_\ell, \cdots, \tau_L) - \tilde{x}(\mathbf{r}, \tau_1, \tau_2, \cdots, \tau_\ell - \Delta\tau_\ell, \cdots, \tau_L), \tag{9.42}$$

which can be sparser than the difference between successive timepoints

$$\nabla_t\{x\} = x(\mathbf{r}, t) - x(\mathbf{r}, t - \Delta t)$$

$$= \tilde{x}(\mathbf{r}, \tau_1(t), \tau_2(t), \cdots, \tau_L(t)) - \tilde{x}(\mathbf{r}, \tau_1(t-\Delta t), \tau_2(t-\Delta t), \cdots, \tau_L(t-\Delta t)), \qquad (9.43)$$

which represents changes from multiple processes.

The XD-GRASP method [81] introduced high-dimensional compressed sensing for dynamic MRI, using $L = 2$ for cardiac- and respiratory-resolved imaging of the heart and for respiratory- and DCE-resolved abdominal imaging. The XD flow method [82] performed high-dimensional compressed sensing with different $L = 3$ combinations of cardiac, respiratory, flow and DCE time dimensions, collapsing the fourth dimension not included in each combination.

9.4.3.2 *Multidynamic low-rank tensor modeling*

A complementary alternative to CS is low-rank tensor image modeling. Revisiting Eq. (9.5) in the context of multidynamic imaging, we have:

$$\tilde{x}(\mathbf{r}, \tau_1, \tau_2, \ldots, \tau_L) = \sum_{r=1}^{R} a_r(\mathbf{r})\tilde{\varphi}_r(\tau_1, \tau_2, \ldots, \tau_L), \qquad (9.44)$$

which simply replaces the $\{\varphi_r(t)\}_{r=1}^R$ with multidimensional functions $\{\tilde{\varphi}_r(\tau_1, \tau_2, \ldots, \tau_L)\}_{r=1}^R$. In discretized form, Eq. (9.44) becomes

$$\mathcal{X} = \Phi \times_1 \mathbf{A}, \qquad (9.45)$$

where $\Phi \in \mathbb{C}^{R\times N_1 \times \cdots \times N_L}$ has elements $[\Phi]_{rn_1n_2\ldots n_L} = \tilde{\varphi}_r(\tau_{1,n_1}, \tau_{2,n_2}, \ldots, \tau_{L,n_L})$.

LRT modeling assumes low-order separability of each $\tilde{\varphi}_r(\tau_1, \tau_2, \ldots, \tau_L)$. More specifically, the CP decomposition of $\mathbf{X}$ implies

$$\tilde{\varphi}_r(\tau_1, \tau_2, \ldots, \tau_L) = \varphi_{1,r}(\tau_1)\varphi_{2,r}(\tau_2)\ldots\varphi_{N,r}(\tau_L), \qquad (9.46)$$

where $\{\varphi_{\ell,r}(\tau_\ell)\}_{r=1}^R$ spans the subspace for the ℓth time dimension. The Tucker decomposition of $\mathcal{X}$ implies

$$\tilde{\varphi}_r(\tau_1, \tau_2, \ldots, \tau_L) = \sum_{r_1=1}^{R_1}\sum_{r_2=1}^{R_2}\cdots\sum_{r_L=1}^{R_L} g_{rr_1r_2\ldots r_L}\varphi_{1,r_1}(\tau_1)\varphi_{2,r_2}(\tau_2)\ldots\varphi_{N,r_L}(\tau_L), \qquad (9.47)$$

where $\{\varphi_{\ell,r_\ell}(\tau_\ell)\}_{r_\ell=1}^{R_\ell}$ spans the subspace for the ℓth time dimension. Fig. 9.8 depicts the Tucker decomposition of a cardiac image with $L = 3$ time dimensions: cardiac motion, respiration, and inversion recovery.

Explicit multidynamic low-rank tensor reconstruction

Explicit low-rank tensor image reconstruction is possible by parameterizing $\mathcal{X}$ according to Eq. (9.45) and either Eq. (9.46) or Eq. (9.47). Image reconstruction can for example proceed according to the CP recovery problem

$$\left\{\hat{\mathbf{A}}, \left\{\hat{\mathbf{\Phi}}_\ell\right\}_{\ell=1}^{L}\right\} = arg\ min_{\mathbf{A},\{\mathbf{\Phi}_\ell\}_{\ell=1}^L} \left\| E\left(\sum_{r=1}^{R} \mathbf{a}_r \circ \boldsymbol{\varphi}_{1,r} \circ \boldsymbol{\varphi}_{2,r} \circ \cdots \circ \boldsymbol{\varphi}_{L,r}\right) - \mathbf{s}\right\|_2^2, \qquad (9.48)$$

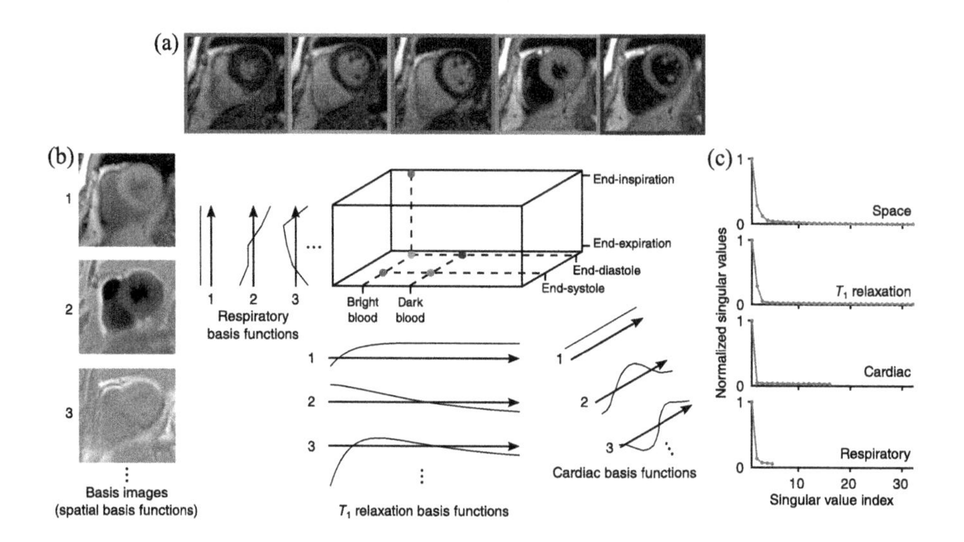

FIGURE 9.8 Illustration and analysis of multiple time dimensions for multidynamic imaging

(**a**) Locations of different images in a space with $L = 3$ time dimensions. Different T_1 weightings lie along the inversion time axis (*horizontal*), different cardiac phases lie along the cardiac time axis (*depth*), and different respiratory phases lie along the respiratory time axis (*vertical*). (**b**) The three most significant orthogonal basis functions in Tucker form, as reconstructed from 1-min worth of data using MR Multitasking [24]. (**c**) Singular value curves from the higher-order SVD of 12.3-min worth of raw training data (i.e., enough data to cover all motion-state and contrast combinations), demonstrating that the singular values decay quickly for all unfoldings of the raw data tensor. *Source: Adapted with permission from Ref. [24].*

where the elements of $\mathbf{a}_r \in \mathbb{C}^{M\times 1}$ are $[\mathbf{a}_r]_m = [\mathbf{A}]_{mr}$ and the elements of $\varphi_{\ell,r} \in \mathbb{C}^{N_\ell \times 1}$ are $[\varphi_{\ell,r}]_n = \varphi_{\ell,r}(\tau_{\ell,n},)$, or according to the Tucker recovery problem

$$\left\{\hat{\mathcal{G}}, \hat{\mathbf{A}}, \left\{\hat{\mathbf{\Phi}}_\ell\right\}_{\ell=1}^{L}\right\} = arg\ min_{\mathcal{G},\mathbf{A},\{\mathbf{\Phi}_\ell\}_{\ell=1}^{L}} \|E(\mathcal{G} \times_1 \mathbf{A} \times_2 \mathbf{\Phi}_1 \times_3 \mathbf{\Phi}_2 \times_4 \cdots \times_{L+1} \mathbf{\Phi}_L) - \mathbf{s}\|_2^2, \quad (9.49)$$

where the elements of $\mathbf{\Phi}_\ell \in \mathbb{C}^{N_\ell \times R_\ell}$ are $[\mathbf{\Phi}_\ell]_{nr} = \varphi_{\ell,r}(\tau_{\ell,n})$.

As in Section 9.3.1.2, fixed-subspace reconstruction is possible with the help of training data. When it is feasible to prospectively control the $(\mathbf{k}, \tau_1, \tau_2, \ldots, \tau_L)$-space sampling in full (e.g., when all of the $\{\tau_\ell(t)\}_{\ell=1}^{L}$ represent sequence parameters), then one can acquire L subsets of training data, the ℓth set of which densely samples across τ_ℓ within a limited region of $(\mathbf{k}, \{\tau_q\}_{q\neq\ell})$-space. The SVD of each training data subset then produces each $\mathbf{\Phi}_\ell$ for the Tucker decomposition [22]. When $(\mathbf{k}, \tau_1, \tau_2, \ldots, \tau_L)$-space sampling cannot be fully controlled (e.g., when some of the $\{\tau_\ell(t)\}_{\ell=1}^{L}$ describe physiological processes such as motion), then training data may sparsely sample $(\mathbf{k}, \tau_1, \tau_2, \ldots, \tau_L)$-space with an unpredictable pattern. In such cases, Φ can instead be calculated by LRT completion of the training data followed by truncated decomposition of the training data tensor, as proposed in the MR multitasking framework [24].

ALS reconstruction can also be used to optimize the cost function in Eqs. (9.48) or (9.49) over all variables without requiring training data. However, in practice training data may already be available in the form of self-gating signals used to determine motion-related $\tau_\ell(t)$'s [24].

Implicit multidynamic low-rank tensor reconstruction

Implicit low-rank tensor modeling can be performed by penalizing a tensor rank surrogate, i.e.,

$$\hat{\mathcal{X}} = arg\ min_{\mathcal{X}} \| E(\mathcal{X}) - \mathbf{s} \|_2^2 + \lambda R_{\text{rank}}(\mathcal{X}), \tag{9.50}$$

with $R_{\text{rank}}(\mathcal{X})$ defined for example as the sum of nuclear norms for each unfolding of $\mathcal{X}$: $\sum_{i=1}^{L+1} \|\mathbf{X}_{(i)}\|_*$ [63]. Note that this approach requires storing and operating upon $\mathcal{X}$ without any parameterization; this $\mathcal{X}$ has $M \prod_{\ell=1}^{L} N_\ell$ complex elements. This can lead to enormous storage requirements that grow geometrically with L, making implicit low-rank tensor image reconstruction impractical for large L.

Additional multidynamic LRT models

The previous discussed extensions to low-rank modeling can be applied to multidynamic imaging as well, e.g., multidynamic LLR tensor modeling [21], multidynamic L&S [22,24], multidynamic L+S [84], spatially separable multidynamic imaging [26], multidynamic patch similarity modeling [25], and multidynamic multiscale LRT image reconstruction. Each of these strategies can support explicit or implicit low-rank reconstruction, with explicit decomposition having the aforementioned memory benefits.

9.5 Summary

Low-rank image reconstruction has enhanced or enabled several MRI applications, and active research continues into more advanced modeling, decompositions, and reconstruction algorithms. Continued development of low-rank image reconstruction for fast dynamic imaging and multidynamic imaging has the potential to change the structure of clinical MR exams from a series of independent scans to a single high-dimensional scan reconstructed as a low-rank tensor. This change would necessitate new modes of storing, viewing, and analyzing high-dimensional images, which, given the utility of low-rank decompositions for these tasks, seems a natural extension of low-rank image reconstruction to a broader low-rank image reconstruction, storage, and analysis pipeline.

References

[1] Liang Z-P. Spatiotemporal imaging with partially separable functions. In: Proc IEEE Int Symp Biomed Imaging; 2007. p. 988–91.
[2] Brinegar C, Wu YJL, Foley LM, Hitchens TK, Ye Q, Ho C, et al. Real-time cardiac MRI without triggering, gating, or breath holding. In: Conf Proc IEEE Eng Med Biol Soc; 2008. p. 4383–6.
[3] Pedersen H, Kozerke S, Ringgaard S, Nehrke K, Kim WY. k-t PCA: temporally constrained k-t BLAST reconstruction using principal component analysis. Magn Reson Med 2009;62(3):706–16.
[4] Lingala S, Hu Y, DiBella E, Jacob M. Accelerated dynamic MRI exploiting sparsity and low-rank structure: k-t SLR. IEEE Trans Med Imaging 2011;30(5):1042–54.

[5] Zhou R, Huang W, Yang Y, Chen X, Weller DS, Kramer CM, et al. Simple motion correction strategy reduces respiratory-induced motion artifacts for kt accelerated and compressed-sensing cardiovascular magnetic resonance perfusion imaging. J Cardiovasc Magn Reson 2018;20(1):1–13.
[6] Huang C, Graff CG, Clarkson EW, Bilgin A, Altbach MI. T2 mapping from highly undersampled data by reconstruction of principal component coefficient maps using compressed sensing. Magn Reson Med 2012;67(5):1355–66.
[7] Zhang T, Pauly JM, Levesque IR. Accelerating parameter mapping with a locally low rank constraint. Magn Reson Med 2015;73(2):655–61.
[8] Zhao B, Lu W, Hitchens TK, Lam F, Ho C, Liang Z-P. Accelerated MR parameter mapping with low-rank and sparsity constraints. Magn Reson Med 2015;74(2):489–98.
[9] Peng X, Ying L, Liu Y, Yuan J, Liu X, Liang D. Accelerated exponential parameterization of T2 relaxation with model-driven low rank and sparsity priors (MORASA). Magn Reson Med 2016;76(6):1865–78.
[10] Tamir JI, Uecker M, Chen W, Lai P, Alley MT, Vasanawala SS, et al. T2 shuffling: sharp, multicontrast, volumetric fast spin-echo imaging. Magn Reson Med 2017;77(1):180–95.
[11] Assländer J, Cloos MA, Knoll F, Sodickson DK, Hennig J, Lattanzi R. Low rank alternating direction method of multipliers reconstruction for MR fingerprinting. Magn Reson Med 2018;79(1):83–96.
[12] Zhao B, Setsompop K, Adalsteinsson E, Gagoski B, Ye H, Ma D, et al. Improved magnetic resonance fingerprinting reconstruction with low-rank and subspace modeling. Magn Reson Med 2018;79(2):933–42.
[13] Doneva M, Amthor T, Koken P, Sommer K, Börnert P. Matrix completion-based reconstruction for undersampled magnetic resonance fingerprinting data. Magn Reson Imaging 2017;41:41–52.
[14] Brinegar C, Schmitter S, Mistry N, Johnson G, Liang Z-P. Improving temporal resolution of pulmonary perfusion imaging in rats using the partially separable functions model. Magn Reson Med 2010;64(4):1162–70.
[15] Feng L, Wen Q, Huang C, Tong A, Liu F, Chandarana H. GRASP-Pro: imProving GRASP DCE-MRI through self-calibrating subspace-modeling and contrast phase automation. Magn Reson Med 2020;83(1):94–108.
[16] Lam F, Liang Z-P. A subspace approach to high-resolution spectroscopic imaging. Magn Reson Med 2014;71(4):1349–57.
[17] Bhattacharya I, Jacob M. Compartmentalized low-rank recovery for high-resolution lipid unsuppressed MRSI. Magn Reson Med 2017;78(4):1267–80.
[18] Klauser A, Courvoisier S, Kasten J, Kocher M, Guerquin-Kern M, Van De Ville D, et al. Fast high-resolution brain metabolite mapping on a clinical 3T MRI by accelerated $_1$H-FID-MRSI and low-rank constrained reconstruction. Magn Reson Med 2019;81(5):2841–57.
[19] Lam F, Zhao B, Liu Y, Liang Z-P, Weiner M, Schuff N. Accelerated fMRI using low-rank model and sparsity constraints. In: Proc Int Soc Magn Reson Med; 2013. p. 2620.
[20] Chiew M, Smith SM, Koopmans PJ, Graedel NN, Blumensath T, Miller KL. k-t FASTER: acceleration of functional MRI data acquisition using low rank constraints. Magn Reson Med 2015;74(2):353–64.
[21] Guo S, Fessler JA, Noll DC. High-resolution oscillating steady-state fMRI using patch-tensor low-rank reconstruction. IEEE Trans Med Imaging 2020;39(12):4357–68.
[22] He J, Liu Q, Christodoulou AG, Ma C, Lam F, Liang Z-P. Accelerated high-dimensional MR imaging with sparse sampling using low-rank tensors. IEEE Trans Med Imaging 2016;35(9):2119–29.
[23] Ma C, Clifford B, Liu Y, Gu Y, Lam F, Yu X, et al. High-resolution dynamic ^{31}P-MRSI using a low-rank tensor model. Magn Reson Med 2017;78(2):419–28.
[24] Christodoulou AG, Shaw JL, Nguyen C, Yang Q, Xie Y, Wang N, et al. Magnetic resonance multitasking for motion-resolved quantitative cardiovascular imaging. Nature Biomed Eng 2018;2(4):215–26.
[25] Bustin A, Lima da Cruz G, Jaubert O, Lopez K, Botnar RM, Prieto C. High-dimensionality undersampled patch-based reconstruction (HD-PROST) for accelerated multi-contrast MRI. Magn Reson Med 2019;81(6):3705–19.
[26] Yaman B, Weingärtner S, Kargas N, Sidiropoulos ND, Akçakaya M. Low-rank tensor models for improved multidimensional MRI: application to dynamic cardiac T_1 mapping. IEEE Trans Comput Imaging 2019;6:194–207.

[27] Candès EJ, Tao T. The power of convex relaxation: near-optimal matrix completion. IEEE Trans Inf Theory 2010;56(5):2053–80.

[28] Jhooti P, Gatehouse P, Keegan J, Bunce N, Taylor A, Firmin D. Phase ordering with automatic window selection (PAWS): a novel motion-resistant technique for 3D coronary imaging. Magn Reson Med 2000;43(3):470–80.

[29] Mardani M, Giannakis GB, Ugurbil K. Tracking tensor subspaces with informative random sampling for real-time MR imaging. Available from: arXiv:1609.04104, 2016.

[30] Zhu Y, Zhang Q, Liu Q, Wang YXJ, Liu X, Zheng H, et al. PANDA-$T_{1\rho}$: integrating principal component analysis and dictionary learning for fast $T_{1\rho}$ mapping. Magn Reson Med 2015;73(1):263–72.

[31] Bieri O, Markl M, Scheffler K. Analysis and compensation of eddy currents in balanced SSFP. Magn Reson Med 2005;54(1):129–37.

[32] Christodoulou AG, Hitchens TK, Wu YL, Ho C, Liang Z-P. Improved subspace estimation for low-rank model-based accelerated cardiac imaging. IEEE Trans Biomed Eng 2014;61(9):2451–7.

[33] Zhao B, Haldar JP, Christodoulou AG, Liang Z-P. Image reconstruction from highly undersampled $(\mathbf{k}, t)$-space data with joint partial separability and sparsity constraints. IEEE Trans Med Imaging 2012;31(9):1809–20.

[34] Biswas S, Poddar S, Dasgupta S, Mudumbai R, Jacob M. Two step recovery of jointly sparse and low-rank matrices: theoretical guarantees. In: Proc IEEE Int Symp Biomed Imaging; 2015. p. 914–7.

[35] Haldar JP, Liang Z-P. Spatiotemporal imaging with partially separable functions: a matrix recovery approach. In: Proc IEEE Int Symp Biomed Imaging; 2010. p. 716–9.

[36] Goud Lingala S, Jacob M. Blind compressive sensing dynamic MRI. IEEE Trans Med Imaging 2013;32(6):1132–45.

[37] Haldar JP, Hernando D. Rank-constrained solutions to linear matrix equations using PowerFactorization. IEEE Signal Process Lett 2009;16(7):584–7.

[38] Majumdar A, Ward RK. An algorithm for sparse MRI reconstruction by Schatten p-norm minimization. Magn Reson Imaging 2011;29(3):408–17.

[39] Mehta BB, Chen X, Bilchick KC, Salerno M, Epstein FH. Accelerated and navigator-gated look-locker imaging for cardiac T1 estimation (ANGIE): development and application to T1 mapping of the right ventricle. Magn Reson Med 2015;73(1):150–60.

[40] Candès EJ, Recht B. Exact matrix completion via convex optimization. Found Comput Math 2009;9(6):717.

[41] Recht B, Fazel M, Parrilo PA. Guaranteed minimum-rank solutions of linear matrix equations via nuclear norm minimization. SIAM Rev 2010;52(3):471–501.

[42] Cai JF, Candès EJ, Shen Z. A singular value thresholding algorithm for matrix completion. SIAM J Optim 2010;20(4):1956–82.

[43] Trzasko J, Manduca A, Borisch E. Local versus global low-rank promotion in dynamic MRI series reconstruction. In: Proc Int Soc Magn Reson Med; 2011. p. 4371.

[44] Saucedo A, Lefkimmiatis S, Rangwala N, Sung K. Improved computational efficiency of locally low rank MRI reconstruction using iterative random patch adjustments. IEEE Trans Med Imaging 2017;36(6):1209–20.

[45] Hu Y, Levine EG, Tian Q, Moran CJ, Wang X, Taviani V, et al. Motion-robust reconstruction of multishot diffusion-weighted images without phase estimation through locally low-rank regularization. Magn Reson Med 2019;81(2):1181–90.

[46] Vitanis V, Manka R, Giese D, Pedersen H, Plein S, Boesiger P, et al. High resolution three-dimensional cardiac perfusion imaging using compartment-based k-t principal component analysis. Magn Reson Med 2011;65(2):575–87.

[47] Ma C, Lam F, Johnson CL, Liang Z-P. Removal of nuisance signals from limited and sparse 1H MRSI data using a union-of-subspaces model. Magn Reson Med 2016;75(2):488–97.

[48] Christodoulou AG, Zhang H, Zhao B, Hitchens TK, Ho C, Liang Z-P. High-resolution cardiovascular MRI by integrating parallel imaging with low-rank and sparse modeling. IEEE Trans Biomed Eng 2013;60(11):3083–92.

[49] Jung H, Sung K, Nayak K, Kim E, Ye J. k-t FOCUSS: a general compressed sensing framework for high resolution dynamic MRI. Magn Reson Med 2009;61(1):103–16.

[50] Gao H, Yu H, Osher S, Wang G. Multi-energy CT based on a prior rank, intensity and sparsity model (PRISM). Inverse Probl 2011;27(11):115012.

[51] Gao H, Rapacchi S, Wang D, Moriarty J, Meehan C, Sayre J, et al. Compressed sensing using prior rank, intensity and sparsity model (PRISM): applications in cardiac cine MRI. In: Proc Int Soc Magn Reson Med; 2012. p. 2242.

[52] Otazo R, Candès E, Sodickson DK. Low-rank plus sparse matrix decomposition for accelerated dynamic MRI with separation of background and dynamic components. Magn Reson Med 2015;73(3):1125–36.

[53] Velikina JV, Samsonov AA. Reconstruction of dynamic image series from undersampled MRI data using data-driven model consistency condition (MOCCO). Magn Reson Med 2015;74(5):1279–90.

[54] Candès EJ, Li X, Ma Y, Wright J. Robust principal component analysis? J ACM 2011;58(3):1–37.

[55] Ong F, Lustig M. Beyond low rank + sparse: multiscale low rank matrix decomposition. IEEE J Sel Top Signal Process 2016;10(4):672–87.

[56] Ong F, Zhu X, Cheng JY, Johnson KM, Larson PE, Vasanawala SS, et al. Extreme MRI: large-scale volumetric dynamic imaging from continuous non-gated acquisitions. Magn Reson Med 2020;84(4):1763–80.

[57] Kolda TG, Bader BW. Tensor decompositions and applications. SIAM Rev 2009;51(3):455–500.

[58] Carroll JD, Chang JJ. Analysis of individual differences in multidimensional scaling via an N-way generalization of "Eckart-Young" decomposition. Psychometrika 1970;35(3):283–319.

[59] Harshman RA, et al. Foundations of the PARAFAC procedure: models and conditions for an "explanatory" multimodal factor analysis. UCLA Working Papers in Phonetics, vol. 16. 1970. p. 1–84.

[60] Tucker LR. Some mathematical notes on three-mode factor analysis. Psychometrika 1966;31(3):279–311.

[61] De Lathauwer L, De Moor B, Vandewalle J. A multilinear singular value decomposition. SIAM J Matrix Anal Appl 2000;21(4):1253–78.

[62] Håstad J. Tensor rank is NP-complete. J Algorithms 1990;11(4):644–54.

[63] Liu J, Musialski P, Wonka P, Ye J. Tensor completion for estimating missing values in visual data. IEEE Trans Pattern Anal Mach Intell 2012;35(1):208–20.

[64] Friedland S, Lim LH. Nuclear norm of higher-order tensors. Math Comput 2018;87(311):1255–81.

[65] Kong H, Xie X, Lin Z. t-Schatten-p norm for low-rank tensor recovery. IEEE J Sel Top Signal Process 2018;12(6):1405–19.

[66] Trzasko JD, Manduca A. CLEAR: calibration-free parallel imaging using locally low-rank encouraging reconstruction. In: Proc Int Soc Magn Reson Med; 2012. p. 517.

[67] Trzasko JD, Manduca A. A unified tensor regression framework for calibrationless dynamic, multi-channel MRI reconstruction. In: Proc Int Soc Magn Reson Med; 2013. p. 21.

[68] Trzasko JD. Exploiting local low-rank structure in higher-dimensional MRI applications. In: Wavelets and sparsity XV, vol. 8858. International Society for Optics and Photonics; 2013. p. 885821.

[69] Akçakaya M, Basha TA, Goddu B, Goepfert LA, Kissinger KV, Tarokh V, et al. Low-dimensional-structure self-learning and thresholding: regularization beyond compressed sensing for MRI reconstruction. Magn Reson Med 2011;66(3):756–67.

[70] Akçakaya M, Basha TA, Pflugi S, Foppa M, Kissinger KV, Hauser TH, et al. Localized spatio-temporal constraints for accelerated CMR perfusion. Magn Reson Med 2014;72(3):629–39.

[71] Chen X, Salerno M, Yang Y, Epstein FH. Motion-compensated compressed sensing for dynamic contrast-enhanced MRI using regional spatiotemporal sparsity and region tracking: block low-rank sparsity with motion-guidance (BLOSM). Magn Reson Med 2014;72(4):1028–38.

[72] Bustin A, Ginami G, Cruz G, Correia T, Ismail TF, Rashid I, et al. Five-minute whole-heart coronary MRA with sub-millimeter isotropic resolution, 100% respiratory scan efficiency, and 3D-PROST reconstruction. Magn Reson Med 2019;81(1):102–15.
[73] Küstner T, Bustin A, Jaubert O, Hajhosseiny R, Masci PG, Neji R, et al. Isotropic 3D Cartesian single breath-hold CINE MRI with multi-bin patch-based low-rank reconstruction. Magn Reson Med 2020;84(4):2018–33.
[74] Yu Y, Jin J, Liu F, Crozier S. Multidimensional compressed sensing MRI using tensor decomposition-based sparsifying transform. PLoS ONE 2014;9(6):e98441.
[75] Yang X, Luo Y, Chen S, Zhen X, Yu Q, Liu K. Dynamic MRI reconstruction from highly undersampled (k, t)-space data using weighted Schatten p-norm regularizer of tensor. Magn Reson Imaging 2017;37:260–72.
[76] Banco D, Aeron S, Hoge WS. Sampling and recovery of MRI data using low rank tensor models. In: Conf Proc IEEE Eng Med Biol Soc; 2016. p. 448–52.
[77] Kanatsoulis CI, Fu X, Sidiropoulos ND, Akçakaya M. Tensor completion from regular sub-Nyquist samples. IEEE Trans Signal Process 2019;68:1–16.
[78] Zhao B, Setsompop K, Salat D, Wald LL. Further development of subspace imaging to magnetic resonance fingerprinting: a low-rank tensor approach. In: Conf Proc IEEE Eng Med Biol Soc; 2020. p. 1662–6.
[79] Liu F, Li D, Jin X, Qiu W, Xia Q, Sun B. Dynamic cardiac MRI reconstruction using motion aligned locally low rank tensor (MALLRT). Magn Reson Imaging 2020;66:104–15.
[80] Roohi SF, Zonoobi D, Kassim AA, Jaremko JL. Multi-dimensional low rank plus sparse decomposition for reconstruction of under-sampled dynamic MRI. Pattern Recognit 2017;63:667–79.
[81] Feng L, Axel L, Chandarana H, Block KT, Sodickson DK, Otazo R. XD-GRASP: golden-angle radial MRI with reconstruction of extra motion-state dimensions using compressed sensing. Magn Reson Med 2016;75(2):775–88.
[82] Cheng JY, Zhang T, Alley MT, Uecker M, Lustig M, Pauly JM, et al. Comprehensive multi-dimensional MRI for the simultaneous assessment of cardiopulmonary anatomy and physiology. Sci Rep 2017;7(1):5330.
[83] Otazo R. Motion-tolerant quantitative cardiovascular MRI. Nature Biomed Eng 2018;2(4):199.
[84] Ramb R, Zenge M, Feng L, Muckley M, Forman C, Axel L, et al. Low-rank plus sparse tensor reconstruction for high-dimensional cardiac MRI. In: Proc Int Soc Magn Reson Med; 2017. p. 1199.

CHAPTER

10 Dictionary, Structured Low-Rank, and Manifold Learning-Based Reconstruction

Mathews Jacob, Sajan Goud Lingala, and Merry Mani
University of Iowa, Iowa City, IA, United States

10.1 Introduction

Efficient image representations are key to reconstructing images from fewer measurements. In particular, low-dimensional representations can represent images efficiently, enabling the robust recovery from sparse and noisy measurements. Many of the early model-based MRI approaches relied on fixed image representations that were often carefully engineered to the data. In contrast, learning the low-dimensional representations from data itself offers improved efficiency. For instance, the compressed sensing methods described in Chapter 8 may be thought of as learning-based algorithms, where the specific basis functions that are best suited to represent a given signal are chosen from a pre-engineered dictionary. The low-rank models reviewed in Chapter 9 go one step further, for instance, to make use of the redundancies of images in a time series that differ in contrasts and/or motion states. This chapter focus on *learned low-dimensional representations* that are generalizations of the low-rank methods discussed in the previous chapter. The objective is to account for image redundancies that are challenging for low-rank methods to capture.

In this chapter, we will review the generalizations of low-rank methods, which can be broadly classified as

1. approaches that capitalize the complex *nonlinear redundancies* with the datasets, which low-rank methods are not capable of capitalizing on. Sparse dictionary learning and smooth manifold representations fall into this category. Unlike low-rank methods that use energy based priors on the factors, dictionary learning methods use sparsity priors on the coefficients and energy-based priors on the dictionary atoms. The smooth manifold models use nonlinear kernel priors, allowing us to account for nonlinear redundancies in the data resulting from motion and contrast changes. Please see Fig. 10.1 for an illustration of the representations in 3D space.
2. approaches that use the above representations to *sub-parts of the image* (e.g., patches of different shape, either in the image domain or k-space) rather than the whole image. We will show that this approach enables one to capitalize on unique signal redundancies that are characteristic of each application (e.g., smoothly varying phase, exponential signal decay in time, uncalibrated parallel MRI), which are often challenging for the traditional low-rank or compressed sensing algorithms. Please see Fig. 10.2 for an illustration of the signals extracted from the images, whose redundancies are leveraged to recover the image.

Advances in Magnetic Resonance Technology and Applications, Volume 7, ISSN 2666-9099. https://doi.org/10.1016/B978-0-12-822726-8.00020-8

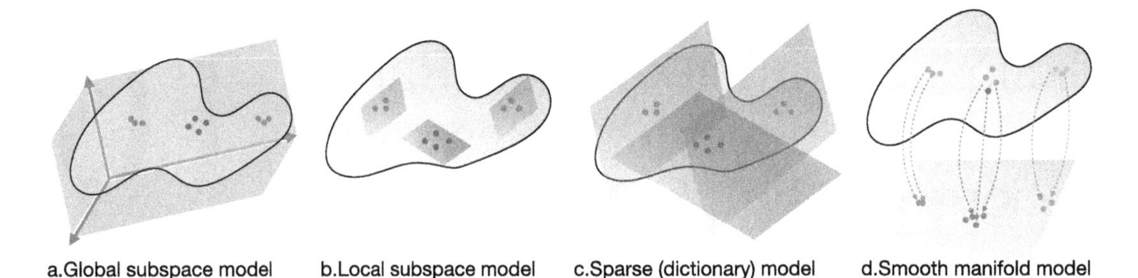

FIGURE 10.1

Overview of signals extracted from the image volumes, whose redundancy is capitalized on in this chapter: The signals (images/patches/timeprofiles) in MRI datasets have extensive redundancy as described in Fig. 10.2; they can be viewed as points lying on a manifold in high-dimensional spaces. The redundancy of the signals are captured by different adaptive models in different ways. Global subspace or global principal component analysis (PCA) models reviewed in Chapter 9 model the signals as a subspace as shown in (**a**); they learn the basis vectors indicated by the red arrows from the data, coming up with a compact representation of the space. Local subspace models cluster the data and learn a subspace for each cluster/neighborhood as shown in (**b**). Sparse dictionary learning methods learn the dictionary basis functions from the data itself; the signal space is modeled by a union of subspaces, where basis functions may be shared by local neighborhoods as shown in (**c**). The main benefit over local PCA approaches is that this approach does not need an explicit clustering step. Smooth manifold and kernel PCA models represent the signals as a smooth manifold; as shown in (**d**), they rely on a mapping that converts the nonlinear manifold to a low-dimensional subspace denoted by the plane; the structure of the low-dimensional subspace is used to recover the signals.

The chapter is organized as follows. Following a brief overview of the background in Section 10.2, we review dictionary-based methods in Section 10.3, where the sparsity of the dictionary coefficients is used to further improve the adaptation of the representation to the specific signal; these approaches use sparsity to bypass the need for clustering that is often needed in the local PCA-based methods reviewed in Chapter 9. The low-rank model in Chapter 9 is then extended to patches in image domain or k-space in Section 10.4. In particular, under specific assumptions, the resulting patch matrix/matrices are highly low-rank, which can be capitalized for acceleration as discussed in Section 10.4. We review smooth manifold models that are efficient in capturing nonlinear redundancies in the dataset in Section 10.5. An overview and broad classification of the methods reviewed in this chapter is given in Table 10.1. In this chapter, the various topics are not presented in the chronological order in which they were introduced in the MRI setting. Rather, our focus is on grouping the various methods into broader themes to facilitate easy comprehension of the links and the generalizations of the various approaches.

10.2 Background

10.2.1 Acquisition scheme

As introduced in Chapter 2, the main goal of image recovery is the estimation of the continuous domain function $\mathbf{x} : \mathbb{R}^n \rightarrow \mathbb{C}$ from a finite number of multichannel k-space measurements $s_{i,j}$, based on the

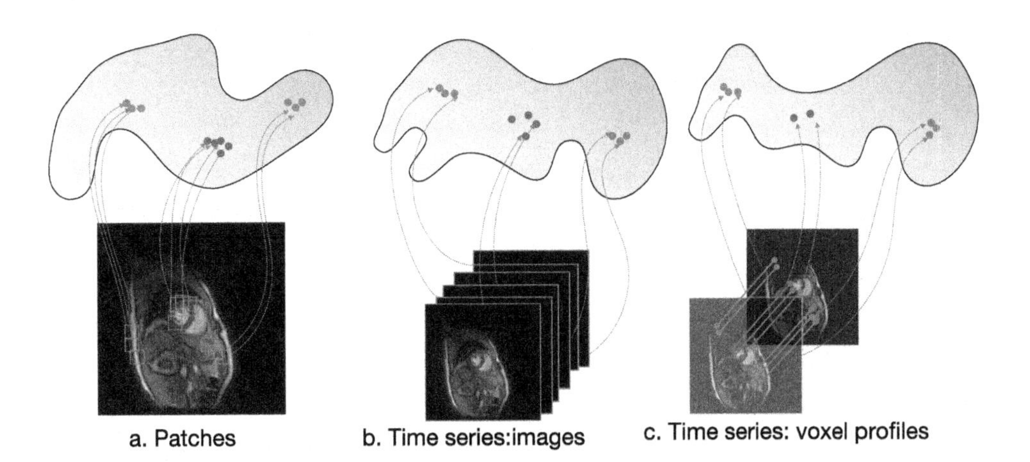

FIGURE 10.2

Types of signals used by different algorithms: The MR methods reviewed in this chapter account for the redundancy within the image volumes, which often manifest as correlations between image sub-parts that we broadly refer to as patches. Depending upon the shape of the patches, they could be cubes in 3D/4D as shown in (**a**), images in a dynamic time series as shown in (**b**), or time profiles of pixels in a time series as shown in (**c**). The images in dynamic imaging or parameter mapping have extensive nonlocal similarity as shown in (**c**). For instance, the images in similar cardiac/respiratory phases are expected to be similar; each image in the time series may be viewed as a mapping of the cardiac/respiratory phases, which are often accounted by self-gating methods. Likewise, the voxel time series in image time series in (c) are also highly correlated. For instance, the time series of pixels from the same organ that experience similar motion pattern or have similar physiology (e.g., myocardium, liver) are expected to have similar intensity profiles. All of these schemes can be seen as patch based methods; the main difference is the shape of the patches. Patch based methods aim to capture the extensive similarity between patches to recover the image dataset from highly under sampled measurements. If the number of pixels in a patch is denoted by p, each of the patches can be viewed as a point in a p dimensional space. However, because of the extensive structure/redundancy between the pixel values, these signals are often localized to low-dimensional structures in this high-dimensional space as shown above.

relationship

$$s_{i,j} = \int x(\mathbf{r})c_j(\mathbf{r})\exp(-j\mathbf{k}_i^T\mathbf{r})d\mathbf{r} + \eta_{i,j}. \tag{10.1}$$

Here, $c_j(\mathbf{r})$ are the sensitivity profiles of the jth coil, and $\mathbf{k}_i$ is the location in k-space, while η denotes noise. The measurement process described in (10.1) can be compactly represented as

$$\mathbf{s} = \mathbf{E}(\mathbf{x}), \tag{10.2}$$

where $\mathbf{E}$ is the multichannel Fourier encoding operator. The function $\mathbf{x}$ may be 2D ($n = 2$) or 3D ($n = 3$), or higher-dimensional, depending on the applications. In 2D/3D + time datasets (e.g., dynamic

Table 10.1 Broad classification of methods reviewed in this chapter.

Method	Signal Type			
	Patch		Images	
	Two-step	Joint	Two-step	Joint
Low-rank image domain	BM3D [16,17]	CLEAR [19], LOST [2], HD-PROST [13] PRICE [42]	- PSF [11,70,31,30]	- k-t SLR [19,33,32]
Low-rank k-space	ESPIRIT [42] GRAPPA [20] SLR [49]	LORAKS [21,23,22] SLR [51,50,48,49] MUSSELS [41,39,40,37]	- - - -	- ALOHA [29,26,25] SLR [4,5]
Dictionary learning	Transform Learning [58,57,60]	BCS [59]	- - -	- BCS [9,8,34,35]
Manifold smoothness manifold	NLM [12] UINTA [3]	Nonlocal Regularization [68,67] Patch-STORM [43]	Kernel PCA [47,66] STORM [54] MLS [64,46] KLR-STORM [56]	Spiral-STORM [1]

imaging applications), the acquisition model correspond to

$$s_{i,j}(t) = \int x(\mathbf{r}, t)\, c_j(\mathbf{r}) \exp(-j\mathbf{k}_i^T \mathbf{r}) d\mathbf{r}. \tag{10.3}$$

In this chapter, we will use the same symbol $\mathbf{E}$ as the multichannel Fourier sampling operator in the dynamic setting. Specifically, $\mathbf{E}$ applied on the 3D/4D volume image $\mathbf{x}$ yields the vector of measurements denoted by $\mathbf{s}$. We denote $\widehat{\mathbf{x}}$ to denote the discrete Fourier coefficients of the signal on a Cartesian grid. Note that the multichannel measurements in (10.1) need not be in the Cartesian domain.

10.2.2 Manifold models of signals

We start with a brief and intuitive illustration of the manifold assumption, with the objective of connecting diverse image models used in the context of MRI. We note that an $n \times n$ image has n^2 pixels and hence can be viewed as a point in n^2 dimensional space. In the absence of any redundancies (e.g., pixels have random values), each image is a random point; the images will fill the n^2 dimensional space. However, natural images have extensive redundancies between the pixels. Hence, the images of interest often lie on low-dimensional structures (e.g., lines, union of subspaces, curves) in n^2 dimensional space.

For example, consider three pixel images whose pixel values are uniformly random. If we plot each of these three pixel images in 3D space, they fill the space. By contrast, if all of the images have their pixel values are linearly increasing with the same slope, all the three pixel images will fall on a straight line passing through the origin. When modeling larger 1D images, one may extract patches consisting of three consecutive pixels from the image, each of which can be viewed as a point on the slanted line. This approach can be viewed as a global subspace model for the image. As discussed in the next section, the patch signals can be collected into a $3 \times P$ matrix, whose columns are the patches. The cited global linear model is restrictive and may not approximate real-world 1D signals. A more general representation is a piecewise linear model, where the slope of the signal is different at different locations. In this case, the three pixel patches from each image may not lie on a single line. Depending on the spatial location of the patch, they would lie on different straight lines; the number of straight lines would depend on the number of piecewise linear regions. This union of lines/subspaces model can be viewed as a generalization of the aforementioned global subspace model.

This representation can be generalized in many different ways to improve the efficiency and approximation power. For instance, one can use piecewise polynomial or exponential signal models, which are more efficient than the piecewise linear model. Note that the dimension of the space depends on the number of pixels in the patch (m). With higher dimensions, one can capitalize on more complex interdependencies between the pixels, beyond the piecewise models just discussed. In general, the patch signals do not fill the m-dimensional space (m being the number of pixels in the patch); they often lie on low-dimensional constructs (e.g., clusters, smooth surfaces, curves), often loosely termed as manifolds, in the m-dimensional space. This property is often referred to as the manifold assumption in machine learning [7,53].

The aforementioned idea can be extended to patches in 2D images, where the pixels within the patches may have nonlinear relationships between them, depending on the type of the image content (e.g., piecewise linear, piecewise polynomial). Depending on the image content, one may consider patches with m pixels of different sizes and shapes (see Fig. 10.2) to capture specific redundancies within the dataset. Generalizing the previous example, the patches can be viewed as points in high-dimensional space. This chapter reviews the several approaches of *learning the compact representation of data matrices*, whose columns are the signals of interest (images, patches, time-profiles of pixels in time series data) that we will capitalize on using advanced algorithms to capitalize on the unique redundancies in each application.

10.2.3 Capitalization of redundancy using structured matrices

A common approach to capitalize the redundancies within the signals of interest (e.g., images in a time-series, patches in an image, voxel profiles) is to create structured matrices from the data and use their properties to recover the images. In the general setting, one can extract patch vectors from images to form a matrix denoted by $\mathcal{T}(\mathbf{x})$:

$$\mathcal{T}(\mathbf{x}) = \begin{bmatrix} P_1(\mathbf{x}) & P_2(\mathbf{x}) & \dots & P_n(\mathbf{x}) \end{bmatrix}. \tag{10.4}$$

Here, $P_1, ..., P_n$ are patch extraction operators that extract a patch from an image and convert it into a column of $\mathcal{T}(\mathbf{x})$. If each patch has m pixels and there are n patches that cover the dataset, the data matrix denoted by $\mathcal{T}(\mathbf{x})$ will be of dimension $m \times n$. The size and shape of the patches could be chosen, depending on the application, to exploit a specific property of the dataset. For instance, if the ith patch extraction operator is chosen as the ith image in a time series $P_i(x(\mathbf{r}, t)) = x(\mathbf{r}, i); i = 1, .., T$, we obtain

$$\mathcal{T}(\mathbf{x}) = \begin{bmatrix} x(\mathbf{r}, 1) & \dots & x(\mathbf{r}, T) \end{bmatrix}, \tag{10.5}$$

which is the standard Casorati matrix reviewed in Chapter 9, whose columns are the reshaped images of the time series from the dynamic data (see Fig. 10.2.(a)). Similarly, if the patch extraction operators extract the time profile of each pixel, we obtain the transpose of a Casorati matrix. These are the two extreme cases. One can choose 2D patch-extraction operators (or 3D patch-extraction operators in the time series) to account for the correlation between patches in the dataset. In these cases, the matrix can have a block convolutional structure. See Fig. 10.6. Hence, the methods that use the low-rank property of these structured matrices are called structured low-rank (SLR) methods [65,21,38,49,23,29,26,51, 41]. One may also create a structured matrix $\mathcal{T}(\hat{\mathbf{x}})$ by choosing the patches from the discrete Fourier samples of the signal $\widehat{\mathbf{x}}$. In fact, several of the structured low-rank algorithms reviewed later in the chapter rely on the low-rank property of structured matrices in the Fourier domain.

The structured matrix $\mathcal{T}(\mathbf{x})$ is often much larger in size than the original dataset $\mathbf{x}$; the operation $\mathbf{x} \to \mathcal{T}(\mathbf{x})$ of creating the structured matrix from the samples is often called as a lifting operation. We term the columns of $\mathcal{T}(\mathbf{x})$ as the signals of interest; the algorithms considered in this chapter will promote the learning and capitalization of redundancies between the columns.

10.2.4 Efficient matrix representation in terms of factors

As discussed in Chapter 9 in the context of low-rank representation, the matrix $\mathcal{T}(\mathbf{x})$ can be efficiently represented in terms of its factors as

$$\mathcal{T}(\mathbf{x}) = \mathbf{\Phi}\mathbf{W}^T, \tag{10.6}$$

where Φ and $\mathbf{W}$ are the factor matrices, of size $m \times R$ and $n \times R$, respectively. In the context of low-rank matrices, R is the rank of the matrix. When the data has high redundancy, R is much smaller than m and n. Most of the algorithms choose an $R > r$, coupled with priors (e.g., ℓ_2 or ℓ_1 norms) on the factors discussed subsequently to make the recovery well-posed.

In this case, the number of free parameters in $\mathbf{\Phi}$ and $\mathbf{W}$ is often much smaller than the size of $\mathcal{T}(\mathbf{x})$. When $\mathcal{T}(\mathbf{x})$ is the Casorati matrix, the columns of $\mathbf{\Phi}$ can be viewed as the spatial factor, while that of $\mathbf{W}$ is the temporal factor. In addition to enabling the recovery from undersampled data in terms of the spatial and temporal factors, the example factor representation can also mitigate the high memory demands of directly working with $\mathcal{T}(\mathbf{x})$, by conserving space and obtain a computationally efficient algorithms. Implicit low-rank methods use the nuclear norm of $\mathcal{T}(\mathbf{x})$, denoted by $\|\mathcal{T}(\mathbf{x})\|_* = \sum_i \sigma_i[\mathcal{T}(\mathbf{x})]$, as a prior in reconstruction problems,

$$\mathbf{x}^* = \arg\min_{\mathbf{x}} \|\mathbf{E}(\mathbf{x}) - \mathbf{s}\|^2 + \lambda\|\mathcal{T}(\mathbf{x})\|_* \tag{10.7}$$

to encourage the recovery of an $\mathbf{x}$ such that the matrix $\mathcal{T}(\mathbf{x})$ is low-rank. The nuclear norm has an alternate form [61]

$$\|\mathcal{T}(\mathbf{x})\|_* = \|\mathbf{\Phi}\|_F^2 + \|\mathbf{W}\|_F^2, \text{ where } \mathcal{T}(\mathbf{x}) = \mathbf{\Phi}\mathbf{W}^T, \tag{10.8}$$

provided R is greater than the rank of the matrix $\mathcal{T}(\mathbf{x})$. This interpretation allows one to implement an implicit low-rank method without storing the large $\mathcal{T}(\mathbf{x})$ matrix

$$\{\mathbf{\Phi}^*, \mathbf{W}^*\}, = \arg\min_{\mathbf{\Phi},\mathbf{W}} \|\mathbf{E}(\mathbf{\Phi}\mathbf{W}^T) - \mathbf{s}\|^2 + \lambda\left(\|\mathbf{\Phi}\|_F^2 + \|\mathbf{W}\|_F^2\right), \tag{10.9}$$

when $\mathcal{T}(\mathbf{x})$ has a Casorati form. In addition to the computational and memory efficiency, the factor interpretation opens the door to the use of other priors on the factors and the matrices, which can offer improved performance. We will now focus on how more general factorization strategies can offer improved performance, compared to the low-rank methods already mentioned.

10.3 Dictionary learning and blind compressed sensing

The global subspace models described in Chapter 9 enables the representation of dynamic datasets. However, when the signals that one is trying to represent is very diverse (e.g., patches in the image), the ability of the global subspace model to represent them is limited (see Fig. 10.1.a). For instance, there may be several groups of patches, each of which may possess a low-rank. However, the global subspace spanned by all the groups may be high-dimensional. Hence, state-of-the art patch-based methods such as BM3D [17,16] cluster the patches into subsets, followed by the application of the subspace model to each subset. These schemes can be viewed as the approximation of the global manifold locally by low-dimensional subspaces, as shown in Fig. 10.1.(b). A challenge for these schemes is the two-step process, involving the identification of the similar subsets, followed by low-rank modeling. These approaches are widely used in the denoising setting. Approaches, such as [2,67–69], extend this approach to image-reconstruction applications. These methods either estimate the clusters from zero-filled MRI data [2,67] or alternate between clustering and recovery of images [68,69]. More information on the alternating scheme is provided in Section 10.5.1.2 in the context of manifold recovery. Dictionary learning and blind compressing can overcome the need for this pre-clustering for the recovery of images from undersampled data.

10.3.1 Subspace selection for each signal of interest using sparse representation

Both dictionary learning and blind compressing schemes rely on a sparse image representation, as shown in Fig. 10.1.(c). The sparse model allows one to choose the specific basis functions needed to represent a specific column of $\mathcal{T}(\mathbf{x})$ (patch) only by allowing only a few coefficients of the representation to be nonzero. For instance, the sparse model [15] represents the signal as

$$x(\mathbf{r}) \approx \sum_{i=1}^{K} w_i\, \varphi_i(\mathbf{r}) = \underbrace{\begin{bmatrix}\varphi_1(\mathbf{r}) & \varphi_2(\mathbf{r}) & \ldots \varphi_K(\mathbf{r})\end{bmatrix}}_{\mathbf{\Phi}} \underbrace{\begin{bmatrix} w_1 \\ w_2 \\ \vdots \\ w_K \end{bmatrix}}_{\mathbf{w}^T}, \text{ where } \|\mathbf{w}\|_{\ell_0} \leq k. \tag{10.10}$$

The matrix Φ is termed as the dictionary, while its columns are the basis vectors and are often referred to as atoms. Here, $\|\mathbf{w}\|_{\ell_0}$ denotes the number of nonzero terms in the coefficient vector $\mathbf{w}$. This model allows the basis functions used to approximate each group of signals to be different and, hence, offer more compact representation of the data. Traditional compressed sensing schemes relies on pre-determined dictionaries (e.g., wavelet transform) $\mathbf{\Phi}$. Rather than using fixed dictionaries, several authors have proposed to adapt or learn the dictionaries or transforms from the data [57,34,35,58,59,8]. The adaptation of the dictionary to the data depending on the specific signal offers a quite significant reduction in the number of measurements. The learning is either performed from several fully-sampled example images [57,58,8,60] or from a single undersampled dataset [34,35,59,8] in a joint manner. The first approach is termed as pre-learning. By contrast, the joint learning of the dictionary and coefficients from the data is termed blind compressed sensing. Since the ℓ_0 norm is not convex, a common approach is to approximate it with the ℓ_1 norm that is convex [15]. Rather than employing dictionaries, the use of analysis operators $\mathbf{\Psi}$ is also a common approach.

10.3.2 Dictionary pre-learning

In pre-learning, the learning of the dictionary from a family of fully sampled signals $\mathbf{X} = [\mathbf{x}_1, .., \mathbf{x}_N]$ is posed as the optimization problem [57,58]

$$\{\boldsymbol{\Phi}, \mathbf{W}\} = \arg\min_{\boldsymbol{\Phi},\mathbf{W}} \|\mathbf{X} - \boldsymbol{\Phi}\mathbf{W}^T\|^2 + \lambda_1 \|\mathbf{W}\|_{\ell_0} + \lambda_2 \ \mathcal{R}(\boldsymbol{\Phi}), \tag{10.11}$$

where $R(\boldsymbol{\Phi})$ is a regularization penalty on the dictionary atoms. A simple choice for $\mathcal{R}(\boldsymbol{\Phi}) = \|\boldsymbol{\Phi}\|_F^2$ is the Frobenius norm, where the energy of the dictionary is restricted. Regularization penalties that encourage the dictionary to be an orthogonal transform to have a low condition number or to be a combination of orthogonal transforms have been introduced by several authors [57,34,35,58,59,8].

Note that solving for $\mathbf{W}$ from (10.11) assuming $\boldsymbol{\Phi}$ to be known is the traditional compressed sensing approach. Likewise, for many of the common choices of $\mathcal{R}$, the optimization of $\boldsymbol{\Phi}$ assuming $\mathbf{W}$ to be known, is also a simple problem. For instance, when $\mathcal{R}(\boldsymbol{\Phi}) = \|\boldsymbol{\Phi}\|_F^2$ is the Frobenius norm, the solution of $\boldsymbol{\Phi}$ is a quadratic problem that has an analytical solution. However, the joint optimization of $\boldsymbol{\Phi}$ and $\mathbf{W}$ is a nonconvex problem. Nevertheless, this problem has been well-studied by several researchers [57,34,35,58,59,8], especially when $\boldsymbol{\Phi}$ is an orthonormal matrix. The learning of the transform to the class of signals results in improved performance over the use of standard transforms such as wavelet transform.

10.3.2.1 *Dictionary pre-learning, applied to static MRI*

In static imaging, a common approach is to assume the image patches of size $p \times p$ to have a sparse dictionary representation. Here, the signal matrix in (10.11) is the patch matrix $\mathbf{X} = \mathcal{T}(\mathbf{x})$ of dimension $p^2 \times n$. The ability to choose the nonzero coefficients for each patch facilitates the use of the same dictionary for the entire image. This formulation can be extended to learning from multiple images by horizontally stacking the patch matrices as $[\mathcal{T}(x_1), ..., \mathcal{T}(x_N)]$. The dictionary learning approach is thus a learning-based alternative for transformations such as wavelets or discrete cosine transform widely used in compressed sensing. It is also an alternative to the patch-based low-rank methods (e.g., BM3D) used in the image domain, where similar patches need to be clustered prior to subspace fitting. Unlike these methods, the patch dictionary-based schemes use the same dictionary for all patches; the sparsity of the coefficients allow the adaptation of the specific basis functions used in the representation to the specific patch. Once the dictionary $\boldsymbol{\Phi}$ is learned from multiple images, it can be used to recover images from undersampled measurements [57] as

$$\mathbf{W}^* = \arg\min_{\mathbf{W}} \alpha \|\mathbf{E}(\mathbf{x}) - \mathbf{s}\|^2 + \|\mathcal{T}(\mathbf{x}) - \boldsymbol{\Phi}\mathbf{W}^T\|^2 + \lambda \|\mathbf{W}\|_{\ell_p}, \tag{10.12}$$

which is the extension of compressive sensing (CS) to image patches. The second and third term encourages the patches in the solution to be a sparse linear combination of atoms in $\boldsymbol{\Phi}$, while the first term encourages $\mathbf{x}$ to satisfy data consistency; the optimal solution is a compromise between the two, where the relative importance of data consistency is controlled by α.

10.3.3 Blind compressed sensing (BCS)

BCS schemes estimate the dictionary and the coefficients directly from the measured under-sampled data, rather than pre-learning the dictionary from exemplar data. By adapting the dictionary to the specific image content, these schemes can offer improved performance.

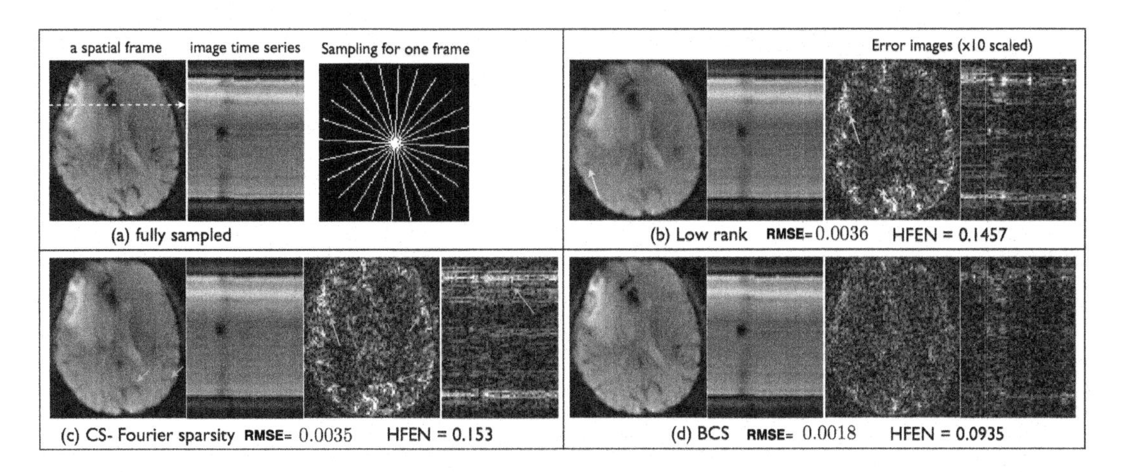

FIGURE 10.3

Comparison of low-rank, Fourier sparsity, and blind compressed sensing (BCS) reconstructions on a brain perfusion MRI dataset retrospectively undersampled at an undersampling factor of 10 fold. BCS is shown to provide superior spatial and temporal fidelity in characterizing the contrast agent temporal dynamics compared to low-rank and CS reconstructions.

10.3.3.1 Application of BCS to dynamic MRI

In the context of dynamic MRI, blind compressed sensing methods learn the bilinear model in (10.11) directly from undersampled data [34,35,59,8]. In the dynamic setting with the transpose of the Casorati matrix, each column of $\mathbf{X}$ corresponds to the temporal profile of a pixel. The factorization $\mathbf{X} = \mathbf{\Phi}\mathbf{W}^T$ amounts to expressing the temporal profiles of each pixel as a linear combination of the columns of the dictionary $\mathbf{\Phi}$. When $\mathbf{W}$ is sparse, the temporal profile of each pixel is expressed as the linear combination of a few atoms, which change from pixel to pixel. Please see Fig. 10.4 for the difference between low-rank and dictionary representation. Unlike the low-rank setting that uses the same basis functions at all pixels, the dictionary learning scheme is able to customize the basis functions for each pixel. In particular, the coefficients that are nonzero (and hence the temporal basis functions that are active) for the heart region with periodic oscillations may be different from that of a static region.

Since the temporal profiles of the pixels change from subject to subject, it is not practical to pre-learn the dictionary from other dataset. The dictionary Φ, which may be overcomplete, and its coefficients $\mathbf{W}$ are hence learned directly from the undersampled dataset itself as

$$\{\mathbf{\Phi}, \mathbf{W}\} = \arg\min_{\Phi, \mathbf{W}} \|\mathbf{E}\left(\mathbf{\Phi}\mathbf{W}^T\right) - \mathbf{s}\|^2 + \lambda_1 \|\mathbf{W}\|_{\ell_0} + \lambda_2 \; \mathcal{R}(\mathbf{\Phi}), \tag{10.13}$$

The use of blind compressed sensing scheme in (10.13) further improves the quality of dynamic MRI reconstructions [34,35,8] compared to linear (low-rank) models. Specifically, low-rank models use the same basis functions for the voxel profiles of each pixel. The projection of the time series to the signal subspace results in nonlocal temporal averaging [9]. Since the basis functions are the same for each pixel, the temporal point spread functions are the same for each pixel. By contrast, the active basis functions (ones corresponding to nonzero coefficients) in sparse models can potentially differ from

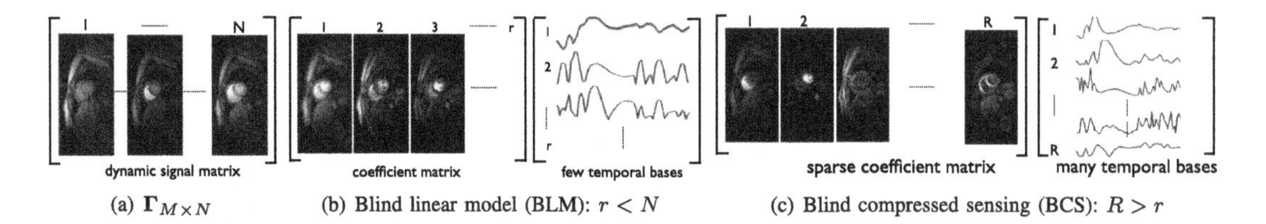

FIGURE 10.4

Comparison of blind compressed sensing (BCS) and low-rank (blind linear model) representations of dynamic imaging data: The Casorati form of the dynamic signal is shown in (**a**). The BLM and BCS decompositions are respectively shown in (**b**) and (**c**). BCS uses a large over-complete dictionary, unlike the orthogonal dictionary with few basis functions in BLM ($R > r$). Note that the coefficients/spatial weights in BCS are sparser than that of BLM. The temporal basis functions in BCS dictionary are representative of specific regions since they are not constrained to be orthogonal. For example, the first, second columns of the temporal basis functions in BCS correspond respectively to the temporal dynamics of the right and left ventricles in this myocardial perfusion data with motion. We observe that only 4-5 coefficients per pixel are sufficient to represent the dataset. This figure adapted from [35] with permission from IEEE.

pixel to pixel and hence, the temporal averaging at each pixel is different. This is a desirable feature in applications where the temporal motion patterns are drastically different from pixel to pixel depending on the organs within the field of view (e.g., heart, lung). An illustration of the blind compressed sensing (BCS) approach is shown in Fig. 10.3, where it is compared against global low-rank methods and approaches that use sparse models with fixed (Fourier) dictionary.

10.3.3.2 *Application of BCS to static imaging*

The blind compressed sensing formulation in (10.13) can also modified to the patch setting. In particular, the dictionary, the coefficient matrix **W**, and the resulting image are all simultaneously learned from the measured data [59] as

$$\{\mathbf{x}, \mathbf{W}, \mathbf{\Phi}\} = \arg\min_{\mathbf{x},\mathbf{W},\mathbf{\Phi}} \|\mathbf{E}(\mathbf{x}) - \mathbf{s}\|^2 + \alpha\|\mathcal{T}(\mathbf{x}) - \mathbf{\Phi}\mathbf{W}^T\|^2 + \lambda_1\|\mathbf{W}\|_{\ell_p} + \lambda_2\mathcal{R}(\mathbf{\Phi}). \tag{10.14}$$

Here, the first term is the data-consistency term that measures the discrepancy of the recovered image $\mathbf{x}$ from the measurements. Ideally, we would like to have the patch matrix $\mathcal{T}(\mathbf{x})$ extracted by the image to have a compact factorization $\mathcal{T}(\mathbf{x}) = \mathbf{\Phi}\mathbf{W}^T$, where the coefficient matrix $\mathbf{W}$ is sparse and the dictionary is compact under a prior $\mathcal{R}(\mathbf{\Phi})$ such as $\|\mathbf{\Phi}\|_{\ell_2}$ or $\|\mathbf{\Phi}\|_{\ell_1}$. Rather than introducing the factorization as a constraint, the formulation in (10.14) relies on a penalty term; when $\alpha \to \infty$, the solution will satisfy $\mathcal{T}(\mathbf{x}) = \mathbf{\Phi}\mathbf{W}^T$.

10.4 Structured low-rank methods

As mentioned in Section 2.2, 2D/3D patch extraction operations can generate data matrices with block convolutional structure. The earlier methods relied on patches exclusively in the image domain. We will now review methods that exploit the similarity of patches in the Fourier domain, or equivalently consider structured matrices $\mathcal{T}(\widehat{\mathbf{x}})$ obtained by lifting the discrete Fourier coefficients of the signal $\widehat{\mathbf{x}}$. It is interesting to note that several image properties result in extensive correlations between the k-space samples, which can be used with global low-rank regularization.

10.4.1 Low-rank structure of patch matrices in k-space

The structured low-rank methods in MRI started with the multichannel methods termed as ESPIRIT [28], simultaneous autocalibrating and k-space estimation (SAKE) [65], followed by single channel approaches termed as Low-Rank Modeling of Local k-Space Neighborhoods (LORAKS) [21,23,22], Annihilating filter based LOw-rank HAnkel matrix ALOHA [29,26,25], structured low-rank (SLR) [49,51,48] and multishot sensitivity-encoded diffusion data recovery using structured low-rank matrix completion (MUSSELS) [38,41]. We now briefly review some of the low-rank relationships resulting from specific signal properties.

10.4.1.1 Low-rank relationships in multichannel MRI

In parallel MRI schemes that acquire multichannel data, the sensitivity-weighted image data are given by

$$x_i(\mathbf{r}) = x(\mathbf{r})\; c_i(\mathbf{r}),\;\; i = 1, .., N_{\text{channels}}, \tag{10.15}$$

where $c_i(\mathbf{r})$ is the sensitivity weighting of the ith receiver coil. The multichannel relations specified by (10.15) results in image domain annihilation relationships [44]

$$\underbrace{x(\mathbf{r})c_1(\mathbf{r})}_{x_1(\mathbf{r})}\, c_2(\mathbf{r}) - \underbrace{x(\mathbf{r})c_2(\mathbf{r})}_{x_2(\mathbf{r})}\, c_1(\mathbf{r}) = 0. \tag{10.16}$$

One can take the Fourier transforms of both sides of the previous equation to obtain [28,65]

$$\widehat{x}_1 * \widehat{c}_2 - \widehat{x}_2 * \widehat{c}_1 = 0. \tag{10.17}$$

When the coil sensitivities c_i are smooth, one can reliably approximate them as bandlimited functions, whose Fourier support is restricted to a $p \times p$ square region. We now focus on the convolution between the signal $\mathbf{x}$ and a finite impulse response filter $\mathbf{c}$ of support $p \times p$. The convolution output at each pixel $\mathbf{r}$ can be thought of as the innerproduct between the flipped version of a $p \times p$ patch of $\mathbf{x}$, centered at $\mathbf{r}$ with $\mathbf{c}$. Thus, the convolution can be expressed in the matrix form as

$$c * x = \mathbf{c}^T\; \mathcal{T}(\mathbf{x}), \tag{10.18}$$

where $\mathcal{T}(\mathbf{x})$ is a lifted matrix, whose columns are flipped versions of $p \times p$ patches from $\mathbf{x}$. The vector $\mathbf{c}$ corresponds to a vectorized version of the filter c. With this property, one can rewrite (10.17) as

$$\mathbf{c}_2^T \mathcal{T}(\widehat{\mathbf{x}}_1) - \mathbf{c}_1^T \mathcal{T}(\widehat{\mathbf{x}}_2) = 0. \tag{10.19}$$

We note that similar annihilation relationships can be found for every pair of channels. We can compactly express these relations in the matrix form as

$$\underbrace{\begin{bmatrix} \widehat{\mathbf{c}}_2^T & -\widehat{\mathbf{c}}_1^T & 0 & \cdots & \\ \widehat{\mathbf{c}}_3^T & 0 & -\widehat{\mathbf{c}}_1^T & \cdots & \\ \vdots & \vdots & \ddots & \cdots & \\ \widehat{\mathbf{c}}_{N_c}^T & 0 & 0 & \cdots & -\widehat{\mathbf{c}}_1^T \end{bmatrix}}_{\mathbf{P}} \underbrace{\begin{bmatrix} \mathcal{T}(\hat{\mathbf{x}}_1) \\ \mathcal{T}(\hat{\mathbf{x}}_2) \\ \vdots \\ \mathcal{T}(\hat{\mathbf{x}}_{N_c}) \end{bmatrix}}_{\mathcal{M}(\widehat{\mathbf{X}})} = \mathbf{0}. \tag{10.20}$$

Note that each of the rows of P are null-vectors of $\mathcal{M}(\widehat{\mathbf{X}})$, which are linearly independent. Hence, the matrix $\mathcal{M}(\widehat{\mathbf{X}})$ is low-rank [28,65]. Here, $\widehat{\mathbf{X}} = [\hat{\mathbf{x}}_1, .., \hat{\mathbf{x}}_{N_c}]$ is the multichannel data in the Fourier domain. The these multichannel convolution relationships can also be rewritten as

$$\widehat{\mathbf{X}} = \underbrace{(\mathcal{I} - \mathbf{P})}_{\mathbf{G}} \ \mathcal{M}(\widehat{\mathbf{X}}), \tag{10.21}$$

where the operator $\mathcal{I}$ in (10.21) extracts $\widehat{\mathbf{X}}$ from $\mathcal{M}(\widehat{\mathbf{X}})$ (*i.e.*, $\mathcal{I}\{\mathcal{M}(\widehat{\mathbf{X}})\} = \widehat{\mathbf{X}}$). This relationship forms the basis of the autocalibrating parallel MRI reconstruction method, SAKE [65], which interpolates the missing k-space samples of the accelerated acquisition, based on the structured low-rank property. The SAKE relation in (10.21) can be viewed as a generalization of GRAPPA reconstruction method for multichannel MRI [20].

The formulations in (10.16)–(10.20) is general enough to be applied to settings beyond multichannel MRI. Researchers have used these relationships in several contexts, such as the calibration-less compensation of phase errors in multichannel diffusion MRI [39,37], the correction of Nyquist ghost artifacts in echo-planar imaging [41,36], and the correction of trajectory errors in radial MRI [40]. In all of these cases, different segments of k-space experience different distortions [39,41,36]. These errors can be modeled as image domain weighting (similar to the coil sensitivity weighting studied previously), allowing the use of the relationships (10.16)-(10.20), to derive the structured data matrix with the low-rank property. Fig. 10.5 shows the application of the patch low-rank idea for the reconstruction of high-resolution diffusion MRI data from multishot acquisitions, where the phase compensation of the multishot data were achieved in a calibration-less manner using patch low-rank.

10.4.1.2 Low-rank structure resulting from finite support and smoothly varying image phase

The LORAKS algorithms introduced in [21,23,22] makes use of the property of finite support of images to derive the patch low-rank relationships, instead of the image domain weightings. Haldar [21] showed that for images possessing finite support, i.e., the signal $x(\mathbf{r})$ is zero within a region $\mathbf{r}_i \in \Omega$, the annihilation relationships of the form

$$x(\mathbf{r}) \cdot f_i(\mathbf{r}) = 0 \tag{10.22}$$

can be derived, where $f_i(\mathbf{r})$ are functions (also referred to as "filters") that are zero at all locations except at $\mathbf{r}_i \in \Omega$. Additionally, when $f_i(\mathbf{r})$ is assumed to be smooth so that it is bandlimited in the

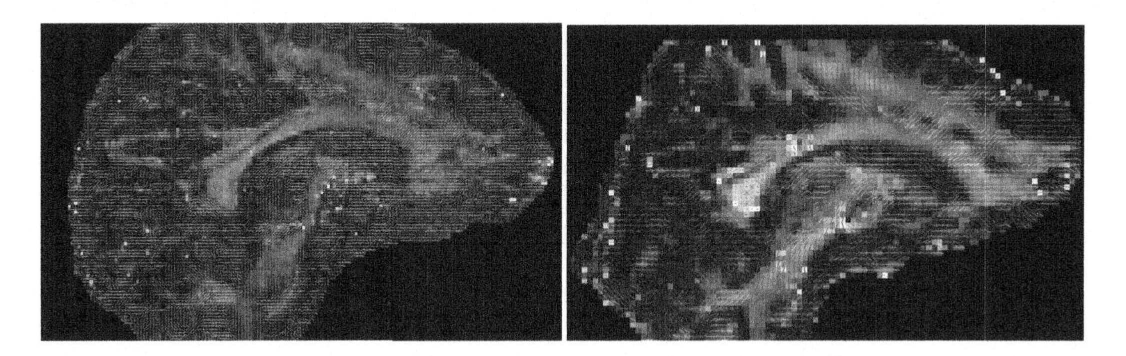

FIGURE 10.5

High-resolution diffusion MRI enabled by the patch low-rank methods in k-space. The image on the left side is acquired using a 4-shot diffusion weighted EPI scan at 1.1 mm isotropic resolution on 7 Tesla MRI with a standard clinical strength gradient of 40 mTesla/m. For comparison, a typical 2 mm-isotropic resolution single-shot diffusion MRI scan slice-matched from the same study is provided on the right. The multishot data on the left is reconstructed using the MUSSELS method that makes use of the patch low-rank in k-space and parallel imaging to recover diffusion weighted images free of phase errors. The iterative reweighted least squares implementation enables efficient reconstruction for such high-dimensional datasets.

Fourier domain, the previous multiplication relationships translate to convolution relations in k-space

$$\widehat{x}(\mathbf{r}) * \widehat{f_i}(\mathbf{r}) = 0, \tag{10.23}$$

resulting in annihilation relationships in the Fourier domain, and results in the reduction of the degrees of freedom. Typically, one can find multiple filters $f_i(\mathbf{r})$ that satisfy the relationships in (10.22). This relationship implies that

$$\mathcal{T}(\widehat{\mathbf{x}})\,\mathbf{F} = 0, \tag{10.24}$$

or equivalently $\mathcal{T}(\mathbf{x})$ is a low-rank matrix. Haldar et al. has empirically indicated that the rank of the lifted convolution matrices, $\mathcal{T}(\mathbf{x})$, corresponding to MR images indeed depends on the support of the signal. Since these patches are formed from single-channel images, it can be applied for single-channel undersampled recovery.

It is well-known that real images exhibit conjugate symmetry in k-space, resulting in annihilation relationships

$$\widehat{x}[k] - \widehat{x}[-k]^* = 0. \tag{10.25}$$

When the phase of the images are smoothly varying, [21] showed that one can construct a specialized convolution matrix using the 2D patches from the conjugate symmetric k-space samples also, that satisfy annihilation relationships and hence is low-rank. These results translate to structured low-rank algorithms that account for the patch low-rank structure.

10.4.1.3 *Low-rank structure resulting from continuous domain sparsity*

Here, we discuss the low-rank relationships for sparse continuous domain functions, which enable super-resolution reconstructions. Traditional CS schemes assume the images to be sparse on a specific grid, which may be an unrealistic assumption. Several researchers have considered the extension of CS for the super-resolution setting, where the sparse samples of the signal may not be localized to a grid [14]. Specifically, we can use an image model using impulse functions

$$x(r) = \sum_{i=1}^{R} \gamma_i \, \delta(r - r_i) \tag{10.26}$$

for sparse images, where γ_i are the weights and r_i are the location of the impulses, which are not necessarily on a uniform grid. The SLR [50] and ALOHA settings [26] extends the Fourier domain annihilation relationships discussed in the previous sections for the recovery of continuous domain sparse signals.

To see how the aforementioned sparse image model can harness the Fourier domain annihilation relationships, we first discuss a simple 1D case. The seminal work by Prony dating back to 1885 showed that 1D exponential signal of the form $\widehat{x}[k] = \alpha^k$ can be annihilated by convolution as follows [27]:

$$\widehat{x}[k] * h[k] = \underbrace{\alpha^k}_{\widehat{x}[k]} - \alpha \underbrace{\alpha^{k-1}}_{\widehat{x}[k-1]} = 0, \tag{10.27}$$

where h[k] is a two tap filter given by $[1, -\alpha]$. This theory is relevant to MR images because, when $\alpha = \exp(jr_0)$, $\widehat{x}[k] = \exp(jr_0 k)$ is the Fourier transform of a sparse signal of the form $\delta(r - r_0)$.

More generally, when the signal is a linear combination of multiple impulses at location $r_0, .., r_k$, its Fourier coefficients can be annihilated by the convolution with a $k+1$ tap filter; the $k+1$ tap filter is obtained by the convolution of the k two-tap filters that annihilates each of the exponentials. Here, the location of the impulses r_i are not required to be localized to a specific grid; this approach may be viewed as the continuous domain extension of discrete compressed sensing methods. The extension of this idea to two dimensions is relatively straightforward.

The convolution-based annihilation relationship in (10.27) for k impulses for the superresolution recovery can thus be compactly expressed as [27,21,49,26]

$$\mathbf{h}^T \underbrace{\begin{bmatrix} \widehat{x}[0] & \widehat{x}[1] & \dots & \widehat{x}[N-k] \\ \widehat{x}[1] & \widehat{x}[2] & \dots & \widehat{x}[N-k+1] \\ \vdots & \vdots & \ddots & \vdots \\ \widehat{x}[k] & \widehat{x}[k+1] & \dots & \widehat{x}[N] \end{bmatrix}}_{\mathcal{T}(\widehat{x})} = \mathbf{0}. \tag{10.28}$$

Note that $\mathcal{T}(\hat{\mathbf{x}})$ is a patch matrix obtained by lifting the 1D signal $\widehat{x}[n]$, which are the Fourier coefficients of $\mathbf{x}$. The columns of $\mathcal{T}(\hat{\mathbf{x}})$ correspond to $(k+1) \times 1$ patches in $\mathbf{x}$. (10.28) implies that the matrix $\mathcal{T}(\hat{\mathbf{x}})$ has a null-space vector. If the number of impulses is $K' < k$, Eq. (10.28) will have $k - K' + 1$ null-space vectors [27,26]. In other words, the rank of the matrix $\mathcal{T}(\widehat{x})$ is a surrogate for the number of impulses in the signal.

Let us also briefly discuss the image domain interpretation of the filters $h[k]$. Taking the inverse Fourier transform of (10.27), we obtain

$$\hat{h}(r) \cdot x(r) = 0, \tag{10.29}$$

where $\hat{h}(r)$ and $x(r)$ the inverse Fourier transforms of $h[k]$ and $\widehat{x}[k]$, respectively. Note that the convolution in (10.27) is translated into the point-by-point multiplication [49,50]. For the example considered in (10.27), we get

$$\mu(r) = 1 - \underbrace{\exp(jr_0)}_{\alpha}\exp(-jr) = 1 - \exp\left(-j(r - r_0)\right). \tag{10.30}$$

Note that this exponential is a first-order polynomial that is nonzero at all locations, except at $r = r_0$, which is the location of the impulse. Likewise, when the signal x consists of k impulses, $\mu(x)$ is a bandlimited function that is zero only on the nonzero locations of $x(r)$. This interpretation will be useful in the next section for the discussion of piecewise smooth images.

10.4.1.4 Low-rank structure of piecewise smooth images

We can extend the annihilation relations for sparse image models to more general settings. The gradients of piecewise constant images are often significantly sparser [49] than the support of the signal considered in [21] or sparse model assumed in [26]. Note that the Fourier transform of the gradients of the 1D signal $\widehat{\partial_{r_1} x} = jk_{r_1}\widehat{x}$. Thus, replacing $\widehat{x}$ by $\widehat{\partial_{r_1} f}$ will result in a matrix with a significantly smaller rank.

The generalization of the idea to multiple dimensions is not straightforward from a theoretical perspective. Specifically, the gradient of a piecewise constant image is nonzero on the edges. Unlike the sparse model considered in the 1D case, where the number of impulses is finite, the gradient cannot be modeled as the sum of a finite number of impulses, which makes the recovery of the 2D images from few measurements is ill-posed. Prony's model and the related theory is only valid when the number of impulses is finite. Nevertheless, the problem can be made well-posed by assuming the edges to be localized to the zero-sets of a 2D bandlimited function $\mu(\mathbf{r})$ [49,50]. This model amounts to stating that the piecewise constant image has smooth edge contours. In this case, we have

$$\mu(\mathbf{r}) \cdot \underbrace{\left[\partial_{r_1} x(\mathbf{r}) \quad \partial_{r_2} x(\mathbf{r})\right]}_{\nabla x(\mathbf{r})} = 0. \tag{10.31}$$

Here, $\mu(\mathbf{r})$ is a bandlimited function that is zero at the edges of the image and nonzero elsewhere. Taking the Fourier transform on both sides, we obtain $h * \mathcal{T}_2(\mathbf{x})$, where

$$\mathcal{T}_2(\hat{\mathbf{x}}) = \left[\mathcal{T}\left(\widehat{\partial_{r_1} x}\right) \quad \mathcal{T}\left(\widehat{\partial_{r_2} x}\right)\right], \tag{10.32}$$

where $\mathcal{T}\left(\widehat{\partial_{r_1} x}\right)$ is the 2D patch matrix of the Fourier coefficients of the partial derivative of $\mathbf{x}$. The same approach can be extended to piecewise polynomials by replacing (10.32) with a matrix with more partial derivatives. The number of rows in equal to p^2, which is dependent on the size of the patch. The number of columns is equal to the number of valid patches in the images, without considering the regions outside the image. Note that, as the size of the patches increase, the number of patches and the number of columns decrease [50]. Theoretical results show that the best performance is obtained when the matrix $\mathcal{T}_2\left(\widehat{\partial_{r_1} x}\right)$ is square shaped, which roughly corresponds to each of the patch dimensions being half the corresponding image dimensions. However, practical algorithmic considerations, such as memory and computational constraints, often force the size of the patches to be smaller.

10.4.1.5 Low-rank relations in parameter mapping

Many parameter mapping applications in MRI consider the imaging of a time series, where the intensity of the pixel values at the spatial location $\mathbf{r}$ and time-point n change in an exponential fashion (e.g., $\rho[\mathbf{r}, n] = \alpha[\mathbf{r}]^n + c$, where c and α are arbitrary constants). Such a signal can be annihilated by a finite difference filter, whose parameters depend on α. The same approach can be readily extended to cases where the signal is the sum of several exponentials, where the size of the filter depends on the number exponentials. For simplicity, we will consider a single exponential signal $x[n] = \alpha^n$. From (10.27), we see that such exponential signals satisfy an annihilation relationship. Thus, the matrices obtained by lifting the time series entries will be low-rank in nature. Moreover, in many cases, the parameter maps vary smoothly in space, i.e., the exponential decay in a given pixel is highly correlated to the exponential decay in the neighboring pixels. In such cases, the coefficients of the exponentials and the parameters of the exponentials, themselves, can be modeled as band-limited functions. In this case, the k-t space samples of the parameter mapping application can be annihilated by multichannel convolution relations. This property was used successfully to recover $T_1\rho$ and T_2 maps in parameter mapping [4] applications as well as B0 field inhomogeneity compensated recovery of EPI images [5].

10.4.2 Algorithms for k-space patch low-rank methods

In Section 4.1, we discussed several properties of the images that result in Fourier domain annihilation relationships that translate to low-rank relationships on the associated structured matrices. The main difference for each problem is the lifting operation that is used to create the structured matrix from the Fourier samples. The lifting operation depends upon the specific image property that is accounted for. Once the structured data matrix is created, the recovery of the images using the low-rank relationships is posed as an optimization problem [22,49,50,26]. We will first discuss the multichannel parallel MRI case where the recovery of the images from the undersampled measurements can be written as the unconstrained optimization

$$\mathbf{x}^* = \arg\min_{\mathbf{x}} \|\mathbf{E}(\mathbf{x}) - \mathbf{s}\|^2 + \lambda \|\mathcal{T}(\hat{\mathbf{x}})\|_*, \tag{10.33}$$

Here, $\mathcal{T}(\hat{\mathbf{x}})$ is the lifted structured matrix formed from the multichannel convolution relationships in (10.20). This problem can be solved as a general low-rank matrix completion problem where the Hankel structure is additionally enforced. The rank minimization can be performed using singular value thresholding schemes and the optimization can be performed in an alternating manner updating the data consistency and rank-minimization. Fig. 10.6 shows a schematic of this approach where the cited problem is applied to the recovery of missing k-space samples as an interpolation problem.

The number of entries of the matrix $\mathcal{T}(\hat{\mathbf{x}})$ is several orders of magnitude larger than the size of the image, as discussed earlier. Because of this, the storage and computation of the matrix is often impossible in high-resolution and multidimensional applications. Moreover, the rank-minimization involves computing the SVD which is also computational demanding. Several algorithms were introduced to solve k-space low-rank problems similar to (10.33), as described in [22,49,50,26].

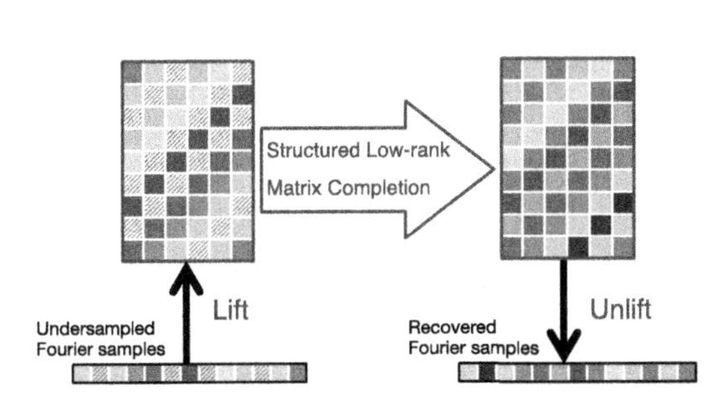

(a) SLR interpolation

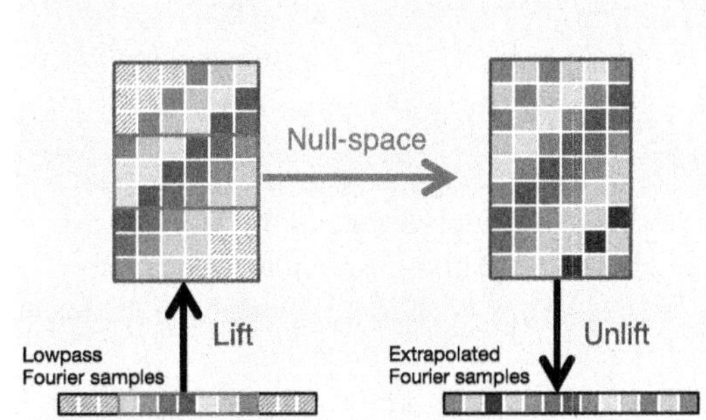

(b) SLR extrapolation

FIGURE 10.6

Illustration of SLR-based interpolation and extrapolation methods in the context of 1D FRI. (**a**) In SLR interpolation, the data is acquired on a nonuniformly subsampled Fourier grid. The SLR interpolation scheme relies on a lifting of the signal samples to a Hankel matrix, which has missing entries indicated by the hashed boxes. The one-to-one relationship between the rank of a matrix and the continuous domain sparsity of the space domain signal is used to pose the recovery of missing samples as a structured low-rank matrix completion (SLRMC) problem in the lifted matrix domain. Specifically, the algorithm determines the matrix with the lowest rank that satisfies the Hankel structure and is consistent with the known matrix entries. Post-recovery, the matrix is unlifted to obtain the Fourier samples of the signal. (**b**) In SLR extrapolation problems, the low-frequency Fourier coefficients of the signal are uniformly sampled. The central fully known matrix region is used to estimate the null space of the matrix, which is used to linear-predict/extrapolate the missing high-frequency samples. The SLR algorithms that exploit the different signal structures differ only in the structure of the lifted matrix; the algorithms are essentially the same. This figure is copied from [24] with permission from IEEE.

10.4.3 Iterative reweighted least square (IRLS) algorithm

The IRLS scheme relies on the approximation of the nuclear norm [51]

$$\|\mathcal{T}_2(\hat{\mathbf{x}})\|_* \approx \|\mathbf{Q}\,\mathcal{T}(\hat{\mathbf{x}})\|^2, \tag{10.34}$$

where $\mathbf{Q}$ is a $p^2 \times p^2$ matrix

$$\mathbf{Q} = \left(\mathcal{T}(\hat{\mathbf{x}})\mathcal{T}(\hat{\mathbf{x}})^T\right)^{-1/4} \tag{10.35}$$

The main benefits of the IRLS algorithm [51] is that it significantly reduces the computational complexity and the memory demand. In addition, we use this algorithm in some of the existing SLR methods in Section 10.4.4 and to connect kernel PCA with manifold methods in Section 10.5. In the perfectly low-rank setting, $\mathbf{Q}$ can be viewed as the projection onto the null-space and hence the right-hand side of (10.34) can be viewed as the energy of the projection of $\mathcal{T}(\mathbf{x})$ on to the null-space.

Using the previous approximation and the structure of the matrix $T(\mathbf{x})$, one can solve (10.28) by alternating between (10.35) and

$$f^* \doteq \arg\min_f \|\mathbf{E}(\mathbf{x}) - \mathbf{s}\|^2 + \lambda \sum_{i=1}^{p^2} \|\widehat{\mathbf{x}} * \mathbf{q}_i\|^2, \tag{10.36}$$

where $\mathbf{q}_i$ are the columns of $\mathbf{Q}$ [51]. The main benefit of (10.36) is that this formulation does not need the computation and storage of the large matrix $\mathcal{T}_2(\mathbf{x})$, which makes it possible to apply the scheme to multidimensional high-resolution applications.

A second approach is to approximate the nuclear norm using UV factorization [26]

$$\|\mathcal{T}_2(\hat{\mathbf{x}})\|_* \approx \arg\min_{\mathbf{U},V}, \|\mathbf{U}\|_F^2 + \|\mathbf{V}\|_F^2, \text{ where } \mathcal{T}_2(\hat{\mathbf{x}}) = \mathbf{U}\mathbf{V}^H, \tag{10.37}$$

as in Chapter 9. This method does not involve SVD computation, thus speeding up the minimization.

10.4.4 Algorithms that rely on calibration data

When a fully sampled center of k-space is available, a calibration based strategy can be employed for solving the minimization problem. Using the known data from the fully sampled region, the null-space filters $\mathbf{q}_i$ can be estimated. Note from (10.34) that the projection to the null-space should be as small as possible, which implies that $\mathbf{Q}\,\mathcal{T}(\mathbf{x}) \approx 0$. Hence, $\mathbf{Q}$ can be estimated from the central k-space regions by solving [49,28]

$$\mathbf{Q}^* = \arg\min_{\mathbf{Q}} \|\mathbf{Q}\,\mathcal{T}(\mathbf{k}_{\text{central}})\|^2 \text{ such that } \|\mathbf{Q}\|_F = 1 \tag{10.38}$$

using eigenvalue decomposition of $\mathcal{T}(\mathbf{k}_{\text{central}})$ corresponding to the matrix constructed from central k-space samples. Once $\mathbf{Q}$ is known, one can solve (10.36) with the knowledge of the filters, resulting in computationally efficient solution. Fig. 10.6 shows a schematic of this approach where the previous problem is applied to the recovery of missing k-space samples as an extrapolation problem.

Before we conclude the patch low-rank methods, we note several existing MRI reconstructions that are related to the patch low-rank methods previously discussed. The popular ESPIRIT reconstruction uses the null-space property to estimate the coil sensitivities using an eigen decomposition [28]. In particular, once $\mathbf{Q}$ or equivalently the signal subspace of $\mathcal{T}_2$ is obtained, it performs a pixel by pixel eigenvalue decomposition to obtain the coil sensitivities. The GRAPPA [20] approach described in Chapter 6 is also related. For instance, as the size of the patch/filter specified by p decreases, the number of columns/rows in $\mathbf{Q}$ will decrease. If there is only column, the equation $\mathbf{q}\ \mathcal{T}_2(\mathbf{k}_{\text{central}}) = 0$ can be rewritten as $\mathbf{q_o}\ \mathcal{T}_2(\mathbf{k_o}) = -\mathbf{q_u}\ \mathcal{T}_2(\mathbf{k_u})$, obtained by partitioning the rows and columns. Here $\mathbf{k_o}$ and $\mathbf{k_u}$ are the k-space samples that can be observed and cannot be unobserved, respectively. These simplifying assumptions can translate to the GRAPPA setting, which is less general than the ESPIRIT setting.

10.5 Smooth manifold models

Smooth manifold models use nonlinear representations, which are more powerful in capturing the nonlinear relationships between signals compared to the linear counterparts. They assume the signals (images/patches/pixel profiles) to be living on a smooth image manifold (see Fig. 10.1). Methods relying on smooth manifold models include nonlocal means [12], nonlocal regularization [42], kernel methods [45,66,47,46,64], STORM [1,43,54,10,56], and recent extensions of STORM using deep generative models [71,72]. Most of the methods rely on modeling/smoothing the signals based on their proximity on the manifold rather than in the original domain. For instance, the patches that may be far apart in space might be similar and hence close on the manifold.

We note that standard Tikhonov regularization penalizes the gradient of the image, which makes use of the fact that the intensities of the adjacent pixels are similar. Most algorithms for Tikhonov regularization rely on the Laplacian of the image. For 1D images with n pixels, the Laplacian is often approximated by the $x \times n$ finite difference matrix

$$\mathbf{L} = \begin{bmatrix} -2 & 1 & 0 & \dots & 1 \\ 1 & -2 & 1 & \dots & 0 \\ 0 & 1 & -2 & 1 & \dots \\ 1 & 0 & \dots & 1 & -2 \end{bmatrix}. \tag{10.39}$$

This matrix is block diagonal and captures the neighbor structure of the pixels in the image, thus facilitating the smoothing of the image. The sum of the off diagonal entries of each row is equal to the negative of the diagonal entry. In manifold methods, the smoothing is enabled by a custom Laplacian matrix, which captures the neighborhood structure of the images in the manifold, and is estimated from the data. The entries of the Laplacian matrix are chosen based on the proximity of the signals on the manifold (see Fig. 10.7). This approach has strong ties to the kernel methods [7] used in machine learning. In particular, one performs the smoothing in a nonlinearly transformed feature space, defined by the nonlinear mapping function $\varphi(\mathbf{x})$. The neighborhood structure or Laplacian is determined based on the distances between the nonlinearly mapped features $d_{i,j} = \|\varphi(\mathbf{x}_i) - \varphi(\mathbf{x}_j)\|^2$, rather than the conventional distance measure $\|\mathbf{x}_i - \mathbf{x}_j\|^2$. The cost functions to solve for the images using these models only depend on inner products between the features $\langle\varphi(\mathbf{x}_i), \varphi(\mathbf{x}_j)\rangle$; the *kernel-trick* that is widely used in machine learning can be used to come up with computationally efficient algorithms that does not require the explicit computation of the nonlinear mapping φ that is expensive to compute.

10.5.1 Analysis manifold methods

The recovery of a smooth multidimensional function $\mathbf{f} : \mathbb{R}^m \to \mathbb{R}^n$ has been considered in machine learning [7]. For example, when one is considering patches, $n = p^2$ is the number of pixels in the patch. Because of the redundancy of the patches, they can be viewed as a function of low-dimensional latent vectors denoted by $\mathbf{r}_k$: $\mathbf{x}_k = f(\mathbf{r}_k)$. Here, $\mathbf{f}$ is a nonlinear function, and $\mathbf{r}_k$ are as the coordinates that are unknown. We note that as the latent coordinates $\mathbf{r}$ is varying in a m-dimensional space (domain), the function values vary on a smooth surface in high-dimensional space ($n \gg m$). The recovery of the function $\mathbf{f}$ from the fully sampled data points $\mathbf{x}_k$ is posed as the Tikhonov regularized problem [7]

$$\mathbf{f}^* = \arg \min_{f,\{\mathbf{r}_k\}} \|\mathbf{f}(\mathbf{r}_k) - \mathbf{s}_k\|^2 + \eta \int_{\mathcal{M}} \|\nabla \mathbf{f}\|^2 d\mathbf{r}, \tag{10.40}$$

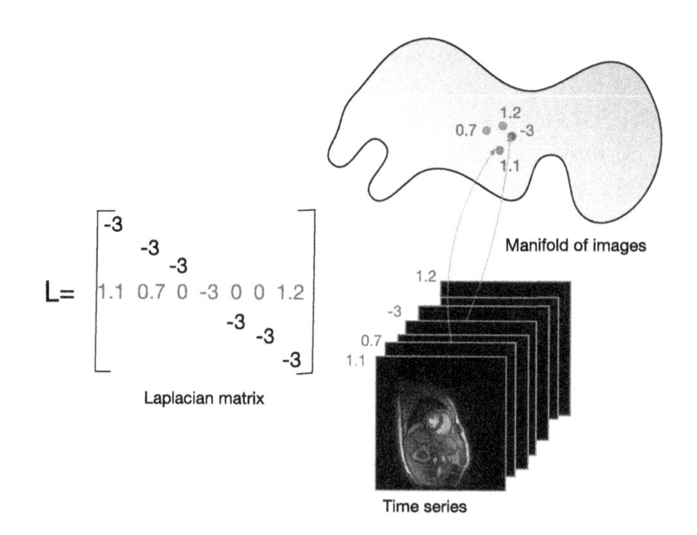

FIGURE 10.7

Illustration of the Laplacian matrix used in manifold methods. Each row of the Laplacian matrix may be thought of a second-order derivative operator in the manifold domain. In this example, the fourth row of the matrix correspond to the finite difference operator for the fourth image in the time series. Note that its neighbors on the manifold are not necessarily its temporal neighbors. The weights denote how close the images are to the fourth image in the time series. We illustrated an example with three neighbors, but the number of neighbors could be chosen arbitrarily and may vary from frame to frame.

where the second term is the smoothness of the function on $\mathcal{M}$. In the discrete setting, the regularization term is approximated as a weighted sum of differences between the points [7]

$$\mathbf{f}^* = \arg\min_{\mathbf{f},\{\mathbf{r}_j\}} \|\mathbf{f}(\mathbf{r}_k) - \mathbf{s}_k\|^2 + \eta \sum_{i=1}^{N}\sum_{i=1}^{N} \mathbf{W}_{i,j} \|\mathbf{f}(\mathbf{r}_i) - \mathbf{f}(\mathbf{r}_j)\|^2, \tag{10.41}$$

where the weights are selected based on the proximity of the points on the manifold. A simple choice of weights is specified by [7]

$$\mathbf{W}_{i,j} = \exp\left(-\frac{\|\mathbf{f}(\mathbf{r}_i) - \mathbf{f}(\mathbf{r}_j)\|^2}{\sigma^2}\right). \tag{10.42}$$

The weights capture the geometry of the manifold. Specifically, closer point pairs on $\mathcal{M}$ will have greater weights, while distant point pairs will have lesser weights. We will discuss more sophisticated approaches for the estimation of weights in Section 10.5.1.2. We note that the weighted sum can be expressed in a compact form as

$$\sum_{i=1}^{N}\sum_{i=1}^{N} \mathbf{W}_{i,j} \|\mathbf{f}(\mathbf{r}_i) - \mathbf{f}(\mathbf{r}_j)\|^2 = \text{trace}\left(\mathbf{X}\mathbf{L}\mathbf{X}^T\right).$$

Here, $\mathbf{X} = \begin{bmatrix}\mathbf{x}_1 & \ldots & \mathbf{x}_N\end{bmatrix} = \begin{bmatrix}\mathbf{f}(\mathbf{r}_1) & \ldots & \mathbf{f}(\mathbf{r}_N)\end{bmatrix}$ and $\mathbf{L}$ is the Laplacian matrix $\mathbf{L} = \mathbf{D} - \mathbf{W}$, which captures the structure of the manifold and $\mathbf{D}$ is a diagonal matrix $\mathbf{D} = \text{diag}(\sum_j \mathbf{W}_{i,j})$. See Fig. 10.7 for an illustration of the structure of this matrix. Thus, the optimization scheme in (10.41) can also be written as

$$\mathbf{X} = \arg\min_{\mathbf{X}} \|\mathbf{X} - \mathbf{S}\|^2 + \eta \text{ trace}\left(\mathbf{X}\mathbf{L}\mathbf{X}^{\mathrm{T}}\right) \tag{10.43}$$

and $\mathbf{S} = \begin{bmatrix}\mathbf{s}_1 & \ldots & \mathbf{s}_N\end{bmatrix}$. The discrete approximation of the manifold can be viewed as a graph, where the structure of the graph is captured by the graph Laplacian $\mathbf{L}$. Signal processing on graphs is extensively studied, and the Laplacian matrix is central to most of the methods [52].

10.5.1.1 Relationship to factor models and binning based approaches

One can perform the eigendecomposition of the known Laplacian matrix $\mathbf{L}$ as

$$\mathbf{L} = \mathbf{\Phi}\mathbf{\Lambda}\mathbf{\Phi}^T \tag{10.44}$$

It is well-known that the eigenfunctions (columns of $\mathbf{\Phi}$) are basis functions of functions on $\mathcal{M}$, analogous to Fourier exponentials being eigenfunctions of Laplacian operator in Euclidean space [52]. In particular, eigenfunctions corresponding to smaller eigen values of $\mathbf{L}$ correspond to smooth functions on $\mathcal{M}$; the eigenvalues are analogous to the frequency or roughness of the function. Since $\mathbf{\Phi}$ is an orthogonal basis analogous to Fourier transform, one can express the signal matrix as

$$\mathbf{X} = \mathbf{\Phi}\mathbf{W}^T, \tag{10.45}$$

where $\mathbf{W}$ can be viewed as the coefficients. Using this property, one can rewrite (10.43) as

$$\mathbf{W}^* = \arg\min_{\mathbf{W}} \|\mathbf{\Phi}\mathbf{W}^T - \mathbf{Z}\|^2 + \eta \underbrace{\text{trace}\left(\mathbf{W}\Lambda\mathbf{W}^T\right)}_{\sum_{i=1}^N \lambda_i \|\mathbf{w}_i\|^2} \tag{10.46}$$

Note that (10.45) is similar to the representation of the signal using dictionaries; each signal $\mathbf{k}$ is expressed as the linear combination of $\mathbf{u}_i$ with the weights specified by the kth row of $\mathbf{W}$. The weights are expected to be similar for points closer on the manifold; the active Φ basis vectors in each manifold neighborhood provides a local linear representation (similar to local PCA) on the manifold. Since the eigenvalues λ_i can be viewed as the *frequency* or the measure of roughness, one can attenuate the high-frequency components on the manifold by increasing η, thus obtaining smoother signals on the manifold. For computational efficiency, one may also truncate the representation by ignoring the basis functions corresponding to higher eigenvalues.

When the images can be grouped into r distinct clusters with minimal inter-group similarity, the Laplacian matrix can be thought of as a block diagonal matrix. In particular, the off-diagonal entries of the matrix corresponding to the images from two different clusters are zero. In this case, it is well-known that the matrix will have r zero eigenvalues. The eigenvectors φ_i corresponding to the zero eigenvalues will be the indicator vectors of the clusters. If the remaining eigenvalues are much higher, one can approximate (10.46) as the independent recovery of each cluster from the measured data. We note that the hard binning approach pursued in GRASP or XD-GRASP framework [18] bins the data to different clusters, followed by the recovery of the bins. Thus, the hard binning-based approaches may be viewed as a special case of the manifold method where the inter-cluster similarities can be ignored. When the nonzero eigenvalues are not ignored in the reconstruction, the additional eigenfunctions (corresponding to nonzero eigenvalues) capture the variability of the images within each cluster. The reconstruction can thus be viewed as a local PCA approach, where each cluster is represented independently by its basis set. As discussed before, the BM3D approach pursues a similar approach, where the patches in the image are clustered into different groups.

10.5.1.2 *Estimation of manifold Laplacian*

As discussed previously, $\mathbf{L}$ captures the manifold structure. The recovery heavily depends on the specific choice of the Laplacian matrix. Several methods were introduced to estimate the Laplacian from its noisy and possibly undersampled data.

Proximity based methods: Early methods directly estimated the weight matrix based on the proximity of the function values [7,12,3,54]. For instance, they are chosen as

$$\mathbf{W}_{i,j} \approx \exp\left(-\frac{\|\mathbf{s}_i - \mathbf{s}_j\|^2}{\sigma_s^2}\right). \tag{10.47}$$

We note that $\mathbf{W}$ is also termed as the kernel matrix. The BM3D approach [16,17] can be viewed as a hard-clustering setting

$$\mathbf{W}_{i,j} = \begin{cases} 1 & \text{if} \quad \|\mathbf{s}_i - \mathbf{s}_j\|^2 \leq \sigma_s^2 \\ 0 & \text{else} \end{cases}. \tag{10.48}$$

Note that the Gaussian choice in (10.47) will be equivalent to hard-clustering if the clusters are well-separated. In this case, the $\mathbf{L}$ matrix will have a block structure with no interactions between clusters. When the points are well-distributed in the manifold, the Gaussian choice will promote the interaction between the points, thus facilitating data-sharing between the corresponding images.

Alternating minimization schemes: In many cases (e.g., signals are patches in an image), the signals $\mathbf{x}_i$ are either noisy or jointly measured using a single-rank deficient linear operator $\mathbf{E}$. In this case, the estimation of the weights from aliased data using (10.47) often results in poor results. An approach to overcome this challenge is to pose the recovery as the minimization of the cost function [67,68]

$$\mathbf{X}^* = \arg\min_{\mathbf{X}} \|\mathbf{E}(\mathbf{X}) - \mathbf{S}\|^2 + \lambda \sum_{i=1}^{N} \sum_{i=1}^{N} \eta\left(\|\mathbf{x}_i - \mathbf{x}_j\|\right). \tag{10.49}$$

Here, $\eta(\cdot)$ is a nonconvex function of its argument (e.g., ℓ_p; $p < 1$ norm). The nonconvexity of the regularization term will encourage each i to be influenced by signals in the immediate proximity, while being minimally impacted by far away points. This criterion can be minimized by alternating between the estimation of the weights (10.47) and the recovery of the signals (10.41). Continuation schemes that start with a quadratic or convex η, and gradually change it to the desired nonconvex cost function during iterations have been introduced to encourage the convergence to the global minimum with improved results in compressed sensing applications.

Sparse optimization: The work in [46,64] proposes to estimate the Laplacian by assuming the weight matrix (and equivalently the Laplacian entries) to be sparse. Specifically, it aims to express each signal

i as a sparse linear combination of the other signals. The intuition is that each signal on the manifold can be expressed as a sparse linear combination of its neighbors

$$\mathbf{W}^*_{i,j} = \arg \min_{\mathbf{W}_{i,j}; \sum_j \mathbf{W}_{i,j}=1} \|\mathbf{s}_i - \sum_j \mathbf{W}_{i,j}\mathbf{s}_j\|^2 + \lambda \sum_j \|\mathbf{W}_{i,j}\|_{\ell_1}. \tag{10.50}$$

This approach is reported to yield improved results over proximity-based methods [7,12,3,54] in (10.47). This approach has similarities to local linear embedding [62], where each signal is expressed as a weighted linear combination of its neighbors.

Kernel based projection: Note that the approach in (10.46) approximates the signals using the eigen vectors corresponding to the lowest eigenvalues of the Laplacian matrix to approximate/denoise them. This is equivalent to approximating the signals using eigenvectors corresponding to the highest eigenvalues of the normalized kernel matrix. Kernel PCA is widely used in machine learning to approximate signals living on manifolds. The eigenvalues of kernel matrix is observed to decay rapidly when the signals are living on smooth manifolds or clustered [6], which is used to denoise the signals.

Early manifold approaches in MRI relied on explicit polynomial features [66,47]. Specifically, polynomial features of the signals were computed, followed by performing PCA in the feature space. Once the features are projected onto a lower dimensional subspace, these methods use the explicit inversion formula to obtain the corresponding signals in the original space; the corresponding signals are termed as pre-images of the projections derived by PCA. This approach demonstrated improved performance over PCA. A challenge for this direct approach is the difficulty applying it to large images, where the explicit lifting is not possible.

Kernel PCA regularization: The low-dimensional structure of the weight matrix is explicitly used for the joint estimation of the Laplacian matrix and the signals from undersampled measurements in [56]. We now show a simple example to illustrate that the kernel matrix can be exactly low-rank if the original points lie on simple constructs such as surfaces or manifolds. For example, the points on a unit circle in two dimensions live on the zero set of $x^2 + y^2 - 1$. Hence, one can consider a nonlinear lifting of the points

$$\begin{bmatrix} x \\ y \end{bmatrix} \rightarrow \begin{bmatrix} 1 \\ x^2 \\ y^2 \end{bmatrix} = \phi\left(\begin{bmatrix} x \\ y \end{bmatrix}\right). \tag{10.51}$$

Because the high-dimensional features of every point on the circle satisfies

$$\begin{bmatrix} -1 & 1 & 1 \end{bmatrix} \underbrace{\begin{bmatrix} 1 & 1 & \dots & 1 \\ x_1^2 & x_2^2 & \dots & x_N^2 \\ y_1^2 & y_2^2 & \dots & y_N^2 \end{bmatrix}}_{\Phi(\mathbf{X})} = 0, \tag{10.52}$$

we can conclude that $\Phi(\mathbf{X})$ is rank deficient by one. If the points lie on the intersection of two polynomials, one would have more null space vectors, further reducing the rank of $\Phi(\mathbf{X})$.

Generalizing this simple illustration, we can show that the kernel matrix $\mathbf{W}$ exactly low-rank when the manifold $\mathcal{M} \in \mathbb{R}^N <$ can be expressed as the zero-level set of a finite linear combination of basis functions $\psi(\mathbf{x}) = \sum_{k=1}^{B} c_k\, \varphi_k(\mathbf{x})$ as follows:

$$\mathcal{M} = \{x | \sum c_k\, \varphi_k(\mathbf{x}) = 0\}. \tag{10.53}$$

Here, $\varphi_k(\mathbf{r})$ are basis functions[1] (e.g., polynomials, exponentials) that span the high-dimensional space. In this case, the feature matrix

$$\Phi(\mathbf{X}) = \left[\underbrace{\begin{bmatrix} \varphi_1(\mathbf{x}_1) \\ \vdots \\ \varphi_S(\mathbf{x}_1) \end{bmatrix}}_{\phi(\mathbf{r}_1)} \cdots \underbrace{\begin{bmatrix} \varphi_1(\mathbf{x}_N) \\ \vdots \\ \varphi_S(\mathbf{x}_N) \end{bmatrix}}_{\phi(\mathbf{x}_N)} \right] \tag{10.54}$$

is low-rank.

The mapping from the original points to the feature vectors $\phi(x)$ can be viewed as a nonlinear lifting. The low-rank structure of this matrix implies that the lifted points lie in a low-dimensional subspace. The lifted points can hence be viewed as low-dimensional latent vectors that compactly represent the signals. The algorithms that use the low-rank property of the feature vectors may be viewed as structured low-rank algorithms with the nonlinear mapping ϕ. Following the approach in Section 10.4, we recover the signals on the manifold from its linear measurements as

$$\mathbf{X}^* = \arg\min_{\mathbf{X}} \|\mathbf{E}(\mathbf{X}) - \mathbf{S}\|^2 + \lambda \|\Phi(\mathbf{X})\|_*. \tag{10.55}$$

This problem cannot be solved in practical applications since the feature vectors $\phi(x)$ are high-dimensional. Hence, one can use the *kernel-trick* that allows the direct computation of the inner-product of the high-dimensional feature maps

$$\langle \phi(\mathbf{r}_i), \phi(\mathbf{r}_j) \rangle = \kappa(\mathbf{x}_i - \mathbf{x}_j) \tag{10.56}$$

as nonlinear functions of the image differences $\mathbf{x}_i - \mathbf{x}_j$ without the direct computation of the feature maps, which is computationally challenging. Here, κ is a function that is dependent on the specific feature maps. This approach is often referred to as the *kernel-trick*, which is widely used in machine learning to avoid explicit lifting to higher-dimensional space. We note that the IRLS approach described in Section 10.4.2 only depends on the inner products of the features, unlike many of the nuclear norm-minimization algorithms. This property can be made use of to solve (10.55), which alternates between

$$\mathbf{X}^* = \arg\min_{\mathbf{X}} \|\mathbf{E}(\mathbf{X}) - \mathbf{S}\|^2 + \lambda \|\Phi(\mathbf{X})\, \kappa(\mathbf{X})^{-1}\|_F^2 \tag{10.57}$$

and the evaluation of the kernel matrix $\kappa(\mathbf{X})$

$$(\kappa(\mathbf{X}))_{i,j} = \langle \phi(\mathbf{r}_i), \phi(\mathbf{r}_j) \rangle = \kappa(\mathbf{r}_i, \mathbf{r}_j). \tag{10.58}$$

[1] The basis functions may also be chosen as a subset of the Mercer decomposition of the kernel function.

This approach eliminates the need for the explicit evaluation of the high-dimensional feature maps $\phi(f_i)$. Since the regularization functional in (10.57) is nonquadratic, this term is solved using steepest descent; this approach is shown to be similar to the alternating strategy in (10.5.1.2) (i.e., alternation between (10.47) and the recovery of the signals (10.41)), where (10.47) is replaced by

$$\mathbf{L} = \mathcal{K}(\mathbf{X}) \odot \mathcal{K}(\mathbf{X})^{-\frac{1}{2}}, \tag{10.59}$$

where $\odot$ denotes point-wise multiplication of the matrices.

10.5.1.3 Image recovery assuming smooth patch manifold

Nonlocal means is an early and powerful algorithm for patch-based image denoising [12,3]. It estimates the Laplacian matrix from noisy data as in (10.47), followed by (10.41) to recover the denoised signals. Note that the evaluation of the $p^2 \times p^2$ patch matrix, as well as its use in (10.41), is computationally expensive. Several assumptions on the structure of the weight matrix (e.g., block diagonal assuming that the similar patches are in the spatial neighborhood) have been introduced to speed up the computations. The BM3D approach is also related, when the structure is determined by the Euclidean proximity of the patches.

Recently, some researchers have proposed to use the decomposition in (10.45) to further improve the denoising performance [69]. Each of the columns of $\mathbf{V}$ are termed as nonlocal basis functions, which shows the similarity between the regions. The corresponding $\mathbf{U}$ basis functions are termed as local basis functions. The spatial variation of the nonlocal basis functions (coefficients of the expansion) allows the subspace to be adapted to each patch, depending on the local neighborhood on the manifold.

When the recovery of image from undersampled Fourier measurements are considered, an alternating minimization scheme that minimizes (10.49) is adopted [68,67]. By making use of the redundancy between the patches, this approach is observed to offer good image quality during reconstruction. A similar strategy, where similarity between image patches in dynamic MRI reconstruction is used in PRICE [42], which offers implicit motion compensation. This approach has conceptual similarities to [13] and the kernel PCA approach in the patch setting [63]. Considering the improved performance offered by kernel PCA methods used in the dynamic MRI setting, we expect better performance with this scheme in the patch setting.

10.5.2 Application to dynamic MRI

The manifold structure of images in a dynamic time series is used to recover them from undersampled data [55,56,54,43]. In particular, the images in a free-breathing cardiac dataset can be viewed as nonlinear functions of the cardiac and respiratory phases; the images can be assumed as points on a low-dimensional surface.

A navigated strategy was used in [54]. In particular, each image is sampled by a sampling pattern that includes a common set of k-space locations, termed as navigators. Specifically, one would obtain $\mathbf{Z} = \mathbf{B}\mathbf{X}$, where $\mathbf{B}$ is the sampling operator corresponding to the k-space navigators. In this case, one can approximate the $\mathbf{W}$ matrix as

$$\mathbf{W}_{i,j} \approx \exp\left(-\frac{\|\mathbf{z}_i - \mathbf{z}_j\|^2}{\sigma_z^2}\right). \tag{10.60}$$

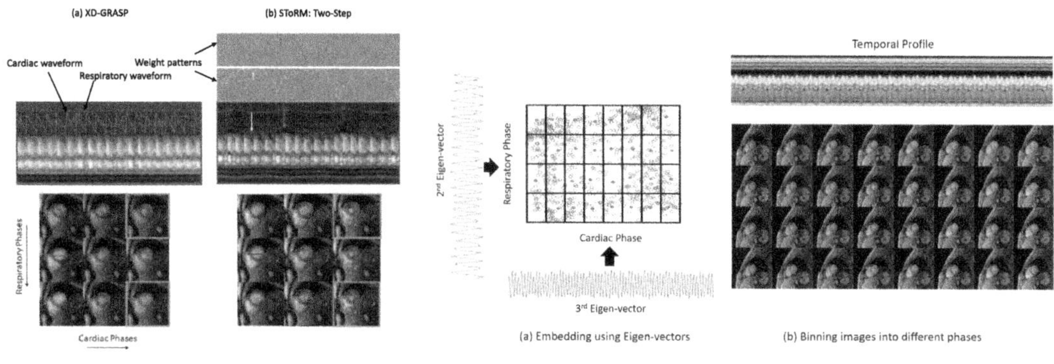

(a) Similarity to XD-GRASP (b) Cardiac and Motion states using Manifold embedding

FIGURE 10.8

Illustration of Smoothness Regularization on Manifolds (STORM) framework and its similarity to XD GRASP for the recovery of free-breathing and ungated MRI data. XD-GRASP bins the data to different cardiac/respiratory phases, followed by the joint recovery of the images. By contrast, STORM estimates a Laplacian matrix that has conceptual similarities to the XD-GRASP approach. The manifold Laplacian is estimated from the k-space navigators using (10.57) and (10.59). The reconstructed data is compared against self-gated XD-GRASP reconstruction of the same data. (**b**) uses the eigenvectors of the Laplacian matrix to bin the reconstructed data into cardiac and respiratory phases. This figure is copied from [56] with permission from IEEE.

Specifically, the navigators are expected to indicate the structure of the manifold.

A challenge with the direct implementation of the manifold-aware recovery in (10.43) in the multidimensional setting is its high computational complexity. The factorization approach is considered in (10.46) enables significant reduction in the computational complexity and memory demand of the algorithm [56]. In practice, 20–30 basis functions were observed to recover the dataset, which offers a 20–30 fold reduction in the memory demand. This approach also estimated the Laplacian matrix from the navigators using the kernel low-rank algorithm was used to estimate the Laplacian in [56], which offered improved performance over [54].

When navigators are not available, the formulation in (10.49) makes it possible to jointly estimate the Laplacian and the signals from the data itself [43]. A key benefit of this approach is that one can customize the Laplacian to different spatial regions. In particular, [43] split the images to patches, each with a different Laplacian. The kernel low-rank algorithm (10.55) is used instead of (10.49) to estimate the Laplacian in [1]. The iterative strategies including (10.49) and (10.55) are reported to yield far superior results compared to the proximity based methods. More studies are needed for the comparison of the iterative methods for the Laplacian matrix [1]. See Fig. 10.8 for an illustration of the manifold recovery. The results show that the generative STORM approach facilitates a significant reduction in acquisition time compared to the analysis counterparts.

10.6 Software

The MATLAB® software associated with this paper is available at https://github.com/sajanglingala/data_adaptive_recon_MRI.

10.7 Summary

This chapter reviewed several learning-based models that are used in MR image reconstruction, building upon compressed sensing methods in Chapter 6 and low-rank methods in Chapter 9. All the approaches reviewed in this chapter make use of the manifold structure of substructures (e.g., patches, pixel time series, images in the time series) of the dataset. The models differ in the representation of the data manifold, resulting in algorithms that rely on matrix factorization. In particular, the subregions are used to populate the columns of a structured matrix, which is factorized into two submatrices. Even though these approaches resemble low-rank methods reviewed extensively in Chapter 9, the main distinction is the nature of the priors used on the factors. The priors encourage the learning of basis functions and coefficients with specific properties, which often provide improved approximations of the data manifold. The matrix structure and basis functions promote the sharing of information between subregions of the dataset, thus facilitating the recovery of the dataset from highly undersampled data.

References

[1] Ahmed AH, Zhou R, Yang Y, Nagpal P, Salerno M, Jacob M. Free-breathing and ungated dynamic MRI using navigator-less spiral SToRM. IEEE Trans Med Imaging 2020:1. https://doi.org/10.1109/TMI.2020.3008329.

[2] Akçakaya M, Basha TA, Goddu B, Goepfert LA, Kissinger KV, Tarokh V, et al. Low-dimensional-structure self-learning and thresholding: regularization beyond compressed sensing for MRI Reconstruction. Magn Reson Med 2011;66. https://doi.org/10.1002/mrm.22841.

[3] Awate SP, Whitaker RT. Unsupervised, information-theoretic, adaptive image filtering for image restoration. IEEE Trans Pattern Anal Mach Intell 2006;28:364–76. https://doi.org/10.1109/TPAMI.2006.64. Available from: http://ieeexplore.ieee.org/document/1580482/.

[4] Balachandrasekaran A, Magnotta V, Jacob M. Recovery of damped exponentials using structured low rank matrix completion. IEEE Trans Med Imaging 2017. https://doi.org/10.1109/TMI.2017.2726995.

[5] Balachandrasekaran A, Mani M, Jacob M. Calibration-free B0 correction of EPI data using structured low rank matrix recovery. IEEE Trans Med Imaging 2019;38. https://doi.org/10.1109/TMI.2018.2876423.

[6] Belkin M, Niyogi P. Laplacian eigenmaps for dimensionality reduction and data representation. Neural Comput 2003. https://doi.org/10.1162/089976603321780317.

[7] Belkin M, Niyogi P, Sindhwani V. Manifold regularization: a geometric framework for learning from labeled and unlabeled examples. J Mach Learn Res 2006;7:2399–434. https://doi.org/10.1016/j.neuropsychologia.2009.02.028.

[8] Bhave S, Lingala SG, Johnson CP, Magnotta VA, Jacob M. Accelerated whole-brain multi-parameter mapping using blind compressed sensing. Magn Reson Med 2016. https://doi.org/10.1002/mrm.25722.

[9] Bhave S, Lingala SG, Newell JD, Nagle SK, Jacob M. Blind compressed sensing enables 3-dimensional dynamic free breathing magnetic resonance imaging of lung volumes and diaphragm motion. Invest Radiol 2016;51:387–99. https://doi.org/10.1097/RLI.0000000000000253.

[10] Biswas S, Aggarwal HK, Jacob M. Dynamic MRI using model-based deep learning and SToRM priors: MoDL-SToRM. Magn Reson Med 2019. https://doi.org/10.1002/mrm.27706.
[11] Brinegar C, Schmitter SS, Mistry NN, Johnson GA, Liang ZP. Improving temporal resolution of pulmonary perfusion imaging in rats using the partially separable functions model. Magn Reson Med 2010;64. https://doi.org/10.1002/mrm.22500.
[12] Buades A, Coll B, Morel JM. A non-local algorithm for image denoising. In: Proceedings - 2005 IEEE computer society conference on computer vision and pattern recognition, CVPR 2005. IEEE Computer Society; 2005. p. 60–5.
[13] Bustin A, Lima da Cruz G, Jaubert O, Lopez K, Botnar RM, Prieto C. High-dimensionality undersampled patch-based reconstruction (HD-PROST) for accelerated multi-contrast MRI. Magn Reson Med 2019;81(6). https://doi.org/10.1002/mrm.27694.
[14] Candès EJ, Fernandez-Granda C. Towards a mathematical theory of super-resolution. Commun Pure Appl Math 2014;67:906–56.
[15] Candes EJ, Wakin MB. An introduction to compressive sampling: a sensing/sampling paradigm that goes against the common knowledge in data acquisition. IEEE Signal Process Mag 2008;25:21–30. https://doi.org/10.1109/MSP.2007.914731.
[16] Dabov K, Foi A, Katkovnik V, Egiazarian K. Image denoising by sparse 3-D transform-domain collaborative filtering. IEEE Trans Image Process 2007;16:2080–95. https://doi.org/10.1109/TIP.2007.901238.
[17] Danielyan A, Katkovnik V, Egiazarian K. BM3D frames and variational image deblurring. IEEE Trans Image Process 2012. https://doi.org/10.1109/TIP.2011.2176954.
[18] Feng L, Axel L, Chandarana H, Block KT, Sodickson DK, Otazo R. Xd-grasp: golden-angle radial mri with reconstruction of extra motion-state dimensions using compressed sensing. Magn Reson Med 2016;75:775–88.
[19] Goud S, Hu Y, Jacob M. Real-time cardiac MRI using low-rank and sparsity penalties. In: 2010 7th IEEE international symposium on biomedical imaging: from nano to macro, ISBI 2010 - proceedings; 2010.
[20] Griswold MA, Jakob PM, Heidemann RM, Nittka M, Jellus V, Wang J, et al. Generalized autocalibrating partially parallel acquisitions (GRAPPA). Magn Reson Med 2002;47:1202–10.
[21] Haldar JP. Low-rank modeling of local -space neighborhoods (LORAKS) for constrained MRI. IEEE Trans Med Imaging 2014;33:668–81. https://doi.org/10.1109/TMI.2013.2293974.
[22] Haldar JP. Low-rank modeling of local k-space neighborhoods (LORAKS) for constrained MRI. IEEE Trans Med Imaging 2014. https://doi.org/10.1109/TMI.2013.2293974.
[23] Haldar JP, Zhuo J. P-LORAKS: low-rank modeling of local k-space neighborhoods with parallel imaging data. Magn Reson Med 2016. https://doi.org/10.1002/mrm.25717.
[24] Jacob M, Mani MP, Ye JC. Structured low-rank algorithms: theory, magnetic resonance applications, and links to machine learning. IEEE Signal Process Mag 2020;37:54–68. https://doi.org/10.1109/MSP.2019.2950432.
[25] Jin KH, Lee D, Ye JC. A novel k-space annihilating filter method for unification between compressed sensing and parallel MRI. In: Proceedings - international symposium on biomedical imaging; 2015.
[26] Jin KH, Lee D, Ye JC. A general framework for compressed sensing and parallel MRI using annihilating filter based low-rank Hankel matrix. IEEE Trans Comput Imaging 2016;2. https://doi.org/10.1109/tci.2016.2601296.
[27] Kay SM, Marple SL. Spectrum analysis—a modern perspective. Proc IEEE 1981. https://doi.org/10.1109/PROC.1981.12184.
[28] Lai P, Lustig M, Vasanawala SS, Brau AC. ESPIRiT (efficient eigenvector-based L1SPIRiT) for compressed sensing parallel imaging - theoretical interpretation and improved robustness for overlapped FOV prescription. Electr Eng 2011;19.
[29] Lee D, Jin KH, Kim EY, Park SH, Ye JC. Acceleration of MR parameter mapping using annihilating filter-based low rank Hankel matrix (ALOHA). Magn Reson Med 2016. https://doi.org/10.1002/mrm.26081.

[30] Liang ZP. Spatiotemporal imaging with partially separable functions. In: 2007 4th IEEE international symposium on biomedical imaging: from nano to macro - proceedings; 2007. p. 988–91.

[31] Liang ZP, Lauterbur PC. A generalized series approach to MR spectroscopic imaging. IEEE Trans Med Imaging 1991;10. https://doi.org/10.1109/42.79470.

[32] Lingala SG, Dibella E, Adluru G, McGann C, Jacob M. Accelerating free breathing myocardial perfusion MRI using multi coil radial k - T SLR. Phys Med Biol 2013. https://doi.org/10.1088/0031-9155/58/20/7309.

[33] Lingala SG, Hu Y, Dibella E, Jacob M. Accelerated dynamic MRI exploiting sparsity and low-rank structure: K-t SLR. IEEE Trans Med Imaging 2011;30:1042–54. https://doi.org/10.1109/TMI.2010.2100850. Available from: https://pubmed.ncbi.nlm.nih.gov/21292593/.

[34] Lingala SG, Jacob M. Blind compressed sensing with sparse dictionaries for accelerated dynamic MRI. In: Proceedings - international symposium on biomedical imaging; 2013.

[35] Lingala SG, Jacob M. Blind compressive sensing dynamic MRI. IEEE Trans Med Imaging 2013;32:1132–45. https://doi.org/10.1109/TMI.2013.2255133.

[36] Lobos RA, Kim TH, Hoge WS, Haldar JP. Navigator-free EPI ghost correction using low-rank matrix modeling: theoretical insights and practical improvements. In: Proceedings of the 25th annual meeting of ISMRM; 2017.

[37] Mani M, Aggarwal HK, Magnotta V, Jacob M. Improved MUSSELS reconstruction for high-resolution multishot diffusion weighted imaging. Magn Reson Med 2020;83. https://doi.org/10.1002/mrm.28090.

[38] Mani M, Jacob M. Fast iterative algorithm for the reconstruction of multishot non-Cartesian diffusion data. Magn Reson Med 2014. Available from: http://onlinelibrary.wiley.com/doi/10.1002/mrm.25486/full.

[39] Mani M, Jacob M, Kelley D, Magnotta V. Multi-shot sensitivity-encoded diffusion data recovery using structured low-rank matrix completion (MUSSELS). Magn Reson Med 2017;78. https://doi.org/10.1002/mrm.26382.

[40] Mani M, Magnotta V, Jacob M. A general algorithm for compensation of trajectory errors: application to radial imaging; 2018. Available from: http://doi.wiley.com/10.1002/mrm.27148.

[41] Mani M, Magnotta V, Kelley D, Jacob M. Comprehensive reconstruction of multi-shot multi-channel diffusion data using mussels. In: Proceedings of the annual international conference of the IEEE engineering in medicine and biology society, EMBS; 2016.

[42] Mohsin YQ, Lingala SG, DiBella E, Jacob M. Accelerated dynamic MRI using patch regularization for implicit motion compensation. Magn Reson Med 2017;77:1238–48. https://doi.org/10.1002/mrm.26215. Available from: http://www.ncbi.nlm.nih.gov/pubmed/27091812. http://www.pubmedcentral.nih.gov/articlerender.fcgi?artid=PMC5300957. http://doi.wiley.com/10.1002/mrm.26215.

[43] Mohsin YQ, Poddar S, Jacob M. Free-breathing ungated cardiac MRI using iterative SToRM (i-SToRM). IEEE Trans Med Imaging 2019;38. https://doi.org/10.1109/TMI.2019.2908140.

[44] Morrison RL, Jacob M, Do MN. Multichannel estimation of coil sensitivities in parallel MRI. In: 2007 4th IEEE international symposium on biomedical imaging: from nano to macro - proceedings; 2007.

[45] Nakarmi U, Slavakis K, Lyu J, Ying L. M-MRI: a manifold-based framework to highly accelerated dynamic magnetic resonance imaging. In: Proceedings - international symposium on biomedical imaging; 2017.

[46] Nakarmi U, Slavakis K, Ying L. MLS: joint manifold-learning and sparsity-aware framework for highly accelerated dynamic magnetic resonance imaging. In: Proceedings - international symposium on biomedical imaging; 2018.

[47] Nakarmi U, Wang Y, Lyu J, Liang D, Ying L. A kernel-based low-rank (KLR) model for low-dimensional manifold recovery in highly accelerated dynamic MRI. IEEE Trans Med Imaging 2017. https://doi.org/10.1109/TMI.2017.2723871.

[48] Ongie G, Biswas S, Jacob M. Convex recovery of continuous domain piecewise constant images from nonuniform Fourier samples. IEEE Trans Signal Process 2018;66:236–50. https://doi.org/10.1109/TSP.2017.2750111.

[49] Ongie G, Jacob M. Super-resolution MRI using finite rate of innovation curves. In: Proceedings - international symposium on biomedical imaging. IEEE Computer Society; 2015. p. 1248–51.

[50] Ongie G, Jacob M. Off-the-grid recovery of piecewise constant images from few Fourier samples. SIAM J Imaging Sci 2016;9:1004–41. https://doi.org/10.1137/15M1042280.
[51] Ongie G, Jacob M. A fast algorithm for convolutional structured low-rank matrix recovery. IEEE Trans Comput Imaging 2017;3:535–50. https://doi.org/10.1109/tci.2017.2721819.
[52] Ortega A, Frossard P, Kovacevic J, Moura JM, Vandergheynst P. Graph signal processing: overview, challenges, and applications. Proc IEEE 2018. https://doi.org/10.1109/JPROC.2018.2820126.
[53] Peyré G. Manifold models for signals and images. Comput Vis Image Underst 2009. https://doi.org/10.1016/j.cviu.2008.09.003.
[54] Poddar S, Jacob M. Dynamic MRI using SmooThness Regularization on Manifolds (SToRM). IEEE Trans Med Imaging 2016. https://doi.org/10.1109/TMI.2015.2509245.
[55] Poddar S, Mohsin Y, Ansah D, Thattaliyath B, Ashwath R, Jacob M. Free-breathing cardiac MRI using bandlimited manifold modelling. IEEE Trans Comput Imaging 2015;35:1106–15.
[56] Poddar S, Mohsin YQ, Ansah D, Thattaliyath B, Ashwath R, Jacob M. Manifold recovery using kernel low-rank regularization: application to dynamic imaging. IEEE Trans Comput Imaging 2019;3:478–91. https://doi.org/10.1109/tci.2019.2893598. Available from: https://ieeexplore.ieee.org/document/8625515/.
[57] Ravishankar S, Bresler Y. MR image reconstruction from highly undersampled k-space data by dictionary learning. IEEE Trans Med Imaging 2011;30:1028–41. https://doi.org/10.1109/TMI.2010.2090538.
[58] Ravishankar S, Bresler Y. Learning sparsifying transforms. IEEE Trans Signal Process 2013;61:1072–86. https://doi.org/10.1109/TSP.2012.2226449.
[59] Ravishankar S, Bresler Y. Efficient blind compressed sensing using sparsifying transforms with convergence guarantees and application to magnetic resonance imaging. SIAM J Imaging Sci 2015;8:2519–57. https://doi.org/10.1137/141002293.
[60] Ravishankar S, Ye JC, Fessler JA. Image reconstruction: from sparsity to data-adaptive methods and machine learning. Proc IEEE 2020;108. https://doi.org/10.1109/JPROC.2019.2936204.
[61] Recht B, Fazel M, Parrilo PA. Guaranteed minimum-rank solutions of linear matrix equations via nuclear norm minimization. SIAM Rev 2010;52:471–501.
[62] Roweis ST, Saul LK. Nonlinear dimensionality reduction by locally linear embedding. Science 2000. https://doi.org/10.1126/science.290.5500.2323.
[63] Schmidt JF, Santelli C, Kozerke S. Mr image reconstruction using block matching and adaptive kernel methods. PLoS ONE 2016;11:e0153736.
[64] Shetty GN, Slavakis K, Bose A, Nakarmi U, Scutari G, Ying L. Bi-linear modeling of data manifolds for dynamic-MRI recovery. IEEE Trans Med Imaging 2020. https://doi.org/10.1109/TMI.2019.2934125.
[65] Shin PJ, Larson PE, Ohliger MA, Elad M, Pauly JM, Vigneron DB, et al. Calibrationless parallel imaging reconstruction based on structured low-rank matrix completion. Magn Reson Med 2014. https://doi.org/10.1002/mrm.24997.
[66] Wang Y, Ying L. Undersampled dynamic magnetic resonance imaging using kernel principal component analysis. In: 2014 36th annual international conference of the IEEE engineering in medicine and biology society, EMBC 2014; 2014.
[67] Yang Z, Jacob M. Robust non-local regularization framework for motion compensated dynamic imaging without explicit motion estimation. In: Proceedings - international symposium on biomedical imaging. NIH Public Access; 2012. p. 1056–9. Available from: /pmc/articles/PMC3956771/?report=abstract, https://www.ncbi.nlm.nih.gov/pmc/articles/PMC3956771/.
[68] Yang Z, Jacob M. Nonlocal regularization of inverse problems: a unified variational framework. IEEE Trans Image Process 2013;22:3192–203. https://doi.org/10.1109/TIP.2012.2216278.
[69] Yin R, Gao T, Lu YM, Daubechies I. A tale of two bases: local-nonlocal regularization on image patches with convolution framelets. SIAM J Imaging Sci 2017. https://doi.org/10.1137/16M1091447.
[70] Zhao B, Haldar JP, Liang ZP. PSF model-based reconstruction with sparsity constraint algorithm and application to real-time cardiac MRI. In: 2010 annual international conference of the IEEE engineering in

medicine and biology society, EMBC'10. NIH Public Access; 2010. p. 3390–3. Available from: /pmc/articles/PMC3121182/?report=abstract, https://www.ncbi.nlm.nih.gov/pmc/articles/PMC3121182/.

[71] Zou Q, Ahmed AH, Nagpal P, Kruger S, Jacob M. Dynamic imaging using a deep generative SToRM (Gen-SToRM) model. IEEE Trans Med Imaging 2021 Nov;40(11):3102–12. https://doi.org/10.1109/TMI.2021.3065948. Epub 2021 Oct 27. PMID: 33720831; PMCID: PMC8590205.

[72] Zou Q, Ahmed AH, Nagpal P, Priya S, Schulte R, Jacob M. Generative storm: a novel approach for joint alignment and recovery of multi-slice dynamic mri. Available from: https://arxiv.org/abs/2101.08196.

CHAPTER

Machine Learning for MRI Reconstruction

11

Kerstin Hammernik[a,d], **Thomas Küstner**[b,d], **and Daniel Rueckert**[a,c]
[a] *Technical University Munich, AI in Healthcare and Medicine, Klinikum Rechts der Isar, Munich, Germany*
[b] *University Hospital Tübingen, Medical Image and Data Analysis, Department of Radiology, Tübingen, Germany*
[c] *Imperial College London, Department of Computing, London, United Kingdom*

11.1 Introduction

The development of Artificial Intelligence (AI) and Machine Learning (ML) has evolved tremendously during the last decade. In the field of medicine, including in the field of medical imaging, AI and ML have emerged as techniques that are likely to transform clinical practice fundamentally in the coming years. In particular in medical imaging, AI and ML have the potential to impact all stages of the imaging pipeline, from (1) image acquisition and reconstruction to (2) image analysis and interpretation and (3) diagnosis and prognosis. The terms AI, ML, and more recently Deep Learning (DL) are used as synonymous, however, there is a substantial difference between these terms.

Artificial intelligence

In 1956, the term *Artificial Intelligence (AI)* was formulated by John McCarthy as "the science and engineering of making intelligent machines" at the 1956 Dartmouth Conference [1]. Since then, AI has served as an umbrella term to solve any task that usually requires human intelligence. Inspired by the biological neuron, Frank Rosenblatt introduced the perceptron [2], a single artificial neuron.

Machine learning

ML is a subbranch of AI that describes any *data-driven* learning model to predict the output of a particular task and summarizes (statistical) methods that enable machines to improve performance with experience. After the AI winter, ML started to flourish in the 1980s with developments in Boltzmann machines [3], recurrent neural networks [4–7], multilayer perceptron, and back-propagation [8]. The neocognitron [9] reported in 1980 can be seen as the origin of Convolutional Neural Networks (CNNs) [10]. All these publications have formed the basis for later developments in the field of DL.

The basic working principle of ML is summarized in Fig. 11.1. The ML algorithm receives as input *features* or a reduced/combined subset of features that are extracted from input data. These input data include acquired sensor signals (e.g., raw MR *k*-space), (reconstructed) images, clinical and anthropometric meta-parameters, among many others. Feature extraction requires hand-crafting and manual

[d] Equal contribution.

Advances in Magnetic Resonance Technology and Applications, Volume 7, ISSN 2666-9099. https://doi.org/10.1016/B978-0-12-822726-8.00021-X

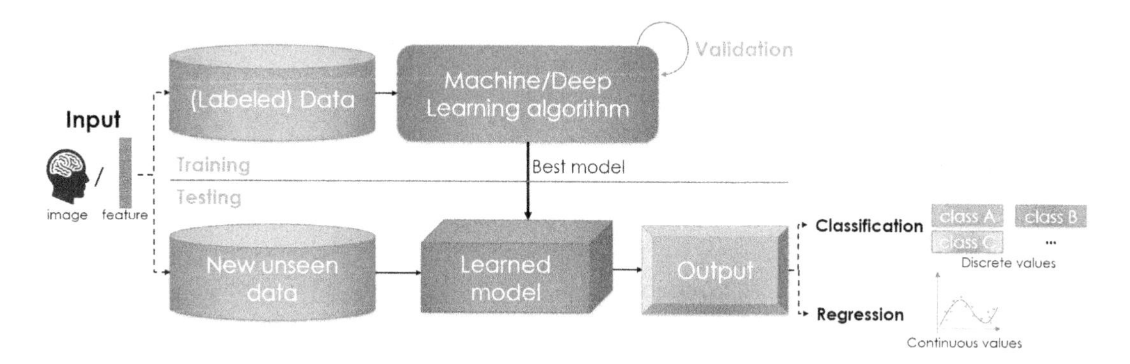

FIGURE 11.1 Processing Steps in Machine Learning (ML)

Processing steps of ML algorithms with corresponding images or feature inputs and task-specific outputs of classification or regression. Algorithms are learned on labeled (supervised) or unlabeled (unsupervised) data in a training step. Validation with independent datasets make possible tracking and optimization of learning. Testing is performed on new unseen data to produce the best trained model.

finetuning of features according to the underlying task and application. Hence, the quality of the predicted output depends not only on the ML algorithm itself but also on the extracted features.

Deep learning

DL [11] is a subbranch of ML that enables automated feature extraction and comprise deep NNs consisting of multiple layers. The basics of these methods were established in the late 1980s but only gained widespread attention with the remarkable performance boost of AlexNet [12] in the ImageNet Challenge 2012 for object recognition. The increasing availability and performance of graphics processing units (GPUs), as well as annotated large-scale databases, helped in the development of novel neural-network building blocks (see Section 11.5) and deep architectures (see Section 11.6).

11.2 Organization of this chapter

In this chapter we focus on the application of ML in the context of Magnetic Resonance Imaging (MRI) reconstruction. ML enables to overcome common challenges in regularized undersampled reconstruction approaches, such as Compressed Sensing (CS) approaches (Chapter 8) that require long reconstruction times and the selection of a hand-crafted sparsifying transform by using data-driven features. To introduce the important elements of building a NN for MR reconstruction, we introduce a general notation to deal with both the complex-valued nature of MR signals, real-valued magnitude images, and the dimensionality of input data in Appendix 11.A. In Section 11.3, we introduce various types of learning and provide insights into training, validating, and testing NNs. To study the application potential of ML in MRI reconstruction, we introduce various tasks in Section 11.4. Basic layers of a NN are presented in Section 11.5 that can be bundled to larger networks outlined in Section 11.6. Finally, we use the introduced concepts to build own reconstruction networks in the Tensorflow[1]/Keras[2] framework. Practical information regarding the implementation in the various frameworks are highlighted in gray boxes throughout this chapter.

11.3 Machine learning definitions

In the following, we introduce important ML definitions. Here, we build on the notation for both real-valued and complex-valued data, introduced in Appendix 11.A.

11.3.1 Learning models

The learning model represents any (non)linear and parametric model

$$\mathbf{x}_{\mathrm{NN}} = f_{\boldsymbol{\theta}}(\mathbf{x}) \tag{11.1}$$

that maps the inputs $\mathbf{x}$ to the neural network outputs $\mathbf{x}_{\mathrm{NN}}$. The mapping function $f_{\boldsymbol{\theta}}$ represents a NN whose parameters $\boldsymbol{\theta}$ can be optimized under some given cost function $\mathcal{L}$ (see Section 11.3.3). During a training process, the model sees different data samples such that the parameters $\boldsymbol{\theta}$ of the NN can be learned to predict a reasonable output $\mathbf{x}_{\mathrm{NN}}$ for any new input $\mathbf{x}$. During testing, the model is fixed and the trained parameters generate the output from new unseen test data. The models perform a sequence of operations on the inputs to yield a task-specific output. The most commonly used learning models and their basic building blocks are summarized in Section 11.5. The reader is referred to textbooks [13–15] for more details on learning models.

11.3.2 Types of learning

Types of learning can be differentiated based on the availability of label information and the type of label integration during training. In principle this relates to how closely a human observer is involved during training.

In *supervised learning*, data samples along with their task-specific labels exist in the database. However, labeling is very time- and cost-intensive and often requires human interaction, ranging from data sorting to curating to annotating structures within the image. For example, in undersampled MR image reconstruction, a labeled ground-truth corresponds to a fully sampled dataset or a dataset acquired with mild undersampling.

In *semi-supervised* learning, both labeled and unlabeled data are included in the learning process. The unlabeled data provides additional information about the underlying data distribution.

Self-supervised learning circumvents the problem of external labels. The input data itself is used to guide the learning. In a similar sense, in *reinforcement learning* the model receives feedback, i.e., rewards or penalties, based on its current prediction, that drives the training procedure.

---

[1] https://www.tensorflow.org/.

[2] https://keras.io/.

Active learning integrates an oracle into the training procedure, which is periodically queried to either label or select the next most meaningful samples for training. The oracle is in most cases a human observer, but can also be another algorithm [16].

In *unsupervised learning*, no labeled data or any type of a-priori information about the data is available that could be leveraged to guide the training. The network learns to identify patterns in the data. Common approaches are clustering [13] (e.g., k-means or Gaussian mixture models), Principal Component Analysis (PCA), (variational) autoencoders [17,18], Deep belief networks [19] or Generative Adversarial Networks (GANs) [20].

Transfer learning investigates the possibility to transfer knowledge between models or tasks. It can involve sharing information from simpler to more complex tasks, or from a source domain to another (but similar) target domain [21–23].

Federated learning [24–26] trains a model across multiple decentralized devices, where each device holds its own set of training data and only the model weights are shared across devices. This allows for training across highly heterogeneous datasets. Furthermore, federated learning has a vast potential in medicine, where sharing of data across multiple centers is challenging due to data protection and data privacy [27].

11.3.3 Cost function, optimization and backpropagation

Cost function

The *cost function* (*error*, *loss*) $\mathcal{L}$ is a quantitative measure that describes how well the neural network $f_{\boldsymbol{\theta}}(\mathbf{x})$ predicts the output $\mathbf{x}_{\mathrm{NN}}$ from the input $\mathbf{x}$. In supervised learning, the aim is to minimize the loss between prediction and label/ground-truth. In general, the ML optimization task can be formulated as

$$\min_{\boldsymbol{\theta}} \mathcal{L}(\boldsymbol{\theta}) = \min_{\boldsymbol{\theta}} \sum_j \lambda_j \mathcal{L}_j(\boldsymbol{\theta}), \tag{11.2}$$

which is composed of several λ_j weighted sub-losses $\mathcal{L}_j(\boldsymbol{\theta})$. This can include, e.g., regularization of weights or penalties on intermediate model outputs. Note that the result of a loss function is a real-valued scalar, even if the input data $\mathbf{x}$ and model output $\mathbf{x}_{\mathrm{NN}}$ are in the complex domain as in MRI reconstruction.

Depending on the chosen architecture (see Section 11.6) and defined task (see Section 11.4), the appropriate losses are selected. Common cost functions in linear-regression problems are the mean-squared-error (MSE, L_2 distance)

$$\mathcal{L}_{MSE}(\boldsymbol{\theta}) = \frac{1}{2N_s} \sum_{i=1}^{N_s} \| f_{\boldsymbol{\theta}}(\mathbf{x}_i) - \mathbf{x}_{\mathrm{GT},i} \|_2^2 \tag{11.3}$$

or the mean-absolute error (MAE, L_1 distance)

$$\mathcal{L}_{MAE}(\boldsymbol{\theta}) = \frac{1}{N_s} \sum_{i=1}^{N_s} \| f_{\boldsymbol{\theta}}(\mathbf{x}_i) - \mathbf{x}_{\mathrm{GT},i} \|_1, \tag{11.4}$$

to minimize the error between predicted output $f_{\boldsymbol{\theta}}(\mathbf{x})$ and ground-truth $\mathbf{x}_{\mathrm{GT}}$ for all N_s data samples. Differences between commonly used loss functions, including MSE, MAE, or Structural Similarity

Index (SSIM) [28], can be found in [29]. Practically, any piece-wise differentiable function can be used as a loss function.

Perceptual or style-transfer losses are often derived from feature maps of pre-trained networks and try to mimic the human visual system by capturing differences between high-level image feature representations [30]. Adversarial losses (minimax loss [20] or Wasserstein loss [31]) of GANs try to replicate the probability distribution of the input data and incorporate this information into training. Losses can also directly operate on the latent space, i.e., space in which feature maps lie, such as Kullback–Leibler divergence [18] to model feature distributions in deeper layers.

Box 11.1 Implementation of losses in ML frameworks

Common loss functions are provided in `tf.keras.losses`. A loss function needs to be defined in order to compile a model. Additional losses on, e.g., intermediate outputs of the model, can be defined in the model using the `add_loss()` method in the `call` method of a layer or sub-classed model. Weight regularization applies penalty on the network parameters. Weight regularization is defined in `tf.keras.regularizers` and can be added when creating a `tf.keras.layer`. The Keras API collects the defined losses in the model.

Optimization

Once a cost function is defined, the model parameters $\boldsymbol{\theta}$ need to be found that minimize the cost function $\mathcal{L}$. Gradient descent algorithms are commonly used to perform this optimization. While the negative gradient of the loss function $-\frac{\partial \mathcal{L}}{\partial \boldsymbol{\theta}}$ points towards the local minimum, optimal parameters $\boldsymbol{\theta}$ can be found by iteratively going along this direction with a pre-defined step size τ, known as learning rate. A simple gradient descent algorithm iteratively solves for

$$\boldsymbol{\theta}^{[it+1]} = \boldsymbol{\theta}^{[it]} - \tau \frac{\partial \mathcal{L}}{\partial \boldsymbol{\theta}}. \tag{11.5}$$

Key challenges for gradient-based optimizations are the proper selection of the learning rate τ to reach the (global) minimum and to avoid getting trapped in numerous suboptimal local minima for the commonly nonconvex optimization problem. Many other *optimizers* exist to toggle these challenges in various ways. They differ in the way how much data is incorporated to update the parameters if gradient momentum is included and/or if the learning rate is adapted, known as learning-rate scheduling [32].

The most common optimizers in ML are stochastic gradient descent (SGD) [33], Adaptive moment estimation (ADAM) [34], RMSProp [35], Nesterov accelerated gradient (NAG) [36], Adagrad [37], Adadelta [38], AdaMax [34], or Nesterov-accelerated Adaptive Moment Estimation (NADAM) [39]. An overview and comparison of these optimizers can be found in [40].

Box 11.2 Implementation of optimizers in ML frameworks

Common optimizers are provided in `tf.keras.optimizers`. An optimizer needs to be defined in order to compile a model. A learning-rate scheduler (see `tf.keras.optimizers.schedules`) can be added to the optimizer call.

Back-propagation

In general, the network $f_{\boldsymbol{\theta}}$ consists of L connected layers

$$\mathbf{x}_{\mathrm{NN}} = f_{\boldsymbol{\theta}}(\mathbf{x}) = f_L \circ f_{L-1} \circ \cdots \circ f_1(\mathbf{x}) = \mathbf{x}_L. \tag{11.6}$$

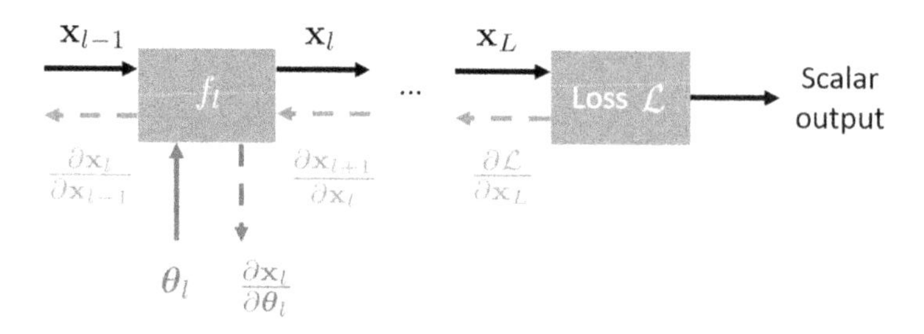

FIGURE 11.2 Back-propagation

The network consists of L connected layers with $f_l = \mathbf{x}_l$ being the output of the l-th layer, $1 \leq l \leq L$. The output prediction $\mathbf{x}_{\text{NN}} = \mathbf{x}_L$ is passed to a loss function $\mathcal{L}$. This gives an error that is back propagated using the chain rule of differentiation to update the network parameters $\boldsymbol{\theta}_l$.

The output of the l-th layer is denoted as $\mathbf{x}_l = f_l(\cdot)$, $1 \leq l \leq L$. In training, the parameters $\boldsymbol{\theta}$ of the network $f_{\boldsymbol{\theta}}$ are learned to minimize the loss function Eq. (11.2). This involves two steps as depicted in Fig. 11.2. First a feed-forward prediction from the input $\mathbf{x}$ to the output $\mathbf{x}_{\text{NN}} = f_{\boldsymbol{\theta}}(\mathbf{x})$ is required to evaluate the loss $\mathcal{L}$ for the current set of parameters $\boldsymbol{\theta}$. Second, the error between the predicted output $f_{\boldsymbol{\theta}}(\mathbf{x})$ and the ground-truth labels $\mathbf{x}_{\text{GT}}$ is back propagated to adapt the parameters $\boldsymbol{\theta}$. Back propagation [41] can be realized by applying the chain-rule of differentiation

$$\frac{\partial \mathcal{L}}{\partial \boldsymbol{\theta}} = \frac{\partial \mathcal{L}}{\partial \mathbf{x}_{\text{NN}}} \frac{\partial \mathbf{x}_{\text{NN}}}{\partial \boldsymbol{\theta}}. \tag{11.7}$$

More specifically, the gradient needs to be back propagated from the output through the single layers to update the layer parameters $\boldsymbol{\theta}_l$. Together with Eq. (11.5), the update rule for the parameters $\boldsymbol{\theta}_l$ of layer f_l in the $[it]$-th training iteration becomes

$$\begin{aligned} \boldsymbol{\theta}_l^{[it+1]} &= \boldsymbol{\theta}_l^{[it]} - \tau \frac{\partial \mathcal{L}}{\partial \boldsymbol{\theta}_l} \\ &= \boldsymbol{\theta}_l^{[it]} - \tau \frac{\partial \mathcal{L}}{\partial \mathbf{x}_L} \frac{\partial \mathbf{x}_L}{\partial \mathbf{x}_{L-1}} \cdots \frac{\partial \mathbf{x}_l}{\partial \boldsymbol{\theta}_l}. \end{aligned} \tag{11.8}$$

If complex-valued layers and network outputs $\mathbf{x}_{\text{NN}}$ are involved, complex back propagation following Wirtinger calculus (see Appendix 11.B) has to be considered.

Box 11.3 Back-propagation in practice

Automatic differentiation is supported in ML frameworks. Complex back propagation according to Wirtinger calculus is supported in Tensorflow/Keras. The most recent versions of PyTorch now also support complex-valued processing.

11.3.4 Training, validation, and testing

Training describes the process of updating the model parameters $\boldsymbol{\theta}$ in the ML model $f_{\boldsymbol{\theta}}(\mathbf{x})$ optimizing a cost function. The training database is the composition of N_{train} training samples. In case of supervised learning, the cost function involves the labels / reference output $\mathbf{x}_{GT}$, hence, the training database contains both training samples and their associated annotated labels. The gradient descent in Eq. (11.8) would require all training samples for the update, however, this requires a huge computing effort especially for large datasets. In ML, stochastic optimization is used to overcome this computational burden. SGD (the stochastic version of GD) computes a single gradient update only on a subset of training samples, provided in a single batch. The *batch size* N_b is set a priori. For N_{train} training samples and a batch size N_b, we obtain $\lfloor N_{train}/N_b \rfloor$ iterations per epoch. An epoch refers to one pass over the full training set. The amount of epochs determines the training duration. Careful consideration needs to be taken in terms of the amount of training data, model capacity, and problem complexity to avoid under- or overfitting which can be observed by comparing discrepancies between the *training* and *validation loss* [15]. During training, dropout layers (see Section 11.5.7) can be included in NNs, weights can be regularized or early stopping can be deployed to reduce the risk of overfitting. The model behavior can be observed by the training loss. However, it is advisable to monitor the training progress with a separate validation set.

Validation is performed during model training to track the model performance or to optimize model hyperparameters, but not the model parameters $\boldsymbol{\theta}$ themselves. Therefore, a separate validation set N_{val} data samples with annotated labels (in case of supervised learning) is used to evaluate the performance of the currently learned model following a predetermined validation cycle. Evaluation metrics are, on one hand, the loss function itself (validation loss), but can also include other quantitative measures (MSE, MAE, SSIM) [29], on the other.

In *testing/inference/prediction*, the trained model is tested with N_{test} newly unseen data samples. Test data is usually not accompanied by annotated labels (in case of supervised learning). In case annotated labels or ground truth references are available, nonblinded similarity measures (e.g., MSE) can be applied to assess the model performance.

11.3.5 Database splitting

Training, validation, and test sets are disjoint. Available data samples N_s can be split up into training N_{train}, validation N_{val}, and test set N_{test} according to different inclusion/exclusion criteria or completely random. Database creation and curation is an important step towards a reliable ML method. The outline of the database specifies the structure of the database pipeline needed to load the data into the ML application (see Section 11.7.3).

Selection of representative training samples N_{train} is crucial to provide enough diversity for training, but with sufficient consistency to converge. The validation set size N_{val} is usually much smaller than N_{train}, e.g., 10%–20% of the number of training samples, depending on the underlying application, task, and hyper-parameter optimization. Test samples N_{test} are never seen by the method during training and ideally (but not necessarily) reflect the same underlying trained application that enables inference of performance and robustness and/or generalization of the model.

Random splitting

While randomly splitting data samples into training, validation, and test set (or between training and validation set) is feasible and justified for some applications, it is strongly discouraged in the field of medical imaging. A random splitting ignores the information from which patient the data sample is derived and hence samples from a specific patient can be assigned to training, validation, and test set that in the end potentially boosts performance, but definitively biases the results.

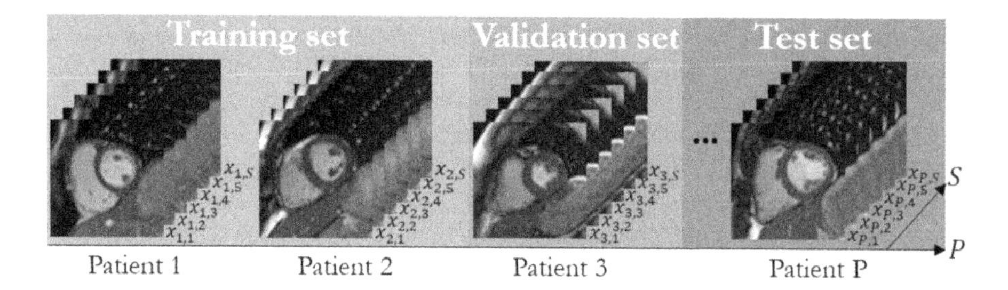

FIGURE 11.3 Database splitting with patient leave-out

Splitting of available data in training, validation, and test set. Each of the P patients has S samples (slices in a volume, imaging contrasts, ...). A patient leave-out approach is chosen, i.e., unique patients are assigned for each set.

Patient leave-out approach

Therefore, common splitting methods consider a patient leave-out approach, i.e., only unique patients are fed through the model in training, validation, and testing. This avoids any bias of the model towards specific anatomies or pathologies. A database of P patients with each patient having S data samples (e.g., slices in a volume), consists of $N_s = P \cdot S$ samples

$$\Big\{\underbrace{x_{1,1}, \dots, x_{1,S}}_{S}, \underbrace{x_{2,1}, \dots, x_{2,S}}_{S}, \underbrace{x_{3,1}, \dots, x_{3,S}}_{S}, \dots, \underbrace{x_{P,1}, \dots, x_{P,S}}_{S}\Big\}. \tag{11.9}$$

In a patient leave-out approach, unique patients $\{x_{i,1}, \dots, x_{i,S}\}$ are assigned for the training, validation, and test set as illustrated in Fig. 11.3. In a similar fashion, one could also define other and/or additional means to sort and split the data along S, e.g., imaging contrast, multi-parametric information, trajectory, etc., depending on the specific reconstruction problem.

Cross-validation

For smaller databases, it is advisable to perform a cross-validation in order to reduce bias towards selected samples and investigate the method's robustness. In a k-fold cross-validation, the available database of N_s samples is split into k folds—considering unique patients are sorted into each fold–with each having N_s/k samples. Assignment of samples to these folds can either be random (after patient consideration) or follow specific criteria. In total, k networks are then trained, one for each of the folds. Deviation in evaluation metrics between folds gives an indicator on the method's robustness.

11.4 Task definition for MR reconstruction

ML has the ability to learn expressive feature extractors from data and enables us to solve tasks that are challenging to solve with fixed, hand-crafted models. Various types of tasks can be solved using ML such as image segmentation, image classification, and regression tasks. For *image classification*, the task of the NN is to learn a function $f_\theta(\mathbf{x}) : \mathbb{K}^N \to \{1, \dots, L\} \triangleq [l]$ to assign a label l to the input $\mathbf{x}$,

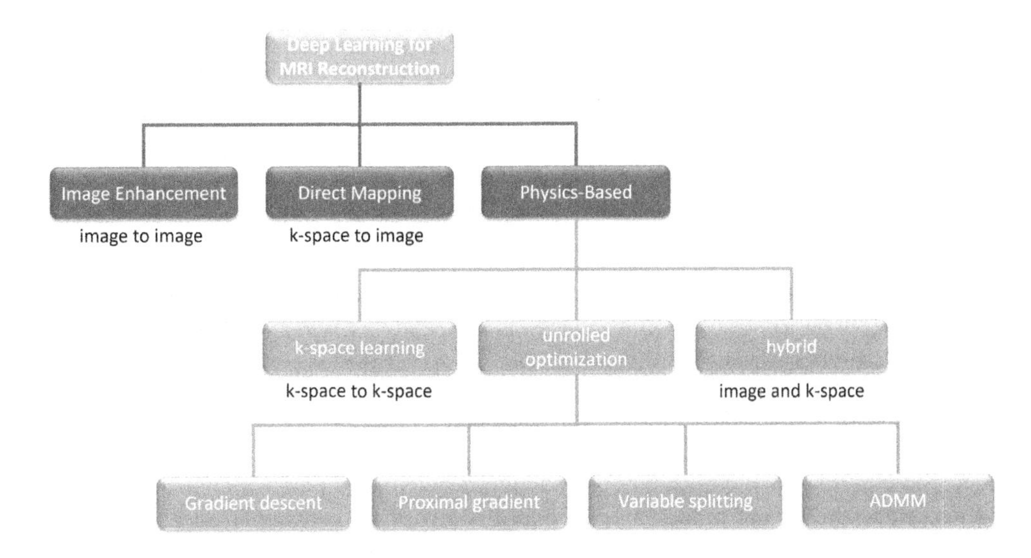

FIGURE 11.4 MR reconstruction task definition

Classification of ML reconstruction tasks that differ in terms of input and targeted application output.

where $\mathbb{K}$ is an arbitrary field. An example is tumor classification with the labels *benign* and *malign*. For *image segmentation* tasks, the network $f_{\boldsymbol{\theta}}(\mathbf{x}) : \mathbb{K}^N \rightarrow [l]^N$ assigns a label l to every pixel of the input vector $\mathbf{x}$, hence, it partitions the image $\mathbf{x}$ into a set of L distinct classes/objects, e.g., organs.

Image classification and segmentation assign a global or local label to the input vector $\mathbf{x}$. In contrast, the function $f_{\boldsymbol{\theta}}(\mathbf{x}) : \mathbb{K}^N \rightarrow \mathbb{K}^{N_d}$ predicts a numerical value for a given input vector $\mathbf{x}$ in an *(image) regression* task, where N_d is the output dimension. Image-to-image translation describes an image regression task if continuous predictions are assigned to every pixel in the image. MRI reconstruction can be viewed as sensor-to-image translation task.

As previously discussed in Chapter 2, in MR image reconstruction, we aim at recovering an image $\hat{\mathbf{x}}$ from the k-space signal $\mathbf{s}$ that is corrupted by measurement noise $\boldsymbol{\epsilon}$ following

$$\mathbf{s} = \mathbf{E}\mathbf{x} + \boldsymbol{\epsilon}, \tag{11.10}$$

where $\mathbf{E} \in \mathbb{C}^{K \times N}$ is the linear forward operator describing the MR acquisition model and K denotes the dimensionality of the underlying k-space data. Depending on the imaging application and signal modeling, the operator $\mathbf{E}$ involves Fourier transforms, sampling trajectories, and coil sensitivity maps. Field inhomogeneities, relaxation effects, motion, and diffusion can be also considered (Chapters 12–16), which may lead to a nonlinear measurement operator.

In ML frameworks, mapping functions $f_{\boldsymbol{\theta}}$ can be used in various ways to reconstruct an image $\mathbf{x}$ from the measured data. All tasks have an image $\mathbf{x}$ and/or k-space $\mathbf{s}$ as input to the NN function $f_{\boldsymbol{\theta}}$, but can also include further MR specific information as meta parameters or other tensors, e.g., trajectories and coil sensitivity maps.

An overview of common ML reconstruction tasks for MRI, differing in terms of input and targeted application output is shown in Fig. 11.4. These reconstruction tasks are further described hereafter.

A summary of recently proposed ML-based MR reconstruction approaches is depicted in Table 11.1, along with details about the underlying task, data processing, input dimensionality, and handling of a complex domain. MRI data can be processed in a coil-combined/single-coil setting or as multicoil data. While multicoil data are usually acquired in practice and provide a real-time scenario, single-coil/coil-combined data reduce the computational burden substantially.

11.4.1 Image enhancement

Certain types of undersampling, such as pseudorandom, radial, or spiral sampling, introduce incoherent noise-like aliasing in the zero-filled reconstructed images, while we observe coherent undersampling artifacts from regular Cartesian sampling. Thus, an *image enhancement* task can be used to reduce the aliasing in the images. The function $f_\theta(\mathbf{x}) : \mathbb{K}^N \rightarrow \mathbb{K}^N$ performs image-to-image regression by predicting the output value $\mathbf{y}$ based on the corrupted input image $\mathbf{x}$. The input to the enhancement task can be the zero-filled (and noise-affected) MR images or reconstructed MR images that present remaining aliasing or noise amplification for large undersampling factors, e.g., images reconstructed with parallel imaging. Instead of learning the enhanced image, some approaches learn the residual noise to be removed from the noisy input [42,43]. The mapping f_θ only acts on the image $\mathbf{x}$ and has no information of the acquired rawdata. Hence, consistency to the measured k-space signal $\mathbf{s}$ cannot be guaranteed. Approaches exist that add additional k-space consistency to the cost function [44] or enforce k-space consistency after image enhancement [45].

11.4.2 Direct k-space to image mapping

Another ML-based approach is to reconstruct the MR image directly from the acquired k-space data. With this approach, the so-called direct k-space to image mapping, the k-space data are fed directly into the mapping function to achieve $\mathbf{y} = f_\theta(\mathbf{s})$. Consequently, the mapping function f_θ approximates the forward model. Learning a direct mapping is especially useful if the forward model or parts of the forward model are not exactly known. In the case of fully sampled MRI under ideal conditions, the learned mapping approximates the Fourier transform [46]. However, this becomes computationally very demanding due to fully connected layers (see Section 11.5.4) that are involved here. Furthermore, consistency to the acquired k-space data cannot be guaranteed.

11.4.3 Physics-based reconstruction

Another family of ML-based MR reconstruction methods is referred as physics-based reconstruction. These approaches integrate the traditional physics-based modeling of the MR encoding with ML, ensuring consistency to the acquired data. We can distinguish two classes of problems: (1) learning in k-space domain and (2) iterative optimization in image domain with interleaved data consistency steps. The first approaches are referred to as k-space learning, whereas the latter one are known as unrolled optimization methods. These two approaches can be combined to *hybrid* approaches that learn both a NN in k-space domain and image domain.

Unrolled optimization

Physics-based learning, which is modeled as iterative optimization, can be viewed as generalization of iterative SENSE [47,48] (Chapter 6) with a learned regularization in image domain. To explore this in more detail, let us recap the basic variational image reconstruction problem defined in (Chapter 2)

$$\mathbf{x}^* \in \arg\min_{\mathbf{x}\in\mathbb{K}^n} \frac{\lambda}{2} \|\mathbf{Ex} - \mathbf{s}\|_2^2 + R[\mathbf{x}], \tag{11.11}$$

which contains a data-consistency term $D[\mathbf{Ex}, \mathbf{s}] = \frac{1}{2}\|\mathbf{Ex} - \mathbf{s}\|_2^2$ and a regularization term $R[\mathbf{x}]$, that imposes prior knowledge on the reconstruction $\mathbf{x}$. As discussed in Chapters 2 and 3, the easiest way to solve Eq. (11.11) is to use a gradient descent scheme to optimize for the reconstruction $\mathbf{x}$. This results in

$$\mathbf{x}^{[it+\frac{1}{2}]} = \mathbf{x}^{[it]} - \tau^{[it]}\mathbf{E}^{\mathbf{H}}(\mathbf{Ex}^{[it]} - \mathbf{s}) \tag{11.12}$$

$$\mathbf{x}^{[it+1]} = \mathbf{x}^{[it+\frac{1}{2}]} - \tau^{[it]}\nabla_{\mathbf{x}} R[\mathbf{x}] \tag{11.13}$$

where τ is the (iteration-dependent) step size. Assuming $R[\mathbf{x}]$ is a fixed, convex function, the gradient descent scheme would be run for a large number of iterations. In learning algorithms, we unroll this scheme for a fixed number of T iterations to obtain a solution for $\mathbf{x}$. An NN replaces the gradient of the handcrafted regularizer by a data-driven mapping function $f_{\boldsymbol{\theta}}$. Thus, training several iterations with alternating mapping functions $f_{\boldsymbol{\theta}}$ and intermittent data consistencies reflect *unrolled optimizations* [49]. For a GD scheme, this results in

$$\mathbf{x}^{[it+\frac{1}{2}]} = \mathbf{x}^{[it]} - \tau^{[it]}\mathbf{E}^{\mathbf{H}}(\mathbf{Ex}^{[it]} - \mathbf{s}) \tag{11.14}$$

$$\mathbf{x}^{[it+1]} = \mathbf{x}^{[it+\frac{1}{2}]} - \tau^{[it]} f_{\boldsymbol{\theta}}(\mathbf{x}). \tag{11.15}$$

Another example would be to learn a proximal gradient scheme (Chapter 3). A common way in MRI reconstruction is to first take a step in the direction of the regularization term and then perform the proximal mapping with respect to the data term [50].

$$\mathbf{x}^{[it+\frac{1}{2}]} = \mathbf{x}^{[it]} - \tau^{[it]}\nabla_{\mathbf{x}} R\left[\mathbf{x}^{[it]}\right], \tag{11.16}$$

$$\mathbf{x}^{[it+1]} = \mathrm{prox}_{\tau^{[it]}\frac{\lambda}{2}(\mathbf{E}\cdot,\mathbf{s})}\left(\mathbf{x}^{[it+\frac{1}{2}]}\right) = \arg\min_{\mathbf{x}\in\mathbb{K}^n} \frac{1}{2\tau^{[it]}}\left\|\mathbf{x} - \mathbf{x}^{[it+\frac{1}{2}]}\right\|_2^2 + \frac{\lambda}{2}\|\mathbf{Ex} - \mathbf{s}\|_2^2. \tag{11.17}$$

Here, again the gradient, with respect to the regularization term, can be replaced by a mapping function $f_{\boldsymbol{\theta}}$ yielding

$$\mathbf{x}^{[it+\frac{1}{2}]} = \mathbf{x}^{[it]} - \tau^{[it]} f_{\boldsymbol{\theta}}(\mathbf{x}^{[it]}), \tag{11.18}$$

$$\mathbf{x}^{[it+1]} = \mathrm{prox}_{\tau^{[it]}\frac{\lambda}{2}(\mathbf{E}\cdot,\mathbf{s})}\left(\mathbf{x}^{[it+\frac{1}{2}]}\right) = \arg\min_{\mathbf{x}\in\mathbb{K}^n} \frac{1}{2\tau^{[it]}}\left\|\mathbf{x} - \mathbf{x}^{[it+\frac{1}{2}]}\right\|_2^2 + \frac{\lambda}{2}\|\mathbf{Ex} - \mathbf{s}\|_2^2. \tag{11.19}$$

While for single-coil imaging a closed-form solution for the proximal mapping exists [51], the proximal mapping for multicoil imaging can be solved using a conjugate gradient optimizer [50]. In practice, any other optimization can be used for algorithm unrolling, such as variable splitting [52] or primal-dual optimization [53]. While this section gave a very general mathematical overview, the implementation of MRI-specific data-consistency layers are described in Section 11.5.11.

Plug-and-play priors

Trained image denoisers can be also combined with physics-based learning (see Section 11.4.3) or conventional reconstructions (see Chapter 2) and serve thus as an advanced regularization for traditional optimization problems. Specifically, it was noticed by Heide et al. [54] and Venkatakrishnan et al. [55] that advanced denoising schemes such as BM3D [56] and Nonlocal means [57] can be used efficiently with multiple forward models. Inspired by [54,55], the fixed denoiser has been replaced by data-driven CNNs in computer vision [58] and MRI [59,60]. The advantage of plug-and-play priors is that the denoiser can be trained off-line and does not have to involve any forward model. The trained denoiser can than be matched with any optimization scheme, e.g., ADMM [55] or Proximal gradient optimization [58].

k-space learning

A prominent approach for physics-based learning in k-space domain [61] can be viewed as an extension of the linear kernel estimation in GRAPPA (Chapter 6). Here, a nonlinear kernel modeled by the mapping function f_θ is learned from the ACS. The missing k-space lines can then be filled using this estimated, nonlinear kernel and the data is then transformed to the image space using an inverse Fourier transform. The final image is obtained by root-sum-of-squares reconstruction of the individual coil images.

Hybrid learning

The last type of approaches we present here are *hybrid* approaches [62] that combine the advantages of learning in k-space domain and learning in image domain. These networks are applied in an alternating manner to obtain the final reconstruction $\mathbf{x}$. When designing hybrid approaches, it is important to keep the basic theorems of the Fourier transform in mind: local changes in image domain result in global changes in k-space domain and vice versa, to avoid unexpected behavior.

11.5 Core concepts: layers

NNs are composed of several layers. Each layer operates on received input variables and output variables, which are denoted by $\hat{\mathbf{x}}$ in the following. A layer may have trainable and non-trainable parameters. Various types of layers are commonly used in ML-based MR reconstruction, including convolution, normalization, activation, among others. These layers are further described hereafter.

11.5.1 Convolution layer

Convolutions are linear operations that extract *features* of an image. While shallow convolution layers extract low-level features such as edges, deeper layers extract more abstract, high-level features such as shapes or objects. A convolution layer maps $N_{f,\text{in}}$ input feature channels to $N_{f,\text{out}}$ output feature channels

The discrete convolution of an image $\mathbf{x}$ with filter kernels $\mathbf{k}_{i,j}$ reads as

$$\hat{\mathbf{x}}_j = \sum_{i=1}^{N_{f,\text{in}}} \mathbf{x}_i * \mathbf{k}_{i,j} = \mathbf{K}_{i,j}\mathbf{x}_i, \quad j = 1, \ldots, N_{f,\text{out}}, \tag{11.20}$$

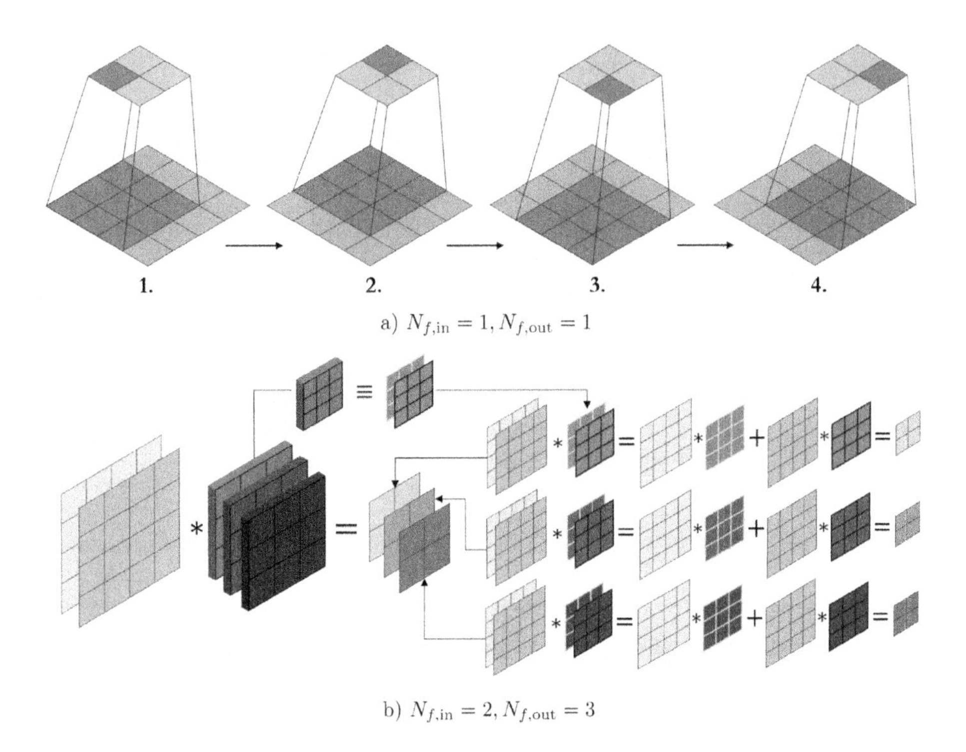

FIGURE 11.5 Convolution operation

The input $\mathbf{x}$ (light blue (light gray in print version)) is convolved with the filter kernel $\mathbf{k}$ (dark blue (dark gray in print version)) producing the output feature map (green (gray in print version)). The filter kernel of size 3×3 is slid over the entire input of size 5×5, yielding a scalar estimate (dark green (mid gray in print version)) at each of the four positions of the 2×2 output feature map. No padding ($p = 0$), a stride of $s = 1$ with dilation rate $d = 1$, was applied in all directions. Convolution operation for learning **a)** a single filter, i.e., output channel $N_{f,\text{out}} = 1$ from a single channel input $N_{f,\text{in}} = 1$ and b) three filters, i.e., output channel $N_{f,\text{out}} = 3$ from a two channel input $N_{f,\text{in}} = 2$. In this case for each input and output channel combination, a kernel is learned, i.e., $2 \cdot 3 = 6$ kernels of size 3×3 are trained. Input channels are convolved with the respective kernel and summed up to form the 2×2 output for each of the three output channels.

where the subscripts denote the feature channels. This produces $N_{f,\text{out}}$ feature maps in $\mathbf{x}_{\text{NN}}$. The convolution with a filter kernel $\mathbf{k}_{i,j}$ can be represented as a matrix–vector multiplication, where $\mathbf{K}_{i,j} : \mathbb{K}^N \rightarrow \mathbb{K}^N$ is a sparse matrix. Note that convolution filters are local feature extractors and invariant to translation. The convolution operation is depicted in Fig. 11.5.

For real-valued convolutions, it is $\mathbb{K} = \mathbb{R}$. For complex convolutions, $\mathbb{K} = \mathbb{C}$, the convolution operation is extended to

$$\begin{aligned}
\mathbf{x}_i * \mathbf{k}_{i,j} &= (Re(\mathbf{x}_i) * Re(\mathbf{k}_{i,j}) - Im(\mathbf{x}_i) * Im(\mathbf{k}_{i,j})) \\
&\quad + \mathrm{i} \cdot (Im(\mathbf{x}_i) * Re(\mathbf{k}_{i,j}) + Re(\mathbf{x}_i) * Im(\mathbf{k}_{i,j})) \\
&= \mathbf{K}_{i,j}\mathbf{x}_i
\end{aligned}$$

$$= (Re(\mathbf{K}_{i,j}) \cdot Re(\mathbf{x}_i) - Im(\mathbf{K}_{i,j}) \cdot Im(\mathbf{x}_i)) \\ + \mathrm{i} \cdot (Im(\mathbf{K}_{i,j}) \cdot Re(\mathbf{x}_i) + Re(\mathbf{K}_{i,j}) \cdot Im(\mathbf{x}_i)). \tag{11.21}$$

If real and imaginary parts are represented as a vector, the following notation can be used:

$$\begin{bmatrix} Re(\mathbf{K}_{i,j}\mathbf{x}_i) \\ Im(\mathbf{K}_{i,j}\mathbf{x}_i) \end{bmatrix} = \begin{bmatrix} Re(\mathbf{K}_{i,j}) & -Im(\mathbf{K}_{i,j}) \\ Im(\mathbf{K}_{i,j}) & Re(\mathbf{K}_{i,j}) \end{bmatrix} \begin{bmatrix} Re(\mathbf{x}_i) \\ Im(\mathbf{x}_i) \end{bmatrix}. \tag{11.22}$$

The number of trainable parameters is determined by the kernel size k along each direction and the amount of input and output features $N_{f,\text{in}}$ and $N_{f,\text{out}}$. The number of trainable parameters in a convolution layer can be calculated as

$$\#\text{parameters} = N_{f,\text{out}} \cdot \left(1 + N_{f,\text{in}} \cdot \prod_p k_p\right), \tag{11.23}$$

where the kernel sizes k_p along each direction $p \in \{x, y, z, t\}$ are multiplied and where a bias term per filter is considered.

Box 11.4 Implementation of a convolution layer

A convolution layer is characterized by the following items:

- `in_channels` $N_{f,\text{in}}$: number of channels in the input image
- `out_channels` $N_{f,\text{out}}$: Number of output channels respectively feature maps in the result of the convolution, equals the number of filters to be trained
- `kernel_size` k: Size of the convolution kernel, integer, or tuple
- `padding` p: Boundary handling, impacts the image size, string, integer, or tuple
- `strides` s (optional): Spacing between pixel elements considered in the convolution, integer, or tuple
- `dilation_rate` d (optional): Spacing between kernel elements, integer, or tuple
- `bias` (optional): Adds a trainable offset to the output of the convolution, bool

A simple Keras 2D convolution layer can be created using

```
tf.keras.layers.Conv2D(out_channels, kernel_size, use_bias=True)
```

The input channels are automatically estimated in a separate `build` function.

All parameters need to be *initialized.* Commonly used initializers for filter kernels follow a Glorot Uniform [63] or a He Normal distribution [64]. The initializer can be passed directly to the Keras `Conv2D` by passing, e.g., `kernel_initializer="glorot_uniform"`.

If complex-valued input data are considered, two processing modes (see Section 11.4) are possible: 1) Real/Imaginary are considered in two input channels, hence, the output for each convolution filter is real-valued. However, the complex relationship between real and imaginary parts is lost; 2) Complex-valued convolutions as stated in Eq. (11.21) can be performed, where the output for each convolution filter is complex-valued. The drawback of this approach is that twice the number of trainable parameters is required as a real and imaginary filter kernel needs to be learned. Additionally, the number of convolution operations doubles compared to 1). In the context of MRI reconstruction, complex-valued processing is conducted both ways, as depicted in Table 11.1.

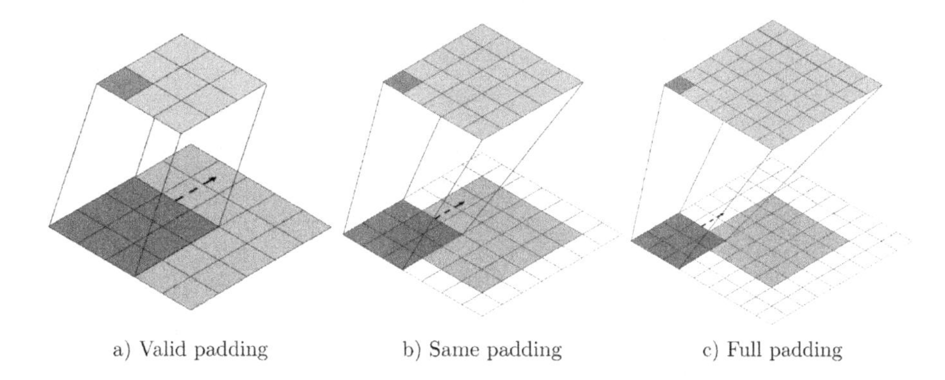

FIGURE 11.6 Convolution operation: Padding

Padding operations of the input. The input $\mathbf{x}$ (light blue (light gray in print version)) is padded with zeros (white) and then convolved with the filter kernel $\mathbf{k}$ (dark blue (dark gray in print version)) producing the output feature map (green (gray in print version)). The filter kernel of size 3×3 is slid over the entire input of size 5×5 yielding a scalar estimate (dark green (mid gray in print version)) at each position of the output feature map of size **a**) 3×3 for valid padding, i.e., $p = 0$, **b**) 5×5 for same padding with $p = 1$ in each direction, i.e., 1×1 padding so that output size is equal to input size, or **c**) 7×7 for full padding with $p = 2$ in each direction, i.e., 2×2 padding to enlarge output size. A stride of $s = 1$ and dilation rate $d = 1$ was used in all cases and all directions. Only a single filter is learned, i.e., output channel $N_{f,\text{out}} = 1$ from a single input channel $N_{f,\text{in}} = 1$.

Padding

Padding appends values at the boundary of the input, which for the majority of frameworks, is a zero-padding, i.e., a constant value of zero, but in general can follow periodic or symmetric padding as well. Three special cases for padding are of importance: 1) no or *valid padding* uses the input without padding ($p = 0$); 2) half or *same padding* appends zeroes so that the output has the same size as the input; and 3) *full padding* appends zeroes such that every pixel in the input contributes equally to the output, but the output size is enlarged. The number of padding p in these cases depends on the kernel size k and stride s. Fig. 11.6 illustrates these three special cases.

Stride

The *stride* describes the step size with which the filter kernel is moved over the input. A stride can be used to perform a down-sampling operation similar to a pooling layer (see Section 11.5.5). Fig. 11.7 shows a convolution operation with stride.

Receptive field

The receptive field describes the spatial extent of the input that a particular feature map within the network observes or is affected by. It is determined by the `kernel_size` k, `padding` p, `stride` s and `dilation_rate` d (see Section 11.5.1.1). These values are prescribed for each input direction and consequently influence the output size respectively receptive field. The output size along each dimension

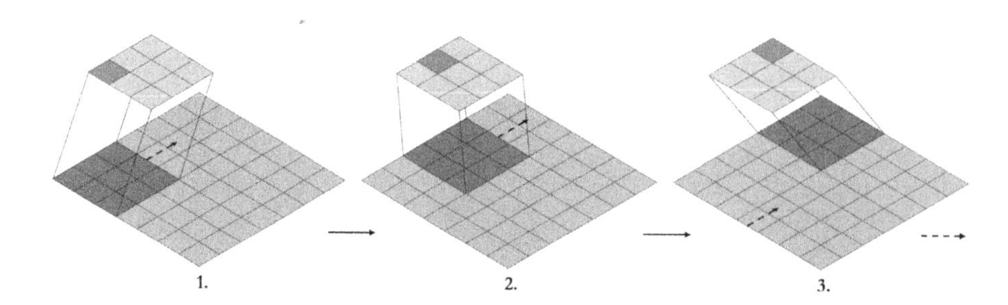

FIGURE 11.7 Convolution operation: Stride

Stride of the filer kernel moving over the input. The input $\mathbf{x}$ (light blue (light gray in print version)) is convolved with the filter kernel $\mathbf{k}$ (dark blue (dark gray in print version)) producing the output feature map (green (gray in print version)). The filter kernel of size 3×3 is slid over the entire input of size 7×7 with a stride of $s = 2$ yielding a scalar estimate (dark green (mid gray in print version)) at each of the output feature map of size 3×3. No padding ($p = 0$) and a dilation rate $d = 1$ was applied in all directions. Only a single filter is learned, i.e., output channel $N_{f,\text{out}} = 1$ from a single input channel $N_{f,\text{in}} = 1$.

is for an input of size R_{in} given by

$$R_{\text{out}} = \left\lfloor \frac{R_{\text{in}} + 2p - d \cdot (k-1) - 1)}{s} \right\rfloor + 1. \tag{11.24}$$

For N-dimensional filters, these values can be stated as $R_{x,\text{out}} \times R_{y,\text{out}} \times \ldots$. For an input $R_{x,\text{in}} \times R_{y,\text{in}} \times \cdots \times N_{f,\text{in}}$, the convolution with $N_{f,\text{out}}$ filters of size $k_x \times k_y \times \ldots$ (with padding $p_x \times p_y \times \ldots$ and stride $s_x \times s_y \times \ldots$) yields an output of size $R_{x,\text{out}} \times R_{y,\text{out}} \times \cdots \times N_{f,\text{out}}$. For a single layer, the receptive field is the same as the size of the output from this layer. If several layers are applied sequentially, the receptive field can be retrieved by consecutive calculation of Eq. (11.24).

11.5.1.1 *Dilated convolution*

Dilated convolution, also known as atrous convolution, introduces a dilation rate along each direction to spread out the filter kernel. With this concept, the same receptive field can be achieved but with less trainable parameters. Fig. 11.8 depicts an example for a convolutional layer with dilation rate of $d = 1$, i.e., no dilation, and $d = 2$. In order to achieve the same receptive field, a filter kernel of 5×5 for $d = 1$ and of 3×3 for $d = 2$ can be used, yielding 25 ($d = 1$) versus 9 ($d = 2$) trainable parameters.

11.5.1.2 *Separable convolution*

While 2D and 3D implementations exist in ML frameworks, spatio-temporal and higher-dimensional convolutions might be designed in a different way. Separable convolution split the N-dimensional kernel operation into multiple steps. *Spatial separable convolution* splits the convolution along the spatial direction if they can be factored into independent directions. While N-dimensional convolutions can be used, various work have shown that separable convolutions can perform superior to N-dimensional convolutions [65,66]. Furthermore, less trainable parameters are required.

For example, in a separable 2D+t convolution, the 3D convolution kernel $\mathbf{k}$ along x, y and t can be separated into a 2D spatial and 1D temporal convolution filter if $\mathbf{k} = \mathbf{k}_{x,y} \otimes \mathbf{k}_t$ applies. Various other

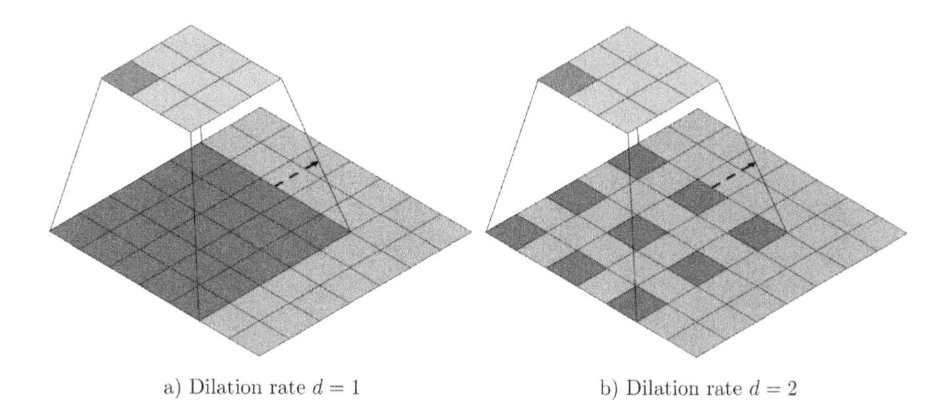

FIGURE 11.8 Dilated convolution

Dilated convolutions on the input $\mathbf{x}$ (light blue (light gray in print version)) which is convolved with the filter kernel $\mathbf{k}$ (dark blue (dark gray in print version)) producing the output feature map (green (gray in print version)). The filter kernel of size **a)** 5×5 for dilation rate of $d = 1$, i.e., no dilation, and b) 3×3 for a dilation rate of $d = 2$, i.e., dilation of 2×2, is slid over the entire input of size 7×7 yielding a scalar estimate (dark green (mid gray in print version)) at each position of the output feature map of size 3×3. A stride of $s = 1$ and valid padding $p = 0$ was used in all cases. Only a single filter is learned, i.e., output channel $N_{f,\text{out}} = 1$ from a single input channel $N_{f,\text{in}} = 1$.

variants exist to further improve the performance [67] or can be extended towards higher dimensions as shown in [66,68].

A *depthwise separable convolution* splits the spatial component from the depth/output channel component. An example calculation for the reduction in trainable parameters is shown in Appendix 11.C.

11.5.1.3 Transposed convolution

Transposed convolutions generally arise from the desire to use a transformation going in the opposite direction of a normal convolution, i.e., from something that has the shape of the output of some convolution to something that has the shape of its input, while maintaining a connectivity pattern that is compatible with convolutions. Transposed convolutions can thus be used for upsampling operations (see Section 11.5.6). Following Eq. (11.20), this yields the transposed filter $\mathbf{K}^T$ for $\hat{\mathbf{x}} = \mathbf{K}^T\mathbf{x}$.

11.5.2 Normalization layer

Adding normalization layers directly after convolution layers is a common way to enable faster and more stable training of NNs. Statistics are estimated from the input and used to reparametrize the input. Hence, the subsequent layers are less tolerant to changes in previous layers. The selection of the normalization layer is task dependent and there is no general recipe which layer should be selected. Although normalization layers are often important to make the NN train, they might lead to unwanted artifacts for image restoration tasks [69].

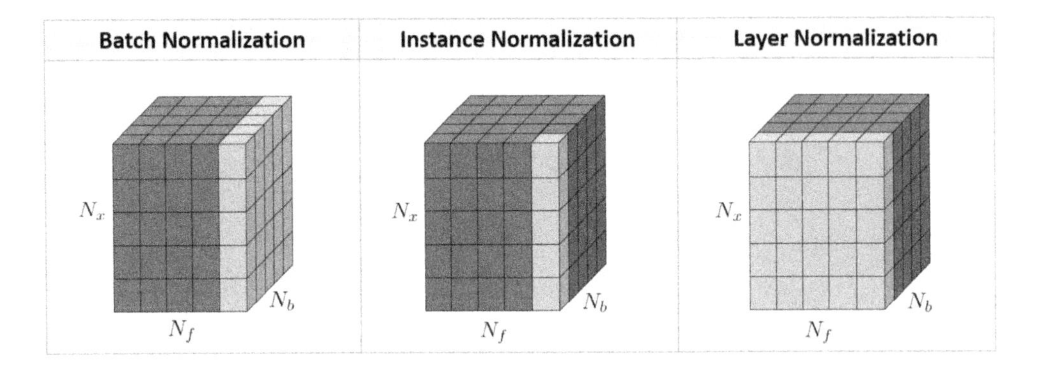

FIGURE 11.9 Normalization techniques

Visualization of common normalization techniques. The green (gray in print version) area denotes the dimensions that are used to calculate the normalization statistics, i.e., mean and standard deviation, over the input batch with dimensions $[N_b, N_x, N_f]$.

In this section, we define normalization layers n for the 1D case of a feature vector $\mathbf{x}$ with the dimensions $[N_b, N_x, N_f]$. We introduce the notation $E_{d=\cdot}[\mathbf{x}]$ and $Var_{d=\cdot}[\mathbf{x}]$ to compute the mean and variance values over the dimension $d \in \{b, x, f\}$. Common normalization methods are depicted in Fig. 11.9.

The normalization layer is defined as

$$\hat{\mathbf{x}} = \frac{\mathbf{x} - E_d[\mathbf{x}]}{\sqrt{Var_d[\mathbf{x}] + \epsilon}} \alpha + \beta, \tag{11.25}$$

where $E[\mathbf{x}]$ and $Var[\mathbf{x}]$ are the estimated mean and variances of the input $\mathbf{x}$. The scaling parameter $\alpha \in \mathbb{R}$ and bias $\beta \in \mathbb{R}$ are optional parameters that can be learned.

Trabelsi et al. [70] introduced complex-valued batch normalization that ensures a circular distribution, i.e., the same variance for real and imaginary part. The scaling terms are replaced by covariance matrices, hence, complex batch-normalization reads as

$$\hat{\mathbf{x}} = \boldsymbol{A}\boldsymbol{V}^{-\frac{1}{2}}(\mathbf{x} - E_d[\mathbf{x}]) + \beta. \tag{11.26}$$

The matrix $\boldsymbol{V}$ defines the covariance matrix with real-valued components V_{rr}, V_{ii} and V_{ri} following

$$\boldsymbol{V} = \begin{bmatrix} E_d(Re(\mathbf{x})^2) & E_d(Re(\mathbf{x})Im(\mathbf{x})) \\ E_d(Im(\mathbf{x})Re(\mathbf{x})) & E_d(Im(\mathbf{x})^2) \end{bmatrix} = \begin{bmatrix} V_{rr} & V_{ri} \\ V_{ri} & V_{ii} \end{bmatrix}. \tag{11.27}$$

The scaling operator A contains (trainable) real-valued parameters α_{rr}, α_{ii} and α_{ri} following

$$\mathbf{A} = \begin{bmatrix} \alpha_{rr} & \alpha_{ri} \\ \alpha_{ri} & \alpha_{ii} \end{bmatrix}. \tag{11.28}$$

The bias $\beta \in \mathbb{C}$ is complex-valued.

Batch normalization

Ioffe et al. [71] introduced batch normalization to reduce the internal covariance shift. Statistics are computed over a mini-batch $d = b, x$, with batch size $N_b > 1$. During training, mini-batch statistics are accumulated over time. The accumulated running means $E[\bar{\mathbf{x}}]$ and variances $Var[\bar{\mathbf{x}}]$ are computed using a running average scheme with momentum ν

$$E[\bar{\mathbf{x}}] = \nu \cdot E[\mathbf{x}] + (1 - \nu) \cdot E[\bar{\mathbf{x}}]$$
$$Var[\bar{\mathbf{x}}] = \nu \cdot Var[\mathbf{x}] + (1 - \nu) \cdot Var[\bar{\mathbf{x}}],$$

where $E[\mathbf{x}]$ and $Var[\mathbf{x}]$ are the estimated mean and variance of the current mini-batch $\mathbf{x}$. It follows that in training mode the running means and averages are only updated, while the real batch statistics are used to compute Eq. (11.25). In the inference mode, the running means and averages are used to compute Eq. (11.25).

Instance normalization

Initially introduced for style transfer, instance normalization is another widely used normalization technique for neural networks [72]. Here, $E[\mathbf{x}]$ and $Var[\mathbf{x}]$ are computed during training and tested individually for each sample and feature independently, i.e., $d = x$. The normalization is computed in the training and inference mode directly from the batch, hence, no moving average and moving variance have to be tracked during training. It is optional to learn additional scaling α and bias β parameters. The complex instance normalization follows Eq. (11.26).

Layer and group normalization

In contrast to batch normalization, *layer normalization* [73] computes the statistics over a whole layer, i.e., $d = x, f$. This is a common way to stabilize training of recurrent neural networks. *Group normalization* [74] divides the feature channels N_f into groups, and statistics are calculated within each individual group.

11.5.3 Activation layer

The activation layer decides if a neuron should be "fired" by analyzing the values of the convolution layer. This is realized by applying a nonlinear activation function in an element-wise way.

Rectified Linear Units (ReLUs) are one of the most commonly used activation functions in ML. These piece-wise linear functions $\phi_{\text{ReLU}} : [-\infty, \infty] \rightarrow [0, \infty]$ set negative values to 0 and keep positive values following the rule

$$\phi_{\text{ReLU}}(\mathbf{x}) = \max(\mathbf{x}, 0).$$

Although ReLUs are not differentiable at $\mathbf{x} = 0$, which might cause issues in gradient-based learning, gradient-based optimizers use the subgradient for back-propagation and have shown to be efficient for a variety of ML tasks [14].

A common problem in training networks are *dying ReLUs* which output all zeros independent of the input values, making parts of the network inactive. To avoid this problem, *leaky ReLUs* are used that apply a small gradient for negative values scaled by $\alpha < 0$, i.e.,

$$\phi_{\text{Leaky ReLU}}(\mathbf{x}) = \begin{cases} \mathbf{x} & \text{if} \quad \mathbf{x} \geq 0 \\ \alpha\mathbf{x} & \text{if} \quad \mathbf{x} < 0 \end{cases}.$$

This extends the output range $\phi_{\text{Leaky ReLU}} : [-\infty, \infty] \to [-\infty, \infty]$. A further extension is *parametrized ReLUs (PReLUs)* that allow the scale α to be learned for the individual feature channels.

When applying nonlinear activation functions ϕ to complex values, several things have to be considered. One possibility is to apply the activation function to the real and imaginary part separately, however, the natural correlation between real and imaginary channels are not considered in this case. Furthermore, the phase information is mapped to the first quadrant, i.e., the interval $[0, \frac{\pi}{2}]$. Another variant is to keep the phase information fixed and only alter the magnitude information. An example is the *ModReLU* [75]

$$\phi_{\text{ModReLU}}(\mathbf{x}) = \max(0, |\mathbf{x}| + \beta)\frac{\mathbf{x}}{|\mathbf{x}|},$$

where β is the bias that is trainable.

Virtue et al. [76] proposed a new complex activation function called *cardioid*

$$\phi_{\text{cardioid}}(\mathbf{x}) = \frac{1}{2}(1 + \cos(\angle\mathbf{x} + \beta))\,\mathbf{x}.$$

The complex cardioid can be seen as a generalization of ReLU activation functions to the complex plane. Compared to other complex activation functions, the complex cardioid acts on the input phase rather than the input magnitude. A bias β can be additionally learned.

11.5.4 Fully connected layer

Fully connected (FC) (or dense) layers connect every input unit to every output unit and are formulated as dense matrix–vector multiplication so

$$\hat{\mathbf{x}} = \mathbf{W}\mathbf{x} + \beta, \tag{11.29}$$

with an additional bias β. Trainable parameters are the dense matrix $\mathbf{W}$ and the bias β. FC layers are global feature extractors and are core components in multi-layer perceptrons (MLPs), object recognition, classification, and global transform learning [46]. In contrast to convolution layers, the number of trainable parameters depend on the amount of input nodes N_i and output nodes N_o (including bias β), yielding #parameters $= N_i \cdot N_o + N_o$. This becomes inefficient to compute when learning global transforms of large image sizes.

11.5.5 Down-sampling layer

Down sampling layers reduce the spatial resolution and make NNs more efficient to train as the number of operations are reduced substantially in subsequent layers. *Pooling layers* are used as down-sampling operations to reduce the spatial resolution in the image and to introduce approximate invariance to small translations [14]. Small patches of size k are analyzed in the individual features maps to keep important information about extracted features. Common pooling layers are *average pooling* and *max pooling*. For complex-valued images, the maximum operation does not exist. Instead, the pooling layer is modified such that it keeps values with, e.g., the maximum magnitude response. A pooling layer does not have any trainable parameters.

Convolutional Downsampling layers are convolution layers (see Section 11.5.1) with a stride greater than 1, resulting in an output with decreased spatial resolution. Down sampling with convolution layers can be seen as a learning the pooling operation. However, the number of trainable parameters are increased when using convolutional downsampling.

11.5.6 Up-sampling layer

Up-sampling layers increase the spatial resolution and are important to revert from coarser scales to the original image resolution in image-to-image networks. *Convolutional Upsampling* with transposed convolutions (see Section 11.5.1) allow to recover the original input size, for strides $s > 1$. However, this kind of upsampling is known to introduce checkerboard artifacts [77]. The number of trainable parameters increases with convolutional upsampling. To overcome checkerboard artifacts in convolutional upsampling, nearest neighbor or bilinear *interpolation* provide an alternative. Interpolation is commonly followed by a convolution layer for further computation. Bilinear interpolation in the complex domain can be conducted by either interpolating real/imaginary part or magnitude/phase separately.

11.5.7 Dropout layer

Dropout [78] is a regularization technique during training to prevent overfitting of the network. At each training iteration, a random selection of nodes and connections to and from the layer are temporarily set to zero with probability $1 - p$ or kept with probability p. The result of dropout is scaled with a factor $1/p$ to maintain the scale of the output layer. The complex dropout is realized the same way to the complex number, but not to real and imaginary parts separately.

11.5.8 Merging layers

Concatenation stacks tensors together along a certain dimension, mostly the feature dimension. The aim is to increase feature diversity or to gather feature information from multiple parts of the network. It is also possible to merge further inputs into the network at deeper layers. Note that the spatial dimensions have to match when using merging layers at various scales. If concatenation is followed by a convolution layer, this can be interpreted as weighted combination of multiple inputs. In a similar sense, other feature map combinations, such as *addition*, *subtraction*, *multiplication*, *averaging*, *minimum* or *maximum*, can be performed.

11.5.9 Recursive layer

Recursive layers enable processing of spatiotemporal sequences of arbitrary length. The temporal frames are processed sequentially, but the temporal features are captured and stored in hidden units

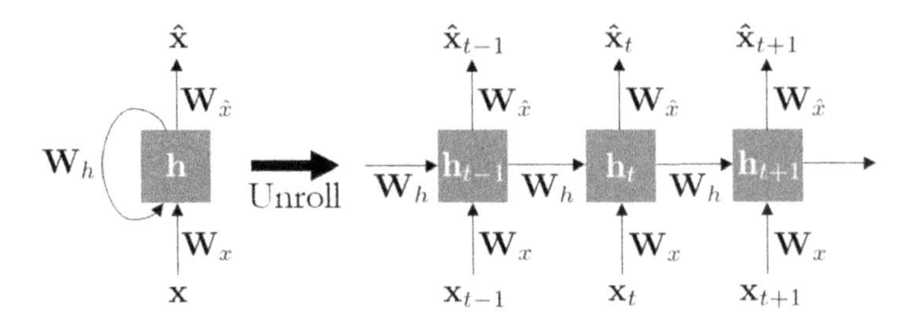

FIGURE 11.10 Recursive layer

Recursive layer with input $\boldsymbol{x}$, hidden unit h, and output $\hat{x}$ to process a sequence of time frames t. Inputs and outputs are weighted by learned filters $\mathbf{W}_x$ and $\mathbf{W}_{\hat{x}}$, respectively, as well as the recursive feedback $\mathbf{W}_h$. Unrolling the recursive unit reveals its temporally sequential processing.

that act as memory. Later frames have access to this memory. A basic recursive layer reads as

$$\begin{aligned} \boldsymbol{h}_t &= \phi_h(\mathbf{W}_h \boldsymbol{h}_{t-1} + \mathbf{W}_x \mathbf{x}_t + b_x) \\ \hat{\mathbf{x}}_t &= \phi_{\hat{x}}(\mathbf{W}_{\hat{x}} \boldsymbol{h}_t + b_{\hat{x}}), \end{aligned} \tag{11.30}$$

where: $\mathbf{W}_h$, $\mathbf{W}_x$, and $\mathbf{W}_{\hat{x}}$ are weight matrices; b_x and $b_{\hat{x}}$ are the bias; ϕ_h and $\phi_{\hat{x}}$ are the activation functions; and $\boldsymbol{h}$ are the hidden units receiving input $\boldsymbol{x}$ and output $\hat{\boldsymbol{x}}$ for the processed temporal frame t. Fig. 11.10 depicts the recursive layer together with its unrolled version in which the temporally sequential processing can be appreciated. The formulation in Eq. (11.30) might result in vanishing gradients during back propagation as each step recurrently depends on all previous steps. This might be avoided by introducing additional *gates* such as in Long Short Term Memory (LSTMs) [6] and Gated Recurrent Units (GRUs) [79]. Note that the weight matrices are defined by convolutions in the case of image reconstruction tasks. While Eq. (11.30) can be straightforwardly translated to complex-valued operations, LSTMs and GRUs require real-valued gating functions as defined in, e.g., [80].

11.5.10 Building blocks

Several layers can be bundled to form larger building blocks as shown in Fig. 11.11. *Convolution blocks* in NNs commonly consist of a sequence of convolution, normalization (optional) and activation. *Residual blocks* [81] learn a residual that is added to the input of the block. The direct identity connections are also known as skip or residual connections. Furthermore, this avoids the vanishing gradient problem and makes the network easier to train. Depending on the kernel sizes and channels, these residual blocks can be also referred to as *bottleneck residuals* (including 1×1 convolutions) or *wide residuals* (increased channel depth) [82]. *Dense blocks* [83] follow a similar principle, but they concatenate feature maps from several previous layers, determined by the growth rate g, to build a collective knowledge. It is also possible to train a convolution filter for the residual/dense path which acts then as a weighted combination or attention focusing depending on if also acting on spatial dimension, also known as *fire blocks* [84], or spatial/channel squeezing/excitation, known as *squeeze-and-excite* [85]. Several convolution blocks acting in parallel to a residual connection form an *assembly block*. An alternative assembling can be achieved with *merge-and-run mapping* [86] which averages the inputs of the

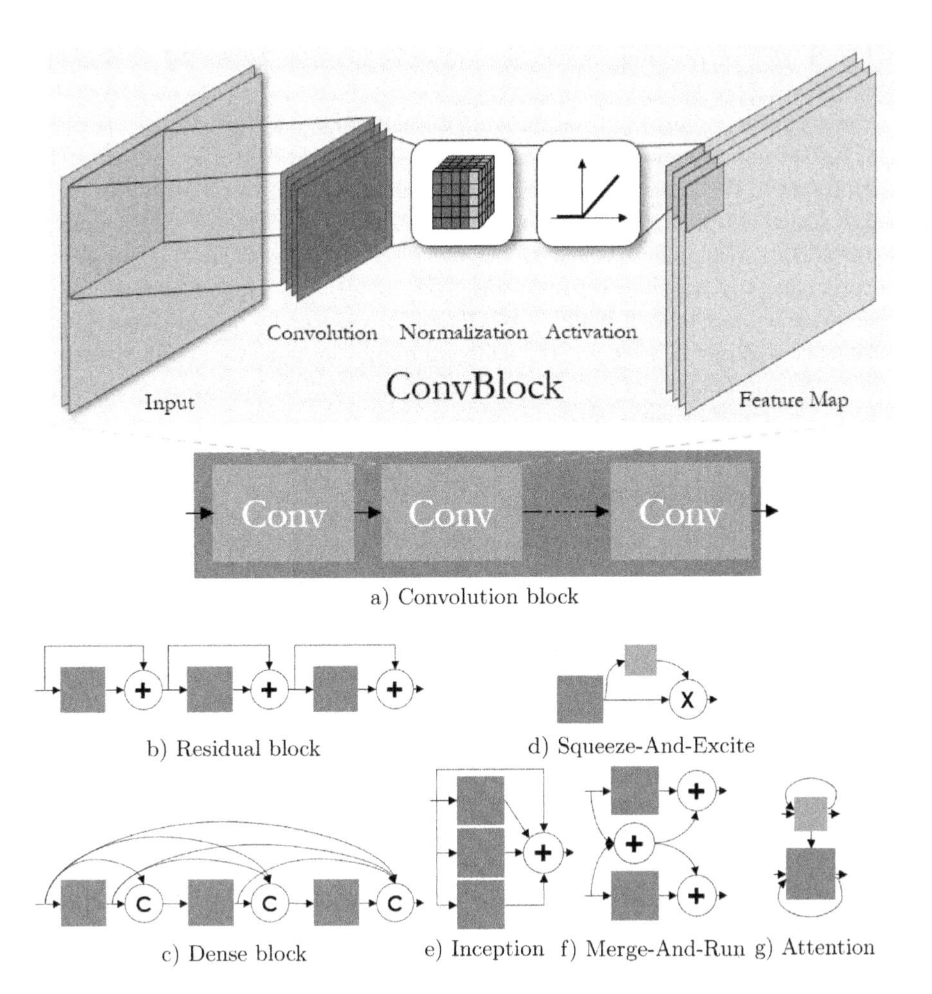

FIGURE 11.11 Building blocks

Common building blocks in neural networks. **a**) Convolution blocks are used for processing image input via convolution layer, followed by normalization (optional) and activation function to form the feature-map output. Several consecutive convolution blocks are usually taken inside other building blocks. **b**) Residual block with residual/skip connection and addition of feature map at output. **c**) Dense block with skip connections over several stages and concatenation of feature maps. **d**) Squeeze-and-excitation by learning a spatial/channel weighting to scale the output feature map. **e**) Inception with multiple parallel convolution blocks of various kernel sizes to allow for a multiresolution approach. **f**) Merge-and-run mapping for merging information between parallel branches. **g**) Attention mechanisms via recursive layers to focus the layer output.

residual branches (merge) and adds the average to the output of each residual branch as the input of the subsequent residual branch (run). A multiresolutional approach is achieved if convolution layers/blocks of different kernel sizes are operating in parallel, known as *inception blocks* [87]. Several versions of this have been proposed [88,89].

Attention mechanisms originated from the field of natural language processing and help to learn dependencies between source and target, consequently focusing the network's attention on the most significant information. The methods differ in the used alignment scores (content [90], additive [91], location [92] or scaled dot-product [93]) and information sharing (self-attention [94], global/soft [95], and local/hard [92,95]). Within these concepts, *recurrent Neural Networks (RNNs)* (see Section 11.5.9) are used to extract the sequence of information based on their relative temporal spreading.

RNNs thus form another type of building blocks that are especially advantageous to process spatiotemporal sequences of arbitrary length. The temporal frames are processed sequentially with temporal information stored in hidden units which later frames can access. Applications of RNNs for MRI reconstruction can be found in, e.g., [96].

11.5.11 Data consistency layers

DC layers are the core layers for physics-based learning Section 11.4.3 and are responsible to induce similarity to the acquired k-space data $\mathbf{s}$. A DC layer receives as input the currently reconstructed image $\mathbf{x}$, its corresponding k-space $\mathbf{s}$, and information about the sampling pattern, e.g., a sampling mask in the Cartesian case or a sampling trajectory in the non-Cartesian case (Chapter 4). The implementation of the DC layers follows the definitions of the encoding operator $\mathbf{E}$. The implementation of various DC layers follows directly the mathematical derivation. In the following, we focus on the most commonly used DC layers that impose DC by gradient descent [97], null-space projection [98], or proximal mapping [51,96,50]. For details on other realizations of DC layers, such as variable splitting and primal–dual optimization, we refer the interested reader to the original publications [52,53].

Gradient descent (GD-DC) layer

Let us consider a GD scheme presented in Eq. (11.15). This involves the encoding operator $\mathbf{E}$ and its adjoint operator $\mathbf{E}^{\mathbf{H}}$. In the Cartesian single-coil case, the forward operator is defined as $\mathbf{E} = \mathbf{BF}$, and its adjoint is defined as $\mathbf{E}^{\mathbf{H}} = \mathbf{F}^{\mathbf{H}}\mathbf{B}^{\mathbf{H}}$. Since the sampling mask $\mathbf{B}$ is a diagonal matrix consisting of ones and zeros, it follows $\mathbf{B}^{\mathbf{H}} = \mathbf{B}$. The adjoint FT equals the inverse FT, i.e., $\mathbf{F}^{\mathbf{H}} = \mathbf{F}^{-1}$. Using these definitions, the GD-DC layer can be easily implemented as

$$\hat{\mathbf{x}} = \mathbf{x} - \lambda \mathbf{E}^{\mathbf{H}}(\mathbf{E}\mathbf{x} - \mathbf{s}), \tag{11.31}$$

where $\lambda > 0$ impacts the weight of DC and is fixed or trainable. This scheme works also for the multicoil as presented in [97].

Null-space projection (NSP-DC) layer

Mardani et al. [98] introduced a layer to project onto the nullspace of $\mathbf{E}$. The NSP-DC layer reads as

$$\hat{\mathbf{x}} = \mathbf{E}^{\dagger}\mathbf{s} + (\mathbf{I} - \mathbf{E}^{\dagger}\mathbf{E})\mathbf{x}, \tag{11.32}$$

where $\mathbf{E}^{\dagger}$ denotes the pseudo inverse.

Proximal mapping (PM-DC) layer

This DC layer is shaped by computing the proximal mapping from Eq. (11.17). For single-coil PM-DC, this yields

$$\hat{\mathbf{x}} = \mathbf{F}^{\mathbf{H}}\mathbf{\Lambda}\mathbf{F}\mathbf{x} + \frac{\lambda}{1+\lambda}\mathbf{F}^{\mathbf{H}}\mathbf{B}^{\mathbf{H}}\mathbf{s}, \quad \mathbf{\Lambda}_{ii} = \begin{cases} 1 & \text{if } i \notin \Omega \\ \frac{1}{1+\lambda} & \text{if } i \in \Omega, \end{cases} \tag{11.33}$$

where $\mathbf{\Lambda}$ is a diagonal matrix and Ω defines the index set of acquired k-space samples. For the multicoil, no closed-form solution to Eq. (11.17) exists, and Aggarwal et al. proposed to solve this problem using a conjugate gradient (CG) optimization. Therefore, Eq. (11.17) is computed and rearranged to form $\mathbf{A}\mathbf{x} = \mathbf{b}$, to get the system matrix $\mathbf{A}$ and the right-hand side term $\mathbf{b}$ for the CG optimizer.

11.6 Network architectures for MRI reconstruction

All the presented layers and building blocks can be used to form a full network. For ML-based MR reconstructions, most commonly CNNs are used that are a sequence of convolution blocks with the aforementioned interconnections. There is, however, no unique definition of networks, and numerous variants exist, depending on the underlying task [45,99–102]. The targeted application, dimensionality, and data availability mainly determine the task definitions (see Section 11.4) and, subsequently, the architectural choices. An overview of various recent network architectures for MRI reconstruction with distinction by task definition are stated in Table 11.1.

A UNet [103] processes the input at various scales, involving convolution blocks, global and local skip connections, down sampling, up sampling and merging layers. Various variants for MRI reconstruction can be found in, e.g., [66,100,104]. ResNet [81] form a large backbone of existing reconstruction networks [51,96,105]. In Variational Networks (VNs) [97,106,107], the main building blocks are motivated by fields-of-experts regularization [108] and have a specific structure defined by the gradient of an energy model. This also involves trainable activation functions, realized with, e.g., Gaussian radial basis functions, linear interpolation, or spline interpolation. Generative adversarial networks (GANs) have been studied to model human-perceived image quality or quality control within the reconstruction process by the adversarial training of generator and discriminator [44,109,110]. Reconstructions can be formulated, as unrolled optimizations [51,66,96,97,111], direct mappings [46], k-space interpolations [61], plug-and-play priors [60,112], or super-resolution task [113], depending on the sampling trajectory and application. In Fig. 11.12, three networks with their targeted application of cardiac, neuro, and MSK imaging are depicted.

11.7 How to build an ML model for MR reconstruction

In this section, we provide an overview of the fundamental study and design choices when building a ML model for MR reconstruction.

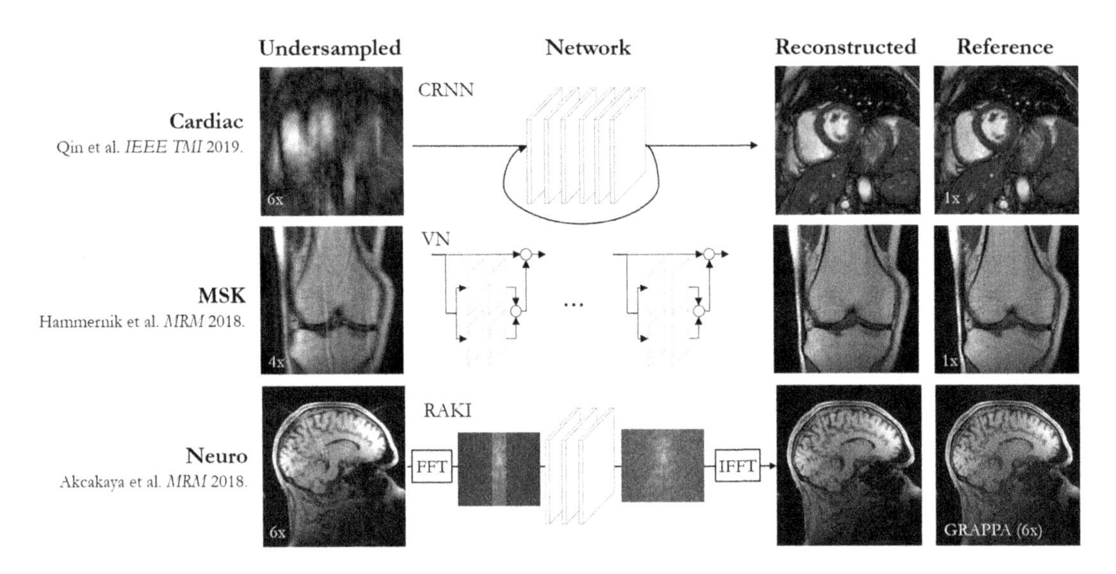

FIGURE 11.12 Machine Learning for MRI reconstruction

Three applications of cardiac, neuro, and musculoskeletal imaging with their proposed machine learning networks for image reconstruction. Details of the networks can be found in Table 11.1.

11.7.1 Checklist to build an ML model

In Table 11.2, we provide an example checklist of important and interdependent points that need to be considered to decide on and design an appropriate architecture.

11.7.2 Database

For training an ML method, a database is required that provides sufficient diversity, but with substantial consistency among datasets targeted to the underlying application. The amount of data required to train an application depends on the complexity of the problem, the complexity of the learning algorithm, and the expected level of performance.

When confronted with a database, you might consider the following:

- Does the database actually resemble the problem?
- Is your database suitable for self-supervised or supervised learning schemes (see Section 11.3.2)?
- Can you use the database as a pre-training step and fine-tune on more specified data?
- Can you collect/acquire more (task-specific) data?
- Can you use data augmentation to increase sample size?
- Can you use phantom data (numerical or acquired) or natural images to mimic the application?

Acquiring new data samples for ML-based MR reconstructions is often not trivial because it requires acquisition of fully sampled (or at least aliasing-free) ground-truth data. Furthermore, simply increasing the sample size does not guarantee a better performance [120,121].

Table 11.1 Network Architectures for MRI Reconstruction. Comparison of network architectures for MRI reconstruction in relation to task definitions (see Section 11.4), processing dimensionality, and input. The processing dimensionality also considers how temporal information (if available) is handled (see Appendix 11.A and Section 11.5.1). Multicoil indicates that information from several MR receiver coils via coil sensitivities were considered inside the network or data consistency. Methods either deploy real-valued processing considering the complex nature of the data in two channels or two networks or perform complex operations (see Appendix 11.A). The input indicates which data (mag: magnitude, pha: phase, real: real, imag: imaginary) is sent into the denoising network (*) deployed in the underlying task.

Task Definition	Method	Dimensionality	Single-/Multicoil?	Real(R)-/Complex(C)-valued?	Input*
Image denoising	Deep Residual [114]	2D	single-coil	R, 2 networks	image (mag/pha)
Image denoising	DAGAN [44]	2D	single-coil	R, 1 channel	image (real)
Image denoising	GANCS [109]	2D	single-coil	R, 1 channel	image (mag)
Image denoising	CNN-SR [113]	2D	single-coil	R, 1 channel	image (mag)
Image denoising	SANTIS [110]	2D	multicoil	R, 2 channels	image (real/imag)
Image denoising	Residual learning [43]	3D	single-coil	R, 1 channel	image (magnitude)
Direct mapping	AUTOMAP [46]	2D	single-coil	R, 2 channels	k-space (real/imag)
PnP	[115]	2D	single-coil	R, 2 channels	image (real/imag)
PnP	RARE [112]	3D	single-coil	R, 2 channels	image (real/imag)
PnP	PnP-CNN [60]	3D	multicoil	R, 2 channels	image (real/imag)
Hybrid	KIKI-net [62]	2D	single-coil	R, 2 channels	image and k-space (real/imag)
k-space learning	ALOHA [116]	2D	single-coil	C	k-space (real/imag)
k-space learning	RAKI [61]	2D	multicoil	R, 2 channels	k-space (real/imag)
k-space learning	sRAKI [117]	2D	multicoil	R, 2 channels	k-space (real/imag)
k-space learning	DeepSpirit [118]	2D	multicoil	R, 2 channels	k-space (real/imag)
Unrolled optimization	ADMM-Net [119]	2D	single-coil	C	k-space (real/imag)
Unrolled optimization	MoDL [111]	2D	multicoil	R, 2 channels	image (real/imag)
Unrolled optimization	DeepCascade [51]	2D+t	single-coil	C	image (real/imag)
Unrolled optimization	CRNN [96]	2D+t	single-coil	C	image (real/imag)
Unrolled optimization	VN [97]	2D	multicoil	C	image (real/imag)
Unrolled optimization	VN [106]	2D	multicoil	C	image (real/imag)
Unrolled optimization	VN [107]	2D	multicoil	C	image (mag/pha)
Unrolled optimization	DL-ESPIRiT [105]	2D+t	multicoil	C	image (real/imag)
Unrolled optimization	CINENet [66]	3D+t	multicoil	C	image (real/imag)

Publicly available databases are rare in the field of medical imaging because of privacy, data confidentiality, and ethical considerations. However, some funding sources request/provide access to the data used in scientific studies (IXI [122], ADNI [123], OASIS [124], etc.) and an increasing trend towards data sharing can be observed (e.g., OpenfMRI [125]). Some databases exist in the form of challenges (fastMRI [126], CHAOS [127], etc.) that were hosted at conferences (MICCAI, ISBI, ISMRM, etc.) and are summarized here: http://grand-challenge.org.

Another challenge in medical imaging is to define the ground-truth in the case of supervised reconstruction approaches. Ideally, fully sampled data serves as reference from which in a retrospective undersampling, input data is created that has a 1:1 voxel-wise correspondence to the ground truth and, hence, enables the usage of voxel-wise loss functions for optimization. However, ground-truth data can

Table 11.2 Checklist for build an ML model. Checklist for designing a learning-based MRI reconstruction architecture.

Targeted application and task definition (see Section 11.4)
☐ Unrolled MRI reconstruction for 2D/3D static/dynamic imaging
☐ Post-processing of pre-reconstructed images
☐ *k*-space processing
☐ Hybrid reconstruction
☐ Plug'n'Play Prior
☐ other:
Data availability
☐ Acquired or simulated measurement data
☐ Pre-processed data
☐ Online database
☐ Existence of ground-truth data, i.e., type of learning, see Section 11.3.2
Data handling (see Appendix 11.A)
☐ Data format: ☐ image and/or ☐ rawdata
☐ Data type: ☐ real-valued or ☐ complex-valued
☐ Data dimensionality: ☐ 2D, ☐ 2D ☐ 2D+t, ☐ 3D, ☐ 3D+t, ☐ *N*D
☐ Pre-processing
☐ other:
Data splitting
☐ Creation of training, validation and test set
☐ Cross-validation needed
Model design
☐ Pre-trained or similar conceptual model available, see Section 11.6
☐ Consistency to measured *k*-space required, see Section 11.5.11
☐ Complex-valued handling, see Appendix 11.B
☐ Single- or multicoil handling, see Section 11.4
☐ Temporal or high-dimensional data handling, see Appendix 11.A
☐ Network architecture and building blocks, see Section 11.5.10
☐ Network input: ☐ corrupted image, ☐ *k*-space, ☐ sampling masks / trajectories ☐ coil-sensitivity maps, ☐ other
☐ Network output: ☐ reconstructed image, ☐ reconstructed *k*-space
Network training
☐ Losses, see Section 11.3.3
☐ Optimizer, see Section 11.3.3
☐ Monitoring of training and hyperparameter optimization, see Section 11.3.4
Evaluation
☐ Quantitative evaluation: ☐ PSNR, ☐ SSIM, ☐ MSE, ☐ other
☐ Qualitative evaluation: ☐ Images, ☐ difference images, ☐ blinded expert reading, ☐ other

in some cases be challenging or impossible to acquire. Even fully sampled data might not be ideal and may suffer from measurement noise and additional artifacts due to the longer acquisition. Additionally, clinical studies to collect patient data are expensive and time consuming. If feasible, phantom data (numerical or acquired), *data augmentation,* or *transfer learning* from other databases can be conducted, also helping to cope with small sample sizes, as detailed in the following.

Pre-training and transfer learning

Transfer learning describes the process of pre-training a network on a dataset with a large amount of samples and then *transferring* and fine-tuning the trained model parameters on a specific tasks where fewer data samples are available. However, pre-trained models for medical images, especially for complex-valued MRI reconstruction, so far don't exist, and leveraging models trained on natural scene images [128] can only be performed to a certain extent, e.g., for perceptual losses or discriminator networks that are applied to the magnitude image. For MRI reconstruction, transfer learning has been used for domain adaptation from CT to radial MRI reconstruction [104] or between imaging sequences [129]. Transfer learning on natural images serves as training database in [46,130,131]. Synthetic phase information, e.g., sinusoidal phase at random spatial frequencies, is added to simulate the complex domain. Multicoil data might be simulated with coil sensitivity maps estimated from real MRI scans.

Data augmentation

Data augmentation is another popular pre-processing step to generate a more versatile dataset. In general, transformations such as rotation, translation, shearing, scaling and nonrigid transformations are used. However, experience from natural image processing cannot be transferred easily to obtain realistic data augmentation for MRI reconstruction. Realistic data augmentation require physics-aware transformations. This can be realized by, e.g., simulating motion [132] or augmentation of the subsampling pattern (beyond changing acceleration factors). Realistic adversarial data augmentation has been introduced, particularly for MRI segmentation, so far [133].

11.7.3 Database pipeline

Data loading pipelines are an essential and integral part of the development. Pipelines are usually specific to the underlying data source, layout, application, and selected ML framework. Some efforts have been made to unify processing in training and testing such as NiftyNet [134] for Tensorflow or TorchIO [135] for PyTorch. Some considerations need to be taken with respect to data loading and format: Data can be pre-processed and stored in an efficient (containerized) format (e.g. TFRecords, HDF5) to enable fast loading of samples into framework during training. The order of the tensor dimensions in the stored file and how they are processed/accessed within the framework also needs to be considered to minimize extensive reshaping operations. Depending on the application, further pre-processing steps (scaling, intensive augmentation, etc.) can be performed as well to minimize the workload during training. If a smaller sample size is used, the database may even be loaded directly into the RAM, instead of being fetched from the hard drive. If fetching from the hard drive is conducted, pre-fetching of data can be considered, e.g., the pipeline loads batches of samples ahead (operated on CPU) while training on previously loaded ones (operated on GPU). This operation can be parallelized and improves overall performance. Core functionalities for this are provided in the respective frameworks.

For testing, access to the trained ML infrastructures through plugins connected to a graphical user interface such as DeepInfer [136] for Slicer3D [137], NiftyLink for NIFTK [138], NvidiaAnnotationPlugin for MITK [139], or Nora [140] have been established. They facilitate the labeling and inspection of test results.

11.7.4 Frameworks

Building your own ML application from scratch is hardly possible nowadays. There exist a number of frameworks that support tremendously the development and production of ML applications. These frameworks take care of core functionalities, automatic differentiation, have CPU/GPU support, provide the user with a numerous amount of tools to process data, build models, and train/evaluate the models to make the start into ML easier. These frameworks are open-source and can be used with python.

- TensorFlow[3] is a ML library developed by Google. Complex-valued support is available. The core of TensorFlow is the computation of graphs. Graphs consist of nodes, describing operations, and edges, which are tensors. The graphs need to be *compiled* before training/evaluation the ML model. While these static graphs have several advantages, e.g., computational optimization and distributed computing, debugging is cumbersome if the eager mode is disabled.
- Keras[4] is a simple API that builds on top of Tensorflow (and Theano). It encapsulates a number of steps in the development of ML algorithms such that only minimal interaction and coding is required from the user.
- PyTorch,[5] developed by Facebook Inc. Pytorch, is a more lightweight ML library, and intuitive to use for Python/NumPy users. Graphs are built dynamically and default execution mode is eager, which allows for easy debugging. PyTorch offers checkpoint processing, which allows for memory-efficient training. Complex-valued support is available for recent versions.

11.8 Summary

This chapter introduced the basic concepts of performing MR image reconstruction with ML frameworks. First, the fundamentals of ML were discussed, including types of learning, cost function, optimization, training/validation/test set composition and their usage. The underlying MR application, including applied body region, imaging sequence, data dimensionality, etc., determine the MR reconstruction task. The various reconstruction tasks differ in their input and targeted application output. Image denoising, direct mapping, and physics-based reconstructions have been introduced and their advantages and disadvantages discussed. Basic building blocks form the backbone of the reconstruction networks and are composed depending on the task and application specific reconstruction. Exemplary ML networks for MR image reconstructions were shown and discussed. A checklist was provided that can guide the initial setup of ML reconstruction frameworks. The chapter concludes with examples and tutorial codes for practice.

11.9 Further resources and tutorials

In the following, a list of resources for further reading is provided. This (incomplete) list includes online resources, tutorials, books, book chapters and survey papers.

[3] https://www.tensorflow.org/.
[4] https://keras.io/.
[5] https://pytorch.org/.

- Keras: https://keras.io/
- Tensorflow tutorials: https://www.tensorflow.org/tutorials
- Tensorflow playground: https://playground.tensorflow.org
- PyTorch tutorials: https://pytorch.org/tutorials
- Peter Chang's DL tutorials: https://github.com/peterchang77/dl_tutor
- Neural network zoo: https://www.asimovinstitute.org/neural-network-zoo/
- Interactive deep learning tutorial: https://d2l.ai/
- Udacity AI courses: https://www.udacity.com/courses/school-of-ai
- Bishop, *Pattern Recognition and Machine Learning* [13]
- Vapnik, *The Nature of Statistical Learning Theory* [120]
- Murphy, *Machine Learning: A Probabilistic Perspective* [15]
- Goodfellow et al., *Deep Learning* [14]
- Survey papers [99,141–143]

11.10 Exercises

11.1 Derive the proximal mapping for single-coil reconstruction

$$\operatorname{prox}_{\tau g}\left(\hat{\mathbf{x}}\right) = \arg\min_{\mathbf{x}\in\mathbb{K}^n} \frac{1}{2\tau}\left\|\mathbf{x}-\hat{\mathbf{x}}\right\|_2^2 + g(\mathbf{x}),$$

where

$$g(x) = \frac{\lambda}{2}\left\|\mathbf{BFx}-\mathbf{s}\right\|_2^2.$$

11.2 No closed-form solution of the proximal mapping exist for multicoil reconstruction

$$\operatorname{prox}_{\tau g}\left(\hat{\mathbf{x}}\right) = \arg\min_{\mathbf{x}\in\mathbb{K}^n} \frac{1}{2\tau}\left\|\mathbf{x}-\hat{\mathbf{x}}\right\|_2^2 + g(\mathbf{x}),$$

where

$$g(x) = \frac{\lambda}{2}\left\|\mathbf{Ex}-\mathbf{s}\right\|_2^2.$$

A CG optimization can be used to solve this problem. First, derive the equation that has to be solved using CG in the form $\mathbf{A}\hat{\mathbf{x}} = \mathbf{x}$. Second, derive the equation to back-propagate the error $\mathbf{e} = \frac{\partial\mathcal{L}}{\partial\mathbf{x}}$ with respect to the input $\hat{\mathbf{x}}$. How can the error be back propagated with CG practically?

Hint. Use following rule to compute the derivative of the matrix inverse $\mathbf{X}^{-1}$

$$\partial\mathbf{X}^{-1} = -\mathbf{X}^{-1}\left(\partial\mathbf{X}\right)\mathbf{X}^{-1}.$$

11.3 Implement a 3D convolution layer with 1 input channel and 32 output channels in the complex domain and ModReLU activation functions. First, use 3D convolutions with a filter size of 5×5×3. How many trainable parameters does this layer have? Second, design a 2D+t convolution layer that uses $5\times5\times1$ spatial filters, followed by $1\times1\times3$ temporal filters. How many intermediate features, i.e., output channels of the spatial convolutions do you need to reach the same model complexity as the 3D convolution?

11.4 Calculate the number of network parameters and the perceptive field for a 5-layer CNN with, 64 features in the first 4 layers, and one output feature in the last layer. The kernel size is set to 3, bias is applied, and ReLU activations are used. The input is a single-channel real-valued image. How does the number of parameters change if this is performed in the complex-valued domain, using complex-valued convolutions and cReLU activations?

11.5 Global transform learning as presented in Zhu et al. [46] is based on fully connected layers. Assume, you have a 64×64 input image, yielding a $N = 64^2$ input. You build a network as follows: FC layer with N hidden neurons, followed by a FC layer with N hidden neurons. This output is reshaped to $n\times n$ and followed by two convolution layers with 32 feature maps. The final convolution maps to a single-channel output image. Compute the number of parameters for the fully complex-valued case and the 2-channel real case. How does the number of parameters change for a 64×64, 128×128 and 512×512, assuming the single-coil case?

Hint. For the 2-channel real case, the first FC layer has $2N$ neurons.

11.10.1 Hands-on examples

The following four exercises correspond to hands-on examples to experiment with basic building blocks, data consistency layers, and network architectures on toy examples in Tensorflow/Keras. Material is provided online,[6] where an overview of the tutorials is given in `merlin.ipynb`. The examples are run on the MNIST [144] dataset with complex-valued data and k-space data being simulated. All examples are provided as Google Colab notebooks,[7] which allow the user to access GPUs to run the code with no payment.

11.6 Image-denoising based MR reconstruction on real-valued images:
This example performs a basic CNN and a residual CNN for an image denoising MR reconstruction task on real-valued images. Follow instructions and steps in `tutorial_denoising_real.ipynb` for this example.

11.7 Image-denoising based MR reconstruction on complex-valued images with complex-valued processing:
This example performs a basic CNN and a residual CNN for an image denoising MR reconstruction task on complex-valued images with complex-valued processing. Follow instructions and steps in `tutorial_denoising_complex.ipynb` for this example.

11.8 Image-denoising based MR reconstruction on complex-valued images with real-valued processing:
This example performs a basic CNN and residual CNN for an image denoising task on complex-valued images with real-valued processing as complex data sent in as 2-channel input. Follow instructions and steps in `tutorial_denoising_2chreal.ipynb` for this example.

---

[6] https://github.com/midas-tum/merlin/blob/master/notebooks/.

[7] https://colab.research.google.com/.

11.9 Physics-based MR reconstruction on complex-valued images:
This example performs and unrolled reconstruction network with denoising regularizer and data consistency blocks for complex-valued input. Follow instructions and steps in `tutorial_physicsbased_complex.ipynb` for this example.

Appendix 11.A ML-specific notation

The acquired k-space data $\mathbf{s} \in \mathbb{C}^K$ are a complex-valued signal usually acquired with multiple receiver coils. Here, we here use a common notation for the input signal $\mathbf{x} \in \mathbb{K}^N$ (image or k-space, depending on the task definition Section 11.4) to represent both real and complex signals where $\mathbb{K}$ denotes a set. In the real-valued case, $\mathbf{x} \in \mathbb{R}^N$ is element of the real-valued numbers $\mathbb{R}$. In the complex-valued case, $\mathbf{x} \in \mathbb{C}^N$ is element of the complex-valued numbers $\mathbb{C}$ and can be represented by its real and imaginary components as $\mathbf{x} = Re(\mathbf{x}) + iIm(\mathbf{x})$.

Dimensionality

N denotes the dimensionality of the image $\mathbf{x}$ and is defined by the product of its image dimensions. In ML frameworks, images are represented by tensors. The characteristic dimensions include the *batch size* N_b, defining the number of image examples in the tensor, spatial dimensions N_x, N_y, N_z, temporal dimension N_t, and *feature channels* N_f, which describe the input characteristics of the data. For the multiple building blocks in Section 11.5, we only denote the 1D case if not stated otherwise.

The rawdata $\mathbf{s}$ can easily become high-dimensional according to the dimensionality (2D, ND) of the problem and the underlying imaging application, e.g., motion-resolved imaging, diffusion imaging, and dynamic contrast-enhanced imaging. For example, for a 3D static imaging sequence, the k-space $\mathbf{s}$ would be a tensor of rank 5, i.e., three spatial dimensions, number of MR receiver channels, and two channels to model the complex domain.

Box 11.5 Data formats in ML frameworks

The location of the feature channels vary in different ML frameworks, leading to different data formats. For a 2D tensor, we find following formats:

- Channel first / `NCHW`: Tensors of size $[N_b, N_f, N_y, N_x]$
- Channel last / `NHWC`: Tensors of size $[N_b, N_y, N_x, N_f]$

The 3D format can be used to represent both 3D and spatiotemporal 2D+t data:

- Channel first / `NCDHW`: Tensors of size $[N_b, N_f, N_z, N_y, N_x]$ (3D) or $[N_b, N_f, N_t, N_y, N_x]$ (2D+t)
- Channel last / `NDHWC`: Tensors of size $[N_b, N_z, N_y, N_x, N_f]$ (3D) or $[N_b, N_t, N_y, N_x, N_f]$ (2D+t)

In Tensorflow / Keras, channel last is the default behavior.

Representation of complex numbers

An interesting aspect in ML MR reconstruction is the choice of how complex-valued MR data is handled. The choice is mainly influenced by the underlying application and architectural design of the specific reconstruction method (see Section 11.4). One way is to view complex-valued MR images as two-channel real image, which is similar to the processing of RGB images. This mode of operation relies on real-valued calculations, i.e., combines the output of two real-valued operations to a single real-valued output. However, this discards the mathematical relationship between real and imaginary part of the MR signal and thus complex-valued processing is performed only implicitly. On the other hand explicit complex-valued operations on either magnitude/phase or real/imaginary tensors require the consideration of complex calculus.

ML algorithms need to be trained using gradient-based optimization, which requires correct processing of the gradients if complex-valued operations are considered. This requires computing complex derivatives using Wirtinger calculus [76], see Appendix 11.B. For an in-depth explanation on this topic, we refer the interested reader to [76,145,146].

Box 11.6 Implementation of complex-valued input in ML frameworks

Not all layers of NNs do yet support complex data type inputs/outputs. Common practice is therefore to stack the real-valued tensors $\boldsymbol{x}_i$

- real: $\boldsymbol{x}_a$, imaginary: $\boldsymbol{x}_b$
- magnitude: $\boldsymbol{x}_a$, phase: $\boldsymbol{x}_b$

into a separate dimension. The choice of the dimension is decided by how the network should process it:

a) Real-valued operations: Stack $\boldsymbol{x}_a$, $\boldsymbol{x}_b$ tensors into the feature channel dimension N_f, discarding mathematical complex-valued relationship between real and imaginary parts.
b) Complex-valued operations (see Appendix 11.B): Stack $\boldsymbol{x}_a$, $\boldsymbol{x}_b$ tensors into a separate dimension, requiring consideration of complex-valued layers and complex calculus

Tensorflow ($\geq$ v1.0) supports complex-valued processing of the datatypes `tf.complex64` and `tf.complex128`. New complex tensors can be created using `tf.complex(real, imag)`.

Pytorch ($\leq$ v1.6.0) does not support complex-valued datatypes. Complex-valued images can be only stored as a two-channel floating-point image. All operations, e.g., complex-valued multiplication, have to be defined manually.

Appendix 11.B Complex calculus

Wirtinger calculus can be leveraged for complex differentiation. Let us assume we have a complex function $f(\mathbf{x}): \mathbb{C} \rightarrow \mathbb{C}$ where $\mathbf{x} = Re(\mathbf{x}) + \mathrm{i}Im(\mathbf{x}) \in \mathbb{C}$ is a scalar. The Wirtinger derivatives are given by a pair of partial derivatives, which are associated with their real and imaginary parts [145]

$$\frac{\partial f}{\partial \mathbf{x}} = \frac{1}{2}\left(\frac{\partial f}{\partial Re(\mathbf{x})} + \mathrm{i}\frac{\partial f}{\partial Im(\mathbf{x})}\right) \tag{11.34}$$

$$\frac{\partial f}{\partial \mathbf{x}^{\mathbf{H}}} = \frac{1}{2}\left(\frac{\partial f}{\partial Re(\mathbf{x})} - \mathrm{i}\frac{\partial f}{\partial Im(\mathbf{x})}\right). \tag{11.35}$$

The advantage of Wirtinger calculus is that derivatives can be computed directly with respect to $\mathbf{x}$ while $\mathbf{x}^{\mathbf{H}}$ is kept constant, and vice versa. Further important identities of complex derivatives [145] are given as

$$\frac{\partial f^{\mathbf{H}}}{\partial \mathbf{x}^{\mathbf{H}}} = \left(\frac{\partial f}{\partial \mathbf{x}}\right)^{\mathbf{H}} \tag{11.36}$$

$$\frac{\partial f^{\mathbf{H}}}{\partial \mathbf{x}} = \left(\frac{\partial f}{\partial \mathbf{x}^{\mathbf{H}}}\right)^{\mathbf{H}}. \tag{11.37}$$

Following these identities, the chain rule to compute the Wirtinger derivatives of a composite function $f(g(\mathbf{x})) : \mathbb{C} \to \mathbb{C}$ is given as

$$\frac{\partial f}{\partial \mathbf{x}} = \frac{\partial f}{\partial g}\frac{\partial g}{\partial \mathbf{x}} + \frac{\partial f}{\partial g^{\mathbf{H}}}\left(\frac{\partial g}{\partial \mathbf{x}^{\mathbf{H}}}\right)^{\mathbf{H}} \tag{11.38}$$

$$\frac{\partial f}{\partial \mathbf{x}^{\mathbf{H}}} = \frac{\partial f}{\partial g}\frac{\partial g}{\partial \mathbf{x}^{\mathbf{H}}} + \frac{\partial f}{\partial g^{\mathbf{H}}}\left(\frac{\partial g}{\partial \mathbf{x}}\right)^{\mathbf{H}}. \tag{11.39}$$

For functions that have a real-valued output, i.e., $f(\mathbf{x}) : \mathbb{C} \to \mathbb{R}$, additional simplifications are given as

$$\frac{\partial f}{\partial \mathbf{x}} = \left(\frac{\partial f}{\partial \mathbf{x}^{\mathbf{H}}}\right)^{\mathbf{H}} \tag{11.40}$$

$$\frac{\partial f}{\partial \mathbf{x}^{\mathbf{H}}} = \left(\frac{\partial f}{\partial \mathbf{x}}\right)^{\mathbf{H}}. \tag{11.41}$$

This results in the simplified chain rule

$$\frac{\partial f}{\partial \mathbf{x}} = \frac{\partial f}{\partial g}\frac{\partial g}{\partial \mathbf{x}} + \left(\frac{\partial f}{\partial g}\right)^{\mathbf{H}}\left(\frac{\partial g}{\partial \mathbf{x}^{\mathbf{H}}}\right)^{\mathbf{H}} \tag{11.42}$$

$$\frac{\partial f}{\partial \mathbf{x}^{\mathbf{H}}} = \left(\frac{\partial f}{\partial g^{\mathbf{H}}}\right)^{\mathbf{H}}\frac{\partial g}{\partial \mathbf{x}^{\mathbf{H}}} + \frac{\partial f}{\partial g^{\mathbf{H}}}\left(\frac{\partial g}{\partial \mathbf{x}}\right)^{\mathbf{H}}, \tag{11.43}$$

and depicts the complex back propagation for optimization of a real-valued loss function $\mathcal{L}$ that is computed on the output of an l layer network $f_1 \circ f_2 \circ \ldots \circ f_L$, where $\mathbf{x}_l = f_l(\mathbf{x}_{l-1})$, see Fig. 11.13. The update rule for the parameters $\boldsymbol{\theta}_l$ in layer f_l following a gradient descent scheme (see Eq. (11.8)) reads as

$$\begin{aligned}\boldsymbol{\theta}_l^{[it+1]} &= \boldsymbol{\theta}_l^{[it]} - \tau\frac{\partial \mathcal{L}}{\partial \boldsymbol{\theta}_l^{\mathbf{H}}} \\ &= \boldsymbol{\theta}_l^{[it]} - \tau\frac{\partial \mathcal{L}}{\partial \mathbf{x}_L^{\mathbf{H}}}\frac{\partial \mathbf{x}_L}{\partial \mathbf{x}_{L-1}^{\mathbf{H}}}\cdots\frac{\partial \mathbf{x}_l}{\partial \boldsymbol{\theta}_l^{\mathbf{H}}}.\end{aligned} \tag{11.44}$$

Wirtinger derivatives for common operations and network layers presented in Section 11.5 and are listed in Table 11.3.

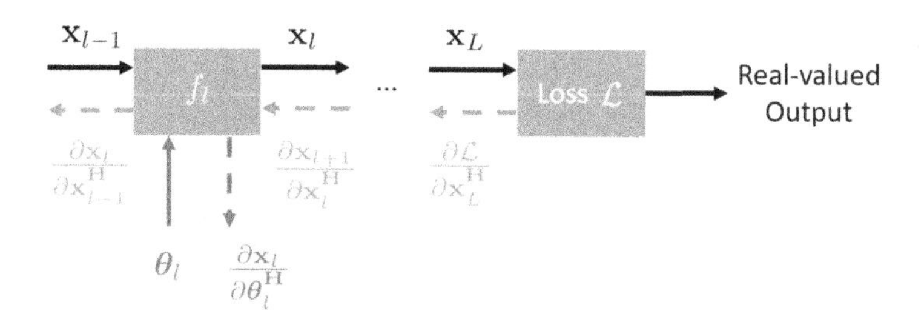

FIGURE 11.13 Complex back-propagation regarding a real-valued loss

To update the network parameters $\boldsymbol{\theta}$, only the adjoint derivatives of the real-valued loss function are required.

Table 11.3 Wirtinger derivatives. Overview of important functions along with their pair of Wirtinger derivatives.

Function	$f(\mathbf{x})$	$\frac{\partial f}{\partial \mathbf{x}}$	$\frac{\partial f}{\partial \mathbf{x}^{\mathrm{H}}}$
Magnitude	$(\mathbf{x}^{\mathrm{H}}\mathbf{x})^{0.5}$	$\frac{\mathbf{x}^{\mathrm{H}}}{2f(\mathbf{x})}$	$\frac{\mathbf{x}}{2f(\mathbf{x})}$
Normalization	$\frac{\mathbf{x}}{(\mathbf{x}^{\mathrm{H}}\mathbf{x})^{0.5}}$	$\frac{1}{2(\mathbf{x}^{\mathrm{H}}\mathbf{x})^{0.5}}$	$-\frac{z^2}{2(\mathbf{x}^{\mathrm{H}}\mathbf{x})^{1.5}}$
Phase	$-\mathrm{i}\log\frac{\mathbf{x}}{\lvert\mathbf{x}\rvert}$	$-\frac{\mathrm{i}}{2\mathbf{x}}$	$\frac{\mathrm{i}}{2\mathbf{x}^{\mathrm{H}}}$
Real Component	$\frac{1}{2}(\mathbf{x}+\mathbf{x}^{\mathrm{H}})$	$\frac{1}{2}$	$\frac{1}{2}$
Imaginary Component	$\frac{1}{2\mathrm{i}}(\mathbf{x}+\mathbf{x}^{\mathrm{H}})$	$\frac{1}{2\mathrm{i}}$	$\frac{\mathrm{i}}{2}$
Scalar product	$\boldsymbol{w}^{\mathrm{H}}\mathbf{x}$	$\boldsymbol{w}^{\mathrm{H}}$	0
Max Pooling	$\mathbf{x}_n,\ n=\arg\max_k\lvert\mathbf{x}_k\rvert$	$\begin{cases}1 & \text{if } n=\arg\max_k\lvert\mathbf{x}_k\rvert\\ 0 & \text{else}\end{cases}$	0
Dropout	$\begin{cases}\frac{1}{p}\mathbf{x}_n & \text{if } n\in\Omega\\ 0 & \text{else}\end{cases}$	$\begin{cases}\frac{1}{p} & \text{if } n\in\Omega\\ 0 & \text{else}\end{cases}$	0
Cardioid [76]	$\frac{1}{2}(1+\cos(\angle\mathbf{x}))\,\mathbf{x}$	$\frac{1}{2}+\frac{1}{2}\cos(\angle\mathbf{x})+\frac{\mathrm{i}}{4}\sin(\angle\mathbf{x})$	$-\frac{\mathrm{i}}{4}\sin(\angle\mathbf{x})\frac{\mathbf{x}}{\mathbf{x}^{\mathrm{H}}}$

Appendix 11.C Trainable parameters of separable convolutions

Depthwise separable convolution is best illustrated in an example. For a normal convolution, let us assume the input of size $32x32$ with 3 input channels is convolved with $f=64$ kernels of size 7×7 (no padding $p=0$ and stride of $s=1$) yielding an output of size 26×26 with 64 feature maps. The operation can be split up by first depthwise convolving each input channel independently with a 7×7 convolutional filter ($p=0, s=1, f=1$), i.e., requiring in total 3 2D filters. The output feature maps are stacked along the channel direction giving a tensor of size $26\times 26\times 3$ (in channels last notation), which is then convolved pointwise with 64 filters of size $1\times 1\times 3$ ($p=0, s=1$) producing the final output tensor of size $26\times 26\times 64$. The advantage is that fewer multiplications are required in a depthwise separable convolution (3 kernels with $7\times 7\times 1$ sliding 32×32 times followed by 64 kernels with $1\times 1\times 3$ sliding 32×32 times, yielding $3\cdot 7\cdot 7\cdot 1\cdot 32\cdot 32+64\cdot 1\cdot 1\cdot 3\cdot 32\cdot 32=216{,}064$ multiplications) than in a normal convolution (64 kernels with $7\times 7\times 3$ sliding 32×32 times, yielding $64\cdot 7\cdot 7\cdot 3\cdot 32\cdot 32=9{,}633{,}792$ multiplications).

References

[1] McCarthy J, Minsky ML, Rochester N, Shannon CE. A proposal for the Dartmouth summer research project on artificial intelligence; 2006.

[2] Rosenblatt F. The perceptron: a probabilistic model for information storage and organization in the brain. Tech. Rep. 6. 1958.

[3] Ackley DH, Hinton GE, Sejnowski J. A learning algorithm for Boltzmann machines. In: Proceedings of the IEEE conference on decision and control, vol. 169. IEEE; 1985. p. 147–69.

[4] Hopfield JJ. Neural networks and physical systems with emergent collective computational abilities. Proc Natl Acad Sci USA 1982;79(8):2554–8. https://doi.org/10.1073/pnas.79.8.2554.

[5] Jordan MI. Attractor dynamics and parallelism in a connectionist sequential machine. In: Proceedings of the eighth annual conference of the cognitive science society. ISBN 0-8186-2015-3, 1986. p. 531–46.

[6] Hochreiter S, Schmidhuber J. Long short-term memory. Neural Comput 1997;9(8):1735–80. https://doi.org/10.1162/neco.1997.9.8.1735.

[7] Williams RJ, Zipser D. A learning algorithm for continually running fully recurrent neural networks. Neural Comput 1989;1(2):270–80. https://doi.org/10.1162/neco.1989.1.2.270.

[8] Rumelhart DE, Hinton GE, Williams RJ. Learning representations by back-propagating errors. Nature 1986;323(6088):533–6. https://doi.org/10.1038/323533a0.

[9] Fukushima K. Neocognitron: a self-organizing neural network model for a mechanism of pattern recognition unaffected by shift in position. Biol Cybern 1980;36(4):193–202. https://doi.org/10.1007/BF00344251.

[10] LeCun Y, Haffner P, Bottou L, Bengio Y. Object recognition with gradient-based learning. Lecture notes in computer science (including subseries lecture notes in artificial intelligence and lecture notes in bioinformatics), vol. 1681. ISBN 3540667229, 1999. p. 319–45.

[11] Lecun Y, Bengio Y, Hinton G. Deep learning. Nature 2015;521(7553):436–44. https://doi.org/10.1038/nature14539.

[12] Krizhevsky A, Sutskever I, Geoffrey EH, Hinton GE. ImageNet classification with deep convolutional neural networks. In: Advances in neural information processing systems; 2012. p. 1097–105. Available from: http://code.google.com/p/cuda-convnet/.

[13] Bishop C. Pattern recognition and machine learning, vol. 9. Springer; 2006.

[14] Goodfellow I, Bengio Y, Courville A. Deep Learning. MIT Press; 2016.

[15] Murphy KP. Machine learning: a probabilistic perspective. The MIT Press. ISBN 0262018020, 2012.

[16] Burr S. Active learning literature survey. Computer sciences technical report. Tech. Rep. January 2009.

[17] Kramer MA. Nonlinear principal component analysis using autoassociative neural networks. AIChE J 1991;37(2):233–43. https://doi.org/10.1002/aic.690370209.

[18] Kingma DP, Welling M. An introduction to variational autoencoders; 2019.

[19] Hinton GE, Salakhutdinov RR. Reducing the dimensionality of data with neural networks. Science 2006;313(5786):504–7. https://doi.org/10.1126/science.1127647.

[20] Goodfellow IJ, Pouget-Abadie J, Mirza M, Xu B, Warde-Farley D, Ozair S, et al. Generative adversarial nets. In: Advances in neural information processing systems, vol. 3; 2014. p. 2672–80.

[21] Pan SJ, Yang Q. A survey on transfer learning; 2010.

[22] Tan C, Sun F, Kong T, Zhang W, Yang C, Liu C. A survey on deep transfer learning. Lecture notes in computer science (including subseries lecture notes in artificial intelligence and lecture notes in bioinformatics), vol. 11141. Springer Verlag. ISBN 9783030014230, 2018. p. 270–9.

[23] Raghu M, Zhang C, Kleinberg J, Bengio S. Transfusion: understanding transfer learning for medical imaging. Tech. Rep. 2019.

[24] Brendan McMahan H, Moore E, Ramage D, Hampson S, Agüera y Arcas B. Communication-efficient learning of deep networks from decentralized data. Tech. Rep. 2017.

[25] Yang Q, Liu Y, Chen T, Tong Y. Federated machine learning: concept and applications. ACM Trans Intell Syst Technol 2019;10(2):1–19. https://doi.org/10.1145/3298981.

[26] Li T, Sahu AK, Talwalkar A, Smith V. Federated learning: challenges, methods, and future directions. IEEE Signal Process Mag 2020;37(3):50–60. https://doi.org/10.1109/MSP.2020.2975749.
[27] Rieke N, Hancox J, Li W, Milletarì F, Roth HR, Albarqouni S, et al. The future of digital health with federated learning. npj Digit Med 2020;3(1):1–7. https://doi.org/10.1038/s41746-020-00323-1.
[28] Wang Z, Bovik AC, Sheikh HR, Simoncelli EP. Image quality assessment: from error visibility to structural similarity. IEEE Trans Image Process 2004;13(4):600–12.
[29] Zhao H, Gallo O, Frosio I, Kautz J. Loss functions for image restoration with neural networks. IEEE Trans Comput Imaging 2016;3(1):47–57.
[30] Johnson J, Alahi A, Fei-Fei L, Assari SM, Idrees H, Shah M, et al. Perceptual losses for real-time style transfer and super-resolution. In: Proceedings of the European conference on computer vision. LNCS, vol. 9906. Springer Verlag. ISBN 9783319464749, 2016. p. 694–711.
[31] Arjovsky M, Chintala S, Bottou L. Wasserstein generative adversarial networks. In: Proceedings of the international conference on machine learning; 2017. p. 214–23.
[32] Robbins H, Monro S. A stochastic approximation method. Ann Math Stat 1951;22(3):400–7. https://doi.org/10.1214/aoms/1177729586.
[33] Richard SS. Two problems with backpropagation and other steepest-descent learning. In: Procedures for networks In Proc 8th annual conf cognitive science society, Erlbaum; 1986. p. 823–31.
[34] Kingma DP, Ba J. Adam: a method for stochastic optimization; 2014. Available from: arXiv:1412.6980.
[35] Hinton GE, Srivastava NN, Swersky K. Neural networks for machine learning lecture 6a overview of mini-batch gradient descent. Tech. Rep. 2012. Available from: https://www.cs.toronto.edu/~tijmen/csc321/slides/lecture_slides_lec6.pdf.
[36] Nesterov Y. A method for unconstrained convex minimization problem with the rate of convergence o(1/k^2). Dokl Akad Nauk USSR 1983;269:543–7. Available from: https://ci.nii.ac.jp/naid/20001173129/en/.
[37] Duchi J, Hazan E, Singer Y. Adaptive subgradient methods for online learning and stochastic optimization. Tech. Rep. 2010.
[38] Zeiler MD. ADADELTA: an adaptive learning rate method. Available from: arXiv:1212.5701, 2012.
[39] Dozat T. Workshop track-ICLR 2016 incorporating Nesterov momentum into ADAM. Tech. Rep. 2016.
[40] Ruder S. An overview of gradient descent optimization algorithms. Available from: arXiv:1609.04747, 2016.
[41] LeCun Y, Bottou L, Orr G, Müller KR. Efficient BackProp. Lecture notes in computer science, vol. 1524. 1998. p. 5–50.
[42] Han YS, Yoo J, Ye JC. Deep residual learning for compressed sensing CT reconstruction via persistent homology analysis. Available from: arXiv:1611.06391, 2016.
[43] Kofler A, Dewey M, Schaeffter T, Wald C, Kolbitsch C. Spatio-temporal deep learning-based undersampling artefact reduction for 2D radial cine MRI with limited training data. IEEE Trans Med Imaging 2020;39(3):703–17. https://doi.org/10.1109/TMI.2019.2930318.
[44] Yang G, Yu S, Dong H, Slabaugh G, Dragotti PL, Ye X, et al. DAGAN: deep de-aliasing generative adversarial networks for fast compressed sensing MRI reconstruction. IEEE Trans Med Imaging 2017;37(6):1310–21. https://doi.org/10.1109/tmi.2017.2785879.
[45] Hyun CM, Kim HP, Lee SM, Lee SM, Seo JK. Deep learning for undersampled MRI reconstruction. Phys Med Biol 2018;63(13):135007. https://doi.org/10.1088/1361-6560/aac71a. Available from: https://github.com/hpkim0512/Deep_MRI_Unet.
[46] Zhu B, Liu JZ, Cauley SF, Rosen BR, Rosen MS. Image reconstruction by domain-transform manifold learning. Nature 2018;555(7697):487–92.
[47] Pruessmann KP, Weiger M, Scheidegger MB, Boesiger P. SENSE: sensitivity encoding for fast MRI. Magn Reson Med 1999;42(5):952–62.
[48] Pruessmann KP, Weiger M, Boernert P, Boesiger P. Advances in sensitivity encoding with arbitrary k-space trajectories. Magn Reson Med 2001;46(4):638–51.

[49] Gregor K, Lecun Y. Learning fast approximations of sparse coding. In: Proceedings of the international conference on machine learning; 2010. p. 399–406.

[50] Aggarwal HK, Mani MP, Jacob M. Model based image reconstruction using deep learned priors (Modl). In: IEEE international symposium on biomedical imaging; 2018. p. 671–4.

[51] Schlemper J, Caballero J, Hajnal JV, Price AN, Rueckert D. A deep cascade of convolutional neural networks for dynamic MR image reconstruction. IEEE Trans Med Imaging 2018;37(2):491–503. https://doi.org/10.1109/TMI.2017.2760978. Available from: https://github.com/js3611/Deep-MRI-Reconstruction.

[52] Duan J, Schlemper J, Qin C, Ouyang C, Bai W, Biffi C, et al. Vs-net: variable splitting network for accelerated parallel MRI reconstruction. Lecture notes in computer science (including subseries lecture notes in artificial intelligence and lecture notes in bioinformatics), vol. 11767. ISBN 9783030322502, 2019. p. 713–22.

[53] Adler J, Öktem O. Learned primal-dual reconstruction. IEEE Trans Med Imaging 2018;37(6):1322–32. https://doi.org/10.1109/TMI.2018.2799231. Available from: https://github.com/adler-j/learned_primal_dual.

[54] Heide F, Steinberger M, Tsai Yt, Gallo O, Heidrich W, Egiazarian K, et al. FlexISP: a flexible camera image processing framework. ACM Trans Graph 2014;33(6):1–13.

[55] Venkatakrishnan SV, Bouman CA, Wohlberg B. Plug-and-play priors for model based reconstruction. In: IEEE global conference on signal and information processing; 2013. p. 945–8.

[56] Dabov K, Foi A, Katkovnik V, Egiazarian K. Image denoising with block-matching and 3D filtering. SPIE electronic imaging; 2006. p. 606414.

[57] Buades A, Coll B, Morel JM. Non-local means denoising. Image Process On Line 2011;1:208–12.

[58] Meinhardt T, Moeller M, Hazirbas C, Cremers D. Learning proximal operators: using denoising networks for regularizing inverse imaging problems. In: IEEE international conference on computer vision, vol. 2017-Octob. Institute of Electrical and Electronics Engineers Inc.; 2017. p. 1799–808.

[59] Kofler A, Haltmeier M, Schaeffter T, Kachelrieß M, Dewey M, Wald C, et al. Neural networks-based regularization for large-scale medical image reconstruction. Phys Med Biol 2020;65(13):135003. https://doi.org/10.1088/1361-6560/ab990e.

[60] Ahmad R, Bouman CA, Buzzard GT, Chan S, Liu S, Reehorst ET, et al. Plug-and-play methods for magnetic resonance imaging: using denoisers for image recovery. IEEE Signal Process Mag 2020;37(1):105–16. https://doi.org/10.1109/MSP.2019.2949470.

[61] Akçakaya M, Moeller S, Weingärtner S, Uğurbil K. Scan-specific robust artificial-neural-networks for k-space interpolation (RAKI) reconstruction: database-free deep learning for fast imaging. Magn Reson Med 2019;81(1):439–53. https://doi.org/10.1002/mrm.27420.

[62] Eo T, Jun Y, Kim T, Jang J, Lee HJ, Hwang D. KIKI-net: cross-domain convolutional neural networks for reconstructing undersampled magnetic resonance images. Magn Reson Med 2018;80(5):2188–201. https://doi.org/10.1002/mrm.27201.

[63] Glorot X, Bengio Y. Understanding the difficulty of training deep feedforward neural networks. Tech. Rep. 2010.

[64] He K, Zhang X, Ren S, Sun J. Delving deep into rectifiers: surpassing human-level performance on imagenet classification. In: Proceedings of the IEEE international conference on computer vision, vol. 2015 Inter. ISBN 9781467383912, 2015. p. 1026–34.

[65] Sandino CM, Cheng JY, Chen F, Mardani M, Pauly JM, Vasanawala SS. Compressed sensing: from research to clinical practice with deep neural networks: shortening scan times for magnetic resonance imaging. IEEE Signal Process Mag 2020;37(1):117–27. https://doi.org/10.1109/MSP.2019.2950433.

[66] Kustner T, Fuin N, Hammernik K, Bustin A, Qi H, Hajhosseiny R, et al. CINENet: deep learning-based 3D cardiac CINE MRI reconstruction with multi-coil complex-valued 4D spatio-temporal convolutions. Sci Rep 2020;10(1):1–13. https://doi.org/10.1038/s41598-020-70551-8.

[67] Stergiou A, Poppe R. Spatio-temporal FAST 3D convolutions for human action recognition. In: Proceedings - 18th IEEE international conference on machine learning and applications, ICMLA 2019. Institute of Electrical and Electronics Engineers Inc. ISBN 9781728145495, 2019. p. 183–90.

[68] Choy C, Gwak J, Savarese S. 4D spatio-temporal convnets: Minkowski convolutional neural networks. In: Proceedings of the IEEE computer society conference on computer vision and pattern recognition, vol. 2019-June; 2019. p. 3070–9.

[69] Wang Y, Song Y, Xie H, Li W, Hu B, Yang G. Reduction of Gibbs artifacts in magnetic resonance imaging based on convolutional neural network. In: Proceedings - 2017 10th international congress on image and signal processing, BioMedical engineering and informatics, CISP-BMEI 2017, vol. 2018-Janua(2); 2018.

[70] Trabelsi C, Bilaniuk O, Zhang Y, Serdyuk D, Subramanian S, Santos JF, et al. Deep complex networks. In: International conference on learning representations; 2018. Available from: https://github.com/ChihebTrabelsi/deep_complex_networks.

[71] Ioffe S, Szegedy C. Batch normalization: accelerating deep network training by reducing internal covariate shift. In: Proceedings of the international conference on machine learning; 2015. p. 448–56.

[72] Ulyanov D, Vedaldi A, Lempitsky V. Instance normalization: the missing ingredient for fast stylization. Available from: arXiv:1607.08022, 2016.

[73] Ba JL, Kiros JR, Hinton GE. Layer normalization. Available from: arXiv:1607.06450, 2016.

[74] Wu Y, He K. Group normalization. Tech. Rep. 3. 2020.

[75] Arjovsky M, Shah A, Bengio Y. Unitary evolution recurrent neural networks. In: Proc. of the international conference on machine learning (ICML); 2016. p. 1120–8.

[76] Virtue P, Yu SX, Lustig M. Better than real: complex-valued neural nets for MRI fingerprinting. In: IEEE international conference on image processing, vol. 2017-Septe; 2018. p. 3953–7.

[77] Odena A, Dumoulin V, Olah C. Deconvolution and checkerboard artifacts. Distill 2016. https://doi.org/10.23915/distill.00003.

[78] Srivastava N, Hinton G, Krizhevsky A, Sutskever I, Salakhutdinov R. Dropout: s simple way to prevent neural networks from overfitting. Tech. Rep. 2014.

[79] Cho K, Van Merriënboer B, Gulcehre C, Bahdanau D, Bougares F, Schwenk H, et al. Learning phrase representations using RNN encoder-decoder for statistical machine translation. In: EMNLP 2014 - 2014 conference on empirical methods in natural language processing, proceedings of the conference. Association for Computational Linguistics (ACL). ISBN 9781937284961, 2014. p. 1724–34. Available from: https://www.aclweb.org/anthology/D14-1179.

[80] Wolter M, Yao A. Complex gated recurrent neural networks. In: Advances in neural information processing systems; 2018. p. 10536–46. Available from: https://github.com/v0lta/Complex-gated-recurrent-neural-networks.

[81] He K, Zhang X, Ren S, Sun J. Deep residual learning for image recognition. In: IEEE conference on computer vision and pattern recognition; 2016. p. 770–8.

[82] Zagoruyko S, Komodakis N. Wide residual networks. In: British machine vision conference 2016, BMVC 2016, vol. 2016-Septe; 2016. p. 1–87.

[83] Huang G, Liu Z, Van Der Maaten L, Weinberger KQ. Densely connected convolutional networks. In: Proceedings - 30th IEEE conference on computer vision and pattern recognition, CVPR 2017, vol. 2017-Janua; 2017. p. 2261–9.

[84] Iandola FN, Han S, Moskewicz MW, Ashraf K, Dally WJ, Keutzer K. SqueezeNet: AlexNet-level accuracy with 50x fewer parameters and <0.5MB model size. Available from: arXiv:1602.07360, 2016.

[85] Hu J, Shen L, Albanie S, Sun G, Wu E. Squeeze-and-excitation networks. IEEE Trans Pattern Anal Mach Intell 2020;42(8):2011–23. https://doi.org/10.1109/TPAMI.2019.2913372.

[86] Zhao L, Li M, Meng D, Li X, Zhang Z, Zhuang Y, et al. Deep convolutional neural networks with merge-and-run mappings. In: IJCAI international joint conference on artificial intelligence, vol. 2018-July. International Joint Conferences on Artificial Intelligence. ISBN 9780999241127, 2018. p. 3170–6.

[87] Szegedy C, Liu W, Jia Y, Sermanet P, Reed S, Anguelov D, et al. Going deeper with convolutions. In: Proceedings of the IEEE computer society conference on computer vision and pattern recognition, vol. 07-12-June. IEEE Computer Society. ISBN 9781467369640, 2015. p. 1–9.

[88] Szegedy C, Vanhoucke V, Ioffe S, Shlens J, Wojna Z. Rethinking the inception architecture for computer vision. In: Proceedings of the IEEE computer society conference on computer vision and pattern recognition, vol. 2016-Decem. IEEE Computer Society. ISBN 9781467388504, 2016. p. 2818–26.

[89] Szegedy C, Ioffe S, Vanhoucke V, Alemi AA. Inception-v4, inception-ResNet and the impact of residual connections on learning. In: 31st AAAI conference on artificial intelligence, AAAI 2017. AAAI Press; 2017. p. 4278–84.

[90] Graves A, Wayne G, Danihelka I. Neural Turing machines. Available from: arXiv:1410.5401, 2014.

[91] Bahdanau D, Cho KH, Bengio Y. Neural machine translation by jointly learning to align and translate. In: 3rd international conference on learning representations, ICLR 2015 - conference track proceedings. International Conference on Learning Representations, ICLR; 2015. Available from: https://arxiv.org/abs/1409.0473v7.

[92] Luong MT, Pham H, Manning CD. Effective approaches to attention-based neural machine translation. In: Conference proceedings - EMNLP 2015: conference on empirical methods in natural language processing. Association for Computational Linguistics (ACL). ISBN 9781941643327, 2015. p. 1412–21.

[93] Vaswani A, Shazeer N, Parmar N, Uszkoreit J, Jones L, Gomez AN, et al. Attention is all you need. Tech. Rep. 2017.

[94] Cheng J, Dong L, Lapata M. Long short-term memory-networks for machine reading. In: EMNLP 2016 - conference on empirical methods in natural language processing, proceedings, vol. 2(3); 2016. p. 551–61.

[95] Xu K, Ba JL, Kiros R, Cho K, Courville A, Salakhutdinov R, et al. Show, attend and tell: neural image caption generation with visual attention. In: 32nd international conference on machine learning, ICML 2015, vol. 3. International Machine Learning Society (IMLS). ISBN 9781510810587, 2015. p. 2048–57.

[96] Qin C, Schlemper J, Caballero J, Price AN, Hajnal JV, Rueckert D. Convolutional recurrent neural networks for dynamic MR image reconstruction. IEEE Trans Med Imaging 2019;38(1):280–90. https://doi.org/10.1109/TMI.2018.2863670.

[97] Hammernik K, Klatzer T, Kobler E, Recht MP, Sodickson DK, Pock T, et al. Learning a variational network for reconstruction of accelerated MRI data. Magn Reson Med 2018;79(6):3055–71. https://doi.org/10.1002/mrm.26977. Available from: https://github.com/VLOGroup/mri-variationalnetwork.

[98] Mardani M, Gong E, Cheng JY, Pauly J, Xing L. Recurrent generative adversarial neural networks for compressive imaging. In: IEEE international workshop on computational advances in multi-sensor adaptive processing (CAMSAP); 2017. p. 1–5.

[99] Lundervold AS, Lundervold A. An overview of deep learning in medical imaging focusing on MRI; 2019.

[100] Knoll F, Zbontar J, Sriram A, Muckley MJ, Bruno M, Defazio A, et al. fastMRI: a publicly available raw k-space and DICOM dataset of knee images for accelerated MR image reconstruction using machine learning. Radiology Artif Intell 2020;2(1):e190007.

[101] Lin DJ, Johnson PM, Knoll F, Lui YW. Artificial intelligence for MR image reconstruction: an overview for clinicians; 2020.

[102] Hammernik K, Knoll F. Machine learning for image reconstruction. In: Rueckert D, Fichtinger G, Zhou SK, editors. Handbook of medical image computing and computer assisted intervention. Elsevier. ISBN 9780128161760, 2020. p. 25–64.

[103] Ronneberger O, Fischer P, Brox T. U-Net: convolutional networks for biomedical image segmentation. In: International conference on medical image computing and computer assisted intervention; 2015. p. 234–41.

[104] Han Y, Yoo J, Kim HH, Shin HJ, Sung K, Ye JC. Deep learning with domain adaptation for accelerated projection-reconstruction MR. Magn Reson Med 2018;80(3):1189–205. https://doi.org/10.1002/mrm.27106.

[105] Sandino CM, Lai P, Vasanawala SS, Cheng JY. Accelerating cardiac cine MRI using a deep learning-based ESPIRiT reconstruction. Magn Reson Med 2020:mrm.28420. https://doi.org/10.1002/mrm.28420. Available from: http://arxiv.org/abs/1911.05845. https://onlinelibrary.wiley.com/doi/abs/10.1002/mrm.28420.

[106] Chen F, Taviani V, Malkiel I, Cheng JY, Tamir JI, Shaikh J, et al. Variable-density single-shot fast spin-echo MRI with deep learning reconstruction by using variational networks. Radiology 2018;289(2):1–8. https://doi.org/10.1148/radiol.2018180445.

[107] Fuin N, Bustin A, Küstner T, Oksuz I, Clough J, King AP, et al. A multi-scale variational neural network for accelerating motion-compensated whole-heart 3D coronary MR angiography. Magn Reson Imaging 2020;70:155–67. https://doi.org/10.1016/j.mri.2020.04.007.

[108] Roth S, Black MJ. Fields of experts. Int J Comput Vis 2009;82(2):205–29.

[109] Mardani M, Gong E, Cheng JY, Vasanawala SS, Zaharchuk G, Xing L, et al. Deep generative adversarial neural networks for compressive sensing (GANCS) MRI. IEEE Trans Med Imaging 2019;38(1):167–79. https://doi.org/10.1109/TMI.2018.2858752. Available from: https://github.com/gongenhao/GANCS.

[110] Liu F, Samsonov A, Chen L, Kijowski R, Feng L. SANTIS: Sampling-Augmented Neural neTwork with Incoherent Structure for MR image reconstruction. Magn Reson Med 2019;82(5):1890–904. https://doi.org/10.1002/mrm.27827.

[111] Aggarwal HK, Mani MP, Jacob M. MoDL: model based deep learning architecture for inverse problems. IEEE Trans Med Imaging 2019;38(2):394–405. https://doi.org/10.1109/TMI.2018.2865356. Available from: https://github.com/hkaggarwal/modl.

[112] Liu J, Sun Y, Eldeniz C, Gan W, An H, Kamilov US. RARE: image reconstruction using deep priors learned without groundtruth. IEEE J Sel Top Signal Process 2020;14(6):1088–99. https://doi.org/10.1109/jstsp.2020.2998402.

[113] Shi J, Liu Q, Wang C, Zhang Q, Ying S, Xu H. Super-resolution reconstruction of MR image with a novel residual learning network algorithm. Phys Med Biol 2018;63(8):085011. https://doi.org/10.1088/1361-6560/aab9e9.

[114] Lee D, Yoo J, Tak S, Ye JC. Deep residual learning for accelerated MRI using magnitude and phase networks. IEEE Trans Biomed Eng 2018;65(9):1985–95. https://doi.org/10.1109/TBME.2018.2821699.

[115] Wang S, Su Z, Ying L, Peng X, Zhu S, Liang F, et al. Accelerating magnetic resonance imaging via deep learning. In: IEEE international symposium on biomedical imaging, vol. 2016-June. IEEE Computer Society. ISBN 9781479923502, 2016. p. 514–7.

[116] Han Y, Sunwoo L, Ye JC. K-space deep learning for accelerated MRI. IEEE Trans Med Imaging 2020;39(2):377–86. https://doi.org/10.1109/TMI.2019.2927101.

[117] Hosseini SAH, Zhang C, Weingärtner S, Moeller S, Stuber M, Ugurbil K, et al. Accelerated coronary MRI with sRAKI: a database-free self-consistent neural network k-space reconstruction for arbitrary undersampling. PLoS ONE 2020;15(2):e0229418. https://doi.org/10.1371/journal.pone.0229418.

[118] Cheng J, Mardani M, Alley M, Pauly J, Vasanawala S. DeepSPIRiT: generalized parallel imaging using deep convolutional neural networks. In: Proceedings of ISMRM 26th annual meeting; 2018. p. 0570.

[119] Yang Y, Sun J, Li H, Xu Z. ADMM-Net: a deep learning approach for compressive sensing MRI. In: Advances in neural information processing systems; 2017. p. 10–8. Available from: https://github.com/yangyan92/Deep-ADMM-Net.

[120] Vapnik VN. The nature of statistical learning theory, No. 4. Springer Science & Business Media; 2013. 409 pp.

[121] Uv Luxburg, Schölkopf B. Statistical learning theory: models, concepts, and results, vol. 10. North-Holland; 2011. p. 651–706.

[122] IXI dataset. Available from: https://brain-development.org/ixi-dataset/.

[123] ADNI. Available from: http://adni.loni.usc.edu/.

[124] OASIS. Available from: www.oasis-brains.org.

[125] OpenfMRI. Available from: http://openfmri.org/.

[126] fastMRI. Available from: https://fastmri.org/dataset/.
[127] CHAOS. Available from: https://chaos.grand-challenge.org/.
[128] Deng Jia, Dong Wei, Socher R, Li Li-Jia, Li Kai, Fei-Fei Li. ImageNet: a large-scale hierarchical image database. In: IEEE conference on computer vision and pattern recognition; 2009. p. 248–55.
[129] Küstner T, Hepp T, Fischer M, Schwartz M, Fritsche A, Häring HU, et al. Fully automated and standardized segmentation of adipose tissue compartments by deep learning in three-dimensional whole-body MRI of epidemiological cohort studies. Available from: arXiv:2008.02251, 2020.
[130] Dar SUH, Özbey M, Çatlı AB, Çukur T. A transfer-learning approach for accelerated MRI using deep neural networks. Magn Reson Med 2020;84(2):663–85. https://doi.org/10.1002/mrm.28148.
[131] Knoll F, Hammernik K, Kobler E, Pock T, Recht MP, Sodickson DK. Assessment of the generalization of learned image reconstruction and the potential for transfer learning. Magn Reson Med 2019;81(1):116–28. https://doi.org/10.1002/mrm.27355.
[132] Oksuz I, Ruijsink B, Puyol-Antón E, Clough JR, Cruz G, Bustin A, et al. Automatic CNN-based detection of cardiac MR motion artefacts using k-space data augmentation and curriculum learning. Med Image Anal 2019;55:136–47. https://doi.org/10.1016/j.media.2019.04.009.
[133] Chen C, Qin C, Qiu H, Ouyang C, Wang S, Chen L, et al. Realistic adversarial data augmentation for MR image segmentation. In: International conference on medical image computing and computer assisted intervention; 2020. Available from: http://arxiv.org/abs/2006.13322.
[134] Gibson E, Li W, Sudre C, Fidon L, Shakir DI, Wang G, et al. NiftyNet: a deep-learning platform for medical imaging. Comput Methods Programs Biomed 2018;158:113–22. https://doi.org/10.1016/j.cmpb.2018.01.025.
[135] Pérez-García F, Sparks R, Ourselin S. TorchIO: a Python library for efficient loading, preprocessing, augmentation and patch-based sampling of medical images in deep learning. Available from: arXiv:2003.04696, 2020.
[136] Mehrtash A, Pesteie M, Hetherington J, Behringer PA, Kapur T, Wells WM, et al. DeepInfer: open-source deep learning deployment toolkit for image-guided therapy. In: Webster RJ, Fei B, editors. Medical imaging 2017: image-guided procedures, robotic interventions, and modeling, vol. 10135. SPIE. ISBN 9781510607156, 2017. p. 101351K.
[137] Pieper S, Lorensen B, Schroeder W, Kikinis R. The NA-MIC Kit: ITK, VTK, pipelines, grids and 3D slicer as an open platform for the medical image computing community. In: 2006 3rd IEEE international symposium on biomedical imaging: from nano to macro - proceedings, vol. 2006. ISBN 0780395778, 2006. p. 698–701.
[138] Clarkson MJ, Zombori G, Thompson S, Totz J, Song Y, Espak M, et al. The NifTK software platform for image-guided interventions: platform overview and NiftyLink messaging. Int J Comput Assisted Radiol Surg 2015;10(3):301–16. https://doi.org/10.1007/s11548-014-1124-7.
[139] Wolf I, Vetter M, Wegner I, Böttger T, Nolden M, Schöbinger M, et al. The medical imaging interaction toolkit. Med Image Anal 2005;9(6):594–604. https://doi.org/10.1016/j.media.2005.04.005.
[140] Nora - the medical imaging platform. Available from: http://www.nora-imaging.com/.
[141] Knoll F, Hammernik K, Zhang C, Moeller S, Pock T, Sodickson DK, et al. Deep-learning methods for parallel magnetic resonance imaging reconstruction: a survey of the current approaches, trends, and issues. IEEE Signal Process Mag 2020;37(1):128–40. https://doi.org/10.1109/MSP.2019.2950640.
[142] Wang G, Ye JC, Mueller K, Fessler JA. Image reconstruction is a new frontier of machine learning. IEEE Trans Med Imaging 2018;37(6):1289–96.
[143] Fessler JA. Optimization methods for MR image reconstruction (long version). Available from: arXiv:1903.03510, 2019.
[144] LeCun Y, Cortes C, Burges C. MNIST handwritten digit database. ATT Labs [Online] 2010;2. Available from: http://yann.lecun.com/exdb/mnist.
[145] Kreutz-Delgado K. The complex gradient operator and the CR-calculus. Available from: arXiv:0906.4835, 2009.
[146] Wirtinger W. Zur formalen Theorie der Funktionen von mehr komplexen Veränderlichen. Math Ann 1927;97(1):357–75. https://doi.org/10.1007/BF01447872.

PART 3

Reconstruction Methods for Nonlinear Forward Models in MRI

CHAPTER

Imaging in the Presence of Magnetic Field Inhomogeneities

12

Bradley P. Sutton[a,b] and Fan Lam[a,b]

[a]*Department of Bioengineering, University of Illinois Urbana Champaign, Urbana, IL, United States*
[b]*Beckman Institute for Advanced Science and Technology, University of Illinois Urbana Champaign, Urbana, IL, United States*

12.1 Introduction

As seen in Chapter 1, accurate spatial localization of spins with MRI relies on a known relationship between the location of a spin in the three-dimensional (3D) space and the magnetic field strength at that location. In the case of Fourier imaging, the linear relationship between spatial position and the magnetic field would result in a one-to-one mapping between the frequency or phase evolution of a spin's signal and its location, ensuring that a Fourier transform can be performed to resolve the signals into the correct locations. For this mapping to be accurate, the only variations in the magnetic field that the spins in an object should experience would be due to the uniform B_0 main magnetic field and the linear magnetic-field gradients. This uniform background magnetic field is essential for generating an accurate, undistorted image using a Fourier transform reconstruction.

However, several physical effects exist that can disrupt the uniformity of the background magnetic field and, therefore, disrupt the mapping relationship. This can result in misplacement of the signals by the Fourier transform, referred to as image distortion. It can also result in several other artifacts which we will discuss later, including signal loss and deviations in the planned k-space trajectory. Without mitigation and correction, these can result in significant errors in interpretation of an image. Although advances in imaging hardware, sampling strategies, and reconstruction approaches have enabled the reduction of the impact of some of these artifacts, the move to higher magnetic field strengths will increase the importance of considering field inhomogeneity impacts on data acquisitions and image reconstructions. This chapter provides a timely review of the sources and effects of B_0 field inhomogeneity in MRI, as well as existing solutions to address these effects for improved image reconstruction. Although many of the examples presented in this chapter are from brain imaging, the physics models and correction methodology are applicable to other body parts [1–3].

12.2 Disruptions to the homogeneity of the magnetic field

Efforts to create a uniform magnetic field are limited by the magnetic properties of the sample in the magnet. Materials, including tissues in the body, possess a property called magnetic susceptibility, χ. Magnetic susceptibility refers to the magnetizability of a material, and it affects the magnetic field that

Advances in Magnetic Resonance Technology and Applications, Volume 7, ISSN 2666-9099. https://doi.org/10.1016/B978-0-12-822726-8.00023-3

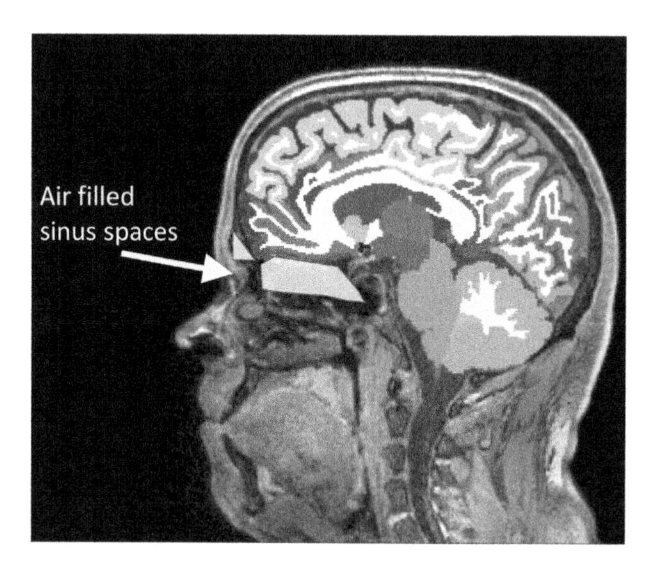

FIGURE 12.1

Many areas of functional significance in the brain are near sinuses, resulting in significant field homogeneity disruptions due to large magnetic-susceptibility differences between sinuses and the surrounding brain tissues. Examples of the problematic air–tissue interfaces are shown as a triangle and box at the anterior part of the brain.

a spin experiences in that material, and, consequently, the signals generated. Materials with a positive magnetic susceptibility ($\chi > 0$) are called paramagnetic, and materials with a negative magnetic susceptibility ($\chi < 0$) are called diamagnetic. Magnetic susceptibility shapes the magnetic field around that tissue based on differences between the magnetic susceptibility of surrounding tissues. Of particular importance for brain and body MR imaging is the significant magnetic susceptibility differences between soft tissues ($\chi_{tissue} = -9 \times 10^{-6}$) and air ($\chi_{air} = 0.4 \times 10^{-6}$) [4]. Such a difference leads to large field inhomogeneity around the air–tissue interfaces. For example, the sinus spaces around the brain are air-filled and close to interesting regions that are implicated in various pathologies and cognitive functions, as shown in Fig. 12.1. The disruption to the magnetic field's uniformity can make detection of signals and depiction of accurate functional anatomy in these regions challenging.

12.3 Field inhomogeneity effects on imaging

Given this large magnetic susceptibility difference between air and tissue of almost 10 ppm (parts per million), the magnetic-field uniformity will be significantly disrupted around air–tissue interfaces in the body and the brain. This is shown in Fig. 12.2 for an axial and sagittal image of the brain at a magnetic field strength of 3 T. Although magnetic fields are measured in Tesla, often magnetic field inhomogeneity (or nonuniformity) will be given with units of Hz. Magnetic field and frequency are related through the Larmor equation and the gyromagnetic ratio. If we consider magnetic field

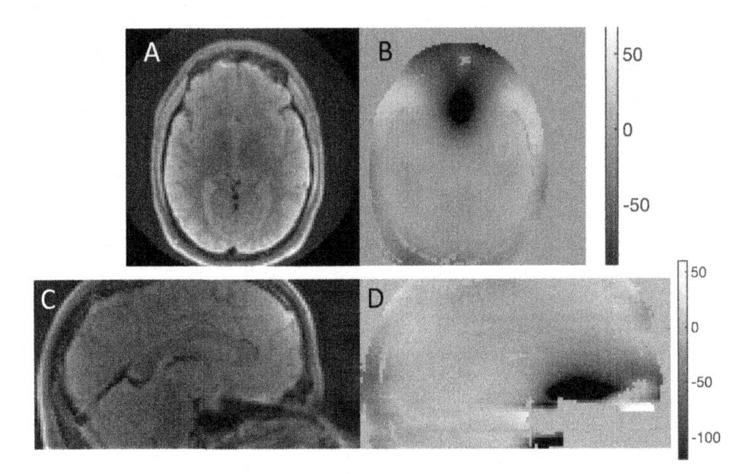

FIGURE 12.2

Structural images of the brain and the magnetic field maps corresponding to the same anatomical locations. **A**) An axial brain image showing anatomical structures. **B**) Magnetic field map in Hz measured for the same slice as in A. **C**) A sagittal brain image. **D**) Magnetic field map in Hz at the same slice as in C. Note the large inhomogeneities around the sinuses.

offset (B_{FM}, magnetic field map) to be the deviation from the uniform magnetic field (B_0), such that $B = B_0 + B_{FM}$, then the off-resonance frequency (deviation from the Larmor frequency) can be given as

$$\omega_{FM} = 2\pi f_{FM} = \gamma B_{FM}, \tag{12.1}$$

where ω_{FM} is in rad/sec and f_{FM} is in Hz. When there is a 10-ppm offset in the magnetic field at an air–tissue interface at 3 T (where f_0 is 128 MHz), then we can end up with an off-resonance frequency of 1.28 kHz. For functional MRI (fMRI) applications where the bandwidth per pixel can be on the order of tens of Hz, this can create significant image distortions. An example of the image distortions that can be experienced due to the magnetic field inhomogeneity is shown in Fig. 12.3.

12.3.1 Three types of effects disrupting the image and its information

As already mentioned, there are several impacts of magnetic field inhomogeneity on the imaging process. We will structure our discussion by considering three distinct effects: image distortion, signal loss, and k-space trajectory distortions. Each of these impacts a different part of the image acquisition and reconstruction process and will be discussed separately in this chapter, although, in practice, they are not fully independent. In this section, we will introduce the concept and implications of each of these disruptions. In subsequent sections, we will focus on ways to correct for these artifacts. In addition to these direct effects on imaging and image encoding, which is the focus of this chapter, there are additional impacts of magnetic field inhomogeneity on RF excitation [5,6], spectroscopy [7,8], and other aspects of MRI that will not be discussed (e.g., susceptibility weighted imaging and relaxation effects

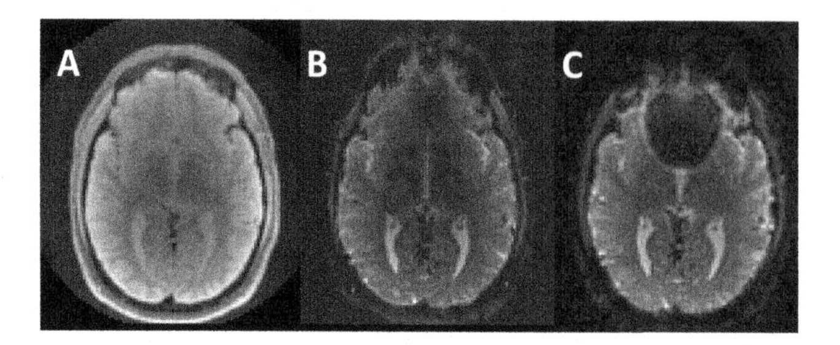

FIGURE 12.3

Magnetic field inhomogeneity induced image distortion for the same axial slice in Fig. 12.2. **A**) Structural image showing the true geometry. **B**) EPI image showing the geometric distortion in the phase encode (y-)direction in the region of large field inhomogeneity in the anterior portion of the brain. **C**) Image reconstructed from a spiral acquisition for the same slice showing circular image distortion.

to name a few) [9,10]. In addition, the presence of metal in the body can cause significant and highly localized field inhomogeneity, requiring different approaches [11] that will not be addressed here.

The most commonly encountered and addressed artifact from magnetic field inhomogeneity is image distortion. This artifact results from the disruption to the linear relationship between the magnetic field and spatial coordinates (introduced by linear gradients), resulting in Fourier transform reconstructions to place the signal intensity for a pixel in the wrong place in the reconstructed image. The extent of the image displacement depends on the amount of off-resonance frequency caused by the magnetic-field inhomogeneity, the timing of the acquisition, and the trajectory through k-space [12]. For Cartesian acquisitions, such as echo planar imaging (EPI), the distortion will be primarily in the slow encoding direction, i.e., the phase encode (PE) axis [13]. For non-Cartesian acquisitions, the distortions will be primarily in the direction of the slow traversal through k-space, such as the radial direction for spiral acquisitions [14,15]. It is important to note that this type of distortion exists for all types of imaging acquisitions, including both gradient echo (GRE) and spin echo (SE) acquisitions, and can cause significant errors in subsequent image analysis (e.g., for functional brain activation and network analysis). An example of the image distortion experienced during an EPI acquisition is shown in Fig. 12.3, showing distortion in primarily the PE direction, i.e., the y-direction in the image. The impact and strategies to correct for field inhomogeneity induced image distortions will be addressed in Section 12.4. It is also important to note that chemical shift is another source of off-resonance frequency and will also result in spatial distortions (more discussion can be found in Chapter 14).

A second artifact resulting from field inhomogeneity is intravoxel signal dephasing leading to signal loss since spins within a voxel accumulate phase relative to each other due to different rates of precession from the different magnetic fields experienced. This type of signal loss is prevalent in GRE imaging, where no refocusing pulse is used. The longer the echo time (TE) is used, the more time there is for accumulating phase differences among spins in a voxel and the more signal loss will result, over and above T_2 losses, as shown in Fig. 12.4. In contrast, for SE imaging, the phase that accumulates due to magnetic field inhomogeneity before the refocusing 180° pulse refocuses from the static field inhomogeneity after the refocusing pulse. In the absence of motion of the spins, the field inhomogeneity

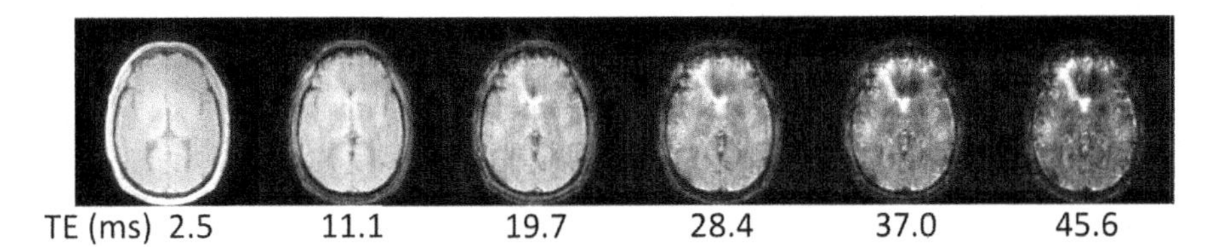

FIGURE 12.4

Signal dephasing occurs as spins within a voxel are allowed to accumulate different phases, such as when gradient echo images are acquired at longer echo times (TE). Shown is an axial acquisition of a gradient echo image at TE's ranging from 2.5 ms to 45.6 ms with a 4-mm slice thickness at 3 T. Signal loss due to within voxel dephasing can be clearly observed for images at longer TE's.

does not have a large impact. There are a few ways to address the within-voxel signal dephasing effect on the GRE image formation process, and these will be addressed in Section 12.5.

The last artifact due to magnetic field inhomogeneity that will be discussed relates to the magnetic-field gradients that will exist across an object due to the distributions in the susceptibility-induced variations in the background magnetic field. Just as the applied linear magnetic-field gradients are used to traverse k-space in a planned manner for adequate sampling, the background gradients of the magnetic field result in unintended spatial encoding and cause deviations in the planned sampling trajectory, as shown in Fig. 12.5. These deviations are spatially varying and depend on the magnetic field inhomogeneity gradients at each position in the image. These can cause deviations in sample density and field of view that can have significant impacts on the ability to reconstruct an image. K-space trajectory distortions will be discussed further in Section 12.6.

12.3.2 Field inhomogeneity and the signal equation

To rigorously describe the impact of magnetic-field inhomogeneity on the MR imaging process, first we will examine the impact on the signal equation for MR. This formal treatment will provide insights into the various effects, which we will explore in more detail later. The magnetic field inhomogeneity at a voxel will cause spins in that voxel to precess at a different off-set frequency relative to the Larmor frequency defined on the nominal magnetic-field strength. This results in a modification of the signal equation to include this background magnetic-field frequency, $f_{FM}(\mathbf{r})$ in Hz or $\omega_{FM}(\mathbf{r})$ in rad/sec, as

$$s(t_m) = \int_{FOV} x(\mathbf{r}) e^{-i2\pi \mathbf{k}_m \cdot \mathbf{r}} e^{-i2\pi f_{FM}(\mathbf{r}) t_m} d\mathbf{r} \tag{12.2}$$

$$s(t_m) = \int_{FOV} x(\mathbf{r}) e^{-i2\pi \mathbf{k}_m \cdot \mathbf{r}} e^{-i\omega_{FM}(\mathbf{r}) t_m} d\mathbf{r}, \tag{12.3}$$

where $x(\mathbf{r})$ is the continuous transverse magnetization that we are imaging, FOV is the spatial field of view, $\mathbf{r}$ is the spatial variable in 2- or 3-dimensions, $\mathbf{k}$ is the k-space variable in 2- or 3-dimensions, t is the time since the RF pulse for GRE or the center of the refocused echo for SE, m indexes the readout dimension of the data sampling event, and $s(t_m)$ denotes the k-space data acquired at t_m. Note that the time in the readout for a GRE acquisition would be $t_m = \text{TE} + m\Delta t$, where TE is the echo time and

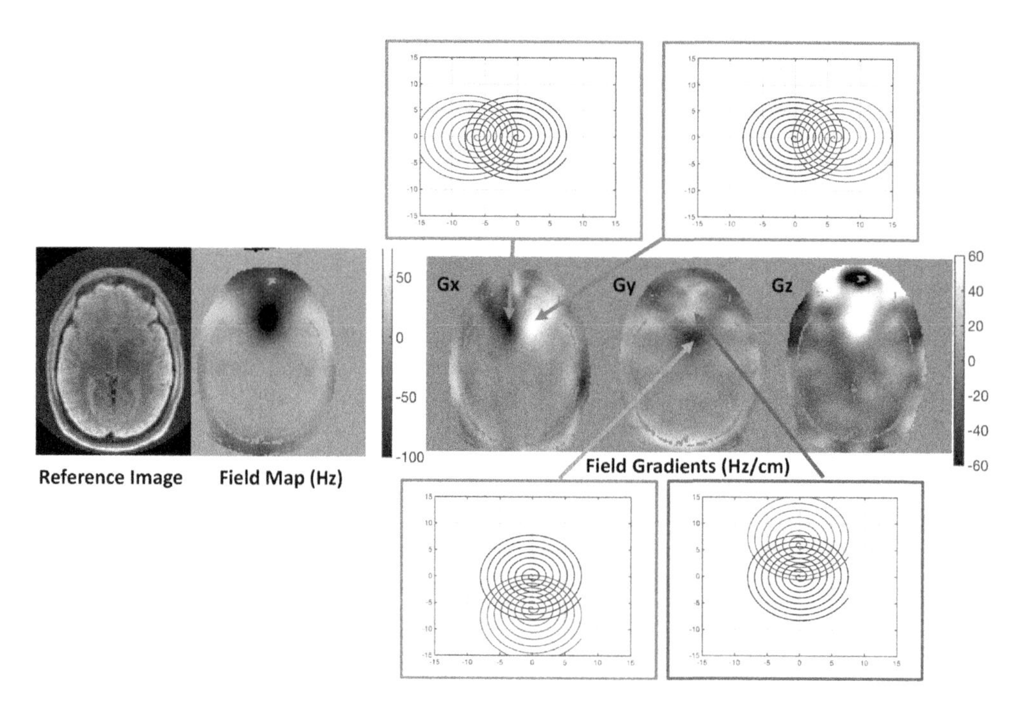

FIGURE 12.5

Anatomical image and magnetic field map in Hz, along with magnetic-field gradients in the x (G_x), y (G_y), and z (G_z) directions. Simulations of spiral trajectories for particular voxels with x- and y-magnetic field gradients are shown in blue (dark gray in print version), compared to the nominal k-space trajectory shown in black. Simulations were performed for ± 30 Hz/cm in x- and y-directions individually.

Δt is the time spacing during the acquisition, and where the bandwidth, BW, of the data acquisition readout is such that $BW = \frac{1}{\Delta t}$.

This equation is formulated as continuous in space, $\mathbf{r}$, but our images will be represented as pixels of constant intensity so we instead approximate our object as a collection of square pixels using the voxel-indicator basis functions, $b_v(\mathbf{r})$, that are centered at each location $\mathbf{r_n}$, for $n = 1, \ldots, N$ where N is the number of voxels in the image. The voxel indicator basis functions are small rectangular functions giving a value of 1 within a voxel and 0 elsewhere, as

$$b_v(\mathbf{r} - \mathbf{r_n}) = \begin{cases} 1, & \text{if } \frac{\mathbf{r}-\mathbf{r_n}}{\Delta_r} \leq \frac{1}{2} \\ 0, & \text{else,} \end{cases} \tag{12.4}$$

with Δ_r being the voxel size. Using the voxel indicator (or other suitable basis function), we can express the continuous image as an approximation of coefficients and basis functions, as

$$x(\mathbf{r}) \approxeq \sum_{n=0}^{N-1} x_n b_v(\mathbf{r} - \mathbf{r_n}). \tag{12.5}$$

In order to get a discrete space model for the signal equation in Eq. (12.3), we must also use a basis expansion of the field inhomogeneity map or field map term. Although representing the image as pixels is straightforward, the expansion basis for the field map will have a significant impact on the signal that is generated by the model. For now, we will expand the field map with the same voxel indicator function as

$$\omega_{FM}(\mathbf{r}) \approxeq \sum_{n=0}^{N-1} \omega_n b_v(\mathbf{r} - \mathbf{r_n}). \tag{12.6}$$

Using this discrete space model for the spatial basis functions and the field map, we can rewrite Eq. (12.3) as

$$s(t_m) = B_v(\mathbf{k}_m) \sum_{n=0}^{N-1} x_n e^{-i2\pi \mathbf{k}_m \cdot \mathbf{r_n}} e^{-i\omega_n t_m}, \tag{12.7}$$

where $B_v(\mathbf{k})$ is the Fourier transform of the voxel indicator basis function that we used in Eq. (12.4), which is just $B_v(\mathbf{k}) = \mathrm{sinc}(k_x \Delta_x)\mathrm{sinc}(k_y \Delta_y)\mathrm{sinc}(k_z \Delta_z)$ [16].

Now that we have a signal model that incorporates the magnetic-field inhomogeneity or field map, we can evaluate the impact of field inhomogeneity on the image. We can also use our model to enable a variety of reconstruction techniques to estimate images that are free from this distortion. For incorporation of this model into iterative reconstructions, we can embed the field-inhomogeneity terms into the MR encoding operator, $\mathbf{E}$, where the m-th time point and the n-th spatial location impact the encoding operator as follows:

$$\mathbf{E}_{m,n} = B_v(\mathbf{k}_m) e^{-i2\pi \mathbf{k}_m \cdot \mathbf{r_n}} e^{-i\omega_n t_m}, \tag{12.8}$$

such that our signal model becomes

$$\mathbf{s} = \mathbf{E}\mathbf{x} + \eta, \tag{12.9}$$

where $\mathbf{s} = [s(t_1), \ldots, s(t_M)]^T$, for M time points in the data acquisition, $\mathbf{x} = [x_1, \ldots, x_N]^T$, for N pixels in the image, and $\eta = [\eta_1, \ldots, \eta_M]^T$, for the noise at each sampled time point.

From Eq. (12.7), it is easy to see how image distortion results from unmodeled field inhomogeneity during an acquisition through the accumulation of phase due to field inhomogeneity and the progression of k-space encoding during the acquisition. Consider an EPI trajectory where the k-space trajectory travels consistently through the k_y phase encode axis during the acquisition. In this case, we can find an approximate relationship between k_y and time, such that $t_m \approx c k_m$ in this direction. For a voxel that has a field inhomogeneity value of ω_n, we can look at the encoding kernel which is

$$e^{-i2\pi k_m y_n} e^{-i\omega_n t_m} = e^{-i2\pi k_m y_n} e^{-i\omega_n c k_m} \tag{12.10}$$

$$= e^{-i2\pi k_m y_n} e^{-i2\pi \frac{\omega_n}{2\pi} c k_m} \tag{12.11}$$

$$= e^{-i2\pi k_m \left(y_n + \frac{c\omega_n}{2\pi}\right)}. \tag{12.12}$$

If a Fourier transform, without field inhomogeneity correction, is used for reconstruction, the spatial representation of the object will be distorted. Image intensity that should have showed up at pixel y_n will

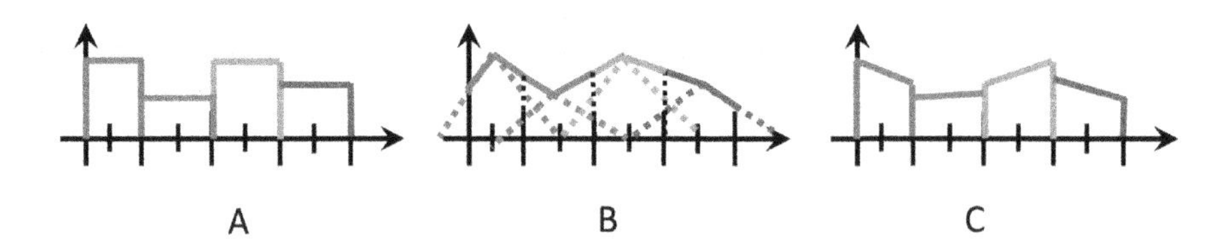

FIGURE 12.6

Basis functions that can be used for representing the continuous image and field map for discretization: **A)** voxel indicator basis, **B)** triangle basis functions that create piecewise continuous and piecewise linear representation, **C)** the piecewise linear basis function for which it is computationally convenient within voxel dephasing.

instead appear at a location $y_n + \frac{c\omega_n}{2\pi}$. The magnitude of the spatial misplacement will be $\frac{c\omega_n}{2\pi} = cf_{FM}$. Recall that the constant c is the scaling of a step in sampling time to a step in k-space. Consider the case where c is small, meaning that we are traveling through k-space quickly, the size of the shift in image space will be small. However, for high-resolution EPI acquisitions, c would be larger since it takes more time to sample each line of k-space (one k_y line), and the distortions would also be larger.

The signal model in Eq. (12.7) allows one to achieve effective field inhomogeneity-corrected image reconstruction. But the evaluations of this model is computationally expensive because the Fast Fourier Transform (FFT) can not be directly applied. Methods have been developed to address this computational challenge, and we will discuss some of those in Section 12.4.4.1. These methods use time segmentation or frequency segmentation to break up the computation of the signal equation into several components and add the results back together to approximate the full signal equation. By fixing time or frequency during these sub-evaluations of the signal equation, they can leverage the FFT to achieve faster computations.

12.3.2.1 *Other basis expansions enable the modeling of additional artifacts*

In Eq. (12.4), we used a voxel indicator's basis function to transform our continuous space signal equation into a discrete space signal equation focused on reconstructing an image of voxels with one coefficient per uniform voxel. Although this makes sense for a representation of the image, magnetic-field inhomogeneity can cause signal dephasing due to a distribution of the magnetic field within a voxel. A voxel indicator basis function does not enable us to model variations in the magnetic field within a voxel. To accommodate this within-voxel dephasing, we need to represent the magnetic-field inhomogeneity with a basis function that varies within a voxel.

Ideal expansions would result in continuous functions of the magnetic-field distribution to have a realistic approximation. However, the overlapping basis functions, such as triangle basis functions, can cause mathematical difficulties in trying to incorporate such models into efficient image-reconstruction routines [16,17]. So, to maintain convenience of computation, we can choose piece-wise linear basis functions that have a mean value and linear slope for each voxel. This results in piecewise linear approximations of the field distribution, but not a continuous function. The various bases discussed are shown in Fig. 12.6, including the voxel indicator function, triangle basis functions, and piecewise linear basis functions.

The piecewise linear basis expansion for the field map requires both a mean field-inhomogeneity value, ω_n for each of the N pixels, but it also requires a slope of the field map in each direction, $\mathbf{G}_n = [X_n, Y_n, Z_n]^T$ which are measured in the x-, y-, and z-direction in rad/s/pixel [17,18]. This gives the following approximations for the image and the field map, where $b_v(\mathbf{r})$ is the same voxel indicator function we used before,

$$x(\mathbf{r}) \approxeq \sum_{n=0}^{N-1} x_n b_v(\mathbf{r} - \mathbf{r_n}); \tag{12.13}$$

$$\omega_{FM}(\mathbf{r}) \approxeq \sum_{n=0}^{N-1} \left(\omega_n + \mathbf{G}_n^T \cdot \frac{\mathbf{r} - \mathbf{r_n}}{\Delta_r} \right) b_v(\mathbf{r} - \mathbf{r_n}). \tag{12.14}$$

If we integrate this approximation into the continuous signal equation in Eq. (12.3), we end up with the following signal model:

$$s(t_m) = \sum_{n=0}^{N-1} B_v(\mathbf{k}_m, \mathbf{G_n}, t_m) x_n e^{-i2\pi \mathbf{k}_m \cdot \mathbf{r_n}} e^{-i\omega_n t_m}, \tag{12.15}$$

where the function B_v is now a more complicated Fourier transform of the voxel indicator function that includes the field inhomogeneity gradient terms. This is given as

$$B_v(\mathbf{k}_m, \mathbf{G_n}, t_m) = \\ \operatorname{sinc}\left(k_{x,m}\Delta_x + \frac{X_n t_m}{2\pi}\right) \operatorname{sinc}\left(k_{y,m}\Delta_y + \frac{Y_n t_m}{2\pi}\right) \operatorname{sinc}\left(k_{z,m}\Delta_z + \frac{Z_n t_m}{2\pi}\right). \tag{12.16}$$

It is important to consider the units of all the terms in this signal equation model when implementing it, including the units of k-space, the field map, the field map gradients, and the voxel size.

We can update our encoding operator, $\mathbf{E}$, to include this advanced signal model. Entries of $\mathbf{E}$ now become

$$\mathbf{E}_{m,n} = B_v(\mathbf{k}_m, \mathbf{G_n}, t_m) e^{-i2\pi \mathbf{k}_m \cdot \mathbf{r_n}} e^{-i\omega_n t_m}. \tag{12.17}$$

Note that the $B_W(\mathbf{k}_m, \mathbf{G_n}, t_m)$ term is computationally expensive to compute as it now depends on both spatial position and time (k-space), hence, we may ask: Why is it still inside the summation in Eq. (12.15) compared to Eq. (12.7). This formulation again precludes the use of FFT for evaluation and requires the use of explicit calculation of the signal equation. Recently, methods have been developed to approximate the full encoding matrix with a low-rank approximation [19,20]. These approximations enable speedups in computation of several orders of magnitude, making this more complete signal model feasible to use in iterative reconstruction schemes. We will discuss the image reconstruction approach that incorporates this model further in Section 12.5.1.

By modeling the linear components of the variations in magnetic field within a voxel, this signal model enables us to address some of the signal dephasing that occurs during imaging. As an example of that, consider 2D gradient echo imaging. For this case, there is no k_z encoding and the $B_v(\mathbf{k}_m, \mathbf{G_n}, t_m)$

term reduces to

$$B_v(\mathbf{k}_m, \mathbf{G}_\mathbf{n}, t_m) = \text{sinc}\left(k_{x,m}\Delta_x + \frac{X_n t_m}{2\pi}\right)\text{sinc}\left(k_{y,m}\Delta_y + \frac{Y_n t_m}{2\pi}\right)\text{sinc}\left(\frac{Z_n t_m}{2\pi}\right). \tag{12.18}$$

Focusing only on the z-direction, i.e., the direction through the imaging slice and the through-plane direction, we can see that the signal will decay as a sinc() function with increasing decay as the through-plane gradient and time increase. If there were no in-plane components (x- and y-directed gradients in the magnetic field map), then the signal would exhibit the through-plane signal dephasing causing signal loss that can not be recovered with a Fourier reconstruction, especially for gradient echo acquisitions with long echo times.

12.3.3 Field inhomogeneity mitigation methods

Before leaving the description of the basic physics of magnetic-field inhomogeneity, we should note that there are several ways to decrease the impact of field inhomogeneity on the image. These can be through hardware-based approaches that minimize the magnitude of the field inhomogeneity distribution over the sample to be imaged. Or they can recognize the dependencies in the signal equation and choose imaging parameters that result in smaller artifacts.

From the early days of functional MRI of the brain, MR engineers realized that the best way to minimize the impact of field inhomogeneity was to minimize the amount of field inhomogeneity in the first place. This can be helped by careful shimming procedures and the use of higher-order shims that are common in modern systems [21–23]. Additionally, the magnetic field distribution can be impacted by placing items in and around the head to reduce the field variations, especially near the orbitofrontal region of the brain. These items included the development of a dielectric mouth insert [24–26], a shim coil that was placed in the mouth [27], or dielectric foam [28]. This has progressed more recently to the use of multichannel receiver coils that put DC currents through the localized loops to create magnetic fields that counteract the local magnetic-field inhomogeneities induced in the sample due to magnetic susceptibility [29]. As receiver coils with higher channels are built, these enable a customized shimming for every sample in the scanner. A review of shimming methods, targeting 7 T MRI systems and the brain, describes several other approaches [30].

Once the magnetic field has been made as uniform as possible, there are several imaging parameters that can have an impact on the amount of distortion or artifacts that will be experienced. As demonstrated in Section 12.3.2, the timing of the acquisition and its relationship to the k-space trajectory can be an important determinant of the size of the artifact. Faster imaging due to higher-bandwidth acquisitions, larger gradient amplitudes, and parallel imaging can reduce the size of the image distortion experienced. These improvements in image fidelity often come with a cost of signal-to-noise ratio, resulting in a tradeoff. For through-plane dephasing, it is the distribution of magnetic fields across the voxel that result in signal cancellation. Thus, thinner imaging slices (smaller voxels) will directly reduce this range of frequencies, and hence, the dephasing in the slice [31–33]. Fig. 12.7 shows a 2D EPI functional MRI acquisition performed at 3 T with various slice thicknesses to illustrate this effect.

In addition to optimizing the parameters, some simple modifications to a pulse sequence can also result in reducing the impact of susceptibility-induced artifacts. One of these is to dynamically update the shim gradients and receive frequency on a slice-by-slice basis to counteract linear components of the field inhomogeneity distribution [34]. Similarly, Z-shimming is the use of an extra gradient

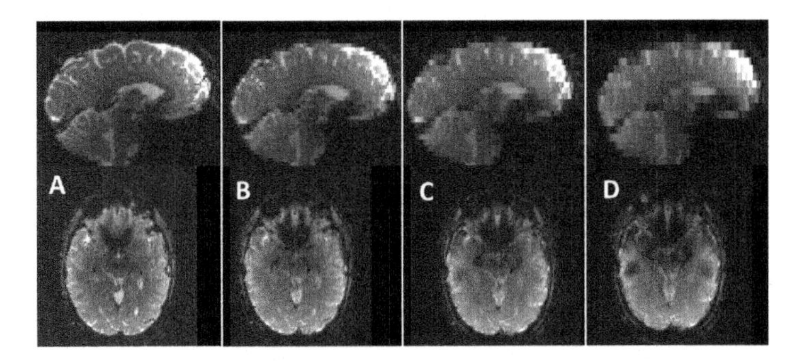

FIGURE 12.7

Decreasing the slice thickness reduces the within voxel dephasing effect. EPI acquisitions from an fMRI scan with slice thickness of **A**) 2 mm, **B**) 4 mm, **C**) 6 mm and **D**) 8 mm.

across a slice (Z-direction) to counteract the dephasing that occurs due to magnetic-field inhomogeneity gradients through the slice direction (which is usually the direction with stronger field gradients or larger dimension) [35–38]. Early methods were limited to only one z-shimming per slice, although several single-shot methods were developed to perform a few z-shimmings in a single-shot acquisition [39–42]. Modern sequences overcome these limitations by designing fully shaped RF pulses that apply a spatial map of through-plane phase, providing compensatory phase to every region of the brain as needed [43,5,44,45]. These RF pulses may become quite long due to the need to provide high-resolution spatial phase to a slice, but the development of multiple channel transmit systems enables time-efficient RF pulses to be used.

Despite these mitigation approaches, field-inhomogeneity effects persist in the image acquisition. Left uncorrected, these will lead to imaging artifacts that can misplace signals, to improper contrast, or to inadequate k-space sampling. In the following sections, we will focus on modeling and correcting these field-inhomogeneity effects.

12.4 Image distortions and correction approaches

As previously noted, image distortion affects nearly all types of image acquisition approaches including both spin echo and gradient echo. Therefore, it is important to understand the impact, how to minimize the distortion, and how to correct for it during or after image reconstruction.

12.4.1 Distortions depend on trajectory and sample timing

As we saw previously, the amount of distortion depends on the magnitude of the magnetic-field offset and the timing of the acquisition trajectory. A useful concept to enable easy interpretation of the impact of field inhomogeneity on an acquisition is the concept of bandwidth per pixel (BWPP). When imaging, if a gradient is applied across a certain direction, for example the x-direction, then we get a distribution of frequencies of spins across this direction. When sampling with the Nyquist criteria, the individual pixels are represented as bins of frequencies within a certain range. Assuming that the correct sampling rate and the correct gradient amplitude have been chosen, the total bandwidth of the acquisition is determined by the inverse of the sample time spacing, $BW = \frac{1}{\Delta t}$. If we consider that the range of frequencies will be spaced across the N pixels of the field of view, then we can calculate the BWPP as

$$BWPP = \frac{BW}{N} = \frac{1}{N\Delta t} \tag{12.19}$$

in Hz per pixel. This interpretation makes it easy to figure out how many pixels of distortion that a particular field inhomogeneity, measured in Hz, will create with a Fourier transform reconstruction. It also makes it easy to determine how choices in the acquisition will impact this level of distortion. A faster time sampling will result in a larger BWPP and lower image distortions (i.e., a smaller c in Eq. (12.12)).

The previous example was for a simple gradient readout event across a single axis, the readout axis, which would be the situation for a Cartesian acquisition with a single readout per TR. In this scenario, typically experienced magnetic field-inhomogeneity levels would not result in measurable image distortion due to the high BW in the data acquisition. If the data sampling of an MRI scanner is 200 KHz, i.e., a time spacing of 5 μs, then, for an image matrix size of $N = 200$, the BWPP would be 1 KHz. Given that, at 3 T, the maximum magnetic susceptibility difference between air and tissue is about 1 KHz, we would expect to get a distortion of, at most, 1 pixel across the 200 pixel field of view. This would have a minor impact on the spatial integrity of the image. However, this concept of BWPP works in the other encoding directions, too. For an EPI acquisition, the data sampling along the phase encode direction is not continuous, but instead has a sampling that is spaced by a larger amount of time, called the echospacing (T_{Echosp}). The echospacing is slightly larger than the time required to measure a whole line in the readout direction. We can think of the sampling bandwidth in this PE direction as $BW_{PE} = 1/T_{Echosp}$, and get the BWPP in the PE direction as

$$BWPP_{PE} = \frac{BW_{PE}}{N_{PE}} = \frac{1}{T_{Echosp}N_{PE}}, \tag{12.20}$$

where N_{PE} is the number of pixels in the phase encode direction. For an EPI acquisition with an echospacing of ≈0.4 ms and 64 pixels, the $BWPP_{PE}$ would be 39 Hz/pixel. In this case, the fat–water frequency difference at 3 T (≈440 Hz) would result in more than 11 pixels displacement in a 64-pixel image that would be a very significant image distortion.

To demonstrate the dependence of the spatial shift in EPI on the magnetic-field inhomogeneity, in Fig. 12.8 a single vertical line of the reconstruction of a brain image is shown with increasing magnetic-field inhomogeneity. This more general formulation of BWPP can be used to approximate the size of the spatial distortion with other k-space trajectories. However, the distortions from magnetic-field inhomogeneity may not be simple geometric shifts when trajectories are used in which the k-space and the time axis are not just scaled versions of each other. For example, in spiral trajectories, such as the spiral-out trajectory, the k-space trajectory starts at the center of k-space and increases its radius as time progresses in the sampling. This means that the azimuthal sampling is very fast and would have a high bandwidth, but the radial sampling is slower, requiring the sampling trajectory to make a full turn through k-space to get to the next sample in the radial direction. The spiral trajectory and its distortion are demonstrated in Fig. 12.9. The radial spacing may not be exactly evenly spaced in time as the spiral

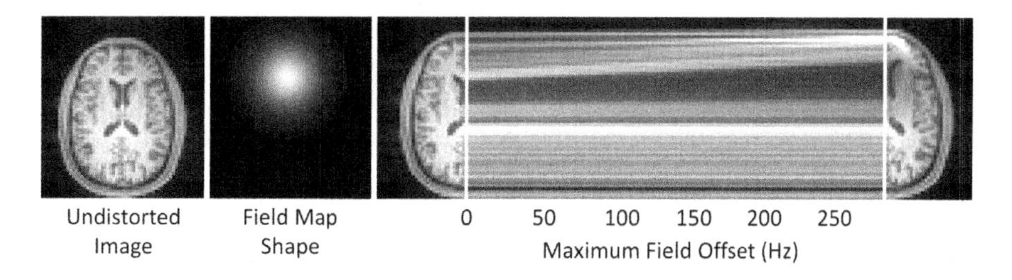

FIGURE 12.8

Distortion of EPI acquisition as a function of size of the magnetic-field inhomogeneity for an acquisition with 64 pixels and an echospacing of 0.4 ms. Undistorted image and the magnetic-field map shape along with an image showing the distortion of the images since the maximum magnetic field map value goes from 0 Hz to 280 Hz. The x-axis shows increasing field inhomogeneity in Hz. The y-axis shows one line along the y-axis (PE axis) in the image, depicting the distortion.

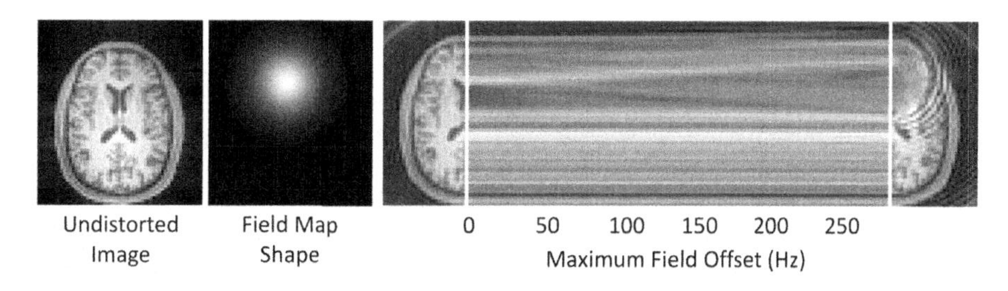

FIGURE 12.9

Distortion of spiral out acquisition with increasing magnetic-field inhomogeneity for a single-shot spiral acquisition with a 20-ms readout for a 64 matrix size acquisition. Similar to Fig. 12.8, the right-most image shows the distortion for the midline of the image as the maximum-field inhomogeneity goes from 0 to 280 Hz.

progresses from smaller inner rungs to outer rungs, so the approximation involved in a $BWPP_{radial}$ is only approximately true. If we assume an $N = 64$ matrix size spiral takes $T_{dur} = 20$ ms to acquire and includes 32 ($N/2$) rungs, then we can approximate the radial bandwidth as, $BW_{radial} = 1/T_{rung}$, where $T_{rung} = T_{dur}/(N/2)) = 2T_{dur}/N$. So, $BWPP_{radial}$ is given by

$$BWPP_{radial} = \frac{BW_{radial}}{N} = \frac{1}{T_{rung}N} = \frac{N}{2T_{dur}N} = \frac{1}{2T_{dur}}. \tag{12.21}$$

So, the $BWPP_{radial}$ is approximately the inverse of twice the spiral readout duration. In this case, that would be 25 Hz/pixel. So, the fat–water frequency shift at 3 T would result in a 17.6 pixel radial blur. This radial blur would extend in all radial directions from the original location of the signal experiencing this field inhomogeneity, resulting in an overall spread of the point spread function of a signal to 35 pixels out of 64. A signal distortion this large would create significant artifacts in images without fat saturation pulses. Partial mitigation of this effect is a necessity and can be attempted through fat saturation pulses, water–fat separation techniques, multishot acquisitions (reducing spiral readout duration per shot through the use of multiple shots) and parallel imaging (which also decreases spiral readout duration). Fig. 12.9 shows the image distortion for a pixel placed in a magnetic-field inhomogeneity of varying intensity. The magnitude of field inhomogeneity is shown on the x-axis, while a single line through the image is shown along the y-axis. Note that the impact of distortions would be similar along both the x- and y-axis of the image for a spiral-out acquisition.

12.4.2 Image correction: image warping approaches

Given that even moderate magnetic-field inhomogeneity can result in substantial spatial distortion in the image, there have been many methods developed to address this distortion during image reconstruction. We start with a simple approach that is commonly used for EPI, where the distortion is primarily in one direction and represented as a spatial shift. We saw previously that field inhomogeneity results in spatial shifts in the encoding axis that is slow in time (hence, low in BW), i.e., the phase-encode direction. Correction for the geometric shift induced by this shift can be corrected through image registration-based approaches that acquire two acquisitions with the phase-encoding direction (and, hence, the axis of the spatial shift) flipped [46,47]. With one acquisition acquired with the PE direction anterior–posterior and another acquired with the PE direction posterior–anterior, the spatial shifts between the two images will be in opposite directions. Nonlinear spatial warping can be done to align one of the images to the other. By applying half the spatial warp to each image, the resulting image will be undistorted. The spatial warp field can be converted to a traditional magnetic field-inhomogeneity map if desired by using the sequence timing to convert the warp field for each voxel back to magnetic-field offset. The disadvantage of this approach would be the need to acquire two images to estimate one undistorted image. However, both fMRI and DTI acquire many EPI images that are subject to the relatively constant magnetic-field inhomogeneity. So, typically, an extra acquisition is acquired before the fMRI or DTI sequence (thus, with minimal time penalty), usually with a spin echo EPI to minimize through-plane signal loss, with the flipped phase-encode images. The distortion map or field map estimated from this flipped PE axis pair is used to correct all subsequent images. Fig. 12.10 shows an example of flipping the phase-encode axis to estimate the magnetic field map and undistorted images. This is the approach that was taken in the Human Connectome Project to provide time-efficient field-inhomogeneity correction to the EPI-based scans (fMRI and DTI) [48].

This method works only for acquisitions, such as EPI where the image distortion manifests as a pixel shift. There are many tools that enable this to be accomplished in a straightforward manner, such as *topup* in FSL [49,50], with just a few parameters that are needed from the acquisition of the flipped PE axis scans and the fMRI or DTI run that is to be corrected. Further advances to this method incorporate spatial atlasing or other anatomically constrained distortion estimation based on other images acquired during a full imaging study and not necessarily a differentially distorted (flipped phase-encode direction) EPI image [51,52].

12.4.3 Image correction: conjugate phase

The image warping approach just explained is limited to trajectories, such as EPI, where there is a linear relationship between the slow k-space axis and time evolution during the acquisition. This results in distortions that are simple geometric shifts. For k-space trajectories that do not fit this acquisition framework, a more direct incorporation of the magnetic field-inhomogeneity map is required in the

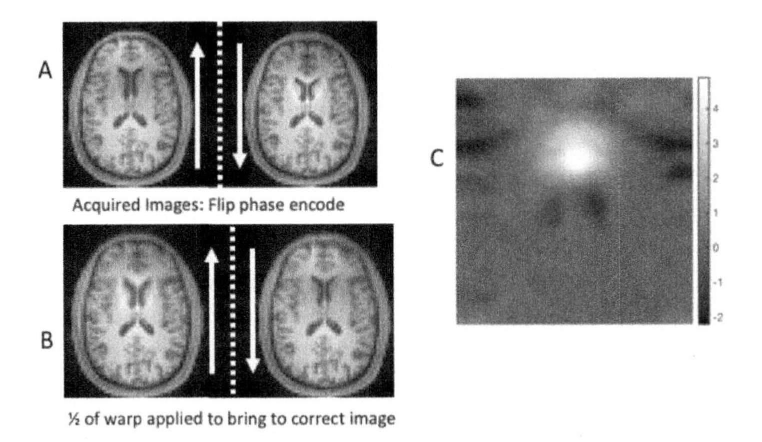

FIGURE 12.10

Flipping the phase encode axis to find the field inhomogeneity-induced distortion. **A**) Images acquired with EPI with opposite PE directions (y-axis), showing image distortions in opposite directions. **B**) After estimating the warping field, half of the warp is applied to each image to bring them back together within an undistorted image. **C**) The warping field is displayed as a map of pixel shifts in the PE direction (simulated).

image reconstruction step. This approach is required for spiral, radial, and rosette trajectories, but also for EPI that do not have the simple linear relationship, such as multishot and centric-reordered EPI.

When incorporating the magnetic field-inhomogeneity map into the image reconstruction, you must first have a measurement of the magnetic-field map, i.e., the spatial distribution of the magnetic field inside the sample that you are imaging. Methods for measuring the field map will be discussed in Section 12.7. For now, we assume that an accurate field map, $\omega_{FM}(\mathbf{r})$, is available.

Once we have an estimate of the magnetic field-inhomogeneity map, we can proceed to invert (or approximately invert) Eq. (12.3) to create a field-corrected image. Since the signal in a standard MRI acquisition is just the Fourier transform of the image, at a base level, the image is just the inverse Fourier transform of the data. However, we must refocus the phase effects of the magnetic-field inhomogeneity in order to bring the net encoding (k-space and field inhomogeneity) of the signal from a pixel back into focus. This approach is called the conjugate-phase method since we apply the conjugate of the phase accumulation due to the magnetic-field inhomogeneity back to the acquired data as we reconstruct a pixel [53–55]. The continuous formulation of this is given as

$$\hat{x}(\mathbf{r}) = \int_{Wk} s(\mathbf{k}(t))e^{i2\pi\mathbf{k}\cdot\mathbf{r}}e^{i\omega_{FM}(\mathbf{r})t}d\mathbf{k}, \tag{12.22}$$

where Wk indicates that the integral is performed over all of continuous k-space. Converting this equation into our case of sampled k-space data requires several considerations, including the basis function expansion of the image discussed in Section 12.3.2 and consideration of the sample density-compensation function, $H(\mathbf{k})$, that takes into account the nonuniformly sampled k-space during an acquisition such as a spiral trajectory. If we assume a voxel-indicator basis function, as was used in Eq. (12.4), and we lump the k-space scaling from the basis function into the sample density compensation function, we get the following image reconstruction equation that corresponds to the signal equation in Eq. (12.7), as

$$\hat{x}(\mathbf{r}_n) = \sum_{m=0}^{M-1} H_m s(t_m) e^{i2\pi \mathbf{k}_m \cdot \mathbf{r_n}} e^{i\omega_n t_m}, \tag{12.23}$$

where H_m includes sample density compensation and division by $B_\upsilon(\mathbf{k})$. If we let $\mathbf{H} = [H_0 \ldots H_{M-1}]^T$, then we can reformulate the conjugate-phase reconstruction in a similar way to the encoding operator in Eq. (12.9) as

$$\hat{x} = \mathbf{E}^H (\mathbf{H} \circ \mathbf{s}), \tag{12.24}$$

where $\circ$ is the Hadamard or element-wise product of the data and weighting vectors and $[\ ]^H$ is the Hermitian transpose, which takes the complex conjugate transpose of the encoding matrix, including the k-space encoding and the field-inhomogeneity term.

Calculation of Eq. (12.23) can be computationally demanding since the kernel that needs to be computed depends on both spatial position, $\mathbf{r}_n$, and time, t_m. This cannot be computed by a FFT directly, and, instead, direct calculation of this reconstruction would require full (and slow) computation of the matrix–vector product of $\mathbf{E}^H$ and $\mathbf{s}$. It should be noted that explicit storing of the large matrix $\mathbf{E}$ must also be avoided since it can be too large to store in computer memory for even moderately sized problems. Instead of direct calculation of the matrix–vector product, we can leverage several FFT's by decomposing the problem into several components by performing either time segmentation [54] or frequency segmentation [55]. With time segmentation and the use of a small time interval, we can drop the time dependence out of the kernel term: $e^{i\omega_n t_m}$ and incorporate field inhomogeneity-induced phase only in the end points of the time interval. If we break up the readout into L time segments, each τ seconds long, we get the following for the reconstruction using an interpolating window function $\alpha(t)$ such as a Hanning window [54]:

$$\hat{x}_n = \sum_{l=0}^{L} e^{i\omega_n l\tau} \sum_{m=0}^{M-1} \alpha(t_m - l\tau) H_m s(t_m) e^{i2\pi \mathbf{k}_m \cdot \mathbf{r_n}}. \tag{12.25}$$

The second summation is just a weighted reconstruction of the k-space data and can be accomplished quickly with gridding for non-Cartesian k-space trajectories [56].

12.4.4 Image correction: inverse problem approach

The conjugate-phase approach provides a one-step, direct reconstruction of magnetic field inhomogeneity-corrected imaging data from arbitrary k-space trajectory acquisitions. This approach works well for fully sampled k-space conditions with a smooth magnetic field-inhomogeneity map. We will discuss comparisons and limitations later in Section 12.4.5. However, in many imaging situations, the data are undersampled and in need of parallel imaging, and additional prior information is available and/or compressed sensing reconstructions are needed in order to produce a high-quality image. For these cases, it is necessary to formulate the MRI image-reconstruction problem in an inverse problem approach. In addition to these cases, the inverse problem approach is not limited by some of the approximations required for a one-step direct reconstruction like the conjugate-phase reconstruction. Conjugate-phase approximations include the need for accurate sample density compensation, $H(\mathbf{k})$, which we will show is potentially complicated by magnetic-field inhomogeneity in Section 12.6. Inverse problem approaches, by simulating the acquired signal, also provide energy-preserving reconstructions in the presence of regions with a high slope in the magnetic field-inhomogeneity map, which will be discussed further in Section 12.4.5.

For the inverse problem approach, additional basis functions can be easily incorporated into the signal model, as was discussed in Section 12.3.2.1. We formulate the image reconstruction using the signal equation model that was given in Eq. (12.9). Then the image reconstruction progresses by finding the image $\mathbf{x}$ that results in a signal that best matches the acquired data, $\mathbf{s}$, as discussed previously in Chapter 2. This optimization problem can be formulated in its simplest form as a least squares problem as

$$\hat{\mathbf{x}} = \arg\min_{\mathbf{x}} \|\mathbf{s} - \mathbf{E}\mathbf{x}\|^2, \tag{12.26}$$

where we have only included a data fit term. Or, it could incorporate additional prior constraints on the image $\mathbf{x}$, along with coil sensitivity information for a SENSE-type parallel imaging [57] (see also Chapter 6). Several open-source implementations of this type of image reconstruction are readily available.[1] This model can be easily incorporated into algorithms covered in other chapters of this book, including constrained, low-rank, and dictionary-based approaches.

12.4.4.1 Computational considerations

The inverse problem approach results in an iterative image-reconstruction algorithm that can take significant time to compute due to the term $e^{-i\omega_n t_m}$ that depends on both spatial position and time. Significant computational speedups can be achieved using similar approaches to the time segmentation that was mentioned for conjugate phase. However, efficient computation is even more critical in iterative reconstruction since the field inhomogeneity-map term will be calculated many times. Additional optimization has been performed to reduce the number of time segments required to get an accurate approximation using a min–max optimal interpolator [16].

Iterative reconstructions must simulate the complicated signal models of $\mathbf{E}$ and $\mathbf{E}^H$ multiple times during image reconstruction. For field inhomogeneity-corrected image reconstructions, due to time segmentation, this usually results in an order-of-magnitude increase in computation time. In order to make field inhomogeneity corrected reconstruction feasible, several other approaches to speedup reconstruction have been developed, including those leveraging graphics processing units (GPUs) and other high-performance computing approaches [58–61]. Through these speedups, work is being done to try to make the field inhomogeneity-corrected reconstructions work in clinically feasible times, which would enable advanced non-Cartesian trajectories with time-efficient, long data readouts usable for clinical imaging purposes.

[1] Please see [16] and associated code included in the Michigan Image Reconstruction Toolbox (MIRT, https://web.eecs.umich.edu/~fessler/code/) or the Reconstruction code from Illinois (RecoIL, https://github.com/mrfil/RecoIL) as examples.

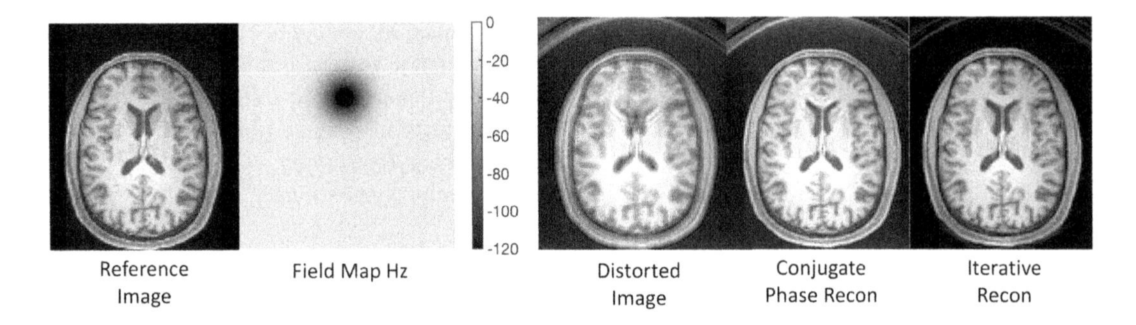

FIGURE 12.11

An axial slice of the brain (reference image), a corresponding field map (in Hz), and reconstructions obtained from a simulated single-shot spiral acquisition with a matrix size of 128 acquired with 2-shot acquisition (equivalent to a single-shot parallel imaging factor of 2 in terms of distortion). The distorted image from a Fourier transform reconstruction shows blurring, but the images reconstructed with both time-segmented conjugate phase and time-segmented NUFFT-based iterative reconstruction approach (10 iterations) recover a sharp image.

12.4.5 Comparing performance of image correction approaches

An example image simulated with magnetic field inhomogeneity and the two reconstruction methods of conjugate-phase and iterative reconstruction are shown in Fig. 12.11. As demonstrated in this example, magnetic field correction methods drastically improve image quality. However, the iterative reconstruction method provides a more complete correction for the field-inhomogeneity effects.

In principle, the conjugate-phase approach refocuses signal intensity to the region where it originated, as indicated by the encoding of k-space and the magnetic-field inhomogeneity it experiences. However, there are many imaging situations where there may be ambiguity about the origin of signal. Such a situation would arise with an EPI-like trajectory, causing a geometric shift of a pixel with high-field inhomogeneity (pixel 1) onto a different location with very low-field inhomogeneity (pixel 2). This situation is demonstrated schematically in Fig. 12.12 for a theoretical 1D imaging experiment. If an uncorrected image is formed, all signal intensity shows up in the location of pixel 2, i.e., it appears that all the signal is coming from pixel 2. The signal encoding (k-space/gradients) combined with the field-inhomogeneity information still does not provide sufficient information to distinguish whether the signal intensity originated from pixel 1 or pixel 2. Conjugate-phase and iterative reconstructions handle this situation differently, as demonstrated in the figure. For conjugate-phase reconstruction, the signal will refocus at both positions, and the conjugate phase reconstruction will put the full amount of signal (sum of pixel 1 and pixel 2) into both locations. The conjugate-phase reconstruction is not energy preserving since it will refocus signal in every location from which it could have originated. In contrast, the iterative reconstruction simulates the signal and tries to match the simulated signal to the received signal. This must be energy preserving in the image and signal domain. So, in the absence of other prior information, it will split the signal between both locations. Although this is closer to the correct image in terms of error, it does not contain any additional information about the actual signal at pixel 1 and pixel 2. However, if prior information is available from surrounding pixels and a regularization

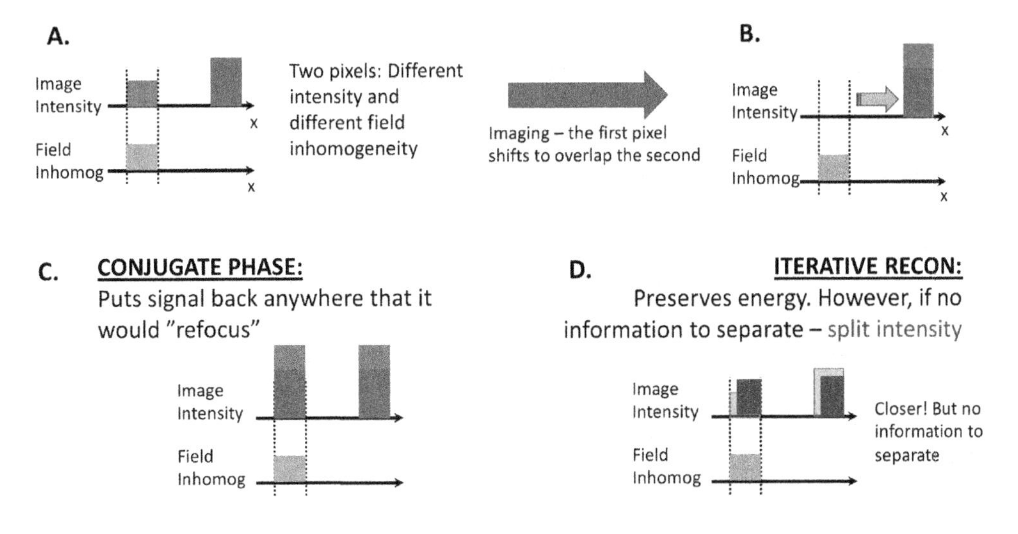

FIGURE 12.12

A simplified 1D example of the different behavior of the conjugate phase and iterative reconstruction methods with respect to placing signal intensity with ambiguity due to magnetic-field inhomogeneity. **A)**. The setup of the problem where two pixels (red and blue (mid gray and dark gray in print version)) have different intensity, but the red pixel also suffers from field inhomogeneity shifting it, in the distorted image (panel **B**), onto the same location as the blue pixel. **C**). Since the signal intensity refocuses to both locations, conjugate phase will place the sum of both pixel intensities at both locations. **D**). The iterative reconstruction is energy preserving and will split the signal between the two locations if no other prior information is available to help determine how much of the signal came from each location.

approach is used in the iterative reconstruction method, then a more accurate image can be formed by finding the best fit image that follows the prior information.

12.5 Phase and signal dephasing correction approaches

In addition to causing geometric distortion, magnetic-field inhomogeneity can lead to phase accumulation due to off-resonant spins. This occurs mainly in gradient echo imaging where the phase is allowed to accumulate over time before or during the acquisition. This phase accumulation can lead to errors in extracting information from images that are dependent on the phase, such as in magnetic resonance elastography [62,63]. Additionally, it can lead to signal loss within a voxel if there are variations in the magnetic field across that voxel such that spins experience a range of fields within that voxel. These spins within a voxel have various frequencies and, over time, they will start to cancel one another, leading to signal loss. This signal dephasing can result in images with large regions of holes or signal voids, especially in functional MRI near the orbitofrontal cortex or around the ear canals. Although other regions may not show complete signal voids, the loss of signal due to magnetic field inhomogeneity-induced dephasing can still cause problems, especially in quantitative imaging methods.

There are several ways to mitigate this effect prior to the image-reconstruction step. This includes the z-shimming and shaped RF pulses that were mentioned in Section 12.3.3. In addition, these gradients through a voxel can be interpreted as additional linear gradient-encoding functions resulting in a unique k-space trajectory for each voxel. This will be discussed in Section 12.6 and can lead to additional mitigation strategies, such as using full 3D k-space sampling or sampling additional locations in k-space to ensure adequate coverage.

12.5.1 Image reconstruction based approaches for within voxel dephasing

The signal loss caused by within voxel dephasing can produce significant artifacts in the image. This is particularly important for gradient echo imaging, multi-echo imaging, spectroscopic imaging, and functional MRI if the voxel size is not small enough to ignore intravoxel B_0 variations. In order to address these effects during the image reconstruction step, we must use a signal model that includes the field gradients. As mentioned in Section 12.3.2.1, we can impose a piecewise linear model of these within-voxel gradients as in Eq. (12.14), leading to a signal model as in Eq. (12.15). This model has a term, $B_v(\mathbf{k}_m, \mathbf{G}_\mathbf{n}, t_m)$, that depends on both spatial position and k-space sampling locations that causes significant computational and memory challenges for incorporating this correction into reconstruction algorithms. To this end, post-processing methods have been described to correct the effects of intravoxel B_0 [64–67,9,1,68]. However, these methods require explicit knowledge of the point spread functions (PSFs) and can become intractable if advanced reconstruction methods are used.

Fortunately, recent methods have been developed to make computation related to $B_v(\cdot)$ more efficient through low-rank approximations [19,20] (see Chapter 9 for more discussions on low-rank modeling). Specifically, $B_v(\mathbf{k}_m, \mathbf{G}_\mathbf{n}, t_m)$ can yield an accurate partially separable functions-based approximation, i.e.,

$$B_v(\mathbf{k}_m, \mathbf{G}_\mathbf{n}, t_m) \approx \sum_{l=1}^{L} \mathbf{u}_l(\mathbf{k}_m)\mathbf{v}_l^T(\mathbf{G}_\mathbf{n}), \quad L \ll M, N. \tag{12.27}$$

If we express the entire operator using a matrix form $\boldsymbol{\Omega}_v \in \mathbb{R}^{M\times N}$, Eq. (12.27) implies that $\boldsymbol{\Omega}_v$ has a low-rank representation $\boldsymbol{\Omega}_v = \mathbf{UV}$, with $\mathbf{U} \in \mathbb{R}^{M\times L}$ and $\mathbf{V} \in \mathbb{R}^{L\times N}$ being two rank-L matrices. Importantly, $\mathbf{U}$ and $\mathbf{V}$ can be estimated without explicitly forming $\boldsymbol{\Omega}_v$ in the process. Specifically, a subset of the columns in $\boldsymbol{\Omega}_v$ can be formed by evaluating Eq. (12.16) at the sampled k-space coordinates and a random subset of voxel field gradients calculated from a pre-estimated B_0 map. SVD can then be applied to this column subset to estimate $\mathbf{U}$. With $\mathbf{U}$ estimated, $\mathbf{V}$ can be efficiently estimated by a least-squares fitting using a row subset of $\boldsymbol{\Omega}_v$ formed similarly as previously described, but with a subset of k-space coordinates instead. This allows us to decompose the complicated encoding operator $\mathbf{E}$ into L components as

$$\mathbf{Ex} \approx \mathbf{B} \sum_{l=1}^{L} \mathbf{u}_l \circ (\mathbf{F}\mathbf{v}_l \circ \mathbf{x}), \tag{12.28}$$

each of which contains a voxel-wise multiplication between the image of interest and $\mathbf{v}_l$, a Fourier transform operator $\mathbf{F}$ that can be easily implemented using FFT's, a k-space point-wise multiplication

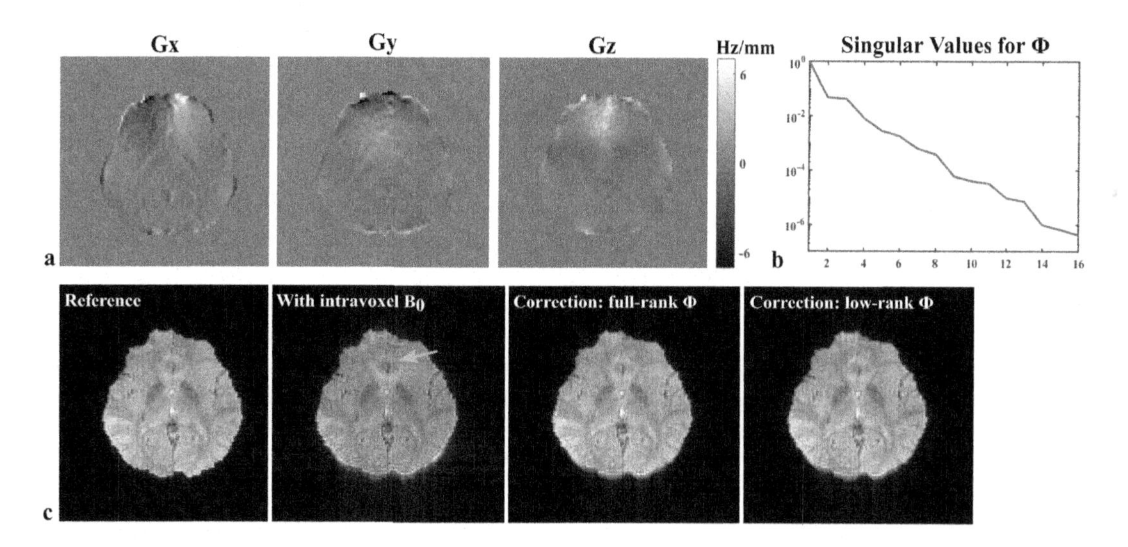

FIGURE 12.13

Illustration of the low-rank property of the encoding operator that incorporates intravoxel B_0 inhomogeneity effects: (**a**) voxel-field gradients along x, y, and z directions for a particular slice of a practical in vivo B_0 map; (**b**) the singular values for a $\mathbf{\Omega}_v$ matrix in log scale (Eqs. (12.15) and (12.27)) for a 2D reconstruction. As can be seen, the singular value decay very rapidly, validating the low-rank assumption. Images in (**c**) compare images from a direct Fourier reconstruction of the low-resolution data that contains intravoxel B_0-induced artifacts (second column), corrected reconstruction using the full B_v (third column), and rank-8 approximated $\mathbf{\Omega}_v$ (last column). The low-rank operator yielded almost the same corrected reconstruction as using the full $\mathbf{\Omega}_v$.

with $\mathbf{u}_l$ and a final arbitrary sampling masking matrix $\mathbf{B}$. Such an approximation results in a reduction of computational complexity from $\mathcal{O}(M \times N)$ to $\mathcal{O}(M \times L + N \times L + L \times N \log N)$, compared to evaluating $\mathbf{E}$ directly [19]. Since L is typically orders-of-magnitude smaller than M and N, the reduction is dramatic (Fig. 12.13). It also allows for flexible uses of any regularization functional in the reconstruction formulation for B_0 correction, and is not limited by the PSF as required by the image post-processing based methods.

Fig. 12.13 shows the intra-voxel field gradients from a practical field map (Fig. 12.13a), the low-rank property of B_v constructed from these maps (Fig. 12.13b); and also a set of simulated intravoxel B_0 inhomogeneity corrected reconstructions (Fig. 12.13c). The low-rank approximated operator produced reconstruction with negligible differences from reconstruction using the full $\mathbf{\Omega}_v$ matrix. However, the low-rank based implementation made the reconstruction 1000 times faster.

12.6 K-space trajectory distortions

The linear portion of the gradients in magnetic-field inhomogeneity, addressed by modeling in the previous section, have an alternative interpretation that helps us to understand their impact. Just like the linear imaging gradients applied to navigate planned sampling trajectories in k-space, the linear field inhomogeneity terms can be thought of as providing additional k-space encoding. Since the field inhomogeneity gradients are varying across the object, they can result in a unique k-space trajectory for each individual voxel in an image. When the field inhomogeneity gradients are added in, as in Fig. 12.5, the k-space trajectory can deviate significantly from its intended trajectory resulting in sampling errors and timing errors that can have a large impact on images.

Trajectories can be significantly shifted in k-space away from the origin, especially with a long echo time that allows the magnetic field-inhomogeneity gradients, which are constant during the entire acquisition, to accumulate significant encoding even before the planned k-space trajectory starts. If the shift is far enough, then the center of k-space may not even be sampled by the effective trajectory (field-inhomogeneity trajectory plus the planned trajectory). This is an additional interpretation for the signal-loss artifact due to dephasing. It also suggests another approach to address the dephasing artifact—sampling further in k-space than is needed for the imaging requirements. This would also be equivalent to the approach of increasing the spatial resolution (making the voxels smaller) to decrease the dephasing.

An additional issue with the effective trajectory is the timing at which the center of k-space is sampled, which is called the echo time, TE. The TE determines the base contrast for an image and it is based on where the signals, in the absence of magnetic field-inhomogeneity gradients, would refocus from the applied imaging gradients. For a spiral-out acquisition, the TE is at the start of the application of the imaging gradients. However, if the magnetic field inhomogeneity gradients have been allowed to travel outward in k-space for some time, such as in a gradient echo acquisition for functional MRI, then the time at which the effective k-space trajectory hits the center of k-space will be delayed to a later time in the readout. For functional MRI, the signal level is dependent on the echo time, with larger echo times resulting in larger BOLD weighting [69–71]. This can potentially result in differential BOLD signals due to differences in magnetic-field inhomogeneity that could be sensitive to acquisition timing, head orientation, or even anatomical changes across subjects [71].

Before closing this section, we note that sources other than magnetic-field inhomogeneity, such as gradient response functions (e.g. [72,73]) and nonlinearity (e.g. [74]), can also lead to k-space distortion and are important to consider for non-Cartesian trajectories. However, they are beyond the scope of this chapter which focuses on the spatially varying k-space trajectories induced by gradients in the magnetic-field inhomogeneity.

12.7 Measuring the field map

When incorporating the magnetic field-inhomogeneity map into the image reconstruction, there must first be a measurement of the magnetic field map, i.e., the spatial distribution of magnetic field inside the sample that you are imaging. This could be done, as was done earlier in the chapter, by estimating it with flipped PE-axis acquisitions to measure distortion with known sequence parameters. However, this only works for situations with field-inhomogeneity artifacts resulting in simple geometric shifts. More generally, the magnetic field-inhomogeneity map is measured by acquiring a multi-echo acquisition and examining the phase evolution at each voxel to infer its off-resonant frequency [75]. This can be done at the beginning of the scan session for a particular slice prescription because it is fairly stable throughout the acquisition. Depending on the sensitivity of the acquisition to magnetic field-inhomogeneity effects, a dynamic estimation of the field map may be necessary to track field changes during the scan.

In [76], a joint estimation framework through the iterative reconstruction scheme was used, resulting in dynamically estimated magnetic-field maps during a functional MRI scan. These dynamic field maps show the oscillations in the magnetic field in the brain during breathing, due to the movement of the chest wall, with a few Hz at 3 T and stronger at ultrahigh fields (e.g., 7 T) [77].

Sequences to measure the field map should be set up so as to not experience significant image distortions or other artifacts due to magnetic-field inhomogeneity. This would mean multishot, high-bandwidth acquisitions that use short echo times and thin slices. In order for the magnetic field to cause an accumulation of phase, gradient echo acquisitions are needed. But, often this is done with an asymmetric spin echo where a reference image is acquired at the spin echo and the other echos are acquired with slight (1–2 ms) shifts away from the echo. At least two images with different phase accumulation are required, but additional echoes can help in obtaining more accurate field maps. Additionally, the field map-estimation approach could be a simple fit of the linear phase accumulation, or it could incorporate prior constraints on the smoothness of the field maps, such as in [78].

Measuring the field maps repeatedly to account for the dynamic B_0 alterations can be time consuming and may not accurately capture the intra-scan field variations (e.g., due to eddy currents or physiological motions). Recently, specialized field probes have emerged as a promising approach to monitor and estimate these spatiotemporal B_0 field changes in real-time [79–81]. The key assumptions for using these field probes are that the spatiotemporal B_0 variations follow a partially separable function representation (with space and time partial separability [82]) and that the spatial components are spherical harmonics [79,83]. Therefore, the temporally varying coefficients can be estimated using limited FIDs from a small number of probes placed around the imaging objects. These field estimates, when incorporated into the image reconstruction process, can offer substantially improved image quality, especially for gradient-echo-based acquisitions at ultrahigh fields [83–85].

12.8 Summary

The standard Fourier imaging for MRI assumes a uniform background magnetic field, which is disrupted in practice by the magnetic properties of the sample that is being imaged. This disruption does not require large ferrous objects to be in the imaging object, and it occurs in every region of the body with large variations of magnetic susceptibility, such as those containing both air and tissues. Uncorrected magnetic-field inhomogeneity can result in image distortions, signal loss, and inadequate k-space sampling due to trajectory deviations. Careful consideration of the physics of magnetic-field inhomogeneity is required in order to minimize its impact through proper choices of imaging sequences and parameters. Several image reconstruction approaches are available that model the effects of magnetic-field inhomogeneity and produce corrected images, as long as the disruptions are not severe. Careful consideration of field inhomogeneity will be of even greater importance for ultrahigh field imaging applications.

References

[1] Hernando D, Vigen K, Shimakawa A, Reeder S. R_2^* mapping in the presence of macroscopic B_0 field variations. Magn Reson Med 2012;68:830–40. https://doi.org/10.1002/mrm.23306.

[2] Diefenbach MN, Ruschke S, Eggers H, Meineke J, Rummeny EJ, Karampinos DC. Improving chemical shift encoding-based water–fat separation based on a detailed consideration of magnetic field contributions. Magn Reson Med 2018;80(3):990–1004.

[3] Tan Z, Voit D, Kollmeier JM, Uecker M, Frahm J. Dynamic water/fat separation and inhomogeneity mapping—joint estimation using undersampled triple-echo multi-spoke radial FLASH. Magn Reson Med 2019;82(3):1000–11.

[4] Yoder DA, Zhao Y, Paschal CB, Fitzpatrick JM. MRI simulator with object-specific field map calculations. Magn Reson Imaging 2004;22(3):315–28.

[5] Stenger VA, Boada FE, Noll DC. Three-dimensional tailored RF pulses for the reduction of susceptibility artifacts in T_2^*-weighted functional MRI. Magn Reson Med 2000;44(4):525–31. https://doi.org/10.1002/1522-2594(200010)44:4<525::AID-MRM5>3.0.CO;2-L [pii]. Available from: http://www.ncbi.nlm.nih.gov/entrez/query.fcgi?cmd=Retrieve&db=PubMed&dopt=Citation&list_uids=11025507.

[6] Wang J, Mao W, Qiu M, Smith MB, Constable RT. Factors influencing flip angle mapping in mri: Rf pulse shape, slice-select gradients, off-resonance excitation, and b0 inhomogeneities. Magn Reson Med 2006;56(2):463–8. https://doi.org/10.1002/mrm.20947. Available from: https://www.ncbi.nlm.nih.gov/pubmed/16773653.

[7] de Graaf RA. In vivo NMR spectroscopy: principles and techniques. 2nd edition. West Sussex, England: John Wiley & Sons, Ltd. ISBN 978-0-470-02670-0, 2007.

[8] Juchem C, de Graaf RA. B_0 magnetic field homogeneity and shimming for in vivo magnetic resonance spectroscopy. Anal Biochem 2017;529:17–29. https://doi.org/10.1016/j.ab.2016.06.003. Available from: https://www.ncbi.nlm.nih.gov/pubmed/27293215.

[9] Fernandez-Seara MA, Wehrli FW. Postprocessing technique to correct for background gradients in image-based R_2^* measurements. Magn Reson Med 2000;44(3):358–66.

[10] Wang Y, Liu T. Quantitative susceptibility mapping (QSM): decoding MRI data for a tissue magnetic biomarker. Magn Reson Med 2015;73(1):82–101.

[11] Hargreaves BA, Worters PW, Pauly KB, Pauly JM, Koch KM, Gold GE. Metal-induced artifacts in MRI. AJR Am J Roentgenol 2011;197(3):547–55. https://doi.org/10.2214/AJR.11.7364. Available from: https://www.ncbi.nlm.nih.gov/pubmed/21862795.

[12] Jezzard P, Clare S. Sources of distortion in functional MRI data. Hum Brain Mapp 1999;8(2–3):80–5. https://doi.org/10.1002/(SICI)1097-0193(1999)8:2/3<80::AID-HBM2>3.0.CO;2-C [pii]. Available from: http://www.ncbi.nlm.nih.gov/entrez/query.fcgi?cmd=Retrieve&db=PubMed&dopt=Citation&list_uids=10524596.

[13] Sekihara K, Kuroda M, Kohno H. Image restoration from non-uniform magnetic field influence for direct Fourier NMR imaging. Phys Med Biol 1984;29(1):15–24.

[14] Yudilevich E, Stark H. Spiral sampling in magnetic resonance imaging - the effect of inhomogeneities. IEEE Trans Med Imaging 1987;6(4):337–45.

[15] Jezzard P, Balaban RS. Correction for geometric distortion in echo planar images from B0 field variations. Magn Reson Med 1995;34:65–73.

[16] Sutton BP, Noll DC, Fessler JA. Fast, iterative image reconstruction for MRI in the presence of field inhomogeneities. IEEE Trans Med Imaging 2003;22(2):178–88. Available from: http://www.ncbi.nlm.nih.gov/pubmed/12715994.

[17] Sutton BP. Physics based iterative reconstruction for MRI: compensating and estimating field inhomogeneity and T_2^* relaxation. Thesis. The University of Michigan; 2003.

[18] Sutton BP, Noll DC, Fessler JA. Compensating for within-voxel susceptibility gradients in BOLD fMRI. In: Proc. 12th Intl. Soc. Mag. Res. Med.; 2004. p. 349.

[19] Lam F, Sutton BP. Intravoxel B0 inhomogeneity corrected reconstruction using a low-rank encoding operator. Magn Reson Med 2020;84(2):885–94. https://doi.org/10.1002/mrm.28182. Available from: https://www.ncbi.nlm.nih.gov/pubmed/32020661. https://onlinelibrary.wiley.com/doi/pdfdirect/10.1002/mrm.28182?download=true.

[20] Fessler JA, Noll DC. Model-based MR image reconstruction with compensation for through-plane field inhomogeneity. In: 4th IEEE Intl Symp Biom Imaging (ISBI); 2007. p. 920–3.

[21] Clare S, Evans J, Jezzard P. Requirements for room temperature shimming of the human brain. Magn Reson Med 2006;55(1):210–4. https://doi.org/10.1002/mrm.20735. Available from: https://www.ncbi.nlm.nih.gov/pubmed/16315227. https://onlinelibrary.wiley.com/doi/pdfdirect/10.1002/mrm.20735?download=true.

[22] Jesmanowicz A, Starewicz P, Hyde JS. Determination of shims needed for correction of tissue susceptibility effects in fMRI. In: Intl Soc Magn Reson Med; 2000. p. 1378.

[23] Kim DH, Adalsteinsson E, Glover GH, Spielman DM. Regularized higher-order in vivo shimming. Magn Reson Med 2002;48(4):715–22. https://doi.org/10.1002/mrm.10267. Available from: https://www.ncbi.nlm.nih.gov/pubmed/12353290.

[24] Wilson JL, Jenkinson M, Jezzard P. Protocol to determine the optimal intraoral passive shim for minimisation of susceptibility artifact in human inferior frontal cortex. NeuroImage 2003;19(4):1802–11.

[25] Wilson JL, Jezzard P. Utilization of an intra-oral diamagnetic passive shim in functional MRI of the inferior frontal cortex. Magn Reson Med 2003;50(5):1089–94.

[26] Cusack R, Russell B, Cox SM, De Panfilis C, Schwarzbauer C, Ansorge R. An evaluation of the use of passive shimming to improve frontal sensitivity in fMRI. NeuroImage 2005;24(1):82–91. https://doi.org/10.1016/j.neuroimage.2004.08.029. Available from: http://www.ncbi.nlm.nih.gov/pubmed/15588599. http://ac.els-cdn.com/S1053811904004902/1-s2.0-S1053811904004902-main.pdf?_tid=2336385e-2eca-11e3-bfe4-00000aab0f6b&acdnat=1381093326_5db05a65309a471aea74316da5754a2e.

[27] Hsu JJ, Glover GH. Mitigation of susceptibility-induced signal loss in neuroimaging using localized shim coils. Magn Reson Med 2005;53(2):243–8.

[28] Lee GC, Goodwill PW, Phuong K, Inglis BA, Scott GC, Hargreaves BA, et al. Pyrolytic graphite foam: a passive magnetic susceptibility matching material. J Magn Reson Imaging 2010;32(3):684–91. https://doi.org/10.1002/jmri.22270. Available from: https://www.ncbi.nlm.nih.gov/pubmed/20815067.

[29] Juchem C, Nixon TW, McIntyre S, Rothman DL, de Graaf RA. Magnetic field modeling with a set of individual localized coils. J Magn Reson 2010;204(2):281–9. https://doi.org/10.1016/j.jmr.2010.03.008. Available from: https://www.ncbi.nlm.nih.gov/pubmed/20347360.

[30] Stockmann JP, Wald LL. In vivo b_0 field shimming methods for MRI at 7 t. NeuroImage 2018;168:71–87. https://doi.org/10.1016/j.neuroimage.2017.06.013. Available from: https://www.ncbi.nlm.nih.gov/pubmed/28602943.

[31] Wadghiri YZ, Johnson G, Turnbull DH. Sensitivity and performance time in MRI dephasing artifact reduction methods. Magn Reson Med 2001;45(3):470–6.

[32] Bellgowan PS, Bandettini PA, van Gelderen P, Martin A, Bodurka J. Improved BOLD detection in the medial temporal region using parallel imaging and voxel volume reduction. NeuroImage 2006;29(4):1244–51.

[33] Merboldt KD, Finsterbusch J, Frahm J. Reducing inhomogeneity artifacts in functional MRI of human brain activation-thin sections vs gradient compensation. J Magn Reson 2000;145(2):184–91.

[34] Blamire AM, Rothman DL, Nixon T. Dynamic shim updating: a new approach towards optimized whole brain shimming. Magn Reson Med 1996;36(1):159–65. https://doi.org/10.1002/mrm.1910360125. Available from: https://www.ncbi.nlm.nih.gov/pubmed/8795035. https://onlinelibrary.wiley.com/doi/pdfdirect/10.1002/mrm.1910360125?download=true.

[35] Frahm J, Merboldt KD, Hanicke W. Direct FLASH MR imaging of magnetic field inhomogeneities by gradient compensation. Magn Reson Med 1988;6(4):474–80.

[36] Glover GH. 3d z-shim method for reduction of susceptibility effects in BOLD fMRI. Magn Reson Med 1999;42(2):290–9

[37] Yang QX, Dardzinski BJ, Li S, Eslinger PJ, Smith MB. Multi-gradient echo with susceptibility inhomogeneity compensation (MGESIC): demonstration of fMRI in the olfactory cortex at 3.0 T. Magn Reson Med 1997;37(3):331–5.

[38] Yang QX, Williams GD, Demeure RJ, Mosher TJ, Smith MB. Removal of local field gradient artifacts in T_2^*-weighted images at high fields by gradient-echo slice excitation profile imaging. Magn Reson Med 1998;39(3):402–9.

[39] Heberlein KA, Hu X. Simultaneous acquisition of gradient-echo and asymmetric spin-echo for single-shot z-shim: Z-SAGA. Magn Reson Med 2004;51(1):212–6.

[40] Song AW. Single-shot EPI with signal recovery from the susceptibility-induced losses. Magn Reson Med 2001;46(2):407–11.

[41] Guo H, Song AW. Single-shot spiral image acquisition with embedded z-shimming for susceptibility signal recovery. J Magn Reson Imaging 2003;18(3):389–95. Available from: http://www.ncbi.nlm.nih.gov/entrez/query.fcgi?cmd=Retrieve&db=PubMed&dopt=Citation&list_uids=12938139.

[42] Truong TK, Song AW. Single-shot dual-z-shimmed sensitivity-encoded spiral-in/out imaging for functional MRI with reduced susceptibility artifacts. Magn Reson Med 2008;59(1):221–7. https://doi.org/10.1002/mrm.21473. Available from: http://www.ncbi.nlm.nih.gov/entrez/query.fcgi?cmd=Retrieve&db=PubMed&dopt=Citation&list_uids=18050341.

[43] Ro YM, Cho ZH. A new frontier of blood imaging using susceptibility effect and tailored RF pulses. Magn Reson Med 1992;28(2):237–48.

[44] Stenger VA, Boada FE, Noll DC. Multishot 3d slice-select tailored RF pulses for MRI. Magn Reson Med 2002;48(1):157–65. https://doi.org/10.1002/mrm.10194. Available from: http://www.ncbi.nlm.nih.gov/entrez/query.fcgi?cmd=Retrieve&db=PubMed&dopt=Citation&list_uids=12111943.

[45] Katscher U, Bornert P, Leussler C, Brink JSvd. Transmit SENSE. Magn Reson Med 2003;49:144–50.

[46] Chang H, Fitzpatrick JM. A technique for accurate magnetic resonance imaging in the presence of field inhomogeneities. IEEE Trans Med Imaging 1992;11(3):319–29. https://doi.org/10.1109/42.158935. Available from: http://www.ncbi.nlm.nih.gov/pubmed/18222873.

[47] Holland D, Kuperman JM, Dale AM. Efficient correction of inhomogeneous static magnetic field-induced distortion in echo planar imaging. NeuroImage 2010;50(1):175–83. https://doi.org/10.1016/j.neuroimage.2009.11.044. S1053-8119(09)01229-4 [pii].

[48] Glasser MF, Smith SM, Marcus DS, Andersson JL, Auerbach EJ, Behrens TE, et al. The Human Connectome Project's neuroimaging approach. Nat Neurosci 2016;19(9):1175–87. https://doi.org/10.1038/nn.4361. Available from: https://www.ncbi.nlm.nih.gov/pubmed/27571196.

[49] Andersson JL, Skare S, Ashburner J. How to correct susceptibility distortions in spin-echo echo-planar images: application to diffusion tensor imaging. NeuroImage 2003;20(2):870–88. https://doi.org/10.1016/S1053-8119(03)00336-7. Available from: https://www.ncbi.nlm.nih.gov/pubmed/14568458.

[50] Smith SM, Jenkinson M, Woolrich MW, Beckmann CF, Behrens TE, Johansen-Berg H, et al. Advances in functional and structural mr image analysis and implementation as FSL. NeuroImage 2004;23(Suppl 1):S208–19. https://doi.org/10.1016/j.neuroimage.2004.07.051. S1053-8119(04)00393-3 [pii]. Available from: http://www.ncbi.nlm.nih.gov/entrez/query.fcgi?cmd=Retrieve&db=PubMed&dopt=Citation&list_uids=15501092.

[51] Kybic J, Thevenaz P, Nirkko A, Unser M. Unwarping of unidirectionally distorted EPI images. IEEE Trans Med Imaging 2000;19(2):80–93. https://doi.org/10.1109/42.836368. Available from: https://www.ncbi.nlm.nih.gov/pubmed/10784280. https://ieeexplore.ieee.org/document/836368/.

[52] Poynton C, Jenkinson M, Wells Wr. Atlas-based improved prediction of magnetic field inhomogeneity for distortion correction of EPI data. Med Image Comput Comput Assist Interv 2009;12(Pt 2):951–9. https://doi.org/10.1007/978-3-642-04271-3_115. Available from: https://www.ncbi.nlm.nih.gov/pubmed/20426203.

[53] Schomberg H. Off-resonance correction of MR images. IEEE Trans Med Imaging 1999;18(6):481–95. Available from: http://ieeexplore.ieee.org/document/781014/.

[54] Noll DC, Meyer CH, Pauly JM, Nishimura DG, Macovski A. A homogeneity correction method for magnetic resonance imaging with time-varying gradients. IEEE Trans Med Imaging 1991;10(4):629–37.

[55] Man LC, Pauly JM, Macovski A. Multifrequency interpolation for fast off-resonance correction. Magn Reson Med 1997;37:785–92.

[56] Jackson JI, Meyer CH, Nishimura DG, Macovski A. Selection of a convolution function for Fourier inversion using gridding. IEEE Trans Med Imaging 1991;10(3):473–8.

[57] Pruessmann KP, Weiger M, Scheidegger MB, Boesiger P. SENSE: sensitivity encoding for fast MRI. Magn Reson Med 1999;42:952–62.

[58] Stone SS, Haldar JP, Tsao SC, Hwu WM, Sutton BP, Liang ZP. Accelerating advanced MRI reconstructions on GPUs. J Parallel Distrib Comput 2008;68(10):1307–18. https://doi.org/10.1016/j.jpdc.2008.05.013. Available from: http://www.ncbi.nlm.nih.gov/pubmed/21796230.

[59] Wu XL, Gai J, Lam F, Fu M, Haldar J, Zhuo Y, et al. IMPATIENT MRI: Illinois massively parallel acceleration toolkit for image reconstruction with ENhanced throughput in MRI. In: Intl Soc Magn Reson Med; 2011. p. 4396.

[60] Gai J, Obeid N, Holtrop JL, Wu XL, Lam F, Fu M, et al. More IMPATIENT: a gridding-accelerated Toeplitz-based strategy for non-Cartesian high-resolution 3D MRI on GPUs. J Parallel Distrib Comput 2013;73(5):686–97. https://doi.org/10.1016/j.jpdc.2013.01.001. Available from: http://www.ncbi.nlm.nih.gov/pubmed/23682203.

[61] Cerjanic A, Holtrop J, Ngo GC, Leback B, Arnold G, Van Moer M, et al. PowerGrid: a open source library for accelerated iterative magnetic resonance image reconstruction. In: Intl Soc Magn Reson Med; 2016. p. 525.

[62] Muthupillai R, Lomas DJ, Rossman PJ, Greenleaf JF, Manduca A, Ehman RL. Magnetic resonance elastography by direct visualization of propagating acoustic strain waves. Science 1995;269(5232):1854–7. Available from: http://www.ncbi.nlm.nih.gov/pubmed/7569924.

[63] Johnson CL, McGarry MD, Van Houten EE, Weaver JB, Paulsen KD, Sutton BP, et al. Magnetic resonance elastography of the brain using multishot spiral readouts with self-navigated motion correction. Magn Reson Med 2013;70(2):404–12. https://doi.org/10.1002/mrm.24473. Available from: http://www.ncbi.nlm.nih.gov/pubmed/23001771.

[64] Yablonskiy DA. Quantitation of intrinsic magnetic susceptibility-related effects in a tissue matrix. Phantom study. Magn Reson Med 1998;39(3):417–28.

[65] An H, Lin W. Cerebral oxygen extraction fraction and cerebral venous blood volume measurements using MRI: effects of magnetic field variation. Magn Reson Med 2002;47(5):958–66.

[66] Dahnke H, Schaeffter T. Limits of detection of SPIO at 3.0 T using T_2^* relaxometry. Magn Reson Med 2005;53(5):1202–6.

[67] Yang X, Sammet S, Schmalbrock P, Knopp MV. Postprocessing correction for distortions in T_2^* decay caused by quadratic cross-slice B_0 inhomogeneity. Magn Reson Med 2010;63(5):1258–68.

[68] Yablonskiy DA, Sukstanskii AL, Luo J, Wang X. Voxel spread function method for correction of magnetic field inhomogeneity effects in quantitative gradient-echo-based MRI. Magn Reson Med 2013;70(5):1283–92.

[69] Deichmann R, Josephs O, Hutton C, Corfield DR, Turner R. Compensation of susceptibility-induced BOLD sensitivity losses in echo-planar fMRI imaging. NeuroImage 2002;15(1):120–35. https://doi.org/10.1006/nimg.2001.0985. S1053811901909851 [pii]. Available from: http://www.ncbi.nlm.nih.gov/entrez/query.fcgi?cmd=Retrieve&db=PubMed&dopt=Citation&list_uids=11771980.

[70] Liu G, Ogawa S. EPI image reconstruction with correction of distortion and signal losses. J Magn Reson Imaging 2006;24(3):683–9. https://doi.org/10.1002/jmri.20672. Available from: http://www.ncbi.nlm.nih.gov/entrez/query.fcgi?cmd=Retrieve&db=PubMed&dopt=Citation&list_uids=16892198. http://onlinelibrary.wiley.com/store/10.1002/jmri.20672/asset/20672_ftp.pdf?v=1&t=h4fyzh53&s=14417f88f5780ad3f8485f782920764ae7238a54.

[71] Ngo GC, Wong CN, Guo S, Paine T, Kramer AF, Sutton BP. Magnetic susceptibility-induced echo-time shifts: is there a bias in age-related fMRI studies? J Magn Reson Imaging 2017;45(1):207–14. https://doi.org/10.1002/jmri.25347. Available from: https://www.ncbi.nlm.nih.gov/pubmed/27299727.

[72] Tan H, Meyer CH. Estimation of k-space trajectories in spiral mri. Magn Reson Med 2009;61(6):1396–404. https://doi.org/10.1002/mrm.21813. Available from: https://www.ncbi.nlm.nih.gov/pubmed/19353671. https://onlinelibrary.wiley.com/doi/pdfdirect/10.1002/mrm.21813?download=true.

[73] Cauley SF, Setsompop K, Bilgic B, Bhat H, Gagoski B, Wald LL. Autocalibrated wave-caipi reconstruction; joint optimization of k-space trajectory and parallel imaging reconstruction. Magn Reson Med 2017;78(3):1093–9. https://doi.org/10.1002/mrm.26499. Available from: https://www.ncbi.nlm.nih.gov/pubmed/27770457. https://onlinelibrary.wiley.com/doi/pdfdirect/10.1002/mrm.26499?download=true.

[74] Gallichan D, Cocosco CA, Dewdney A, Schultz G, Welz A, Hennig J, et al. Simultaneously driven linear and nonlinear spatial encoding fields in mri. Magn Reson Med 2011;65(3):702–14. https://doi.org/10.1002/mrm.22672. Available from: https://www.ncbi.nlm.nih.gov/pubmed/21337403. https://onlinelibrary.wiley.com/doi/pdfdirect/10.1002/mrm.22672?download=true.

[75] Glover GH, Schneider E. Three-point Dixon technique for true water/fat decomposition with B_0 inhomogeneity correction. Magn Reson Med 1991;18:371–83.

[76] Sutton BP, Noll DC, Fessler JA. Dynamic field map estimation using a spiral-in/spiral-out acquisition. Magn Reson Med 2004;51(6):1194–204. https://doi.org/10.1002/mrm.20079. Available from: http://www.ncbi.nlm.nih.gov/pubmed/15170840.

[77] Liu J, de Zwart JA, van Gelderen P, Murphy-Boesch J, Duyn JH. Effect of head motion on MRI B_0 field distribution. Magn Reson Med 2018;80(6):2538–48.

[78] Funai AK, Fessler JA, Yeo DT, Olafsson VT, Noll DC. Regularized field map estimation in MRI. IEEE Trans Med Imaging 2008;27(10):1484–94. https://doi.org/10.1109/TMI.2008.923956. Available from: http://www.ncbi.nlm.nih.gov/entrez/query.fcgi?cmd=Retrieve&db=PubMed&dopt=Citation&list_uids=18815100.

[79] Barmet C, Zanche ND, Pruessmann KP. Spatiotemporal magnetic field monitoring for MR. Magn Reson Med 2008;60(1):187–97.

[80] Barmet C, De Zanche N, Wilm BJ, Pruessmann KP. A transmit/receive system for magnetic field monitoring of in vivo MRI. Magn Reson Med 2009;62(1):269–76.

[81] De Zanche N, Barmet C, Nordmeyer-Massner JA, Pruessmann KP. NMR probes for measuring magnetic fields and field dynamics in MR systems. Magn Reson Med 2008;60(1):176–86.

[82] Liang ZP. Spatiotemporal imaging with partially separable functions. In: Proc IEEE Int Symp Biomed Imag; 2007. p. 988–91.

[83] Vannesjo SJ, Wilm BJ, Duerst Y, Gross S, Brunner DO, Dietrich BE, et al. Retrospective correction of physiological field fluctuations in high-field brain MRI using concurrent field monitoring. Magn Reson Med 2015;73(5):1833–43.

[84] Wilm BJ, Nagy Z, Barmet C, Vannesjo SJ, Kasper L, Haeberlin M, et al. Diffusion MRI with concurrent magnetic field monitoring. Magn Reson Med 2015;74(4):925–33.

[85] Ma R, Akçakaya M, Moeller S, Auerbach E, Ugurbil K, Van de Moortele PF. A field-monitoring-based approach for correcting eddy-current-induced artifacts of up to the 2nd spatial order in human-connectome-project-style multiband diffusion MRI experiment at 7 T: a pilot study. NeuroImage 2020;216:116861.

CHAPTER

Motion-Corrected Reconstruction 13

Freddy Odille[a,b]

[a]*IADI U1254, Inserm and University of Lorraine, Nancy, France*
[b]*CIC-IT 1433, Inserm, Université de Lorraine and CHRU Nancy, Nancy, France*

13.1 Introduction

Several methods are described in previous chapters that can be applied in the context of patient motion, and these are often referred to as dynamic imaging methods. Firstly, highly undersampled acquisition of the k space or k-t space can be used, in combination with sparse reconstruction methods, in order to make the acquisition time short with regard to patient motion (this is also referred to as imaging with high temporal resolution). In that case, "real-time imaging" is achieved, and the impact of motion may be neglected. However, a high temporal resolution is generally achieved at the cost of reduced spatial resolution. Secondly, temporally resolved imaging is a common alternative that consists of synchronizing the MRI data to certain physiological signals, such as the electrocardiogram or a respiratory motion signal. Such synchronization can be performed either at the acquisition stage, by prospectively selecting which k-space samples are acquired depending on which motion state we are in. It can also be performed at the reconstruction stage, by retrospectively sorting out the k-space data using extra motion information that was acquired simultaneously with the k-space data. The drawback is that such methods assume a high level of reproducibility of motion throughout the acquisition.

Motion correction methods have been further developed to enable k-space data acquired from multiple motion states to be combined. Similar to synchronization methods, motion correction in MRI can be prospective or retrospective. Prospective correction consists of modulating, in real time, the pulse sequence waveforms, i.e., the radiofrequency (RF) pulses and the magnetic field gradient pulses, so that the spatial encoding of the image accounts for the deformation of the imaged object or subject, thus prospectively "inverting" the effect of motion [65]. However, since the spatial encoding uses linear magnetic-field gradients, only linear changes of the coordinate system can be performed with such techniques, which correspond to affine transformations (i.e., a global translation, rotation, and scaling). Retrospective correction consists of "inverting" the effect of motion at the image reconstruction stage, and it is the focus of this chapter. One of the main advantages of retrospective correction over prospective methods is that it is not limited to affine motion, i.e., deformable motion can also be addressed, as we will see.

Whether we consider the acquisition of a static image or a dynamic scene, the reconstruction methods that we have seen so far implicitly assume that the imaged organs are static during the acquisition of the k-space data of each image to be reconstructed. To ensure this assumption holds in the clinical practice, the patient is asked to remain still during the sequence, for a period ranging from a few seconds to several minutes. Specifically, in cardiac or abdominal MRI protocols, patients are asked to hold their

Advances in Magnetic Resonance Technology and Applications, Volume 7, ISSN 2666-9099. https://doi.org/10.1016/B978-0-12-822726-8.00024-5

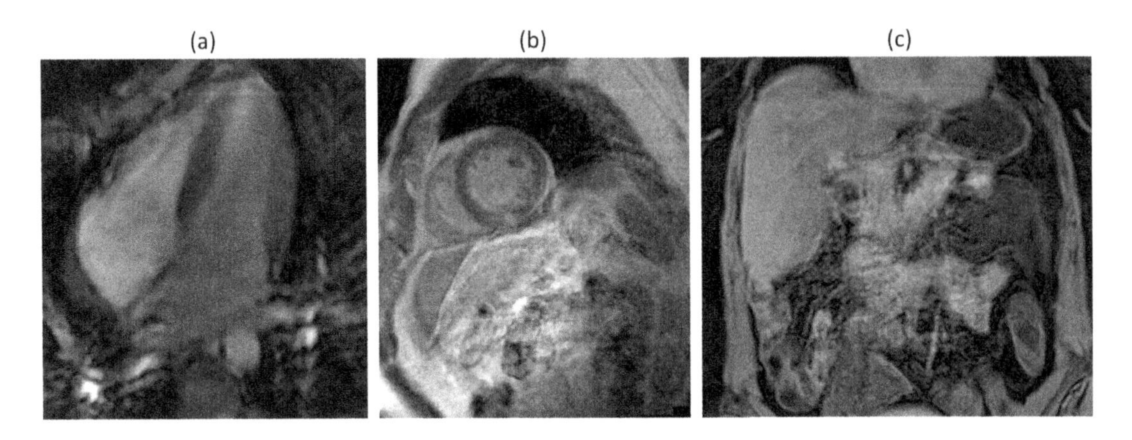

FIGURE 13.1

Example motion artifacts in patients: cardiac cine MRI (balanced SSFP sequence, systolic phase) from a eight-year-old Duchenne muscular dystrophy patient unable to hold his breath (**a**), showing blurring throughout the image and ghosts near the ventricular apex; 3D late gadolinium enhancement sequence (inversion-recovery fast spoiled gradient echo) from an adult cardiac patient with poor breath-holding (**b**), showing blurred myocardium hindering the precise delineation of the myocardial scar; abdominal slice from an Crohn disease patient referred for MR enterography (3D late gadolinium enhanced VIBE sequence) obtained after poor breath-holding (**c**), showing blurring/ghosting of all abdominal structures, including liver and bowel.

breath repeatedly for periods of 10 to 20 s. Some patients, including elderly, pediatric, or psychiatric patients, are less likely to comply with those instructions. Moreover, there are some types of motion that simply cannot be avoided (e.g., heart beating). Motion can occur at random times (e.g., involuntary head motion, swallowing, coughing) or it can be pseudo-periodic (heart beating, breathing, peristalsis). The most obvious manifestation of motion is a degradation of the image quality (see Fig. 13.1). As we know, the MRI data is sampled in the Fourier domain, thus the resulting artifacts in the image domain due to motion in k-space are quite complex, in that they are not only localized in the area of the moving structure. This can alter the quality and diagnostic value of the image. As a result, it can also lengthen the examination due to the need to repeat corrupted scans, and ultimately degrade the patient experience with MRI. Besides this direct impact, motion also has an indirect impact on clinical MRI. Indeed, since most patients cannot hold their breath for much longer than 20 s, cardiac and abdominal imaging protocols are designed with a compromised image quality, typically using a much lower spatial resolution (especially in the slice direction) than in static organs. Thus, reconstruction methods from free-breathing data can therefore enable new applications as shown by several approaches that we will discuss in this chapter.

In this chapter, we will focus on estimating and integrating motion into the forward acquisition model E, introduced in Chapter 2. To achieve that goal, we will need to make a number of assumptions pertaining to MRI physics. First, we will assume that motion between the RF excitation and the sampling of k-space data—also termed intraview motion—can be neglected.[1] This assumption is often reasonable in practice when motion-induced changes are slow compared to the echo time of the imaging sequence. As a consequence, motion can be considered to occur between successive phase encoding steps—this is also known as interview motion—and only the spatial encoding process is affected. Due to inconsistencies in the imaged content during the acquisition of the different sets of k-space samples, a conventional reconstruction by inverse Fourier transform results in a combination of blurring (in all directions of motion) and ghosting artifacts (in the phase-encoding direction(s)) [92]. Such artifacts are illustrated in Figs. 13.1 and 13.2, and are precisely what we seek to suppress, or at least minimize, with the motion corrected reconstruction techniques described in this chapter. Furthermore, it will be assumed that a motion-corrected image does exist, which means that there exists a reference motion state from which any motion state can be derived by a spatial deformation of the imaged organs/tissues, and only by a spatial deformation. In terms of physics, this assumption can be thought of as a signal conservation rule—the signal being what is "seen" by the imaging system, analogous to mass conservation. Therefore, large through-plane or out-of-volume motion, contrast changes during the course of the sequence, or changes of the magnetization due to spin history effects should be treated carefully. This assumption should also be viewed from a physiological standpoint. For instance, the reconstruction of phase-contrast cardiac MRI data acquired during free breathing uses data from multiple respiratory motion states that may correspond to different velocities.

With that in mind, we will first consider the problem of MRI reconstruction in the presence of known motion. Here "known motion" means that we assume motion has already been estimated or guessed by any means. This can be, for instance, from real-time motion measurements of bulk head motion, from a low-resolution image navigator, or from a predictive model combining real-time motion signals and prior imaging data. Because this motion estimation step is generally not perfect, we will next consider motion as an additional optimization variable for the reconstruction problem. Provided motion can be modeled and parameterized, the reconstruction can then be reformulated as a joint estimation of a static image and some motion parameters. In the methods section we will mainly review the strategies that can be used in practice to provide estimates of motion. Then a few applications will be described, with the objective to illustrate how these methods can be fit to various settings, i.e. using either extra hardware for motion sensing, or MR navigator data, or nothing but the raw k-space data. Finally, we will discuss some of the remaining challenges, and, in particular, we will come back to the validity of the various assumptions and possible directions for future research.

[1] This assumption is invalidated, for instance, when fast motion occurs during a conventional fast spin echo or a diffusion-weighted sequence. Certain types of motion (e.g., pulsatile motion in the brain, cardiac contraction) may cause intravoxel dephasing, leading to signal dropouts. In some cases, a sequence design using gradient moment-nulling techniques can minimize such effects.

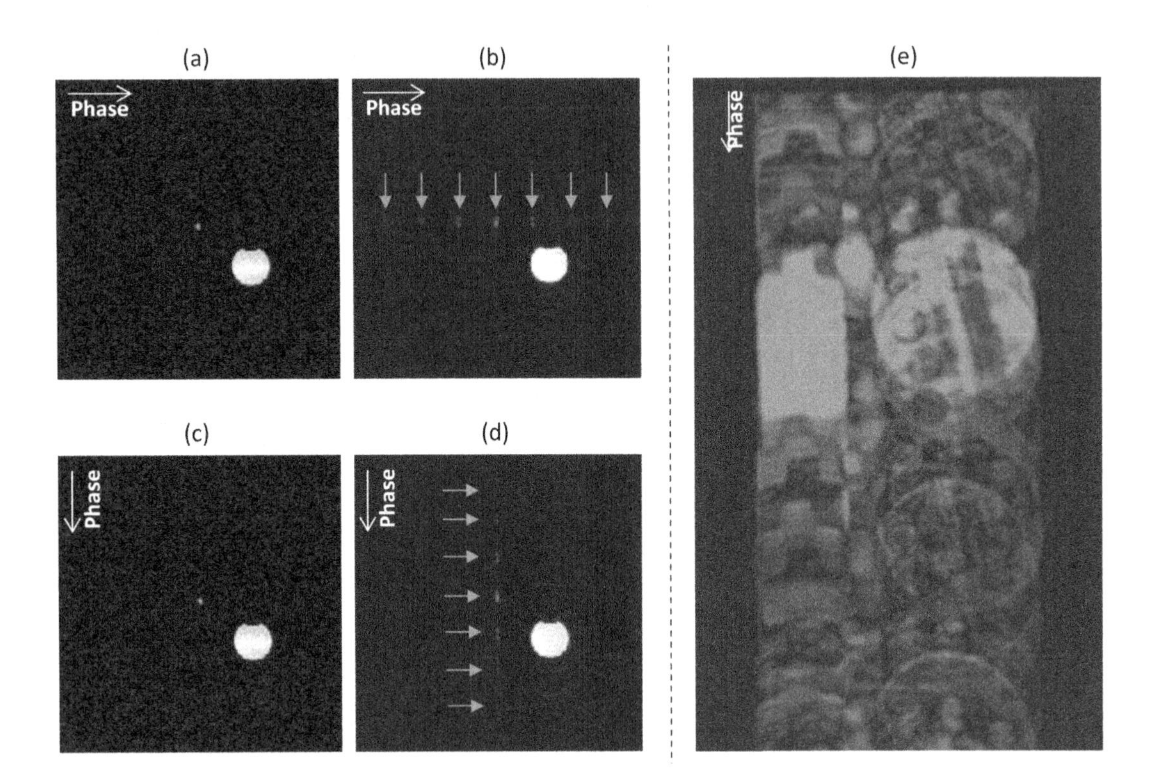

FIGURE 13.2

Example motion artifacts in phantoms subjected to a controlled, periodic, translational motion in the vertical direction. The left-hand figures show spin-echo images without (respectively with) motion of a small gadolinium sphere (at the center of the image), when the phase encoding direction is horizontal (**a** and **b** respectively) or vertical (**c** and **d**). Artifacts appear as replicas (ghosts) in phase direction and localized blurring in the direction of motion. The right-hand figure shows another moving phantom, during a fast spin-echo sequence with interleaved k-space sampling.

13.2 Theory

13.2.1 Reconstruction with known motion: the particular case of translational motion

In the particular case of a bulk translational motion of the imaged object/subject, the motion correction problem may be simplified using properties of the Fourier transform. Let $\boldsymbol{\tau}_{\boldsymbol{k}} = \begin{bmatrix} \tau_{k,x} & \tau_{k,y} & \tau_{k,z} \end{bmatrix}$ be a translation vector affecting the image during the acquisition of a given k-space sample of coordinate $\boldsymbol{k} = \begin{bmatrix} k_x & k_y & k_z \end{bmatrix}$. The MRI signal for that k-space sample, for a given receiving coil c (assumed to be static), is the Fourier transform of the shifted object, weighted by the coil sensitivity (defined in

Chapters 2 and 6 for single and multiple coils, respectively)

$$s_c(\boldsymbol{k}) = \iiint C_c(\boldsymbol{r})\, x(\boldsymbol{r} - \boldsymbol{\tau}_{\boldsymbol{k}})\, e^{-2i\pi \boldsymbol{k}\cdot\boldsymbol{r}} d\boldsymbol{r}. \tag{13.1}$$

Using the substitution rule $\boldsymbol{r}' = \boldsymbol{r} - \boldsymbol{\tau}_{\boldsymbol{k}}$ in the integral,

$$s_c(\boldsymbol{k}) = e^{-2i\pi \boldsymbol{k}\cdot\boldsymbol{\tau}_{\boldsymbol{k}}} \iiint C_c\left(\boldsymbol{r}' + \boldsymbol{\tau}_{\boldsymbol{k}}\right) x\left(\boldsymbol{r}'\right) e^{-2i\pi \boldsymbol{k}\cdot\boldsymbol{r}'} d\boldsymbol{r}'. \tag{13.2}$$

If we further assume that spatial variations in the coil sensitivity are small for the given motion, we can make the approximation that $C_j\left(\boldsymbol{r}' + \boldsymbol{\tau}_{\boldsymbol{k}}\right) \approx C_j\left(\boldsymbol{r}'\right)$, leading to a simplified expression

$$s_c(\boldsymbol{k}) \approx e^{-2i\pi \boldsymbol{k}\cdot\boldsymbol{\tau}_k} F\left(C_c x\right). \tag{13.3}$$

It follows that each coil image $C_c x$, rewritten as x_c, can then be reconstructed directly using the inverse Fourier transform of the k-space data, multiplied by phase-correction terms for each k-space sample

$$x_c(\boldsymbol{r}) \approx \iiint s_c(\boldsymbol{k})\, e^{2i\pi \boldsymbol{k}\cdot\boldsymbol{\tau}_{\boldsymbol{k}}} e^{2i\pi \boldsymbol{r}\cdot\boldsymbol{k}} d\boldsymbol{k}. \tag{13.4}$$

Finally, the image is reconstructed by a coil combination technique. Conventional methods include the sum-of-squares reconstruction, or the SENSE reconstruction with an acceleration factor of 1 [78], i.e., $x = \sum_{c=1}^{N_c} C_c^* x_c / \sum_{c=1}^{N_c} |C_c|^2$.

This approach can be extended to rotational motion or affine motion [73]. Affine motion can be defined by a transformation of the form $\boldsymbol{r} \mapsto \mathcal{A}\boldsymbol{r} + \boldsymbol{\tau}$, with $\mathcal{A}$ a 3×3 matrix (which can describe rotation, shearing, scaling), and the translation vector $\boldsymbol{\tau} = \begin{bmatrix} \tau_x & \tau_y & \tau_z \end{bmatrix}$. In particular, a rotation in image space is equivalent to a rotation by the same angle in k-space. A regridding may be necessary in that case, as well as a density compensation (see Chapter 4). However the general formulation described in the next section will be preferred, even in the case of simple motion (translations, rotations), for several reasons: (i) This k-space correction method makes an additional assumption about coil sensitivities as we have seen, which may lead to errors; (ii) the proposed reconstruction formula in Eq. (13.4) is an empirical method rather than a mathematically well-defined inversion procedure; and as a result it does not provide an optimal solution to the motion correction problem, e.g., in the least-squares sense. We will come back to this point in the next section, and the differences between the two approaches are illustrated in the practical tutorial provided with this chapter.

13.2.2 Reconstruction with known motion: the general case

In this section we seek to integrate "arbitrary" patient motion in the forward acquisition model E of the linear inverse problem formulation of MRI reconstruction. By "arbitrary", we mean that it can be a simple translation or a rigid transformation of the image, but it can also consist of free local deformations, which will be termed elastic or nonrigid. Physically though, motion is not completely arbitrary. The displacement fields describing subject motion in the field-of-view can be assumed to be continuous and smoothly varying in space, except potentially at the interface between certain organs exhibiting a sliding motion (e.g., at the interface between the liver and abdominal wall), or at the interface between cavity and blood for instance. Importantly, we assume that the imaged organs remain mostly within the imaged slice or volume during the acquisition, which means that, for 2D imaging, only in-plane motion will be considered, and through-plane motion will be disregarded. In the particular case of 2D imaging, through-plane motion may also result in an *apparent* in-plane displacement of the structures that is different for the true 3D motion, e.g., in a short axis slice the heart ventricles may appear to be a little bit larger or smaller due to the component of the 3D heart motion that lies in the longitudinal direction of the heart. Therefore, an apparent 2D displacement field may be used to model motion in that short-axis slice, which may actually combine structures coming from closely adjacent regions of the heart in the through-plane direction. Since relatively thick slices are generally used in cardiac imaging (8 mm, typically) the framework described here can also be useful in practice, but it should be kept in mind that what is corrected for is this apparent 2D motion.

13.2.2.1 Motion operators

In the mechanics of continuous media, motion is described by a displacement vector field $\boldsymbol{u}(\boldsymbol{r}, t)$, which means that, for each point $\boldsymbol{r}$ in space, its displacement vector at time t is $\left[u_x(\boldsymbol{r}, t) \quad u_y(\boldsymbol{r}, t) \quad u_z(\boldsymbol{r}, t)\right]$. The intensity of the MR image at a given time t is therefore given by $x(\varphi(\boldsymbol{r}, t))$, with the mathematical transformation $\varphi : \boldsymbol{r} \mapsto \varphi(\boldsymbol{r}, t) = \boldsymbol{r} + \boldsymbol{u}(\boldsymbol{r}, t)$. The question that follows naturally is: How can we move on from this mechanics formulation of the spatial transformation φ to a matrix formalism that is better suited for image reconstruction problems as we have seen in previous chapters, and which would allow us to apply some numerical solvers? We want to formulate the operator $M_u(x)$ that takes the image matrix x as an input, applies the displacement field u to it, and returns the transformed image matrix. If we try, "naively", to apply u to the spatial coordinates of each voxel of a digital image x, we see that M_u has to affect the intensity value at the reference motion state, $x(\boldsymbol{r})$, to a voxel near the coordinate $\boldsymbol{r} + \boldsymbol{u}(\boldsymbol{r}, t)$. This approach has a major drawback: For certain types of displacement fields, e.g., a local expansion, it will leave some "holes" in the transformed image matrix, i.e., there is no guarantee that each voxel of the transformed image will be filled in with a proper value.

To solve this problem, a more convenient way is generally used in the field of digital image processing, in particular for image registration, and that is illustrated in Fig. 13.3. The transformation is described by the inverse displacement field instead, i.e., we will now consider that u is the displacement field from the target motion state to the reference motion state [68]. By doing so, the transformed image $M_{\boldsymbol{u}}(x)$ can be formed by searching, for each of its voxels r, the corresponding intensity value which is $x(\boldsymbol{r} + \boldsymbol{u}(\boldsymbol{r}, t))$. The intensity values in image x are only known at the nodes of a regular grid so, in general, $\boldsymbol{r} + \boldsymbol{u}(\boldsymbol{r}, t)$ does not lie exactly on one of these particular nodes. Therefore, $x(\boldsymbol{r} + \boldsymbol{u}(\boldsymbol{r}, t))$ is estimated by interpolation, which means that it is estimated by a weighted sum of the intensity values in x at the closest nodes. As a result, the operator $M_{\boldsymbol{u}}$ is a linear function of x and can be described by a matrix. Therefore, in the remainder, we will note the transformed image as the matrix–vector product $\boldsymbol{M}_{\boldsymbol{u}}\boldsymbol{x}$. Each row of the matrix $\boldsymbol{M}_{\boldsymbol{u}}$ is responsible for calculating the intensity value of one particular voxel of the transformed image. In the case a 2D linear interpolation kernel is used, the closest voxels form a 2×2 square so this row of $\boldsymbol{M}_{\boldsymbol{u}}$ will comprise four nonzeros elements, which are the four interpolation weights. These four elements will be located in the columns of $\boldsymbol{M}_{\boldsymbol{u}}$ corresponding to the indices of the four surrounding voxels in x (see Fig. 13.3).

Using this explicit matrix formulation of the motion operators, its transpose can be computed exactly, without any approximation (the transpose operators are also required for solving the inverse problem). Indeed, some authors have used the approximation $\boldsymbol{M}_{\boldsymbol{u}}^T \approx \boldsymbol{M}_{-\boldsymbol{u}}$, but it is not always valid.

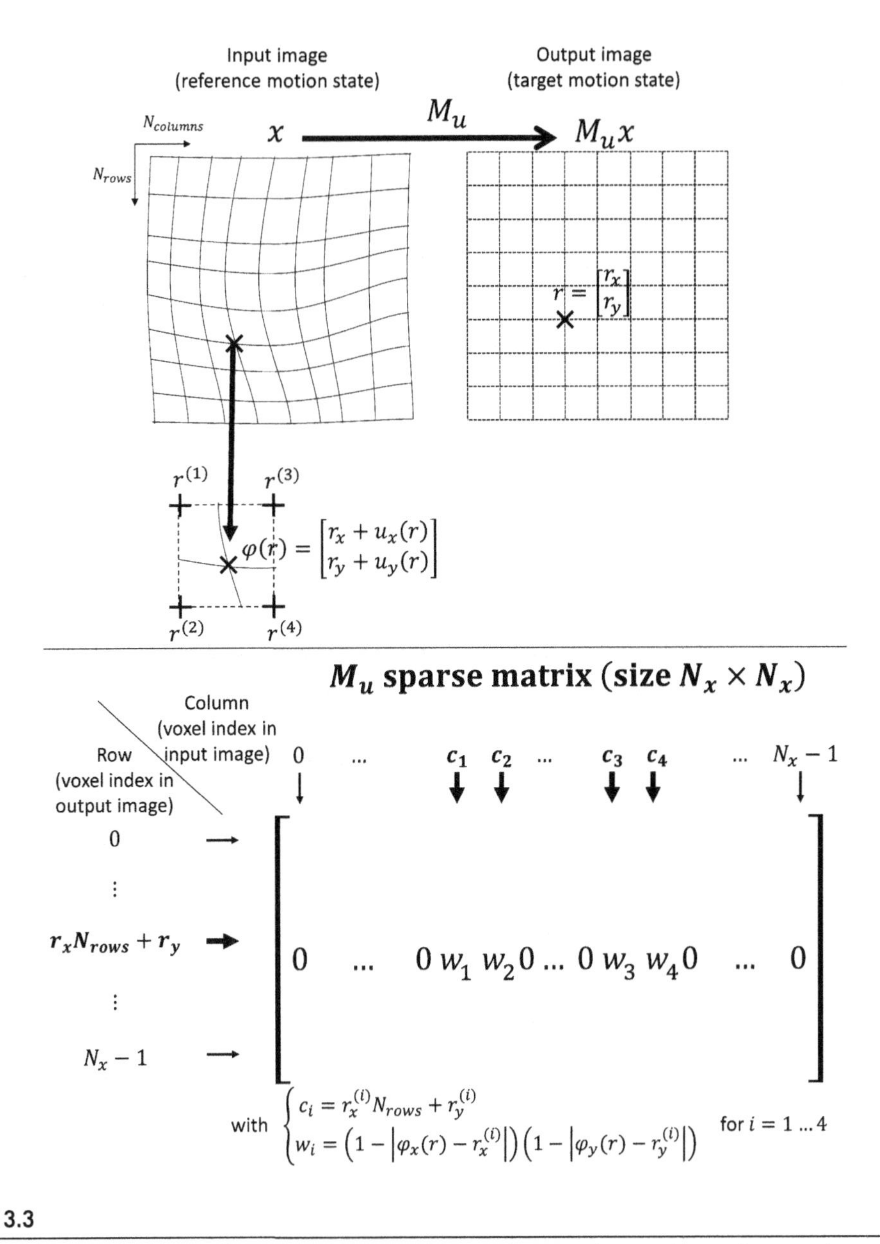

FIGURE 13.3

Definition and implementation of the motion operator $\boldsymbol{M_u}$ used to transform an image from a reference to a target motion state, knowing the underlying displacement field $\boldsymbol{u}$.

13.2.2.2 *Forward acquisition model including motion operators*

Once the motion operators have been determined, forming the forward acquisition model is relatively straightforward.

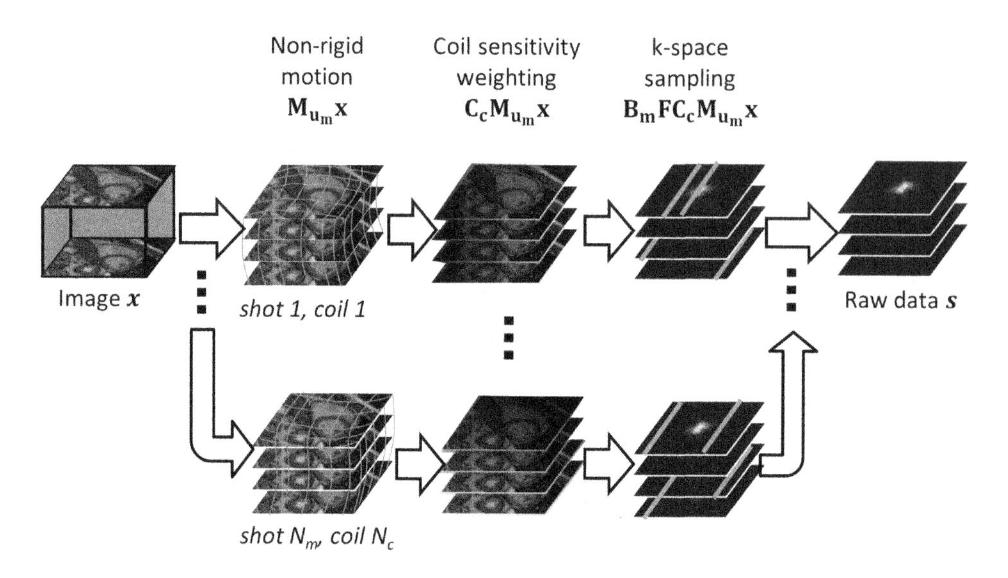

FIGURE 13.4

Forward model of MRI acquisition in the presence of known motion between successive phase-encoding steps or shots, including the motion operators.

Let us assume that the MR acquisition comprises N_m shots (i.e., N_m subsets of the full k-space scan) and that motion occurs between each of these shots, as described by a motion operator $\boldsymbol{M}_{\boldsymbol{u}_m}$ ($m = 1 \ldots N_m$) that takes the imaged subject from a reference motion state to its motion state at shot m. The reference motion state can be chosen arbitrarily, but it is often convenient to define it as an "average" position, or possibly a position where data are more frequently acquired (e.g. the end-expiratory plateau of a free-breathing scan). As we have just discussed, the underlying displacement field $\boldsymbol{u}_m$ defines the mapping of physical coordinates from the mth motion state to the reference motion state.

In the most general case, each shot contains only one k-space readout (e.g., one line in a Cartesian acquisition) and N_m is the number of k-space lines. A shot can also contain several lines of k-space acquired within a very short time period, during which motion is assumed to be negligible. In a more computationally efficient implementation, a shot can be a grouping of k-space lines acquired in a similar motion state (e.g., same phase as the respiratory cycle). In the latter case, N_m is the number of distinct motion states and motion within each shot is disregarded.

Given these definitions, the forward acquisition model is constructed in a way to simulate the motion-corrupted scan, given a static image $\boldsymbol{x}$ (in the reference motion state), and to return the motion-corrupted k-space data s

$$\mathrm{s} = \boldsymbol{E}\boldsymbol{x}. \tag{13.5}$$

Therefore, the encoding operator $\boldsymbol{E}$ is, as illustrated in Fig. 13.4, the concatenation of k-space samples acquired by each shot or motion state (with $\mathbf{B}_m$ the sampling matrix at the motion state/shot m) and by each coil receiver, which are the Fourier transforms ($\boldsymbol{F}$ being the Fourier transform operator) of the deformed image $\mathbf{M}_{u_n}\boldsymbol{x}$, weighted by the sensitivity of the coil receiver $\boldsymbol{C}_c$ [9,68]

$$\mathbf{E} = \begin{bmatrix} \mathbf{B}_1\mathbf{F}\mathbf{C}_1\mathbf{M}_{u_1} \\ \vdots \\ \mathbf{B}_{N_m}\mathbf{F}\mathbf{C}_1\mathbf{M}_{u_{N_m}} \\ \hline \vdots \\ \hline \mathbf{B}_1\mathbf{F}\mathbf{C}_{N_c}\mathbf{M}_{u_1} \\ \vdots \\ \mathbf{B}_{N_m}\mathbf{F}\mathbf{C}_{N_c}\mathbf{M}_{u_{N_m}} \end{bmatrix}. \tag{13.6}$$

Let us explicitly give the size of the matrices composing $\boldsymbol{E}$. The motion operators $\mathbf{M}_{u_m}$ $(m = 1 \ldots N_m)$ are sparse matrices as described in the previous section, of size $N_x \times N_x$, where N_x is the number of voxels is the image. The coil sensitivity operators $\mathbf{C}_c$ $(c = 1 \ldots N_c)$ are diagonal matrices of size $N_x \times N_x$, and the diagonal elements are the complex sensitivity values of coil c. The Fourier transform operator $\boldsymbol{F}$ (which can be either 2D or 3D) is a dense matrix of size $N_x \times N_x$. In a Cartesian acquisition, it is the classic discrete Fourier transform, but in a non-Cartesian acquisition it can be defined as the nonuniform fast Fourier transform (NuFFT) operator. The sampling operators $\mathbf{B}_m$ $(m = 1 \ldots N_m)$ are sparse matrices with values 0 or 1. Each $\mathbf{B}_m$ matrix has N_x columns, and its number of rows is the number of k-space samples acquired at shot m. Finally, the size of $\boldsymbol{E}$ is $N_c N_s \times N_x$, with N_s the total number of acquired k-space samples for each coil.

13.2.2.3 Solving the inverse problem

A motion-corrected image $\hat{\boldsymbol{x}}$ can be obtained by solving the inverse problem, using a classic least-squares optimization scheme

$$\hat{\boldsymbol{x}} = \arg\min_{\boldsymbol{x}} \| \boldsymbol{E}\boldsymbol{x} - \boldsymbol{s} \|_2^2 + \lambda \, \|\boldsymbol{x}\|_2^2 . \tag{13.7}$$

Here, for simplicity, a Tikhonov regularizer is used, but the framework can be easily adapted to include more advanced regularizers [83], such as those used to enforce sparse models (L_1 norms, low-rank models, etc.) which have been more thoroughly described in the previous chapters. As shown in Chapter 2, Problem (13.7) has the following solution:

$$\hat{\boldsymbol{x}} = \left(\boldsymbol{E}^H \boldsymbol{E} + \lambda \boldsymbol{I}\right)^{-1} \boldsymbol{E}^H s. \tag{13.8}$$

The inversion can be performed using a matrix-free solver, i.e. it does not require the full $\boldsymbol{E}$ matrix to be formed and stored explicitly, such as the conjugate-gradient algorithm. Algorithms to solve a problem of this type are discussed in Chapter 3.

The solution provided by Eq. (13.8) is a mathematical inversion of a large composite operator, which can be seen as the sum of several individual operators corresponding to the various motion states or shots in the acquisition. It is interesting to compare this approach to the k-space phase correction given in Section 13.2.1 in the case of a translational motion. The phase-correction method consists of applying an inverse motion operator for each shot individually (since it is equivalent to applying the inverse translation in image space for each shot), and then gathering the data from all shots. Therefore, it is the sum of motion-inverted k-space data rather than the invert of the sum of motion-corrupted k-space data. In general, the sum of inverses is different from the inverse of the sum, as observed in [9], and therefore the phase-correction method provides a suboptimal solution in the least squares sense.

In the particular case where no motion occurs, the motion operators are all equal to the identity matrix, and the reconstruction in Eq. (13.8) is a classic regularized SENSE reconstruction, as seen in Chapter 6. Similar to parallel imaging reconstruction problems, noise-correlation matrices can be incorporated to provide an SNR-optimal solution [79]. This can be done in a pre-processing step, prior to image reconstruction, by forming virtual receiver channels—with their associated virtual coil sensitivities and k-spaces—which have ideal noise decorrelation across channels [76]. To keep notations simple, in this chapter we will therefore assume receiver channels are ideally decorrelated.

Conditioning of the system

The condition number of the $\boldsymbol{E}^H\boldsymbol{E}$ matrix is larger than that a classic inverse Fourier transform or SENSE reconstruction because the introduction of the motion operators makes the linear system of equations deviate from an ideal set of linearly independent equations. The actual value of the condition number depends on the type of motion and on the k-space sampling pattern. The effect of these elements on conditioning can be illustrated by a simple thought experiment, as described in [7]. Let us consider an acquisition made of two shots or motion states: The first half of k-space is sampled during the first shot, the second half during the second shot, and there is a 90° rotation of the imaged object between the two shots. Because the 90° rotation in image space is equivalent to a 90° rotation in k-space, this means that we have effectively collected samples corresponding to three out of four tiles of the Fourier plane, and the fourth tile is left unsampled. This means there is not enough information to reconstruct the image exactly according to the Nyquist limit.

Apart from using regularization, one should also consider another convenient way to improve the conditioning of the system. If we acquire the k-space with several repetitions (i.e., with a number of excitations Nex $>$ 1), provided that the same k-space samples are collected several times but in different motion states, they will provide new linearly independent data to the linear system of equations [68]. Unlike classic reconstructions, the motion-corrected reconstruction does not average the k-space data in the dimension of Nex, but instead it considers multiple Nex data as separate shots or motion states.

As a conclusion, in the ideal case of known motion, using a combination of repeated sampling (Nex $>$ 1) and regularization, Eq. (13.8) provides a mathematically rigorous framework for motion correction and can provide very accurate solutions. The main difficulty that remains for its practical implementation is to accurately estimate motion. Several approaches will be described in Section 13.3 but they generally provide imperfect estimates. Therefore, one will also consider the case where motion is treated as an additional unknown of the reconstruction problem.

13.2.3 Joint reconstruction of image and motion

As discussed in Chapter 2 for nonlinear reconstructions in MRI, in this section we rewrite the motion-corrupted acquisition model $\boldsymbol{E}$ as $\boldsymbol{E}(\boldsymbol{u})$ in order to show explicitly its dependency on the displacement field $\boldsymbol{u}$. For the moment we make no assumption on $\boldsymbol{u}$, which can vary in space and time (the time dimension corresponding to the shots or motion states, as defined in the previous section). We now consider the following optimization problem to solve for the motion-corrected image and the motion

Table 13.1 Autofocus approach versus joint estimation of image and motion.

	"Autofocus" reconstruction	Joint optimization of image and motion
Main cost function to be minimized	$\mathcal{R}\left(E(u)^{-1}s\right)$ e.g., $\mathcal{R}(x)$ = image entropy	$\|E(u)x - s\|_2^2$
Variables to be optimized explicitly	Motion parameters: u	- Motion parameters: u - Image: x
Variables to be optimized implicitly	Solution image $\hat{x} = E(\hat{u})^{-1} s$ with $\hat{u}$ the solution of the motion parameter optimization problem	None
Optional regularizers to be added to the cost function	On motion parameters: $+\mu\mathcal{S}(u)$ e.g., $\mathcal{S}(u)$ = spatial/temporal smoothing of motion parameters	- On the solution image: $+\lambda\mathcal{R}(x)$ e.g. $\mathcal{R}(x)$ = L2 norm, L1 norm, image entropy... - On motion parameters: $+\mu\mathcal{S}(u)$
Approximations for practical solving	$E(u)^{-1}s \approx$ sum of motion-inverted k-space data (e.g., phase correction method described in Section 13.2.1.)	None

field:

$$\left(\hat{\boldsymbol{x}}, \hat{\boldsymbol{u}}\right) = \underset{\boldsymbol{x},\boldsymbol{u}}{\arg\min} \|\boldsymbol{E}(\boldsymbol{u})\boldsymbol{x} - \boldsymbol{s}\|_2^2 + \lambda \|\boldsymbol{x}\|_2^2 + \mu \|\boldsymbol{S}\boldsymbol{u}\|_2^2. \tag{13.9}$$

Here, we have introduced a second regularization term in order to impose a constraint on the displacement field $\boldsymbol{u}$. This term includes an operator $\boldsymbol{S}$, which will be specified later according to the additional assumptions that will be made on motion. Indeed, in the present form, the number of unknowns is extremely large. The image $\boldsymbol{x}$ has N_x elements, the displacement field $\boldsymbol{u}$ has $N_{dims}N_m N_x$ elements, with N_m the number motion states (or shots), and N_{dims} the number of dimensions of the image (2D or 3D). The number of collected k-space samples is $N_c N_s$. For example, if we consider a 256 × 256 image, acquired with 8 shots, 1 Nex, 32 coils, we have about 1.1 million unknowns and 2.1 million acquired data samples. However, there is a lot of redundancy within the 32 coils so, although the system appears overdetermined, it is likely to be severely ill-conditioned.

Relation to "autofocus" method

It should be noted that another formulation has been proposed in the literature, called autofocus that consists of optimizing motion parameters that minimize an image-quality metric, such as the image entropy [5,6,15,54,56]. Table 13.1 summarizes the main differences between the "autofocus" approach and the joint optimization approach. Autofocus methods also result in a joint optimization problem for the image and motion parameters, but the image reconstruction step is hidden, or embedded, in the motion estimation step (e.g., optimizing the image entropy of the motion-corrected image). Mathematically, optimizing the reconstructed image quality with respect to the motion parameters is a more difficult problem because the associated forward model (relating the motion parameters to the cost function) requires a model for the motion-corrected image reconstruction. However, this involves an inversion procedure, i.e., $\boldsymbol{E}(\boldsymbol{u})^{-1}s$, so a direct model of the reconstruction is not easily available. In practice authors used an estimate of the "exact" reconstruction, e.g., a phase-corrected k-space reconstruction in case of translational motion, as described in Section 13.2.1. This is conceptually close to assuming that $\boldsymbol{E}(\boldsymbol{u})^{-1} \approx \boldsymbol{E}(\boldsymbol{u})^T$. Though this approach can be effective in the rigid case, it does involve an approximation, and it is difficult (or computationally inefficient) to generalize it to nonrigid motion. Therefore, Eq. (13.9) will be preferred. Note that the image-entropy metric $H(x) = \sum_{n=1}^{N_x} |x_n|/x_{max} \log(|x_n|/x_{max})$ [5], with x_{max} the maximal intensity in the image, could be used in Eq. (13.9) instead of the classic regularizer $\|\boldsymbol{x}\|_2^2$. Actually, if we ignore the normalization by x_{max}, the image entropy can be thought of as an "intermediate" between the more familiar L_1 and L_2 norms, respectively $\|\boldsymbol{x}\|_1 = \sum_{n=1}^{N_x} |x_n|$ and $\|\boldsymbol{x}\|_2^2 = \sum_{n=1}^{N_x} |x_n|^2$.

Parameterizing motion

Instead of directly solving the general problem in Eq. (13.9), in this chapter we will seek to parameterize motion in order to reduce the number of unknowns significantly. If we choose a particular model for motion, the displacement field $\boldsymbol{u}$ can be described by a reduced number of parameters $\boldsymbol{\alpha}$, and therefore only those parameters need to be optimized. We will consider two motion models: The first one will be a spatially uniform displacement field, i.e., only global translations of the object/subject will be considered (reducing the number of motion parameters to $N_{dims} N_m$); the second one will allow spatially varying (i.e., nonrigid) displacements, under a temporal constraint that time variations of $\boldsymbol{u}$ are linear functions of a number N_k of known motion signals (reducing the number of motion parameters to $N_{dims} N_x N_k$ with $N_k \ll N_m$). Therefore, the joint reconstruction is reformulated as a joint optimization of the static image and the motion model parameters $\boldsymbol{\alpha}$

$$(\hat{\boldsymbol{x}}, \hat{\boldsymbol{\alpha}}) = \underset{x,\alpha}{\arg\min} \, \|\boldsymbol{E}(\boldsymbol{u}(\boldsymbol{\alpha}))\boldsymbol{x} - \boldsymbol{s}\|_2^2 + \lambda \|\boldsymbol{x}\|_2^2 + \mu \|\boldsymbol{S}\boldsymbol{\alpha}\|_2^2 . \quad (13.10)$$

In case of a translational motion model, the regularization on $\boldsymbol{\alpha}$ may be omitted since the number of added parameters remains small. Alternatively, $\boldsymbol{S}$ can be chosen to be the identity operator or a temporal smoothing if there are good reasons to assume motion cannot vary too abruptly between two successive shots or motion states. In case of a temporally constrained nonrigid motion model, $\boldsymbol{S}$ can be the spatial gradient operator to ensure $\boldsymbol{\alpha}$, and thereby $\boldsymbol{u}$ in this case is spatially smooth.

Alternating optimization scheme

In order to solve the problem in Eq. (13.10), an alternating optimization scheme can be used. The main difficulty is that the cost function is a nonlinear function of $\boldsymbol{u}$ (and thereby of $\boldsymbol{\alpha}$). This problem belongs to the class of nonlinear least squares problems. Methods for solving this type of problems have been described in Chapter 3 and, specifically for motion-compensated reconstruction, have been described in [69,71] in the case of a nonrigid motion model and in [21,38] in the case of rigid motion. To keep the formalism consistent for motion-compensated reconstruction in the following sections, we will describe the Gauss–Newton strategy initially proposed for nonrigid motion models, and briefly discussed in Chapter 3, but we will also adapt it to the rigid case (translations only). The Gauss–Newton method consists of linearizing the cost function around the current estimate of motion. This turns the nonlinear least-squares problem into a sequence of linear least-squares problems, which we are more familiar with. This linearization step is described mathematically by the so-called Jacobian matrix. This matrix may seem a bit abstract so it may be useful to give a physical interpretation for it. It actually relates to how a small error in the motion model will propagate into the forward model that we have described in the previous section. In other words, it tells us how much our motion-corrupted acquisition model deviates from the actual k-space data measurements when a small perturbation of the motion model is applied.

13.2.3.1 Propagation of motion errors

Let us assume an estimate of motion $\boldsymbol{u}$ is available that is related to the true displacement field $\boldsymbol{u}_{true}$ as: $\boldsymbol{u}_{true} = \boldsymbol{u} + \delta \boldsymbol{u}$. We define the residual reconstruction error ε as the error in the k-space domain due to the error in motion estimates

$$\boldsymbol{\varepsilon} = \boldsymbol{E}\left(\boldsymbol{u}_{true}\right)\boldsymbol{x} - \boldsymbol{E}\left(\boldsymbol{u}\right)\boldsymbol{x} = \boldsymbol{E}\left(\boldsymbol{u} + \delta\boldsymbol{u}\right)\boldsymbol{x} - \boldsymbol{E}\left(\boldsymbol{u}\right)\boldsymbol{x}. \tag{13.11}$$

This expression can be expanded to

$$\boldsymbol{\varepsilon} = \left[\begin{array}{c} \mathbf{B}_1\mathbf{F}\mathbf{C}_1\left(\mathbf{M}_{u_1+\delta u_1}\boldsymbol{x} - \mathbf{M}_{u_1}\boldsymbol{x}\right) \\ \vdots \\ \mathbf{B}_{\mathrm{N}_m}\mathbf{F}\mathbf{C}_1\left(\mathbf{M}_{u_{\mathrm{N}m}+\delta u_{\mathrm{N}m}}\boldsymbol{x} - \mathbf{M}_{u_{\mathrm{N}m}}\boldsymbol{x}\right) \\ \hline \vdots \\ \hline \mathbf{B}_1\mathbf{F}\mathbf{C}_{\mathrm{N_c}}\left(\mathbf{M}_{u_1+\delta u_1}\boldsymbol{x} - \mathbf{M}_{u_1}\boldsymbol{x}\right) \\ \vdots \\ \mathbf{B}_{\mathrm{N}_m}\mathbf{F}\mathbf{C}_{\mathrm{N_c}}\left(\mathbf{M}_{u_{\mathrm{N}m}+\delta u_{\mathrm{N}m}}\boldsymbol{x} - \mathbf{M}_{u_{\mathrm{N}m}}\boldsymbol{x}\right) \end{array}\right]. \tag{13.12}$$

We can further expand this expression using the optical flow equation: Given an image $\boldsymbol{I}$ and a small displacement $\delta\boldsymbol{u}$, one can write $\boldsymbol{I}\left(\boldsymbol{u}+\delta\boldsymbol{u}\right) - \boldsymbol{I}\left(\boldsymbol{u}\right) \approx \nabla\boldsymbol{I}^T \cdot \delta\boldsymbol{u}$. Noting the motion-transformed images $\boldsymbol{x}_m = \mathbf{M}_{u_m}\boldsymbol{x}$ $(m = 1 \ldots N_m)$, we end up with

$$\boldsymbol{\varepsilon}\left(\delta\boldsymbol{u}\right) \approx \left[\begin{array}{c} \mathbf{B}_1\mathbf{F}\mathbf{C}_1(\nabla\boldsymbol{x}_1^T \cdot \delta\boldsymbol{u}_1) \\ \vdots \\ \mathbf{B}_{\mathrm{N}_m}\mathbf{F}\mathbf{C}_1(\nabla\boldsymbol{x}_{\mathrm{N}_m}^T \cdot \delta\boldsymbol{u}_{\mathrm{N}_m}) \\ \hline \vdots \\ \hline \mathbf{B}_1\mathbf{F}\mathbf{C}_{\mathrm{N_c}}(\nabla\boldsymbol{x}_1^T \cdot \delta\boldsymbol{u}_1) \\ \vdots \\ \mathbf{B}_{\mathrm{N}_m}\mathbf{F}\mathbf{C}_{\mathrm{N_c}}\nabla\boldsymbol{x}_{\mathrm{N}_m}^T \cdot \delta\boldsymbol{u}_{\mathrm{N}_m} \end{array}\right] = \left[\begin{array}{c} \mathbf{B}_1\mathbf{F}\mathbf{C}_1\boldsymbol{G}_1 \\ \vdots \\ \mathbf{B}_{\mathrm{N}_m}\mathbf{F}\mathbf{C}_1\boldsymbol{G}_{\mathrm{N}_m} \\ \hline \vdots \\ \hline \mathbf{B}_1\mathbf{F}\mathbf{C}_{\mathrm{N_c}}\boldsymbol{G}_1 \\ \vdots \\ \mathbf{B}_{\mathrm{N}_m}\mathbf{F}\mathbf{C}_{\mathrm{N_c}}\boldsymbol{G}_{\mathrm{N}_m} \end{array}\right] \delta\boldsymbol{u} = J_u\left(\boldsymbol{x}, \boldsymbol{u}\right)\delta\boldsymbol{u}. \tag{13.13}$$

We have introduced sparse matrices $\boldsymbol{G}_m$ $(m = 1 \ldots N_m)$, of size $N_x \times N_x N_m N_{dims}$, that select within $\delta\boldsymbol{u}$ the displacement field corresponding to the mth shot and apply the dot product with the image gradients $\nabla\boldsymbol{x}_m$.

Eq. (13.13) enables the residual reconstruction error to be predicted as a linear function of the error in the displacement fields $\delta\boldsymbol{u}$. The linear operator $J_u\left(\boldsymbol{x}, \boldsymbol{u}\right)$ is actually the Jacobian matrix of the error function that we used in the data fidelity term $f\left(\boldsymbol{x}, \boldsymbol{u}\right) = \boldsymbol{E}\left(\boldsymbol{u}\right)\boldsymbol{x} - \boldsymbol{s}$. In other words, $J_u\left(\boldsymbol{x}, \boldsymbol{u}\right) =$

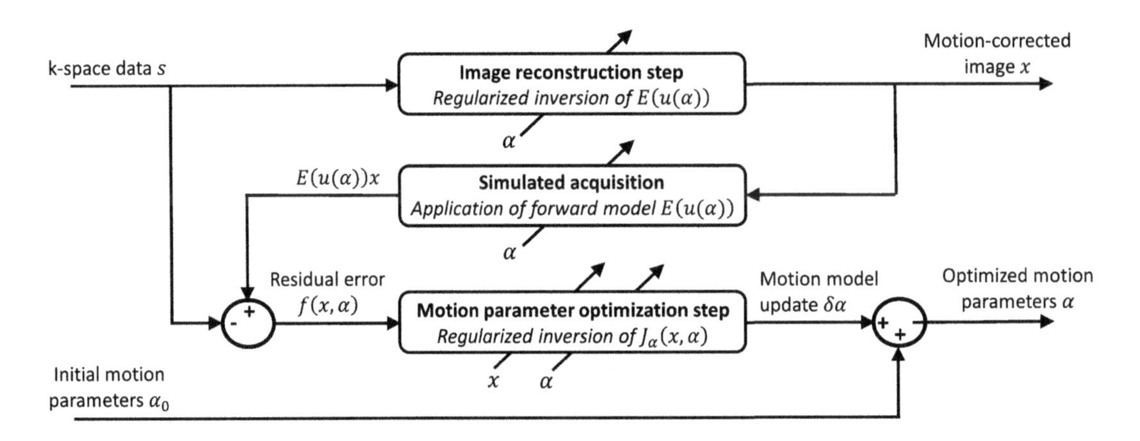

FIGURE 13.5

Graphical representation of the algorithm described for joint optimization of image and motion. From the motion-corrupted k-space data and an initial set of motion parameters (e.g., a rigid or a nonrigid motion model), a first image reconstruction step is performed that provides a first estimate of the image $\boldsymbol{x}$. Then the forward model is applied and the simulated k-space data are compared with the measured ones. The residual error in k-space domain is the input for the motion model optimization step that provides an update of the motion parameters minimizing the residual error. Then the motion model is updated and the process is repeated until a convergence criterion is satisfied.

$\nabla_u f(\boldsymbol{x}, \boldsymbol{u})$, i.e. it represents the partial derivatives of f with respect to the elements of the displacement fields.

If we further introduce a motion model $\boldsymbol{\alpha}$ that is a linear function of $\boldsymbol{u}$ (since that is the case for the two examples taken in this chapter), then we can write similarly

$$\varepsilon(\delta\boldsymbol{\alpha}) = J_\alpha(\boldsymbol{x}, \boldsymbol{\alpha})\,\delta\boldsymbol{\alpha}. \tag{13.14}$$

Detailed expressions for J_α will be given in Sections 13.2.3.3 and 13.2.3.4 in full-matrix formalism, which is not necessarily the preferred way for implementing these expressions, but which is extremely useful for understanding how the Hermitian transpose operators can be computed in a mathematically rigorous way.

13.2.3.2 *Alternating Gauss–Newton optimization*

Now we can formulate a general algorithm for solving Eq. (13.10). Considering an initial guess of motion defined by the motion model parameters $\boldsymbol{\alpha}$, we can proceed iteratively using the following scheme [69,71]: (i) Solve the optimization problem with respect to $\boldsymbol{x}$, considering $\boldsymbol{\alpha}$ fixed (using Eq. (13.8)); (ii) compute the residual reconstruction error $f(\boldsymbol{x}, \boldsymbol{\alpha}) = \boldsymbol{E}(\boldsymbol{u}(\boldsymbol{\alpha}))\boldsymbol{x} - s$; and (iii) solve the optimization problem with respect to the motion model parameters $\boldsymbol{\alpha}$, considering $\boldsymbol{x}$ fixed. Then we repeat step (i) to (iii) until some stopping condition has been reached. A graphical representation of this optimization scheme is given in Fig. 13.5.

Step (iii) is the "Gauss–Newton" step, which involves the linearization around $\boldsymbol{u}(\boldsymbol{\alpha})$. Indeed, since the error function f is a nonlinear function of $\boldsymbol{\alpha}$, instead of directly searching for $\boldsymbol{\alpha}$ minimizing $f(\boldsymbol{x}, \boldsymbol{\alpha})$, the idea is to search for an optimal refinement $\delta\boldsymbol{\alpha}$ that minimizes $f(\boldsymbol{x}, \boldsymbol{\alpha} + \delta\boldsymbol{\alpha})$. The optimization with respect to the motion model is then formulated as

$$\begin{aligned} \hat{\delta\boldsymbol{\alpha}} &= \underset{\delta\boldsymbol{\alpha}}{\arg\min} \| f(\boldsymbol{x}, \boldsymbol{\alpha} + \delta\boldsymbol{\alpha}) \|_2^2 + \mu \| \boldsymbol{S}(\boldsymbol{\alpha} + \delta\boldsymbol{\alpha}) \|_2^2 \\ \hat{\delta\boldsymbol{\alpha}} &\approx \underset{\delta\boldsymbol{\alpha}}{\arg\min} \| f(\boldsymbol{x}, \boldsymbol{\alpha}) + J_\alpha(\boldsymbol{x}, \boldsymbol{\alpha})\, \delta\boldsymbol{\alpha} \|_2^2 + \mu \| \boldsymbol{S}(\boldsymbol{\alpha} + \delta\boldsymbol{\alpha}) \|_2^2 . \end{aligned} \tag{13.15}$$

The least-squares solution is given by

$$\hat{\delta\boldsymbol{\alpha}} \approx \left(J_\alpha(\boldsymbol{x}, \boldsymbol{\alpha})^H J_\alpha(\boldsymbol{x}, \boldsymbol{\alpha}) + \mu \boldsymbol{S}^H \boldsymbol{S} \right)^{-1} \left(J_\alpha(\boldsymbol{x}, \boldsymbol{\alpha})^H \left(s - \boldsymbol{E}(\boldsymbol{u}(\boldsymbol{\alpha}))\, x\right) - \mu \boldsymbol{S}^H \boldsymbol{S}\boldsymbol{\alpha} \right). \tag{13.16}$$

It should be noted that the motion parameters must be real numbers. However, the variables and operators on the right-hand side of Eq. (13.13) are complex, and therefore there is no guarantee that the solution $\hat{\delta\boldsymbol{\alpha}}$ will be purely real. Therefore, the motion model is updated using the formula $\boldsymbol{\alpha} := \boldsymbol{\alpha} + Re\left(\hat{\delta\boldsymbol{\alpha}}\right)$.

Since the linearization with the optical flow equation is only valid for small motion errors, in order to handle large displacements (especially when no initial guess of motion is available), a multiresolution strategy can be implemented. This means that the low-resolution k-space data (i.e., a central square) can be used first to reconstruct a low-resolution image (assuming motion is null), and then a low-resolution estimate of the motion parameters can be obtained. At the end of each resolution level, the motion model solution is scaled/interpolated to the next resolution level, providing an adequate initial guess for the higher resolution estimate.

13.2.3.3 Case of translational motion

In the case of translational motion, the motion parameters are defined as the translations occurring at each shot (in x and y dimensions, considering a 2D case for simplicity): $\boldsymbol{\alpha} = \left[\tau_1^{(x)} \ldots \tau_{N_m}^{(x)} \quad \cdots \quad \tau_1^{(y)} \ldots \tau_{N_m}^{(y)} \right]^T$. In matrix formalism, the relationship between the displacement fields and the motion model parameters is expressed as

$$\boldsymbol{u} = \left[u_1^{(x)} \ldots u_{N_m}^{(x)} \quad \cdots \quad u_1^{(y)} \ldots u_{N_m}^{(y)} \right]^T = \boldsymbol{P}\boldsymbol{\alpha}, \tag{13.17}$$

where $\boldsymbol{P}$ is a sparse matrix of size $N_x N_m N_{dims} \times N_m N_{dims}$ that duplicates the scalar translation value at each shot onto each pixel of the corresponding displacement field. Finally, the Jacobian matrix becomes

$$J_{\alpha}(\boldsymbol{x},\boldsymbol{\alpha}) = \begin{bmatrix} \mathbf{B}_1\mathbf{F}\mathbf{C}_1\boldsymbol{G}_1\boldsymbol{P} \\ \vdots \\ \mathbf{B}_{\mathrm{N}_m}\mathbf{F}\mathbf{C}_1\boldsymbol{G}_{\mathrm{N}_m}\boldsymbol{P} \\ \hline \vdots \\ \hline \mathbf{B}_1\mathbf{F}\mathbf{C}_{\mathrm{N_c}}\boldsymbol{G}_1\boldsymbol{P} \\ \vdots \\ \mathbf{B}_{\mathrm{N}_m}\mathbf{F}\mathbf{C}_{\mathrm{N_c}}\boldsymbol{G}_{\mathrm{N}_m}\boldsymbol{P} \end{bmatrix}. \tag{13.18}$$

This approach can be extended to include rotations as well; for more details, refer to Cordero-Grande et al. [21].

13.2.3.4 *Case of a temporally constrained, nonrigid motion model*

In the case of nonrigid motion, it has been proposed to use a separable motion model of the form $\boldsymbol{u}(r,t) = \sum_{k=1}^{N_k} v_k(r) w_k(t)$, where w_k $(k = 1 \ldots N_k)$ are known motion signals obtained from external sensors (e.g., respiratory sensors) or MR navigator data, and v_k are spatial coefficient maps controlling the local amplitude and direction of motion [68]. We also refer to Chapter 9 where the concept of partial separability has been discussed. We define $\boldsymbol{\alpha}$, to be the concatenation of all v_k maps, which have an x, y (and z) component like $\boldsymbol{u}$: $\boldsymbol{\alpha} = \begin{bmatrix} v_1^{(x)} \ldots v_{N_k}^{(x)} & \ldots & v_1^{(y)} \ldots v_{N_k}^{(y)} \end{bmatrix}^T$. Hence, $\boldsymbol{\alpha}$ is a vector of $N_x N_k N_{dims}$ elements, and, if the number of motion signals is small compared to the number of motion states ($N_k \ll N_m$), and preferably if we have a certain degree of overdetermination (Nex > 1), this model can reduce the number of unknowns sufficiently to make the optimization tractable. In matrix formalism, the displacement fields can be written as $\boldsymbol{u} = \boldsymbol{Q}\boldsymbol{\alpha}$, with $\boldsymbol{Q}$ a sparse matrix of size $N_x N_m N_{dims} \times N_x N_k N_{dims}$, composed of diagonal block matrices, each diagonal having a constant value given by the motion signals w_k at the corresponding motion state m. Finally, the Jacobian matrix reads:

$$J_{\alpha}(\boldsymbol{x},\boldsymbol{\alpha}) = \begin{bmatrix} \mathbf{B}_1\mathbf{F}\mathbf{C}_1\boldsymbol{G}_1 \\ \vdots \\ \mathbf{B}_{\mathrm{N}_m}\mathbf{F}\mathbf{C}_1\boldsymbol{G}_{\mathrm{N}_m} \\ \hline \vdots \\ \hline \mathbf{B}_1\mathbf{F}\mathbf{C}_{\mathrm{N_c}}\boldsymbol{G}_1 \\ \vdots \\ \mathbf{B}_{\mathrm{N}_m}\mathbf{F}\mathbf{C}_{\mathrm{N_c}}\boldsymbol{G}_{\mathrm{N}_m} \end{bmatrix} \boldsymbol{Q}. \tag{13.19}$$

13.3 Methods

We will now focus on how the three families of motion correction techniques that we have described can be implemented in practice. The k-space phase correction technique described in Section 13.2.1

(or its extension to rigid motion) has been widely used, despite its limitations, because it is the simplest and most computationally efficient. As for the general motion-corrected reconstruction described in Section 13.2.2, efforts need to be made in order to estimate the motion parameters corresponding to each k-space sample, shot or motion state. Even for the joint reconstruction of image and motion, in the nonrigid case, some prior information about motion is necessary, in the form of motion signals. Therefore, we will review some of the strategies that can be used to provide motion estimates. Generally speaking, motion information can be obtained in the form of time-varying scalar quantities (e.g., translation parameters given by a navigator measurement) or time-varying vector fields of one, two or three spatial dimensions (e.g., from the registration of image navigators). Some of these motion signals may provide a direct motion measurement of the organ of interest, while others may only provide a surrogate measurement (e.g., an external sensor measurement or motion from a remote body region). In the latter case, they need to be related to the organ motion through a motion model. The choice of the k-space sampling scheme will also be discussed since it is particularly important for the motion-estimation step, in particular, for the joint reconstruction of image and motion. Finally, we will also discuss the connection between motion estimation/correction and accelerated imaging, i.e., how it can help or improve the solving of highly undersampled MRI data for applications in dynamic MRI with high temporal resolution or even real-time MRI.

13.3.1 Strategies for motion sensing

13.3.1.1 External sensor measurements

External sensors are routinely used in clinical cardiac MRI. The electrocardiogram (ECG) is most commonly used (an alternative is the peripheral pulse oximetry) as a cardiac motion signal. From this signal, a cardiac phase can be derived that is a temporal position in the cardiac cycle, assumed to be periodic, up to a stretching of the systole and diastole periods. The ECG is generally used only to synchronize the acquisition and/or the reconstruction with the heart beating, but the cardiac-phase signal can also be used as a surrogate signal for cardiac motion in the vicinity of a given cardiac phase [88].

Respiratory belts or bellows are also provided by the MRI manufacturers. The associated signals are approximately proportional to the amplitude of chest surface or abdominal surface deformation, according to their placement on the patient. Either the amplitude or the phase of the respiratory signal can be used as an input signal for a predictive-motion model. Accelerometer-based sensors (actually used as inclinometers) have also been proposed for obtaining respiratory signals [14].

For tracking head motion in MRI, optical systems have also been extensively developed [31,58], but they have been mostly used with prospective correction rather than retrospective correction. Such systems aim to track the displacement of reflective markers placed on the head with infrared cameras. Multiple cameras can be used with a stereoscopic reconstruction. More recently, single-camera systems have been proposed, with the camera located inside the bore, close to the patient. They use particular markers such as encoded checkerboards or moiré patterns in order to resolve motion. These systems aim to provide rigid head-motion parameters with six degrees of freedom (three translations and three rotations).

13.3.1.2 Extracting motion from MR data

Motion can be extracted from the MR data in multiple ways. In MRI, an image of interest is acquired by repeating many pulse sequence blocks. Separate pulse-sequence blocks can be added and interleaved with the imaging pulse-sequence blocks. In that case, the added blocks are dedicated to motion estimation. This strategy will be termed "separate navigation" in the remainder of the section. Another strategy is to use the data from the imaging pulse-sequence blocks themselves to extract motion information. This strategy will be termed self-navigation. For both strategies, the extracted motion exformation can be of low spatial dimension: It can be dimensionless (a point-like particle with no spatial dimension), i.e., represented by a time-varying scalar term such as the k-space center, or it can be 1D (one spatial dimension), i.e., a vector of time-varying parameters or a time-varying line through the k-space or through the image. It can also be of higher spatial dimension (2D or 3D), and in that case it makes it possible for a 2D or 3D image to be acquired and post-processed in order to estimate more complex motion information, such as motion in multiple directions or local deformations. Finally, we will also describe alternative ways using built-in or additional MRI hardware or those based on the analysis of the acquired noise. An illustration of these methods is given in Fig. 13.6 for cardiovascular applications.

Separate navigation signals

Dedicated MR sequence blocks interleaved in the imaging sequence can provide motion tracking data. Such navigator blocks are available on clinical scanners for cardiac or abdominal applications. Typically, they consist of exciting a pencil-beam region placed on the diaphragm and aim to track the displacement of the interface between lung and liver (see Fig. 13.6). Navigators have also been extensively developed for head applications to track rigid motion [31,58], including orbital, spherical, and cloverleaf navigators, in the effort to identify the six motion parameters with the minimal amount of extra k-space data.

Navigator data can also be obtained by a small modification of the imaging sequence block, which results in a minor increase of the repetition time. The so-called DC navigator (or FID navigator) consists of opening the MR data acquisition window between the slice selection and the phase-encoding gradient, which corresponds to sampling the k-space center point ($k_x = k_y = k_z = 0$) [12]. This sample corresponds to the sum of all pixel intensities in the image. With a receiving surface coil, it is typically modulated by anatomical changes such as bulk or breathing motion. Another strategy consists of sampling MR data during the prewinder and/or rewinder gradients. This amounts to sampling the k-space with a butterfly-shaped trajectory, which can be tuned to retrieve directional motion-information (bulk motion in x, y, z) [15].

All these navigator data come as multicoil k-space data so, even if they provide a global information from the whole image, it is inherently localized to a body region close to the receiving coil, due to the coil sensitivity weighting. Therefore, the analysis of such data often comes with either a coil selection criterion or more generally, a principal component analysis combined with a frequency analysis, in order to extract signals of cardiac or respiratory origin.

Separate image navigation for 2D/3D motion estimation

Navigators providing low spatial-resolution images have also been proposed for tracking head motion, e.g., using fast spiral scans in the plane of interest [8] or in three orthogonal planes [89]. Similarly, for cardiovascular or abdominal applications, 2D or 3D image navigators have been proposed [1,39,47,57]. A specificity of cardiac imaging is that, often, image acquisition is prospectively synchronized to the

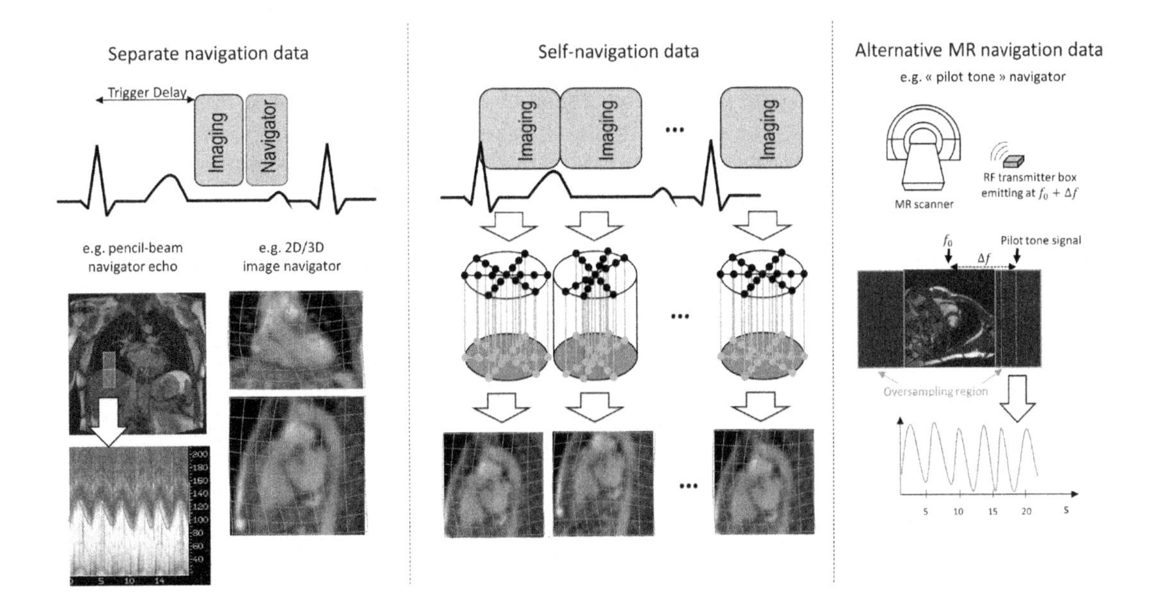

FIGURE 13.6

Illustration of a few methods available for extracting motion information from MR data for cardiovascular applications. On the left, separate MR navigation data are obtained from separate pulse-sequence blocks, dedicated to motion monitoring. They can provide a 1D image (e.g., a pencil beam navigator placed on the diaphragm), or a low-resolution 2D or 3D image that can be used to track more complex motion. In the middle, self-navigated imaging typically uses radial/spiral sampling (or radial/spiral phase encoding in 3D Cartesian sequences). They can also provide a rough motion signal (using the k-space center, or center line, which is sampled frequently), or a low-resolution 2D or 3D image reconstructed from undersampled k-space data acquired in similar motion states (e.g., consecutive spokes or respiratory-binned spokes). On the right, the pilot tone navigator is an example alternative MR method. An RF transmitter emits at a frequency intended to be outside the imaging bandwidth, corresponding to the field of view of interest, but still within the measured MRI data (oversampling region). The pilot tone signal is modulated by patient motion, which allows a motion signal to be extracted.

heart rate so that the imaging data are acquired in the end-diastolic rest period (see Fig. 13.6). This means other periods of the cardiac cycle can be filled with motion tracking-sequence blocks to get heart-motion information in all three directions, as well as nonrigid components of motion.

Self-navigation signals

The imaging sequence can be designed to acquire the central k-space data of the imaging experiment repeatedly, at a sufficiently fast rate, in order to get motion tracking data. This feature is intrinsic to radial or spiral sequences since the center k-space point is acquired at every repetition. With Cartesian trajectories, in particular (but also with radial trajectories), the center k-space line can be interleaved in the sequence [51,80,82], providing, after using a Fourier transform, a projection of the image onto the frequency encoding axis, from which cardiac/respiration signals can be derived. Trajectories made of rotated Cartesian tiles, as used by the PROPELLER technique [73], allow a full disk to be acquired at the center of k-space, i.e., providing a low-resolution image at each shot.

Furthermore, in some cases, the imaging data themselves allow more complex 2D or 3D motion information to be extracted, based on smart sampling strategies in the k-space. The radial trajectory has been very popular for motion-corrected reconstruction. Indeed, it is generally designed with a specific angular step between successive radii based on the golden ratio (111.246 degrees) [90]. This scheme allows a flexible choice for the reconstruction of dynamic data since k-space data can be grouped in multiple ways, each way providing a relatively uniform k-space sampling: either small blocks of consecutive radii can be formed, providing low-spatial, high-temporal resolution images; or large blocks can be formed, providing high-spatial, low-temporal resolution images (see Fig. 13.6). The high spatial resolution images are typically corrupted by motion due to the large temporal footprint. The fast, low spatial resolution images can therefore be used to estimate motion fields, by image registration, and this enables a motion-corrected reconstruction of the high spatial-resolution data. In 3D applications, similar properties are obtained by using a Cartesian sampling with a pseudo-radial or pseudo-spiral sampling of the k_y-k_z plane. Such sampling strategies have been very popular for respiratory and/or cardiac motion correction [24,36,43,72,75,79,83]. Often, in such applications, a fixed number of motion states is determined based on various types of motion signals (ECG, respiration sensors, self-navigation data). The k-space data can then be split into different sets, each corresponding to one motion state. For instance, a respiration signal may allow a number N_m of motion states to be defined by binning the data according to the amplitude of the motion signal. Then an image reconstruction technique suited for (possibly highly) undersampled data is employed to form images, which can be of moderate quality but sufficient to estimate motion by image registration.

Alternative MR navigation data

Small dedicated radiofrequency receiver coils, also termed pick-up coils or NMR field probes, can also be placed on the patient to provide real-time positional information [31]. While initial systems required additional gradients to measure the NMR probe position, simultaneous imaging and field-probe measurements have been demonstrated with ^{19}F probes and with carefully designed gradient waveforms [3,33]. It has also been proposed to probe the magnetic field gradients of the imaging sequence with a three-axis Hall magnetometer, in order to derive its location in the scanner, and therefore to estimate patient motion [86].

Another strategy consists of placing a small radiofrequency transmitter, called pilot tone, emitting at a frequency outside the imaging bandwidth but still within the receiving coil bandwidth [85]. Due to the interaction between the pilot-tone transmitter and the patient, the pilot-tone signals measured by each coil are modulated by patient motion (see Fig. 13.6).

Motion also modulates the variance of the thermal noise associated with the k-space data measurements, and these changes can be measured [2].

13.3.2 Image registration

Image registration is the process by which motion parameters can be estimated from images in two or more motion states. There is a rich literature in digital image processing applied to medical image and many freely available software tools [35,48,62,63]. Since we aim to form the forward model described in Section 13.2.2.2, it is the image in the reference motion state that should be chosen as the floating image, i.e., which should be registered onto the image at each other motion states (i.e., the target images). When choosing a particular registration technique, several aspects should be considered. Firstly, we should consider how motion is parameterized, e.g., with a rigid/affine transformation, with a dense displacement field, or with free local displacements at some control points with spline interpolation. Then, we need to choose a similarity metric to quantify to spatial alignment of the images. The sum of squared differences (i.e., the L_2 norm of the difference image) is the natural choice (leading to a computationally efficient optimization) when there are no (or minor) intensity changes between the images. Other metrics, such as the correlation coefficient or the normalized mutual information, are popular choices otherwise.

13.3.3 Motion models

When the motion parameters of interest are available (e.g., six parameters of rigid body motion or deformable motion field), they can be used directly as inputs of a motion-corrected reconstruction. Otherwise, surrogate motion data can be used in order to estimate the motion parameters or the deformable motion fields. We have seen indeed many ways of collecting motion signals, but often these signals do not provide a quantitative measure of motion in mm units (e.g., respiratory belt, DC navigator, pilot tone, etc.) and/or they provide a measure at the body surface or in another organ such as the diaphragm. So they do not provide a direct measure of motion for the organ of interest. However, it may be assumed that they are temporally correlated with the motion of the organ of interest. Let us consider that we seek to estimate N_p motion parameters at a given time t, i.e., a vector $p(t)$, which can be, e.g., rigid motion parameters ($N_p = 6$) or a displacement vector field ($N_p = N_x N_{dims}$, i.e., the displacement in x, y, z for each voxel). A linear model can be introduced to relate $p_i(t)$ $(i = 1 \dots N_p)$ to the measured physiologic motion signals $\varphi_k(t)$ $(k = 1 \dots N_k)$ [59,65]

$$p_i(t) = \sum_{k=1}^{N_k} a_{i,k} \varphi_k(t). \tag{13.20}$$

The $a_{i,k}$ are some coefficients to be determined. A calibration scan can be implemented to acquire the motion signals $\varphi^{calib}(t)$ simultaneously with fast (possibly low-resolution) imaging data. Then the fast imaging data are registered with a rigid or nonrigid registration technique according to the desired type of motion model, thus providing $p^{calib}(t)$. The $a_{i,k}$ are estimated by linear regression. Finally, the motion model can be used during the sequence of interest to predict the motion parameters from the new motion signals, using Eq. (13.20). This linear model has been the most widely used in MRI, but more general models can be designed [60].

13.3.4 Optimal k-space sampling for motion correction

The joint reconstruction of image and translational (or rigid) motion has been shown to be feasible without any additional motion measurement. However, the k-space sampling scheme has an impact on the accuracy of motion estimation and thereby the performance of the reconstruction. A uniformly distributed, and even incoherent, sampling of k-space should be preferred to a linear sampling since it results in a much more accurate reconstruction of the motion parameters [19]. A similar observation has already been made in Section 13.3.1.2 in the case of a radial sampling strategy, with the golden angular step ensuring a relatively uniform, but also incoherent, sampling of k-space in each motion state. In that case it is well understood that such a sampling scheme is a necessary condition for the reconstruction of images from each motion state using a compressed sensing or low-rank technique. The quality of those images indeed impacts the accuracy of the subsequent motion estimation (i.e., the registration of images from each motion state).

13.3.5 Motion correction to improve dynamic MRI

When imaging a moving patient, reconstructing a static, motion-corrected image is one possible approach to the problem, but another approach is to reconstruct a dynamic imaging dataset by thinking of the acquisition domain as a k-t space, rather than a k-space. When is it better to use one approach or the other? Is there any connection between them, and can we combine them?

The answer to the first question may depend on the target application. On the one hand, if we need some quantitative information from the static image only, motion correction may be more relevant since the combination of all motion-corrupted data will lead to a higher SNR. On the other hand, if we seek to quantify the dynamic information, as in cardiac cine imaging (contractile function assessment), and if many parameters are required to describe motion, one may think it is more efficient to encode the whole k-t space (and possibly several temporal dimensions for cardiac and respiratory motion as in XD-GRASP [27]) than a static k-space and the motion parameters. However, it was observed in [74] that, when the displacement fields of the moving structures were relatively smooth in time, the temporal bandwidth associated with the motion parameters was lower than that associated with the pixel intensities. Indeed, pixels at the edge of a moving organ underwent a sharp intensity transition that resulted in high temporal frequencies in the k-t space. Thus, it was shown in Prieto et al. [74] that a motion model could be fit to undersampled k-t space data, which allowed the full cardiac cycle to be modeled. The motion model was then applied to fill in the missing k-space data. This approach allowed high undersampling factors.

It was further proposed to integrate motion estimates into a compressed sensing reconstruction of undersampled data in the k-t space, namely k-t FOCUSS, for cardiac cine imaging [45,46]. The authors noted that the concept of motion estimation and compensation was present in video compression, such as MPEG, and could improve the encoding of the dynamic information. The method in Jung et al. [46] generalized the classic k-t FOCUSS by introducing a prediction of the various imaging frames, using motion estimates between certain key frames of the series (obtained by a block matching technique). Therefore, only the residual dynamic imaging data needed to be encoded in the k-t space. As a result, the encoded data in k-t space were sparser, and higher acceleration factors could be achieved. This concept has been used with different combinations of algorithms for the image reconstruction step and for the motion estimation step [4,83].

A similar concept was proposed in real-time radial MRI [52], with applications in cardiac and speech imaging. In that work, images were first reconstructed from highly undersampled radial data using a specific technique (so-called "nonlinear inversion" that reconstructs the image and coil sensitivities jointly). Then data from five consecutive frames were combined: Motion between these frames was estimated by an optic flow registration technique; then motion was integrated into the image reconstruction step. Motion correction was thus shown to improve the temporal fidelity of the real-time reconstruction.

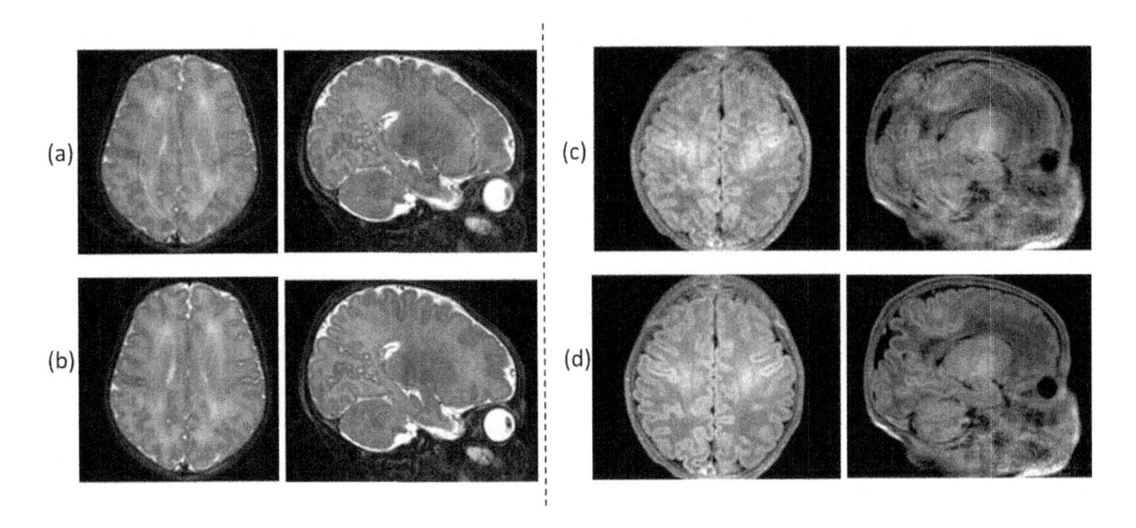

FIGURE 13.7

Example neonatal brain images without (respectively with) motion correction by joint reconstruction of image and rigid motion parameters: T_2-weighted (**a** and **b** respectively) and T_1-weighted (**c** and **d** respectively) multishot multislice FSE sequences. Extracted from Cordero-Grande et al. [20].

13.4 Clinical application examples

We will now illustrate the described theory and method with some clinical applications. Rather than providing an exhaustive list, we aim to show some representative examples in various organs and using various types of motion inputs: no motion prior at all, motion estimates from low-resolution or self-navigation images, or motion prior from external sensors or navigators.

13.4.1 Brain

PROPELLER, thanks to its self-navigated k-space trajectory, is one the earliest motion-corrected reconstruction techniques and is currently available on clinical scanners (also called BLADE). It was shown to improve the quality of clinical images in FLAIR, T_2-, T_1- and/or contrast-enhanced T_1-weighted imaging compared to uncorrected, conventional scans. Various populations were studied including adults/elderly patients [66,91] and children [87]. Applications to high-resolution diffusion imaging have also been demonstrated in acute cerebral infarction [29] and diffusion tensor imaging [18].

The joint reconstruction of image and rigid motion proposed in Cordero-Grande et al. [20] has the great advantage of being applicable to conventional sequences so it does not lengthen the imaging protocol. It was successfully applied to a very large database of neonatal brain scans (>1800 volumes from >500 babies, gestational age between 32 and 45 weeks). The babies were scanned during natural sleep, and the 3D motion correction framework was combined with outlier rejection in order to discard shots with extreme artifacts. This database included T_1- and T_2-weighted multishot, multislice, fast spin-echo sequences (see examples in Fig. 13.7).

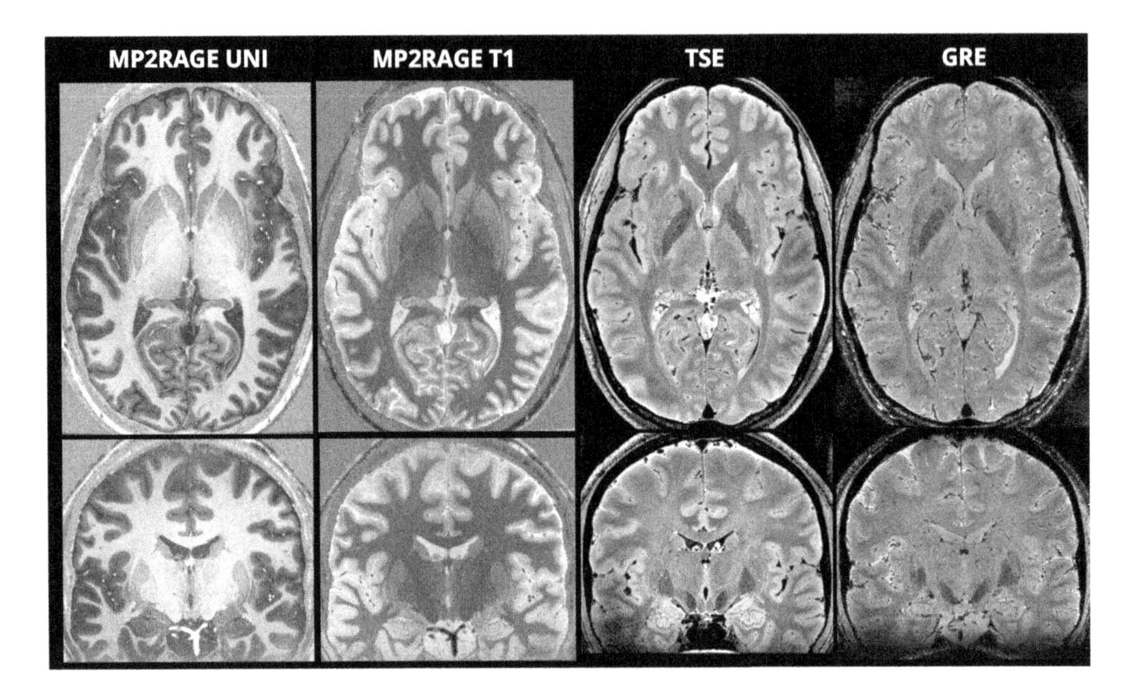

FIGURE 13.8

Ultra-high resolution brain MRI at 7T with motion-corrected reconstruction: T1-weighted image (MP2RAGE UNI), quantitative T_1 map (MP2RAGE T_1), Turbo-Spin Echo (TSE), and Gradient-Recalled Echo (GRE). Extracted from Federau and Gallichan [26].

In conventional multishot diffusion MRI sequences, an image navigator is generally acquired in order to obtain a shot-dependent phase map, required to combine the multiple shots. This low resolution navigator image has been further used for rigid motion estimation and allowed motion-corrected reconstruction of the multishot diffusion images [32,81].

Motion-corrected reconstruction also allowed ultra-high resolution human brain images at 7T with T_1-weighted, T_2-weighted and T_2^*-weighted sequences at 350 to 380 μm nominal isotropic resolution [26] (see Fig. 13.8).

13.4.2 Cardiovascular

Motion-corrected reconstruction has been an application of choice for high-resolution, 3D whole-heart coronary MR angiography [10,13,22,25,30,42,72,75,79]. Indeed, visualizing the coronaries typically requires lengthy 3D cardiac-gated scans and respiratory correction to achieve, ideally, submillimeter spatial resolution. These studies typically use radial/spiral sampling, or 3D Cartesian sampling with a radial/spiral sampling order, respiratory binning to define the motion states, and image registration to estimate motion. Throughout these studies, nonrigid correction was shown to improve vessel sharpness compared to rigid correction or conventional respiratory-gated solutions (see examples in Fig. 13.9).

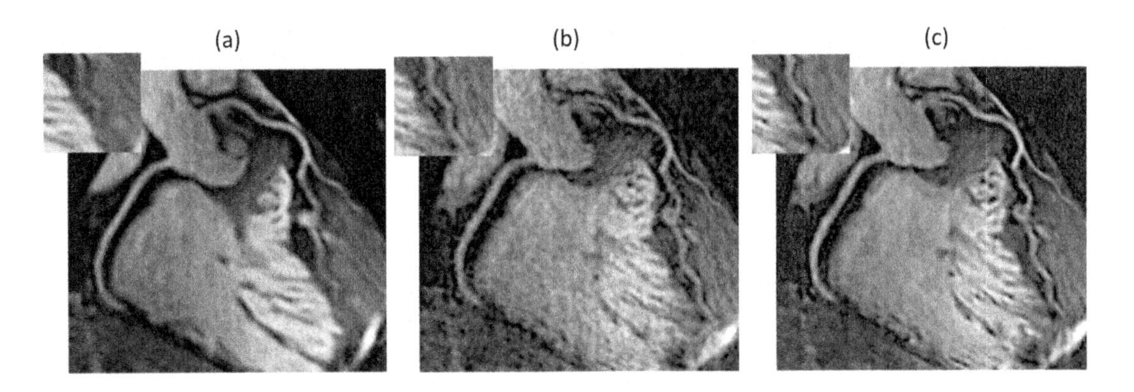

FIGURE 13.9

Example reformatted view of a 3D whole heart coronary MR angiography, with 0.9-mm isotropic resolution, obtained from free-breathing, ECG-gated data (3D balanced SSFP sequence with pseudo-spiral sampling). Various motion-corrected reconstruction techniques are shown: translational motion correction with a patch-based image regularization technique (PROST) (**a**); nonrigid motion correction (**b**); nonrigid correction with patch-based image regularization (**c**). Extracted from [13].

Motion correction also allows highly efficient acquisition, i.e., using 100% of the acquired data, as opposed to conventional gating approaches that reject data from undesired motion states [34]. This also makes the scan time predictable (i.e., not depending on how regular the breathing motion is).

Cardiac function assessment by cine imaging has also been a popular application. It is generally achieved clinically by a multi-slice coverage of the ventricles with a cardiac-gated balanced SSFP sequence. Nonrigid motion-corrected reconstruction has enabled free-breathing protocols using either radial sampling schemes and self-navigation (registration of extracted low resolution images) [36,83], or motion signals as a prior for a joint reconstruction of image and a nonrigid motion model (GRICS technique), including external sensors [88] and center k-space line navigators [70]. Clinical studies in adult patients have shown that such free-breathing, motion-corrected images are in good agreement with breath-held protocols in terms of volumetric parameters, i.e., end-diastolic and end-systolic volumes, stroke volume, and ejection fraction [23,88]. Free-breathing cardiac MRI assessment (function and tissue characterization) was also shown in Duchenne muscular dystrophy patients (from children to young adults), a population that has severely impaired respiratory function and therefore can only perform very limited breath-holds [11,67] (see example images in Fig. 13.10). Application to 4D flow MRI was also shown in pediatric congenital heart disease patients [16].

Extensions of the nonrigid motion correction framework to multicontrast datasets have also been proposed for free-breathing dynamic contrast enhanced (DCE) cardiac imaging [28] and for T_1 and T_2 mapping [61,69].

Multimodal cardiac imaging by combined PET/MRI systems is another appealing application. Indeed, the high spatial resolution of MRI and its soft tissue contrast makes it a better modality for motion tracking, which allows the simultaneously acquired PET images to be corrected using similar motion-corrected reconstruction techniques [49]. Nonrigid correction made it possible for coronary MR

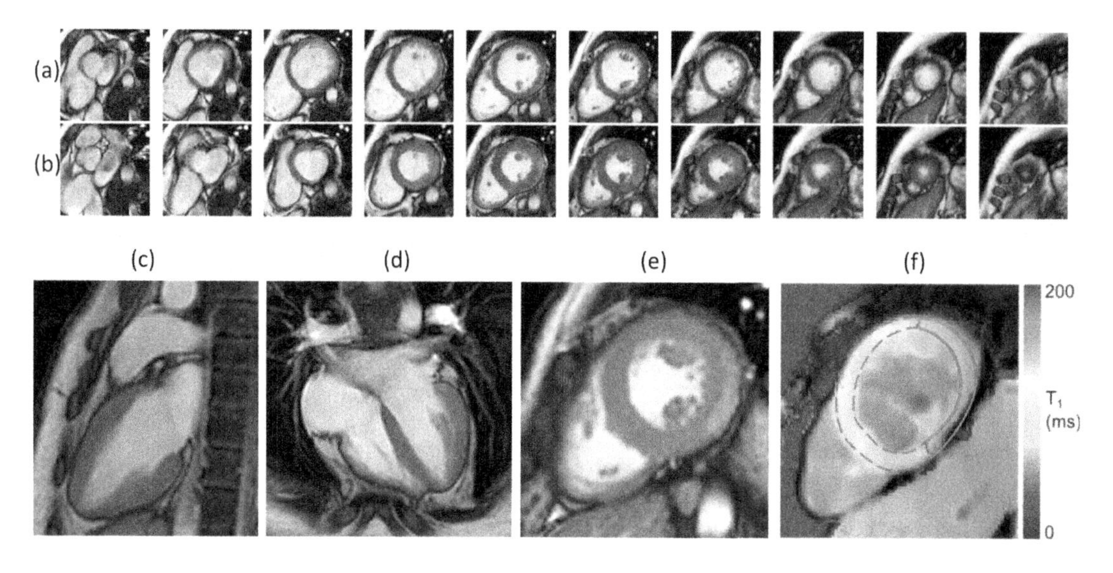

FIGURE 13.10

Example cardiac MR images from a sixteen-year-old Duchenne muscular dystrophy patient, using free-breathing and nonrigid motion correction (GRICS technique): 2D multislice cine (balanced SSFP sequence) in diastole (**a**) and systole (**b**); post-contrast 2D cine imaging in vertical, horizontal, and short axis views (**c**, **d**, **e** respectively); short axis T_1 map from a saturation recovery sequence (**f**). Data from clinical studies in Bonnemains et al. [11] and Odille et al. [67].

angiography to be obtained together with myocardial viability assessment by ^{18}F-FDG PET in patients with chronic total occlusion [64].

13.4.3 Body imaging (other than brain and heart)

Other body regions have also benefited from motion-corrected reconstruction, in particular, when high resolution 3D imaging and/or DCE imaging is necessary. Applications to 3D liver DCE imaging [53,43,44] have been shown, with the respiratory motion correction allowing quantitative perfusion maps to be calculated using Tofts or extended Tofts models (see example in Fig. 13.11). Nonrigid correction in 3D liver scans was also shown to improve the image quality in healthy subjects [24] and pediatric patients [17]. Application to free-breathing UTE imaging of the lung was shown in adults and pediatric patients, and nonrigid correction enabled better depiction of the lesions compared to conventional motion management techniques [93,94].

13.5 Current challenges and future directions

Despite significant advances over the last years, there are still several remaining challenges and unanswered questions in motion-corrected MRI reconstruction. One of the main challenges remains the

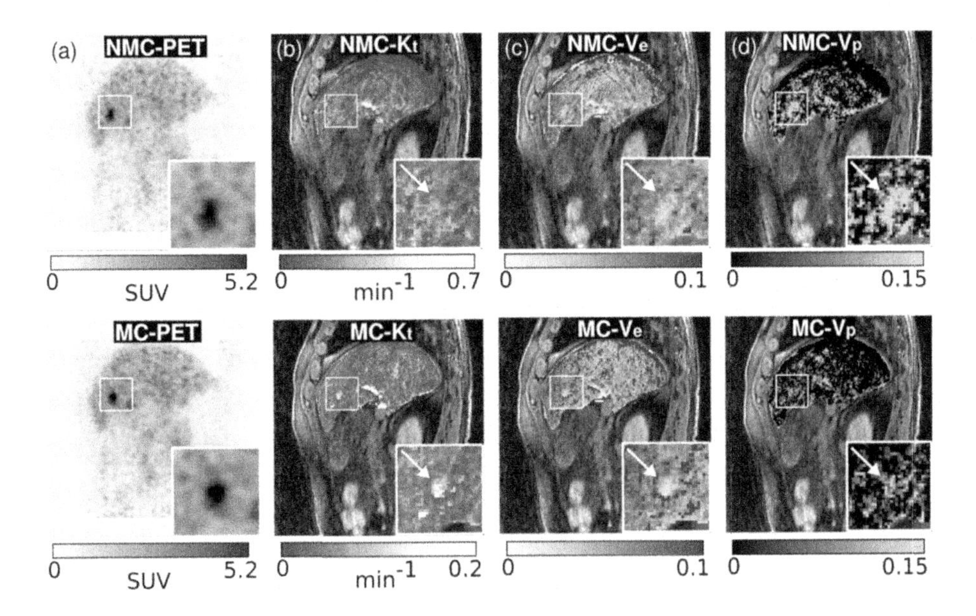

FIGURE 13.11

Example simultaneous acquisition of PET and DCE-MRI of the liver in a patient (lesion highlighted with the white arrow), injected with gadoxetate disodium. The authors used a 3D fat-suppressed T_1-weighted gradient-echo sequence with pseudo-radial sampling. PET images and parametric permeability maps (K_t (**b**), V_e (**c**) and V_p (**d**) fits from an extended Tofts model) are shown without motion correction (NMC, top row) and with nonrigid motion correction (MC, bottom row). Extracted from Ippoliti et al. [44].

increased computational complexity of motion-corrected reconstruction. Even in the case of known motion, like for other iterative MR reconstruction techniques, the computation time is mainly determined by the number of calls to the $\boldsymbol{E}^H \boldsymbol{E}$ matrix, and more precisely by the number of Fourier transformations. Compared to uncorrected reconstruction, this number is roughly multiplied by the number of motion states. In the case of a joint motion estimation, there is another added computational burden due to the motion estimation step.

Deep-learning methods have the potential to speed up different aspects of the motion-corrected reconstruction, and initial work has been done in that direction. A convolutional neural network (CNN) can be designed and trained to speedup the motion estimation step. Such CNNs have been proposed to estimate motion between pairs of patches from two images to be registered [77] or to estimate motion from a motion-corrupted patch [37]. Methods have also been proposed to model the reconstruction step as well [50]. An end-to-end deep-learning framework, consisting of a diffeomorphic registration network and a motion-informed model-based deep-learning reconstruction network, has been also recently proposed for nonrigid motion-corrected reconstruction of nine-fold undersampled, free-breathing, whole-heart coronary MR angiography [77]. Alternatively, a generative adversarial network (GAN) has been proposed to model the motion-corrected reconstruction problem without explicitly modeling motion [84]. The idea underlying GAN is that it includes two neural networks: a generator, here designed to produce motion-corrected images from corrupted input images, and a discriminator,

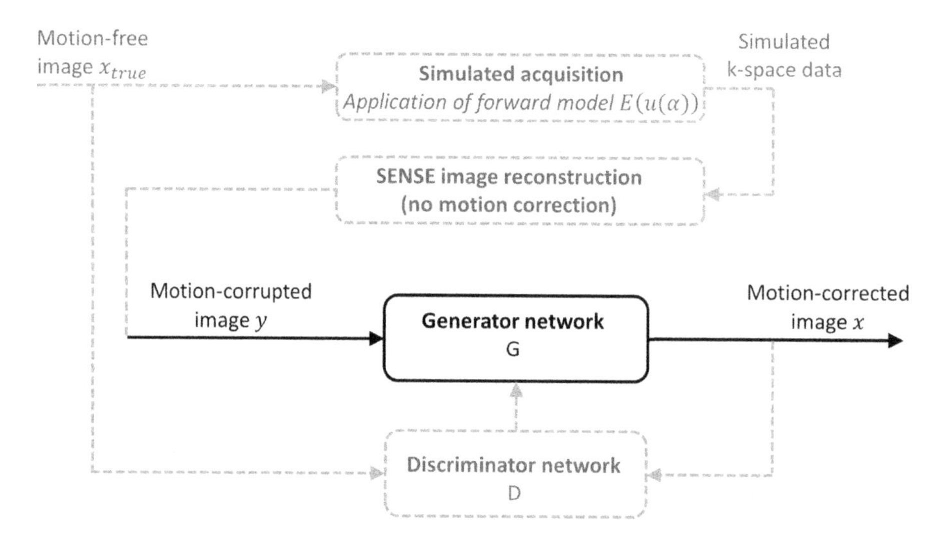

FIGURE 13.12

Example strategy for modeling the whole motion-corrected reconstruction problem with deep-learning, using a generative adversarial network (GAN), as proposed in Usman et al. [84]. The schema shows the motion-corrected image reconstruction, as modeled by a generator network, in solid black lines. The dashed gray lines indicate the elements needed for training the generator. This involves: a database of reference, motion-free images; a forward model of the motion-corrupted scan, including random motion parameters, together with a conventional (uncorrected) image reconstruction, that are used to synthetize corrupted images; these synthetic, corrupted images are the inputs of the generator, which is designed to produce motion-corrected images; finally, a discriminator network attempts to recognize which images are "true" (i.e., similar to those from the reference database) or "fake" (i.e., still motion-corrupted).

designed to distinguish between the generated images and some reference images without motion artifacts. In Usman et al. [84], the following strategy was proposed to train the GAN: a database of brain images, free of motion artifacts, was selected; motion-corrupted images were synthesized using a forward acquisition model similar to that described in Section 13.2.2.2, using a wide range of motion parameters (rigid motion); the synthetized images were used as input of the generator; and both the synthetized and reference (motion-free) images were used as inputs as the discriminator. See Fig. 13.12.

The joint estimation of image and motion can be solved very accurately in the case of rigid motion, possibly with no sequence modification (though optimizing the k-space trajectory is preferable). In the nonrigid case, authors have all used motion models, either implicitly (e.g., respiratory binning) or explicitly (e.g., as in GRICS). Such models only provide an approximate of the true motion, which thereby limits the achievable accuracy of the reconstruction, however they render the optimization tractable. A more general parameterization of motion, using a low-rank model for the full spatiotemporal displacement fields, was recently proposed in Huttinga et al. [41,40] and showed very promising results. How accurate the motion model can be remains an open question since a ground truth for the motion is difficult to obtain.

As mentioned in the introduction of this chapter, the impact of motion on MR physics is complex. We have assumed that the spatial encoding inconsistencies were the main effect, but this may not be the case in all applications. Patient motion also induces changes in B_0 field maps, especially at high field levels. This can be a concern with T_2^*-weighted imaging at 7T [55]. It is also a problem at 3T with balanced SSFP sequences in cardiac or abdominal applications since off-resonance artifacts (so-called banding artifacts) appear as a local distortion of the complex image and do not strictly follow the displacement of the neighboring organs, making motion estimation challenging. We have also implicitly assumed that the receiving coils were static and that the associated B_1^- field did not change with motion. The latter point means coil-loading changes were disregarded, however changes in sensitivity "seen" by moving tissues were intrinsically modeled in $\boldsymbol{E}$. Through-plane motion is another concern, but either 3D-encoded or stacks of 2D sequences can be used in order to model the actual 3D displacement. Besides artifacts in contrast-weighted images, motion can also lead to a bias in the quantitative MRI parameters. More generally, spin-history effects should always be kept in mind. Sequences with motion-compensated gradients can be used to mitigate those effects. The balanced SSFP sequence actually has a good motion-robustness in that regard, and it provides a relatively stable signal in presence of motion (due to its balanced gradients), which is part of the reasons why it is widely used in clinical cardiac protocols. Conversely, some sequences like fast spin-echo or diffusion-weighted imaging are known to be particularly sensitive to motion since it can create signal drops due to intravoxel dephasing. To tackle all these issues, should our future efforts be focused on combining prospective and retrospective motion correction techniques? Or should we rather aim at measuring all these motion-induced effects and modeling them into the reconstruction?

13.6 Summary

General motion (rigid or nonrigid) can be integrated in the MRI reconstruction in a mathematically rigorous way by embedding linear motion operators in the forward model. Various strategies can be used to estimate motion and thereby implement motion-corrected reconstruction. In the case of rigid motion, a blind reconstruction of the motion-free image and motion parameters is feasible in brain imaging and particularly effective when the k-space sampling scheme is uniform and incoherent. In the case of nonrigid motion, prior information is still needed in order to determine the number of motion states and to estimate motion between them. This prior can be obtained from separate image navigators or from self-navigation strategies (particularly with radial or 3D pseudo-radial/spiral trajectories). Alternatively, the joint reconstruction of a motion-free image and a time-constrained motion model is possible, using prior from 1D motion signals (e.g., external sensors or navigators). Applications in cardiac and body imaging include free-breathing imaging of lengthy 3D or dynamic sequences.

13.7 Practical tutorial

A demo code (Matlab®, The MathWorks, Inc., Natick, MA) illustrating the various types of motion-corrected reconstruction techniques, in synthetic data, is provided with this chapter, including: phase correction of translational motion in k-space; motion-corrected reconstruction using motion operators (rigid and nonrigid case); joint reconstruction of image and motion (translational motion and temporally constrained nonrigid motion). To run the tutorial, open the file "motion_corrected_reconstruction_demo.m", edit the simulation options in the first lines (number of shots, number of excitations, type of motion, type of k-space sampling), and run.

References

[1] Addy NO, Ingle RR, Luo J, Baron CA, Yang PC, Hu BS, et al. 3D image-based navigators for coronary MR angiography. Magn Reson Med 2017;77:1874–83. https://doi.org/10.1002/mrm.26269.

[2] Andreychenko A, Raaijmakers AJE, Sbrizzi A, Crijns SPM, Lagendijk JJW, Luijten PR, et al. Thermal noise variance of a receive radiofrequency coil as a respiratory motion sensor. Magn Reson Med 2017;77:221–8. https://doi.org/10.1002/mrm.26108.

[3] Aranovitch A, Haeberlin M, Gross S, Dietrich BE, Wilm BJ, Brunner DO, et al. Prospective motion correction with NMR markers using only native sequence elements. Magn Reson Med 2018;79:2046–56. https://doi.org/10.1002/mrm.26877.

[4] Asif MS, Hamilton L, Brummer M, Romberg J. Motion-adaptive spatio-temporal regularization for accelerated dynamic MRI. Magn Reson Med 2013;70:800–12. https://doi.org/10.1002/mrm.24524.

[5] Atkinson D, Hill DL, Stoyle PN, Summers PE, Clare S, Bowtell R, et al. Automatic compensation of motion artifacts in MRI. Magn Reson Med 1999;41:163–70.

[6] Atkinson D, Hill DL, Stoyle PN, Summers PE, Keevil SF. Automatic correction of motion artifacts in magnetic resonance images using an entropy focus criterion. IEEE Trans Med Imaging 1997;16:903–10. https://doi.org/10.1109/42.650886.

[7] Atkinson D, Hill DLG. Reconstruction after rotational motion. Magn Reson Med 2003;49:183–7.

[8] Bammer R, Aksoy M, Liu C. Augmented generalized SENSE reconstruction to correct for rigid body motion. Magn Reson Med 2007;57:90–102.

[9] Batchelor PG, Atkinson D, Irarrazaval P, Hill DLG, Hajnal J, Larkman D. Matrix description of general motion correction applied to multishot images. Magn Reson Med 2005;54:1273–80. https://doi.org/10.1002/mrm.20656.

[10] Bhat H, Ge L, Nielles-Vallespin S, Zuehlsdorff S, Li D. 3D radial sampling and 3D affine transform-based respiratory motion correction technique for free-breathing whole-heart coronary MRA with 100% imaging efficiency. Magn Reson Med 2011;65:1269–77. https://doi.org/10.1002/mrm.22717.

[11] Bonnemains L, Odille F, Cherifi A, Marie P-Y, Pasquier C, Felblinger J. Free-breathing with motion-correction and video projection during cardiac MRI: a paediatric design! J Cardiovasc Magn Reson 2014;16:P319. https://doi.org/10.1186/1532-429X-16-S1-P319.

[12] Brau ACS, Brittain JH. Generalized self-navigated motion detection technique: preliminary investigation in abdominal imaging. Magn Reson Med 2006;55:263–70. https://doi.org/10.1002/mrm.20785.

[13] Bustin A, Rashid I, Cruz G, Hajhosseiny R, Correia T, Neji R, et al. 3D whole-heart isotropic sub-millimeter resolution coronary magnetic resonance angiography with non-rigid motion-compensated PROST. J Cardiovasc Magn Reson 2020;22:24. https://doi.org/10.1186/s12968-020-00611-5.

[14] Chen B, Weber N, Odille F, Large-Dessale C, Delmas A, Bonnemains L, et al. Design and validation of a novel MR-compatible sensor for respiratory motion modeling and correction. IEEE Trans Biomed Eng 2017;64. https://doi.org/10.1109/TBME.2016.2549272.

[15] Cheng JY, Alley MT, Cunningham CH, Vasanawala SS, Pauly JM, Lustig M. Nonrigid motion correction in 3D using autofocusing with localized linear translations. Magn Reson Med 2012;68:1785–97. https://doi.org/10.1002/mrm.24189.

[16] Cheng JY, Hanneman K, Zhang T, Alley MT, Lai P, Tamir JI, et al. Comprehensive motion-compensated highly accelerated 4D flow MRI with ferumoxytol enhancement for pediatric congenital heart disease. J Magn Reson Imaging 2016;43:1355–68. https://doi.org/10.1002/jmri.25106.

[17] Cheng JY, Zhang T, Ruangwattanapaisarn N, Alley MT, Uecker M, Pauly JM, et al. Free-breathing pediatric MRI with nonrigid motion correction and acceleration. J Magn Reson Imaging 2015;42:407–20. https://doi.org/10.1002/jmri.24785.
[18] Chuang T-C, Huang T-Y, Lin F-H, Wang F-N, Juan C-J, Chung H-W, et al. PROPELLER-EPI with parallel imaging using a circularly symmetric phased-array RF coil at 3.0 T: application to high-resolution diffusion tensor imaging. Magn Reson Med 2006;56:1352–8. https://doi.org/10.1002/mrm.21064.
[19] Cordero-Grande L, Ferrazzi G, Teixeira RPAG, O'Muircheartaigh J, Price AN, Hajnal JV. Motion-corrected MRI with DISORDER: Distributed and incoherent sample orders for reconstruction deblurring using encoding redundancy. Magn Reson Med 2020;84:713–26. https://doi.org/10.1002/mrm.28157.
[20] Cordero-Grande L, Hughes EJ, Hutter J, Price AN, Hajnal JV. Three-dimensional motion corrected sensitivity encoding reconstruction for multi-shot multi-slice MRI: application to neonatal brain imaging. Magn Reson Med 2018;79:1365–76. https://doi.org/10.1002/mrm.26796.
[21] Cordero-Grande L, Teixeira RPAG, Hughes EJ, Hutter J, Price AN, Hajnal JV. Sensitivity encoding for aligned multishot magnetic resonance reconstruction. IEEE Trans Comput Imaging 2016;2:266–80. https://doi.org/10.1109/TCI.2016.2557069.
[22] Correia T, Cruz G, Schneider T, Botnar RM, Prieto C. Technical note: accelerated nonrigid motion-compensated isotropic 3D coronary MR angiography. Med Phys 2018;45:214–22. https://doi.org/10.1002/mp.12663.
[23] Cross R, Olivieri L, O'Brien K, Kellman P, Xue H, Hansen M. Improved workflow for quantification of left ventricular volumes and mass using free-breathing motion corrected cine imaging. J Cardiovasc Magn Reson 2016;18:10. https://doi.org/10.1186/s12968-016-0231-8.
[24] Cruz G, Atkinson D, Buerger C, Schaeffter T, Prieto C. Accelerated motion corrected three-dimensional abdominal MRI using total variation regularized SENSE reconstruction. Magn Reson Med 2016;75:1484–98. https://doi.org/10.1002/mrm.25708.
[25] Cruz G, Atkinson D, Henningsson M, Botnar RM, Prieto C. Highly efficient nonrigid motion-corrected 3D whole-heart coronary vessel wall imaging. Magn Reson Med 2017;77:1894–908. https://doi.org/10.1002/mrm.26274.
[26] Federau C, Gallichan D. Motion-correction enabled ultra-high resolution in-vivo 7T-MRI of the brain. PLoS ONE 2016;11:e0154974. https://doi.org/10.1371/journal.pone.0154974.
[27] Feng L, Axel L, Chandarana H, Block KT, Sodickson DK, Otazo R. XD-GRASP: golden-angle radial MRI with reconstruction of extra motion-state dimensions using compressed sensing. Magn Reson Med 2016;75:775–88. https://doi.org/10.1002/mrm.25665.
[28] Filipovic M, Vuissoz P-A, Codreanu A, Claudon M, Felblinger J. Motion compensated generalized reconstruction for free-breathing dynamic contrast-enhanced MRI. Magn Reson Med 2011;65:812–22. https://doi.org/10.1002/mrm.22644.
[29] Forbes KP, Pipe JG, Karis JP, Heiserman JE. Improved image quality and detection of acute cerebral infarction with PROPELLER diffusion-weighted MR imaging. Radiology 2002;225:551–5. https://doi.org/10.1148/radiol.2252011479.
[30] Forman C, Piccini D, Grimm R, Hutter J, Hornegger J, Zenge MO. Reduction of respiratory motion artifacts for free-breathing whole-heart coronary MRA by weighted iterative reconstruction. Magn Reson Med 2015;73:1885–95. https://doi.org/10.1002/mrm.25321.
[31] Godenschweger F, Kägebein U, Stucht D, Yarach U, Sciarra A, Yakupov R, et al. Motion correction in MRI of the brain. Phys Med Biol 2016;61:R32–56. https://doi.org/10.1088/0031-9155/61/5/R32.
[32] Guhaniyogi S, Chu M-L, Chang H-C, Song AW, Chen N. Motion immune diffusion imaging using augmented MUSE for high-resolution multi-shot EPI. Magn Reson Med 2016;75:639–52. https://doi.org/10.1002/mrm.25624.
[33] Haeberlin M, Kasper L, Barmet C, Brunner DO, Dietrich BE, Gross S, et al. Real-time motion correction using gradient tones and head-mounted NMR field probes. Magn Reson Med 2015;74:647–60. https://doi.org/10.1002/mrm.25432.

[34] Hajhosseiny Reza, Bustin Aurelien, Munoz Camila, Rashid Imran, Cruz Gastao, Manning Warren J, et al. Coronary magnetic resonance angiography. JACC: Cardiovasc Imaging 2020;13:2653–72. https://doi.org/10.1016/j.jcmg.2020.01.006.

[35] Hajnal JV, Hill DLG. Medical image registration, biomedical engineering. CRC Press; 2001.

[36] Hansen MS, Sørensen TS, Arai AE, Kellman P. Retrospective reconstruction of high temporal resolution cine images from real-time MRI using iterative motion correction. Magn Reson Med 2012;68:741–50. https://doi.org/10.1002/mrm.23284.

[37] Haskell MW, Cauley SF, Bilgic B, Hossbach J, Splitthoff DN, Pfeuffer J, et al. Network Accelerated Motion Estimation and Reduction (NAMER): convolutional neural network guided retrospective motion correction using a separable motion model. Magn Reson Med 2019;82:1452–61. https://doi.org/10.1002/mrm.27771.

[38] Haskell MW, Cauley SF, Wald LL. Targeted motion estimation and reduction (TAMER): data consistency based motion mitigation for MRI using a reduced model joint optimization. IEEE Trans Med Imaging 2018;37:1253–65. https://doi.org/10.1109/TMI.2018.2791482.

[39] Henningsson M, Koken P, Stehning C, Razavi R, Prieto C, Botnar RM. Whole-heart coronary MR angiography with 2D self-navigated image reconstruction. Magn Reson Med 2012;67:437–45. https://doi.org/10.1002/mrm.23027.

[40] Huttinga NRF, Bruijnen T, van den Berg CAT, Sbrizzi A. Nonrigid 3D motion estimation at high temporal resolution from prospectively undersampled k-space data using low-rank MR-MOTUS. Magn Reson Med 2020;65:2309–26. https://doi.org/10.1002/mrm.28562.

[41] Huttinga NRF, van den Berg CAT, Luijten PR, Sbrizzi A. MR-MOTUS: model-based non-rigid motion estimation for MR-guided radiotherapy using a reference image and minimal k-space data. Phys Med Biol 2020;65:015004. https://doi.org/10.1088/1361-6560/ab554a.

[42] Ingle RR, Wu HH, Addy NO, Cheng JY, Yang PC, Hu BS, et al. Nonrigid autofocus motion correction for coronary MR angiography with a 3D cones trajectory. Magn Reson Med 2014;72:347–61. https://doi.org/10.1002/mrm.24924.

[43] Ippoliti M, Lukas M, Brenner W, Schaeffter T, Makowski MR, Kolbitsch C. 3D nonrigid motion correction for quantitative assessment of hepatic lesions in DCE-MRI. Magn Reson Med 2019;82:1753–66. https://doi.org/10.1002/mrm.27867.

[44] Ippoliti M, Lukas M, Brenner W, Schatka I, Furth C, Schaeffter T, et al. Respiratory motion correction for enhanced quantification of hepatic lesions in simultaneous PET and DCE-MR imaging. Phys Med Biol 2021;66:095012. https://doi.org/10.1088/1361-6560/abf51e.

[45] Jung H, Park J, Yoo J, Ye JC. Radial k-t FOCUSS for high-resolution cardiac cine MRI. Magn Reson Med 2010;63:68–78. https://doi.org/10.1002/mrm.22172.

[46] Jung H, Sung K, Nayak KS, Kim EY, Ye JC. k-t FOCUSS: a general compressed sensing framework for high resolution dynamic MRI. Magn Reson Med 2009;61:103–16. https://doi.org/10.1002/mrm.21757.

[47] Kawaji K, Spincemaille P, Nguyen TD, Thimmappa N, Cooper MA, Prince MR, et al. Direct coronary motion extraction from a 2D fat image navigator for prospectively gated coronary MR angiography. Magn Reson Med 2014;71:599–607. https://doi.org/10.1002/mrm.24698.

[48] Keszei AP, Berkels B, Deserno TM. Survey of non-rigid registration tools in medicine. J Digit Imaging 2017;30:102–16. https://doi.org/10.1007/s10278-016-9915-8.

[49] Kolbitsch C, Neji R, Fenchel M, Schuh A, Mallia A, Marsden P, et al. Joint cardiac and respiratory motion estimation for motion-corrected cardiac PET-MR. Phys Med Biol 2018;64:015007. https://doi.org/10.1088/1361-6560/aaf246.

[50] Küstner T, Pan J, Gilliam C, Qi H, Cruz G, Hammernik K, et al. Deep-learning based motion-corrected image reconstruction in 4D magnetic resonance imaging of the body trunk. In: 2020 Asia-Pacific signal and information processing association annual summit and conference (APSIPA ASC). Presented at the 2020 Asia-Pacific signal and information processing association annual summit and conference (APSIPA ASC); 2020. p. 976–85.

[51] Larson AC, White RD, Laub G, McVeigh ER, Li D, Simonetti OP. Self-gated cardiac cine MRI. Magn Reson Med 2004;51:93–102. https://doi.org/10.1002/mrm.10664.
[52] Li H, Haltmeier M, Zhang S, Frahm J, Munk A. Aggregated motion estimation for real-time MRI reconstruction. Magn Reson Med 2014;72:1039–48. https://doi.org/10.1002/mrm.25020.
[53] Lin W, Guo J, Rosen MA, Song HK. Respiratory motion-compensated radial dynamic contrast-enhanced (DCE)-MRI of chest and abdominal lesions. Magn Reson Med 2008;60:1135–46. https://doi.org/10.1002/mrm.21740.
[54] Lin W, Ladinsky GA, Wehrli FW, Song HK. Image metric-based correction (autofocusing) of motion artifacts in high-resolution trabecular bone imaging. J Magn Reson Imaging 2007;26:191–7. https://doi.org/10.1002/jmri.20958.
[55] Liu J, Zwart JA, de Gelderen P van, Murphy-Boesch J, Duyn JH. Effect of head motion on MRI B0 field distribution. Magn Reson Med 2018;80:2538–48. https://doi.org/10.1002/mrm.27339.
[56] Loktyushin A, Nickisch H, Pohmann R, Schölkopf B. Blind retrospective motion correction of MR images. In: Proceedings 20th scientific meeting. Melbourne: International Society for Magnetic Resonance in Medicine; 2012.
[57] Luo J, Addy NO, Ingle RR, Baron CA, Cheng JY, Hu BS, et al. Nonrigid motion correction with 3D image-based navigators for coronary MR angiography. Magn Reson Med 2017;77:1884–93. https://doi.org/10.1002/mrm.26273.
[58] Maclaren J, Herbst M, Speck O, Zaitsev M. Prospective motion correction in brain imaging: a review. Magn Reson Med 2013;69:621–36. https://doi.org/10.1002/mrm.24314.
[59] Manke D, Rosch P, Nehrke K, Bornert P, Dossel O. Model evaluation and calibration for prospective respiratory motion correction in coronary MR angiography based on 3-D image registration. IEEE Trans Med Imaging 2002;21:1132–41. https://doi.org/10.1109/TMI.2002.804428.
[60] McClelland JR, Hawkes DJ, Schaeffter T, King AP. Respiratory motion models: a review. Med Image Anal 2013;17:19–42. https://doi.org/10.1016/j.media.2012.09.005.
[61] Menini A, Slavin GS, Stainsby JA, Ferry P, Felblinger J, Odille F. Motion correction of multi-contrast images applied to T_1 and T_2 quantification in cardiac MRI. Magn Reson Mater Phys Biol Med 2014;28. https://doi.org/10.1007/s10334-014-0440-9.
[62] Modersitzki J. Fair: flexible algorithms for image registration. USA: Society for Industrial and Applied Mathematics; 2009.
[63] Modersitzki J. Numerical methods for image registration. 1st ed. USA: Oxford University Press; 2004.
[64] Munoz C, Kunze KP, Neji R, Vitadello T, Rischpler C, Botnar RM, et al. Motion-corrected whole-heart PET-MR for the simultaneous visualisation of coronary artery integrity and myocardial viability: an initial clinical validation. Eur J Nucl Med Mol Imaging 2018;45:1975–86. https://doi.org/10.1007/s00259-018-4047-7.
[65] Nehrke K, Boernert P. Prospective correction of affine motion for arbitrary MR sequences on a clinical scanner. Magn Reson Med 2005;54:1130–8.
[66] Nyberg E, Sandhu GS, Jesberger J, Blackham KA, Hsu DP, Griswold MA, et al. Comparison of brain MR images at 1.5T using BLADE and rectilinear techniques for patients who move during data acquisition. Am J Neuroradiol 2012;33:77–82. https://doi.org/10.3174/ajnr.A2737.
[67] Odille F, Bustin A, Liu S, Chen B, Vuissoz P-A, Felblinger J, et al. Isotropic 3D cardiac cine MRI allows efficient sparse segmentation strategies based on 3D surface reconstruction. Magn Reson Med 2018;79:2665–75. https://doi.org/10.1002/mrm.26923.
[68] Odille F, Cindea N, Mandry D, Pasquier C, Vuissoz P-A, Felblinger J. Generalized MRI reconstruction including elastic physiological motion and coil sensitivity encoding. Magn Reson Med 2008;59:1401–11. https://doi.org/10.1002/mrm.21520.
[69] Odille F, Menini A, Escanyé J-M, Vuissoz P-A, Marie P-Y, Beaumont M, et al. Joint reconstruction of multiple images and motion in MRI: application to free-breathing myocardial quantification. IEEE Trans Med Imaging 2016;35. https://doi.org/10.1109/TMI.2015.2463088.

[70] Odille F, Uribe S, Batchelor PG, Prieto C, Schaeffter T, Atkinson D. Model-based reconstruction for cardiac cine MRI without ECG or breath holding. Magn Reson Med 2010;63:1247–57. https://doi.org/10.1002/mrm.22312.
[71] Odille F, Vuissoz P-A, Marie P-Y, Felblinger J. Generalized reconstruction by inversion of coupled systems (GRICS) applied to free-breathing MRI. Magn Reson Med 2008;60:146–57. https://doi.org/10.1002/mrm.21623.
[72] Pang J, Sharif B, Arsanjani R, Bi X, Fan Z, Yang Q, et al. Accelerated whole-heart coronary MRA using motion-corrected sensitivity encoding with three-dimensional projection reconstruction. Magn Reson Med 2015;73:284–91. https://doi.org/10.1002/mrm.25097.
[73] Pipe JG. Motion correction with PROPELLER MRI: application to head motion and free-breathing cardiac imaging. Magn Reson Med 1999;42:963–9.
[74] Prieto C, Batchelor PG, Hill DLG, Hajnal JV, Guarini M, Irarrazaval P. Reconstruction of undersampled dynamic images by modeling the motion of object elements. Magn Reson Med 2007;57:939–49. https://doi.org/10.1002/mrm.21222.
[75] Prieto C, Doneva M, Usman M, Henningsson M, Greil G, Schaeffter T, et al. Highly efficient respiratory motion compensated free-breathing coronary mra using golden-step Cartesian acquisition. J Magn Reson Imaging 2015;41:738–46. https://doi.org/10.1002/jmri.24602.
[76] Pruessmann KP, Weiger M, Boernert P, Boesiger P. Advances in sensitivity encoding with arbitrary k-space trajectories. Magn Reson Med 2001;46:638–51.
[77] Qi H, Fuin N, Cruz G, Pan J, Kuestner T, Bustin A, et al. Non-rigid respiratory motion estimation of whole-heart coronary MR images using unsupervised deep learning. IEEE Trans Med Imaging 2021;40:444–54. https://doi.org/10.1109/TMI.2020.3029205.
[78] Roemer PB, Edelstein WA, Hayes CE, Souza SP, Mueller OM. The NMR phased array. Magn Reson Med 1990;16:192–225. https://doi.org/10.1002/mrm.1910160203.
[79] Schmidt JFM, Buehrer M, Boesiger P, Kozerke S. Nonrigid retrospective respiratory motion correction in whole-heart coronary MRA. Magn Reson Med 2011;66:1541–9. https://doi.org/10.1002/mrm.22939.
[80] Stehning C, Börnert P, Nehrke K, Eggers H, Stuber M. Free-breathing whole-heart coronary MRA with 3D radial SSFP and self-navigated image reconstruction. Magn Reson Med 2005;54:476–80. https://doi.org/10.1002/mrm.20557.
[81] Steinhoff M, Nehrke K, Mertins A, Börnert P. Segmented diffusion imaging with iterative motion-corrected reconstruction (SEDIMENT) for brain echo-planar imaging. NMR Biomed 2020;33(12):e4185. https://doi.org/10.1002/nbm.4185.
[82] Uribe S, Muthurangu V, Boubertakh R, Schaeffter T, Razavi R, Hill DLG, et al. Whole-heart cine MRI using real-time respiratory self-gating. Magn Reson Med 2007;57:606–13. https://doi.org/10.1002/mrm.21156.
[83] Usman M, Atkinson D, Odille F, Kolbitsch C, Vaillant G, Schaeffter T, et al. Motion corrected compressed sensing for free-breathing dynamic cardiac MRI. Magn Reson Med 2013;70:504–16. https://doi.org/10.1002/mrm.24463.
[84] Usman M, Latif S, Asim M, Lee B-D, Qadir J. Retrospective motion correction in multishot MRI using generative adversarial network. Sci Rep 2020;10:4786. https://doi.org/10.1038/s41598-020-61705-9.
[85] Vahle T, Bacher M, Rigie D, Fenchel M, Speier P, Bollenbeck J, et al. Respiratory motion detection and correction for MR using the pilot tone: applications for MR and simultaneous PET/MR examinations. Invest Radiol 2020;55:153–9. https://doi.org/10.1097/RLI.0000000000000619.
[86] van Niekerk A, Berglund J, Sprenger T, Norbeck O, Avventi E, Rydén H, et al. Control of a wireless sensor using the pulse sequence for prospective motion correction in brain MRI. Magn Reson Med 2022;87(2):1046–61. https://doi.org/10.1002/mrm.28994.
[87] von Kalle T, Blank B, Fabig-Moritz C, Müller-Abt P, Zieger M, Wohlfarth K, et al. Reduced artefacts and improved assessment of hyperintense brain lesions with BLADE MR imaging in patients with neurofibromatosis type 1. Pediatr Radiol 2009;39:1216. https://doi.org/10.1007/s00247-009-1370-y.

[88] Vuissoz P-A, Odille F, Fernandez B, Lohezic M, Benhadid A, Mandry D, et al. Free-breathing imaging of the heart using 2D cine-GRICS (generalized reconstruction by inversion of coupled systems) with assessment of ventricular volumes and function. J Magn Reson Imaging 2012;35:340–51. https://doi.org/10.1002/jmri.22818.

[89] White N, Roddey C, Shankaranarayanan A, Han E, Rettmann D, Santos J, et al. PROMO: real-time prospective motion correction in MRI using image-based tracking. Magn Reson Med 2010;63:91–105. https://doi.org/10.1002/mrm.22176.

[90] Winkelmann S, Schaeffter T, Koehler T, Eggers H, Doessel O. An optimal radial profile order based on the golden ratio for time-resolved MRI. IEEE Trans Med Imaging 2007;26:68–76. https://doi.org/10.1109/TMI.2006.885337.

[91] Wintersperger BJ, Runge VM, Biswas J, Nelson CB, Stemmer A, Simonetta AB, et al. Brain magnetic resonance imaging at 3 Tesla using BLADE compared with standard rectilinear data sampling. Invest Radiol 2006;41:586–92. https://doi.org/10.1097/01.rli.0000223742.35655.24.

[92] Wood M, Henkelman R. Mr image artifacts from periodic motion. Med Phys 1985;12:143–51. https://doi.org/10.1118/1.595782.

[93] Zhu X, Chan M, Lustig M, Johnson KM, Larson PEZ. Iterative motion-compensation reconstruction ultra-short TE (iMoCo UTE) for high-resolution free-breathing pulmonary MRI. Magn Reson Med 2020;83:1208–21. https://doi.org/10.1002/mrm.27998.

[94] Zucker EJ, Cheng JY, Haldipur A, Carl M, Vasanawala SS. Free-breathing pediatric chest MRI: performance of self-navigated golden-angle ordered conical ultrashort echo time acquisition. J Magn Reson Imaging 2018;47:200–9. https://doi.org/10.1002/jmri.25776.

CHAPTER

14 Chemical Shift Encoding-Based Water-Fat Separation

Stefan Ruschke, Christoph Zoellner, Christof Boehm, Maximilian N. Diefenbach, and Dimitrios C. Karampinos

Klinikum rechts der Isar, Technical University of Munich, Department of Diagnostic and Interventional Radiology, School of Medicine, Munich, Germany

14.1 Introduction

Chemical shift encoding-based imaging (CSI) tries to differentiate chemical species based on their chemical shift property that originates from the shielding and deshielding effects caused by the electron density around the nucleus. In the context of proton-based water–fat imaging (WFI), this means that most protons of the fat component experience higher shielding and thus precess at lower frequencies. The frequency is measured in ppm units as a fraction of the absolute resonance frequency. In WFI at body temperature, the reference frequency of water is at about 4.67 ppm, while the reference frequency of the methylene peak (the main fat peak) is at about 1.30 ppm. This difference in chemical shift that corresponds to a frequency difference depending on the actual experienced field strength is encoded in the echo time dimension. For example, at 3T, the main frequencies of water and fat are 430 Hz (3.37 ppm) apart, corresponding to a phase cycling period of approximately 2.3 ms. Consequently, echoes acquired at TEs of n * 2.3 ms are called in-phase and at TEs of (n * 2.3 + 1.15) ms are called out-of-phase. The signal can then be acquired at different echo times and decomposed into its water and fat signal using an appropriate model (Fig. 14.1).

The idea of CSI was first introduced in a publication by eponymous W. Thomas Dixon in 1984 [1], and, since then, the use of WFI has significantly grown over the years both in clinical diagnostic imaging and in the research setting. WFI enables efficient fat suppression especially in experimental settings and anatomical locations where significant transmit B_1 and B_0 inhomogeneities can be observed. WFI can thus overcome some of the challenges faced by traditional chemical selective-fat suppression and inversion-based fat suppression [2]. In addition, WFI enables the simultaneous acquisition of coregistered nonfat-suppressed and fat-suppressed images in a single scan that can be beneficial in diagnostic imaging (albeit at the cost of prolonged scan time). Finally, WFI has recently gained considerable attention by enabling the quantification of the proton density fat fraction (PDFF), which constitutes a standardized metric of tissue fat concentration [3]. PDFF mapping is currently used in multiple organs (liver, pancreas, skeletal muscle, bone marrow) with a wide range of applications in fatty liver disease, metabolic disorders (obesity, diabetes), and musculoskeletal diseases [4].

WFI requires estimating the water and fat signal components in the presence of field-map variations. In addition, if non-Cartesian imaging is employed or quantitative parameter maps are of interest, this simple concept often requires to be extended to take additional effects into account and correct for

Advances in Magnetic Resonance Technology and Applications, Volume 7, ISSN 2666-9099. https://doi.org/10.1016/B978-0-12-822726-8.00025-7

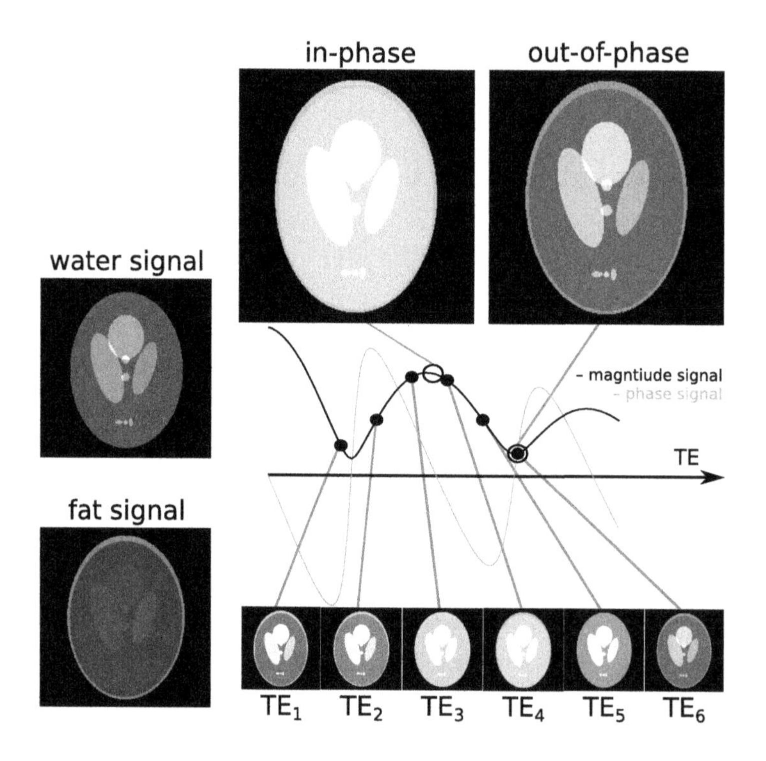

FIGURE 14.1 Schematic concept of chemical shift encoding-based signal encoding

On the left: The column on the left shows the ground truth water (top) and fat (bottom) signal of a simple numerical water–fat phantom. Right top row depicts characteristics contrasts for in-phase and out-of-phase images of the phantom that can be acquired at distinct echo times. Right middle row: A conceptional signal evolution for a water–fat signal (magnitude and phase-signal components) with a proton density fat fraction (PDFF) of 50%. Right bottom row: Example contrast of the same phantom sampled at six different echo times with equidistant echo time spacing. In order to separate the water from the fat signal, multiple sampling points (traditionally six points) are used to fit a water–fat model to the signal.

various sources of errors. Therefore, WFI requires multiple considerations in the data acquisition and signal modeling, which essentially include the whole chain of image reconstruction. In the following, the most important aspects in the reconstruction of WFI are presented.

14.2 Theory on chemical species separation

We first explain the chemical shift property, discuss the chemical shift property of fat, and then introduce the general formulation for chemical species separation.

14.2.1 The chemical shift property

In many quantitative MRI parameter-estimation techniques, the time evolution of the net magnetization of an ensemble of protons in a gradient echo or a free induction decay signal is the central focus. In human tissue these protons are always bound in molecules and therefore surrounded by an electron cloud defining the protons' chemical environment. Electron clouds are subject to polarization by the main magnetic field and therefore generate their own magnetic field based on the magnetic susceptibility of the molecule. The polarization of electrons during an MRI experiment, among others, gives rise to the effect of chemical shift affecting the MR signal dynamics [5].

For observer protons inside a molecule, the field created by the polarized electron cloud is known as the demagnetization field [6], which is dependent on the chemical surrounding and the local geometry of the proton's location inside the molecule (electron configuration, angle and length of chemical bonds between nuclei, etc.). According to Lenz's law [7], the demagnetization field is opposed to the polarizing field, the main magnetic field B_0 in MRI. Protons in a specific chemical environment, a so-called chemical species denoted by subscript p, therefore experience a reduced field strength and precess with a specific Larmor precession frequency defined by their shielding constant δ_p

$$\omega_p = -\gamma B_0(1 - \sigma_p) = \omega_0(1 + \delta_p). \tag{14.1}$$

By convention the chemical shift in MRI (and magnetic resonance spectroscopy) is often given in units of the deshielding constant δ_p [8]. Compared to water, most molecules in human tissue have a higher electron density and therefore a negative deshielding constant with respect to the assumed center frequency ω_0 tuned to the protons in water molecules. In the MR signal evolution, the presence of a single chemical species denoted p adds an additional magnetic-field term to the precession of an observer proton

$$S(t) \sim \int_V m(\mathbf{r}, t) \exp(-i\omega_0(1 + \delta_p)t) d\mathbf{r}. \tag{14.2}$$

The factor $\omega_0\delta_p$ leads to a modulation of the periodic precession of the magnetization around the main magnetic field with frequency ω_0. Therefore, the dynamics of the signal $S(t)$ over multiple echoes effectively encodes the presence of species p in the modulated frequency of the signal phase.

14.2.2 The chemical shift of fat

In body MRI, the most relevant molecules besides water are lipids due to their high abundance and chemical shifts properties. The lipid signal has a more complex spectral appearance compared to the water signal, which is usually well approximated with a single resonance frequency (Fig. 14.2) In MRI, all fat signals are typically assumed to arise exclusively from triglycerides due to their dominant abundance. Thus, the spectral model of the fat signal is based on the chemical structure of triglycerides. Triglycerides hold a number of different chemical species, each with their own specific chemical shifts. The most abundant chemical species in a triglyceride is the methylene group with its specific chemical shift of $\delta_{CH_2} = 1.3$ ppm.

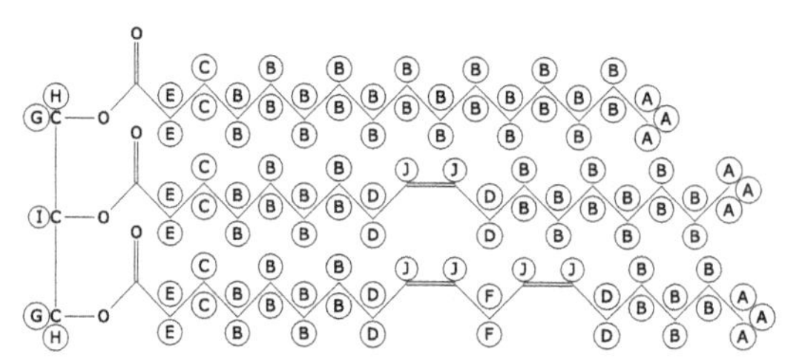

spectral representation

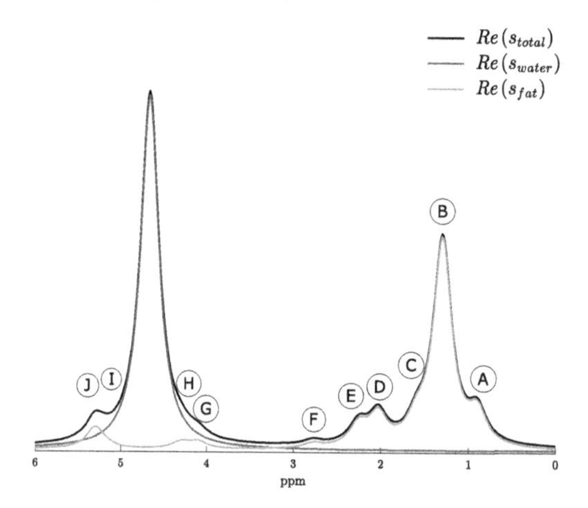

FIGURE 14.2 Chemical structure of a triglyceride and corresponding ^{1}H proton spectral appearance

The chemical structure of a triglyceride is characterized by a glycerol (on the left side) esterified with three fatty acids (palmitic acid, oleic acid, and linoleic acid from top to bottom on the right side). The circled letters (**A-J**) label and group the protons according to their specific resonance frequency. The corresponding spectral appearance of a mixture (black line) of triglycerides (red (mid gray in print version) line) and water (blue (dark gray in print version) line, peak at 4.67 ppm) is displayed below and shows the characteristic peaks that can be assigned to a corresponding proton environment in the chemical structure (see labels **A-J**).

14.2.3 Signal model for water–fat separation

Chemical shift encoding-based imaging uses acquisitions of multiple echoes at increasing echo times to estimate the contributions of possibly several chemical species to the MR signal, yielding important biomarkers like, e.g., the proton density fat fraction and other specific parameters like R_2^*.

The time evolution of the complex MR signal inside a single voxel, $s_n \equiv s(t_n)$, can be written in general as

$$s_n = \sum_{p=1}^{P} \varrho_p e^{i\phi_p} e^{(i\omega_p - r_p)t_n}, \qquad (14.3)$$

where the contribution to the signal from chemical species, $p = 1, ..., P$, is characterized by their magnitude ϱ_p, phase after the RF-excitation (at $t = 0$) ϕ_p, resonance frequency ω_p, and transverse relaxation rate r_p.

Since the estimation of the whole set of model parameters $\{\varrho_p, \phi_p, \omega_p, r_p\}$, $p = 1, ..., P$ in Eq. (14.3) would require an impractical amount of echoes s_n, it is necessary to reduce the number of parameters on the right-hand side of Eq. (14.3). A typical example in WFI is the reduction of Eq. (14.3) to the widely used multipeak single-R_2^* model by fixing the chemical shifts of all $p \in 2, ..., 10$ fat peaks—assuming the first peak $p = 1$ corresponds to the water peak—to a priori values normally obtained from MR spectroscopy by setting $\omega_m = \omega_1 + \delta\omega_p$, $p \in 2, ..., P$. Furthermore, all peaks, including water, are assumed to decay with the same relaxation time R_2^*, which finally yields

$$\hat{s}_n = (W + c_n F)\, e^{(i2\pi f_B - r_2^*)t_n}, c_n = \sum_{p=1}^{P} a_p e^{i2\pi\delta f_p t_n}, \quad \text{with} \quad \sum_{p=1}^{P} a_p = 1, \tag{14.4}$$

where we substituted $W = \varrho_1 \exp(i\phi_1)$ and $f_B = \omega_1/2\pi$ for the water peak and $F\alpha_p = \varrho_p \exp(i\phi_p)$, $p \in \{1, ..., P\}$ for the fat peaks, introducing f_B for the field-map and α_p for the weight of the p-th fat peak in the fat spectrum model.

The signal model given by Eq. (14.4) describes the signal in a single observation at acquired echo time t_n. The definition of the following model matrix [9,10]:

$$\mathbf{A}(f_B, R_2^*) = \text{diag}\left[e^{(i2\pi f_B - R_2^*)t_1}, \ldots, e^{(i2\pi f_B - R_2^*)t_n}\right] \begin{pmatrix} 1 & c_1 \\ \vdots & \vdots \\ 1 & c_N \end{pmatrix}, \tag{14.5}$$

enables us to write the signal model in a multi-observation matrix formulation

$$\hat{\mathbf{s}} = [s_1, \ldots, s_n]^\mathsf{T} = \mathbf{A}(f_B, R_2^*)[W, F]^\mathsf{T}.$$

14.3 Solving the water–fat separation problem

14.3.1 Parameter estimation in water–fat separation

The least-squares parameter estimation problem can then be stated as

$$W, F, f_B, R_2^* = \underset{W', F', f_B', R_2^{*\prime}}{\text{argmin}} \left\|\mathbf{s} - \hat{\mathbf{s}}(W', F', f_B', R_2^{*\prime})\right\|_2^2. \tag{14.6}$$

By means of the VARPRO [11], the linear parameters in Eq. (14.6) can be substituted by

$$[W, F]^\mathsf{T} = \mathbf{A}^+(f_B, R_2^*)\hat{\mathbf{s}}, \tag{14.7}$$

where $\mathbf{A}^{+} = \mathbf{A}^{\dagger}(\mathbf{A}^{\dagger}\mathbf{A})^{-1}$ is the Moore–Penrose pseudo-inverse of matrix in Eq. (14.5), resulting in the optimization

$$f_B, R_2^* = \underset{f_B', R_2^{*\prime}}{\operatorname{argmin}} \left|\left|\left(1 - \mathbf{A}\mathbf{A}^{+}\right)\hat{\mathbf{s}}\right|\right|_2^2, \tag{14.8}$$

which only minimizes the nonlinear parameters f_B, R_2^* in the signal model of Eq. (14.4) since the linear parameters W and F are determined by Eq. (14.7).

The variable projection (VARPRO) minimization problem of Eq. (14.8) is specific to the single-R_2^* water–fat signal model of Eq. (14.4), however it can be generalized to the broader class of weighted sums of complex exponentials with phase terms linearly varying in the echo time [12].

The minimization in Eq. (14.8) can in general be solved by standard nonlinear fitting but requires two additional considerations: the field map estimation and the analysis of the noise performance of the estimate parameters.

14.3.2 The field-map estimation problem

The field-map parameter f_B describes local frequency shifts due to static field inhomogeneity. The static field inhomogeneity mainly originates from the large magnetic susceptibility difference between air and tissue and the geometry of the subject placed inside the scanner. Additionally, different magnetic susceptibility of different tissues contributes to the field inhomogeneity.

The field-map parameter f_B is subject to phase wrapping, and its value can only be determined up to modulo 2π. Traditionally, WFI primarily focuses on the extraction of the water and the fat-signal parameters, where a wrapped field-map is not of importance. However, in cases where further processing of the field-map is needed, the wrapping needs to be taken into account. In addition, and also connected to the complex exponential with the time-dependent argument scaled by the field-map parameter, the least-squares cost function in Eq. (14.8) is nonconvex in the field-map parameter.

The nonlinearity of the field-map parameter often leads to the occurrence of multiple local minima of the cost function that can correspond to unphysical parameter combinations. A convergence of the field-map to such local minima can lead to the infamous water–fat swaps, where the water and the fat signals in a voxel are incorrectly assigned [13]. Water–fat swaps can in particular occur in regions with a rapid spatial variation of the field-map and/or low signal to noise ratio, where the number of local minima of the cost function increases. Therefore, iterative nonlinear optimization schemes require an initialization of the field-map close to the true parameter value to ensure convergence to the global minimum. Finding a good initialization of the field-map often necessitates the incorporation of additional information to the single-voxel minimization problem (14.8). Common approaches involve the incorporation of prior knowledge from several field-map contributions to the problem [14,15] or the incorporation of neighborhood information of each voxel by setting up a global cost function for all voxels similar to

$$f_B, R_2^* = \underset{f_B', R_2^{*\prime}}{\operatorname{argmin}} \sum_{\mathbf{r}} \left[\left|\left|(1 - \mathbf{A}(f_B', R_2^{*\prime})\mathbf{A}^{+}(f_B', R_2^{*\prime}))\mathbf{s}\right|\right|_2 + \sum_{\mathbf{r}' \in \mathcal{N}(\mathbf{r})} U(|f_B'(\mathbf{r}) - f_B'(\mathbf{r}')|^2) \right], \tag{14.9}$$

where the least-squares terms for all voxels are summed up and an "interaction term" U between each voxel (at discrete location $\mathbf{r}$) and its neighborhood voxels at $\mathbf{r}' \in \mathcal{N}(\mathbf{r})$ is added. Global minimization problems like Eq. (14.9) can successfully be solved with graph-cut algorithms [16], first demonstrated for WFI in [17] and further developed and employed in specific applications in [18–20]. We will be describing next the most important methods employed for estimating the field-map in water–fat separation.

Assuming a simple single-fat-peak water–fat voxel signal model, the complex signal at the n-th echo is

$$\mathbf{s}_{\text{model}}(t_n) = \left(\rho_W + \rho_F e^{i2\pi\Delta f_p t_n}\right) e^{i2\pi f_B t_n}, \quad (14.10)$$

with $t_1, t_2, ..., t_N$ the different echo times, Δf_p the chemical shift of fat, ρ_W, and ρ_F the complex signal of the water and fat components. To estimate the static field inhomogeneity f_B in Eq. (14.10), Glover et al. proposed a method where three echoes are recorded with phase shifts of $-\pi$, 0, and π between the fat and water resonances, simplifying the maximum likelihood estimation to a closed-form problem [21]. However, this choice of phase shifts restricts the echo times to be fixed. The fixed echo times are accompanied by limitations such as the restriction in the selection of other imaging parameter, e.g., the size of the FOV in a multi-echo acquisition or increased total scan time. Furthermore, it has been shown that these phase shifts are not the optimal candidates to reach the maximum possible noise performance [22,12]. Additionally, the signal model in Eq. (14.10) accounts for one chemical shift specimen only and does not account for R_2^*-decay effects. These limitations can significantly reduce the quantitative performance of the parameter estimation [10,23].

Therefore, there has been considerable interest in alternative choices of echo shifts, the inclusion of a multi-fat-peak model, and R_2^* [10,23]. In the case of arbitrary echo-time shifts, the estimation of the local frequency shift can not be decoupled from the estimation of water–fat contribution due to the nonlinearity of the frequency shift within the signal model. To estimate the local frequency in case of arbitrary echo-time shifts and a nonlinear maximum-likelihood estimate, Reeder et al. introduced a method for iterative decomposition of water and fat with echo asymmetry and least-squares estimates (IDEAL, [22]). The IDEAL method consists of repeated linearizations of the original nonlinear problem, alternatingly estimating the water–fat signals and the field-map using a gradient descent scheme. However, a gradient descent-based method with a simple initialization of $f_B = 0$ in all voxels has problems dealing with large-field inhomogeneity. The problem with large-field inhomogeneity arises since the maximum-likelihood estimate is not only nonlinear but also nonconvex and potentially contains several local and global minima [24,25] as depicted in Fig. 14.3. The assumption of a moderate field inhomogeneity is not generally true (especially at higher field strengths) and can lead to water–fat swaps associated with a significant error in the field-map estimation as depicted in the first row of Fig. 14.5.

Several methods have been proposed to address the problem with large-field inhomogeneities. Yu et al. [24] proposed an extension to the original gradient descent-based method, where a slowly varying field-map is imposed by a region-growing scheme. Lu et al. [25] proposed a coarse-to-fine grid approach in addition to the region-growing algorithm to further improve robustness, and Hernando et al. [26] proposed a method using variable projection (VARPRO) to resolve the large-field inhomogeneities by calculating the cost function Eq. (14.8) over its period length. The VARPRO approach reveals all local minima, and the extraction of the global minima is possible (see Fig. 14.3). The extraction of the global minimum from the VARPRO residual will subsequently be referred to as gmVARPRO. Field-map and water–fat separated images obtained by the gmVARPRO method are depicted in the second row of Fig. 14.5. However, all the aforementioned methods still can fail to correctly estimate the field-map in challenging anatomies or are prone to noise due to their voxel-independent fitting scheme.

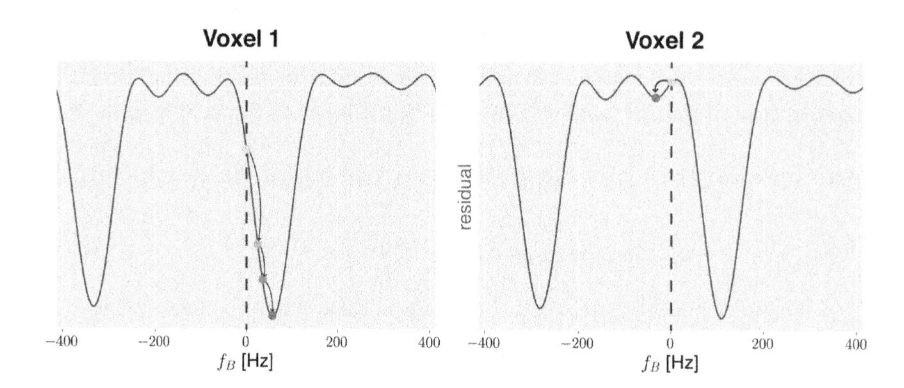

FIGURE 14.3 The voxel residual space of Eq. (14.8) of two voxels

The figure shows the VARPRO residual space of Eq. (14.8) in dependence of the field-map parameter f_B of two voxels from the knee data set shown in Fig. 14.5. Both voxel reveal the nonconvex (multiple minima) and periodic nature of the field-map estimate. In voxel 1, a gradient descent-based optimization is able to find the global minimum because the 0 initialization (dashed gray line) is within the correct valley. However, in voxel 2 the field-map shift is too large, and the 0 initialization is within a valley adjacent to the global minimum.

In the past decade, algorithms formulating the field-map estimation as a graph search have emerged and have proven to be particularly successful in solving the constrained optimization problem, using either min-cuts iteratively as proposed by Hernando et al. [17], single-min-cut approaches proposed by Cui et al. [20,27] and Boehm et al. [28], or a iterative fine to coarse-grid approach proposed by Berglund et al. [19]. These methods either add neighborhood regularization to enforce the field-map to be slowly varying [17,20,28] or use a threshold value for the maximal allowed field-map variation between adjacent voxels [27] or smoothing by a fine-to-coarse grid approach [19].

While significantly improving field-map estimation and water–fat separation, most proposed graph-cut methods still have intrinsic limitations. The iterative graph cut in [17] does not necessarily converge to the global minimum of its defined cost function, smooths the field-map by construction, and allows for adopting only two-dimensional neighborhood information in the field-map smoothness-constraint term (inter-slice regularization is not possible). The iterative fine-to-coarse grid approach in [19] is restricted to compare only two candidates per voxel and can consequently miss the correct solution. The single-min-cut in [27] labeled as *rapid Globally Optimal Surface Estimation* (rGOOSE) tried to address these problems. The rGOOSE method: i) Does not smooth the field map since it restricts the field-map candidates to be only local minima of the voxel-wise field-map estimate; ii) it also allows for 3D neighborhood regularization; and iii) it theoretically yields the exact solution of its defined cost function. However, the rGOOSE method is associated with long computation times and can still sometimes miss the optimal solution due to limitations in its underlying graph construction. The most recently proposed method by Boehm et al. [28] based on a so-called variable-layer, single-min-cut graph cut: i) Reportedly overcomes the limitations in graph construction in the rGOOSE method; ii) uses a novel data consistency formulation in its global problem formulation; iii) shows excellent field-mapping and water–fat separation results in a plethora of imaging situations; and iv) yields wrap-free field-maps. The exemplary illustration of the graph construction based on VARPRO residuals for the

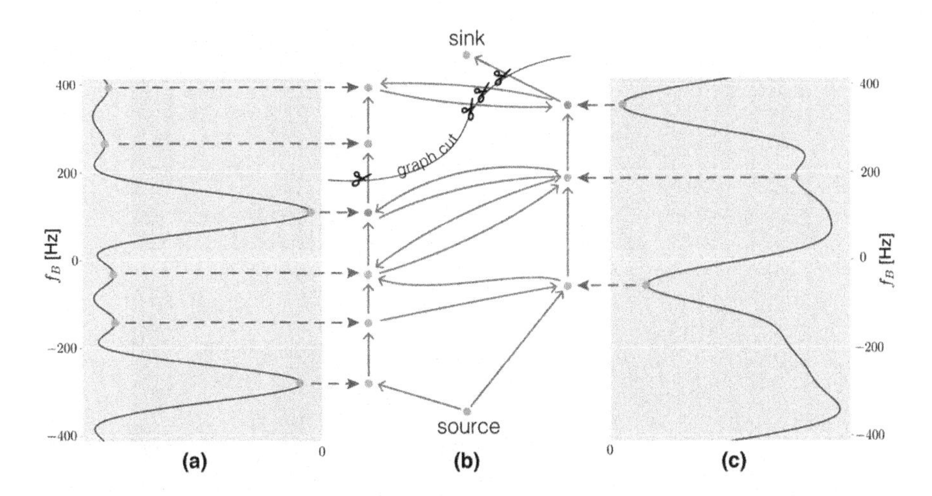

FIGURE 14.4 Exemplary illustration of the minima extraction and graph construction of two adjacent voxels in the graph cut based method field-mapping method by Boehm et al. [28]

The figure shows the residual space of two randomly selected adjacent voxels from the knee data set shown in Fig. 14.5. From the first voxel **(a)** 6 minima and from the second voxel **(c)** 3 minima are extracted, respectively. Based on the extracted minima a graph is constructed **(b)** that encodes the cost function (14.9) into its graph representation. Subsequently, the cost function is minimized by searching for the minimum cut in graph and the global optimal minima (green (light gray in print version) node) are identified. Please be note that the horizontal axis direction is flipped between (a) and (c).

method by Boehm et al. is shown in Fig. 14.4. Please note that the graph construction is always tailored to the specifics of the cost function to be minimized. Hence the graph for the different graph cut-based methods can be significantly different. The characteristic of the method by Boehm et al. to directly yield wrap-free field-maps is especially of high importance for field-map-based parameter estimation, such as quantitative susceptibility mapping (QSM) (Chapter 16), where wraps are expected to translate in strong susceptibility artifacts [29]. Field-map and water–fat separated images obtained by the state-of-the-art variable-layer single-min-cut graph cut method are depicted in the last row of Fig. 14.5.

14.3.3 Noise performance analysis

The second important consideration in solving (14.8) (or the first term in the global estimation (14.9)) is the noise propagation in the optimization, which depends on the echo-time selections in combination with the physical tissue parameters. At certain unfavorable combinations of echo times and model parameters, the noise in the input data $\boldsymbol{s}$ amplifies in the parameter estimates, which renders them useless for later diagnostic or analytic purposes. It is therefore important to choose well-suited echo times for the expected tissue parameters to optimally design the experiment before the data acquisition. The standard method for such an optimal experimental design in terms of echo-time selection and model parameters is the Cramér–Rao analysis [30,31], based on the computation of the Fisher information matrix (FIM) [32] for the chosen signal model. The FIM is defined as the expectation value ($\mathbb{E}[\ldots]$) of

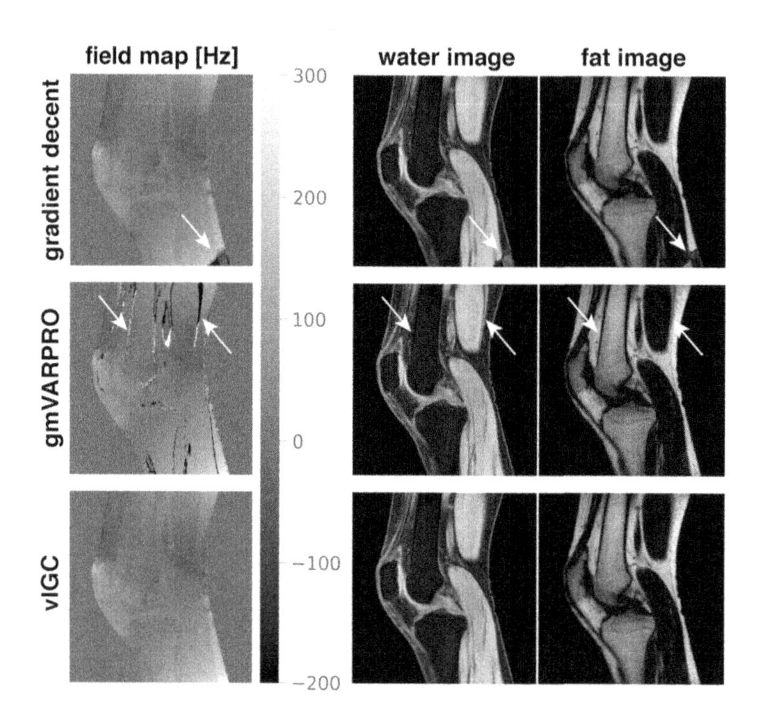

FIGURE 14.5 Field-map and water–fat separated images obtained by a voxel-wise gradient descent method [22], the gmVARPRO method [26] and a state-of-the-art variable layer graph cut (vlGC) [28]

The voxel-wise gradient-descent method (first row) is able to estimate field-map and water–fat separated images to a large extent correctly. However, at the bottom right corner of the FOV, the field-map variation is too large, and a field-map wrap and water–fat swap occurs in the corresponding maps. In contrast to the gradient descent method, the gmVARPRO method (second row) is able to resolve the large field-map variation at right left bottom of the FOV by extracting the global minimum of the residual space (compare with Fig. 14.3) per voxel. However, at regions with low signals such as the cortical bone shell, noise-like artifacts are visible in field-map indicating water–fat swaps. Furthermore, partial swaps occur in the whole FOV where water- or fat-only regions are falsely estimated to contain both water and fat. Field-mapping based on a vlGC method (last row) is able to correctly resolve the large field-map variation at the bottom of the FOV, has no noise-like estimates, and can correctly separate water and fat contributions throughout the whole volume.

the second derivative of the log-likelihood $\ln \mathcal{L}$

$$\boldsymbol{I}_{kl} = \mathbb{E}\left[\frac{\partial}{\partial \beta_l}\frac{\partial}{\partial \beta_k} \ln \mathcal{L}\right], \tag{14.11}$$

where the likelihood $\mathcal{L}$ in the case of additive white Gaussian noise with variance σ^2 follows

$$\mathcal{L} \sim \exp\left(\frac{1}{2\sigma^2}||\boldsymbol{s} - \mathbf{A}\boldsymbol{\rho}||_2^2\right), \boldsymbol{\rho} = [W, F]^{\mathsf{T}}. \tag{14.12}$$

In general, the FIM is then given by

$$\mathbf{I} = \frac{1}{\sigma^2} \Re\{\mathbf{J}^\dagger \mathbf{J}\}, \tag{14.13}$$

where $\mathbf{J}$ is the Jacobian of the chosen signal model [33]. In the case of the widely used single-R_2^* signal model (14.4), the Jacobian that defines the FIM by (14.13) is

$$\mathbf{J} = \left[\frac{\partial \hat{s}}{\partial |W|}, \frac{\partial \hat{s}}{\partial |F|}, \frac{\partial \hat{s}}{\partial \angle W}, \frac{\partial \hat{s}}{\partial \angle F}, \frac{\partial \hat{s}}{\partial f_B}, \frac{\partial \hat{s}}{\partial R_2^*} \right], \tag{14.14}$$

where $\angle W$ and $\angle F$ refers to the angle of W and F, respectively.

The Cramér–Rao lower bound (CRLB), which gives the theoretically minimal variances of the parameter estimates [33], is defined as

$$\text{CRLB} = \text{diag}\,\mathbf{I}^{-1} = [\text{Var}|W|, \text{Var}|F|, \text{Var}\angle W, \text{Var}\angle F, \text{Var}\, f_B, \text{Var} R_2^*]^\mathsf{T}. \tag{14.15}$$

Minimizing the CRLB for the model parameters of interest and/or different echo samplings can be used for optimal design [34] of the experimentally selected echo times. Again, the CRLB's in (14.15) are specific for the single-R_2^* water–fat signal model (14.4), but can be generalized for the class of summed cisoid (i.e., complex exponential) signal models [12].

The noise efficiency can also be quantified by the effective number of signals averages (NSA) defined as the variance in each of the measured images over the variance of a given parameter [34]. See Fig. 14.6 for an example.

14.4 Water–fat separation in non-Cartesian imaging

Non-Cartesian trajectories such as radial or spiral are more robust towards motion artifacts compared to Cartesian trajectories due to their dispersed distribution of motion artifacts [35–37]. For example, radial sampling is able to eliminate k-space gaps that are caused by motion-induced phase shifts, by repeated sampling of the k-space center. However, substantial motion still remains a challenge for non-Cartesian imaging and can result in blurring and aliasing artifacts, which appear as streaks for example when using radial trajectories [38,39]. An advantage of many non-Cartesian acquisitions, such as radial Stack-of-Stars (SoS) or spirals, is the continuous passage of the radial lines through the k-space center, allowing for retrospective self-gating and thus eliminating the need to use navigator signals or external devices for motion correction [40,41]. Especially, the use of golden-angle-ordered 3D SoS trajectories has been investigated for free-breathing water–fat imaging [42–44].

While being less sensitive to motion artifacts, non-Cartesian trajectories are less robust to gradient errors and delays than Cartesian ones (see Chapter 4). Different non-Cartesian trajectories, such as radial [42,45,46,43], spiral [47–49], PROPELLER [50,51], and concentric rings [52], have been investigated for water–fat separation. Besides gradient errors and delays, off-resonance effects affect the use of non-Cartesian trajectories for water–fat imaging. The origination of the blurring of the fat signal and its implications for non-Cartesian trajectories will be explained briefly in this section.

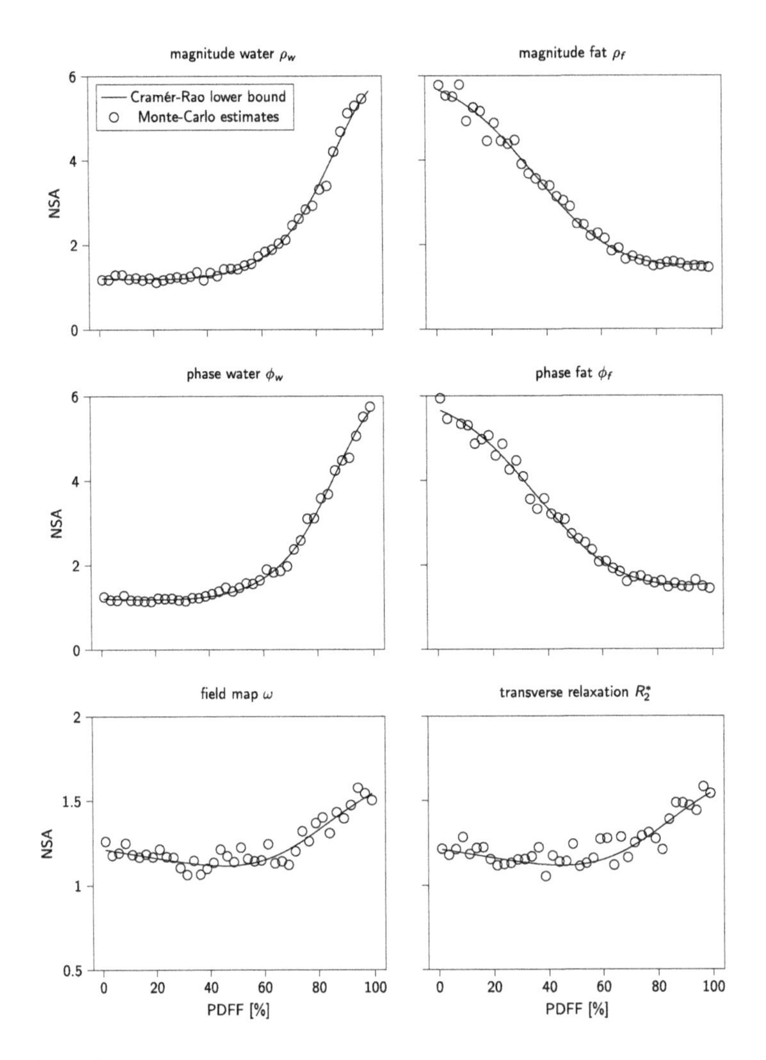

FIGURE 14.6 WFI simulation

Cramér–Rao lower bounds with Monte-Carlo estimates of parameter noise in a single-R_2^* water–fat model with a multi-peak fat model. Experimental setting is defined by: number of echoes $N_{TE} = 6$, first echo time $TE_1 = 1$ ms, echo spacing $\Delta TE = 1.2$ ms, $B_0 = 3$ T, $R_2^* = 5\ \mathrm{s}^{-1}$, and field-map $\omega/2\pi = 10$ Hz. The Monte-Carlo simulation was performed with 1000 independent noise realizations with signal-to-noise-ratio (SNR) of 100.

14.4.1 Water–fat shift artifact

The different Larmor frequencies of water and fat cause a misregistration of fatty tissue along the frequency-encoding direction (read or slice selection).

As discussed in Section 14.2, the signal from lipid protons will have a lower frequency compared to that of water protons. When the resonance frequency of the system is set to water, the signal from the lipid protons will appear to originate from water protons from a voxel from a lower part of the field.

The difference between the precession frequencies $\Delta f_{fw}(x)$ is given by

$$\Delta f_{fw}(x) = f_f - f_w = -\sigma_{fw} \frac{\gamma}{2\pi} B_0, \tag{14.16}$$

where the suffixes w and f stand for water and fat respectively σ_{fw} is the chemical shift between water and fat as a fraction of the field B_0 and γ is the gyromagnetic ratio [53]. If the bandwidth per voxel Δf_{voxel} is not much greater than $\Delta f_{fw}(x)$, this shift of the fat signal becomes visible in the resulting image. The fat signal is then spatially misregistered and invades into neighbored voxels in the read-and-slice selection directions. The bandwidth per voxel is defined as

$$\Delta f_{voxel} = \frac{\gamma}{2\pi} G_R \Delta x, \tag{14.17}$$

with Δx being the voxel size an G_R the readout gradient strength. The spatial shift of fatty tissue can then be calculated with

$$\Delta x_{shift} = \frac{\Delta f_{fw}}{\Delta f_{voxel}} \Delta x = N_{shift} \Delta x. \tag{14.18}$$

As long as the readout gradient is sufficiently large so that $N_{shift} < 0.5$, most of the fat signal remains within the voxel. Eq. (14.18) illustrates that, the lower the bandwidth per voxel, the larger the spatial misregistration artifact, and thus the fat-shift artifact.

For conventional Cartesian spin-warp imaging, the bulk shift of fat due to the chemical shift artifacts is well-understood and to some degree clinically acceptable. Specifically, in conventional Cartesian spin-warp imaging, the phase accumulation is identical in each echo and varies linearly across k-space in the frequency-encoding direction. However, in non-Cartesian imaging, the phase accumulated by off-resonant spins leads to distortion and highly degraded image quality. The blurring from the off-resonance fat signal has long challenged the clinical adoption of non-Cartesian acquisitions, especially in protocols employing low bandwidths and long readouts.

14.4.2 Fat blurring in non-Cartesian acquisitions

For most non-Cartesian acquisitions, the frequency-encoding direction varies during the acquisition and can cover all possible directions in k-space. While for most Cartesian trajectories the frequency-encoding dimension is fixed to one k-space direction, it changes, e.g., step-wise with every spoke in a radial stack-of-stars trajectory or even continuously for a spiral trajectory. Consequently the off-resonant fat signal is not bulk shifted any more in one direction in image space with regards to the water signal as in the Cartesian case, but instead the fat signal is shifted in all frequency-encoding directions that are used for the non-Cartesian trajectory (Fig. 14.7). This leads to blurring and distortions of the fat signal when using non-Cartesian trajectories.

The blurred fat signal can obscure underlying or neighboring pathologies including tumors, inflammation, and edema [46]. Especially, the longer readouts that are needed for spiral trajectories and low bandwidths are heavily affected by blurring and distortions of the fat signal since the longer the readout, the more off-resonant phase is accumulated. Consequently, short readouts or fat suppression are

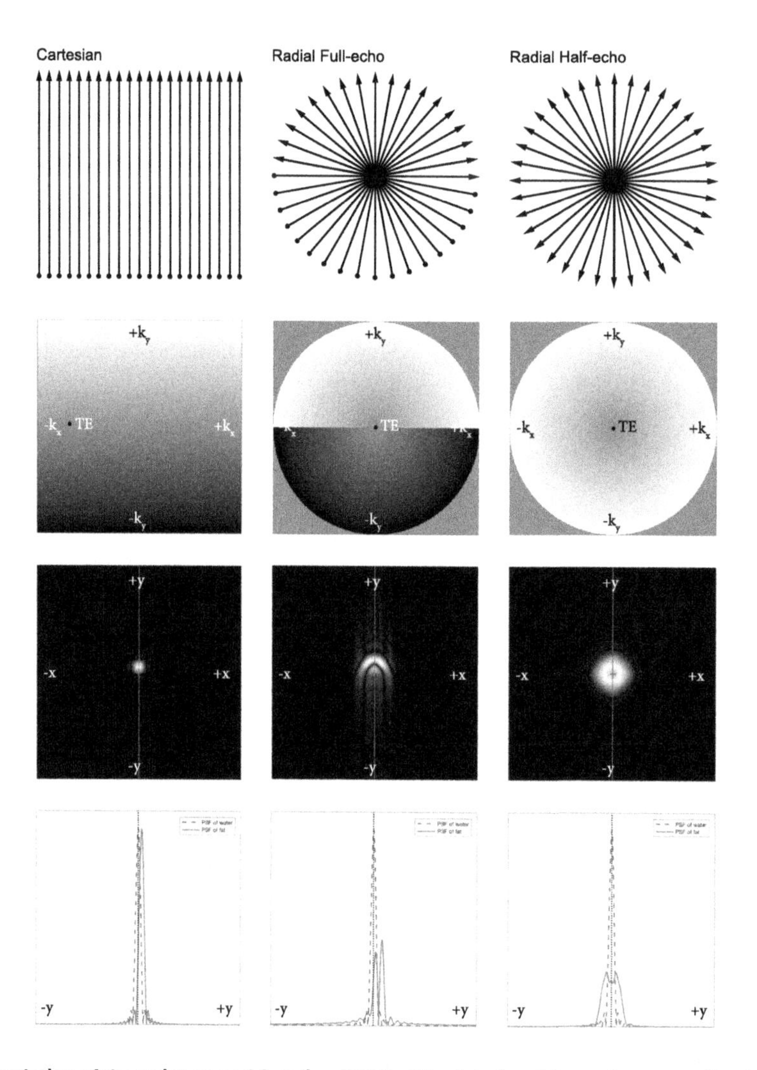

FIGURE 14.7 Simulation of the point spread function (PSF) of the fat signal for various sampling trajectories

Exact knowledge of the acquisition time for each k-space point is necessary to remove phase errors due to the off-resonance of fat. Graphical acquisition time maps in k-space are shown in the second row under the corresponding trajectories. Dark values represent a negative time shift respectively to TE, which means that those k-space points were sampled earlier than the actual TE time, while brighter values were sampled later than TE. The third row shows the PSF for the off-resonant fat components. For the Cartesian case, the PSF is bulk shifted in the frequency-encoding direction, while it is spread out for the two radial trajectories. Note that, for the unsymmetric full-radial case, the PSF also shows a clear directional dependence. 1D profiles of the PSF plotted along the blue (dark gray in print version) line in y-dimension are shown in the bottom row. The PSF of the water signal is added as a reference (blue (dark gray in print version) dotted line). Adapted from [46].

often needed for non-Cartesian imaging techniques to reduce the confounding blurred fat. On the other hand, fat suppression using inversion or spectrally selective pulses [54] can reduce the signal-to-noise ratio (SNR) or perform poorly in regions with B_0 inhomogeneities. To increase SNR, lower receiver bandwidths may be used that increase the water–fat shift and thus the blurring of the fat signal. Chemical shift encoding-based water–fat separation methods can be effective in removing the fat blurring in non-Cartesian acquisitions.

14.4.3 k-space-based water–fat separation

A direct approach to correct for the off-resonance of the various fat spectrum peaks is therefore to already include the phase accumulation in the k-space signal model. Historically IDEAL [22]-based methods and other image-based separation methods are performed in the image domain and treat each echo as if it was sampled instantaneously at the echo-time point t_n at which the k-space center is sampled. But, since the readout time is a finite process, it can be added to the signal model in k-space for a more accurate water–fat separation. This is done by extending t_n by a second time parameter $\tau_{\mathbf{k},n}$, which represents the relative time between the acquisition of sample point $\mathbf{k}$ and the center of k-space [55]. Brodsky et al. [46] assume a signal model that ignores the effects of T_2^* decay and changes in the field-map variation between echoes. In this case, the B_0 can be demodulated from the signal, and it can be expressed in k-space as

$$s_n(\tau_{\mathbf{k},n}, \mathbf{k}) = \sum_{m=1}^{M} \rho_m(\mathbf{k}) e^{i2\pi \Delta f m (t_n - \tau_{\mathbf{k},n})}. \tag{14.19}$$

$\tau_{\mathbf{k},n}$ models the finite readout and specifies the exact sampling time point $t_n - \tau_{\mathbf{k},n}$ of each k-space point $\mathbf{k}$. Similar to the VARPRO approach explained in Section 14.3 in image space, the signal can then be decomposed into water and fat maps in k-space directly. The reconstructed fat maps from this water–fat separation do not show the typical non-Cartesian fat blurring due to off-resonance of the fat spectrum peaks [46].

A more analytical but therefore computationally demanding approach to correct for the water–fat shift using the same signal model idea is presented by Benkert et al. [42]. Water (W), fat (F), and B_0 field-map (Φ) are directly fitted to the radial k-space data y in one single step by solving the minimization problem for each coil element c and echo time t

$$\arg\min \sum_{c,t} ||\mathbf{E}(W, F, \Phi)_{c,t} - y_{c,t}||_2^2 + \lambda_W ||\mathbf{S}(W)||_1 + \lambda_F ||\mathbf{S}(F)||_1. \tag{14.20}$$

With $\mathbf{S}$ being the sparsifying transform, the last two terms are the compressed sensing [56] regularizer for the water and the fat signal. The desired parameters are transformed to k-space data via the forward operator $\mathbf{E}$

$$\mathbf{E}(W, F, \Phi)_{c,t} = \mathbf{FT}(C_c \cdot \exp(2\pi i \cdot \Phi \cdot t_n) \cdot W) + D(t) \cdot \mathbf{FT}(C_c \cdot \exp(2\pi i \cdot \Phi \cdot t_n) \cdot F). \tag{14.21}$$

The parameter $\tau_{\mathbf{k},n}$ is included here in the forward operator to model the various chemical shifts of the fat peaks and to account for the off-resonant blurring of fat due to the radial readout

$$D(t) = \sum_{m=1}^{M} \alpha_m \cdot e^{2\pi i \cdot \Delta f_m \cdot (t_n + \tau_{\mathbf{k},n})}. \tag{14.22}$$

Both presented methods were primarily used to perform water–fat separation for non-Cartesian trajectories. However, the modification of the k-space signal by modeling the off-resonances of the various fat model peaks makes it, in general, possible to correct for chemical shift artifacts in any acquisition. Finally, while the deblurring of fat may be more convenient in k-space, depending on the employed model, the water–fat separation may be performed in image space or in k-space.

14.5 Confounding factors in quantitative water–fat imaging

Quantitative water–fat separation requires the consideration of potential confounding factors, which are traditionally characterized by either their origin as physical effects and hardware characteristics or by their appearance as magnitude and phase errors [57–61].

Physical effects typically include B_0 inhomogeneities [21,26], T_2^* decay [62,23], T_1^* relaxation differences between the water and fat [63,64], spectral complexity of the fat model [23,65], susceptibility-induced resonance shifts [66], temperature-induced resonance shifts of water [67], noise bias [63,64], and concomitant gradient effects [68,69]. Hardware imperfections include gradient delays [68] and non-flat frequency response characteristics of the receiver coils [58].

Gradient delays and concomitant gradients are two confounders that can be corrected during the reconstruction process and are described in the following section for Cartesian imaging trajectories.

14.5.1 Correction of hardware imperfections: gradient delays

In multi-echo gradient–echo-based water–fat imaging, multiple subsequent echoes are acquired in a train of echoes at different echo times. Due to hardware limitations and imperfections, the actual played out gradient waveform may deviate from the nominal defined gradient waveform. Consequently, this may lead to echo misalignments (and also higher-order effects which are not discussed further here), which may also differ across the echo train. For Cartesian trajectories, these echo misalignments often appear as an additional linear phase in the image domain along the readout direction (Fig. 14.8) according to the Fourier shift theorem

$$s(k - k_0) \longleftrightarrow S(r)e^{-i2\pi r k_0}. \tag{14.23}$$

This type of gradient delay can be estimated under the assumption that delays are symmetric with respect to the readout gradient polarity. Therefore, a reference measurement can be acquired by flipping the readout direction that corresponds to inverting the gradient polarity. Usually, it is sufficient to estimate the delay for the center k-space line only and to assume that the delay is independent of the phase-encoding gradients.

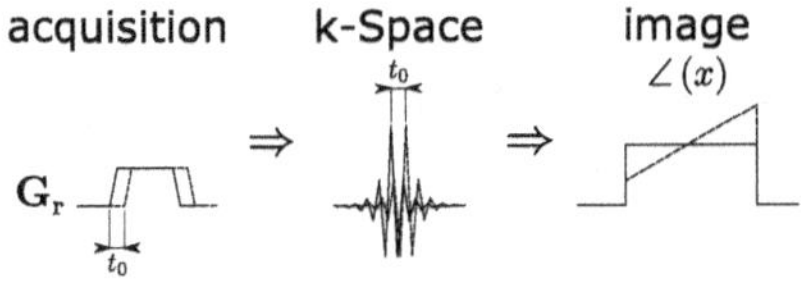

FIGURE 14.8 Effect of gradient delay in Cartesian imaging: k-space shifts propagate as a linear phase term into the image space according to the Fourier theorem

A gradient delay t_0 of the readout gradient G_r leads to an echo shift k_0 in k-space. According to the Fourier shift theorem, an echo shift k_0 in k-space leads to a linear phase along the readout direction in image space.

Therefore, the echo misalignment between opposite readout polarities can be estimated in k-space and corrected using the following optimization problem [68]:

$$k_0 = \arg\min_{k_0^*} \left\| \frac{\partial \left| \mathbf{F}\left(p^+ e^{(-i\pi k_0^* x)}\right) \right|}{\partial k} - \frac{\partial \left| \mathbf{F}\left(p^- e^{(+i\pi k_0^* x)}\right) \right|}{\partial k} \right\|_2, \qquad (14.24)$$

where p^+ and p^- are the reference measurements (complex 1D image space profiles) acquired with opposite readout polarities and $\mathbf{F}$ is the Fourier transform. In some cases, the precision of the estimation can be further improved by zero padding the reference measurements in image space before the Fourier transform is applied.

The corresponding linear phase $\Delta\Phi k_0^n$ that is a function of x in image space and the nth echo can be computed as

$$\Delta\Phi k_0^n(x) = -\pi k_0^n x \qquad (14.25)$$

and demodulated from the complex images.

Noticeable gradient delays typically occur when the immediate gradient waveform history of relevance for short-time eddy currents varies when comparing echoes. Typically, this can be the case when the first echo of a multi-echo gradient-echo train is compared with subsequent echoes of the train. An example of the effect on water–fat parameter estimation is given in Fig. 14.9.

14.5.2 Correction of concomitant gradients

According to Gauss's law of magnetism, a magnetic-field gradient played out along a spatial axis is accompanied by concomitant gradients in the other two axes. Historically, the concomitant gradient field was first described as a confounding factor at low field strengths [70,71] and was then also described at higher field strengths in the context of phase contrast angiography [72], echo planar imaging [73,74] fast spin-echo imaging [75], balanced steady-state free-precession imaging [76], spiral imaging [77], and diffusion tensor imaging [78]. The correction of the concomitant gradient field-induced phase accumulation $\Delta\Phi_C$ was also investigated in CSI evaluating the influence on the estimation of PDFF [68,69], T_2^* [69,79] and B_0 mapping [69]. The effect of the concomitant gradient field on the water–fat signal model differs depending on the employed acquisition scheme: the additional phase term adds up

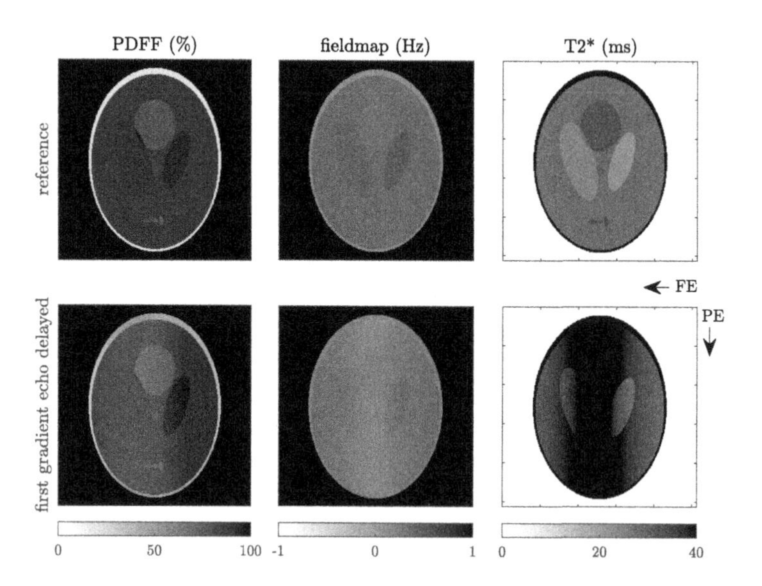

FIGURE 14.9 Exemplary parameter bias caused by a gradient delay of the first (out of six) gradient echo in Cartesian imaging

The upper row shows the reference parameter maps for PDFF, field-map, and T_2^* for a numerical phantom. The lower row shows the introduced bias (when compared to the upper row) by an echo delay of one k-space sample. Simulations were performed for a six-echo gradient-echo readout with a delta TE of 1 ms and first TE of 1 ms. This effect can be observed due to short-time eddy-current effects that are differently affecting the first gradient-echo compared to subsequent gradient-echoes in a multi-echo gradient-echo train. Frequency (FE) and phase encoding (PE) directions are indicated by black arrows.

to the field-map term for single TR measurements (where all echoes are acquired in a single TR) and causes a PDFF bias for interleaved acquisition schemes (where echoes are acquired over multiple TRs).

The phase term $\Delta\Phi_C$ can be approximated by an expression introduced by Bernstein et al. [72]

$$\Delta\Phi_C(x, y, z, t_n) = A(t_n) z^2 + B(t_n)\left(x^2 + y^2\right) + C(t_n) xz + D(t_n) yz, \tag{14.26}$$

where $A(t_n)$, $B(t_n)$, $C(t_n)$, and $D(t_n)$ are defined as

$$A(t_n) = \frac{\gamma}{2B_0}\int_0^{t_n} G_x(t)^2 + G_y(t)^2 dt \tag{14.27a}$$

$$B(t_n) = \frac{\gamma}{8B_0}\int_0^{t_n} G_z(t)^2 dt \tag{14.27b}$$

$$C(t_n) = -\frac{\gamma}{2B_0}\int_0^{t_n} G_x(t)G_z(t) dt \tag{14.27c}$$

$$D(t_n) = -\frac{\gamma}{2B_0}\int_0^{t_n} G_y(t)G_z(t) dt, \tag{14.27d}$$

where t is the time after excitation, γ is the gyromagnetic ratio, B_0 is the main magnetic field, and $G_x(t)$, $G_y(t)$, $G_z(t)$ denote the played out gradient waveforms in the physical coordinate system of the gradient system. As can be depicted from Eq. (14.27), the coefficients A and B arise from gradients G_x and G_y and G_z, respectively, whereas C and D are cross terms of the G_x and G_y with G_z, respectively. Therefore, the more gradient axes are used, the more complex the phase term $\Delta\Phi_C$ becomes.

Simplified examples of concomitant gradient-induced spatially dependent dephasing $\Delta\Phi_C$ for multi-echo gradient-echo readout in X, XZ and XYZ direction are given in Fig. 14.10, 14.11 and 14.12, respectively.

14.5.3 Proton density fat-fraction determination

After an optimal design of the experimental scenario based on CRLB analysis, an adequate field-map initialization and consideration of confounding factors, clinically important biomarkers can then be determined, e.g. tissue R_2^* or the proton density-fat fraction (PDFF). The PDFF can be estimated using either magnitude-valued or complex-valued signal-based techniques. While magnitude-based techniques are insensitive to phase errors (including gradient delays and concomitant gradients effects), complex-based methods are favored due to their superior noise performance.

The PDFF is typically computed by the magnitude-discrimination method [80] for complex-valued signal-based techniques as

$$\text{PDFF} = \begin{cases} 1 - \frac{|W|}{|W+F|} & \text{for } |W| > |F| \\ \frac{|F|}{|W+F|} & \text{for } |W| \leq |F|, \end{cases} \tag{14.28}$$

which is generally more noise robust against phase errors in the input images than simply computing the PDFF as $|F|/(|W| + |F|)$.

14.6 Current challenges and future directions

Despite the large growth of water–fat separation techniques over the last two decades, some exciting future directions exist for combining the method with many of the concepts emerging especially within the arena of quantitative imaging.

First, efficient fat suppression remains a challenge in acquisitions employing non-steady-state pulse sequences and requiring simulation of spin-history effects. The adoption of chemical shift encoding-based water–fat separation in such acquisitions is an emerging field. Techniques that employ nonsteady-state sequences and rely on undersampled non-Cartesian acquisitions, like Magnetic Resonance Fingerprinting (MRF), can be prone to fat blurring and can particularly benefit from Dixon imaging in achieving efficient fat suppression in body applications [81,82].

Second, water–fat separation has also been increasingly applied in quantitative imaging of multiple body tissues. Specifically, a quantitative imaging approach that aims to quantify the properties of the water component requires the application of efficient water–fat separation approaches to assure complete suppression of the fat signal [83–85].

Third, quantitative imaging of lipids and adipose tissue remains a major area of application of water–fat separation subject to growing interest. After the consideration of confounding effects, fat

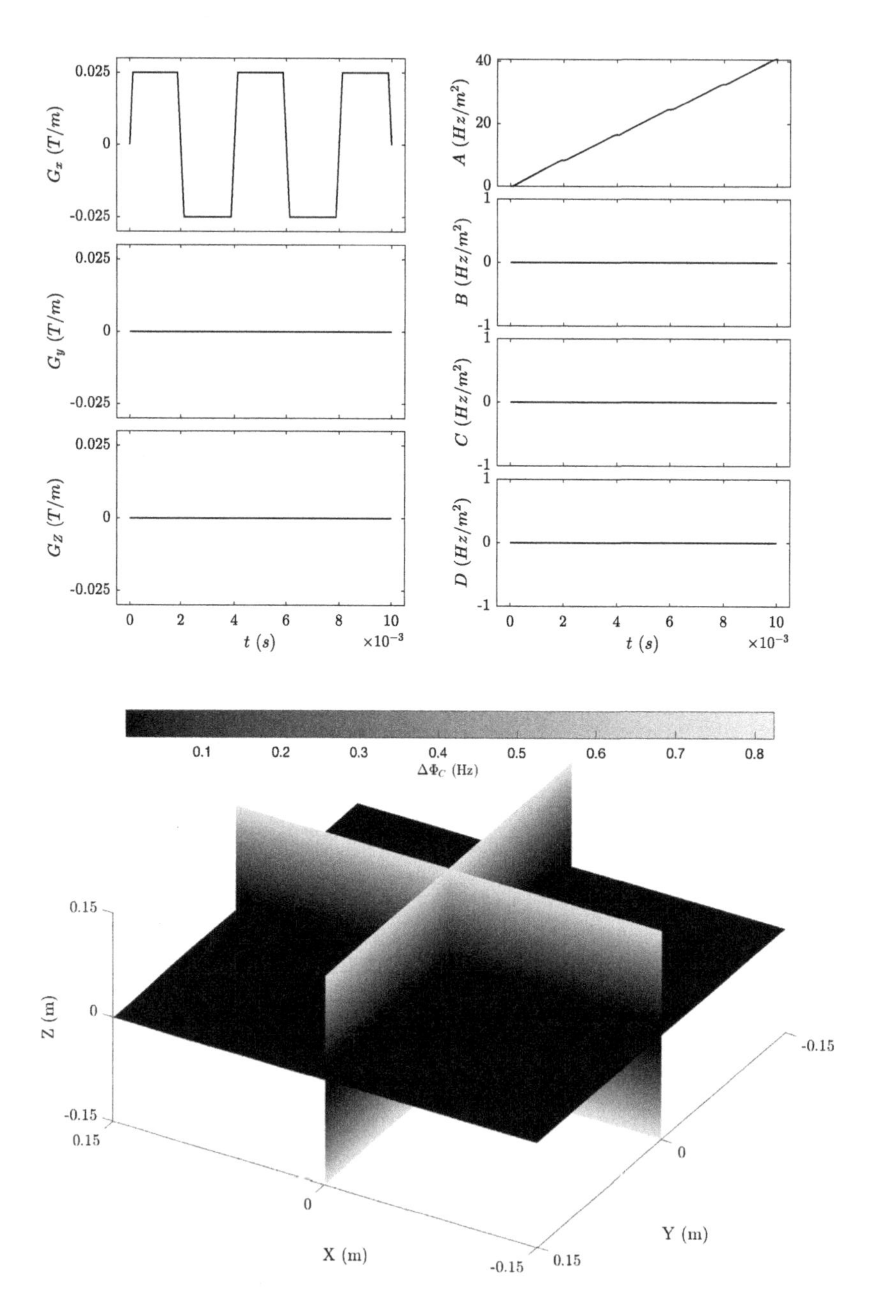

FIGURE 14.10 Spatial dependent dephasing due to concomitant gradients for readout gradients played out in X

Gradient waveforms G_x, G_y and G_z (represented in the scanner coordinate system XYZ), derived concomitant gradient coefficients A, B, C and D and resulting spatially dependant dephasing $\Delta\Phi_C$. The concomitant gradient coefficients A, B, C and D, and $\Delta\Phi_C$ were calculated assuming $B_0 = 3$ Tesla. $\Delta\Phi_C$ is displayed for the 5th echo.

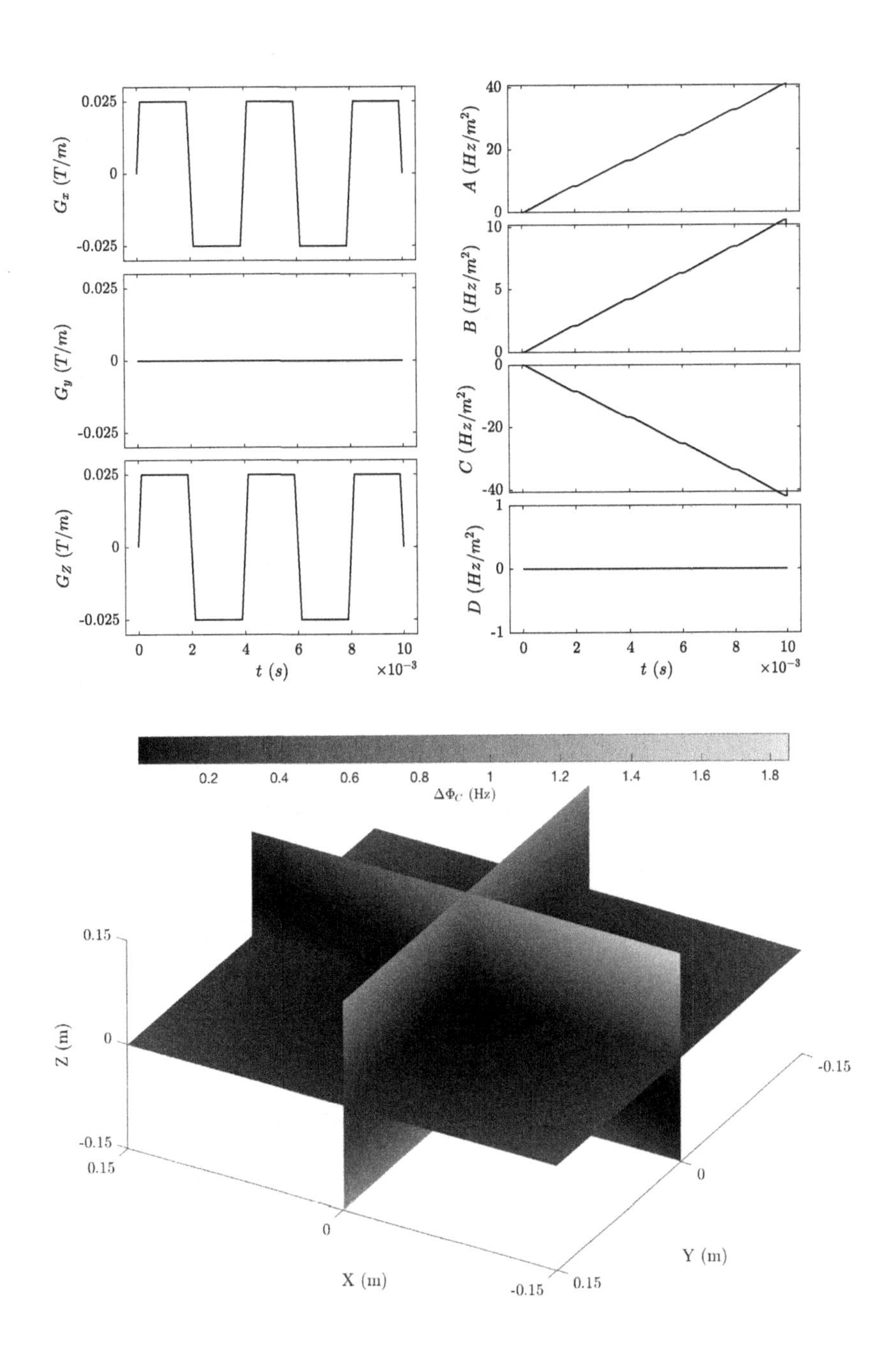

FIGURE 14.11 Spatially dependent dephasing due to concomitant gradients for readout gradients played out in XZ

Gradient waveforms G_x, G_y, and G_z (represented in the scanner coordiante system XYZ), derived concomitant gradient coefficients A, B, C and D and resulting spatially dependent dephasing $\Delta\Phi_C$. The concomitant gradient coefficients A, B, C, D and $\Delta\Phi_C$ were calculated assuming $B_0 = 3$ Tesla. $\Delta\Phi_C$ is displayed for the 5th echo.

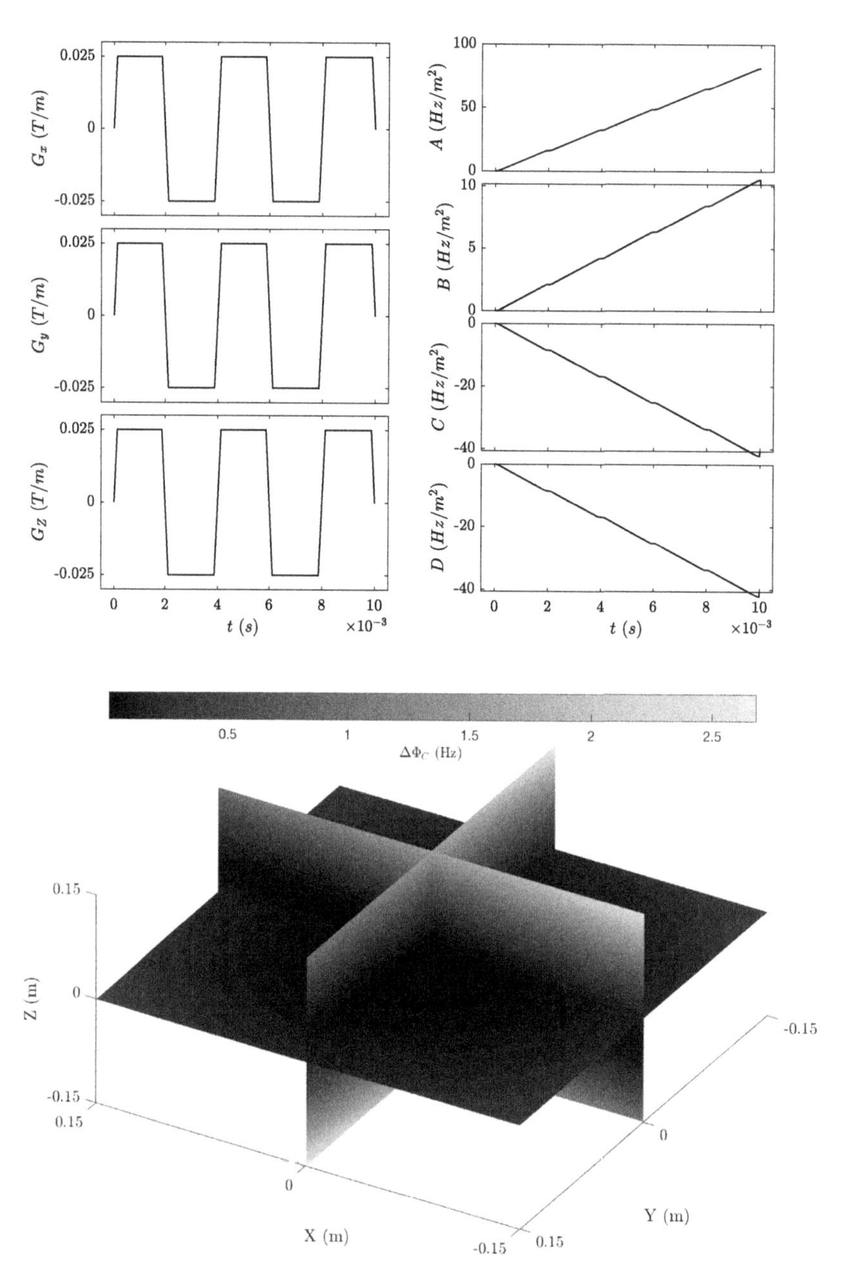

FIGURE 14.12 Spatially dependent dephasing due to concomitant gradients for readout gradients played out in XYZ

Gradient waveforms G_x, G_y, and G_z (represented in the scanner coordiante system XYZ), derived concomitant gradient coefficients A, B, C and D and resulting spatially dependent dephasing $\Delta\Phi_C$. The concomitant gradient coefficients A, B, C, D, and $\Delta\Phi_C$ were calculated assuming $B_0 = 3$ Tesla. $\Delta\Phi_C$ is displayed for the 5th echo.

quantification is nowadays a well-established technique for the liver [86]. However, a consensus on best practices and protocols for fat quantification beyond the liver is needed for the community [4].

14.7 Summary

Solving the water–fat separation problem requires primarily the treatment of a nonconvex optimization problem and the reliable estimation of the field-map. However, water–fat separation in general constitutes an image reconstruction problem that relies on a signal model that has to consider both the underlying data acquisition and spatial encoding process. A variety of signal models have been proposed for water–fat separation. Their complexity increases for the needs of PDFF quantification, as well as with the need for fat deblurring when applying water–separation in non-Cartesian acquisitions.

14.8 Further reading

In 2012, the International Society of Magnetic Resonance in Medicine (ISMRM) organized a scientific workshop on water–fat separation [87] where W. Thomas Dixon gave an enjoyable talk on how it all began. In addition, the workshop website is a good source to review some more detailed concepts and to get some hands-on practice with Matlab® code that was made available (ISMRM water–fat toolbox), together with example data.[1]

References

[1] Dixon WT. Simple proton spectroscopic imaging. Radiology 1984;153(1):189–94. https://doi.org/10.1148/radiology.153.1.6089263.

[2] Eggers H, Börnert P. Chemical shift encoding-based water–fat separation methods. J Magn Reson Imaging 2014;40(2):251–68. https://doi.org/10.1002/jmri.24568.

[3] Reeder SB, Hu HH, Sirlin CB. Proton density fat-fraction: a standardized mr-based biomarker of tissue fat concentration. J Magn Reson Imaging 2012;36(5):1011–4. https://doi.org/10.1002/jmri.23741.

[4] Hu HH, Branca RT, Hernando D, Karampinos DC, Machann J, McKenzie CA, et al. Magnetic resonance imaging of obesity and metabolic disorders: summary from the 2019 ISMRM Workshop. Magn Reson Med 2020;83:1565–76. https://doi.org/10.1002/mrm.28103.

[5] Schenck JF. The role of magnetic susceptibility in magnetic resonance imaging: Mri magnetic compatibility of the first and second kinds. Med Phys 1996;23(6):815–50. https://doi.org/10.1118/1.597854.

[6] Levitt MH. Demagnetization field effects in two-dimensional solution nmr. Concepts Magn Reson 1996;8(2):77–103. https://doi.org/10.1002/(sici)1099-0534(1996)8:2<77::aid-cmr1>3.0.co;2-l.

[7] Lenz E. Ueber die bestimmung der richtung der durch elektrodynamische vertheilung erregten galvanischen ströme. Ann Phys Chem 1834;107(31):483–94. https://doi.org/10.1002/andp.18341073103.

[8] Levitt MH. The signs of frequencies and phases in nmr. J Magn Res 1997;126(2):164–82. https://doi.org/10.1006/jmre.1997.1161.

[1] https://www.ismrm.org/workshops/FatWater12/data.htm.

[9] Yu H, Shimakawa A, Hines CDG, McKenzie CA, Hamilton G, Sirlin CB, et al. Combination of complex-based and magnitude-based multiecho water–fat separation for accurate quantification of fat-fraction. Magn Reson Med 2011;66(1):199–206. https://doi.org/10.1002/mrm.22840.
[10] Yu H, Shimakawa A, McKenzie CA, Brodsky E, Brittain JH, Reeder SB. Multiecho water–fat separation and simultaneous r2* estimation with multifrequency fat spectrum modeling. Magn Reson Med 2008;60(5):1122–34. https://doi.org/10.1002/mrm.21737.
[11] Golub G, Pereyra V. Separable nonlinear least squares: the variable projection method and its applications. Inverse Probl 2003;19(2):R1–26. https://doi.org/10.1088/0266-5611/19/2/201.
[12] Diefenbach MN, Liu C, Karampinos DC. Generalized parameter estimation in multi-echo gradient-echo-based chemical species separation. Quant Imaging Med Surg 2020;10(3):554–67. https://doi.org/10.21037/qims.2020.02.07.
[13] Ma J. Dixon techniques for water and fat imaging. J Magn Reson Imaging 2008;28(3):543–58. https://doi.org/10.1002/jmri.21492.
[14] Sharma SD, Artz NS, Hernando D, Horng DE, Reeder SB. Improving chemical shift encoded water–fat separation using object-based information of the magnetic field inhomogeneity. Magn Reson Med 2014;73(2):597–604. https://doi.org/10.1002/mrm.25163.
[15] Diefenbach MN, Ruschke S, Eggers H, Meineke J, Rummeny EJ, Karampinos DC. Improving chemical shift encoding-based water–fat separation based on a detailed consideration of magnetic field contributions. Magn Reson Med 2018;80(3):990–1004. https://doi.org/10.1002/mrm.27097.
[16] Kolmogorov V, Rother C. Minimizing nonsubmodular functions with graph cuts-a review. IEEE Trans Pattern Anal Mach Intell 2007;29(7):1274–9. https://doi.org/10.1109/tpami.2007.1031.
[17] Hernando D, Kellman P, Haldar JP, Liang ZP. Robust water/fat separation in the presence of large field inhomogeneities using a graph cut algorithm. Magn Reson Med 2010;63(1):79–90. https://doi.org/10.1002/mrm.22177.
[18] Dong J, Liu T, Chen F, Zhou D, Dimov A, Raj A, et al. Simultaneous phase unwrapping and removal of chemical shift (spurs) using graph cuts: application in quantitative susceptibility mapping. IEEE Trans Med Imaging 2015;34(2):531–40. https://doi.org/10.1109/tmi.2014.2361764.
[19] Berglund J, Skorpil M. Multi-scale graph-cut algorithm for efficient water–fat separation. Magn Reson Med 2016;78(3):941–9. https://doi.org/10.1002/mrm.26479.
[20] Cui C, Wu X, Newell JD, Jacob M. Fat water decomposition using globally optimal surface estimation (goose) algorithm. Magn Reson Med 2014;73(3):1289–99. https://doi.org/10.1002/mrm.25193.
[21] Glover GH, Schneider E. Three-point Dixon technique for true water/fat decomposition with B0 inhomogeneity correction. Magn Reson Med 1991;18(2):371–83. https://doi.org/10.1002/mrm.1910180211.
[22] Reeder SB, Pineda AR, Wen Z, Shimakawa A, Yu H, Brittain JH, et al. Iterative decomposition of water and fat with echo asymmetry and least-squares estimation (IDEAL): Application with fast spin-echo imaging. Magn Reson Med 2005;54(3):636–44. https://doi.org/10.1002/mrm.20624.
[23] Bydder M, Yokoo T, Hamilton G, Middleton MS, Chavez AD, Schwimmer JB, et al. Relaxation effects in the quantification of fat using gradient echo imaging. Magn Reson Imaging 2008;26(3):347–59. https://doi.org/10.1016/j.mri.2007.08.012.
[24] Yu H, Reeder SB, Shimakawa A, Brittain JH, Pelc NJ. Field map estimation with a region growing scheme for iterative 3-point water–fat decomposition. Magn Reson Med 2005;54(4):1032–9. https://doi.org/10.1002/mrm.20654.
[25] Lu W, Hargreaves BA. Multiresolution field map estimation using golden section search for water–fat separation. Magn Reson Med 2008;60(1):236–44. https://doi.org/10.1002/mrm.21544.
[26] Hernando D, Haldar JP, Sutton BP, Ma J, Kellman P, Liang ZP. Joint estimation of water/fat images and field inhomogeneity map. Magn Reson Med 2008;59(3):571–80. https://doi.org/10.1002/mrm.21522.
[27] Cui C, Shah A, Wu X, Jacob M. A rapid 3d fat-water decomposition method using globally optimal surface estimation (r-goose). Magn Reson Med 2017;79(4):2401–7. https://doi.org/10.1002/mrm.26843.

[28] Boehm C, Diefenbach MN, Makowski MR, Karampinos DC. Improved body quantitative susceptibility mapping by using a variable-layer single-min-cut graph-cut for field-mapping. Magn Reson Med 2021;85(3):1697–712. https://doi.org/10.1002/mrm.28515.

[29] Bechler E, Stabinska J, Wittsack H. Analysis of different phase unwrapping methods to optimize quantitative susceptibility mapping in the abdomen. Magn Reson Med 2019;82(6):2077–89. https://doi.org/10.1002/mrm.27891.

[30] Cavassila S, Deval S, Huegen C, van Ormondt D, Graveron-Demilly D. Cramér-Rao bounds: an evaluation tool for quantitation. NMR Biomed 2001;14(4):278–83. https://doi.org/10.1002/nbm.701.

[31] van den Bos A. A Cramér-Rao lower bound for complex parameters. IEEE Trans Signal Process 1994;42(10):2859. https://doi.org/10.1109/78.324755.

[32] Jauffret C. Observability and Fisher information matrix in nonlinear regression. IEEE Trans Aerosp Electron Syst 2007;43(2):756–9. https://doi.org/10.1109/taes.2007.4285368.

[33] Scharf L, McWhorter L. Geometry of the Cramer-Rao bound. Signal Processing 1993;31(3):301–11. https://doi.org/10.1016/0165-1684(93)90088-R.

[34] Pineda AR, Reeder SB, Wen Z, Pelc NJ. Cramér-Rao bounds for three-point decomposition of water and fat. Magn Reson Med 2005;54(3):625–35. https://doi.org/10.1002/mrm.20623.

[35] Glover GH, Pauly JM. Projection reconstruction techniques for reduction of motion effects in MRI. Magn Reson Med 1992;28(2):275–89. https://doi.org/10.1002/mrm.1910280209.

[36] Chandarana H, Block TK, Rosenkrantz AB, Lim RP, Kim D, Mossa DJ, et al. Free-breathing radial 3D fat-suppressed T1-weighted gradient echo sequence: a viable alternative for contrast-enhanced liver imaging in patients unable to suspend respiration. Invest Radiol 2011;46(10):648–53. https://doi.org/10.1097/RLI.0b013e31821eea45.

[37] Liao JR, Pauly JM, Brosnan TJ, Pelc NJ. Reduction of motion artifacts in cine MRI using variable-density spiral trajectories. Magn Reson Med 1997;37(4):569–75. https://doi.org/10.1002/mrm.1910370416.

[38] Lauzon ML, Rutt BK. Effects of polar sampling in k-space. Magn Reson Med 1996;36(6):940–9. https://doi.org/10.1002/mrm.1910360617.

[39] Lauzon ML, Rutt BK. Polar sampling in k-space: reconstruction effects. Magn Reson Med 1998;40(5):769–82. https://doi.org/10.1002/mrm.1910400519.

[40] Larson AC, Kellman P, Arai A, Hirsch GA, McVeigh E, Li D, et al. Preliminary investigation of respiratory self-gating for free-breathing segmented cine MRI. Magn Reson Med 2005;53(1):159–68. https://doi.org/10.1002/mrm.20331.

[41] Liu J, Spincemaille P, Codella NC, Nguyen TD, Prince MR, Wang Y. Respiratory and cardiac self-gated free-breathing cardiac CINE imaging with multiecho 3D hybrid radial SSFP acquisition. Magn Reson Med 2010;63(5):1230–7. https://doi.org/10.1002/mrm.22306.

[42] Benkert T, Feng L, Sodickson DK, Chandarana H, Block KT. Free-breathing volumetric fat/water separation by combining radial sampling, compressed sensing, and parallel imaging. Magn Reson Med 2017;78(2):565–76. https://doi.org/10.1002/mrm.26392.

[43] Armstrong T, Dregely I, Stemmer A, Han F, Natsuaki Y, Sung K, et al. Free-breathing liver fat quantification using a multiecho 3D stack-of-radial technique. Magn Reson Med 2018;79(1):370–82. https://doi.org/10.1002/mrm.26693.

[44] Feng L, Axel L, Chandarana H, Block KT, Sodickson DK, Otazo R. XD-GRASP: golden-angle radial MRI with reconstruction of extra motion-state dimensions using compressed sensing. Magn Reson Med 2016;75(2):775–88. https://doi.org/10.1002/mrm.25665.

[45] Moran CJ, Brodsky EK, Bancroft LH, Reeder SB, Yu H, Kijowski R, et al. High-resolution 3D radial bSSFP with IDEAL. Magn Reson Med 2014;71(1):95–104. https://doi.org/10.1002/mrm.24633.

[46] Brodsky EK, Holmes JH, Yu H, Reeder SB. Generalized K-space decomposition with chemical shift correction for non-Cartesian water-fat imaging. Magn Reson Med 2008;59(5):1151–64. https://doi.org/10.1002/mrm.21580.

[47] Börnert P, Koken P, Eggers H. Spiral water-fat imaging with integrated off-resonance correction on a clinical scanner. J Magn Reson Imaging 2010;32(5):1262–7. https://doi.org/10.1002/jmri.22336.
[48] Moriguchi H, Lewin JS, Duerk JL. Dixon techniques in spiral trajectories with off-resonance correction: a new approach for fat signal suppression without spatial-spectral RF pulses. Magn Reson Med 2003;50(5):915–24. https://doi.org/10.1002/mrm.10629.
[49] Wang D, Zwart NR, Li Z, Schär M, Pipe JG. Analytical three-point Dixon method: with applications for spiral water-fat imaging. Magn Reson Med 2016;75(2):627–38. https://doi.org/10.1002/mrm.25620.
[50] Weng D, Pan Y, Zhong X, Zhuo Y. Water-fat separation with parallel imaging based on BLADE. Magn Reson Imaging 2013;31(5):656–63. https://doi.org/10.1016/j.mri.2012.10.018.
[51] Huo D, Li Z, Aboussouan E, Karis JP, Pipe JG. Turboprop IDEAL: a motion-resistant fat-water separation technique. Magn Reson Med 2009;61(1):188–95. https://doi.org/10.1002/mrm.21825.
[52] Wu HH, Jin HL, Nishimura DG. Fat/water separation using a concentric rings trajectory. Magn Reson Med 2009;61(3):639–49. https://doi.org/10.1002/mrm.21865.
[53] Brown RW, Cheng YCN, Haacke EM, Thompson MR, Venkatesan R. Magnetic resonance imaging: physical principles and sequence design. second edition. ISBN 9781118633953, 2014.
[54] Meyer CH, Pauly JM, Macovskiand A, Nishimura DG. Simultaneous spatial and spectral selective excitation. Magn Reson Med 1990;15(2):287–304. https://doi.org/10.1002/mrm.1910150211.
[55] Levin YS, Mayer D, Yen YF, Hurd RE, Spielman DM. Optimization of fast spiral chemical shift imaging using least squares reconstruction: application for hyperpolarized 13C metabolic imaging. Magn Reson Med 2007;58(2):245–52. https://doi.org/10.1002/mrm.21327.
[56] Lustig M, Donoho DL, Santos JM, Pauly JM. Compressed sensing MRI. IEEE Signal Process Mag 2008;25(2):72–82. https://doi.org/10.1109/MSP.2007.914728.
[57] Lu W, Yu H, Shimakawa A, Alley M, Reeder SB, Hargreaves BA. Water–fat separation with bipolar multi-echo sequences. Magn Reson Med 2008;60(1):198–209. https://doi.org/10.1002/mrm.21583.
[58] Yu H, Shimakawa A, McKenzie CA, Lu W, Reeder SB, Hinks RS, et al. Phase and amplitude correction for multi-echo water-fat separation with bipolar acquisitions. J Magn Reson Imaging 2010;31(5):1264–71. https://doi.org/10.1002/jmri.22111.
[59] Yu H, Shimakawa A, Hines CDG, McKenzie CA, Hamilton G, Sirlin CB, et al. Combination of complex-based and magnitude-based multiecho water-fat separation for accurate quantification of fat-fraction. Magn Reson Med 2011;66(1):199–206. https://doi.org/10.1002/mrm.22840.
[60] Hernando D, Hines CDG, Yu H, Reeder SB. Addressing phase errors in fat-water imaging using a mixed magnitude/complex fitting method. Magn Reson Med 2012;67(3):638–44. https://doi.org/10.1002/mrm.23044.
[61] Peterson P, Mansson S. Fat quantification using multiecho sequences with bipolar gradients: investigation of accuracy and noise performance. Magn Reson Med 2014;71(1):219–29. https://doi.org/10.1002/mrm.24657.
[62] Yu H, Yu H, McKenzie CA, Shimakawa A, Shimakawa A, Vu AT, et al. Multiecho reconstruction for simultaneous water-fat decomposition and T2* estimation. J Magn Reson Imaging 2007;26(4):1153–61. https://doi.org/10.1002/jmri.21090.
[63] Liu CY, McKenzie CA, Yu H, Brittain JH, Reeder SB. Fat quantification with IDEAL gradient echo imaging: correction of bias from T1 and noise. Magn Reson Med 2007;58(2):354–64. https://doi.org/10.1002/mrm.21301.
[64] Karampinos DC, Yu H, Yu H, Shimakawa A, Shimakawa A, Link TM, et al. T_1-corrected fat quantification using chemical shift-based water/fat separation: application to skeletal muscle. Magn Reson Med 2011;66(5):1312–26. https://doi.org/10.1002/mrm.22925.
[65] Yu H, Shimakawa A, McKenzie CA, Brodsky E, Brittain JH, Reeder SB. Multiecho water-fat separation and simultaneous R2* estimation with multifrequency fat spectrum modeling. Magn Reson Med 2008;60(5):1122–34. https://doi.org/10.1002/mrm.21737.
[66] Karampinos DC, Yu H, Shimakawa A, Link TM, Majumdar S. Chemical shift-based water/fat separation in the presence of susceptibility-induced fat resonance shift. Magn Reson Med 2012;68(5):1495–505. https://doi.org/10.1002/mrm.24157.

[67] Hernando D, Sharma SD, Kramer H, Reeder SB. On the confounding effect of temperature on chemical shift-encoded fat quantification. Magn Reson Med 2014;72(2):464–70. https://doi.org/10.1002/mrm.24951.

[68] Ruschke S, Eggers H, Kooijman-Kurfuerst H, Diefenbach MN, Baum T, Haase A, et al. Correction of phase errors in quantitative water–fat imaging using a monopolar time-interleaved multi-echo gradient echo sequence. Magn Reson Med 2017;78(3):984–96. https://doi.org/10.1002/mrm.26485.

[69] Colgan TJ, Hernando D, Sharma SD, Reeder SB. The effects of concomitant gradients on chemical shift encoded MRI. Magn Reson Med 2017;78(2):730–8. https://doi.org/10.1002/mrm.26461.

[70] Norris DG, Hutchison JMS. Concomitant magnetic field gradients and their effects on imaging at low magnetic field strengths. Magn Reson Imaging 1990;8(1):33–7. https://doi.org/10.1016/0730-725X(90)90209-K.

[71] Volegov PL, Mosher JC, Espy MA, Kraus RH. On concomitant gradients in low-field MRI. J Magn Res 2005;175(1):103–13. https://doi.org/10.1016/j.jmr.2005.03.015.

[72] Bernstein MA, Zhou XJ, Polzin JA, King KF, Ganin A, Pelc NJ, et al. Concomitant gradient terms in phase contrast MR: analysis and correction. Magn Reson Med 1998;39(2):300–8. https://doi.org/10.1002/mrm.1910390218.

[73] Zhou XJ, Du YP, Bernstein MA, Reynolds HG, Maier JK, Polzin JA. Concomitant magnetic-field-induced artifacts in axial echo planar imaging. Magn Reson Med 1998;39(4):596–605. https://doi.org/10.1002/mrm.1910390413.

[74] Du YP, Joe Zhou X, Bernstein MA. Correction of concomitant magnetic field-induced image artifacts in nonaxial echo-planar imaging. Magn Reson Med 2002;48(3):509–15. https://doi.org/10.1002/mrm.10249.

[75] Zhou XJ, Tan SG, Bernstein MA. Artifacts induced by concomitant magnetic field in fast spin-echo imaging. Magn Reson Med 1998;40(4):582–91. https://doi.org/10.1002/mrm.1910400411.

[76] Sica CT, Meyer CH. Concomitant gradient field effects in balanced steady-state free precession. Magn Reson Med 2007;57(4):721–30. https://doi.org/10.1002/mrm.21183.

[77] King KF, Ganin A, Zhou XJ, Bernstein MA. Concomitant gradient field effects in spiral scans. Magn Reson Med 1999;41(1):103–12. https://doi.org/10.1002/(sici)1522-2594(199901)41:1<103::aid-mrm15>3.3.co;2-d.

[78] Baron CA, Lebel RM, Wilman AH, Beaulieu C. The effect of concomitant gradient fields on diffusion tensor imaging. Magn Reson Med 2012;68(4):1190–201. https://doi.org/10.1002/mrm.24120.

[79] Hofstetter LW, Morrell G, Kaggie J, Kim D, Carlston K, Lee VS. T2* measurement bias due to concomitant gradient fields. Magn Reson Med 2017;77(4):1562–72. https://doi.org/10.1002/mrm.26240.

[80] Liu CY, McKenzie CA, Yu H, Brittain JH, Reeder SB. Fat quantification with ideal gradient echo imaging: correction of bias from t1 and noise. Magn Reson Med 2007;58(2):354–64. https://doi.org/10.1002/mrm.21301.

[81] Koolstra K, Webb AG, Veeger TTJ, Kan HE, Koken P, Börnert P. Water–fat separation in spiral magnetic resonance fingerprinting for high temporal resolution tissue relaxation time quantification in muscle. Magn Reson Med 2020;84(2):646–62. https://doi.org/10.1002/mrm.28143.

[82] Nolte T, Gross-Weege N, Doneva M, Koken P, Elevelt A, Truhn D, et al. Spiral blurring correction with water–fat separation for magnetic resonance fingerprinting in the breast. Magn Reson Med 2020;83(4):1192–207. https://doi.org/10.1002/mrm.27994.

[83] Hernando D, Karampinos D, King KF, Haldar JP, Majumdar S, Georgiadis JG, et al. Removal of olefinic fat chemical shift artifact in diffusion MRI. Magn Reson Med 2011;65(3):692–701. https://doi.org/10.1002/mrm.22670.

[84] Janiczek RL, Gambarota G, Sinclair CDJ, Yousry TA, Thornton JS, Golay X, et al. Simultaneous T2 and lipid quantitation using IDEAL-CPMG. Magn Reson Med 2011;66(5):1293–302. https://doi.org/10.1002/mrm.22916.

[85] Dieckmeyer M, Ruschke S, Eggers H, Kooijman-Kurfuerst H, Rummeny EJ, Kirschke JS, et al. ADC quantification of the vertebral bone marrow water component: removing the confounding effect of residual fat. Magn Reson Med 2017;78(4):1432–41. https://doi.org/10.1002/mrm.26550.

[86] Yokoo T, Serai SD, Pirasteh A, Bashir MR, Hamilton G, Hernando D, et al. Linearity, bias, and precision of hepatic proton density fat fraction measurements by using MR imaging: a meta-analysis. Radiology 2018;286(2):486–98. https://doi.org/10.1148/radiol.2017170550.
[87] Hu HH, Börnert P, Hernando D, Kellman P, Ma J, Jingfei, et al. ISMRM workshop on fat–water separation: insights, applications and progress in MRI. Magn Reson Med 2012;68(2):378–88. https://doi.org/10.1002/mrm.24369.

CHAPTER

15 Model-Based Parametric Mapping Reconstruction

Christoph Kolbitsch, Kirsten Kerkering, and Tobias Schaeffter
Physikalisch-Technische Bundesanstalt (PTB), Braunschweig and Berlin, Germany

15.1 Introduction

The aim of parameter mapping is to determine quantitative physical and/or biophysical parameters ($\boldsymbol{\nu}$) that provide diagnostic information. This chapter will focus on the estimation of MR relaxation times (T_1, T_2, T_2^*...), but the methods discussed here can in principle also be applied to other approaches for quantitative parameter estimation such as model-based fat–water estimation (see Chapter 14).

Relaxation times are determined by quantum mechanical properties of nuclear spins. Even before the first MR images were acquired, it was shown that pathological changes of tissue can lead to changes of these relaxation times [1,2]. Nowadays, mapping of relaxation times is used in a wide range of various clinical applications. T_1 can be used to detect and characterize tumors or to assess changes in the brain due to epilepsy or multiple sclerosis. Changes of T_1 times in the myocardium can be an indication for the presence of fibrosis or fat in the heart muscle. Inflammatory processes can increase the amount of water retained in tissue and thus lead to changes of T_2. T_2^* mapping can be used to detect iron depositions in tissue. In addition to the parameters that are of diagnostic interest, often additional parameters such as the spin density ρ also need to be estimated. Therefore, $\boldsymbol{\nu} = [\boldsymbol{\nu}_1, ...\boldsymbol{\nu}_M]$ where each $\boldsymbol{\nu}_i$ represents a spatial map of one parameter such as T_1 or ρ.

The main advantage of using mapping approaches for clinical applications is that they provide quantitative parameters that are easy to compare between different scans, scanners, patients, and institutions. The standard qualitative images (e.g., T_1 or T_2 weighed images), on the other hand, strongly depend on hardware and sequence specific parameters, which can make a comparison very challenging. As already mentioned, relaxation times are not biophysical parameters such as blood flow but are determined by the nuclear spins. Therefore, there can be a dependency of $\boldsymbol{\nu}$ on the field strength of the MR scanner. Nevertheless, clinical scanners operate at well-defined field strengths of 1.5T or 3T, making this less of a problem.

The main challenges of parameter mapping are long acquisition times and more complex signal processing compared to qualitative MRI. Section 15.2 introduces some basic aspects of MR data acquisition for parameter mapping, but the main focus of this chapter will be on image reconstruction. For more information on data acquisition, please refer, e.g., to [3].

In order to be able to obtain parameters ν_i from acquired k-space data $\mathbf{s}$, the data acquisition has to be sensitive to changes in ν_i, i.e., different values for ν_i have to lead to different k-space data $\mathbf{s}$. This is commonly achieved by using preparation pulses (e.g., inversion, saturation or T_2-sensitive pulses) that prepare the nuclear spin system for the subsequent data acquisition. Another option is to vary

Advances in Magnetic Resonance Technology and Applications, Volume 7, ISSN 2666-9099. https://doi.org/10.1016/B978-0-12-822726-8.00026-9

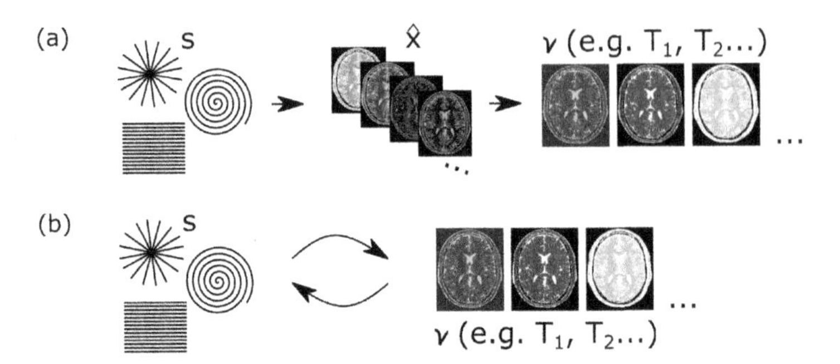

FIGURE 15.1 Parameter mapping

(**a**) For image-based parameter mapping, a set of qualitative images $\hat{\mathbf{x}}$ are reconstructed, and in a second step the quantitative parameters $\boldsymbol{\nu}$ are estimated, which can represent T_1, T_2 or other parameters that are unknowns in the applied model q such as the flip angle. (**b**) For reconstruction-based mapping, $\boldsymbol{\nu}$ is obtained iteratively from the acquired k-space data $\mathbf{s}$. This ensures that the obtained $\boldsymbol{\nu}$ are consistent with $\mathbf{s}$.

MR sequence parameters (e.g., flip angle (α), echo time (T_E) or repetition time (T_R)) during data acquisition. The effect of the MR sequence on the image signal $\mathbf{x}$ given a certain set of parameters $\boldsymbol{\nu}$ can be described by a signal model q. The acquired k-space data is therefore given by

$$\mathbf{s} = \mathbf{E}q(\boldsymbol{\nu}) \quad \text{with} \quad q : \boldsymbol{\nu} \mapsto q(\boldsymbol{\nu}), \tag{15.1}$$

and, where the encoding operator $\mathbf{E}$ describes the entire MR acquisition process including coil sensitivity weighting, Fourier encoding and k-space sampling (see Chapter 2).

Parameter mapping, i.e., the estimation of $\boldsymbol{\nu}$ given $\mathbf{s}$, can be carried out in two different ways (Fig. 15.1). The first approach, which we will refer to here as *image-based mapping*, separates Eq. (15.1) into two subproblems: image reconstruction and parameter estimation. First, images $\hat{\mathbf{x}}$ are reconstructed from $\mathbf{s}$. Each image shows a different image contrast due to the underlying differences of the parameter of interest $\boldsymbol{\nu}$, e.g., to estimate T_1, $\hat{\mathbf{x}}$ can be images at different time points after an inversion pulse or for T_2 mapping, $\hat{\mathbf{x}}$ can be images acquired after different T_2-sensitive preparation pulses. In an independent second step $\boldsymbol{\nu}$ is estimated from the reconstructed images $\hat{\mathbf{x}}$ using an appropriate signal model q. The second class of techniques, solves Eq. (15.1) directly (*reconstruction-based mapping*). These two methods are discussed in Sections 15.3 and 15.4 in more detail.

15.2 MR mapping sequences

In order to be able to estimate $\boldsymbol{\nu}$ from the acquired MR data, the data acquisition has to be sensitive to $\boldsymbol{\nu}$, i.e., different combinations of $\boldsymbol{\nu}$ need to result in different MR signals. Due to the flexibility of MRI, there is a wide range of approaches to how the MR data acquisition can be made sensitive to T_1, T_2... or a combination of these parameters.

Fig. 15.2 shows simulated signal curves (i.e., longitudinal magnetization) of three different species (i.e., T_1 - T_2 combinations) for a gradient-echo sequence. For constant sequence parameters (i.e., T_R,

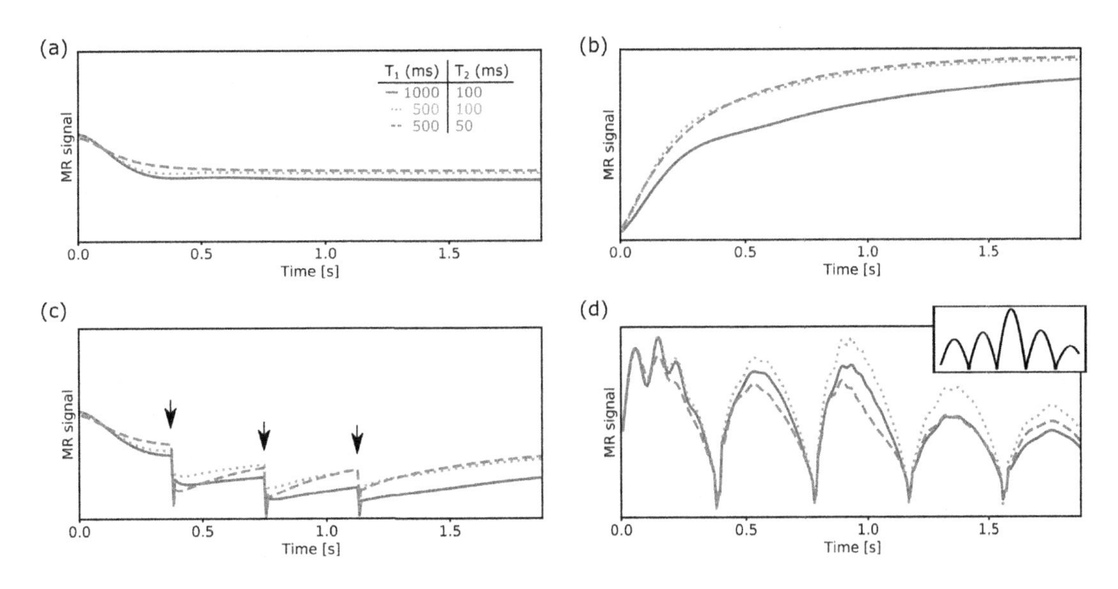

FIGURE 15.2 Signal curves for different MR sequences

Signal curves were simulated for three different species (i.e., T_1 - T_2 combinations). (**a**) Without any pre-pulses or sequence parameter modulation, the species yield very similar signals. (**b**) An inversion pulse prior to the data acquisition makes the sequence dependent on T_1. (**c**) T_2-sensitive preparation pulses (black arrows) lead to T_2 dependency. (**d**) Magnetic resonance fingerprinting uses, e.g., a varying flip-angle pattern (small figure insert) to achieve both T_1 and T_2 dependency. All simulations were carried out using an extended phase graph algorithm assuming an unspoiled gradient-echo sequence.

T_E and α), all species yield very similar signal curves and cannot be distinguished (Fig. 15.2a). For T_1 mapping, commonly an inversion pre-pulse or saturation pre-pulse is applied and the signal is then acquired during T_1 relaxation (Fig. 15.2b). Another option is to use different flip angles (variable flip-angle approach) in order to achieve signal variation as a function of T_1. Fig. 15.2c shows T_2 dependency achieved by multiple T_2-sensitive preparation pulses, e.g., a combination of 90°–180°–180°–90° pulses, which lead to T_2 weighting of transversal magnetization which is then flipped back to the longitudinal direction and can be measured. T_2 mapping can also be carried out using spin-echo acquisitions at different echo times and similarly multi-echo gradient-echo acquisitions can be used for T_2^* mapping.

The examples discussed so far aim at achieving MR signal dependency mainly on a single parameter. A multi-echo spin-echo sequence with appropriately set T_R depends only on T_2 and a spoiled low-flip angle gradient echo sequence is highly T_1 dependent. The advantage of this approach is that simple signal models (e.g., mono-exponential functions) can be used to describe the MR signal as a function of $\boldsymbol{\nu}$, which allows for robust parameter mapping. Nevertheless, this requires different acquisitions for different parameters and hence can make a comprehensive assessment of multiple mapping parameters very time consuming.

A wide range of approaches has been proposed that overcome this challenge by obtaining multiple parameters $\boldsymbol{\nu}_i$ from a single scan (see, e.g., [4–11]). These methods provide different diagnostic parameters such as T_1 or T_2 that are perfectly spatially coregistered allowing for a direct voxel-wise comparison of both parameters. In order to be able to estimate multiple ν_i, the MR sequence has to be sensitive to all these parameters, which is often achieved by varying multiple sequence parameters. An example is shown in Fig. 15.2d for Magnetic Resonance Fingerprinting (MRF) [6]. A varying flip angle (small figure insert in Fig. 15.2d) leads to signal dependency on both T_1 and T_2. Commonly, this is also combined with an inversion preparation pulse to achieve a strong T_1 effect. The main challenge here is that such signal variations cannot be described by a single exponential model, but the evolution of the MR signal has to be calculated using more complex models such as Bloch equations or Extended Phase Graph (EPG) approaches [12]. These methods ensure that the entire spin system is traced over time, and the recorded signal contains the appropriate contributions from gradient echoes, spin echoes, and stimulated echoes.

In order to estimate $\boldsymbol{\nu}$, it is assumed that the sequence parameters are well known. This is in general true for the sequence timings (e.g., T_R, T_E). Parameters, such as the flip angle α on the other hand, can deviate strongly from the theoretical value. This of course means that the model used to describe the MR data is inaccurate, leading to errors in the parameter estimation. In order to minimize these errors, the sequence design can be optimized (e.g., using longer RF excitation pulses to achieve a more accurate α), or these sequence parameters have to be included as unknowns in the signal model and also estimated during the mapping. Therefore, many of the multi-parametric methods already mentioned do not just provide several diagnostic parameters but also additional maps of, e.g., B_0 or B_1 fields or other model parameters.

The flexibility of MR offers a wide range of possibilities to design sequences for parameter mapping. It is therefore crucial that both the MR sequence and the MR signal model are adapted to a given mapping problem. In addition, the limitations of the MR sequence (e.g., inaccurate α) need to be well understood and need to be taken into considerations during mapping.

15.3 Image-based mapping

Image-based parameter mapping can be divided into three steps: sampling of the magnetization over time (Fig. 15.3a), reconstruction of qualitative images with different contrast weighting (Fig. 15.3b), and prediction of the parameter of interest (Fig. 15.3c). Image reconstruction and parameter estimation are usually independent of each other, with parameter estimation implemented as a post-processing step, performed voxel-wise based on the qualitative images.

The data acquisition is chosen such that $\mathbf{s}$ is sensitive to the parameter to be resolved, with preferably low sensitivities to other undesired biophysical parameters. Independent of the used method, a series of qualitative images $\hat{\mathbf{x}}_i$ has to be sampled to resolve the evolution of the magnetization ($\mathbf{M}(t)$), e.g., T_1 recovery after an inversion pulse. This is achieved by sampling of $\mathbf{M}(t)$ at multiple predefined time points τ_i leading to multiple sets of k-space data $\mathbf{s}_i$ (Fig. 15.3a).

From $\mathbf{s}_i$, multiple contrast-weighted qualitative images $\hat{\mathbf{x}}_i$ are reconstructed. Ideally, each $\mathbf{s}_i$ is obtained within an infinitesimally short time compared to the temporal changes of $\mathbf{M}(t)$. In practice this is often not possible and hence the changes of $\mathbf{M}(t)$ over the acquired $\mathbf{s}_i$ can also lead to artifacts (e.g., image blurring). Furthermore, restrictions due to clinically feasible scan time can also mean that not a fully sampled k-space can be acquired for each τ_i, but undersampling is necessary. Commonly, uniform Cartesian sampling is used for $\mathbf{s}_i$, but for certain applications non-Cartesian (e.g., radial or spiral trajectories) or nonuniform Cartesian k-space sampling is preferred because it can lead to incoherent

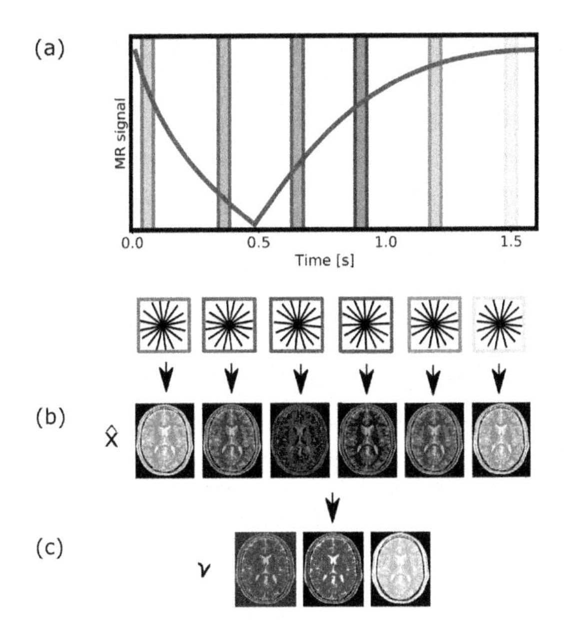

FIGURE 15.3 Overview of image-based parameter mapping

(**a**) After magnetization preparation, $\mathbf{s}$ is acquired at various time points using, e.g., a radial acquisition scheme. The blue (dark gray in print version) curve shows the temporal evolution of the absolute value of the magnetization for an arbitrary T_1 time. (**b**) Images $\hat{\mathbf{x}}_i$ are then reconstructed from $\mathbf{s}_i$. (**c**) From the signal evolution in each voxel of $\hat{\mathbf{x}}_i$, parameters $\boldsymbol{\nu}$ are estimated by, e.g., carrying out a fit of the data to a signal model q. In this example, $\boldsymbol{\nu}$ consists of three parametric maps showing T_1, the equilibrium magnetization M_0, and the flip angle α. Only T_1 is clinically relevant, but the other two are required to ensure the model describes the signal evolution well. There are, of course, other approaches, that provide multiple diagnostic parameters as discussed in Section 15.2.

undersampling artifacts. Incoherence of the undersampling artifacts not just within each $\hat{\mathbf{x}}_i$ but also between different $\hat{\mathbf{x}}_i$ is advantageous because these artifacts propagate less into the parameter maps $\boldsymbol{\nu}$. This requires a different k-space sampling pattern for each $\hat{\mathbf{x}}_i$. For imaging affected by physiological motion (e.g., breathing or heartbeat), triggering and gating, or motion correction (see Chapter 13) is needed, which can make it challenging to cover the entire temporal evolution of $\mathbf{M}(t)$.

The qualitative images $\hat{\mathbf{x}}_i$ can be reconstructed using multiple reconstruction methods that affect the image quality of $\hat{\mathbf{x}}_i$ and therefore $\boldsymbol{\nu}$. Fig. 15.4 compares various image reconstruction schemes for undersampled $\mathbf{s}_i$ and the effect on the obtained T_1 map. The simplest approach is to use FFT or NUFFT (see Chapter 4) to reconstruct $\hat{\mathbf{x}}_i$ (Fig. 15.4a). Nevertheless, because $\mathbf{s}_i$ is undersampled, undersampling artifacts are present that also affect the T_1 map. An improvement of $\hat{\mathbf{x}}_i$ can be achieved by using parallel imaging (see Chapter 6), which also increases the quality of the T_1 map (Fig. 15.4b). Both approaches treat each $\hat{\mathbf{x}}_i$ separately and do not utilize the fact that each image shows the same anatomy, just with a different image contrast. Image reconstruction can thus be further improved by applying regularization along the temporal dimension of $\hat{\mathbf{x}}_i$. This can be carried out independently of the underlying physical

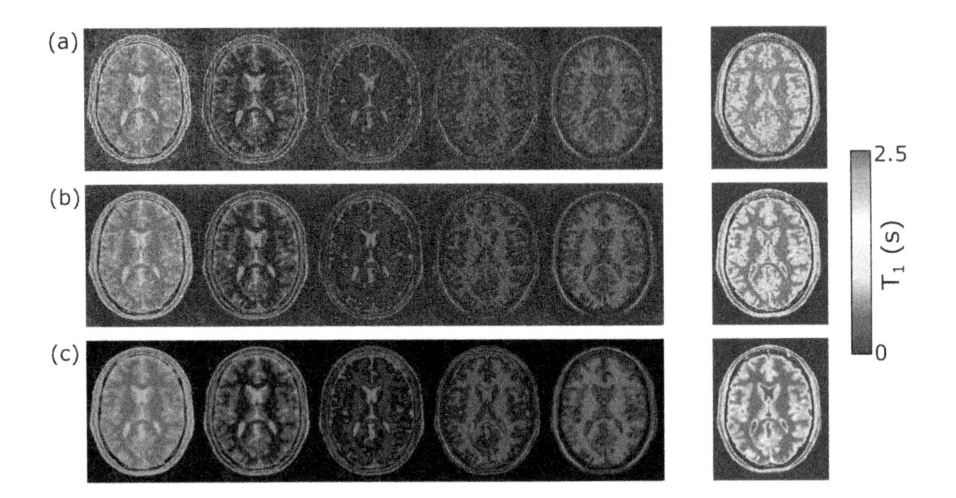

FIGURE 15.4 Image reconstruction approaches for direct T_1 mapping

2D radial data acquisition after an inversion pulse was simulated. Images at multiple time points after the inversion pulse were reconstructed from highly undersampled data using (**a**) nonuniform FFT, (**b**) iterative non-Cartesian SENSE and (**c**) iterative reconstruction with TV-based regularization in space and over time. T_1 maps were determined using a voxel-wise fit with a Levenberg–Marquardt approach.

model q utilizing, e.g., TV-based regularization (Fig. 15.4c) or enforcing sparsity along the temporal dimension. Temporal acceleration techniques such as kt-SENSE or kt-PCA can also be used [13]. Low-rank based-reconstruction can also be applied (see Chapter 9) which assumes that groups of voxels follow similar temporal behavior without having to explicitly specify the temporal model [14,15,10,16]. Another option is to already utilize q during image reconstruction in order to calculate a dictionary-based sparsifying transform [17,18].

As shown in Fig. 15.4, the image quality of $\hat{\mathbf{x}}_i$ affects the quality of the obtained parameter map. Adding regularization can strongly improve $\hat{\mathbf{x}}_i$, especially for highly undersampled $\mathbf{s}_i$. Nevertheless, regularization needs to be carefully chosen for the reconstruction task at hand, and the regularization strength also needs to be optimized. If the regularization is not well chosen, regularization artifacts (e.g., staircasing artifacts for regularization using total variation [19]) can appear in $\hat{\mathbf{x}}_i$ and can thus also affect the final $\boldsymbol{\nu}$, which needs to be taken into consideration when interpreting $\boldsymbol{\nu}$. For more details on regularized image reconstruction, please also see Chapter 2.

Based on $\hat{\mathbf{x}}_i$, $\boldsymbol{\nu}$ is predicted. This is done voxel-wise, e.g., by L2-norm minimization of $\hat{\mathbf{x}}_i$ and q of the specific acquisition. For example, for T_1 mapping, an inversion preparation could be used, where the magnetization recovery can be derived from the Bloch equations

$$\frac{dM_z}{dt} = \frac{M_0 - M_z}{T_1}, \tag{15.2}$$

resulting in

$$M_z(t) = M_0 + (M_z(t_0) - M_0)e^{-t/T_1}, \tag{15.3}$$

with $M_z(t_0)$ the initial magnetization at $t = 0$. Another approach is a variable flip-angle method, where at least two gradient-echo acquisitions with different flip angles are acquired. These can be described by the signal model

$$M_z(T_R, \alpha) = M_0 \frac{1 - e^{-T_R/T_1}}{1 - \cos\, \alpha\, e^{-T_R/T_1}} \sin \alpha, \tag{15.4}$$

and T_1 can be calculated based on the signal from both acquisitions.

T_2 can be estimated using a spin-echo sequence that can be described by

$$M_{xy}(t) = M_0\, e^{-t/T_2}. \tag{15.5}$$

Next to these monoexponential models, the Bloch model or EPG can be used to obtain q.

Multiple relaxation parameters can be important for an accurate classification of a disease so more recent approaches employ the combination of $\boldsymbol{\nu}$ in a single acquisition (see also Section 15.2). One of these approaches is MRF. Here, data acquisition is performed using pseudorandomly chosen TR and flip angles instead of keeping these values constant during a MR scan. This leads to a signal intensity sensitive to T_1 as well as T_2, with a specific signal evolution for each combination of these parameters, a so-called "fingerprint".

Instead of conventional fitting, dictionary matching is performed to obtain $\boldsymbol{\nu}$. For this, a dictionary $\mathbf{D}$ is generated for this specific acquisition, with signal time curves $\mathbf{d_i}$ of a wide range of physiologically possible combinations of $\boldsymbol{\nu}$. The Bloch equations (see also Chapter 1) or EPG [12] are commonly used to calculate the dictionary entries. Usually, this step has to be performed only once for a specific MRF sequence. Eventually, $\hat{\mathbf{x}}$ is matched voxel-wise to $\mathbf{D}$ to obtain the best $\boldsymbol{\nu}$. This is done by computing the dot product between the normalized $\hat{\mathbf{x}}$ and all entries $\mathbf{d_i}$ of $\mathbf{D}$ ($\mathbf{d_i} \cdot \hat{\mathbf{x}}$), as a correlation-based metric. The best $\boldsymbol{\nu}$ is given by the dictionary entry with the highest dot product (i_{max}). M_0, proportional to the proton density, is calculated by the scaling factor between $\hat{\mathbf{x}}$ and $\mathbf{d_{i_{max}}}$. One thing to keep in mind is that the dictionary $\mathbf{D}$ is always only a discrete representation of all possible values of $\boldsymbol{\nu}$. If $\mathbf{D}$ only coarsely resolves, e.g., T_1, then the same will be true for the obtained map $\boldsymbol{\nu}$.

For MRF sequences that have to be synchronized with physiological motion such as the heart beat, $\mathbf{D}$ has to be calculated for each scan because the physiological motion determines the timings of the sequence and hence $\mathbf{M}(t)$ [20]. In order to minimize artifacts due to cardiac motion, the MR scan can be ECG-triggered such that data is only acquired in a specific cardiac phase. This improves the final quality of $\boldsymbol{\nu}$ but also means that the timings of the MR scan depend on the heart rate.

One of the main drawbacks of MRF is the large range of combinations of parameters. This leads to a very large dictionary and computationally demanding pattern recognition. A common approach to speed up the process is using singular value decomposition (SVD)

$$\mathbf{D} = \mathbf{U}\boldsymbol{\Sigma}\mathbf{V}^*. \tag{15.6}$$

Based on this, a low-rank approximation of the dictionary can be calculated

$$\mathbf{D_k} = \mathbf{D}\mathbf{V_k}, \tag{15.7}$$

with $\mathbf{V_k}$ containing the first k right singular vectors of $\mathbf{D}$ [21]. For compression in the time domain, the time evolution of the dictionary entries $\mathbf{d}$ can be projected onto the low-rank space

$$\mathbf{d_k} = \mathbf{d}\mathbf{V_k}, \tag{15.8}$$

and pattern matching can be performed in this subspace. Many other approaches have been proposed to speed up the dictionary matching and reduce its memory requirements using, for example, group matching or linear interpolation [22–25]. For more details on MRF, please refer [26] to and [27]. As a starting point, there are several open-source software packages for fast calculation of $\mathbf{D}$ available on github.[1,2]

Machine learning can also be applied in parameter mapping at any stage of the process. For example, image reconstruction could be performed with a neuronal network to replace iterative reconstructions and thus to speed up image reconstruction (see Chapter 10). Furthermore, it could also be applied for data fitting, fast dictionary generation, or pattern recognition [28–30].

Parameter estimation as part of image postprocessing has the advantage that it is easy to implement because the parameters are obtained separately for each voxel. Furthermore, since the estimation has only to be performed once for each voxel, this method is fast, and the total time of parameter mapping depends largely on the data acquisition and the image reconstruction method used.

One of the main challenges is that artifacts in $\hat{\mathbf{x}}$ (e.g., due to coherent undersampling, mismatch of motion states, or field inhomogeneities) can impair $\boldsymbol{\nu}$ since no data consistency between $\boldsymbol{\nu}$ and $\mathbf{s}$ is included.

15.4 Reconstruction-based mapping

Undersampling of k-space speeds up data acquisition but can lead to residual artifacts in $\hat{\mathbf{x}}$. Depending on the artifacts, this can lead to quantification inaccuracies in $\boldsymbol{\nu}$. No information of the encoding model $\mathbf{E}$ is included in the model operator q, and hence it is not possible to distinguish between true signal and undersampling artifacts.

Reconstruction-based mapping tries to overcome this problem by directly solving Eq. (15.1), i.e., estimating $\boldsymbol{\nu}$ from given k-space data $\mathbf{s}$. The main advantages of this approach are that data consistency with the acquired k-space data can be ensured and regularization can be applied directly on $\boldsymbol{\nu}$ rather than the intermediate $\hat{\mathbf{x}}$. Rather than estimating $\boldsymbol{\nu}$ from image data impaired by undersampling artifacts, the quantitative parameters are directly estimated from $\mathbf{s}$. Assuming we have an image with $N \times N$ voxels and want to estimate N_ν quantitative parameters from N_a quantitative images $\hat{\mathbf{x}}$, then, for image-based mapping, we need to estimate $N \times N \times N_a$ unknowns from $\mathbf{s}$. For the reconstruction-based mapping, we need to estimate $N \times N \times N_\nu$ unknowns and as N_ν is often much smaller than N_a, this is a better determined problem. The main challenge of the model-based approach is that Eq. (15.1) describes a nonlinear problem. In the following we will describe two approaches to solve this problem. Model-based acceleration of parameter mapping (MAP) separates the problem of finding $\boldsymbol{\nu}$ into several individual steps that each on its own can be solved in a straightforward way. Model-based optimization directly solves Eq. (15.1) using a nonlinear optimization.

[1] https://github.com/chixindebaoyu/MRF_CUDA.

[2] https://github.com/utcsilab/mri-sim-py.epg.

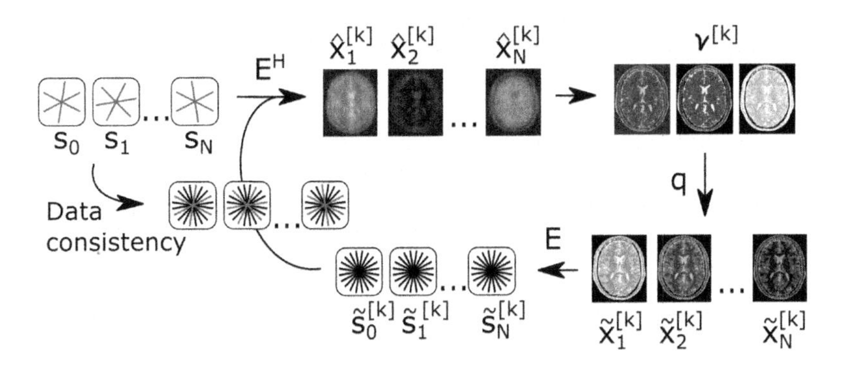

FIGURE 15.5 Overview of MAP

Starting from the acquired k-space $\mathbf{s}_i$, first image estimates $\hat{\mathbf{x}}_i^{[k=1]}$ are reconstructed and $\boldsymbol{\nu}^{[k]}$ is estimated. In each iteration k, k-space data $\tilde{\mathbf{s}}_i^{[k]}$ is calculated from $\boldsymbol{\nu}^{[k]}$ using q and $\mathbf{E}$, and data consistency is ensured. More and more of the missing k-space data is filled in at each iteration, which improves the estimation of $\boldsymbol{\nu}$. Please refer to Section 15.4.1 for further details.

15.4.1 Model-based acceleration of parameter mapping (MAP)

For image-based mapping, images $\hat{\mathbf{x}}$ are reconstructed, and $\boldsymbol{\nu}$ is determined from them. The main idea of MAP is to extend this and calculate the k-space data representing $\boldsymbol{\nu}$ using q and $\mathbf{E}$. This k-space data $\tilde{\mathbf{s}}$ is then compared to the acquired data $\mathbf{s}$ and used to update $\hat{\mathbf{x}}$. This is repeated multiple times until convergence is reached.

MAP assumes a set of k-space data $\mathbf{s}_i$, where i represents the different dynamics (e.g., different inversion times for inversion-recovery based T_1 mapping). The main requirements are that the k-space sampling leads to incoherent artifacts along the dynamic dimension (e.g., spiral or radial acquisitions turned by the golden angle for each i) and that the k-space center is well covered for each dynamic i to ensure contrast changes are captured and hence mainly high k-space frequencies are undersampled. MAP then carries out the following steps for each iteration k, which are also depicted in Fig. 15.5:

1. Reconstruct an initial estimate $\hat{\mathbf{x}}_i^{[k=1]}$ of the acquired k-space data $\mathbf{s}_i$ using $\mathbf{E}^{\mathbf{H}}$.
2. Estimate $\boldsymbol{\nu}^{[k]}$.
3. Calculate the qualitative image sequence $\tilde{\mathbf{x}}_i^{[k]}$ from $\boldsymbol{\nu}^{[k]}$ using q.
4. Calculate the k-space data $\tilde{\mathbf{s}}_i^{[k]}$ representing $\tilde{\mathbf{x}}_i^{[k]}$ using $\mathbf{E}$.
5. Ensure data consistency (e.g., by substituting the acquired k-space positions in $\tilde{\mathbf{s}}_i^{[k]}$ with $\mathbf{s}_i$).
6. Repeat steps 2–5 until convergence has been reached.

In this example, the acquired radial k-space $\mathbf{s}_i$ has a higher sampling density in the center of k-space. Therefore, the initial images $\hat{\mathbf{x}}_i^{[k=1]}$ reconstructed by applying the adjoint of the encoding operator $\mathbf{E}^{\mathbf{H}}$ contain very little high-resolution information. Commonly, this is compensated for using a density compensation function to ensure a balanced contribution of high- and low-k-space frequencies to the reconstructed image (see also Chapter 4). For the example shown in Fig. 15.6, no density compensation function is applied, leading to a higher contribution of low k-space frequencies to the image for the initial iterations. This also means that undersampling artifacts are attenuated allowing for an estimation

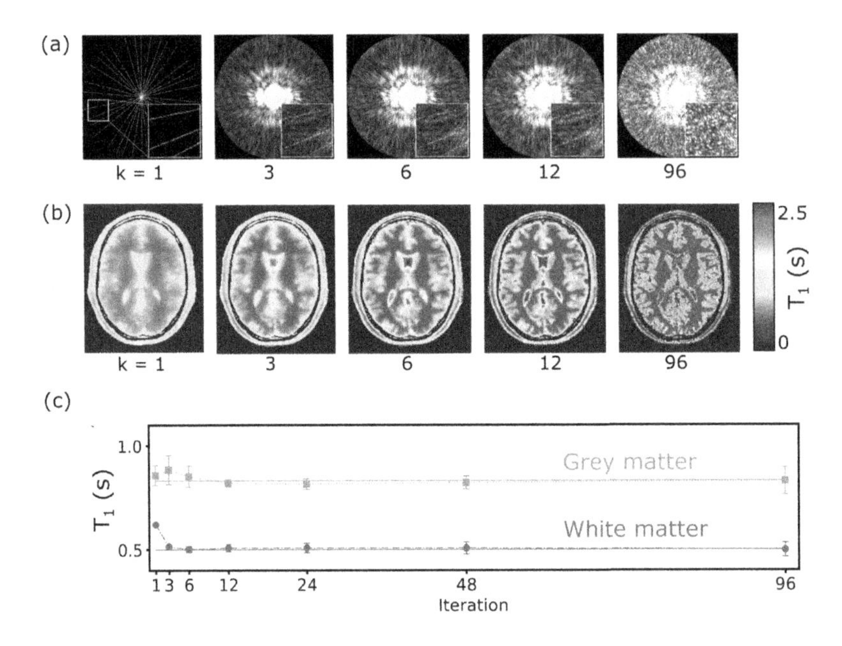

FIGURE 15.6 Convergence of MAP

(**a**) K-space data $\mathbf{s}_i$ and (**b**) T_1 maps at multiple numbers of iterations (k) of the MAP reconstruction. (**c**) Convergence of T_1 values for region-of-interests in gray and white matter to the ground truth values (dashed lines).

of $\boldsymbol{\nu}^{[k]}$ with fewer artifacts but, of course, also fewer image details. If too many artifacts are present in $\hat{\mathbf{x}}_i^{[k=1]}$, then additional low-pass filters can also be applied to $\mathbf{s}_i$ prior to image reconstruction. The low-pass filter strength can then be decreased in each iteration to allow for more and more high-resolution information in the images and in the parametric maps [31].

The estimation of $\boldsymbol{\nu}^{[k]}$ from $\hat{\mathbf{x}}_i^{[k=1]}$ can be carried out using fitting algorithms, dictionary matching-based approaches, or any other parameter-estimation technique discussed in Section 15.3. Despite $\hat{\mathbf{x}}_i^{[k]}$ being low-resolution images, they might still be impaired by undersampling artifacts, and therefore the method used to estimate $\boldsymbol{\nu}^{[k]}$ needs to be robust in the presence of these artifacts.

Fig. 15.6 shows $\boldsymbol{\nu}^{[k]}$ for various iteration numbers. For this reconstruction, a 2D golden-radial acquisition after a single inversion pulse was simulated. Each $\mathbf{s}_i$ consisted of eight radial lines at various time points after the inversion pulse. A Look–Locker model [32] was used as q and, in addition to T_1, also the spin density and the actual flip angle was estimated as $\boldsymbol{\nu}$. The radial sampling leads to a higher density of k-space points in the center of k-space and hence the first reconstruction using $\mathbf{E}^{\mathbf{H}}$ leads to low-resolution images $\hat{\mathbf{x}}_i^{[k]}$ and hence a low-resolution estimation of $\boldsymbol{\nu}^{[k=1]}$. With each iteration, more and more details are visible in $\boldsymbol{\nu}^{[k]}$ as more and more of the missing k-space data at higher k-space frequencies are filled in. Fig. 15.6c depicts T_1 values of a region-of-interest in the white and gray matter as a function of the number of iterations. For higher iteration numbers, the mean values approach the true value (dashed lines) but also the standard deviation in each region increases. Undersampling is higher for higher k-space frequencies, and hence the data consistency step has less of an effect here.

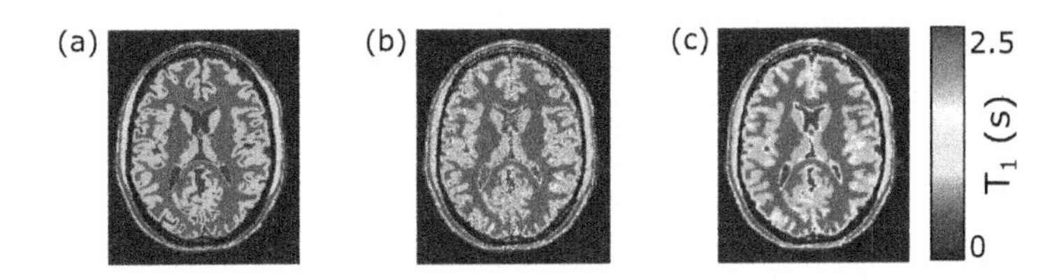

FIGURE 15.7 MAP with de-noising

(**a**) Ground truth T_1 map. (**b**) T_1 map after 96 iterations using MAP without de-noising. (**c**) T_1 map after 96 iterations with TV-based de-noising.

With increasing number of iterations, more of the higher k-space data get filled in but also noise starts to build up. The number of iterations has to be therefore carefully selected to ensure all image features are recovered and an accurate estimation of $\boldsymbol{\nu}$ is achieved before noise starts to dominate the data. In order to further stabilize the reconstruction and avoid the problem of noise amplification for high numbers of iterations, de-noising of $\boldsymbol{\nu}^{[k]}$ can be added as a reconstruction step. Fig. 15.7 compares the ground truth T_1 map (a) to the estimated T_1 maps after 96 iterations without de-noising (b) and with TV-based de-noising (c). With the de-noising step, the reconstruction can be further stabilized, and the problem of noise dominating the reconstruction for higher iteration numbers is avoided. Of course, this requires a suitable selection of a de-noising approach adding further parameters to the reconstruction and leading to a potential bias of the final result.

MAP has been used for T_1 and T_2 estimation and a similar approach has also been proposed for MRF [33,31].

15.4.2 Model-based optimization

As just mentioned, Eq. (15.1) describes a highly nonlinear problem. Nevertheless, a minimization of a likelihood cost function is possible to estimate the quantitative parameters $\hat{\boldsymbol{\nu}}$ from the k-space data $\mathbf{s}$

$$\hat{\boldsymbol{\nu}} = argmin_{\nu}||n(\boldsymbol{\nu}) - \mathbf{s}||_2^2, \tag{15.9}$$

with

$$n : \boldsymbol{\nu} \mapsto \mathbf{E}q(\boldsymbol{\nu}) \tag{15.10}$$

and using the L2-norm. Commonly, this problem is extended by further regularization to stabilize the solution

$$\hat{\boldsymbol{\nu}} = argmin_{\nu}||n(\boldsymbol{\nu}) - \mathbf{s}||_2^2 + \sum_i \lambda_i R_i(\boldsymbol{\nu}) \quad . \tag{15.11}$$

Suitable regularization terms R_i have to be adapted to the requirements of each problem. Examples of proposed regularizations are:

- piecewise constant $\boldsymbol{\nu}$ by minimizing the L1-norm of TV or total generalized variation of $\boldsymbol{\nu}$ [34];
- smoothness of $\boldsymbol{\nu}$ by minimizing the L2-norm of the first-order difference matrix of $\boldsymbol{\nu}$ [35];

- smoothness of the k-space data by minimizing the L2-norm of $\nabla\mathcal{F}\boldsymbol{\nu}$, where ∇ is the spatial finite-difference operator and $\mathcal{F}$ the spatial Fourier transform [36];
- sparsity of $\boldsymbol{\nu}$ and similarity between $\boldsymbol{\nu}_i$ using wavelets as sparsifying transforms and ensuring that wavelet coefficients that are present in all $\boldsymbol{\nu}_i$ are favored [37].

Similar to the choice of regularization, the choice of the solver for (15.11) also depends strongly on $\mathbf{E}$ and q, which is still an active field of research. Olafsoon et al. linearized the problem of T_2^* mapping from a multi-echo gradient-echo acquisition and used a conjugate gradient approach [35]. Sbrizzi et al. used a variable projection method [9], and other groups proposed iteratively regularized Gauss–Newton methods [36,38,34].

In order to solve this problem iteratively, the Jacobian matrix of Eq. (15.9) needs to be calculated

$$\begin{aligned}\frac{\partial}{\partial \nu}||n(\boldsymbol{\nu})-\mathbf{s}||_2^2 &= \frac{\partial}{\partial \nu}||\mathbf{E}q(\boldsymbol{\nu})-\mathbf{s}||_2^2 = \frac{\partial}{\partial \nu}\left((\mathbf{E}q(\boldsymbol{\nu})-\mathbf{s})^H(\mathbf{E}q(\boldsymbol{\nu})-\mathbf{s})\right)\\ &= \frac{\partial}{\partial \nu}(\mathbf{E}q(\boldsymbol{\nu}))^H(\mathbf{E}q(\boldsymbol{\nu})-\mathbf{s}) + (\mathbf{E}q(\boldsymbol{\nu})-\mathbf{s})^H\frac{\partial}{\partial \nu}(\mathbf{E}q(\boldsymbol{\nu}))\\ &= \frac{\partial}{\partial \nu}(\mathbf{E}q(\boldsymbol{\nu}))^H(\mathbf{E}q(\boldsymbol{\nu})-\mathbf{s}) + \left(\frac{\partial}{\partial \nu}(\mathbf{E}q(\boldsymbol{\nu}))^H(\mathbf{E}q(\boldsymbol{\nu})-\mathbf{s})\right)^H\\ &= 2Re\left(\frac{\partial}{\partial \nu}(\mathbf{E}q(\boldsymbol{\nu}))^H(\mathbf{E}q(\boldsymbol{\nu})-\mathbf{s})\right)\\ &= 2Re\left(\left(\frac{\partial q(\boldsymbol{\nu})}{\partial \nu}\right)^H\mathbf{E}^H(\mathbf{E}q(\boldsymbol{\nu})-\mathbf{s})\right),\end{aligned} \tag{15.12}$$

where we used $A + A^H = 2Re(A)$ with $Re(A)$ referring to taking the real part of A.

Approaches have also been proposed that make use of the Hessian using trust-region algorithms to solve Eq. (15.9) [39]. Calculating and storing the Hessian explicitly can be computationally demanding. Therefore, sparse approximations of the Hessian have been proposed instead.

The problem in Eq. (15.11) is nonconvex, and, therefore, it is also important to provide a good starting point for the minimization in order to ensure convergence to the desired solution. The convergence rate of most solvers will strongly depend on the correct scaling of the Jacobian/Hessian matrix. The quantitative parameters $\boldsymbol{\nu}$ can represent physical quantities such as relaxation times, and hence they have units (e.g., ms or s) or they can be unitless, such as the spin density. Therefore, the various partial derivatives need to scaled appropriately [36,37]. The same considerations apply to the strength of regularization λ_i in Eq. (15.11). The regularization type and strength have to be carefully optimized for each physical quantity of $\boldsymbol{\nu}$.

Once the objective function and the Jacobian/Hessian are defined, the problem can be treated similarly to the classical image-reconstruction problems described in Chapters 2 and 3.

Fig. 15.8 shows an example of a T_2 map reconstructed from a multi-echo spin-echo sequence acquired in a human brain using a golden-ratio radial acquisition pattern [40]. Each echo is highly undersampled. In this example, $\boldsymbol{\nu}$ representing a M_0 and T_2 map, and the coil sensitivity maps were jointly estimated using a model-based optimization. L_1-wavelet regularization was applied to $\boldsymbol{\nu}$, and the nonlinear problem was solved using an iteratively regularized Gauss–Newton method. For comparison, a standard image-based mapping approach is also shown that uses iterative non-Cartesian SENSE

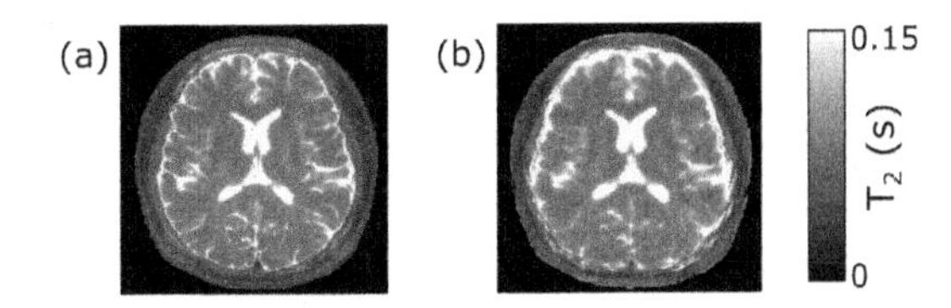

FIGURE 15.8 T_2 **mapping using model-based optimization**

T_2 map obtained from a multi-echo spin-echo in-vivo scan using (**a**) model-based optimization compared to (**b**) image-based mapping. This figure was reproduced from [40] using the code and data that was made publicly available with this publication.

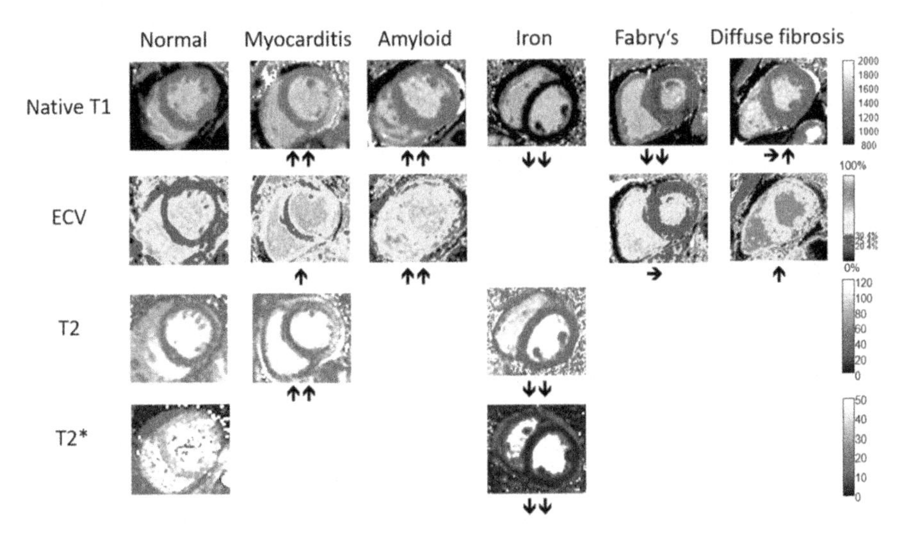

FIGURE 15.9 Parameter mapping for cardiac applications

T1, T2, T2*, and ECV (extracellular volume) mapping are useful for the diagnosis of many different myocardial diseases. The arrows denote changes to the parameter value compared to healthy myocardium. Reprinted from Messroghli et al., JCMR (2017) 19:75, https://dx.doi.org/10.1186/s12968-017-0389-8 [42].

for the reconstruction of $\hat{\mathbf{x}}_i$, followed by a voxel-wise fit of Eq. (15.5) to the data. The model-based optimization shows a better depiction of anatomical details compared to the image-based approach.

15.5 Clinical applications

Parameter mapping can be used to improve the diagnosis and classification of pathologies in various body regions, such as the brain, heart, and abdomen [41]. Pathological processes alter the local molecular environment of tissue, implicating altered relaxation times of this tissue. Fig. 15.9 gives an overview of changes in mapping parameters for multiple pathologies of the heart. The increase of free-water content causes prolonged T_2 times so T_2 is sensitive to edema. Edema is present in acute inflammatory diseases or acute ischemia. Furthermore, absolute changes in T_2 also make it possible to monitore inflammatory processes over time. Another application of T_2 mapping is hemorrhage, which can induce shortened T_2 times.

Iron is important for the production of hemoglobin, usually stored in the liver and spleen, but accumulation of iron can occur, damaging the organs. The superparamagnetic behavior of the iron-stored proteins leads to susceptibility-induces distortion in the local magnetic field, making T_2^* the most suitable parameter for the detection of iron storage diseases, with shortened T_2^* times. Moreover, the T_2^* time correlates with the absolute iron level in the liver, making it possible to replace biopsy for accurate diagnosis.

Infiltrative diseases are also affecting the T_1 times because of the change in tissue composition. For example, specific lipids (Fabry's disease) or insoluble proteins, such as amyloid, could be deposed in the interstitial space of the tissue, severely damaging the tissue if untreated. These changes lead to increased T_1 times (amyloidosis) or shortened T_1 times (Fabry's disease).

Native T_1 times obtained by T_1 mapping without the application of a contrast agent are sensitive to intracellular as well as extracellular changes of the tissue. However, in many diseases, especially, the extracellular space is of great importance. Using both native T_1 mapping and T_1 mapping after the administration of a contrast agent changing the T_1 times, the extracellular volume can be calculated, directly obtaining a physiologically parameter that is important in all diseases involving the remodeling of the tissue. In general, remodeling of the extracellular space is present in many diseases, thus T_1 can be used for the detection and monitoring of many pathophysiological changes, such as myocarditis, chronic diseases (e.g., multiple sclerosis, dementia) or post-transplantation changes of the kidney and liver. The T_1 value itself provides an indication of the underlying pathology and its extent. By comparing the quantitative relaxation times to healthy tissue, overall longer T_1 times could be caused by diffuse scarring fibrosis, and shortened T_1 times could be a marker for diffuse fat infiltration.

The combination of multiple parameters enables a more certain decision about the pathological state of the tissue. Therefore, methods such as MRF, combining T_1, and T_2 mapping in a short scan time could highly impact future clinical routine. As shown in Fig. 15.10, using an MRF approach, metastases in the liver caused by lung adenocarcinoma can be better visualized compared to its T_1-weighted image, as shown in the study of Chen et al. [43].

Whereas local pathological tissue, and thus locally altered parameters, could possibly be also detected in qualitative images, the main advantage of parameter mapping could lie in the detection of diffuse diseases where no contrast to healthy tissue is present. Furthermore, because of the system- and acquisition-independent nature of T_1 and T_2 and T_2^*, parameter mapping facilitates objective interpretations. Comparability across patients or institutes is increased, making large-scale multi-center studies possible and thus enabling broader research and therefore better understanding of diseases. For this, robust and fast techniques have to be investigated to be able to obtain parameter maps during clinically feasible scan times.

15.6 Current challenges and future directions

Mapping is a very powerful technique, providing quantitative diagnostic parameters for a wide range of pathologies. The main challenge are inconsistencies between the acquired data and the applied model

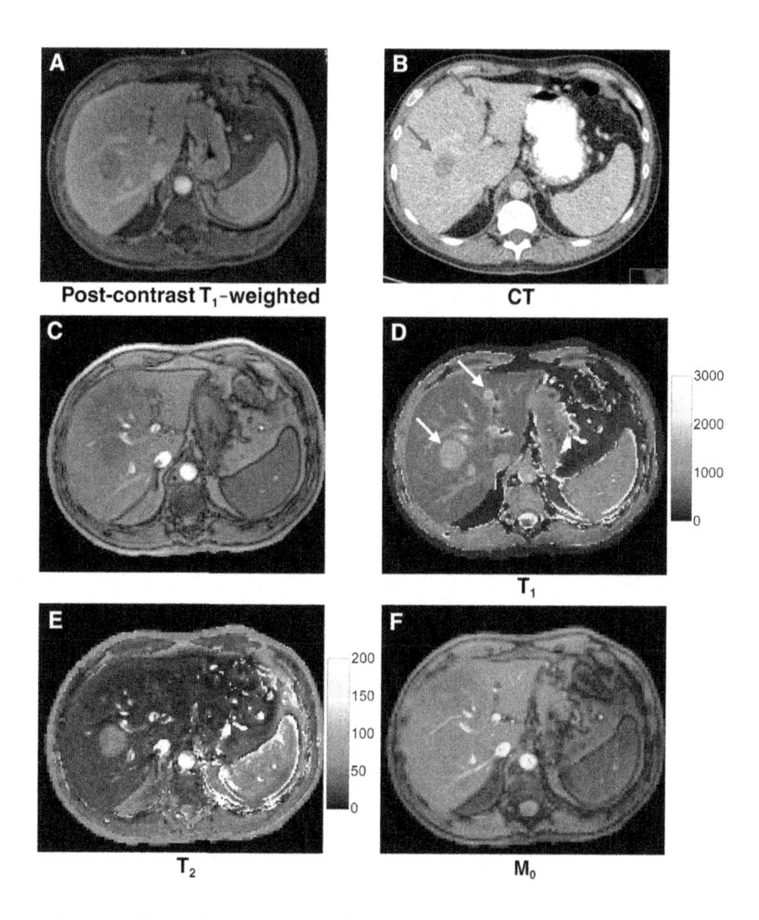

FIGURE 15.10 Parameter mapping of lung adenocarcinoma metastatic to the liver using MRF

Metastases (arrows) are more visible in parameter maps obtained by MRF (C-F) compared to a qualitative contrast-enhanced T1-weighted image (A). Reprinted with permission from Radiology Society of North America (RSNA) from Fig. 5, Chen Y, Jiang Y, Pahwa S, Ma D, Lu L, Twieg MD, Wright KL, Seiberlich N, Griswold MA, Gulani V: MR Fingerprinting for Rapid Quantitative Abdominal Imaging. Radiology 2016, 279:278–286 [43].

operator q leading to errors in the estimated $\boldsymbol{\nu}$. Often a simplified model (e.g., single-exponential signal recovery) is used since it provides fast mapping and is robust towards noise and other artifacts in the data. The model is commonly adapted to the data acquisition and hence different MR mapping sequences (e.g., T_1 mapping using either an inversion or a saturation pulse) lead to different model errors and hence different estimations of $\boldsymbol{\nu}$. In clinical practice, reference values for healthy and pathological tissue therefore have to be determined for each sequence, separately. Future work is required to improve the comparability of various mapping approaches.

MR relaxation times are parameters that describe the behavior of nuclear spins in a magnetic field, hence they can only be determined with MRI. Other quantitative parameters used for clinical diagnosis,

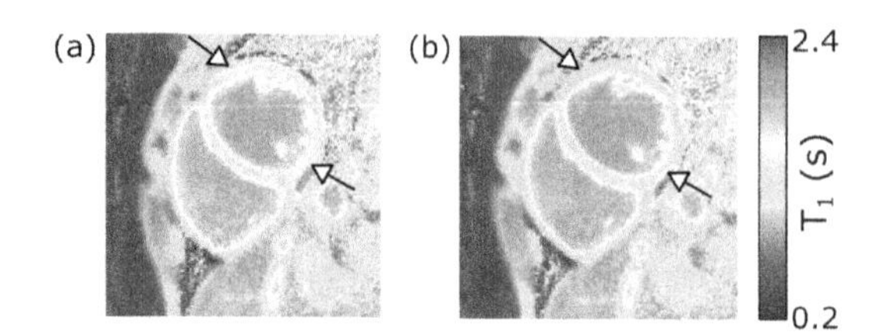

FIGURE 15.11 Fig. T_1 mapping with and without motion correction

(**a**) Residual cardiac motion leads to inaccurate myocardial T_1 times which would suggest the presence of a myocardial pathology (arrows). (b) By correcting for the cardiac motion, these artifacts are removed, showing a homogeneous healthy myocardium.

such as blood flow velocities, can be measured with MRI but also with ultrasound or invasively with catheters, providing reference measurements that are independent to MRI. This is not possible for mapping. MR-based reference measurements for relaxation times are available, but they are commonly very time consuming (longer than one hour) and therefore not applicable in vivo. This makes it challenging to fully evaluate the accuracy of mapping sequences and also to determine the true values for various tissue types.

One approach to minimize differences between the actual MR signal evolution and q is to make q more and more complex (e.g., by including magnetization-transfer effects) to ensure the MR physics is accurately described. Of course, this makes mapping more computationally demanding and more time consuming. In addition, it can also make it less robust in the presence of noise or residual artifacts. Recently, machine learning has also been introduced for parameter mapping. On one hand, this offers the possibility to learn q directly from the data, assuming an accurate reference measurement is available as training data. On the other hand, it can also be combined with a MR physics-based model to enable fast parameter estimation, while ensuring $\boldsymbol{\nu}$ is consistent with the acquired data [29]. Fast parameter estimation is also essential if complex models are to be used in clinical practice.

Inconsistencies between the data and the applied model can have two different effects. On the one hand, model errors can lead to a bias of the estimated $\boldsymbol{\nu}$, but still result in a good match between model and data. An example of this is given in the tutorial Section 15.8 where the effects of an incomplete inversion on the estimation of T_1 is analyzed. Although the estimated value for T_1 is inaccurate, the model fits well to the data. Such errors can only be detected if reference measurements are available. The other type of errors lead to a mismatch between model prediction and acquired data. This can also lead to errors of $\boldsymbol{\nu}$ but not necessarily. An example of this is MRF. The undersampling of each image is not taken into consideration in the signal model. Therefore, there is actually a poor match between the dictionary entries and the acquired data. Because the undersampling artifacts are incoherent compared to the signal model, an accurate estimation of $\boldsymbol{\nu}$ is still possible.

Also, physiological changes during data acquisition can lead to differences between q and the acquired data. One of the most common causes is physiological motion, such as head motion, swallowing, breathing motions, or the beating of the heart. MR scans are carried out to minimize any physiological motion, but this is not always possible. Sometimes, patients cannot hold their breath long enough during data acquisition or they experience heart rate irregularities that lead to residual motion in the data and hence artifacts in the reconstructed images. Fig. 15.11 shows an example for T_1 mapping in a short-axis view of the heart. Respiratory motion is minimized by asking subjects to hold their breath. Nevertheless, cardiac motion is still present and can lead to errors in $\boldsymbol{\nu}$. In this case the T_1 value in some parts of the myocardium are elevated which would suggest the presence of fibrosis but is simply an artifact due to cardiac motion (Fig. 15.11) [44]. Physiological motion has to be taken into consideration to ensure accurate and reliable parameter mapping. For more details, please refer to (Chapter 13).

15.7 Summary

Mapping yields quantitative parameters that make objective diagnosis and treatment monitoring possible. Image-based mapping approaches separate image reconstruction and parameter estimation and enable fast estimation of the parameters. Parameter estimation can be carried out by data fitting, dictionary matching, or machine-learning approaches. The physical model used for parameter estimation can already be incorporated into image reconstruction to enable accurate parameter mapping even from highly undersampled data. Image reconstruction and parameter estimation can also be combined in an iterative process to further improve quantification accuracy by ensuring data consistency between the acquired raw data and the obtained parametric maps. Clinical applications for parametric mapping range from assessing inflammatory processes to detecting iron accumulation and fibrotic tissue. Nevertheless, large clinical trials are still needed to establish parametric mapping in clinical routine.

15.8 Tutorial

15.8.1 Image-based T1 mapping

15.8.1.1 Problem description

In this exercise we will carry out T_1 mapping for four tissue types (fat, white matter, gray matter and cerebrospinal fluid (CSF)), assuming data acquisition carried out with an inversion recovery sequence using the following signal model:

$$M_z(t) = M_0(1 - 2I_{eff}e^{-t/T_1}), \tag{15.13}$$

where $M_z(t)$ is the longitudinal magnetization at time point t, M_0 is the equilibrium magnetization. Ideally, an inversion pulse will lead to a full inversion of the magnetization (i.e., 180° pulse), but, due to hardware limitations and interaction of the magnetic field with the body, the inversion pulse can lead to an excitation less than 180°. This is described with the inversion efficiency $I_{eff} \in [0, 1]$ (e.g., $I_{eff} = 1$ is the ideal case and corresponds to a 180° pulse).

This exercise requires the Curve Fitting Toolbox of Matlab®.

15.8.1.2 Provided material

- **sim_sig = sig_ir(M0, T1, Ieff, t_acq, noise_level)** This function simulates data acquisition after an inversion pulse. Example: `sim_sig = sig_ir(1, 1000, 0.9, [0, 100, 200], 0.05)` simulates a signal with $M_0 = 1$, $T_1 = 1000$ ms, $I_{eff} = 0.9$, a noise level of 0.05 at 0 ms, 100 ms, and 200 ms after the inversion pulse.

- **fit_ir(sig, t_acq)** This function fits the signal model of the inversion recovery sequence to given data. Example: `fit_ir(sig, [0, 100, 200])` fits the model to a 1D signal sig acquired at the time points [0, 100, 200].

15.8.1.3 Questions

1. Simulate signals for fat (T_1: 350 ms), white matter (T_1: 500 ms), gray matter (T_1: 833 ms) and CSF (T_1: 2569 ms) with $M_0 = 1$, $I_{eff} = 1.0$, noise level $= 0.1$ at the time points $t =$ [0, 600, 1200, 1800, 2400, 3000].
2. Fit the signal model to this data and report the error between the estimate and true T_1 times for each tissue type separately.
3. Simulate the signals for fat, white, and gray matter and CSF again, but for the time points $t =$ [0, 1800, 2400, 3000, 3600, 4200]. Repeat the fitting and report the errors for each tissue. Why can T_1 for white and gray matter and CSF still be well estimated, whereas T_1 for fat shows larger errors?
4. Simulate the signals for fat, white, and gray matter and CSF again, with $t =$ [0, 600, 1200, 1800, 2400, 3000] and $I_{eff} = 0.9$. This corresponds to the case when we do not achieve a full 180° inversion with the inversion pulse. Calculate the errors of T_1. Is T_1 over- or underestimated and explain why?

15.8.2 Magnetic resonance fingerprinting

15.8.2.1 Problem description

In this exercise we will use a precalculated dictionary $\mathbf{D}$ to estimate the T_1 and T_2 of four signal curves $\hat{\mathbf{x}}$. The (T_1, T_2) combination used to calculate the entries of the dictionary are listed in $\boldsymbol{\nu}_{Dict}$. As discussed in Section 15.3, for dictionary matching we carry out the following steps:

- Calculate the dot product between $\mathbf{D}$ and $\hat{\mathbf{x}}$ for each (T_1, T_2) combination in the dictionary.
- Find the dictionary entry with the index i_{max} where this dot product is highest.
- The i_{max} entry in $\boldsymbol{\nu}_{Dict}$ then provides the estimated T_1 and T_2 values.

15.8.2.2 Provided material

- **ex_15_2_mrf_data.mat** `Dict_sig` is the precalculated dictionary ($\mathbf{D}$) of 60,000 (T_1, T_2) combinations over 1000 MR acquisition data points, `Dict_par` contains the corresponding T_1 and T_2 values ($\boldsymbol{\nu}_{Dict}$), and `Sig` are the four unknown signal curves $\hat{\mathbf{x}}$ (again, each with 1000 data points) from which we want to estimate the T_1 and T_2 values. In addition, the true T_1 and T_2 values of the unknown signals are also provided as `T1_true` and `T2 true`.

15.8.2.3 Questions

1. Carry out dictionary matching with the provided dictionary $\mathbf{D}$, report the found T_1 and T_2 values, and calculate the error compared to the true T_1 and T_2 values.
2. Carry out a singular value decomposition (SVD) of the dictionary along the 1000 data points.
3. Compress the dictionary $\mathbf{D}$ to only include the first N_k SVD components with $N_k \ll 1000$.
4. Repeat the dictionary matching for different values of N_k and report the errors of the T_1 and T_2 estimation.
5. What is the main benefit of carrying out the SVD compression?

Acknowledgment

The images of the brain simulated in this chapter to illustrate the various reconstruction techniques were created based on data from the BrainWeb: http://www.bic.mni.mcgill.ca/brainweb/ [45]. We would like to thank Dr. Martin Uecker (University Medical Center Göttingen) for his help with reproducing Fig. 15.8.

References

[1] Damadian R. Tumor detection by nuclear magnetic resonance. Science 1971;171(3976):1151–3. https://doi.org/10.1126/science.171.3976.1151. Available from: https://www.sciencemag.org/lookup/doi/10.1126/science.171.3976.1151.

[2] Lauterbur PC. Image formation by induced local interactions: examples employing nuclear magnetic resonance. Nature 1973;242(5394):190–1. https://doi.org/10.1038/242190a0. Available from: http://www.nature.com/doifinder/10.1038/242190a0. http://www.nature.com/articles/242190a0.

[3] Seiberlich N, Gulani V, Campbell-Washburn AE, Sourbron S, Doneva M, Calamante F, et al., editors. Quantitative magnetic resonance imaging. 1st ed. San Diego: Elsevier Science Publishing Co Inc. ISBN 9780128170588, 2020.

[4] Schmitt P, Griswold MA, Jakob PM, Kotas M, Gulani V, Flentje M, et al. Inversion recovery TrueFISP: quantification of T(1), T(2), and spin density. Magn Reson Imaging 2004;51(4):661–7. https://doi.org/10.1002/mrm.20058.

[5] Warntjes J, Leinhard OD, West J, Lundberg P. Rapid magnetic resonance quantification on the brain: optimization for clinical usage. Magn Reson Med 2008;60(2):320–9. https://doi.org/10.1002/mrm.21635. Available from: http://doi.wiley.com/10.1002/mrm.21635.

[6] Ma D, Gulani V, Seiberlich N, Liu K, Sunshine JL, Duerk JL, et al. Magnetic resonance fingerprinting. Nature 2013;495(7440):187–92. https://doi.org/10.1038/nature11971.

[7] Heule R, Ganter C, Bieri O. Triple echo steady-state (TESS) relaxometry. Magn Reson Med 2014;71(1):230–7. https://doi.org/10.1002/mrm.24659. Available from: http://doi.wiley.com/10.1002/mrm.24659.

[8] Metere R, Kober T, Möller HE, Schäfer A. Simultaneous quantitative MRI mapping of T1, T2* and magnetic susceptibility with multi-echo MP2RAGE. PLoS ONE 2017;12(1):e0169265. https://doi.org/10.1371/journal.pone.0169265. Available from: https://dx.plos.org/10.1371/journal.pone.0169265.

[9] Sbrizzi A, van der Heide O, Cloos M, van der Toorn A, Hoogduin H, Luijten PR, et al. Fast quantitative MRI as a nonlinear tomography problem. Magn Reson Imaging 2018;46(June 2017):56–63. https://doi.org/10.1016/j.mri.2017.10.015. Available from: arXiv:1705.03209. https://linkinghub.elsevier.com/retrieve/pii/S0730725X17302400.

[10] Christodoulou AG, Shaw JL, Nguyen C, Yang Q, Xie Y, Wang N, et al. Magnetic resonance multitasking for motion-resolved quantitative cardiovascular imaging. Nat Biomed Eng 2018;2(4):215–26. https://doi.org/10.1038/s41551-018-0217-y. Available from: http://www.nature.com/articles/s41551-018-0217-y.

[11] Cheng C, Preiswerk F, Hoge WS, Kuo T, Madore B. Multipathway multi-echo (MPME) imaging: all main MR parameters mapped based on a single 3D scan. Magn Reson Med 2019;81(3):1699–713. https://doi.org/10.1002/mrm.27525. Available from: https://onlinelibrary.wiley.com/doi/abs/10.1002/mrm.27525.

[12] Weigel M. Extended phase graphs: dephasing, RF pulses, and echoes - pure and simple. J Magn Reson Imaging 2015;41(2):266–95. https://doi.org/10.1002/jmri.24619.

[13] Petzschner FH, Ponce IP, Blaimer M, Jakob PM, Breuer FA. Fast MR parameter mapping using k-t principal component analysis. Magn Reson Med 2011;66(3):706–16. https://doi.org/10.1002/mrm.22826.

[14] Zhang T, Pauly JM, Levesque IR. Accelerating parameter mapping with a locally low rank constraint. Magn Reson Med 2015;73(2):655–61. https://doi.org/10.1002/mrm.25161. Available from: http://doi.wiley.com/10.1002/mrm.25161.

[15] Doneva M, Amthor T, Koken P, Sommer K, Börnert P. Matrix completion-based reconstruction for under-sampled magnetic resonance fingerprinting data. Magn Reson Imaging 2017;41:41–52. https://doi.org/10.1016/j.mri.2017.02.007.

[16] Lima da Cruz G, Bustin A, Jaubert O, Schneider T, Botnar RM, Prieto C. Sparsity and locally low rank regularization for MR fingerprinting. Magn Reson Med 2019;(December 2018):mrm.27665. https://doi.org/10.1002/mrm.27665. Available from: https://onlinelibrary.wiley.com/doi/abs/10.1002/mrm.27665.

[17] Doneva M, Börnert P, Eggers H, Stehning C, Sénégas J, Mertins A. Compressed sensing reconstruction for magnetic resonance parameter mapping. Magn Reson Med 2010;64(4):1114–20. https://doi.org/10.1002/mrm.22483.

[18] Li W, Griswold M, Yu X. Fast cardiac T1 mapping in mice using a model-based compressed sensing method. Magn Reson Med 2012;68:1127–34. https://doi.org/10.1002/mrm.23323. Available from: http://www.pubmedcentral.nih.gov/articlerender.fcgi?artid=3324650&tool=pmcentrez&rendertype=abstract.

[19] Knoll F, Bredies K, Pock T, Stollberger R. Second order total generalized variation (TGV) for MRI. Magn Reson Med 2011;65(2):480–91. https://doi.org/10.1002/mrm.22595. Available from: http://doi.wiley.com/10.1002/mrm.22595.

[20] Hamilton JI, Jiang Y, Chen Y, Ma D, Lo WC, Griswold M, et al. MR fingerprinting for rapid quantification of myocardial T 1, T 2, and proton spin density. Magn Reson Med 2017;77(4):1446–58. https://doi.org/10.1002/mrm.26668. Available from: http://doi.wiley.com/10.1002/mrm.26668.

[21] McGivney DF, Pierre E, Ma D, Jiang Y, Saybasili H, Gulani V, et al. SVD compression for magnetic resonance fingerprinting in the time domain. IEEE Trans Med Imaging 2014;33(12):2311–22. https://doi.org/10.1109/TMI.2014.2337321. Available from: arXiv:1533.4406. http://ieeexplore.ieee.org/lpdocs/epic03/wrapper.htm?arnumber=6851901.

[22] Cauley SF, Setsompop K, Ma D, Jiang Y, Ye H, Adalsteinsson E, et al. Fast group matching for MR fingerprinting reconstruction. Magn Reson Med 2015;74(2):523–8. https://doi.org/10.1002/mrm.25439. Available from: http://doi.wiley.com/10.1002/mrm.25439.

[23] Yang M, Ma D, Jiang Y, Hamilton J, Seiberlich N, Griswold MA, et al. Low rank approximation methods for MR fingerprinting with large scale dictionaries. Magn Reson Med 2018;79(4):2392–400. https://doi.org/10.1002/mrm.26867.

[24] Roeloffs V, Uecker M, Frahm J. Joint T1 and T2 mapping with tiny dictionaries and subspace-constrained reconstruction. IEEE Trans Med Imaging 2019;39(4):1008–14. https://doi.org/10.1109/tmi.2019.2939130. Available from: arXiv:1812.09560.

[25] van Valenberg W, Klein S, Vos FM, Koolstra K, van Vliet LJ, Poot DH. An efficient method for multi-parameter mapping in quantitative MRI using B-spline interpolation. IEEE Trans Med Imaging 2019;39(5):1681–9. https://doi.org/10.1109/tmi.2019.2954751.

[26] Poorman ME, Martin MN, Ma D, McGivney DF, Gulani V, Griswold MA, et al. Magnetic resonance fingerprinting part 1: potential uses, current challenges, and recommendations. J Magn Reson Imaging 2020;51(3):675–92. https://doi.org/10.1002/jmri.26836.

[27] McGivney DF, Boyacıoğlu R, Jiang Y, Poorman ME, Seiberlich N, Gulani V, et al. Magnetic resonance fingerprinting review part 2: technique and directions. J Magn Reson Imaging 2020;51(4):993–1007. https://doi.org/10.1002/jmri.26877. Available from: https://onlinelibrary.wiley.com/doi/abs/10.1002/jmri.26877.

[28] Cohen O, Zhu B, Rosen MS. MR fingerprinting Deep RecOnstruction NEtwork (DRONE). Magn Reson Med 2018;80(3):885–94. https://doi.org/10.1002/mrm.27198. Available from: arXiv:1710.05267.

[29] Liu F, Feng L, Kijowski R. MANTIS: model-augmented neural neTwork with incoherent k-space sampling for efficient MR parameter mapping. Magn Reson Med 2019;82(1):174–88. https://doi.org/10.1002/mrm.27707. Available from: arXiv:1809.03308.

[30] Jeelani H, Yang Y, Zhou R, Kramer CM, Salerno M, Weller DS. A myocardial T1-mapping framework with recurrent and U-net convolutional neural networks. In: 2020 IEEE 17th international symposium on biomedical imaging (ISBI). ISBN 9781538693308, 2020. p. 1941–4.

[31] Pierre EY, Ma D, Chen Y, Badve C, Griswold MA. Multiscale reconstruction for MR fingerprinting. Magn Reson Med 2016;75(6):2481–92. https://doi.org/10.1002/mrm.25776.

[32] Look DC, Locker DR. Time saving in measurement of NMR and EPR relaxation times. Rev Sci Instrum 1970;41:250.

[33] Tran-Gia J, Stäb D, Wech T, Hahn D, Köstler H. Model-based Acceleration of Parameter mapping (MAP) for saturation prepared radially acquired data. Magn Reson Med 2013;70(6):1524–34. https://doi.org/10.1002/mrm.24600. Available from: http://www.ncbi.nlm.nih.gov/pubmed/23315831.

[34] Maier O, Schoormans J, Schloegl M, Strijkers GJ, Lesch A, Benkert T, et al. Rapid T 1 quantification from high resolution 3D data with model-based reconstruction. Magn Reson Med 2019;81(3):2072–89. https://doi.org/10.1002/mrm.27502.

[35] Olafsson V, Noll D, Fessler J. Fast joint reconstruction of dynamic R_2^* and field maps in functional MRI. IEEE Trans Med Imaging 2008;27(9):1177–88. https://doi.org/10.1109/TMI.2008.917247. Available from: http://ieeexplore.ieee.org/document/4446611/.

[36] Block KT, Uecker M, Frahm J. Model-based iterative reconstruction for radial fast spin-echo MRI. IEEE Trans Med Imaging 2009;28(11):1759–69.

[37] Wang X, Roeloffs V, Klosowski J, Tan Z, Voit D, Uecker M, et al. Model-based T 1 mapping with sparsity constraints using single-shot inversion-recovery radial FLASH. Magn Reson Med 2018;79(2):730–40. https://doi.org/10.1002/mrm.26726. Available from: http://doi.wiley.com/10.1002/mrm.26726.

[38] Sumpf TJ, Uecker M, Boretius S, Frahm J. Model-based nonlinear inverse reconstruction for T2 mapping using highly undersampled spin-echo MRI. J Magn Reson Imaging 2011;34:420–8. https://doi.org/10.1002/jmri.22634.

[39] Wübbeler G, Elster C. A large-scale optimization method using a sparse approximation of the Hessian for magnetic resonance fingerprinting. SIAM J Imaging Sci 2017;10(3):979–1004. https://doi.org/10.1137/16M1095032. Available from: https://epubs.siam.org/doi/10.1137/16M1095032.

[40] Wang X, Tan Z, Scholand N, Roeloffs V, Uecker M. Physics-based reconstruction methods for magnetic resonance imaging. Philos Trans R Soc A, Math Phys Eng Sci 2021;379:20200196. https://doi.org/10.1098/rsta.2020.0196.

[41] Margaret Cheng HL, Stikov N, Ghugre NR, Wright GA. Practical medical applications of quantitative MR relaxometry. J Magn Reson Imaging 2012;36:805–24. https://doi.org/10.1002/jmri.23718.

[42] Messroghli DR, Moon JC, Ferreira VM, Grosse-Wortmann L, He T, Kellman P, et al. Clinical recommendations for cardiovascular magnetic resonance mapping of T1, T2, T2 and extracellular volume: A consensus statement by the Society for Cardiovascular Magnetic Resonance (SCMR) endorsed by the European Association for Cardiovascular Imagin. J Cardiovasc Magn Reson 2017;19(75). https://doi.org/10.1186/s12968-017-0389-8.

[43] Chen Y, Jiang Y, Pahwa S, Ma D, Lu L, Twieg MD, et al. MR fingerprinting for rapid quantitative abdominal imaging. Radiology 2016;279(1):278–86.

[44] Becker KM, Blaszczyk E, Funk S, Nuesslein A, Schulz-Menger J, Schaeffter T, et al. Fast myocardial T 1 mapping using cardiac motion correction. Magn Reson Med 2020;83(2):438–51. https://doi.org/10.1002/mrm.27935. Available from: https://onlinelibrary.wiley.com/doi/abs/10.1002/mrm.27935.

[45] Cocosco C, Kollokian V, Kwan R, Evans AC. Brainweb: online interface to a 3d mri simulated brain database; 1997.

CHAPTER

Quantitative Susceptibility-Mapping Reconstruction

16

Berkin Bilgic[a,b,c], Itthi Chatnuntawech[d], and Daniel Polak[e]

[a]*Athinoula A. Martinos Center for Biomedical Imaging, Charlestown, MA, United States*
[b]*Harvard Medical School, Boston, MA, United States*
[c]*Harvard/MIT Health Sciences and Technology, Cambridge, MA, United States*
[d]*National Nanotechnology Center, Pathum Thani, Thailand*
[e]*Siemens Healthcare GmbH, Erlangen, Germany*

16.1 Introduction

Quantitative Susceptibility Mapping (QSM) aims to estimate tissue magnetic susceptibility distribution, χ_{tissue}, that gives rise to subtle changes in the main magnetic field of an MRI scanner. These effects can be in the range of ∓ 10 Hz and thus remain in the parts-per-million (ppm) level relative to the Larmor frequency. Despite being small, these changes in the magnetic field can provide a significant contrast to the noise ratio (CNR) boost over standard GRE magnitude images when certain post-processing steps are applied [1] (Fig. 16.1, top). Unfortunately, this tissue field map, δ_{tissue}, depends on the orientation of the tissue relative to the main magnetic field. This implies that, if the same subject is imaged at different positions, drastically different tissue field-map information may be obtained. Fig. 16.1 (bottom) shows the effect of this in imaging of the brain, the most commonly targeted organ in QSM, where subjects were asked to tilt their head to multiple positions.

Another complication in the interpretation of tissue field maps is the influence of a spatial "blur" imposed by the dipole kernel d (Fig. 16.2). This additional blur stems from dipole physics [2], and can be expressed as

$$\chi_{tissue} \otimes d = \delta_{tissue}. \tag{16.1}$$

Accordingly, there is a convolutional relationship between the unknown susceptibility map χ_{tissue} and the acquired field map. This convolution, denoted as $\otimes$, causes a mismatch between the tissue field map and the underlying anatomical boundaries of tissues. QSM aims to deconvolve this dipole kernel to eliminate the orientation bias and spatial blurring effects, while providing a quantitative and sensitive biomarker. The convolutional relationship can be conveniently expressed in k-space via multiplication with the frequency-domain representation of the dipole kernel, D, as follows

$$F^{-1} D F \chi_{tissue} = \delta_{tissue}, \tag{16.2}$$

where F denotes 3D discrete Fourier transform.

Advances in Magnetic Resonance Technology and Applications, Volume 7, ISSN 2666-9099. https://doi.org/10.1016/B978-0-12-822726-8.00027-0

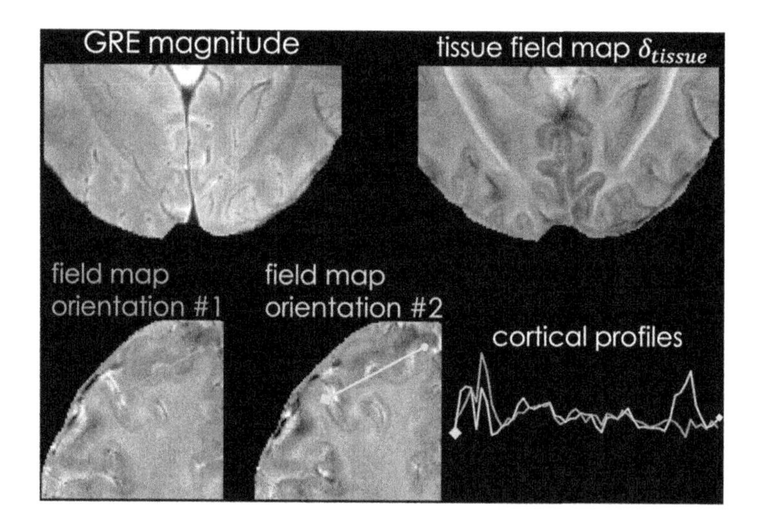

FIGURE 16.1

top: Tissue field-map provides a significant CNR boost of the GRE magnitude signal, especially at ultra-high fields. **bottom**: However, this field map is dependent on the orientation of the head relative to the main magnetic field, which can result in large differences when plotted as a 1D signal profile in the cortex, despite coregistering the two acquisitions.

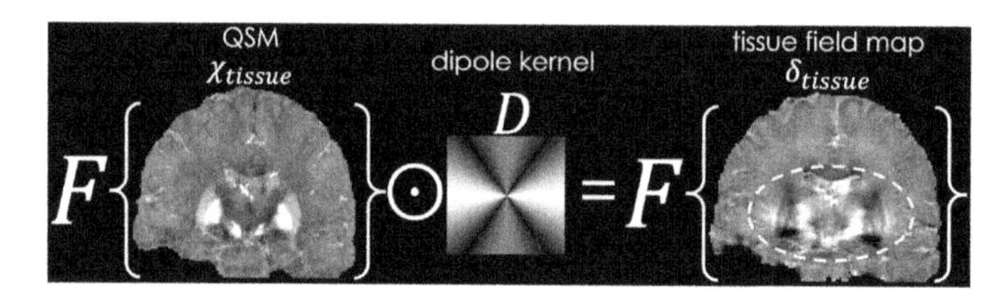

FIGURE 16.2

Tissue susceptibility is related to the acquired field map through a convolution, which can be represented with a voxel-wise multiplication in frequency space (Eq. (16.2)). This convolution operation creates a spatial blur, which complicates the interpretation of the field map since this information does not accurately depict the underlying anatomical boundaries. This can best be visualized in the basal ganglia, as indicated by the yellow (light gray in print version) oval.

This imaging biomarker has been found to be highly correlated with tissue iron concentration, especially in the basal ganglia [3], and has been used to estimate oxygen concentration in the brain vessels [4]. QSM also provides high enough CNR to provide exquisite details at mesoscale spatial resolutions, consistent with the underlying histology [5]. It is important to note that neuroimaging has been the predominant focus of QSM, and, as such, proposed reconstruction techniques have been specialized to the brain. However, recent developments have expanded the application domain of QSM to abdominal [6], cardiac [7], bone [8], and breast imaging [9]. Imaging these organs has been hampered by challenges, including the presence of fat signal, motion, and, in the case of bone imaging, rapid T_2^* decay. These challenges have been mitigated by advanced water–fat signal modeling, breath-held and gated acquisitions, and ultra-short echo times, respectively.

An additional challenge is that, the dipole inversion problem (Eq. (16.1)) is an ill-posed and ill-conditioned linear inverse problem. The equation is not well satisfied due to physical and physiological biases in the data. Moreover, the kernel D undersamples the frequency content of the susceptibility map and is not directly invertible. Before tackling this inverse problem, we first detail data-acquisition aspects and provide a checklist of desirable choices so that the acquired GRE volume lends itself well to reconstruction by the methods presented in the following. We also describe the pre-processing steps that need to be applied on the acquired GRE signal phase to obtain the desired tissue field map. We define and discuss the implementation of these pre-processing steps, detail popular algorithms to solve the inverse problem, and focus on "single-step" approaches that merge the pre-processing and dipole inversion steps. We conclude the chapter by reporting about current challenges, as well as some of the future directions in the field.

16.2 GRE data acquisition

The same GRE acquisition lends itself to both QSM and susceptibility weighted imaging (SWI) [10] processing. Clinical SWI brain-imaging acquisitions are mostly made with a 3D-GRE sequence using high in-plane and relatively low-slice resolution, e.g., $1 \times 1 \times 2$ mm^3. Since a minimum intensity projection (MIP) across the slices is usually performed, such anisotropic voxel sizes can be conveniently used in SWI. For QSM, however, the 3D nature of the dipole convolution often calls for isotropic voxel sizes. A common choice is 1-mm isotropic resolution, which usually has ample SNR at 3T. Although 2D-GRE acquisitions lend themselves to QSM processing as well, they usually fail to match the SNR gain provided by the volumetric noise-averaging ability of 3D-GRE sequences.

A constraint in 3D-GRE acquisitions that may dictate the achievable resolution is the scan time. The magnitude signal in GRE scans decays with T_2^* as a function of the echo time (TE). As such, shorter TEs provide increased magnitude SNR. This is convenient since a short TE would enable a short repetition time (TR) and reduce the overall scan time. The phase SNR, however, is maximized when TE equals to the T_2^* of the tissue of interest [11]. This would require an optimal TE of, e.g., 35 msec at 3T for the brain parenchyma, thus pushing the TR to around 40 msec. Assuming matrix sizes of 240 in phase- and 120 in partition-encoding directions for 1-mm^3 resolution with whole-brain coverage, this would necessitate a ~19-min scan without acceleration. In QSM (and SWI), it is desired to maximize the phase SNR since this is the input data used in subsequent processing. Reaching resolutions such as 1-mm isotropic within clinically feasible scan times of, e.g., 5 min for QSM acquisition, would require additional acceleration. With standard parallel-imaging methods such as SENSE [12] or GRAPPA [13], achieving $R = 2 \times 2$-fold acceleration with modern head coils (e.g., 32 channel) is possible and would render such 5-min QSM acquisition feasible.

Having identified a 5-min 3D-GRE brain-imaging acquisition at 1-mm isotropic resolution at TE/TR = 35–40 msec as a desirable target at 3T, we can next consider boosting its sampling efficiency. One trivial option would be to sample a single echo, and use as low of a readout bandwidth as possible to improve the SNR of this single volume. This would still leave significant unused sequence time prior to this single readout, which can be better utilized by sampling additional, earlier echoes.

FIGURE 16.3

Phase pre-processing steps include coil combination, tissue masking, phase unwrapping (and echo combination in multi-echo acquisitions), and background phase removal. At the completion of these steps, the obtained tissue field map is used in the final dipole-inversion step to estimate the susceptibility map.

Having multiple echoes would both provide the ability to quantify T_2^* maps as an additional quantitative contrast, and also yield an SNR boost via combining the phase information across the echoes. We will touch upon multi-echo combinations in the following section. Finally, we recommend using unipolar echoes to simplify phase combination across the echoes. Although bipolar sampling provides higher efficiency, subsequent echoes suffer from geometric distortions in opposite directions along the readout direction, which need to be corrected before echo combination. The final protocol we recommend is a 5-min, unipolar multi-echo 3D-GRE acquisition using $R = 2 \times 2$ parallel imaging acceleration at TR = 40 msec and sampling multiple echoes ranging from early TEs to the ones closer to the optimal TE = T_2^* point.

16.3 Phase pre-processing

Phase pre-processing pipeline comprises coil combination, tissue masking, phase unwrapping, and background field removal (Fig. 16.3). Each of these steps is discussed hereafter; example links to openly available code are provided.

3.1 Coil combination: is a crucial first step since potential errors may propagate all the way to the final susceptibility map. These errors stem from imperfect estimation and removal of spatially varying phase offsets of each receiver coil. Estimating the coil sensitivities makes it possible to remove the spatially varying coil-phase offsets and obtain an SNR-optimal coil combination. This is referred to as Roemer [14] or SENSE combination [12] and is usually implemented in standard GRE acquisitions. If this is not present in the vendor-provided sequences, either a separate low-resolution reference scan or the fully sampled autocalibration signal (ACS) region of the GRE acquisition can be used in coil sensitivity estimation. Further details can be found in Chapter 6. ESPIRiT algorithm [15] allows for automated sensitivity estimation and is freely available as a part of the BART toolbox (https://mrirecon.github.io/bart/). ESPIRiT provides *relative* coil-phase offsets, and thus needs a coil that can act as the phase reference. This reference coil needs to be free of phase singularities and can be obtained using body coil reception, or with virtual coil computation [16]. At ultra-high fields, there is usually no body coil that would serve as a reference coil, and the simple virtual coil approach may include phase singularities since the sensitivity profiles vary more rapidly in space. To combat this, more involved virtual reference-coil computations, such as block coil compression (BCC), can be used [17] (available at https://bit.ly/370Srl8).

While using an additional GRE calibration scan allows for coil sensitivity mapping, it can also be directly utilized for coil combination without the need for explicit sensitivity estimation. COMPOSER uses a low-resolution, short-TE acquisition for this purpose [18]. This models the acquired coil-phase signal ϕ_{acq} as the summation of a coil phase offset ϕ_{coil} and a time-dependent term:

$$\phi_{acq} = \phi_{coil} + 2\pi \cdot TE \cdot f \tag{16.3}$$

where f denotes the frequency map and TE is the echo time. It can then be seen that using a separate calibration acquisition with $TE \approx 0$ gives a good approximation to the underlying coil-phase offsets due to $\phi_{acq} \approx \phi_{coil}$. Another multi-echo acquisition/reconstruction approach is ASPIRE [19], which is publicly available (http://github.com/korbinian90/ASPIRE) and permits online coil combination at the scanner console.

3.2 Masking: This step becomes necessary for most of the algorithms that are used in the subsequent pre-processing steps. Limiting the computation inside a mask helps improve the processing speed of most phase-unwrapping techniques. During background field removal, large phase contributions due to air–tissue interfaces are estimated and removed. This necessitates masking so that tissue and background regions of interests (ROIs) can be defined. Dipole inversion step that estimates the susceptibility distribution from the tissue phase may also benefit from masking, or from conditioning the inverse problem using an organ-masked weighting matrix.

Current open-source tissue masking algorithms focus on brain imaging. FSL's BET [20] is by far the most commonly used tool for brain imaging (https://fsl.fmrib.ox.ac.uk). It is fast (on the order of seconds) and robust, albeit in ultra-high fields it may benefit from tuning the parameter that controls the mask size. The power of deep learning in segmentation and classification tasks has also been leveraged for masking. An open-source tool for brain imaging is *deepbrain* (https://github.com/iitzco/deepbrain), which aims to improve the robustness and speed of BET.

3.3 Phase unwrapping: Having combined coils (and potentially performed masking), ambiguities caused by multiples of 2π phase jumps need to be removed from the phase image. This process is called phase unwrapping and can be carried out in numerous different ways. Software tools such as FSL's PRELUDE [21], as well as computationally faster SEGUE [22] (https://xip.uclb.com/i/software/SEGUE.html?print) and ROMEO [23] (https://github.com/korbinian90/ROMEO) provide open-source solutions to this problem. An alternative and robust technique is Laplacian unwrapping [24,25]. This can be carried out with or without tissue masking using the following relationship:

$$\Psi(\phi) = \Delta^{-1} \cdot Imag\left(e^{-j\phi} \cdot \Delta e^{j\phi}\right) = \Delta^{-1} \cdot (cos\phi \cdot \Delta sin\phi - sin\phi \cdot \Delta cos\phi), \tag{16.4}$$

where $\Psi(\phi)$ is the unwrapped phase, Δ is the 3D Laplacian operator, and $Imag()$ denotes the imaginary component. Source code and exemplar data that implement Eq. (16.4) are included in *Tutorial 16.A.1*, and are also a part of the STI Suite (https://people.eecs.berkeley.edu/~chunlei.liu/software.html). It should be noted that Laplacian unwrapping is not an unwrapping operator in the sense that the residual $(\Psi(\phi) - \phi)$ is not guaranteed to be an integer multiple of 2π. This, however, may not be crucial because additional background filtering needs to be applied after phase unwrapping, and conventional and Laplacian unwrapping techniques yield very similar results when these filtered tissue field maps are compared [26]. Laplacian unwrapping has emerged as a popular choice since it can relate the wrapped phase to the unwrapped phase through simple trigonometric functions and it readily eliminates some background signals as they satisfy the Laplace equation ($\Delta\phi = 0$).

3.4 Echo combination: In multi-echo acquisitions, a separate image volume is obtained per each echo time. As such, phase information from these multiple images needs to be combined into a single-phase image. One simple way of achieving this is to sum the unwrapped phase from each echo. Since phase unwrapping is not a linear operator, this needs to be performed before summing the echo phase images. The echo-combined phase image ϕ_{sum} is given by

$$\phi_{sum} = \sum_t \phi_t = 2\pi \cdot f \cdot \sum_t TE_t, \tag{16.5}$$

where ϕ_t is the unwrapped phase image from the tth echo and TE_t is the corresponding echo time. Assuming that phase evolves linearly with time and the underlying frequency map is denoted by f, it can be seen that the "effective TE" of this echo-combined phase image is $\sum_t TE_t$. This allows us to compute a frequency map f_{sum} normalized by the effective TE due to $f_{sum} = \frac{\phi_{sum}}{2\pi \cdot \sum_t TE_t}$.

An alternative approach is to normalize the echo-phase images by their TE, and then perform the averaging

$$f_{avg} = \sum_t \frac{\phi_t}{2\pi \cdot TE_t} \tag{16.6}$$

This yields an average frequency map f_{avg}. The main difference between Eqs. (16.5) and (16.6) is the SNR weighting: Eq. (16.5) weights the frequency map f of each echo by its TE, whereas Eq. (16.6) normalizes the effect of TE before summing. Since early echoes have low-phase SNR, it is beneficial to down-weight their contribution. Eq. (16.5) aims to achieve this using TE-weighting, but this may become suboptimal at the very late TEs since the magnitude signal will decay with T_2^* and become negligibly small. A better approach could be using an SNR-optimal combination of echoes as per [11]:

$$f_{weighted} = \sum_t \frac{\phi_t}{2\pi \cdot TE_t} \cdot w_t, \tag{16.7}$$

where the weighting amount of the tth echo w_t is equal to $\frac{TE_t \cdot e^{-TE_t/T_2^*}}{\sum_t TE_t \cdot e^{-TE_t/T_2^*}}$. The T_2^* map that is required in this equation can be readily estimated, e.g., using nonnegative least-squares fitting to the echo-magnitude images. Eqs. (16.5)–(16.7) can be applied after background-field removal as well, but, since background filtering is a linear operation, it is computationally more efficient to perform echo combination first.

Another echo-combination technique is applied on the complex-valued signals directly [27]. This technique estimates a frequency map and a phase offset that match the acquired complex multi-echo images by fitting a complex exponential across the echoes; it is publicly available in the MEDI toolbox (http://pre.weill.cornell.edu/mri/pages/qsm.html). Compared to the preceding echo-combination techniques, this exponential fit is applied prior to phase unwrapping. This estimated frequency map can contain phase wraps and will benefit from unwrapping.

3.5 Background field removal: So far, we have combined coils, performed phase unwrapping, and, for multi-echo acquisitions, combined echoes as well. Using either echo summation, averaging, or weighted combination, these steps yield the frequency map f_{total}. This can be further normalized by the gyromagnetic ratio γ and the field strength B_0 to switch to a unitless scale where the influence of field strength can be eliminated. This normalized field map $\delta_{total} = f_{total}/(\gamma \cdot B_0)$ is expressed in parts per million (ppm), and is approximately the summation of a foreground tissue component δ_{tissue} and an unwanted background signal δ_{back}.

The background-field map δ_{back} stems from several sources including large susceptibility differences between air–tissue and bone–tissue interfaces, the static main magnetic field inhomogeneity, and macroscopic currents [28], and can be more than an order of magnitude stronger than the tissue field map of interest. As such, estimating and eliminating this background component is critical.

Based on underlying assumptions, problem formulation, and characteristics of the solution, background-field removal methods can be rigorously categorized into three types [29]: assumption of no sources close to the boundary (NOS), assumption of no harmonic (NOHA) internal and boundary fields in the boundary region, and minimization of an objective function involving a norm (MOIN). Each type has its own advantages and limitations (summarized in Table 1 of [29]), and, hence, one type may be preferred to the others depending on the situation. According to quantitative comparisons of different background-field removal methods on a numerical model, all the methods performed similarly in deep brain regions, whereas NOS methods resulted in less reliable foreground-tissue component (i.e., internal fields) δ_{tissue} in the cortical regions of the brain [29]. In this chapter, we provide an overview of two of the most widely used background field-removal methods that have repeatedly demonstrated their robustness on *in vivo* data: projection onto dipole fields (PDF) [30] and sophisticated harmonic-artifact reduction for phase data (SHARP) [31].

PDF, a NOHA method, aims to place a background susceptibility distribution χ_{back} outside of the brain that best explains the total field map δ_{total} inside the tissue mask M:

$$\chi_{back}^{pdf} = argmin_{\chi_{back}} \| M(F^{-1} D F \underline{M} \chi_{back} - \delta_{total}) \|_2^2 \tag{16.8}$$

where $\underline{M}$ denotes the complement of the tissue mask (i.e., background mask). Having found this background susceptibility distribution, its effect inside of the tissue foreground is computed by convolving it with the dipole kernel, $\delta_{back}^{pdf} = M F^{-1} D F \underline{M} \chi_{back}^{pdf}$. This estimated background-field map can then be subtracted from the total field to yield the tissue field map estimate due to $\delta_{tissue}^{pdf} = \delta_{total} - \delta_{back}^{pdf}$. An implementation of PDF is included in the MEDI Toolbox.

On the other hand, SHARP, a NOS method, relies on the assumption that the background-field map is harmonic inside the mask. This allows us to invoke the spherical mean-value (SMV) property of harmonic functions, which states that a harmonic function is preserved when convolved with a radially symmetric, nonnegative kernel that sums up to one. As such, when a radial filter r is applied on the total field map, it should eliminate the background component

$$\delta_{total} - r \otimes \delta_{total} = \delta_{back} - r \otimes \delta_{back} + \delta_{tissue} - r \otimes \delta_{tissue} \approx (u - r) \otimes \delta_{tissue} \tag{16.9}$$

where u denotes the Dirac delta function, and we used $\delta_{back} \approx r \otimes \delta_{back}$. We define $h = (u - r)$ to be an SMV filter, which can be applied rapidly using multiplication with its k-space counterpart H. An issue with SHARP arises at the boundary of the tissue foreground mask, where convolution with the

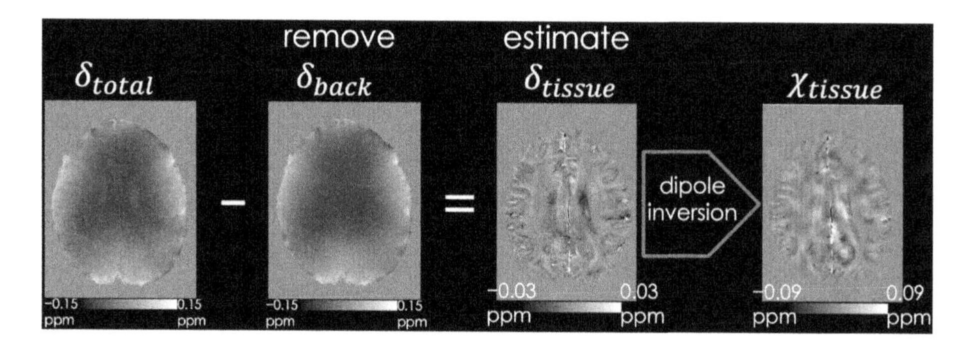

FIGURE 16.4

Background-field removal aims to eliminate contributions from background susceptibility sources that induce slowly varying fields inside of the tissue mask, by which the tissue field map is obtained. This is followed by the dipole inversion step that estimates the tissue susceptibility that gave rise to the tissue field map. Note that the dynamic range of the tissue field map is 5x smaller than those of the total and background fields.

radial filter has contributions from both the tissue and background components due to the extent of the kernel. This ambiguity necessitates eroding several voxels away from the surface of the brain mask and leads to the following linear system:

$$M_{sharp} F^{-1} H F \delta_{total} = M_{sharp} F^{-1} H F \delta_{tissue}, \tag{16.10}$$

where M_{sharp} is the eroded mask. Eq. (16.10) is underdetermined, and its conditioning can be improved using an SMV with a larger radius. However, a larger filter would require more erosion and further reduce the size of the tissue mask. To address this complication, V-SHARP proposes to use multiple filters with varying radii [32]. Large filters provide high-fidelity tissue field-map estimates in the interior of the object, whereas using smaller filter sizes towards the edge of the tissue helps preserve more information at the tissue surface. V-SHARP filtering is governed by the following equation,

$$\sum_i M_i F^{-1} H_i F \delta_{total} = \sum_i M_i F^{-1} H_i F \delta_{tissue} \tag{16.11}$$

where M_i represents the reliable field-map mask belonging to the ith SMV filter H_i. Eq. (16.11) can be solved iteratively with Tikhonov regularization [33], or the impact of SMV filtering can be deconvolved due to

$$\delta_{tissue}^{v-sharp} = F^{-1} H_{inv} F \sum_i M_i F^{-1} H_i F \delta_{total}, \tag{16.12}$$

where H_{inv} denotes the inverse of the largest SMV filter, which needs to be truncated with a certain threshold parameter [34]. This deconvolution step may also be bypassed since large SMV filters have a flat point-spread function in k-space [32]. An implementation of V-SHARP is included in Tutorial 16.A.1, as well as in the STI Suite. The background field removal step is depicted in Fig. 16.4, which is followed by the dipole inversion step that will be covered in the next section.

16.4 Dipole inversion

In the previous section, several pre-processing steps were introduced to convert the raw GRE phase to a tissue field map. In the last step of the reconstruction, an inverse problem known as the dipole inversion needs to be solved to estimate the desired susceptibility. The QSM forward model relates the susceptibility χ_{tissue} to the tissue field δ_{tissue} using the dipole kernel d. This relation can either be expressed as a convolution in the image domain (Eq. (16.1)) or more conveniently as a multiplication in k-space (Eq. (16.2)). Unfortunately, this inverse problem is ill-conditioned because the conical dipole kernel in k-space has diminishing values on its surface (Fig. 16.2) and thus undersamples the frequency space of the susceptibility χ_{tissue}. This impedes solving Eq. (16.2) using a simple voxel-wise kernel division in k-space because this would cause streaking artifacts in the reconstructed susceptibility map.

The inversion of the QSM forward model does not lead to a unique solution and hence multiple susceptibility distributions are in agreement with the acquired MR phase signal. To identify the physically meaningful solution, one can either aim to "fully sample" the susceptibility by acquiring data in multiple orientations with respect to the main magnetic field B_0, or utilize regularization in the reconstruction. Regularized approaches can be divided into image- and k-space methods that either allow a closed-form solution or necessitate iterative optimization. In the following, four common dipole inversion techniques are briefly introduced, and example reconstructions are provided in Fig. 16.6.

16.4.1 COSMOS

The dipole inversion in QSM is ill-posed and ill-conditioned, e.g., without prior constraints, an image reconstruction will assign arbitrary values in the zero-cone neighborhood of the dipole kernel in k-space. Calculation of susceptibility using multiple orientation sampling (COSMOS) [35] recovers the "missing" k-space information from these regions. This is achieved by changing the orientation of the subject's head with respect to the main magnetic field B_0 and reacquiring the MR data. In the COSMOS reconstruction, the tissue field maps from all orientations are coregistered. Then, an overdetermined linear system of equations is solved to estimate the susceptibility map that explains the tissue field-map data from each orientation.

Using the head orientation index $j \in \{1, \ldots, N\}$, where D_j corresponds to the dipole kernel for the jth orientation in k-space, this inverse problem can be formulated in the matrix form

$$\begin{bmatrix} D_1 \\ \vdots \\ D_N \end{bmatrix} F\chi = \begin{bmatrix} F\delta_1 \\ \vdots \\ F\delta_N \end{bmatrix}. \tag{16.13}$$

It also admits the closed-form solution $\chi = F^{-1}(\sum_j D_j^T D_j)^{-1} \sum_j D_j^T F\delta_j$ and thus allows reconstructions with minimal computational footprint.

Reorientation and reacquisition of MR data at different angles with respect to the main magnetic field reduces the ambiguity of the ill-conditioned dipole inversion. This is because the dipole kernel rotates for each orientation and, hence, spatial frequency undersampling due to diminishing values on the surface of the conical dipole kernel in k-space vary in position (Fig. 16.5). This facilitates QSM reconstruction as the dipole inversion problem is no longer underdetermined. Unfortunately, the need for multi-orientation data also leads to longer scan times which have impeded widespread clinical

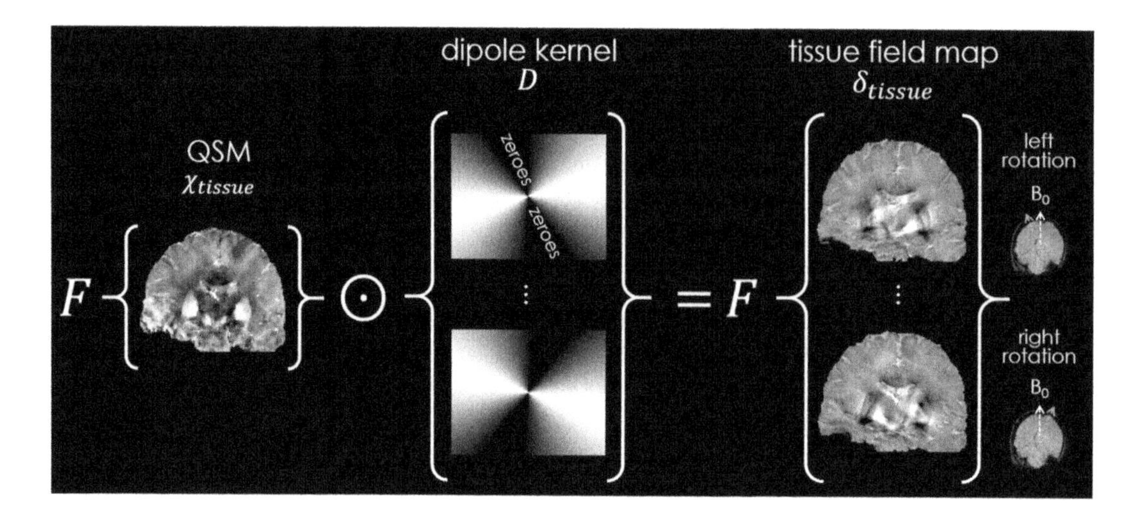

FIGURE 16.5

COSMOS recovers "missing" k-space information in the zero-cone region of the dipole kernel by re-orienting the subject with respect to the main magnetic field B_0 and resampling the MR data. After coregistration of the tissue field maps, an overdetermined linear system of equations is solved. COSMOS reduces the ambiguity of the dipole inversion since the position of the zero-cone area of the dipole kernel varies in each orientation, thus each tissue field map provides complementary k-space information about the desired susceptibility.

usage. However, COSMOS often serves as a ground-truth reference when evaluating and comparing dipole-inversion techniques.

16.4.2 K-space reconstruction with closed-form solution

In the previous section, the conditioning of the dipole inversion was improved by better sampling the susceptibility distribution through the acquisition of multi-orientation data. In this section, regularization is employed to identify a unique solution. Regularization can be applied directly in k-space, and truncated k-space division (TKD) [36] is the most commonly used technique among the k-space methods. It inverts the dipole kernel D directly and replaces small absolute values in D by a constant number ϵ.

$$\hat{D}(k) = \frac{sign(D(k))}{max(|D(k)|, \epsilon)} \tag{16.14}$$

Here, the division is performed in an element-wise fashion, and $D(k)$ denotes the kth diagonal element of the matrix D. The desired susceptibility is then obtained using the closed-form solution that involves Fourier transforms and simple element-wise matrix multiplication, $\chi = F^{-1}\hat{D}F\delta_{tissue}$. Note that closed-form solutions also exist for, e.g., quadratic (L2) gradient penalties [37]. Unfortunately, the modification of the dipole kernel in k-space can lead to systematic underestimation of the tissue susceptibility in k-space, especially around the zero-cone region of the dipole kernel. Corresponding gaps

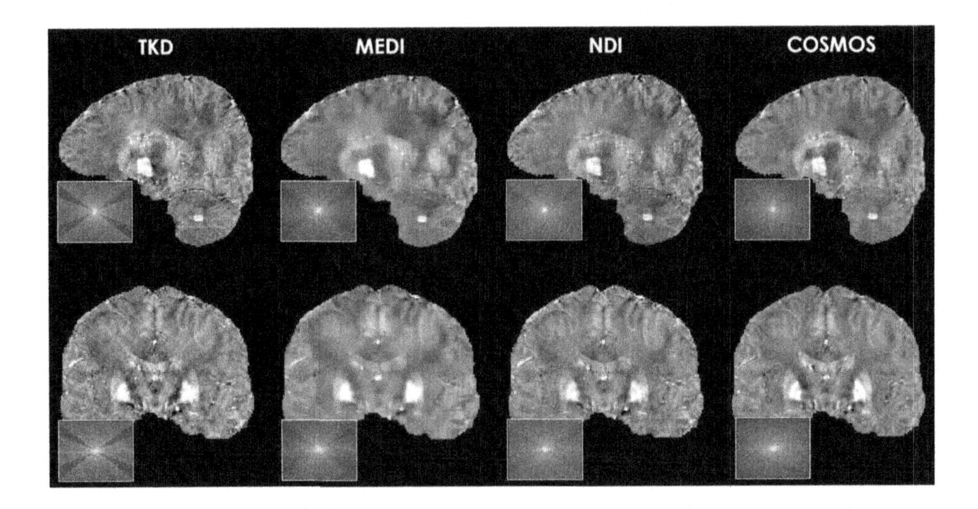

FIGURE 16.6

Comparison of single-orientation TKD, NDI, and MEDI. COSMOS was computed on data from five head orientations and serves as a ground-truth reference. The truncation of the dipole kernel in TKD results in residual streaking artifacts and underestimation of the susceptibility signal in image space. In k-space these artifacts manifest as gaps around the zero-cone of the dipole kernel. These artifacts can be mitigated, albeit at the cost of oversmoothing in MEDI. The gradient-based regularization suppresses streaking artifacts and fills gaps in k-space. Similar image quality is also obtained from NDI, which does not use explicit regularization but relies on the inherent regularization effect from magnitude weighting and nonlinear formulation.

in k-space typically manifest as streaking artifacts and noise amplification in the image domain [38], which is also seen in the example reconstruction in Fig. 16.6.

16.4.3 Iterative reconstructions in image space

Several dipole inversion techniques that exploit prior constraints in the image domain have been proposed. These priors are applied either to the image itself or a representation using a transform. The image reconstruction then involves the minimization of a cost function that typically follows Eq. (16.15).

$$minimize_{\chi} \frac{1}{2} \left\| F^{-1} D F \chi - \delta_{tissue} \right\|_2^2 + \lambda R(\chi) \tag{16.15}$$

The first term is the data-consistency term which enforces agreement with the QSM forward model. The second term is the regularizer that imposes prior constraints on the reconstructed susceptibility map (such as image smoothness). A regularization parameter λ is used to balance data-consistency (model accuracy) and regularization error and needs to be tuned for a specific application.

In the literature, a wide range of regularized dipole-inversion techniques has been reported, which mainly differ in the choice of the regularizer [27,37,39–41]. While the majority of these methods require iterative optimization, some of them also admit closed-form k-space solutions. In this section, we will briefly introduce Morphology-Enabled Dipole Inversion (MEDI) [39], a commonly used iterative technique. In MEDI, anatomical information from the magnitude image is used as a prior for the image reconstruction. This is based on the assumption that edges in the desired susceptibility should be colocated with anatomical edges in the known GRE magnitude image. MEDI (Eq. (16.16)) incorporates this idea into the regularization term where a 3D gradient operator ∇ acts on the susceptibility χ. The gradient is weighted by a diagonal binary matrix M, derived from the magnitude image, to exclude known anatomical edge structures from the gradient penalty. The use of the L1-norm then enforces gradient sparsity and thus suppresses streaking artifacts. Moreover, a diagonal noise-weighting matrix W is included in the data-consistency term to account for the nonuniform phase-noise distribution across the image.

$$minimize_{\chi} \frac{1}{2} \left|\left| W(F^{-1}DF\chi - \delta_{tissue}) \right|\right|_2^2 + \lambda \left|\left| M\nabla\chi \right|\right|_1 \tag{16.16}$$

In practice, the acquired MR data are contaminated by noise, which can lead to unreliable field map values. Especially in low-SNR regions, the phase-noise distribution can deviate from a Gaussian distribution and is hence not governed by the linear susceptibility-to-field relationship shown in Eq. (16.15). This mismatch can lead to artifacts in the reconstructed susceptibility map, particularly in patients who exhibit strong susceptibility signals, e.g., due to hemorrhages. This issue was recognized in the nonlinear-MEDI (NMEDI) approach [27], which employs a nonlinear data-consistency term (Eq. (16.17)) to better model the complex MR signal and noise distribution.

$$minimize_{\chi} \frac{1}{2} \left|\left| W(e^{iF^{-1}DF\chi} - e^{i\delta_{tissue}}) \right|\right|_2^2 + \lambda \left|\left| M\nabla\chi \right|\right|_1 \tag{16.17}$$

Regularized reconstructions, including (N)MEDI, typically require tuning of the regularization parameter λ. This value trades-off model accuracy and image smoothness, and the optimal choice of this parameter can be subject- and application-specific. Moreover, nonlinear regularizers, as used in (N)MEDI, often require complicated iterative optimization techniques that can be time-consuming. The recently proposed FANSI algorithm [42] aims to address this issue. It minimizes the same cost function as in NMEDI (Eq. (16.17)) but relies on a different optimization strategy (Split Bregman method [43] and ADMM [44]), which transforms the regularization penalty and nonlinear data consistency into simpler decoupled problems that can be solved more rapidly. In particular, parameter splitting allows for using a voxel-wise Newton–Raphson approach to rapidly solve for the roots of the nonlinear data-consistency function with closed-form iterations. While this can provide up to ten-fold speedup, it comes at the cost of additional regularization parameters that need to be tuned. Implementations of both (N)MEDI and FANSI are publicly available (http://pre.weill.cornell.edu/mri/pages/qsm.html; https://gitlab.com/cmilovic/FANSI-toolbox).

So far, we have briefly introduced popular dipole-inversion techniques that impose prior constraints by explicitly regularizing the reconstructed susceptibility map. In the recently proposed NDI technique (Nonlinear Dipole Inversion, [45]), regularization terms are removed, and the inherent regularization effect from magnitude weighting and nonlinear formulation is exploited. The NDI cost function is a special case of NMEDI and is obtained by choosing $\lambda = 0$ in Eq. (16.17).

$$f(\chi) = \frac{1}{2} \left|\left| W(e^{iF^{-1}DF\chi} - e^{i\delta_{tissue}}) \right|\right|_2^2 \tag{16.18}$$

The removal of gradient-based regularization admits a simple gradient-descent optimization via the analytical derivative $\nabla_{\chi} f(\chi)$ [39]. Using an iterative update rule, the tth image estimate in the NDI reconstruction becomes

$$\chi^{[t+1]} = \chi^{[t]} - D^T W^T W sin(F^{-1} DF\chi^{[t]} - \delta_{tissue}) \ with \ \chi^{[0]} = 0, \tag{16.19}$$

which can be implemented, e.g., in MATLAB®, using just a few lines of code, and is explored in Tutorial 16.A.2.

Gradient-descent optimization (see Chapter 3) is often prone to error propagation/noise amplification in ill-conditioned problems. As previously reported [2], early iterations mainly fill k-space outside of the zero-cone region, while later iterations insert structured noise into this area to further reduce the cost function. Unfortunately, the latter often causes streaking artifacts and noise amplification in the image domain. NDI also suffers from this overfitting problem when too many iterations are performed. However, this issue can be mitigated by the use of early-stopping (implicit regularization) or a small amount of Tikhonov regularization. This improves the convergence of the optimization and typically does not result in oversmoothing, which is often observed in highly regularized techniques.

16.5 Recent advances: single-step QSM and deep-learning-based QSM

The QSM methods that we have discussed so far are made up of a sequence of multiple processing steps including phase unwrapping, background-field removal, and dipole inversion. These multistep QSM methods often suffer from error propagation between these processing steps. For instance, an erroneous phase-unwrapping algorithm that gives rise to a corrupted unwrapped phase adversely affects the result obtained from the subsequent background field-removal step. The errors from both the phase unwrapping and background-field removal steps could then get propagated to the dipole-inversion step.

In order to tackle such an error propagation, QSM methods that perform the processing steps simultaneously have been proposed [16,40,46–53] (Fig. 16.7). Several optimization problems, each with the objective function that simultaneously incorporates the phase unwrapping, background-field removal based on the mean-value property of harmonic functions, and regularized dipole inversion components have been formulated [16,40,46–51]. Despite sharing the same principle, these proposed formulations differ in how each of the processing steps is incorporated. They all use Laplacian-based methods to perform phase unwrapping and background-field removal. However, a single Laplacian operator is used in [40,46,48,51], whereas multiple operators are used in [16,47,49,50]. The regularization part of the objective functions also differs: the l_2-norm of the weighted gradients of the tissue magnetic susceptibility used in [46,47], weighted total variation used in [16,48–50], and total generalized variation used in [16,40,50,51]. Some of these methods have been made publicly available (http://martinos.org/~berkin/software.html and http://qsm.neuroimaging.at). The general optimization problem solved in single-step methods is of the form

$$minimize_{\chi} \frac{1}{2} \left\| \sum_i M_i F^{-1} H_i DF\chi - M_i F^{-1} H_i F\delta_{tissue} \right\|_2^2 + R(\chi), \tag{16.20}$$

where H_i and M_i again denote the ith SMV kernel and reliable mask region and $R(.)$ is a regularizer. The solution to these single-step optimization problems is an estimate of the underlying magnetic susceptibility map in the region of interest, χ_{tissue}. It has been shown that these single-step QSM

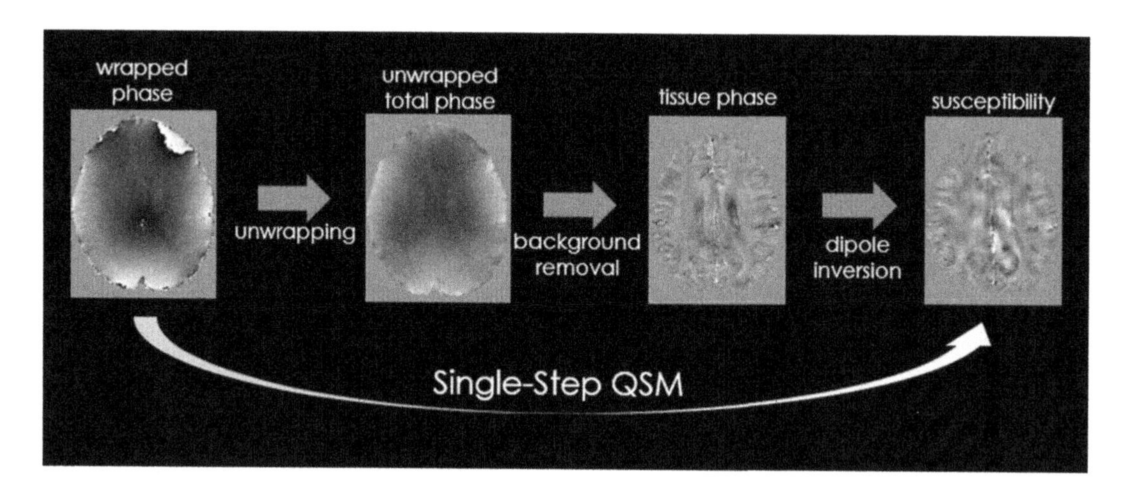

FIGURE 16.7

As opposed to a standard multistep QSM pipeline, which performs phase unwrapping, background field removal, and dipole inversion sequentially (the blue (dark gray in print version) arrows), single-step QSM methods perform these operations simultaneously (the yellow (light gray in print version) arrow).

methods resulted in magnetic susceptibility estimates with fewer artifacts than those of the multistep QSM methods [16,40,46–53]. In Tutorial 16.A.3 at the end of this chapter, we will implement the total variation-regularized single-step QSM method proposed in [16,50] and compare it to some of the multistep QSM methods.

The Laplacian-based single-step QSM methods mentioned in the preceding paragraph, especially those that use a single Laplacian operator, typically lose the information near regions with a large susceptibility dynamic range. To address this "erosion" problem, the total field inversion (TFI) [52] and its fast version [53] have been proposed. The TFI methods model the total susceptibility map, χ_{total}, as the summation of the susceptibility map inside a region of interest, χ_{tissue}, and the susceptibility map outside the region of interest, χ_{back}, as $\chi_{total} = \chi_{tissue} + \chi_{back}$. Under this model assumption, χ_{tissue} and χ_{back} can be jointly estimated without having to evaluate the Laplacian, shielding the TFI methods from the erosion problem [52]. It has been demonstrated that the TFI methods not only were able to reconstruct a larger region of the susceptibility maps, but also yielded improved susceptibility map estimation near the brain regions with large susceptibility differences such as the brain boundary and the interface between intracerebral hemorrhage (ICH) and surrounding tissue, compared to several Laplacian-based single-step QSM methods [52,53].

While all of these single-step methods yield superior results compared to those obtained from the multistep QSM methods (see Fig. 16.8 for example results), they are time-consuming iterative algorithms that rely on carefully chosen regularization parameters, limiting their practicality in clinical settings. Recently, several deep learning-based QSM methods have been proposed to make QSM become more practical by not only reducing computational workload through the use of non-iterative algorithms without requiring users to select regularization parameters at test time, but also providing

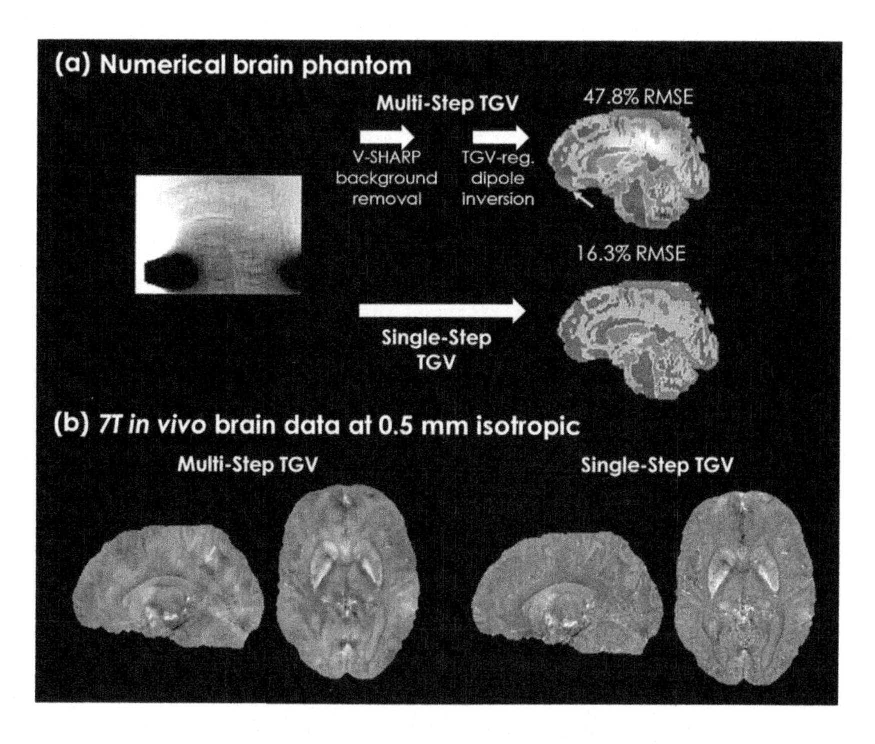

FIGURE 16.8

Comparisons between the TGV-regularized multistep QSM and TGV-regularized single-step QSM methods on (**a**) numerical brain phantom and (textbfb) 7T *in vivo* brain data. Residual artifacts are clearly visible in the reconstructed susceptibility maps obtained from the multistep TGV method (indicated by the yellow (light gray in print version) arrows).

improved reconstruction quality [45,54,63–72,55,73–81,56–62]. These deep learning-based approaches often use high-quality COSMOS images acquired on multiple subjects as reference data.

A deep learning-based method (please see also Chapter 11) typically requires large amounts of data to train its model to perform the given tasks successfully, resulting in a training phase that is computationally expensive due to iterative optimization. Nevertheless, the test phase can be done much more quickly. The fast inference time of deep learning-based methods is an attractive characteristic for the practical aspect of QSM (Fig. 16.9). Developing deep learning-based QSM methods is thus a promising direction to accelerate QSM towards clinical applications.

With the availability of increasing amounts of QSM data, deep learning has rapidly gained success in the QSM community. In particular, deep learning has been used to improve each of the processing steps in QSM: phase unwrapping [54–56], background-field removal [57–59], and dipole inversion [45, 59,68–74,60–67]. Please refer to Table 1 of Ref. [75] for a comprehensive list of these deep learning-based QSM methods with their key features. The authors of [75] have also put together a list of the deep learning-based QSM implementations that are available online on https://github.com/dlQSM/.

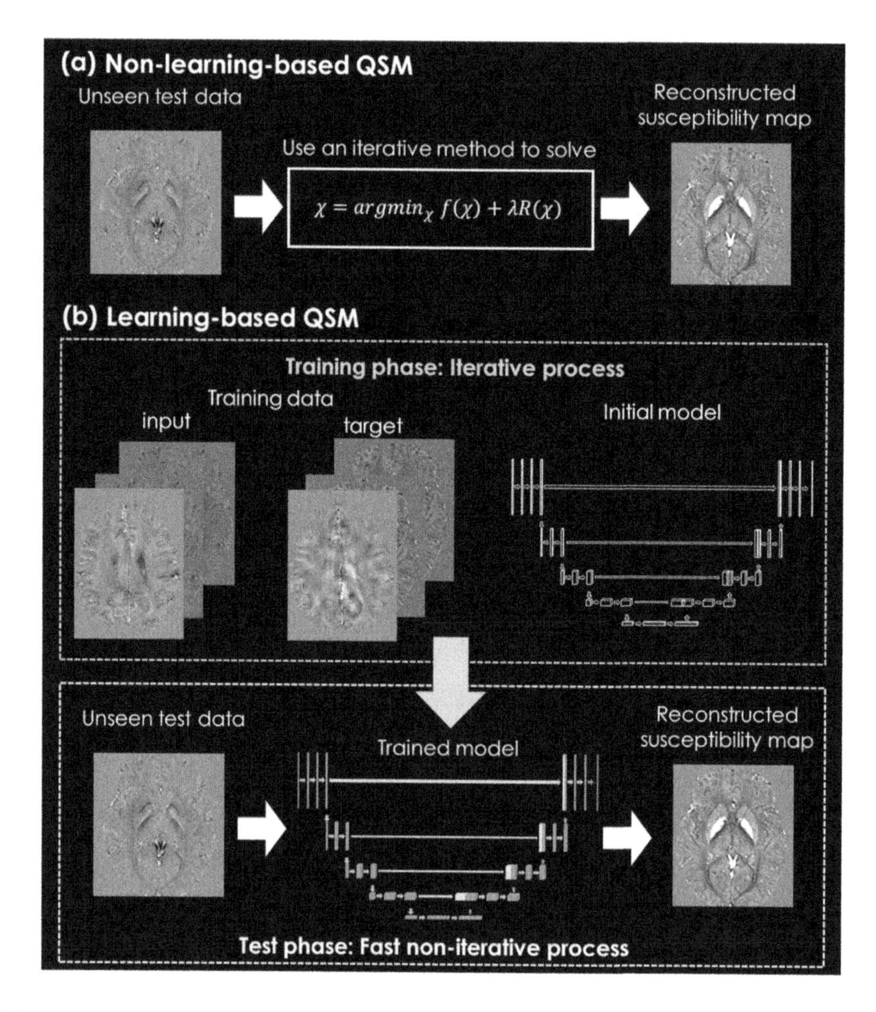

FIGURE 16.9

(**a**) Non-learning-based QSM methods reconstruct the magnetic susceptibility map using an iterative algorithm. (**b**) Learning-based QSM methods usually consist of two phases: training and test phases. The training phase creates an initial model and iteratively updates the model using training data. The test phase reconstructs the susceptibility map using a single forward pass through the trained model. This non-iterative process in the test phase enables fast magnetic susceptibility reconstruction.

In addition to improving each of the QSM processing steps separately, deep learning-based methods that perform some of the processing steps simultaneously have been proposed [76–81]. Heber et al. have proposed a deep learning-based single-step QSM using a single fully convolutional network that is a concatenation of two U-shaped subnetworks: one performing the background field-removal step and the other performing the dipole inversion step [76]. The proposed method uses a simulated total phase map as its input and the corresponding susceptibility map in the region-of-interest as its target (or labels) to train its two subnetworks to perform the two steps simultaneously. This is in contrast to the deep learning-based QSM method proposed in [59] that trains its two subnetworks separately. Liu and Kock have proposed deep learning-based single-step QSM methods that use a single 3D encoder–decoder network [77,78]. The proposed methods use the whole brain total-field map and brain mask as the inputs to the network and reconstruct both the underlying magnetic susceptibility map and the local field map simultaneously in [77], or reconstruct only the susceptibility map in [78]. Similarly, Gessner et al. explored the possibility of reconstructing the susceptibility map directly from the total field map and the mask defining the background regions using a single two-channel-input U-Net [79]. Kames et al. have proposed variants of variational network-based single-step QSM methods that estimate the magnetic susceptibility map from the total phase and an initial estimate of the susceptibility map [80]. Wei et al. have developed the deep learning-based *autoQSM* that reconstructs the magnetic-susceptibility map directly from the total phase map without requiring brain-volume extraction, bypassing the error-prone organ-masking process that frequently causes "erosion" problem [81].

Deep learning-based methods work especially well when training data and test data are drawn from the same underlying probability distribution. However, in most practical applications, the underlying probability distribution is unknown. Consequently, the success of many deep-learning models rely heavily on having access to large amounts of training data with enough variability to ensure that the model can be generalized to unseen test data. Unfortunately, it is costly to acquire extensive QSM data with enough variability. With the limited amounts of training data, the degraded performance of deep learning-based QSM methods can be observed when the unseen data characteristics differ from those of the training data [70]. It has been shown that QSMNet [61], when trained with data from only healthy volunteers, underestimated the susceptibility values in the hemorrhagic lesions of the patient test sets [70]. The different susceptibility dynamic range between the training data (−0.41 ppm–0.61 ppm) and the test data (approximately −0.5 ppm–1.40 ppm) led to the degraded performance of QSMNet. In order to mitigate this out-of-distribution problem, the authors augmented the training data by scaling their susceptibility values and demonstrated improved performance on the test set [70]. In addition to the susceptibility dynamic range mismatch, the resolution mismatch, different noise levels of the tissue phase, and different B_0 orientations also lead to degraded performance [75,79,82].

There are several approaches that can be adopted to mitigate this out-of-distribution problem in QSM, some of which have already been mentioned in the preceding paragraphs. Synthetic QSM datasets can be simulated with a wide range of conditions, such as different spatial resolutions, SNRs, B_0 orientations, and magnetic-susceptibility dynamic ranges. More data can also be generated by applying data augmentation to existing in vivo data such as scaling susceptibility values as done in [70], spatially shifting and rotating the data, and passing the estimated susceptibility map through a physics-based forward model. Moreover, widespread data-, model-, and code sharing within the QSM community would rapidly enable access to large QSM datasets and also better benchmarking of different QSM methods. Combining both synthetic and *in vivo* datasets with data augmentation techniques for training could also help with the generalization capability. Regarding the model training procedures, additional types of loss functions can be incorporated into the main objective function, such as the adversarial loss [63,83], perceptual/feature loss [84], or loss functions specifically designed to address susceptibility anisotropy, to complement the more commonly used Mean-Square-Error (MSE) and Mean-Absolute-Error (MAE) losses, which are pixel-based losses that are not designed for capturing perceptual characteristics. Furthermore, developing semi-supervised, self-supervised, or unsupervised learning-based methods [74,85] (see also Chapter 11) that typically require training data along with fewer or none of the targets (labels) than commonly used supervised-learning-based QSM methods would also be a promising complementary approach toward bringing QSM to clinical applications.

16.6 Summary and outlook

QSM is a novel biomarker that has been linked to several physiological and pathological conditions. However, the required pre-processing and reconstruction steps may seem involved, and this chapter aimed to provide an overview and guideline to these numerous processing steps. In the following, the reconstruction pipeline will be briefly summarized and some recommendations for practical use will be given. We conclude this section with an outlook and information about ongoing communal efforts.

QSM reconstruction typically involves several pre-processing steps that convert the raw GRE data to a tissue field map. This pre-processing typically starts with the coil combination of multichannel data where vendor implementations of Roemer combination help estimate and remove phase offsets between coil channels. In the absence of such implementations, ESPIRiT can be used as a powerful alternative. The resulting coil-combined image is then brain masked to improve the efficiency of subsequent processing steps, and we recommend the use of FSL BET as a rapid and robust tool. As a next step, the image phase needs to be unwrapped to remove 2π phase jumps and obtain a global phase map. This processing step can be carried out in several different ways and established algorithms either rely on path-following or Laplacian techniques, which often result in comparable quality as recently demonstrated by an excellent review article [86]. As part of this contribution, we also provided source code and example data for Laplacian Unwrapping (see Tutorial 16.A.1) which should allow the interested readers to become familiar with this important processing step. Last, any unwanted background-phase contributions need to be removed to obtain a tissue phase map. In this chapter, we discussed two of the most widely used methods for background-phase removal, PDF and SHARP, but we would like to refer the interested readers to the following review article [29] for further information.

Next, the desired quantitative susceptibility is extracted from the tissue phase, and this step involves the solution of a complicated inverse problem. This is because the dipole kernel in k-space undersamples the frequency content of the desired susceptibility map and thus physically meaningful solutions can only be obtained through data oversampling or the use of prior information. The COSMOS algorithm oversamples the dipole inversion by jointly reconstructing data acquired in multiple orientations with respect to the main magnetic field, and this technique is often referred to as a gold standard due to its high-quality reconstructions. However, the need for reacquisitions in multiple orientations (at least three) and the associated long scan times limit its widespread clinical usage. In contrast, regularized algorithms enforce additional constraints, such as image smoothness to facilitate single-direction reconstructions, and MEDI is among the most widely used techniques. However, to trade-off accuracy/smoothing vs. streaking artifacts, regularization parameter(s) need to be well-chosen, and in Tutorial 16.A.2 we demonstrate the need for parameter tuning, which can be a tedious task. Moreover, regularized reconstructions often do not permit closed form solutions but require time-consuming iterative optimization that can be prohibitive for clinical usage.

Given these considerations, our overall recommendation for an easy-to-use, robust pipeline is to unwrap the images with Laplacian unwrapping (sum the unwrapped phase images across the echoes in a multi-echo acquisition), perform V-SHARP background-field removal, and continue with NDI dipole inversion. These steps have been implemented in Tutorials 16.A.1 and 16.A.2.

In this chapter, we also discussed recent developments in QSM. Single-step algorithms combine phase unwrapping, background-field removal, and dipole inversion into a single optimization problem. This was shown to reduce error propagation between the different processing steps thus yielding improved susceptibility-map estimation especially near regions with large susceptibility differences, such as the brain boundary and interfaces between intracerebral hemorrhage (ICH) and surrounding tissue. In Tutorial 16.A.3, we also provided example code that enables the interested readers to explore this novel field of research. However, while single-step methods generally yield superior image quality than classical multistep approaches, they entail an even more complicated reconstruction with increased computational footprint. In addition, regularization parameters still need to be chosen carefully to prevent oversmoothing, which poses yet another challenge for clinical adoption. Therefore, recent advances in deep-learning reconstruction may present a viable alternative to classical physics-based algorithms. These techniques have shown improvement across all QSM processing steps, offering improved image quality and faster reconstruction. Moreover, these methods no longer necessitate manual tuning of regularization parameters and thus offer greater practicability for clinical deployment. However, to achieve robust reconstructions, supervised deep-learning networks require training across large representative sets of training cases, and the availability of such datasets remains challenging and costly. Further, supervised deep-learning approaches are not guaranteed to generalize to unseen patient data and may thus result in undesirable image artifacts and distortions. These issues are often a consequence of limited training data and are increasingly tackled through synthetic/simulated data as well as data sharing.

To foster more widespread clinical usage of QSM, it will be crucial to resolve the remaining implementation challenges, and we believe that the use of deep learning could play an important role in this. In addition, more clinical studies will be needed to identify useful applications and increase the clinical utility of QSM. This may also increase the likelihood of commercial vendors to offer online scanner implementations and thus obviate the current need for offline data processing.

Ever since the advent of the Susceptibility Weighted Imaging (SWI) approach [10], there has been tremendous interest in mapping the phase and susceptibility distributions in the brain, with recent demonstrations in abdominal [6] and cardiac imaging [7]. The growth in the QSM community has been fostered by five ISMRM-sponsored workshops and two reconstruction challenges [87,88] to date. EMTP Hub (https://www.emtphub.org/) aims to bring QSM, as well as electrical properties-mapping communities together, and provide a common platform to share code, data, literature, and tutorials. We look forward to seeing these communal efforts on technical development to help deploy QSM in neuroscientific studies and clinical applications. For further information on the physics and applications of QSM, we refer the interested readers to excellent review articles in [2,89].

Appendix 16.A Tutorials

16.A.1 Phase pre-processing

16.A.1.1 Provided materials

- **main_phase_preprocessing.m** The main file that loads exemplar 3D-GRE data, and performs phase unwrapping and V-SHARP filtering.
- **imagesc3d.m** This function displays 3D data such as a magnetic-susceptibility map.
- **laplacian_unwrap.m** This function performs Laplacian unwrapping.

- **create_SMVkernel.m** This function creates spherical mean value (SMV) kernels.
- **msk.mat** Brain mask obtained from FSL BET.
- **prot.mat** Header information related to the 3D-GRE acquisition.
- **IMG.mat** Complex-valued 3D-GRE data that were acquired with five head orientations. The first orientation is processed in the main function.

16.A.1.2 Exercises

(a) Display and compare V-SHARP reconstructions by varying the maximum SMV radius in *main_phase_preprocessing.m.*

(b) V-SHARP result can be deconvolved with the response function of the largest SMV filter. This deconvolution benefits from additional regularization. Observe the effects of varying the regularization parameter on the final deconvolved tissue field map.

16.A.2 Dipole inversion

16.A.2.1 Provided materials

- **main_dipole_inversion.m** The main file loads the pre-processed tissue phase, GRE magnitude, and brain mask data. TKD and NDI are computed on single-orientation data, while COSMOS utilizes data from five head orientations.
- **imagesc3d.m** This function displays 3D data such as a magnetic-susceptibility map.
- **ndi_gradient_descent.m** This function performs the iterative NDI gradient-descent optimization
- **mag_dir1.mat** Magnitude image (single-orientation)
- **tissue_phase_dir1.mat** Tissue phase image (single-orientation)
- **mask_dir1.mat** Brain mask (single-orientation)
- **tissue_phase_all.mat** Tissue phase images from five head orientations (used for COSMOS only)
- **rotation_matrix_all.mat** Rotation matrices obtained from coregistration of multi-orientation tissue-phase images (used for COSMOS only)

16.A.2.2 Exercises

(a) Display and compare the reconstructions of COSMOS, TKD, and NDI by running *main_dipole_inversion.m.*

(b) COSMOS calculates the susceptibility map from multi-orientation data. Assess the effect of using data from only 2, 3, 4, and 5 head orientations.

(c) TKD utilizes a k-space truncation parameter ϵ. Display and compare TKD for various values of ϵ.

(d) NDI relies on a gradient-descent optimization, which is prone to overfitting when too many iterations are performed. Compare the image quality of NDI after 50, 200, and 1000 iterations. Besides early-stopping, overfitting can also be prevented by the use of Tikhonov regularization. Set $\lambda = 1e-3$, and rerun NDI for a large number of iterations.

16.A.3 Total variation regularized single-step QSM (single-step TV)

16.A.3.1 Provided materials

- **SS_TV_QSM.m** An incomplete implementation of the Single-Step TV method proposed in [16,50].

- **main_single_step.m** The main file for parts (a)–(c) of this tutorial. It loads the relevant data, estimates the magnetic susceptibility map using the method implemented in *SS_TV_QSM.m*, and displays the results.
- **main_multi_step.m** The main file for part (d) of this tutorial. It loads the relevant numerical data, estimates the magnetic susceptibility map using two multistep QSM methods, and displays the results.
- **main_in_vivo.m** The main file for part (e) of this tutorial. It loads the relevant in vivo data, estimates the magnetic susceptibility map using Single-Step TV and two multistep QSM methods, and displays the results.
- **create_dipole_kernel.m** This function creates a dipole kernel in k-space.
- **create_SMVkernel.m** This function creates Spherical-Mean-Value (SMV) kernels.
- **imagesc3d.m** This function displays 3D data such as a magnetic-susceptibility map.
- **MS_L2_QSM.m** An implementation of the multistep QSM denoted as V-SHARP L2 in [16,50].
- **MS_TV_QSM.m** An implementation of the multistep QSM denoted as V-SHARP TV in [16,50].
- **Data/numerical_qsm_phantom.mat** The numerical brain phantom used for this tutorial.
- **Data/Wave_Caipi_7T.mat** The 7T in vivo brain data used for this tutorial.
- **poly_sensitivity_fit_coils.m** This function is used to remove a bias field from the magnitude images in part (e) of this tutorial.

16.A.3.2 Problem description

Conventional QSM methods that consist of multiple processing steps suffer from potential error propagation between the steps. In order to mitigate such errors, several single-step QSM methods have been proposed [16,40,46–53]. In this tutorial, we will implement the Single-Step TV method proposed in [16,50]. Specifically, the magnetic susceptibility map is estimated by solving the following regularized optimization problem:

$$minimize_{\chi} \frac{1}{2} \sum_{i} \left\| M_i F^{-1} H_i D F \chi - M_i F^{-1} H_i F \Psi(\phi) \right\|_2^2 + \alpha \|G\chi\|_1, \quad (16.T3.1)$$

where χ is the underlying magnetic-susceptibility map to be estimated from the acquired raw phase ϕ, Ψ is the Laplacian unwrapping operator divided by $2\pi * B_0 * \gamma * TE$, F is the discrete Fourier transform operator, D is the dipole kernel in k-space, M_i is a binary mask for the ith reliable phase region, H_i is the Fourier transform of an SMV kernel for the ith reliable phase region, $G = [G_x, G_y, G_z]^T$ is the gradient operator with G_x, G_y, and G_z being the gradient operators along the x-, y-, and z-axes, respectively. The first term of Eq. (16.T3.1), commonly called the data consistency/fidelity term, ensures that the reconstructed susceptibility map when passed through the forward model is consistent with the acquired data ϕ. The second term imposes prior information on the reconstructed susceptibility map. α is the regularization parameter balancing the effects of the first and second terms. Different α values result in solutions with different characteristics.

While Eq. (16.T3.1) can be solved directly using standard optimization algorithms, Chatnuntawech et al. proposed to transform the problem into the following form:

$$minimize_{\chi, z_1, z_2} \frac{1}{2} \sum_{i} \left\| M_i F^{-1} z_{2,i} - M_i F^{-1} H_i F \Psi(\phi) \right\|_2^2 + \alpha \|z_1\|_1 \quad (16.T3.2)$$

$$\text{subject to } G\chi = z_1, \text{ and } H_i D F \chi = z_{2,i},$$

where z_1 and z_2 are auxiliary variables. It can then be shown that the problem in Eq. (16.T3.2) can be decomposed into simpler subproblems with the following closed-form updates [50]:

$$F\chi^{[k+1]} := \left(\mu_1 E^* E + \mu_2 D^* \sum_i H_i^* H_i D\right)^{-1} \left(\mu_1 E^* F\left(z_1^{[k]} - s_1^{[k]}\right) + \mu_2 D^* \sum_i H_i^* \left(z_{2,i}^{[k]} - s_{2,i}^{[k]}\right)\right) \tag{16.T3.3}$$

$$z_{2,i}^{[k+1]} = F\left(M_i^* M_i + \mu_2 I\right)^{-1}\left(M_i^* M_i F^{-1} H_i F \Psi(\phi) + \mu_2 F^{-1}\left(H_i D F \chi^{[k+1]} + s_{2,i}^{[k+1]}\right)\right) \tag{16.T3.4}$$

$$z_1^{[k+1]} = \left(\left|G\chi^{[k+1]} + s_1^{[k]}\right| - \frac{\alpha}{\mu_1}, 0\right) \cdot sign\left(G\chi^{[k+1]} + s_1^{[k]}\right) \tag{16.T3.5}$$

$$s_1^{[k+1]} := s_1^{[k]} + G\chi^{[k+1]} - z_1^{[k+1]} \tag{16.T3.6}$$

$$s_{2,i}^{[k+1]} := s_{2,i}^{[k]} + H_i D F \chi^{[k+1]} - z_{2,i}^{[k+1]} \tag{16.T3.7}$$

where s_1 and s_2 are the scaled dual variables, μ_1 and μ_2 are the augmented Lagrangian parameters, and k in the square brackets is an iteration counter. An optimal solution to Eq. (16.T3.2) can be found by iterating Eqs. (16.T3.3)–(16.T3.7) until convergence.

16.A.4 Exercises

(a) Complete the implementation of the Single-Step TV method based on Eqs. (16.T3.3)–(16.T3.7) by modifying the parts marked as "**TO DO**" in *SS_TV_QSM.m*.

(b) Display the reconstructed magnetic-susceptibility maps generated by running *main_single_step.m*.

(c) Investigate the effects of α in Eq. (16.T3.2) on the reconstructed magnetic susceptibility map.

(d) Compare the reconstructed magnetic susceptibility map obtained from running *main_single_step.m* with your chosen α to several multistep QSM methods by running *main_multi_step.m*. Your Single-Step TV implementation should work better than the multistep QSM methods.

(e) In the previous parts, you have tested your implementation on the numerical brain phantom. In this part, you will test it on 7 Tesla in vivo data. Display the reconstructed magnetic-susceptibility maps generated by running *main_in_vivo.m*.

References

[1] Duyn JH, van Gelderen P, Li T-Q, de Zwart JA, Koretsky AP, Fukunaga M. High-field MRI of brain cortical substructure based on signal phase. Proc Natl Acad Sci USA 2007;104:11796–801. https://doi.org/10.1073/pnas.0610821104.

[2] Wang Y, Liu T. Quantitative susceptibility mapping (QSM): decoding MRI data for a tissue magnetic biomarker. Magn Reson Med 2015;73(1):82–101. https://doi.org/10.1002/mrm.25358.

[3] Langkammer C, Schweser F, Krebs N, et al. Quantitative susceptibility mapping (QSM) as a means to measure brain iron? A post mortem validation study. NeuroImage 2012;62:1593–9. https://doi.org/10.1016/J.NEUROIMAGE.2012.05.049.

[4] Fan AP, Bilgic B, Gagnon L, Witzel T, Bhat H, Rosen BR, et al. Quantitative oxygenation venography from MRI phase. Magn Reson Med 2014;72:149–59. https://doi.org/10.1002/MRM.24918.

[5] Deistung A, Schäfer A, Schweser F, Biedermann U, Turner R, Reichenbach JR. Toward in vivo histology: a comparison of quantitative susceptibility mapping (QSM) with magnitude-, phase-, and R2*-imaging at ultra-high magnetic field strength. NeuroImage 2013;65:299–314. https://doi.org/10.1016/J.NEUROIMAGE.2012.09.055.

[6] Sharma SD, Hernando D, Horng DE, Reeder SB. Quantitative susceptibility mapping in the abdomen as an imaging biomarker of hepatic iron overload. Magn Reson Med 2015;74:673–83. https://doi.org/10.1002/MRM.25448.

[7] Wen Y, Nguyen TD, Liu Z, et al. Cardiac quantitative susceptibility mapping (QSM) for heart chamber oxygenation. Magn Reson Med 2018;79:1545–52. https://doi.org/10.1002/MRM.26808.

[8] Dimov AV, Liu Z, Spincemaille P, Prince MR, Du J, Wang Y. Bone quantitative susceptibility mapping using a chemical species–specific signal model with ultrashort and conventional echo data. Magn Reson Med 2018;79:121–8. https://doi.org/10.1002/MRM.26648.

[9] Dimov AV, Liu T, Spincemaille P, Ecanow JS, Tan H, Edelman RR, et al. Joint estimation of chemical shift and quantitative susceptibility mapping (chemical QSM). Magn Reson Med 2015;73:2100–10. https://doi.org/10.1002/MRM.25328.

[10] Haacke EM, Xu Y, Cheng Y-CN, Reichenbach JR. Susceptibility weighted imaging (SWI). Magn Reson Med 2004;52:612–8. https://doi.org/10.1002/mrm.20198.

[11] Wu B, Li W, Avram AV, Gho S-M, Liu C. Fast and tissue-optimized mapping of magnetic susceptibility and T2* with multi-echo and multi-shot spirals. NeuroImage 2012;59:297–305. https://doi.org/10.1016/j.neuroimage.2011.07.019.

[12] Pruessmann KP, Weiger M, Scheidegger MB, Boesiger P. SENSE: sensitivity encoding for fast MRI. Magn Reson Med 1999;42:952–62.

[13] Griswold MA, Jakob PM, Heidemann RM, Nittka M, Jellus V, Wang J, et al. Generalized autocalibrating partially parallel acquisitions (GRAPPA). Magn Reson Med 2002;47:1202–10. https://doi.org/10.1002/mrm.10171.

[14] Roemer PB, Edelstein WA, Hayes CE, Souza SP, Mueller OM. The NMR phased array. Magn Reson Med 1990;16:192–225.

[15] Uecker M, Lai P, Murphy MJ, Virtue P, Elad M, Pauly JM, et al. ESPIRiT-an eigenvalue approach to autocalibrating parallel MRI: where SENSE meets GRAPPA. Magn Reson Med 2014;71:990–1001. https://doi.org/10.1002/mrm.24751.

[16] Chatnuntawech I, McDaniel P, Cauley SF, et al. Single-step quantitative susceptibility mapping with variational penalties. NMR Biomed 2016. https://doi.org/10.1002/nbm.3570.

[17] Bilgic B, Marques JP, Wald LL, Setsompop K. Block coil compression for virtual body coil without phase singularities. In: Fourth international workshop on MRI phase contrast & quantitative susceptibility mapping; 2016.

[18] Robinson SD, Dymerska B, Bogner W, Barth M, Zaric O, Goluch S, et al. Combining phase images from array coils using a short echo time reference scan (COMPOSER). Magn Reson Med 2017;77:318–27. https://doi.org/10.1002/mrm.26093.

[19] Eckstein K, Dymerska B, Bachrata B, Bogner W, Poljanc K, Trattnig S, et al. Efficient combination of multi-channel phase data from multi-echo acquisitions (ASPIRE). Magn Reson Med 2018;79:2996–3006. https://doi.org/10.1002/mrm.26963.

[20] Smith SM. Fast robust automated brain extraction. Hum Brain Mapp 2002;17:143–55. https://doi.org/10.1002/hbm.10062.

[21] Jenkinson M. Fast, automated, N-dimensional phase-unwrapping algorithm. Magn Reson Med 2003;49:193–7. https://doi.org/10.1002/mrm.10354.

[22] Karsa A, Shmueli K. SEGUE: a Speedy rEgion-Growing algorithm for Unwrapping Estimated phase. IEEE Trans Med Imaging 2019;38(6):1347–57. https://doi.org/10.1109/TMI.2018.2884093.

[23] Dymerska B, Eckstein K, Bachrata B, Siow B, Trattnig S, Shmueli K, et al. Phase unwrapping with a rapid opensource minimum spanning tree algorithm (ROMEO). Magn Reson Med 2021;85:2294–308. https://doi.org/10.1002/MRM.28563.

[24] Li W, Wu B, Liu C. Quantitative susceptibility mapping of human brain reflects spatial variation in tissue composition. NeuroImage 2011;55:1645–56. https://doi.org/10.1016/j.neuroimage.2010.11.088.

[25] Schofield MA, Zhu Y. Fast phase unwrapping algorithm for interferometric applications. Opt Lett 2003;28:1194. https://doi.org/10.1364/OL.28.001194.

[26] Li N, Wang W-T, Sati P, Pham DL, Butman JA. Quantitative assessment of susceptibility-weighted imaging processing methods. J Magn Reson Imaging 2014;40:1463–73. https://doi.org/10.1002/jmri.24501.

[27] Liu T, Wisnieff C, Lou M, Chen W, Spincemaille P, Wang Y. Nonlinear formulation of the magnetic field to source relationship for robust quantitative susceptibility mapping. Magn Reson Med 2013;69:467–76. https://doi.org/10.1002/mrm.24272.

[28] Li L. Magnetic susceptibility quantification for arbitrarily shaped objects in inhomogeneous fields. Magn Reson Med 2001;46:907–16. https://doi.org/10.1002/MRM.1276.

[29] Schweser F, Robinson SD, de Rochefort L, Li W, Bredies K. An illustrated comparison of processing methods for phase MRI and QSM: removal of background field contributions from sources outside the region of interest. NMR Biomed 2017;30:e3604. https://doi.org/10.1002/nbm.3604.

[30] Liu T, Khalidov I, de Rochefort L, Spincemaille P, Liu J, Tsiouris AJ, Wang Y. A novel background field removal method for MRI using projection onto dipole fields (PDF). NMR Biomed 2011;24:1129–36. https://doi.org/10.1002/nbm.1670.

[31] Schweser F, Deistung A, Lehr BW, Reichenbach JR. Quantitative imaging of intrinsic magnetic tissue properties using MRI signal phase: an approach to in vivo brain iron metabolism? NeuroImage 2011;54:2789–807. https://doi.org/10.1016/j.neuroimage.2010.10.070.

[32] Wu B, Li W, Guidon A, Liu C. Whole brain susceptibility mapping using compressed sensing. Magn Reson Med 2012;67:137–47. https://doi.org/10.1002/mrm.23000.

[33] Sun H, Wilman AH. Background field removal using spherical mean value filtering and Tikhonov regularization. Magn Reson Med 2014;71:1151–7. https://doi.org/10.1002/MRM.24765.

[34] Özbay PS, Deistung A, Feng X, Nanz D, Reichenbach JR, Schweser F. A comprehensive numerical analysis of background phase correction with V-SHARP. NMR Biomed 2017;30:e3550. https://doi.org/10.1002/NBM.3550.

[35] Liu T, Spincemaille P, de Rochefort L, Kressler B, Wang Y. Calculation of susceptibility through multiple orientation sampling (COSMOS): a method for conditioning the inverse problem from measured magnetic field map to susceptibility source image in MRI. Magn Reson Med 2009;61:196–204. https://doi.org/10.1002/mrm.21828.

[36] Shmueli K, de Zwart JA, van Gelderen P, Li T-Q, Dodd SJ, Duyn JH. Magnetic susceptibility mapping of brain tissue in vivo using MRI phase data. Magn Reson Med 2009;62:1510–22. https://doi.org/10.1002/mrm.22135.

[37] Bilgic B, Chatnuntawech I, Fan AP, Setsompop K, Cauley SF, Wald LL, et al. Fast image reconstruction with L2-regularization. J Magn Reson Imaging 2013;00:1–11. https://doi.org/10.1002/jmri.24365.

[38] Schweser F, Deistung A, Sommer K, Reichenbach JR. Toward online reconstruction of quantitative susceptibility maps: superfast dipole inversion. Magn Reson Med 2013;69:1582–94. https://doi.org/10.1002/mrm.24405.

[39] Liu T, Liu J, de Rochefort L, Spincemaille P, Khalidov I, Ledoux JR, et al. Morphology enabled dipole inversion (MEDI) from a single-angle acquisition: comparison with COSMOS in human brain imaging. Magn Reson Med 2011;66:777–83. https://doi.org/10.1002/mrm.22816.

[40] Langkammer C, Bredies K, Poser BA, Barth M, Reishofer G, Fan AP, et al. Fast quantitative susceptibility mapping using 3D EPI and total generalized variation. NeuroImage 2015;111. https://doi.org/10.1016/j.neuroimage.2015.02.041.

[41] Wei H, Dibb R, Zhou Y, Sun Y, Xu J, Wang N, et al. Streaking artifact reduction for quantitative susceptibility mapping of sources with large dynamic range. NMR Biomed 2015;28:1294–303. https://doi.org/10.1002/NBM.3383.

[42] Milovic C, Bilgic B, Zhao B, Acosta-Cabronero J, Tejos C. Fast nonlinear susceptibility inversion with variational regularization. Magn Reson Med 2018;80:814–21. https://doi.org/10.1002/mrm.27073.

[43] Goldstein T, Osher S. The split Bregman method for L1-regularized problems. SIAM J Imaging Sci 2009;2:323–43. https://doi.org/10.1137/080725891.

[44] Boyd S, Parikh N, Chu E, Peleato B, Eckstein J. Distributed optimization and statistical learning via the alternating direction method of multipliers. Found Trends Mach Learn 2010;3:1–122. https://doi.org/10.1561/2200000016.

[45] Polak D, Chatnuntawech I, Yoon J, Iyer SS, Milovic C, Lee J, et al. Nonlinear dipole inversion (NDI) enables robust quantitative susceptibility mapping (QSM). NMR Biomed 2020:1–13. https://doi.org/10.1002/nbm.4271.

[46] Sharma SD, Hernando D, Horng DE, Reeder SB. A joint background field removal and dipole deconvolution approach for quantitative susceptibility mapping in the liver. In: Proceedings of the 22nd annual meeting of ISMRM; 2014. p. 606.

[47] Bilgic B, Langkammer C, Wald LL, Setsompop K. Single-Step QSM with fast reconstruction. In: Third Int. Work. MRI phase contrast quant. susceptibility mapping; 2014. p. 40.

[48] Liu T, Zhou D, Spincemaille P, Wang Y. Differential approach to quantitative susceptibility mapping without background field removal. In: Proceedings of the 22nd annual meeting ISMRM; 2014. p. 597.

[49] Kan H, Arai N, Takizawa M, Kasai H, Kunitomo H, Hirose Y, et al. Improvement of signal inhomogeneity induced by radio-frequency transmit-related phase error for single-step quantitative susceptibility mapping reconstruction. Magn Reson Med Sci 2019;18. https://doi.org/10.2463/mrms.tn.2018-0066.

[50] Chatnuntawech I. Acquisition and reconstruction methods for magnetic resonance imaging. Massachusetts Institute of Technology; 2016.

[51] Bredies K, Ropele S, Poser BA, Barth M, Langkammer C. Single-step quantitative susceptibility mapping using total generalized variation and 3D EPI. In: Proc. Intl. Soc. Mag. Reson. Med., vol. 22; 2014.

[52] Liu Z, Kee Y, Zhou D, Wang Y, Spincemaille P. Preconditioned total field inversion (TFI) method for quantitative susceptibility mapping. Magn Reson Med 2017;78. https://doi.org/10.1002/mrm.26331.

[53] Zhang L, Chen X, Lin J, Ding X, Bao L, Cai C, et al. Fast quantitative susceptibility reconstruction via total field inversion with improved weighted L0 norm approximation. NMR Biomed 2019;32:e4067.

[54] Johnson KM. 3D velocimetry phase unwrapping using block-wise classification with a shift variant fully 3D convolutional neural network. In: ISMRM workshop on machine learning; 2018.

[55] He JJ, Sandino C, Zeng D, Vasanawala S, Cheng J. Deep spatiotemporal phase unwrapping of phase-contrast MRI data. In: Proc Intl Soc Mag Reson Med, vol. 27; 2019. p. 1962.

[56] Ryu K, Gho S-M, Nam Y, Koch K, Kim D-H. Development of a deep learning method for phase unwrapping MR images. In: Proc Intl Soc Mag Reson Med, vol. 27; 2019. p. 4707.

[57] Bollmann S, Kristensen MH, Larsen MS, Olsen MV, Pedersen MJ, Østergaard LR, et al. SHARQnet–sophisticated harmonic artifact reduction in quantitative susceptibility mapping using a deep convolutional neural network. Z Med Phys 2019;29:139–49.

[58] Liu J, Nencka A, Koch K. Deep residual neural networks for QSM background removal. In: Proc Intl Soc Mag Reson Med, vol. 27; 2019. p. 4852.

[59] Kim H, Yoon J, Lee J. Achieving real-time QSM reconstruction using deep neural network. In: Proc Intl Soc Mag Reson Med, vol. 27; 2019. p. 4029.

[60] Gong E, Bilgic B, Setsompop K, Fan A, Zaharchuk G, Pauly J. Accurate and efficient QSM reconstruction using deep learning. In: Proc Intl Soc Mag Reson Med, vol. 26; 2018. p. 189.

[61] Yoon J, Gong E, Chatnuntawech I, et al. Quantitative susceptibility mapping using deep neural network: QSMnet. NeuroImage 2018;179:199–206. https://doi.org/10.1016/J.NEUROIMAGE.2018.06.030.

[62] Bollmann S, Rasmussen KGB, Kristensen M, Blendal RG, Østergaard LR, Plocharski M, et al. DeepQSM-using deep learning to solve the dipole inversion for quantitative susceptibility mapping. Neuroimage 2019;195:373–83.

[63] Chen Y, Jakary A, Avadiappan S, Hess CP, Lupo JM. QSMGAN: improved quantitative susceptibility mapping using 3D generative adversarial networks with increased receptive field. NeuroImage 2020;207:116389.

[64] Gao Y, Zhu X, Bollmann S. OctQSM - a deep learning QSM method with Octave convolution. In: 5th international QSM workshop; 2019.

[65] Kames C, Doucette J, Rauscher A. Proximal variational networks: generalizable deep networks for solving the dipole-inversion problem. In: 5th international QSM workshop; 2019.

[66] Liu J, Koch KM. Meta-QSM: an image-resolution-arbitrary network for QSM reconstruction. Available from: arXiv:1908.00206, 2019.

[67] Liu J, Koch KM. Non-locally encoder-decoder convolutional network for whole brain QSM inversion. Available from: arXiv:1904.05493, 2019.

[68] Liu J, Nencka AS, Muftuler LT, Swearingen B, Karr R, Koch KM. Quantitative susceptibility inversion through parcellated multiresolution neural networks and k-space substitution. Available from: arXiv:1903.04961, 2019.

[69] Liu Z, Zhang J, Zhang S, Spincemaille P, Nguyen T, Wang Y. Quantitative susceptibility mapping using a deep learning prior. In: Proc Intl Soc Mag Reson Med, vol. 27; 2019. p. 4933.

[70] Jung W, Yoon J, Ji S, Choi JY, Kim JM, Nam Y, et al. Exploring linearity of deep neural network trained QSM: QSMnet+. NeuroImage 2020;211:116619.

[71] Jochmann T, Haueisen J, Schweser F. Physics-aware augmentation, artificial noise, and synthetic samples to train a convolutional neural network for QSM. In: 5th international QSM workshop; 2019.

[72] Jochmann T, Haueisen J, Zivadinov R, Schweser F. U2-Net for DEEPOLE QUASAR–a physics-informed deep convolutional neural network that disentangles MRI phase contrast mechanisms. In: Proc Intl Soc Mag Reson Med, vol. 27; 2019. p. 320.

[73] Zhang H, Bao L. Quantitative susceptibility mapping using a three dimensional enhanced U-Net. In: 5th international QSM workshop; 2019.

[74] Zhang J, Liu Z, Zhang S, Zhang H, Spincemaille P, Nguyen TD, et al. Fidelity imposed network edit (FINE) for solving ill-posed image reconstruction. NeuroImage 2020;211:116579.

[75] Jung W, Bollmann S, Lee J. Overview of quantitative susceptibility mapping using deep learning: current status, challenges and opportunities. NMR Biomed 2020:e4292.

[76] Heber S, Tinauer C, Bollmann S, Ropele S, Langkammer C. Deep quantitative susceptibility mapping by combined background field removal and dipole inversion. In: Proc. Intl. Soc. Mag. Reson. Med.; 2019. p. 4028.

[77] Liu J, Koch KM. Deep quantitative susceptibility mapping for background field removal and total field inversion. Available from: arXiv:1905.13749, 2019.

[78] Liu J, Koch KM. MRI tissue magnetism quantification through total field inversion with deep neural networks. Available from: arXiv:1904.07105, 2019.

[79] Geßner C, Meineke J. Exploring the U-Net for dipole-inversion and combined background-field removal and dipole-inversion for quantitative susceptibility mapping in MRI. In: 5th international workshop on MRI phase contrast & quantitative susceptibility mapping; 2019.

[80] Kames C, Doucette J, Rauscher A. Training a variational network for use on 3D high resolution MRI data in 1 day; 2019.

[81] Wei H, Cao S, Zhang Y, Guan X, Yan F, Yeom KW, et al. Learning-based single-step quantitative susceptibility mapping reconstruction without brain extraction. NeuroImage 2019;202:116064.

[82] Høy PC, Sørensen KS, Østergaard LR, O'Brien K, Barth M, Bollmann S. Deep learning for solving ill-posed problems in quantitative susceptibility mapping: what can possibly go wrong? In: Annual meeting of the international society for magnetic resonance in medicine, ISMRM; 2019. p. 321.

[83] Goodfellow I, Pouget-Abadie J, Mirza M, Xu B, Warde-Farley D, Ozair S, et al. Generative adversarial nets. Adv Neural Inf Process Syst 2014;27.

[84] Johnson J, Alahi A, Fei-Fei L. Perceptual losses for real-time style transfer and super-resolution. In: European conference on computer vision. Springer; 2016. p. 694–711.

[85] Zhang J, Zhang H, Sabuncu M, Spincemaille P, Nguyen T, Wang Y. Probabilistic dipole inversion for adaptive quantitative susceptibility mapping. Available from: arXiv:2009.04251, 2020.

[86] Robinson SD, Bredies K, Khabipova D, Dymerska B, Marques JP, Schweser F. An illustrated comparison of processing methods for MR phase imaging and QSM: combining array coil signals and phase unwrapping. NMR Biomed 2017;30:e3601. https://doi.org/10.1002/nbm.3601.

[87] Langkammer C, Schweser F, Shmueli K, et al. Quantitative susceptibility mapping: report from the 2016 reconstruction challenge. Magn Reson Med 2018;79. https://doi.org/10.1002/mrm.26830.

[88] QSM Challenge 2.0 Organization Committee, Bilgic B, Langkammer C, Marques JP, Meineke J, Milovic C, Schweser F. QSM reconstruction challenge 2.0: design and report of results. Magn Reson Med 2021;86:1241–55. https://doi.org/10.1002/MRM.28754.

[89] Liu C, Li W, Tong KA, Yeom KW, Kuzminski S. Susceptibility-weighted imaging and quantitative susceptibility mapping in the brain. J Magn Reson Imaging 2015;42:23–41. https://doi.org/10.1002/JMRI.24768.

APPENDIX

Linear Algebra Primer

Gastao Cruz[a], **Burhaneddin Yaman**[b,c], **Mehmet Akçakaya**[b,c], **Mariya Doneva**[d], and **Claudia Prieto**[a]

[a]*School of Biomedical Engineering and Imaging Sciences, King's College London, London, United Kingdom*
[b]*Department of Electrical and Computer Engineering, University of Minnesota, Minneapolis, MN, United States*
[c]*Center for Magnetic Resonance Research, University of Minnesota, Minneapolis, MN, United States*
[d]*Philips Research, Hamburg, Germany*

A.1 Vector spaces

Definition (Vector space). A vector space V is a set that is defined over a field F ($\mathbb{R}$ or $\mathbb{C}$) with two operations:

- vector addition: $\mathbf{v}, \mathbf{w} \in V \longrightarrow \mathbf{v} + \mathbf{w} \in V$
- scalar multiplication: $\mathbf{v} \in V, \alpha \in F \longrightarrow \alpha\mathbf{v} \in V$

that satisfy the following properties:

- commutativity of vector addition: $\mathbf{v} + \mathbf{w} = \mathbf{w} + \mathbf{v}$
- associativity of vector addition: $\mathbf{v} + (\mathbf{w} + \mathbf{y}) = (\mathbf{v} + \mathbf{w}) + \mathbf{y}$
- additive identity: $\mathbf{v} + 0 = \mathbf{v}$
- additive inverse: $\mathbf{v} + (-\mathbf{v}) = 0$
- associativity of scalar multiplication: $(\alpha\beta)\mathbf{v} = \alpha(\beta\mathbf{v})$
- distributivity of vector sums: $\alpha(\mathbf{v} + \mathbf{w}) = \alpha\mathbf{v} + \alpha\mathbf{w}$
- distributivity of scalar sums: $(\alpha + \beta)\mathbf{v} = \alpha\mathbf{v} + \beta\mathbf{v}$
- identity for scalar multiplication: $1\mathbf{v} = \mathbf{v}$.

A.1.1 Linear independence

Definition. A subset of vectors $S = \{\mathbf{v_1}, \ldots, \mathbf{v_n}\} \subset V$ is linearly dependent if there exists $\{\alpha_1, \ldots, \alpha_n\} \in F$, not all zero, such that $\sum_{k=1}^{n} \alpha_k \mathbf{v}_k = \mathbf{0}$.

Definition. A subset of vectors $S = \{\mathbf{v_1}, \ldots, \mathbf{v_n}\} \subset V$ is linearly independent if it is not linearly dependent. In other words, the equation $\sum_{k=1}^{n} \alpha_k \mathbf{v}_k = \mathbf{0}$ can only be satisfied if $\alpha_k = 0$ for all $k \in \{1, \ldots, n\}$.

A.1.2 Span

Definition. The span of $S \subset V$ is the set of all finite linear combinations in S. More formally,

$$span(S) = \{\sum_{k=1}^{n} \lambda_k \mathbf{v}_k | n \in \mathbb{Z}^+, \mathbf{v}_k \in S, \lambda_k \in F\}.$$

If $V = span(S)$, S is called a spanning set of V.

A.1.3 Basis

Definition. A basis B of a vector space V is a linearly independent subset of vectors in V that spans V. As a result, $B \subset V$ is called a basis if every element of V can be uniquely written as a linear combination of vectors in B.

A.1.4 Normed space

Definition. A normed space is a vector space in which a norm is defined. Norm is a real-valued function that maps a vector space to real numbers and denoted with $\|.\| : V \longrightarrow \mathbb{R}$ subject to:

(i) $\|\mathbf{v}\| \geq 0$ and $\|\mathbf{v}\| = 0 \Longleftrightarrow \mathbf{v} = 0$
(ii) $\|\alpha\mathbf{v}\| = |\alpha|\|\mathbf{v}\|$
(iii) $\|\mathbf{u} + \mathbf{v}\| \leq \|\mathbf{u}\| + \|\mathbf{v}\|$ (triangle inequality).

Definition (ℓ_p norms). For an n-dimensional vector $\mathbf{x} \in \mathbb{C}^n$, the ℓ_p norm is defined as

$$||\mathbf{x}||_p = \Big(\sum_{k=1}^{n} |x_k|^p\Big)^{\frac{1}{p}}.$$

Remark. Note two special ℓ_p norms that will be used throughout the text:

- ℓ_1 norm: $\|\mathbf{u}\|_1 = \sum_{i=1}^{n} |u_i|$
- ℓ_2 norm: $\|\mathbf{u}\|_2 = \sqrt{\sum_{i=1}^{n} |u_i|^2}$.

A.1.5 Inner product space

Definition. An inner product space is a vector space in which an inner product is defined. An inner product is a map, $< ., . >: V \times V \longrightarrow F$ that satisfies:

(i) $< \mathbf{u}, \mathbf{v} >= \overline{< \mathbf{v}, \mathbf{u} >}$ (conjugate symmetry)
(ii) $< \alpha\mathbf{u}, \mathbf{v} >= \alpha < \mathbf{u}, \mathbf{v} >$
(iii) $< \mathbf{u} + \mathbf{v}, \mathbf{w} >=< \mathbf{u}, \mathbf{w} > + < \mathbf{v}, \mathbf{w} >$
(iv) $< \mathbf{u}, \mathbf{u} >\geq 0$ and $< \mathbf{u}, \mathbf{u} >= 0 \Longleftrightarrow \mathbf{u} = 0$.

Remark. It is worth noting that:

- A norm can be defined based on the inner product as $\|\mathbf{u}\| = \sqrt{< \mathbf{u}, \mathbf{u} >}$
- For $\mathbb{R}^n$, which is a Euclidean space, the dot product is the inner product, defined as $< \mathbf{u}, \mathbf{v} >= \mathbf{u}^T\mathbf{v} = \sum_{i=1}^{n} \mathbf{u}_i\mathbf{v}_i$.

A.2 Matrix theory

An $m \times n$ matrix has m rows and n columns. The $(i, j)^{th}$ entry of a matrix $\mathbf{A}$ is denoted with $\mathbf{A}_{ij}$.

Remark. A matrix can be used to represent a linear mapping/transformation between two vector spaces of finite dimension n and m. Consider a linear mapping T from $\mathbb{C}^n$ to $\mathbb{C}^m$. Then, one can find a matrix

$\mathbf{A} \in \mathbb{C}^{m \times n}$ such that

$$T(\mathbf{x}) = \mathbf{A}\mathbf{x} \qquad \text{for } \mathbf{x} \in \mathbb{C}^n.$$

In order to explicitly write the $\mathbf{A}$ corresponding to the linear mapping T, the canonical basis elements $\mathbf{e}^k \in \mathbb{C}^n$ can be used. Here, $\mathbf{e}^k$ is a vector that takes the value 1 in its kth coordinate and 0 elsewhere. Using the canonical basis, $\mathbf{A}$ can be explicitly written using

$$\mathbf{A} = \begin{bmatrix} T(\mathbf{e}^1) & T(\mathbf{e}^2) & \ldots & T(\mathbf{e}^n) \end{bmatrix}$$

Definition (**Range space**). The linear span of columns of a matrix. Given a matrix $\mathbf{A} \in \mathbb{R}^{m \times n}$, the range (or column) space of matrix $\mathbf{A}$ is denoted as $\text{Ran}(\mathbf{A})$ and defined as

$$\text{Ran}(\mathbf{A}) = \{\mathbf{A}\mathbf{x} \mid \mathbf{x} \in \mathbb{R}^n\} \quad \subseteq \mathbb{R}^m.$$

Definition (**Null space**). Null space of a matrix $\mathbf{A}$ is denoted as $\text{Null}(\mathbf{A})$ (or $N(\mathbf{A})$) and defined as

$$\text{Null}(\mathbf{A}) = \{\mathbf{x} \in \mathbb{R}^n \mid \mathbf{A}\mathbf{x} = 0\} \quad \subseteq \mathbb{R}^n.$$

Definition (**Rank**). Number of linearly independent columns of a matrix. Given a matrix $\mathbf{A} \in \mathbb{R}^{m \times n}$, rank of the matrix $\mathbf{A}$ is denoted as $\text{rank}(\mathbf{A})$ and defined as

$$\text{rank}(\mathbf{A}) = \dim(\text{Ran}(\mathbf{A})) \leq n,$$

where dim(.) denotes the dimension of the matrix.

Remark. The matrix $\mathbf{A}$ is full-rank if $\text{rank}(\mathbf{A}) = \min\{m, n\}$. Otherwise, the matrix $\mathbf{A}$ is rank-deficient.

Definition (**Transpose**). A transpose of a matrix $\mathbf{A}$, denoted by $\mathbf{A}^T$, switches the row and column indices of the matrix $\mathbf{A}$, i.e.,

$$(\mathbf{A}^T)_{ij} = \mathbf{A}_{ji}.$$

Definition (**Conjugate-transpose**). The conjugate-transpose, also known as the Hermitian transpose, of a matrix $\mathbf{A}$ is obtained by taking the transpose and then taking the complex conjugate of each entry. It is often denoted as $\mathbf{A}^H$ or $\mathbf{A}^*$.

Remark. For real matrices, the conjugate transpose is equivalent to the transpose $\mathbf{A}^H = \mathbf{A}^T$.

Definition (**Inverse**). Inverse of a $n \times n$ square matrix $\mathbf{A}$ is $\mathbf{A}^{-1}$ such that

$$\mathbf{A}\mathbf{A}^{-1} = \mathbf{A}^{-1}\mathbf{A} = \mathbf{I}.$$

Definition (**Pseudo-inverse**). The (Moore–Penrose) pseudoinverse, denoted $\mathbf{A}^\dagger$, generalizes the notion of inverse of a square matrix to arbitrary matrices. The pseudoinverse of a matrix $\mathbf{A}$, with linearly independent columns, is defined as

$$\mathbf{A}^\dagger = \left(\mathbf{A}^H\mathbf{A}\right)^{-1}\mathbf{A}^H,$$

which yields a left inverse, i.e., $\mathbf{A}^\dagger\mathbf{A} = \mathbf{I}$.

For a matrix **A**, with linearly independent rows, it is defined as

$$\mathbf{A}^{\dagger} = \mathbf{A}^{H}\left(\mathbf{A}\mathbf{A}^{H}\right)^{-1},$$

yielding a right inverse, i.e., $\mathbf{A}\mathbf{A}^{\dagger} = \mathbf{I}$.

Definition (Eigenvalue and & eigenvector). An eigenvalue and corresponding eigenvector of a square matrix A is a scalar λ and vector $\mathbf{v}$, respectively, such that

$$\mathbf{A}\mathbf{v} = \lambda\mathbf{v}.$$

Definition (Trace). The trace of a square matrix is the sum of all its diagonal elements, i.e.,

$$tr(\mathbf{A}) = \sum_{i=1}^{n} a_{ii}.$$

Remark. Traces and eigenvalues are related, in particular, the trace of matrix **A** is the sum of all its eigenvalues.

A.2.1 Types of matrices

Identity matrix: A square matrix with ones in the diagonal and zeros elsewhere is defined as

$$\mathbf{I}_{ij} = \begin{cases} 1 & \text{if } i = j \\ 0 & \text{if } i \neq j. \end{cases}$$

Diagonal matrix: A square matrix in which all nondiagonal entries are zeros.

Symmetric matrix: A square matrix that is equal to its transpose.

Hermitian matrix: A complex square matrix that is equal to its conjugate-transpose, $\mathbf{A}^{H} = \mathbf{A}$., is a complex square matrix that is equal to its own conjugate transpose—that is, the element in the i-th row and j-th column is equal to the complex conjugate of the element in the j-th row and i-th column, for all indices i and j.

Unitary matrix: A square matrix, **U**, is unitary if $\mathbf{U}\mathbf{U}^{H} = \mathbf{U}^{H}\mathbf{U} = \mathbf{I}$. For a unitary matrix $\mathbf{U}^{H} = \mathbf{U}^{-1}$.

A.2.2 Matrices with special structures

Toeplitz matrix: Entries are constant along diagonals, e.g.,

$$\mathbf{T} = \begin{bmatrix} t_1 & t_2 & t_3 & t_4 & t_5 \\ t_6 & t_1 & t_2 & t_3 & t_4 \\ t_7 & t_6 & t_1 & t_2 & t_3 \\ t_8 & t_7 & t_6 & t_1 & t_2 \\ t_9 & t_8 & t_7 & t_6 & t_1 \end{bmatrix}$$

Remark. The recall the linear convolution of two N-dimensional vectors (or sequences) is defined as

$$(\mathbf{x} * \mathbf{h})_n = \sum_{k=0}^{N-1} x_k h_{n-k},$$

where h_k is defined to be 0 for $k \notin \{0, \ldots, N-1\}$. The linear convolution can be implemented via matrix multiplication with a Toeplitz matrix as follows:

$$\mathbf{x} * \mathbf{h} = \begin{bmatrix} h_0 & 0 & \cdots & 0 & 0 \\ h_1 & h_0 & & \vdots & \vdots \\ h_2 & h_1 & \cdots & 0 & 0 \\ \vdots & \vdots & & \vdots & \vdots \\ h_{N-1} & h_{N-2} & \cdots & \vdots & h_0 \\ \vdots & \vdots & & \vdots & \vdots \\ \vdots & \vdots & & h_{N-1} & h_{N-2} \\ 0 & 0 & 0 & \cdots & h_{N-1} \end{bmatrix} \begin{bmatrix} x_0 \\ x_1 \\ \vdots \\ x_{N-1} \end{bmatrix}.$$

Note here we have indexed the vector coordinates from 0 to $N-1$ for a simpler notation in the convolution output.

Hankel matrix: Entries are constant along antidiagonals, e.g.,

$$\mathbf{H} = \begin{bmatrix} h_1 & h_2 & h_3 & h_4 & h_5 \\ h_2 & h_3 & h_4 & h_5 & h_6 \\ h_3 & h_4 & h_5 & h_6 & h_7 \\ h_4 & h_5 & h_6 & h_7 & h_8 \\ h_5 & h_6 & h_7 & h_8 & h_9 \end{bmatrix}.$$

Circulant matrix: Entries in a row are cyclically right-shifted to form the next row, e.g.,

$$\mathbf{C} = \begin{bmatrix} c_1 & c_2 & c_3 & c_4 & c_5 \\ c_5 & c_1 & c_2 & c_3 & c_4 \\ c_4 & c_5 & c_1 & c_2 & c_3 \\ c_3 & c_4 & c_5 & c_1 & c_2 \\ c_2 & c_3 & c_4 & c_5 & c_1 \end{bmatrix}$$

Remark. The recall the circular convolution of two N-dimensional vectors (or sequences) is defined as

$$(\mathbf{x} *_c \mathbf{h})_n = \sum_{k=0}^{N-1} x_k h_{(n-k) \bmod N}.$$

Circular convolution can be implemented via matrix multiplication with a circulant matrix as follows:

$$\mathbf{x} *_c \mathbf{h} = \begin{bmatrix} h_0 & h_{N-1} & \cdots & h_2 & h_1 \\ h_1 & h_0 & h_{N-1} & \cdots & h_2 \\ h_2 & h_1 & h_0 & \cdots & h_3 \\ \vdots & \vdots & & \vdots & \vdots \\ h_{N-1} & h_{N-2} & \cdots & \vdots & h_0 \end{bmatrix} \begin{bmatrix} x_0 \\ x_1 \\ \vdots \\ x_{N-1} \end{bmatrix}$$

Discrete Fourier transform (DFT) matrix: DFT matrix implements the DFT as a transformation matrix. For the N-point DFT, the unitary DFT matrix, $\mathbf{F}$, is the $N \times N$ matrix defined as

$$\mathbf{F} = \frac{1}{\sqrt{N}} \begin{bmatrix} 1 & 1 & 1 & 1 & \cdots & 1 \\ 1 & \omega & \omega^2 & \omega^3 & \cdots & \omega^{N-1} \\ 1 & \omega^2 & \omega^4 & \omega^6 & \cdots & \omega^{2(N-1)} \\ 1 & \omega^3 & \omega^6 & \omega^9 & \cdots & \omega^{3(N-1)} \\ \vdots & \vdots & \vdots & \vdots & \ddots & \vdots \\ 1 & \omega^{N-1} & \omega^{2(N-1)} & \omega^{3(N-1)} & \cdots & \omega^{(N-1)(N-1)} \end{bmatrix},$$

where $\omega = e^{-\frac{i2\pi}{N}}$.

Remark. Any circulant matrix is diagonalizable by the DFT matrix. In particular, we have

$$\mathbf{C} = \begin{bmatrix} c_0 & c_{N-1} & \cdots & c_2 & c_1 \\ c_1 & c_0 & c_{N-1} & \cdots & c_2 \\ c_2 & c_1 & c_0 & \cdots & c_3 \\ \vdots & \vdots & & \vdots & \vdots \\ c_{N-1} & c_{N-2} & \cdots & \vdots & c_0 \end{bmatrix} = \mathbf{F}^H \mathbf{D}_C \mathbf{F},$$

where $\mathbf{D}_C$ is a diagonal matrix whose m^{th} diagonal entry is the mth DFT coefficient of $\mathbf{c} = [c_0, \ldots, c_{N-1}]^T$. In effect, this implements the convolution theorem, i.e., convolution in image domain as a multiplication in Fourier domain.

Casorati matrix: Given a 3D multidimensional array $\mathcal{X}$ of size $I \times J \times K$, Casorati matrix is formed by vectorizing the frontal slices of the $\mathcal{X}$, i.e., vectorizing $\{\mathcal{X}(:, :, k)\}_{k=1}^{K}$, leading to a matrix with a size of $IJ \times K$.

A.2.3 Special matrix products

Kronecker product: The Kronecker product between two matrices $\mathbf{A} \in \mathbb{R}^{I \times J}$ and $\mathbf{B} \in \mathbb{R}^{K \times L}$, denoted by $\mathbf{A} \otimes \mathbf{B}$, is defined as

$$\mathbf{A} \otimes \mathbf{B} = \begin{bmatrix} a_{11}\mathbf{B} & a_{12}\mathbf{B} & \cdots & a_{1J}\mathbf{B} \\ a_{21}\mathbf{B} & a_{22}\mathbf{B} & \cdots & a_{2J}\mathbf{B} \\ \vdots & \vdots & \ddots & \vdots \\ a_{I1}\mathbf{B} & a_{I2}\mathbf{B} & \cdots & a_{IJ}\mathbf{B} \end{bmatrix}.$$
$$= \begin{bmatrix} \mathbf{a}_1 \otimes \mathbf{b}_1 & \mathbf{a}_1 \otimes \mathbf{b}_2 & \mathbf{a}_1 \otimes \mathbf{b}_3 & \cdots & \mathbf{a}_J \otimes \mathbf{b}_{L-1} & \mathbf{a}_J \otimes \mathbf{b}_L \end{bmatrix}.$$

Hadamard product: The element-wise matrix product. Hadamard product between two matrices $\mathbf{A} \in \mathbb{R}^{I \times K}$ and $\mathbf{B} \in \mathbb{R}^{I \times K}$ is denoted by $\mathbf{A} \circ \mathbf{B}$ and defined as

$$\mathbf{A} \circ \mathbf{B} = \begin{bmatrix} a_{11}b_{11} & a_{12}b_{12} & \cdots & a_{1K}b_{1K} \\ a_{21}b_{21} & a_{22}b_{22} & \cdots & a_{2K}b_{2K} \\ \vdots & \vdots & \ddots & \vdots \\ a_{I1}b_{I1} & a_{I2}b_{I2} & \cdots & a_{IK}b_{IK} \end{bmatrix}.$$

A.2.4 Matrix decompositions

Eigenvalue decomposition (EVD): EVD of a $n \times n$ matrix $\mathbf{A}$ is the factorization of $\mathbf{A}$ into the product of three matrices as

$$\mathbf{A} = \mathbf{Q}\mathbf{\Lambda}\mathbf{Q}^{-1},$$

where $\mathbf{Q}$ is a $n \times n$ square matrix whose columns are eigenvectors of $\mathbf{A}$ and $\mathbf{\Lambda}$ is a $n \times n$ diagonal matrix whose elements are eigenvalues of $\mathbf{A}$.

Singular value decomposition (SVD): SVD of a $m \times n$ matrix $\mathbf{A}$ is the factorization of $\mathbf{A}$ into the product of three matrices as

$$\mathbf{A} = \mathbf{U}\mathbf{D}\mathbf{V}^T,$$

where $\mathbf{U}$ is a $m \times m$ unitary matrix, $\mathbf{D}$ is a $m \times n$ diagonal matrix with nonnegative entries on the diagonal, and $\mathbf{V}$ is a $n \times n$ unitary matrix. SVD of the matrix $\mathbf{A}$ can be also written as

$$\mathbf{A} = \sum_{i=1}^{r} \sigma_i \mathbf{u}_i \mathbf{v}_i^T,$$

where $\sigma_i = \mathbf{D}_{ii}$ are the singular values of $\mathbf{A}$, $\mathbf{u}_i$ and $\mathbf{v}_i$ are the columns of $\mathbf{U}$ and $\mathbf{V}$, respectively, and $r \leq \min\{m, n\}$ is the rank of $\mathbf{A}$.

Definition (Condition number). Measures amount of change in output of a function based on a small change in the input argument. Given a matrix $\mathbf{A}$, the condition number is denoted with $\kappa(\mathbf{A})$ and defined as

$$\kappa(\mathbf{A}) = \frac{\sigma_{\max}(\mathbf{A})}{\sigma_{\min}(\mathbf{A})},$$

where $\sigma_{\max}(\mathbf{A})$ and $\sigma_{\min}(\mathbf{A})$ are maximum and minimum singular values of $\mathbf{A}$, respectively.

Remark. If a problem has a low condition number, it is *well-conditioned.* Otherwise, it is *ill-conditioned.*

Definition (Principal component analysis (PCA)). PCA is a dimensionality reduction approach that finds directions of maximum variance in high-dimensional data and projects it onto a lower-dimensional subspace. In particular, PCA is a unitary linear transformation converting the data into a new coordinate system. These new orthogonal axes or principal components represent directions of maximum variance, where the greatest variance of data lies on the first coordinate, the second greatest variance in the second coordinate, and so on.

A.2.5 Matrix norms

Definition. A matrix norm is a function $f : \mathbb{R}^{m\times n} \longrightarrow \mathbb{R}$ that satisfies the following properties for $\mathbf{A}, \mathbf{B} \in \mathbb{R}^{m\times n}$:

(i) $\|\mathbf{A}\| \geq 0$ and $\|\mathbf{A}\| = 0 \Longleftrightarrow \mathbf{A} = 0_{m\times n}$
(ii) $\|\alpha\mathbf{A}\| = |\alpha|\|\mathbf{A}\|$
(iii) $\|\mathbf{A}+\mathbf{B}\| \leq \|\mathbf{A}\| + \|\mathbf{B}\|$ (triangle inequality).

Definition (p-norm). Given a matrix $\mathbf{A}$ in $\mathbb{C}^{m\times n}$, the set of matrix norms is defined as

$$\|\mathbf{A}\|_p = \max_{x\in\mathbb{C}^n, x\neq 0} \frac{\|\mathbf{Ax}\|_p}{\|\mathbf{x}\|_p}.$$

The matrix norm $\|.\|_p$ is induced by the vector norm $\|.\|_p$. The induced matrix norms for the special cases of $p = 1, 2, \infty$ is

$$\|\mathbf{A}\|_1 = \max_{1\leq j\leq n} \sum_{i=1}^{m} |a_{ij}|,$$

$$\|\mathbf{A}\|_2 = \sqrt{\lambda_{\max}(\mathbf{A}^H\mathbf{A})} = \sigma_{\max}(\mathbf{A}),$$

$$\|\mathbf{A}\|_\infty = \max_{i=1,\ldots,m} \sum_{j=1}^{n} |a_{ij}|,$$

where $\lambda_{max}(.)$ denotes the largest eigenvalue.

Definition (Frobenius norm). Given a matrix $\mathbf{A}$, the Frobenius norm is defined as the square root of the sum of the absolute squares of all the matrix entries, i.e.,

$$\|\mathbf{A}\|_F = \left(\sum_{j=1}^{n}\sum_{i=1}^{m} |a_{ij}|^2\right)^{1/2}$$

$$= \sqrt{\text{trace}\left(\mathbf{A}^H\mathbf{A}\right)} = \sqrt{\sum_{i=1}^{\min\{m,n\}} \sigma_i(\mathbf{A})^2}.$$

Definition (Nuclear norm). The nuclear norm, also known as the trace norm or the Schatten 1-norm, is defined as

$$\|\mathbf{A}\|_* = \text{trace}\left(\sqrt{\mathbf{A}^H\mathbf{A}}\right) = \sum_{i=1}^{\min\{m,n\}} \sigma_i(\mathbf{A}).$$

A.3 Tensors

Definition. Tensors are multidimensional arrays indexed with three or more indices.

A.3.1 Tensor properties

Definition (Order). The order of a tensor is the number of dimensions. An n-order tensor is given as $\mathcal{X} \in \mathbb{R}^{I_1 \times I_2 \times \cdots \times I_N}$. For instance, an entry of a third order tensor $\mathcal{X} \in \mathbb{R}^{I \times J \times K}$ is shown with $\mathcal{X}_{i,j,k} \in \mathbb{R}$ for each $i \in \{1, \ldots, I\}$, $j \in \{1, \ldots, J\}$, $k \in \{1, \ldots, K\}$.

Definition (Slice). Two-dimensional sections defined by fixing all except two indices. For a third-order tensor, horizontal, lateral, and frontal slices are denoted by $\mathcal{X}_{i::}$, $\mathcal{X}_{:j:}$, and $\mathcal{X}_{::k}$, respectively.

Definition (Fiber). Fibers are defined by fixing every index but one. A third-order tensor has column, row and tube fibers, denoted by mode-1 (column) fiber: $\mathcal{X}_{:,j,k}$, mode-2 (row) fiber : $\mathcal{X}_{i,:,k}$, mode-3 (tube) fiber : $\mathcal{X}_{i,j,:}$.

Definition (Matricization). Matricization, also known as unfolding, is the process of transforming a tensor into a matrix by taking all slices along one direction and stacking them together. A third-order tensor has three unfoldings defined by the order of stacking the slices. The three unfolding modes are

$$\mathcal{X}_{(1)} \in \mathbb{R}^{I \times JK}, \mathcal{X}_{(2)} \in \mathbb{X}^{J \times IK}, \mathcal{X}_{(3)} \in \mathbb{R}^{K \times IJ}.$$

Definition (Vectorization). The process of flattening a tensor into a vector, denoted by $\text{vec}(\mathcal{X})$.

A.3.2 Tensor products

Definition (Inner product). An inner product on a tensor is defined by the entry-wise product of the two tensors and summing them up. For third-order tensors, the inner product is defined as

$$\langle \mathcal{X}, \mathcal{Y} \rangle = \sum_{ijk} \mathcal{X}_{ijk} \mathcal{Y}_{ijk}.$$

Definition (n-Mode product). The n-mode (matrix) product of a tensor $\mathcal{X} \in \mathbb{R}^{I_1 \times I_2 \times \cdots \times I_N}$ with a matrix $\mathbf{U} \in \mathbb{R}^{J \times I_n}$ is denoted by $\boldsymbol{X} \times_n \mathbf{U}$. The n-mode product is defined as

$$(\mathcal{X} \times_n \mathbf{U})_{i_1 \cdots i_{n-1} j i_{n+1} \cdots i_N} = \sum_{i_n=1}^{I_n} x_{i_1 i_2 \cdots i_N} u_{j i_n}.$$

Definition (Outer product). Given we have three vectors $\mathbf{a} \in \mathbb{R}^I$, $\mathbf{b} \in \mathbb{R}^J$, and $\mathbf{c} \in \mathbb{R}^K$, their outer product is defined as

$$\mathcal{X} = \mathbf{a} \circledcirc \mathbf{b} \circledcirc \mathbf{c} \in \mathbb{R}^{I \times J \times K}, \text{ where } \mathcal{X}(i, j, k) = a_i b_j c_k,$$

where $\circledcirc$ denotes the outer product.

A.3.3 Tensor ranks

Definition (Rank-one). A rank-one $N-$ order tensor $X \in \mathbb{R}^{I_1 \times I_2 \times \cdots \times I_N}$ is an outer product of N vectors:

$$\mathcal{X} = \mathbf{a}^{(1)} \circledcirc \mathbf{a}^{(2)} \circledcirc \cdots \circledcirc \mathbf{a}^{(N)}.$$

Definition (Rank). The rank of a tensor $\mathcal{X}$ is the minimum number of rank-one tensors that can produce the tensor $\mathcal{X}$ as their summation.

A.3.4 Tensor decompositions

CANDECOMP/PARAFAC decomposition. CP Decomposition factorizes a tensor into a sum of rank-one components. CP decomposition for a third-order tensor, $\mathcal{X} \in \mathbb{R}^{I \times J \times K}$, is defined as

$$\mathcal{X} = \sum_{r=1}^{R} \mathbf{a}_r \circledcirc \mathbf{b}_r \circledcirc \mathbf{c}_r \Longleftrightarrow \mathcal{X}(i, j, k) = \sum_{r=1}^{R} \mathbf{a}_r(i) \mathbf{b}_r(j) \mathbf{c}_r(k),$$

where R is the rank of the tensor and $\mathbf{a}_r \in \mathbb{R}^I$, $\mathbf{b}_r \in \mathbb{R}^J$, and $\mathbf{c}_r \in \mathbb{R}^K$ for $r = 1, \ldots, R$. The element-wise CP decomposition can be written as

$$\begin{aligned} \mathcal{X}(i, j, k) &= \sum_{r=1}^{R} \mathbf{a}_r(i) \mathbf{b}_r(j) \mathbf{c}_r(k) \\ &= \sum_{r=1}^{R} \mathbf{A}(i, r) \mathbf{B}(j, r) \mathbf{C}(k, r), \forall\, i \in 1, \ldots, I, j \in 1, \ldots, J, k \in 1, \ldots, K, \end{aligned}$$

where factor matrices $\mathbf{A} = [\mathbf{a}_1, \cdots, \mathbf{a}_R]$, $\mathbf{B} := [\mathbf{b}_1, \cdots, \mathbf{b}_R]$, and $\mathbf{C} := [\mathbf{c}_1, \cdots, \mathbf{c}_R]$.

Tucker decomposition. Tucker decomposition is a form of higher-order singular-value decomposition method. It decomposes a tensor into a core tensor with a matrix multiplied along each mode. Tucker decomposition for a third-order tensor, $\mathcal{X} \in \mathbb{R}^{I \times J \times K}$, is defined as

$$\mathcal{X} = \sum_{p=1}^{P} \sum_{q=1}^{Q} \sum_{r=1}^{R} \mathbf{G}(p, q, r)\mathbf{U}(:, p) \circledcirc \mathbf{V}(:, q) \circledcirc \mathbf{W}(:, r),$$

where $\mathbf{G} \in \mathbb{R}^{P \times Q \times R}$ is a core tensor, $\mathbf{U} \in \mathbb{R}^{I \times P}$, $\mathbf{V} \in \mathbb{W}^{J \times Q}$, and $\mathbf{C} \in \mathbb{R}^{K \times R}$ are the factor matrices that are obtained through principal components along corresponding modes. The element-wise Tucker decomposition can be written as

$$\mathcal{X}(i, j, k) = \sum_{p=1}^{P} \sum_{q=1}^{Q} \sum_{r=1}^{R} \mathbf{G}(p, q, r)\mathbf{U}(i, p) \circledcirc \mathbf{V}(j, q) \circledcirc \mathbf{W}(k, r).$$

Index

D

M

V

W

Z

Printed in the United States
by Baker & Taylor Publisher Services